# LACTIC ACID BACTERIA – R & D FOR FOOD, HEALTH AND LIVESTOCK PURPOSES

Edited by **Marcelino Kongo**

**Lactic Acid Bacteria – R & D for Food, Health and Livestock Purposes**
http://dx.doi.org/10.5772/2825
Edited by Marcelino Kongo

## Contributors

J. Marcelino Kongo, A.K. Szczepankowska, R.K. Górecki, P. Kołakowski, J.K. Bardowski, F. Martin, B. Ebel, C. Rojas, P. Gervais, N. Cayot, R. Cachon, Jirasak Kongkiattikajorn, Lothar Kröckel, Belal J. Muhialdin, Zaiton Hassan, Nazamid Saari, Lavinia Claudia Buruleanu, Magda Gabriela Bratu, Iuliana Manea, Daniela Avram, Carmen Leane Nicolescu, Daoud Harzallah, Hani Belhadj, Katia Sivieri, Raquel Bedani, Daniela Cardoso Umbelino Cavallini, Elizeu A. Rossi, Gülden Başyiğit Kılıç, Lijuan Yuan, Ke Wen, Fangning Liu, Guohua Li, Nditange Shigwedha, Li Jia, Anne Pihlanto, Edson Mauro Santos, Thiago Carvalho da Silva, Carlos Henrique Oliveira Macedo, Fleming Sena Campos, Yang Cao, Yimin Cai, Toshiyoshi Takahashi, Mahdi Ghanbari, Mansooreh Jami, Masoud Rezaei, Mansoureh Jami, Shirin Tarahomjoo, Junichi Takahashi, Tamara Aleksandrzak-Piekarczyk, Anna Sikora, Mieczysław Błaszczyk, Marcin Jurkowski, Urszula Zielenkiewicz, Tsuda Harutoshi, Diana I. Serrazanetti, Davide Gottardi, Chiara Montanari, Andrea Gianotti, Charina Gracia B. Banaay, Marilen P. Balolong, Francisco B. Elegado, Panagiota Florou-Paneri, Efterpi Christaki, Eleftherios Bonos, Sanna Taskila, Heikki Ojamo, Bertrand Tatsinkou Fossi, Frédéric Tavea

## Published by InTech

Janeza Trdine 9, 51000 Rijeka, Croatia

## Copyright © 2013 InTech



## Notice



**Publishing Process Manager** Dragana Manestar
**Technical Editor** InTech Prepress, Novi Sad
**Cover** InTech Design team

Additional hard copies can be obtained from orders@intechopen.com

Lactic Acid Bacteria – R & D for Food, Health and Livestock Purposes,
Edited by Marcelino Kongo
p. cm.
ISBN 978-953-51-0955-6

For EU product safety concerns:
IN TECH d.o.o., Prolaz Marije Krucifikse Kozulić 3, 51000 Rijeka, Croatia,
info@intechopen.com or visit our website at intechopen.com.

# Meet the editor

Dr. J. Marcelino Kongo is currently a Post-Doc fellow and a Special Researcher Graduate Faculty at the Canadian Research Institute for Food Safety, University of Guelph in Canada, where he pursues research on bacteriocins of Lactic Acid Bacteria (LAB) . He has a PhD in Biotechnology & Food Engineering, Msc in Food Science and Technology from the School of Biotechnology and a BA in Biology of Aquatic Ecosystems from the University of Oporto, Portugal. He is the recipient of "The President Award" from the Society for Applied Microbiology (SFAM), UK and is a Reviewer for the Journal of Dairy Science. His research interests include Systems Biology of Lactic Acid Bacteria, their Antimicrobial Systems (bacterocins) applied in non-thermal processing, as well as the Multiple Socio Economic Roles of Artesan Cheeses.

# Contents

# Preface

The identification of solutions to improve the life and health of consumers, providing safe food of high quality, is the major concern in Food Science. Towards that goal, preservation methods such as salting, drying, high/low temperature application, fermentation, and more recently, pulsed electric field, high pressure and radiation - alone or combined – may be applied. The choice of the preservation process will depend on the food type, availability of the method, cost effectiveness and the degree of change it causes to the flavor and nutritional features of the food product. However, few preservation methods present the advantages of fermentation or biopreservation using lactic acid bacteria (LAB), which is cheap, widely accessible and meets today's increasing consumer's demand for minimally processed/preserved food products. Biopreservation is indeed one of the oldest forms of food preservation technologies and a proved highly efficient non-thermal processing method.

The rapidly changing consumption patterns of the global market and the desire for minimally processed foods is pushing the industry to find processing methods that meet the consumers desire and also increase products safety. LAB fermentation is common in production of dairy, meat, vegetable and fish products, as well as, animal feed. It is based on LAB ability to ferment sugars, especially glucose and galactose, leading to formation of lactic acid, and other metabolites that bring desirable texture and flavors changes to fermented foods, besides increasing their safety via release of bacteriocins. Considerable research is focused on these ribossomaly sintethesized proteins, because of their potential optimized applications in food, pharmaceuticals, nutraceuticals, and veterinary and human medicine. Their presence in foods is in general considered safe for consumers because they are inactivated by pancreatic or gastric enzyme.

New research in the fields of genomics, proteomics and genetic engineering is helping us understand better LAB physiology, pushing further the boundaries of their potential applications. In fact the isolation and characterization of new LAB strains, is one of the most active fields of research in Food Science today.

Many LAB have been identified as suitable for probiotic foods production - defined as a preparations containing viable defined microorganisms in sufficient numbers, which positively alter the GI tract microbiota. In fact LAB probiotic cultures have been

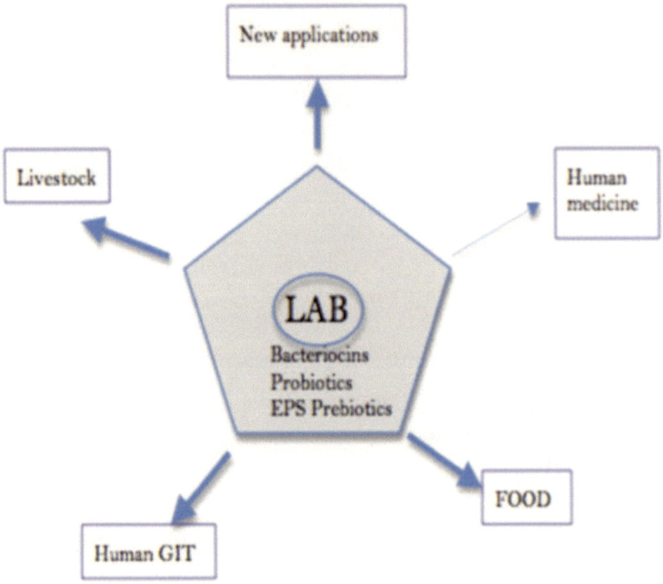


proposed as alternative to antibiotic therapy in patients affected *Clostridium difficile.* LAB are applied in production of prebiotics, which are polyssacharides ingredients that encourage the growth and activity of beneficial bacteria in the GI tract, as well as in the manufacture of enzymes and other pharmaceutical products - namely as potential vehicles for production of new vaccines and new antibiotics.

The compilation and discussion of these many fields of application of LAB is the core of this book. We thank all contributors from around the world that have lay down here their outstanding scientific research and make it available to the vast public of readers - researcher, academics, corporate R & D, students or just curious person interested in the subject. Either way, I hope you enjoy reading it and find the contents of the book useful.

I would like to thank my wife, Karyne for her support of my academic endeavors, which, from time to time have meant, unfortunately, enjoying a little less of our family time.

**J. Marcelino Kongo**
Instituto de Inovação Tecnológica dos Açores (INOVA)
Canadian Research Institute of Food Safety)

Section 1

# Dairy Food Products

Chapter 1

# Lactic Acid Bacteria as Starter-Cultures for Cheese Processing: Past, Present and Future Developments

J. Marcelino Kongo

Additional information is available at the end of the chapter

http://dx.doi.org/10.5772/55937

## 1. Introduction

The identification of solutions to improve the life and health of consumers, providing safe and nutritious foods, is the major concern in Food Science. Toward that goal, preservation methods such as salting, drying, high/low temperature application, fermentation, and more recently, pulsed electric field, high pressure and radiation - alone or in combination – may be applied. The chosen method will depend on the type of raw materials, availability of the method, cost, effectiveness and degree of change it causes to the flavor and nutritional features of the food product. Fermentation, also called biopreservation, is a cheap, widely accessible method that meets today's increasing consumer´s demand for minimally processed/preserved food products. Biopreservation with lactic acid bacteria (LAB) is indeed one of the oldest and highly efficient forms of non-thermal processing method. Cheese production is based on LAB ability to ferment sugars, especially glucose and galactose, so to produce lactic acid and aroma substances that give typical flavors and tastes to fermented products. LAB also release antimicrobial metabolites so called bacteriocins, which are considered safe and natural preservatives, with great potential to be used on their own, or synergistically with other methods in food preservation.

## 2. Lactic acid bacteria in dairy processing

Milk is a highly perishable food raw material, therefore, its transformation in cheese or other form of fermented dairy product, provides an ideal vehicle to preserve its valuable nutrients (Table 1), making them available throughout the year. It is known that while unprocessed milk can be stored for only a few hours at room temperatures, cheeses may reach a shelf-live up to 5 years (depending on variety).

| Animal | Fat | Casein | Lactose | Albumin | Ash | Water |
|---|---|---|---|---|---|---|
| | | | % | | | |
| Cow | 3.75 | 3.0 | 4.75 | 0.4 | 0.75 | 87.3 |
| Goat | 6.0 | 3.3 | 4.6 | 0.7 | 0.84 | 84.5 |
| Ewe | 9.0 | 4.6 | 4.7 | 1.1 | 1.0 | 79.6 |
| Camel | 3.0 | 3.5 | 5.5 | 1.7 | 1.5 | 84.8 |
| Buffalo | 6.0 | 3.8 | 4.5 | 0.7 | 0.75 | 85 |

Note: In cheese, these nutrients will appear at a concentration approximately ten times higher, while the water content decreases.

**Table 1.** Approximate composition of milk from various species of mammals

Fermentation with lactic acid bacteria (LAB) is a cheap and effective food preservation method that can be applied even in more rural/remote places, and leads to improvement in texture, flavor and nutritional value of many food products. LAB have a long and safe history of application and consumption namely in cheese processing (Aquilanti et al., 2006, Caplice & Fitzgerald, 1999, Giraffa et al., 2010, Ray, 1992; Wood, 1997; Wood & Holzapfel, 1995) thus being generally regarded as safe (GRAS). Increasing knowledge of LAB physiology, together with new developments in processing technology, is leading to their application beyond traditional starter culture application, namely in new food safety roles and direct health applications.

## 2.1. LAB as starter-cultures in cheese processing

Cheese-making is based on application of LAB in the form of defined or undefined starter cultures that are expected to cause a rapid acidification of milk through the production of lactic acid, with the consequent decrease in pH, thus affecting a number of aspects of the cheese manufacturing process and ultimately cheese composition and quality (Briggiler-Marco et al., 2007).

The earliest productions of cheeses were based on the spontaneous fermentation, resulting from the development of the microflora naturally present in the raw milk and its environment. The quality of the end product was a reflex of the microbial load and spectrum of the raw material. Spontaneous fermentation was later optimized through backslopping, i.e., inoculation of the raw material with a small quantity of whey from a previously performed successful fermentation, and the resulting product characteristics depended on the best-adapted strains dominance (Leroy & De Vuyest, 2004). Today, backslopping is still used to produce many artisanal raw-milk cheeses, namely those bearing the PDO (Protected Designation of Origin) status, which are considered to be an important source of LAB genetic diversity, as well as being crucial from an economic and even ecologic point of view, since production of said cheeses (usually processed on a small-scale) contributes to local employment and maintains people functioning as "guardians of local environment" in regions that otherwise would be deserted.

The starter-culture applied in this, so-called, natural fermentation, is usually a poorly-known microflora mix that although having a predominance of LAB, may also contain non-LAB microorganisms, and its microbial diversity and load is usually variable over time. In fact, studies directed to characterize traditional cheeses show that those made from raw milk harbor a diversity of LAB (Bernardeau et al., 2008) depending on geographical region, where a few may show particular interesting technological features that upon optimization may have industrial applications (Buckenhiiskes, 1993). For example, because wild strains need to withstand the competition of other microorganisms to survive in their hostile natural environment, they often produce antimicrobials substances called bacteriocins (Ayad et al., 2002), which are natural antibacterial proteins that can be incorporated directly into fermented foods as such (food–grade) or indirectly as starter culture (Bernardeau et al., 2008). Although nisin is today the only bacteriocin that reached commercial status, approved worldwide as a natural food preservative, many other bacteriocins may soon reach similar status. Recently, our work (to be published) with LAB isolates from traditional portuguese raw-milk cheeses, revealed several lactobacilli having antibacterial activity against pathogens such as *Listeria monocytogenes, Staphyloccus aureus, Salmonella newport* and even *E.coli*. Future studies may allow us using these isolates or their metabolites, applied *in situ or ex situ* fashion, in applications where food safety is a concern.

Moreover, traditional cheeses also obtain their flavor intensity also from the non-starter lactic acid bacteria (NSLAB), which are not part of the normal starter flora but develop in the product, particularly during maturation, as a secondary flora (Beresford, et al., 2001). The isolation and optimization of wild-type strains from traditional products, to be used as starter cultures in cheese processing, is indeed a highly active field of research in Food Science today.

## 2.2. LAB food safety and cheese technology

Cheese is made in almost every country of the world and there are more than 2000 varieties, made from milk of several mammals, processed industrially or by traditional methods (Figure 1).

However, despite the large number of varieties, the basic steps required in any cheese processing are essentially the same, and slight variations in any of these steps may result in products of different general quality (Figure 2).

**Milk treatment**. In large-scale cheese processing, the milk is heat-treated, e.g. 73 ºC for 15 seconds, to destroy pathogens and reduce microbial numbers, while in most traditional PDO raw-milk cheeses heat treatment is not applied. Also the milk may be standardized, i.e. the fat content may be increased or reduced, or the casein-to-fat ratio may be adjusted.

**Starter-culture addition**. The type of commercially available starter preparation to be used will be determined by the cheese recipe. As previously stated, large-scale processing relies on using defined, commercially available starters, while for traditional cheeses, a natural fermentation (whey from the previous lot) is often used.

**Figure 1.** (Top) - Brine salting of cheeses in a large-scale plant processing 20 tons of cheese a day. (Bottom) - Small-scale unit processing 50 Kg per day of a traditional PDO cheese

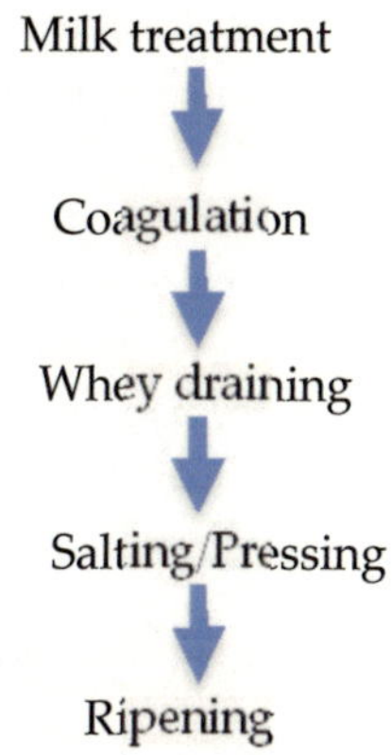


**Figure 2.** Common steps to most cheese making processes

**Coagulation**. During coagulation, modifications on the milk protein complex occur under defined conditions of temperature and by action of a coagulant agent, which changes the physical aspect of milk from liquid to a jelly-like mass. Various coagulants are available, e.g. lemon juice, plant rennet or more commonly a proteolytic enzyme such as chymosin (rennin) or – due to high demand from the cheese industry - proteolytic enzymes from the mould *Rhizomucor miehei* obtained via biotechnology. These enzymes have an acidic nature, meaning they have optimum activity in a slightly acidic environment. Therefore, the action of LAB in this phase is crucial as they are required to rapidly release enough lactic acid, to lower the milk pH from 6.7 to near 6.2, (thus creating an appropriate environment for optimum activity of rennin) and later to pH 4.5 as the processing proceeds, creating an inhospitable environment for many unwanted bacteria, thus increasing the end product safety.

**Cutting the coagulum**. The resulting coagulum may be cut with appropriate knives into curd particles of a defined size, e.g. 1–2 cm, or it may be transferred into containers or cheese moulds. The cutting or ladling of the coagulum is a very important step in the manufacture of some cheese varieties as it determines the rate of acid development and the body (firmness) and texture of the cheese.

**Heating or cooking the curds**. Heating (37–45 ºC, depending on the type of cheese) the curds and whey affects the rate at which whey is expelled from the curd particles and the growth of the starter microorganisms. During heating, the curds and whey are often stirred to maintain the curd in the form of separate particles.

**Whey removal**. After heating and stirring, and when the curd particles have firmed and the correct acid development have taken place, the whey is removed allowing the curd particles to mat together.

**Milling the curd**. In cheeses such as Cheddar, when the curd has reached the desired texture, it is broken up into small pieces to enable it to be salted evenly. Milling the curd can be done either by hand or mechanically. Salting is usually done to enhance the taste of the curd and to increase its safety and shelf life.

**Ripening.** Finally, for most cheeses, the resulting mass is molded and put to ripening for periods that may vary from 15 days to one, two or more years. Ripening is a slow phase, crucial for the development of aroma and flavor, brought about by the action of the many enzymes released by LAB. During ripening the protein in cheese is broken down from casein to low molecular weight peptides and amino acids. Proteolysis is the major – and certainly the most complex of biochemical events that take place during ripening of most cheese varieties and LAB play an important role in it. This happens while the cheeses are stored in the curing cabinets and in some cases in caves, usually with temperature and humidity controlled (Figure 3).

**Figure 3.** Cheese ripening in cabinets with controlled temperature and humidity.

During coagulation, the initial step of casein hydrolysis is performed by chymosin (milk coagulant) and proteinases from starter lactic acid bacteria, starter moulds and other microorganisms. The further degradation of high molecular weight peptides produced at the initial step, is subsequently catalised to low molecular weight peptides by endopeptidases from LAB during ripening (see Fig. 4 and 5).

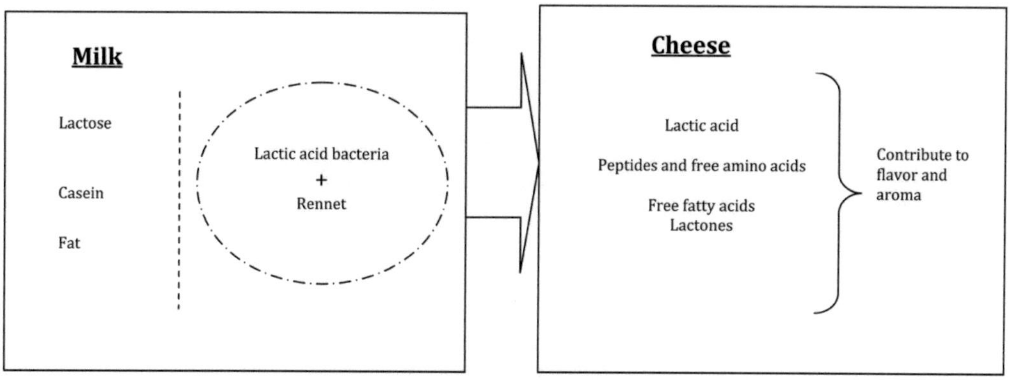


**Figure 4.** Simplified view of the biochemical changes that lead to texture and flavour changes in cheeses.

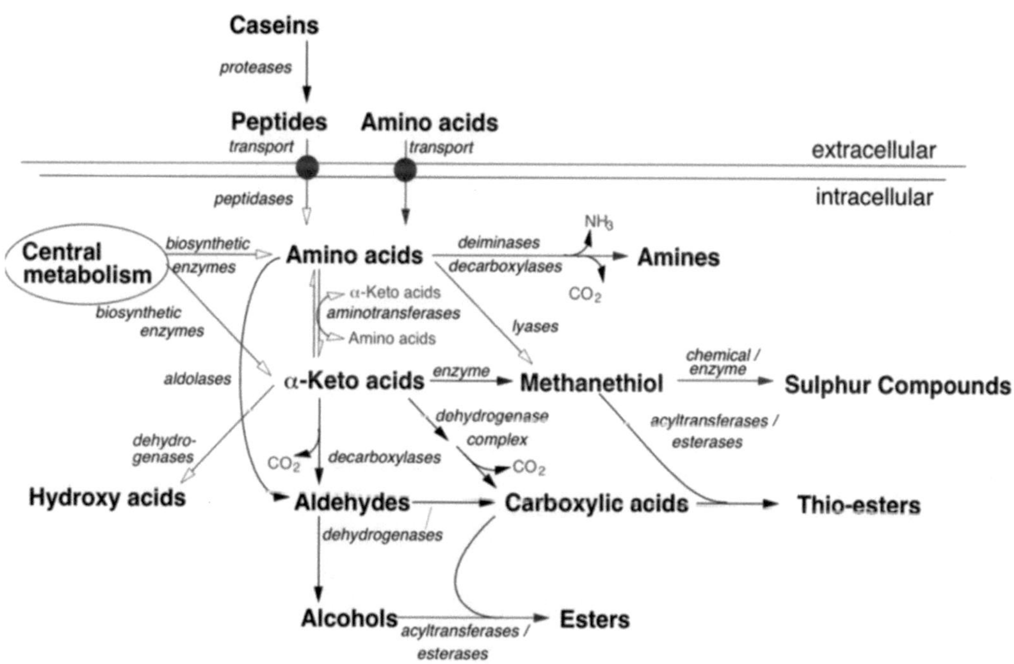


**Figure 5.** General pathways leading to intracellular meatabolites, and their degradation routes to potential flavour compounds. More specifically, pathways from methionine to flavour compounds (methanethiol, thioesters, sulphur compounds) are shown (Adapted from Kranemburg et al., 2002).

Primary proteolysis in cheese is defined as changes in β-, γ-, αs-caseinpeptides, and other minor proteins that are detected by PAGE (Figure 6). Primary proteolysis leads to the formation of large water-insoluble peptides and smaller water-soluble peptides (Fox, 1993, Mooney *et al.*, 1998). Secondary proteolysis products include those peptides, proteins and amino acids soluble in the aqueous phase of cheese and are extractable as the water-soluble nitrogen (WSN) fraction. The WSN fraction is a complex mixture of large, medium, and small peptides and amino acids. These components result from the action of milk clotting enzymes, milk proteases, starter LAB and contaminating microorganisms (Rank *et al.*, 1985).

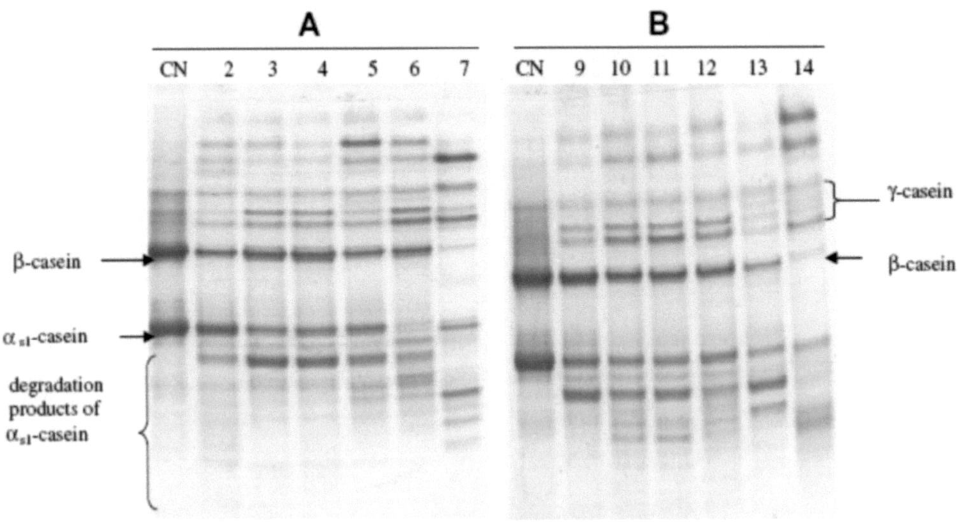


**Figure 6.** Evolution of proteolysis via urea-polyacrylamide gel electrophoresis in São Jorge cheeses from dairies A and B, by 1, 15, 30, 60, 90 or 130 days of ripening. Lanes 1, 8 and 15, Na- caseinate; lanes 2-6: cheese A; lanes 9-14: cheese B (Kongo et al., 2012).

Typical cheese pH values measured at 3–7 days after manufacture are 4.9–5.5 in most firm and hard ripened varieties, and 4.4–4.8 in fresh lactic and most soft ripened varieties (Table 2 and Figures 7 and 8).

| Operations | Swiss type | | Gouda | | Cheddar | | Feta | | Cottage | |
|---|---|---|---|---|---|---|---|---|---|---|
| | Time | pH | Time | pH | Time | pH | Time | pH | Time | pH |
| Add starter | 0 | 6.60 | 0 | 6.60 | 0 | 6.60 | 0 | 6.60 | 0 | 6.60 |
| Add rennet | 15 | 6.60 | 35 | 6.55 | 30 | 6.55 | 75 | 6.50 | 60 | 6.50 |
| Cut | 45 | 6.55 | 70 | 6.50 | 75 | 6.50 | 115 | 6.4 | 300 | 4.80 |
| Drain or dip into forms | 150 | 6.35 | 100 | 6.45 | 195 | 6.3 | 130 | NA | 360 | 5.0 |
| Milling | NA | NA | NA | NA | 315 | 5.45 | NA | NA | NA | NA |
| Pressing | 165 | 6.35 | 130 | | 390 | 5.40 | NA | NA | NA | NA |
| Demoulding | 16 h | 5.30 | 8 h | 5.40 | 10 h | 5.20 | 24 h | 4.6 | NA | NA |
| Minimum pH | 1 wk | 5.20 | 1 wk | 5.20 | 1 wk | 5.10 | 1 wk | 4.4 | NA | NA |
| Retail | 6 mo | 5.6 | 6 mo | 5.6 | 4 mo | 5.3 | 6 wk | 4.4 | 2–14 d | 5.2 |

**Table 2.** Typical pH *vs* time profiles for several cheese varieties (time is in minutes unless otherwise noted).

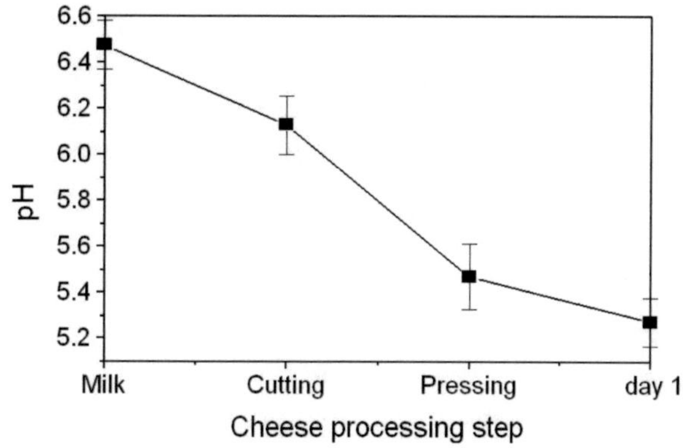


**Figure 7.** Evolution of pH (average ± standard deviation) in experimental cheese made with a starter culture of authoctonous São Jorge cheese LAB isolates.

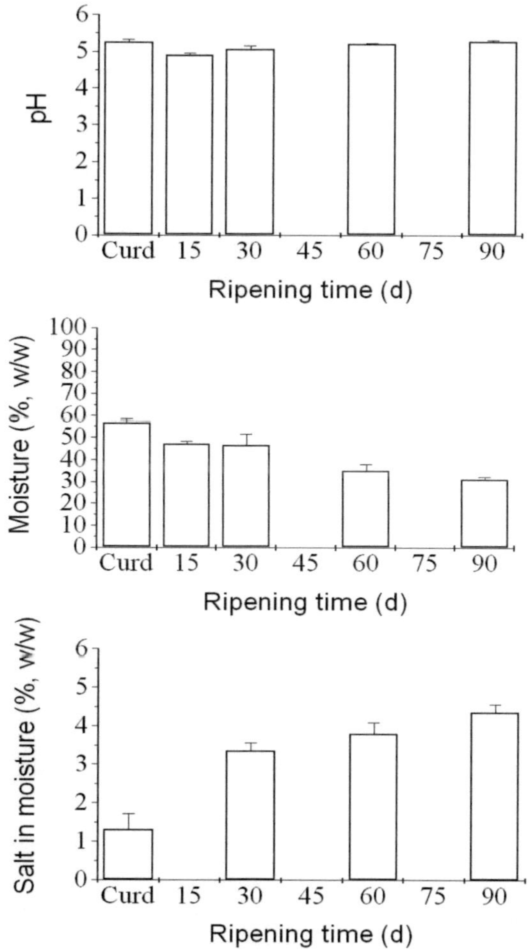


**Figure 8.** Evolution of physicochemical parameters (average ± standard deviation) throughout ripening of cheeses made with an experimental starter culture.

During processing, the pH history of the cheese is a good indicator of the actual product safety. For example a 'slow vat' allows more time at high pH for undesirable bacteria to grow, while during cheese ripening, unwanted bacteria may grow due to an acidity neutralization resulting from secondary microflora growth such as moulds. For most ripened varieties the combination of a low pH and ripening time, which leads to moisture decrease in the cheese, will in general cause a gradual decline of all groups of bacteria due to increasing inhospitable conditions inside the cheese.

The pH history of a cheese and the hygienic practices applied in its manufacture are thus key factors to guarantee safe products. Thus, the isolation of autochthonous LAB intend to be used for development of specific starter cultures with improved acid production and other antimicrobial activities may be an excellent way towards reaching the goals of simultaneously obtaining safe traditional cheeses, still bearing their unique flavors.

Nowadays, western consumers still enjoy artisan cheeses thanks to their outstanding gastronomic qualities; however, in most industrialized countries the large-scale cheese processing is the most important branch of the food industry. In such cases, there is a strong need to control the fermentation process towards maximum efficiency in terms of yields and standardization of the end product. This, and the need to fulfill the safety assurance of the final product, is usually achieved by, among other improvements, adding a high dosage of pure LAB selected starter cultures, commercially available (today's world starter culture market is more than US$1 billion), as well as by heat treating the raw milk, most commonly by pasteurization.

## 3. Development of new starter cultures for cheese processing

Traditional raw-milk cheeses are highly valued for their flavors, while large-scale products are often perceived by the consumer as "boring" (Law, 2001) – a consequence of the elimination by pasteurization, of the flora that has a key role in flavor development; and this puts the food industry under pressure to look for alternative LAB cultures capable of improving products flavor (Leroy & De Vuyest, 2004).

Today, the increased understanding of the genomics and metabolomics of food microbes opens up new perspectives for starter-cultures improvements and through genetic engineering it is now possible to express their desirable properties or suppress undesirable features (Del- cour, De Vuyst, & Shortt, 1999; Law, 2001; Mogensen, 1993).

Originally, starter cultures for the cheese industry were maintained by daily propagation, and later, they became available as frozen concentrates and dried or lyophilised preparations, produced on an industrial scale, some of them allowing direct vat inoculation (Sandine, 1996). Because the original starter cultures were mixtures of several undefined microbes, the daily propagation, eventually led to shifts of the ecosystem resulting in the disappearance of certain strains. Because some important metabolic traits in LAB are plasmid-encoded, there was a risk that they would be lost during propagation (Weerkamp et al., 1996). Lactococci are generally used as starter cultures in the production of industrial cheeses and cultured milk products. In traditional cheeses the natural starter cultures may harbor many different species and strains.

On the other hand, cheeses manufactured in a standard (large-scale) processing manner, are considered as safer because of the application of pasteurization and following the standard hygienic practices, including the HACCP. Traditional cheeses have their own specific processing methods, namely the common use of raw milk, however the hygienic procedures and HACCP approaches adapted to their specificities should be applied as well.

| **Species / subspecies** | **Main uses / Other comments** |
|---|---|
| **Lactococcus** | Mesophilic starter used for many cheese types. |
| *Lc. lactis* subsp. *lactis* | Used in Gouda, Edam, sour cream and lactic butter. Mesophilic starter used for many cheese types. |
| *Lc. lactis* subsp. *lactis* biovar diacetylactis | |
| *Lc. lactis* subsp. *cremoris* | |
| ***Streptococcus*** | Thermophilic starter used for yogurt and many cheese types particularly hard and semi hard high-cook cheeses. |
| *Sc. thermophilus* | |
| ***Lactobacillus*** | |
| *Lb. acidophilus* | Probiotic adjunct culture used in cheese and yogurt. |
| *Lb. delbrueckii* subsp. *bulgaricus* | Thermophilic starter for yogurt. and many cheese types, particularly hard and semi hard high-cook cheeses. |
| *Lb. delbrueckii* subsp. *lactis* | Used in fermented milks and high-cook cheese. |
| *Lb. helveticus* | Thermophilic starter for fermented milks and many cheese types particularly hard and semi hard high-cook cheeses |
| *Lb. casei* | Cheese ripening adjunct culture. |
| *Lb. plantarum* | Cheese ripening adjunct culture. |
| *Lb. rhamnosus* | Cheese ripening adjunct culture. |
| **Leuconostoc** | Mesophilic culture used for Edam, Gouda, fresh cheese, lactic butter and sour cream. |
| *Ln. mesenteroides* subsp. *cremoris* | |
| ***Brevibacterium*** | Used in smear surface-ripened cheeses, Camembert, Stilton and Limburger and as a cheese ripening adjunct culture. |
| *Brev. linens* | |
| ***Propionibacterium*** | Used in Gruyère and Emmental cheeses. |
| *Prop. Acidipropionici* | Used in Gruyère and Emmental cheeses. |
| *Prop. freudenreichii* subsp. *shermanii* | |

**Table 3.** Main bacteria associated with cheeses or other fermented products (From: Broome et al., 2003).

As previously stated, LAB are only a part of the complete microflora of raw milk (Kongo et al, 2007) and this, associated to other technological methods such as pressing, allows the production of a diversity of traditional cheeses (Parguel, 2011). This raw-milk microflora represents the contamination from the environment (air, utensils, the animal skin), and the load and its diversity will thus vary with local, season and livestock type, influenced by temperature.

These microbial mixes have an interdependent activity when together in their ecosystem and therefore their physiological properties may differ when the biodiversity is disrupted. In fact, it has been shown that certain microbial associations reveal a higher protecting effect against pathogens such as listeria, than when their association diversity is disrupted, (Montel 2010) see Figure 9.

Bacteriocinogenic probiotic bacteria could be beneficial when used as starter cultures in cheese, as they may prolong the shelf-life of the products, while simultaneously providing the consumer with a healthy advantage at a low cost (Gomes et al. 1998). The presence of bacteriocins in foods is, in general, seen as safe for consumers because bacteriocins are inactivated by pancreatic or gastric enzymes (Liu et al., 2011).

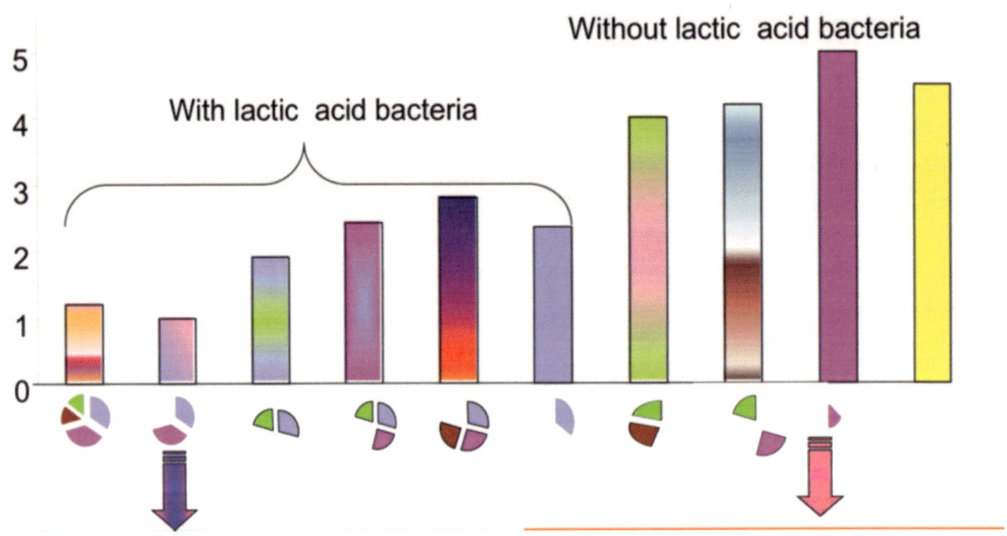


Low level of *L. monocytogenes* in cheeses prepared with consortium associating lactic acid bacteria (species) and non lactic acid bacteria.

Highest level of *L. monocytogenes* in cheeses with *S.thermophilus* and without lactic acid bacteria in the consortium

**Figure 9.** Level of *L. monocytogenes* in the core of Saint-Nectaire type cheese (28d) (Adapted from Montel & Samelis, (2010).

## 3.1. EPS-producing cultures and acceleration of cheese ripening

Many LAB produce exopolysaccharides (EPS), which may provide viscosifying, stabilizing, and water-binding effects in cheeses. The growing demand for all-natural, healthy food products, foods with low fat or sugar content and low levels of additives, as well as cost factors has increased the interest of food industry to use LAB polysaccharides. Research has also shown that EPS+ LAB can enhance the functional properties of low fat cheese and that the excellent water- binding properties and moisture retention of EPS can improve the melting properties of low fat Mozzarella cheese. These properties show that EPS have wide technical potentials for development of novel and improved food products with enhanced texture, mouth-feel, taste perception and stability, representing potential sources for economic gains for the dairy industry.

EPS have also the potential to be used as surface carriers of bacteriocins or bacteriocin-producing LAB, and species such as *Leuconostoc mesenteroides, Streptococcus mutans* and several lactobacilli (Lactobacillus brevis, *Lactococcus lactis* subsp. *lactis, L. lactis* subsp. *cremoris, Lactobacillus casei, Lb. sake, Lb. rhamnosus,*) and thermophilic (*Lb. acidophilus, Lb. delbrueckii* subsp. *bulgaricus, Lb.helveticus* and *S. thermophilus*) are known to produce EPS. The isolation and characterization of EPS from new wild LAB species, which are ubiquitous in traditional cheeses, is a key strategy towards finding strains with optimized production of EPS.

Finally, cheese ripening is a lengthy and costly process. Therefore, attenuated starter cultures with high autolysis are being sought towards increasing the amount of endogenous peptides, thus accelerating the cheese ageing process as well as enhancing flavour and texture. These cultures may be obtained via application of several techniques such as pulsed electric field, heat treatment, freeze–thawing and lysozyme treatment (Briggs, 2003).

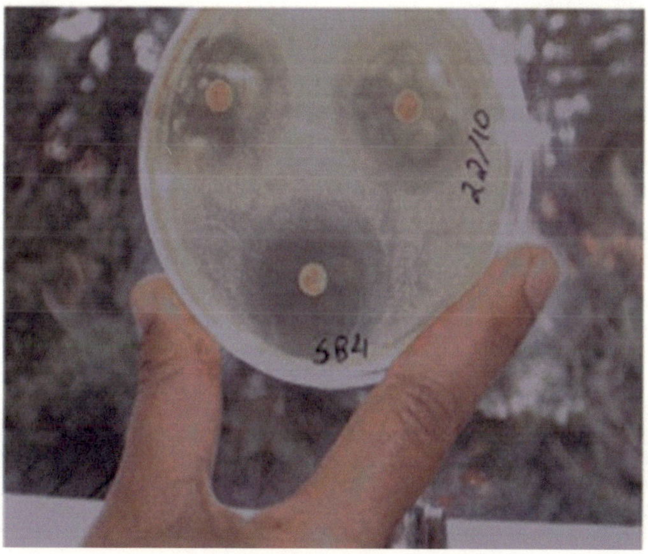


**Figure 10.** Antilisterial activity of LAB isolates from a traditional cheese.

Thus, the cheese industry in looking for new types of LAB starter-cultures bearing several properties: – cultures that increase microbial safety or offer one or more organoleptic, technological, nutritional (enzymes, or polyunsaturated fatty acids - PUFAs) or health advantages such as probiotic properties, starter cultures with increased resistance to bacteriophage, (recall that high product loss, especially in cheese manufacturing, is often associated with bacteriophages (Parente and Cogan, 2004), cultures that produce EPS and cultures that accelerate cheese ripening.

### 3.2. Methods used to characterize LAB for starter cultures development

To characterize new LAB isolates, phenotypic methods relying on physiological or biochemical criteria have been widely applied (Montel, Talon, Fournaud, & Champomier, 1991, Kongo et al., 2007). These phenotypic profiling methods are very important - especially related to finding the technological features, such as the acidification, proteolytic and lipolytic activity, of a new isolate (see Tables 3 and 4, and Figure 11) and have the advantage of requiring less sophisticated equipment. In most of the cases however, these tests are insufficient for accurate species identification due to the great number of different LAB species with similar phenotypic characteristics (Temmerman et al. 2004).

| Characteristics | *Lactobacillus* | *Enterococcus* | *Lactococcus* | *Leuconostoc* | *Pediococcus* | *Streptococcus* |
|---|---|---|---|---|---|---|
| $CO_2$ from glucose | +/- | - | - | + | - | - |
| Growth at 10 ºC | +/- | + | + | + | +/- | - |
| Growth at 45 ºC | +/- | + | - | - | +/ | +/- |
| Growth at 6.5% NaCl | +/- | + | - | +/- | +/ | - |
| Growth at 18% NaCl | - | - | - | - | - | - |
| Growth at pH 4.4. | +/- | + | +/- | +/- | + | - |
| Growth at pH 9.6 | - | + | - | - | - | - |
| Type of lactic acid | D, L, DL | L | L | D | L, DL | L |

**Table 4.** Phenotypic characteristics for discrimination of common LAB for dairy processing (modified from Batt, 1995).

| | Genus | | | | | |
|---|---|---|---|---|---|---|
| | *Lactobacillus* | | | *Enterococcus* | | |
| | Per cent of strains giving indicated reaction* | | | Per cent of strains giving indicated reaction* | | |
| Enzyme | – | + | ++ | – | + | ++ |
| Alkaline phosphatase | 100 | | | 100 | | |
| Esterase | 100 | | | 100 | | |
| Lipase | 100 | | | | 25 | 75 |
| Leucine aminopeptidase | 12 | 88 | | | | 100 |
| Cystine aminopeptidase | 100 | | | | 100 | |
| α-Chymotrypsin | 100 | | | | 100 | |
| Acid-phosphatase | 88 | 24 | | 15 | 75 | |
| Phosphoamidase | | 100 | | | 100 | |
| β-Galactosidase | | 100 | | | 50 | 50 |
| β-Glucuronidase | | 100 | | | 25 | 75 |
| α-Glucosidase | 100 | | | 100 | | |
| β-Glucosaminidase | 100 | | | 100 | | |

*API-ZYM color: grade 0, negative (–); grades 1 through 3, positive (+); grades 4 and 5, plus positive (++).

**Table 5.** Enzyme profiling of 14 representative LAB isolates found in Sao Jorge traditional cheese (from Kongo et al., 2007).

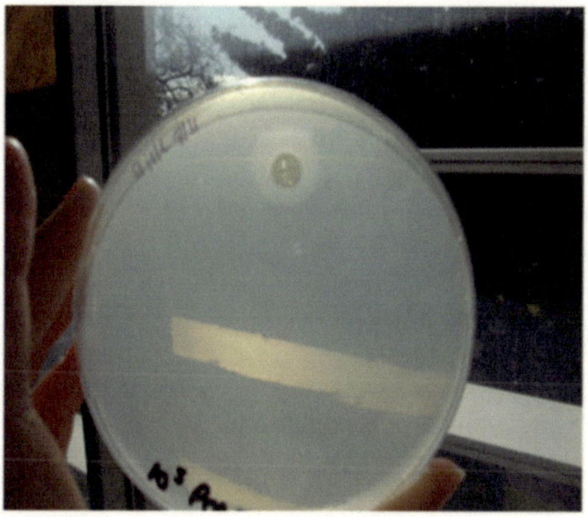

**Figure 11.** Proteolytic acitivity of a Lactobacillus ssp isolate on milk agar.

Molecular biology (genotypic) methods (Figure 12) on the other hand - largely DNA-based techniques - offer much greater discriminatory power, all the way to differentiation of individual strains (Aymerich et al., 2006, Cocolin et al., 2004, Furet et al., 2004, Prabhakar et al., 2011). Thus, a combination of both phenotypic and genotypic identification techniques (so called polyphasic approach) is preferred (Temmerman et al., 2004, Aquilanti et al., 2006).

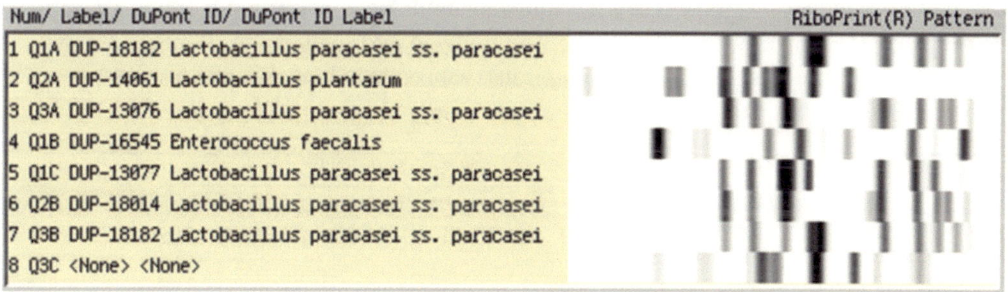


**Figure 12.** Ribotyping as molecular biology technique for identification of LAB to type or strain level (Kongo et al., 2007)

Finally, it should be mentioned that there are concerns today that commensal bacterial populations from food and the gastrointestinal tract (GIT) of humans and animals, such as LAB, could act as a reservoir for antibiotic resistance genes, and therefore, be transferred to possibly pathogenic bacterial species, complicating the treatment of a disease or infection and leading to the spread of antibiotic-resistant bacteria (Ammor et al., 2007). Thus, before using new isolates as starter cultures or as probiotics, the antibiotic resistance must be addressed.

The European Food Safety Agency (EFSA) proposed a system for a pre-market safety assessment of selected groups of microorganisms, leading to granting a "Qualified Presumption of Safety (QPS)". Therefore, EFSA proposed that a safety assessment of a defined taxonomic group, such as a genus or group of related species could be made based on establishing identity, body of knowledge, possible pathogenicity and end use (European Commission 2007). The 33 *Lactobacillus* species shown in Table 6 are the ones that in 2007 EFSA stated could be considered to have QPS-status. In addition to *Lactobacillus* species, also other LAB species have been granted QPS –status. They include three leuconostocs, (*Ln. citreum, Ln. lactis* and *Ln. mesenteroides*), three pediococci (*P. acidilactici, P. dextrinicus* and *P. pentosaceus*), *Lc. lactis* and *Streptococcus thermophilus.*

| | | |
|---|---|---|
| *Lb acidophilus* | *Lb farciminis* | *Lb paracasei* |
| *Lb amylolyticus* | *Lb fermentum* | *Lb paraplantarum* |
| *Lb amylovorus* | *Lb gallinarum* | *Lb pentosus* |
| *Lb alimentarius* | *Lb gasseri* | *Lb plantarum* |
| *Lb aviaries* | *Lb heveticus* | *Lb pontis* |
| *Lb brevis* | *Lb hilgardii* | *Lb reuteri* |
| *Lb bucheneri* | *Lb johnsonii* | *Lb rhamnosus* |
| *Lb casei* | *Lb kefiranofaciens* | *Lb sakei* |
| *Lb crispatus* | *Lb kefiri* | *Lb salivarius* |
| *Lb curvatus* | *Lb mucosae* | *Lb sanfranciscensis* |
| *Lb delbrueckii* | *Lb panis* | *Lb zeae* |

**Table 6.** *Lactobacillus* (Lb) species with QPS- status according to EFSA (from Korhonen, 2010).

Lactobacilli are generally susceptible to antibiotics inhibiting the synthesis of proteins, such as chloramphenicol, erythromycin, clindamycin and tetracycline, and more resistant to aminoglycosides (neomycin, kanamycin, streptomycin and gentamicin. While some species show a high level of resistance to glycopeptides (vancomycin and teicoplanin), susceptibility to bacitracin will vary greatly (Ammor et al, 2007; Coppola et al., 2005).

| | Species | | | | | | | |
|---|---|---|---|---|---|---|---|---|
| Antibiotic | Lactobacillus obligate homofermentative | Lactobacillus heterofermentative | Lactobacillus plantarum | Enterococcus spp | Pediococcus pp | Leuconostoc | Lactococcus lactis | Streptococcus thermophilus |
| Ampicillin | 4 | 4 | 4 | 8 | 4 | 4 | 4 | 4 |
| Vancomycin | 4 | IR | IR | 8 | IR | IR | 4 | 4 |
| Gentamycin | 8 | 8 | 64 | 512 | 4 | 4 | 8 | 8 |
| Kanamycin | 16 | 16 | 64 | 1024 | 4 | 8 | 8 | 8 |
| Streptomycin | 16 | 16 | 64 | 1024 | 4 | 8 | 16 | 16 |
| Neomycin | 16 | 16 | 32 | 1024 | 8 | 8 | 8 | 8 |
| Erythromycin | 4 | 4 | 4 | 4 | 4 | 4 | 4 | 4 |
| Clindamicin | 4 | 4 | 4 | 4 | 4 | 4 | 4 | 4 |
| Tetracycline | 8 | 8 | 32 | 16 | 4 | 4 | 4 | 4 |
| Chloranphenicol | 4 | 4 | 8 | 8 | 4 | 4 | 8 | 8 |
| Trimethoprim | 8 | 8 | 8 | 8 | 8 | 8 | n.r | n.r |

Key: IR, intrinsically resistant

**Table 7.** Microbiological break points (μg $mL^{-1}$) categorizing some LAB species as resistant (Adapted from Ammor et al., 2007)

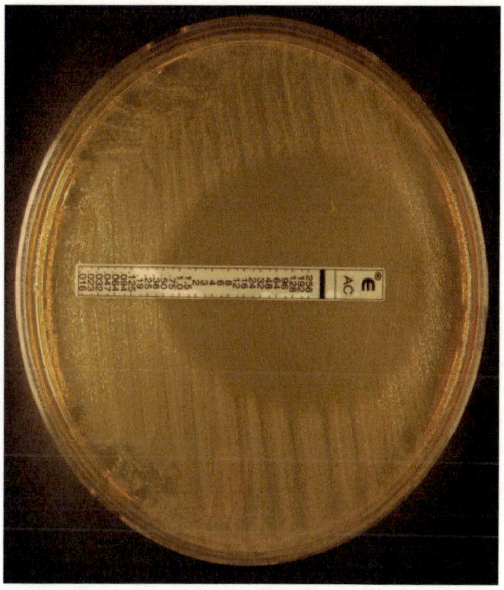

**Figure 13.** Result of a screening for antibiotic resistance of a *Lactobacillus paracasei* isolate

# 4. Concluding remarks

LAB are important in cheese processing because (i) they increase food safety through the release of lactic acid and bacteriocins, (ii) produce aromas and flavor and accelerate the maturation process of cheese via their proteolytic and lipolytic activities, bringing economic advantages to the industry, (iii) bring about desirable food textures via release of polysaccharides that increase the viscosity and firmness, and reduce susceptibility to syneresis, (iv) they may be used to deliver polyunsaturated fatty acids (PUFA) and vitamins, leading to dairy products with increased nutritional value, (v) specific probiotic strains contribute to liberation of health-enhancing bioactive peptides improving absorption in the intestinal tract, stimulating the immune system, exerting antihypertensive, antithrombotic effects, or functioning as carriers for minerals.

Novel insights arising from use of Bioinformatics, Systems Biology and Bioengeneering approaches will offer perspectives for the application of a new generation of starter cultures for cheese-making, having enhanced functional features and offering several health, marketing, and technological advantages, contributing to the development of small and medium sized enterprises on the one hand, and product diversification of large companies on the other.

However, there are still many developments to be achieved towards fully realizing the many foreseen potential of LAB or their products. For example extraction and purification of bacteriocins is still difficult as they form micelles or clumps with the nitrogen sources already in the growth medium. On the other hand while genetic engineering may offer many solutions related to optimal use of LAB, they may not be easily allowed by food legislation.

## Author details

J. Marcelino Kongo
*Instituto de Inovação Tecnológica dos Açores (INOVA)*
*Currently at Canadian Research Institute of Food Safety*

## 5. References

Ammor, M.S., Florez, A.B. & Mayo, B. (2007) Antibiotic resistance in non- enterococcal lactic acid bacteria and bifidobacteria. *Food Microbiol* 24, 559-570.

Aquilanti, L., Dell'Aquila, L., Zannini, E., Zocchetti, A., & Clementi, F. (2006) Resident lactic acid bacteria in raw milk Canestrato Pugliese cheese. Lett. Appl Microbiol 43 (2006) 161–167.

Ayad, E.H.E., A. Verheul, J.T.M. Wouters & G. Smit, (2002). Antimicrobial-producing wild *Lactococci* isolated from artisanal and non-dairy origins. Int. Dairy J., 12: 145-150.

Aymerich, T., Martín, B., Garriga, M., Vidal-Carou, M.C., Bover-Cid, S., & Hugas, M. (2006) Safety properties and molecular strain typing of lactic acid bacteria from slightly fermented sausages. J Appl Microbiol 100, 40-49.

Batt, C. A., Erlandson, K., & Bsat, N. (1995) Design and implementation of a strategy to reduce bacteriophage infection of dairy starter cultures. Int Dairy J 5, 949–962.

Beresford, T.P., Fitzsimons, N.A., Brennan, N.L., & Cogan, T.M. (2001). Recent advances in cheese microbiology. Int. Dairy J., 11: 259-274.

Bernardeau, M., Vernoux, J. P., Henri-Dubernet, S., Guéguen, M. (2008). Safety assessment of dairy microorganisms: The Lactobacillus genus Int J Food Microbiol 126, 278–285.

Briggiler-Marcó, M., Capra, ML., Quiberoni, A., Vinderola, G., Reinheimer, J.A., & Hynes, E. (2007) Nonstarter Lactobacillus strains as adjunct cultures for cheese making: in vitro characterization and performance in two model cheeses. J Dairy Sci 90, 4532-4542.

Briggs, S. S. (2003). Evaluation of lactic acid bacteria for the acceleration of cheese ripening using pulsed electric fields. MSc Thesis, McGill University, Montreal Quebec, Canada.

Broome, M.C., Powel, I. B & Limsowtin, G. K. Y. (2003). Starter cultures: Specific properties. *In* Enyclopedia of Dairy Sciences. Vol I ed. Regisnki, H. Fuquay, J.W. & Fox, P. F. 269 – 275. London: Academic Press.

Buckenhiiskes, H. J. (1993) Selection criteria for lactic acid bacteria to be used as starter cultures for various food commodities. FEMS Microbiol Rev 12, 253-272

Caplice, E., & Fitzgerald, G. F. (1999). Food fermentations: role of microorganisms in food production and preservation. Int J Food Microbiol 50, 131–149.

Carminati, D., Giraffa, G., Quiberoni, A., Binetti, A.,Suarez, V., & Reinhemer, J. (2010). Advances and Trends in Starter cultures for Dairy Fermentation Chapter 10. *In:* Biotechnology of Lactic Acid Bacteria: Novel Applications. Edited by Mozi, F., Raya, R. R., & Vignolo, G. M. Wiley Blackwell Publisher.

Cocolin, L., Rantsiou, K. Iacumin, L., Urso, R., Cantoni, C., & Comi, G. (2004) Study of the ecology of fresh sausages and characterisation of populations of lactic acid bacteria by molecular methods. Appl Environ Microbiol 4, 1883-1894.

Coppola, R., Succi, M., Tremonte, P., Reale, A., Salzano, G. & Sorrentino, E. (2005) Antibiotic susceptibility of *Lactobacillus rhamnosus* strains isolated from Parmigiano Reggiano cheese. *Lait* 85, 193-204.

Delcour, J., de Vuyst, L., & Shortt, C. (1999) "Recombinant dairy starters, probiotics, and prebiotics: Scientific, technological, and regulatory challenges". Int Dairy J (Special issue) 9, 3–80.

European Commission. (2007) Opinion of the Scientific Committee on a request from EFSA on the introduction of a Qualified Presumption of Safety (QPS) approach for assessment of selected microorganisms referred to EFSA. The EFSA J 587, 1-16.

Fox P.F. (1993) Cheese chemistry physics and microbiology. Vol. 1. Chapman and Hall London, 303-340

Furet, J. P., Quenee, P., & Tailliez, P. (2004) Molecular quantification of lactic acid bacteria in fermented milk products using. J Food Microbiol 103, 131–142.

Giraffa, G., Chanishvili, N., & Widyastuti, Y. (2010) Importance of lactobacilli in food and feed biotechnology. Research Microbiol 161, 480-487.

Gomes, A.M., Malcata, F.X. (1998) Development of probiotic cheese manufactured from goat milk: response surface analysis via technological manipulation. J Dairy Sci 81, 1492-507.

Grappin, R., Beuvier, E. (1997) Possible implications of milk pasteurization on the manufacture and sensory quality of ripened cheese. Int Dairy J 7, 751-761.

Hayaloglu, A.A., Guven, M., & Fox, P.F. (2002) Microbiological, biochemical and technological properties of Turkish White cheese, "Beyaz Peynir". Int Dairy J 12, 635-648.

Klaenhammer, T.R., Barrangou, R., Buck, B.L., Azcarate-Peril, M.A., Alterman, E. (2005) Genomic features of lactic acid bacteria affecting bioprocessing and health, FEMS Microbiol Rev 29, 391-409.

Kongo, J.M., Ho, A.J., Malcata, F.X., & Wiedmann, M. (2007) Characterization of dominant lactic acid bacteria isolated from Sao Jorge cheese, using biochemical and ribotyping methods. J. Appl. Microbiology 103, 1838 –1844.

Kongo, J.M., & Malcata, F.X. (2012) Azorean traditional cheesesmaking: a case study pertaining to a unique food chain. In Food Chains: New Research, edited by Melissa A., Jensen & Danielle W. Mueller. Nova Science Publisher Inc, New York.

Korhonen, J. (2010) Antibiotic Resistance of Lactic Acid Bacteria PhD Disseration, University of Eastern Finland.

Kranenburg, R., Kleerebezem, M., Vlieg, J. H., Ursing, B., Boekhorst, J., Smit, B. A., Ayad, E.H.E., Smit, G., & Siezen, R. J. (2002) Flavour formation from amino acids by lactic acid bacteria: predictions from genome sequence analysis. Int Dairy J 12, 111–121.

Law, B. A. (2001) Controlled and accelerated cheese ripening: the research base for new technology. Int Dairy J 11, 383–398.

Leroy, F., & De Vuyst, L. (2004) Lactic acid bacteria as functional starter cultures for the food fermentation industry. Food Sci Technol 15, 67-78.

Liu, G., Wang, H., Griffiths, M.W. & Li, P. (2011) Heterologous extracellular production of enterocin P in Lactococcus lactis by a food-grade expression system. Europ Food Res Technol 233, 123-129.

McSweeney, P.L.H., & Sousa, M.J. (2000) Biochemical pathways for the production of flavour compounds in cheeses during ripening. A review. Lait 80, 293-324.

Mogensen, G. (1993) Starter cultures. *In*: Smith, J. (Ed.), Technology of reduced-additive foods. Blackie Academic and Professional, London, pp. 1-25.

Montel M-C., & Samelis, J.(2010) Microbial Safety of Traditional cheeses. NAGREF WP2A Final TRUEFOOD Conference *New roots for traditional European foods: Possibilities for success and sustainability*. Brussels, 13 April.

Montel, M.C., Talon, R., Fournaud, J., & Champomier, M.C. (1991) A simplified key for identifying homofermentative *Lactobacillus* and *Carnobacterium* spp. from meat. J Appl Bacteriol 70, 469–472

Mooney, J.S., Fox, P.F., Healy, A., & Leaver, J. (1998) Identification of the principal water-solubel peptides in Cheddar cheese. Int Dairy J 8, 813-818.

Parente E., & Cogan T.M. (2004). Starter cultures: General aspects. In: General Chemistry, Physics and Microbiology. Vol I, edited by P.F. Fox., P.J. H. McSweeney, T. M. Cogan and T. P. Guninee. Amsterdam Elsevier.

Parguel P (2004). "Milk flores", group Malbuisson (Doubs), pp. 1-7.

Prabhakar V, Kocaoglu-Vurma N, Harper J, Rodriguez-Saona L. (2011). Classification of Swiss cheese starter and adjunct cultures using Fourier transform infrared microspectroscopy. J. Dairy Sci 94, 4374-4382.

Rank, T.C., Grappin, R., & Olson, N.F. (1985) Secondary Proteolysis of Cheese During Ripening: A Review. J Dairy Sci 68, 801- 805.

Ray, B. (1992) The need for food biopreservation. *In:* B. Ray, & M. Daeschel (Eds.), Food biopreservatives of microbial origin (pp. 1–23). Boca Raton, Florida: CRC Press.

Sandine, W.E., (1996) Commercial Production of Dairy Starter Cultures. *In*: Dairy Starter Cultures, Cogan, T.M. and J.P. Accolas (Eds.). Wiley-VCH, New York.

Smit, G., Smit, B. A., & Engels, W.J.M. (2005) Flavour formation by lactic acid bacteria and biochemical flavour profiling of cheeses products. FEMS Microbiol Rev 29, 591- 610.

Songisepp, E., Kullisaar, T., Hutt, P., Elias, P., Brilene, T., Zilmer, M., & Mikelsaar, M. (2004) A New Probiotic Cheese with Antioxidative and Antimicrobial Activity. J Dairy Sci 87, 2017–2023.

Temmerman, R., Huys, G. and Swings, J. (2004) Identification of lactic acid bacteria: culture-dependent and culture-independent methods. Trends Food Sci Tech 15, 348-359.

Weerkam, A.H., Klijn, N., Neeter, R., & Smit, G. (1996) Properties of mesophilic lactic acid bacteria from raw milk and naturally fermented raw products. Neth Milk Dairy J 50, 319-322.

Wood, B.J.B. (1997) Microbiology of fermented foods. London: Blackie Academic & Professional.

Wood, B.J.B. & W.H. Holzapfel, (1995.) The Genera of Lactic Acid Bacteria. Vol. 2, Blackie Academic and Professional, Glasgow.

Wouters, J. T. M., Ayad, E. H. E., Hugenholtz, J., & Smit, G. (2002) Microbes from raw milk for fermented dairy products. Int Dairy J 12, 91–109.

Yvon, M., & Rijnen, L. (2001) Cheese flavour formation by aminoacid catabolism. Int Dairy J 11, 185–201.

Chapter 2

# Lactic Acid Bacteria Resistance to Bacteriophage and Prevention Techniques to Lower Phage Contamination in Dairy Fermentation

A.K. Szczepankowska, R.K. Górecki,
P. Kołakowski and J.K. Bardowski

Additional information is available at the end of the chapter

http://dx.doi.org/10.5772/51541

## 1. Introduction

The first negative effect of bacteriophages on dairy fermentation was reported in the mid 30s of the XX century [1]. Regardless of sanitary precautions, starter strain rotations and constant development of new phage-resistant bacterial strains, phages remain one of the main and economically most serious sources of fermentation failures. Due to their natural presence in the milk environment, bacteriophages cause problems in industrial dairy fermentations world-wide. Their short latent period, relatively large burst size and/or resistance to pasteurization makes them difficult to eliminate [2]. Phage-induced bacterial cell lysis leads to failed or slow fermentation, decrease in acid production and reduction of milk product quality (e.g. nutritive value, taste, texture, etc.), which in effect cause profound economical losses [3]. An intriguing high number of bacteriophages of *Lactococcus* and *Streptococcus* bacteria reflects the biotechnological interest and engagement of the dairy industry in research on biology of these phages [4].

Since *Lactococcus lactis* strains are widely used as starter cultures for milk fermentation during manufacturing of many types of cheeses, sour cream and buttermilk, bacteriophages virulent against these strains appear commonly in the fermentation environment. It is estimated that 60 – 70% of technological problems in production of cottage and hard cheeses are caused by bacteriophage infection of bacteria from the *Lactococcus* genus [5]. The raise of interest in lactococcal phages due to economical aspects has subsequently led to a more global research on the biology of lactococcal phages, ways of their appearance in dairy environments and means of their elimination as well as characterization of phage resistance mechanisms encoded by bacteria exploited by the industry.

## 2. Lactic acid bacteria used in dairy industry

Lactic acid bacteria (LAB) comprise different groups of microorganisms, such as *Carnobacterium, Enterococcus, Tetragenococcus, Vagococcus, Weissella* as well as species of genera which constitute the "industrial" core of LAB, like *Lactococcus, Lactobacillus, Streptococcus, Pediococcus* and *Leuconostoc* [6]. LAB reside in different natural habitats, including healthy and decaying plants, milk and dairy products, oral cavity and gastrointestinal tract of humans and animals. In addition, lactic acid bacteria can grow on meat and wine. These features are used in the production of fermented sausages (*Lactobacillus, Pediococcus*) and to improve the organoleptic characteristics of wine (*Oenococcus oeni*) [6].

The genus *Lactococcus* is the best characterized food-related LAB. As lactococcal strains are able to grow in milk and transform lactose to lactic acid, they are commonly used as starter cultures in industrial fermentations for cheese production. The ability of LAB to transform raw milk into other products suitable for consumption has been used by man for millenniums. Such long history record of interactions of man with lactic acid bacteria and present knowledge led to assigning these bacteria the GRAS status (generally recognized as safe) [7]. Dairy products and the respective LAB species are gathered in Table 1 based on specifications and recommendations released by the main culture suppliers.

A typical lactococcal mixed starter culture consists of 2-3 well defined strains, which specific properties have significant impact on the texture and flavor of the end product. Nowadays, large dairy plants process up to $10^6$ liters of milk per day, producing annually approximately $10^7$ tons of cheese [8]. Therefore, technological problems in production of cottage and hard cheeses caused by bacteriophage infections have serious economical consequences.

## 3. Lactic acid bacteria phages – history background, morphology, classification

The history of discovery of bacteriophages originates in the research of Felix d'Herelle and Frederick Twort in the beginning of the XX century and further development of phage biology studies spans the fourth quarter of the last century. Bacteriophages (phages) are defined as viruses that exert their activity against prokaryotic cells – both bacterial as well as archeal.

The name "bacteriophage"derives from the Greek word "phagein", meaning "to eat", which points to their destructive action. Bacteriophages exist in two states – extra- and intracellular – which place them half-way between live organisms and non-viable forms. As obligate intracellular parasites their survival is dependent on host organisms. Phage "life functions", such as genome replication and synthesis of capsid components, are restricted to occur within infected cells. Outside of the host phages are regarded as metabolically inert, unable to carry out neither biosynthetic nor respiratory functions.

Phages intrigue by their simplistic organization and submicroscopic sizes. These infectious particles consist of a single- or double-stranded nucleic acid genome (DNA or RNA), enveloped in a protein structure (capsid). Current taxonomy and classification of bacteriophages rely on the type of nucleic acid genome and phage morphology, physiology (temperate and virulent life cycles) and genomics. Taxonomy of viruses is supervised by the International Committee for Taxonomy of Viruses (ICTV) that imposes rules for names and writings.

| **Product** | **LAB species** |
|---|---|
| Yoghurt | *Streptococcus thermophilus, Lactobacillus delbrueckii* subsp. *bulgaricus* |
| Cottage cheese, Cheddar, Pasta Filata | *Lactococcus lactis* subsp. *lactis, Lactococcus lactis* subsp. *cremoris*<br>*Streptococcus thermophilus** |
| Tvarog, blue cheese | *Lactococcus lactis* subsp. *lactis, Lactococcus lactis* subsp. *cremoris*<br>*Lactococcus lactis* subsp. *lactis* var. *diacetylactis, Leuconostoc mesenteroides* subsp. *cremoris*<br>*Leuconostoc mesenteroides* subsp. *mesenteroides* |
| Butter milk, fermented cream, butter | *Lactococcus lactis* subsp. *lactis, Lactococcus lactis* subsp. *cremoris*<br>*Lactococcus lactis* subsp. *lactis* var. *diacetylactis* |
| Ryazanka | *Streptococcus thermophilus*** |
| Cheddar, Feta | *Lactococcus lactis* subsp. *lactis, Lactococcus lactis* subsp. *cremoris* |
| Mozzarella, Pizza cheese | *Streptococcus thermophilus, Lactobacillus delbrueckii subsp. bulgaricus* |
| Masdamer, Gouda, Edam, Tilsitter, soft mould ripened cheese, quark, fermented milk beverages | *Lactococcus lactis* subsp. *lactis, Lactococcus lactis* subsp. *cremoris*<br>*Lactococcus lactis* subsp. *lactis* var. *diacetylactis*<br>*Leuconostoc mesenteroides* subsp. *cremoris* |
| Mozzarella, Swiss, stabilized soft mould ripened cheese | *Streptococcus thermophilus* |
| Swiss, Grana | *Lactobacillus helveticus, Lactobacillus delbrueckii* subsp. *lactis* |
| Fermented cream, fermented milk beverages | *Lactobacillus acidophilus, Streptococcus thermophilus* |
| Actimel®-like products | *Lactobacillus casei, Lactobacillus paracasei* subsp. *paracasei*<br>*Lactobacillus rhamnosus, Lactobacillus acidophilus* |
| Swiss, Italian | *Lactococcus lactis* subsp. *lactis, Lactococcus lactis* subsp. *cremoris*<br>*Lactococcus lactis* subsp. *lactis* var. *diacetylactis*<br>*Lactobacillus helveticus, Lactobacillus delbrueckii* subsp. *lactis* |

* seldom applied in cottage cheese, ** texturizing strains

**Table 1.** Various dairy products and LAB species applied in their production.

The majority of known viruses are bacteriophages, which infect cells of Eubacteria and Archaea. It is also accepted that most phages (96%) isolated so far belong to one taxonomic order of *Caudovirales* [9]. Bacteriophages within this order contain tails and a linear dsDNA genome. They are further classified into three phylogenetically linked families of: *Myoviridae, Siphoviridae, Podoviridae* [9]. *Myoviridae* phages contain a long and contractile tail, while *Siphoviridae* and *Podoviridae* are equipped with a non-contractile tail, long and short, respectively [10]. Isometric heads are dominating (85%) in the morphology of phages from all three families [11]. It is worth to mention that 61% of known phages are classified into the *Siphoviridae* family, of which most of them infect strains of enterobacteria (906 phages), *Lactococcus* (700), *Bacillus* (380) and *Streptococcus* (290) [9]. Apart from the tailed *Caudovirales* phages, there are others demonstrating filamentous, pleomorphic or polyhedral morphology.

Bacteriophages, although simple in organization, are the most diverse life forms in the biosphere. Their apparent heterogeneity is reflected by various features – both morphological as genetic, and their persistence on Earth, estimated as high as $10^{31}$, outnumbers by far their bacterial hosts [12]. Phages inhabit various niches, like oceans [13], thermal waters [14], gastrointestinal tract [15] and superficial ecosystems created by man, including fermentation tanks in dairy industry [16]. Hence, their impact on the microbial world cannot be underestimated.

Bacteriophage genome structure, indicating linear and double-stranded characteristics of the DNA molecule, containing or not cohesive ends and sometimes presenting terminal redundancy and circular permutation, describes the general feature of LAB phage genomes.

## 4. Molecular mechanisms of phage infection of LAB

To enter the host, phages firstly come in contact and adsorb to the bacterial cell wall. The adsorption process has been well studied in Gram-negative bacteria, where it was found that two components are involved in the phage-host interaction. One of them is a receptor located in the bacterial cell envelope (membrane or wall), whereas the second component, called the receptor binding protein (RBP), is presented on the phage surface. RBP is responsible for recognition and binding of the phage particle to the bacterial receptor [17]. In the first stage of phage infection, the RBP protein recognizes and binds to a suitable sugar receptor. However, such binding is reversible and thus, the initial phage-bacteria interaction does not ensure commencement of a successful infection event. In contrast to this, in the second stage, a stable phage attachment to the bacterial cell occurs due to an irreversible binding between proteins located on bacterial and phage surfaces [18]. Both stages of adsorption are observed in Gram-positive bacteria: phages that attack *Lactococcus lactis* cells bind to specific receptors, mainly sugars, located in the cell wall. It is widely known that rhamnose, glucose, galactose, and galactosamine are compounds with which the phage RBP interacts at the initial stage of adsorption [19]. In the case of *Lactococcus* c2-type phages, effective infection requires interaction between phage and the bacterial protein Pip (phage infection protein) [20]. The Pip protein of *L. lactis* is an integral membrane protein [21] and its interaction is crucial both for establishing the reversible and irreversible contact between the phage and the host. In contrast to c2-type phages, phages representing P335 and 936 groups bind to other various bacterial membrane proteins and have been examined in a lesser extent [22]. After establishing a tight connection, they inject their genetic material inside the host cytoplasm, while the capsid remains outside the cell. Then, subsequent steps of phage infection are effectuated which follows either the lytic or lysogenic life cycle. Phages entering the lytic mode immediately redirect the host replication machinery and metabolic functions to replicate its own genetic material and synthesize phage encoded proteins. In effect, abundant amounts of progeny particles are produced. Phages executing only the lytic cycle are designated as virulent and their infection implicates cell death. Yet, certain phages termed as temperate can lead an alternating existence between a dormant state inside the bacterial cell and lytic growth. These phages can exist in the cell in a latent form for generations, replicating in synchrony with the bacterial chromosome. A dormant form of the phage is called a prophage and leads a lysogenic life cycle in a bacterial host strain, which is regarded as a lysogen. Conversion from the lysogenic life cycle to the lytic often occurs spontaneously or can be induced by various mutagens (UV, mitomycin).

## 5. Phage sensitivity of LAB starters used in dairy industry

Virulent phages of *Lactococcus lactis* spp. are the most frequently encountered phages in milk plants during cheese and dairy beverages production. Additionally, phages attacking *Streptococcus thermophilus* are often observed in cheese and less distinct in yoghurt manufacturing. Phages against *Lactobacillus* spp. and *Leuconostoc* spp. starter cultures represent a minor problem [23]. Currently, in production of dairy beverages functional *Lactococcus* and *Streptococcus thermophilus* texturizing strains with ability to produce exopolysaccharides (EPS) are commonly used. In nature it is very difficult to find strains with similar rheological properties differing in resistance to phages. Thus, phage contamination of texturizing strains can lead to serious problems in ensuring quality dairy products.

## 6. Defense mechanisms of lactic acid bacteria

It is well documented that lactic acid bacteria evolved defense systems against bacteriophages, which allow them to survive in an environment full of their predators. These anti-phage systems have been organized into five groups depending on the manner by which they operate: (i) inhibition of phage adsorption, (ii) blocking of phage DNA injection, (iii) restriction modification systems, (iv) phage abortive infection systems, and finally, the most recently described, (v) CRISPR/*cas* systems. The knowledge about natural phage resistance mechanisms together with a set of genetic tools were applied to develop also (vi) engineered defense systems that confer higher levels of resistance and/or broader phage specificity.

### 6.1. Inhibition of phage adsorption

Basic mechanisms of inhibition of phage adsorption to the bacterial cell are associated either with physical masking of the receptor or with changes in its structure, or even with its absence in the cell envelopes [24]. Lack of a functional receptor might be due to spontaneous mutations in the genetic material, leading in turn to bacteriophage insensitive mutants (BIM). A good illustration of the BIM phenomenon is a lactococcal mutant in the chromosomally-encoded *pip* gene. The resultant strain is unable to interact with phages of the c2 group, revealing high level of c2-specific resistance [24] (for further details on BIMs see section 12.2.).

Mechanisms preventing phage adsorption are not only mediated by the bacterial chromosome, but also by acquired plasmids. The best documented plasmid-encoded mechanisms of inhibition of phage adsorption rely on either direct synthesis of cell surface antigens or the production of extracellular carbohydrates. Of the two modes of action, the former reveals phage specificity, whereas the latter seems to restrict access to the bacterial cell for various harmful factors, including bacteriophages [25]. Studies carried by Tuncer and Akcelic demonstrated that a 28.5-kb plasmid, isolated from *L. lactis* subsp. *lactis* MPL56, causes complete inhibition of four lactococcal phages due to the production of a 55.4-kDa protein [25]. The protein exhibits similarity to lectins, a group of proteins that adsorb to specific monosaccharide components of polysaccharides in the cell wall, hence, impairing specific recognition of the phage receptor sites by these four phages. Thus, this plasmid-encoded 55.4-kDa protein shields specifically the galactose-containing receptor rather than interacts with the phage, in other words, the bacterial lectin and the phage RBP compete for the receptor [25]. Another example of physical masking of the receptor is the plasmid-mediated production of extracellular carbohydrates, called exopolysaccharides (EPS) [26]. Such EPS envelope coats the cell surface giving bacteria extra protection, not only against bacteriophages, but also against desiccation. There is some evidence that EPSs contain sugar residues that are similar or even identical to initial phage receptors. Therefore, phage insensitivity of LAB strains that carry EPS-encoding plasmids, for instance, pCI658, might be due to phage immobilization by binding to EPS [26]. On the other hand, polysaccharides have an impact on the properties of dairy products, like: texture, viscosity and smoothness of mouthfeel. Thereby, application of EPS-producing phage-resistant strains might be limited to a narrow range of dairy products [25-26].

## 6.2. Blocking of phage DNA injection

After phage binding to the receptor, phage DNA is introduced into the bacterial cell. In the cytoplasm, phage genetic information is amplified and consequently progeny particles are produced. However, studies of Watanabe on the interaction between phage PL-1 and a *Lactobacillus casei* strain showed no bacterial lysis, despite phage adsorption to cell envelopes [27]. An electron microscopy image indicated that the phage DNA remains intact in the capsid. In contrast to this, a significant increase in the number of empty capsids was observed on the surface of the sensitive strain. In the light of this evidence, it is obvious that phage DNA injection might be interrupted, although the adsorption of phages to the cell surface occurred. Intensive attempts to elucidate the injection blocking phenomenon have allowed identifying different Sie (superinfection exclusion) or Sie-like systems. On the other hand, only few of them have been well characterized [28]. Therefore, the mechanism preventing entry of phage DNA to the cell is still poorly understood, both in LAB and other microorganisms. Surprisingly, it was discovered that most *sie* genes are located within the prophage regions of the bacterial chromosome [28]. However, the first lactococcal injection blocking system was identified on the pNP40 plasmid, which blocks DNA penetration specifically for $\phi$c2 phage of the lactococcal c2 phage group [29]. As it was described in the previous section, the membrane Pip protein is essential for c2 adsorption to *Lactococcus lactis*. It was speculated that the pNP40-encoded protein product might have an impact on the activity, production, or membrane insertion of Pip, thereby affect its biological function and prevent phage DNA entry [29]. The first description of a *sie* system of *Lactococcus* was published in 2002 and referred to the P335-type temperate lactococcal bacteriophage Tuc2009 [30]. After integration of the bacteriophage Tuc2009 genome into the lactococcal chromosome, the prophage protein $Sie_{2009}$ is produced and blocks superinfecting phage DNA entry into the cell. The blocking mechanism has not been fully elucidated; nevertheless, it has been proposed that $Sie_{2009}$ interacts with factor(s) responsible for initiating the phage DNA release from the capsid. Alternatively, the $Sie_{2009}$ protein might interact with cell membrane proteins that are essential for DNA translocation. The effect of $Sie_{2009}$ seems to be analogous to the effect of the lysogenic phage repressor (CI) preventing re-infection. In contrast, the presence of the *sie*$_{2009}$ gene determines resistance to various phages, also to phages from other species [28,30]. Similarly to lactococci, in lactobacilli prophages are also a common phenomenon [31]. Comparative genomics of lactobacilli revealed the presence of genes coding for putative proteins with a close sequence match to a surface-exposed lipoprotein encoded by bacteriophage TP-J34 of *Streptococcus thermophilus*, another bacterial species used in industrial milk fermentation processes. The TP-J34 prophage carries a Sie-like system consisting of the *ltp* (lipoprotein of temperate phage) gene, encoding a surface-exposed lipoprotein of biologically proven phage-resistance functions. In view of the fact that the *sie* genes of lactic acid bacteria are located on lysogeny modules of prophages and confer infection exclusion, they have been termed phage-derived phage resistance systems [32].

### 6.3. Restriction modification systems

Following successful injection of DNA, phage infection might be completed or hindered by the presence of restriction modification systems (RM). RM systems comprise two activities represented by the following enzymes: endonuclease (restriction) and methyltransferase (modification) [33]. Simultaneously, both activities are specific to the same target sequences. The endonucleolytic activity is responsible for degradation of invading foreign DNA, including phage DNA, which lack a unique methylation pattern, while the methyltransferase activity protects the host DNA against degradation by introducing a methyl group into a specific nucleotide of the target site [34]. In detail, phage DNA usually reveal different methylation patterns than those recognized by innate RM systems. Unmethylated target sequences are significantly susceptible to endonucleolytic attack, resulting in DNA degradation [35]. Such mode of action guarantees that the presence of RM systems limits phage proliferation in the cytoplasm, causing no harm to the cell. RM systems are classified into four groups, based on their molecular structure, co-factor requirements, sequence recognition and cleavage position [34-36].

### *6.3.1. Type I RM*

Type I is the most complex RM system in terms of genetic organization and biochemical activity. It is composed of three different *hsd* (host specificity determinant) genes coding for the following subunits: HsdR - responsible for restriction, HsdM - involved in modification and HsdS - responsible for specific sequence recognition. None of them reveals any activity as a single protein [36]. In order for the modification activity to occur, a combination of one HsdS and two HsdM subunits is required. The $M_2S_1$ multifunctional enzyme acts as protective methyltransferase, which modifies DNA through the transfer of the methyl group from S-Adenosyl-methionine (AdoMet) to the specific adenines in the recognition site [36,38]. For restriction activity, all subunits are absolutely required in a stoichiometric ratio of $R_2M_2S_1$. This holoenzyme exhibits both endonucleolytic and helicase activities, and is active only in presence of $Mg^{2+}$, AdoMet and ATP [36].

Besides the complex structure of this multifunctional enzyme, also structure of the recognized sequences and cleavage position are the distinguishing features of type I RM systems. Type I RM enzymes specifically recognize asymmetric and bipartite sequences. These non-palindromic DNA sequences consist of two specific components, one of 3-4 bp and the other of 4-5 bp, separated by a 6-8 bp non-specific sequence [34,36-37]. The innate methylation state of the target sequence determines the activity of the multifunctional $R_2M_2S_1$ enzyme. When the target sequence is methylated or semi-methylated (e.g. just after replication), the enzyme will exhibit activity of a methyltransferase, which completes DNA modification. In contrast, if the holoenzyme binds to an unmethylated recognition site, DNA translocation past the DNA-enzyme complex occurs in an ATP-dependent manner [35,38]. In spite of DNA translocation, the enzyme remains bound to the target site. DNA is cleaved at a position, where either collision with another translocating complex has appeared or translocation is halted due to the topology of the DNA substrate. Consequently, type I restriction enzymes cleave DNA randomly at a nonspecific site, far from the recognition sequence [38].

Interaction between subunits, leading to formation of multifunctional enzymes as well as interaction of resultant enzyme molecules with DNA, are determined by the structure of the HsdS subunits. HsdS subunits consist of regions, which amino acid sequences are conserved within an enzyme family, and two independent target recognition domains (TRD) that share low level of amino acid identity [34,39]. TRDs are involved in target sequence recognition, each TRD recognizes one-half of the split target site and is responsible for DNA binding. Since TRDs are highly variable, they recognize multiple target sequences, and thus, provide a variety of phage resistance types [34,36,39]. The central domain, located between two TRDs, is responsible for interaction with one HsdM subunit. Other conserved regions located at N and C termini have been proposed to form a split domain, which makes contact with a second HsdM subunit [35,37].

Type I systems have been further classified into four families based on genetic and biochemical criteria, such as: gene order, identity at amino acid level, complementation assay and enzymatic properties. RM systems belonging to type IA, IB, and ID are only chromosomally-encoded, while most complete type IC systems are either chromosomal or carried on large conjugative plasmids [36]. Additionally, numerous small plasmids carry the *hsdS* gene alone [34,40]. While all subunits belong to the same subtype, a plasmid-encoded HsdS protein is able to form a multifunctional enzyme with chromosomally-encoded HsdM and HsdR subunits [41]. Thus, acquisition of a new *hsdS*, revealing new sequence specificity, leads to the increase of phage resistance.

Among LAB, type IC systems seem to be most widespread. Type IC RM loci of both *L. lactis* IL1403 and *L. cremoris* MG1363 consist of three genes: *hsdR*, *hsdM*, *hsdS*, and two promoters, one for transcription of *hsdR* and the other for transcription of both *hsdM* and *hsdS* [17,42]. Nevertheless, there is no clear evidence for transcription regulation of type I RM enzymes [42]. Under these circumstances, an unmodified chromosome is exposed to endonucleolytic digestion after acquisition of either a new system or just the subunit specificity genes. It was observed that a delay in the appearance of restriction activity, which ensures the survival of recipient cells in the absence of complete modification of chromosomal target sites, depends on host function [36,43]. Chromosomally-encoded energy-dependent proteases ClpP and ClpX, co-operating in a complex, are implicated in the regulation of restriction activity [36]. The ClpXP complex is responsible for restriction alleviation through proteolytic degradation of HsdR subunits. Based on results of Janscak and colleagues concerning the EcoR124I endonuclease, an alternative mechanism of delay in restriction alleviation has been proposed. As each of the two HsdR subunits interacts differently with HsdM, it has been postulated that the control of restriction activity is implemented at the level of subunit assembly [38]. Formation of a weak $R_2M_2S_1$ restriction complex will be suspended, unless accumulation of HsdR molecules occurs. Excess of HsdR over HsdM is observed in the late stage of establishing of the RM system in a recipient cell; hence, the unmodified chromosome is protected against premature restriction activity [38].

### 6.3.2. *Type II RM*

In contrast to type I, type II RM systems are structurally the simplest of all restriction modification systems. They are generally encoded by two genes, but the key defining feature of this RM type is the independent activity of restriction and modification enzymes [33]. Methyltransferase is active as an asymmetric monomer, requires only AdoMet, and recognizes the same target sequences as the cognate endonuclease. In contrast, restriction endonuclease is a homodimer and requires divalent $Mg^{2+}$ cations for proper activity. Endonucleases generally recognize a palindromic 4-8 bp DNA sequence and cleave within or in a fixed distance of the recognition site. In contrast to type I, ATP has no effect on the cleavage activity of type II endonucleases [44].

As this RM type is more heterogeneous in respect to endonucleolytic activity than originally thought based on their structural simplicity, the described mode of action refers mainly to typical (orthodox) type II endonucleases [45].

Apart from the orthodox type (called IIP ), type II restriction enzymes have been categorized into the following subclasses: IIA, IIB, IIC, IIE, IIF, IIG, IIH, IIM, IIS and IIT. Endonucleases of these subclasses differ in structure of the recognized sequence (asymmetric or symmetric), cleavage positions and cofactor requirements. Type IIA endonucleases behave similarly to the orthodox class, but recognize asymmetric sequences [45]. The unique feature of subclass IIB refers to the cleavage position. These endonucleases cut DNA from both sides, which results in complete extraction of the target sequence from the DNA molecule [46]. Subclasses IIC and IIE have both modification and restriction domains present in one polypeptide. Additionally, class IIE endonucleases interact with two copies of their recognition site, one copy being the target for cleavage, the other serving as an allosteric effector [47]. Similarly to subclass IIE, class IIF restriction enzymes interact with two copies of their recognition sequences, but cleavage occurs at both sequences. Type IIG restriction enzymes seem to combine properties of both IIB and IIC subclasses. The methyltransferase activity of class IIG, like IIB, is stimulated by AdoMet. The main similarity between IIG and IIC is that they both have restriction and modification activities located on one polypeptide chain [45,47]. Subclass IIH, represented by the AhdI system, appears to be a novel RM system due to its genetic organization resembling that of type I. As in type II systems, the AhdI endonuclease is encoded by a single gene; on the other hand, similarly to type I, its cognate methyltransferase forms a complex consisting of two modification and two specificity subunits [44,48]. Subclass IIM is at the opposite extreme from other type II subclasses as it recognizes and cleavages methylated target sequences. The key distinguishing feature of type IIS is the cleavage position outside of the recognition sequence at a defined distance [49]. Subclass IIT is an example of a variation in the typical genetic organization of type II RM systems, as the endonuclease is composed of two different subunits. Moreover, some IIT endonucleases function not only as heterodimers, but also as heterotetramers [44-45].

As enzymes belonging to type II systems are the most abundant and mainly encoded on plasmids, they can be acquired by the bacterial cell through plasmid transfer events. Therefore, a question arises as how to protect the host cell against an incoming endonuclease. In many cases, each gene of the type II RM system has its own promoter. Thus, a delay in appearance of the endonuclease activity is regulated at the transcriptional level. The lactococcal LlaDII RM system is a good example which illustrates this type of regulation [50]. At the initial stage of establishing in the host cell, the LlaDII methyltransferase is overexpressed, whereas the restriction enzyme is produced in small amounts due to the weak constitutive expression of its gene. On the other hand, a permanently high concentration of methylases is an unfavorable circumstance due to possible methylation and therefore protection of the invading phage DNA. The LlaDII methyltransferase contains HTH motifs, which were shown to be engaged in direct interaction with its promoter sequence, causing silencing of its own gene expression [50].

### 6.3.3. *Type III RM*

Unlike types I and II, type III systems are less spread among lactic acid bacteria. The *Lla*FI system identified on the lactococcal pND801 plasmid is the first type III RM system described not just in LAB, but generally in Gram-positive bacteria [51]. Based on computational analyses of genome sequences, type III systems were observed to occur also in lactobacilli (for instance *Lactobacillus johnsonii* and *Lb. rhamnosus*) [52]. On the one hand, type III resemble type II systems in their structural and genetic organization. Type III, like type II systems, consists of two genes, one encoding a methyltransferase (Mod) and the other - an endonuclease (Res). Mod is responsible for binding and methylating the recognition sequences, regardless of the presence of Res. On the other hand, type III systems are similar to type I, in respect to endonuclease activity, as the Res subunit is only active in a complex with Mod. Another basic similarity to type I systems is the fact that they both comprise the helicase domain and require both AdoMet and ATP for full restriction activity. The distinctive features characterizing type III systems concern recognition sequences and cleavage sites. The Mod subunit recognizes asymmetric, opposite-oriented sequences and methylation takes place only on one strand of the DNA [53]. The Res endonuclease cuts both strands of the DNA at the distance of 24-27 nucleotides downstream of the unmethylated specific sites [53].

Lactococci have been found to possess three types of RM systems: type I, II and III. Based on genomic sequence data, it is evident that RM genes are both chromosomally- and plasmid-encoded. However, a variety of RM determinants is generally associated with plasmids [17]. In contrast, very few phage defense mechanisms have been described for *S. thermophilus*. In 2001, Solow and Somkuti reported on the discovery of a complete type I RM system encoded on a streptococcal plasmid pER35 [54]. Further progress in genome sequencing led to finding complete type I and III RM systems in chromosomes of *S. thermophilus* strains. Genome sequence analyses revealed that lactobacilli, like lactococci and streptococci, possess in their chromosomes three types (I-III) of RM systems [55].

### 6.3.4. *Type IV RM*

To date, no type IV RM systems has been distinguished in lactic acid bacteria. It is highly likely that in the future members of this class will be discovered in LAB. For that reason as well as from the evolutionary point of view, the type IV RM system is worth mentioning. A fusion of genes coding for Mod and Res subunits of type III systems was a key step for evolution of type IV RM [56]. The resulting endonuclease (revealing also methyltransferase activity) has an asymmetrical recognition sequence and cleavage occurs at a fixed distance from the recognition site, like for the type IIS enzymes. On the other hand, this endonuclease requires AdoMet, which distinguishes it from type II endonuclease activity. Therefore, taking into account the enzymatic features of model type IV *Eco*57I and *Bse*MII endonucleases, it has been hypothesized that type IV endonucleases are an intermediate between type III and type IIS enzymes.

In summary, it has been well documented that phage restriction-modification systems are widely spread among lactic acid bacteria. Nevertheless, comparative genomics of LAB demonstrated that bacteria representing different niches vary in the presence of restriction-modification genes. The lack of RM systems is a common feature for LAB isolated from the gut, whereas the presence of RM genes is a typical feature for dairy species. Therefore, it was proposed that genes constituting the restriction-modification systems, together with certain genes of sugar metabolism and the proteolytic system, constitute "a barcode" of genes, which can indicate the ability of the microorganism to occupy either dairy or gut niches [57].

## 6.4. Phage abortive infection systems

When the RM systems fail in protecting the bacterium against invading phage DNA, initiation of the phage propagation cycle occurs. However, proliferation of progeny particles might be dramatically limited due to systems that abort the infection at various points of the phage cycle. Abortive infection mechanisms (Abi) have different targets in the cell. They are able to interrupt phage DNA replication, transcription, protein synthesis, phage particle assembly or induce premature cell lysis [17,58]. The Abi mechanisms have been found in many bacterial species, including *Escherichia coli, Bacillus subtilis, Streptococcus pyogenes, Vibrio cholerae* and *Lactococcus lactis* [58]. The most known Abis have been found in the latter species. To date, 22 lactococcal Abi mechanisms have been identified and designated into various groups distinguished by a subsequent letter of the alphabet [58-60]. Most of Abi systems are plasmid-encoded and only three are located on chromosomal DNA (*abi*H, *abi*N, *abi*V) [60]. For instance, *abi*N is located in a prophage region of the *L. lactis* subsp. *cremoris* MG1364 genome and exhibits significant similarity to a corresponding region of the lactococcal temperate phage rlt [61]. Abi systems present simple genetic organization. The Abi phenotype is most frequently encoded by a single gene; however, more complex structures have been identified in six systems. AbiE, AbiG, AbiL, AbiT and AbiU are encoded by two genes, whereas AbiR is the only system identified until now that is encoded by three separate genes [58, 62-63]. Proteins encoded by *abi* genes are cytoplasm-located, where they reveal their activity. In contrast, the AbiP system is represented by a membrane-anchored protein [64].

Abi systems reveal a variety of modes of action. However, in many cases, mechanisms of action of the individual systems were not fully elucidated. Some Abis, like AbiA, AbiD1, AbiF, AbiK, AbiP and AbiT, have been found to interfere with DNA replication, whereas AbiB, AbiG and AbiU arrest mRNA synthesis or have a negative impact on stabilization of transcripts. Haaber and colleagues presented that the AbiV system strongly affects translation of both early and late phage proteins, shortly after infection. Based on this observation, it was concluded that the AbiV system arrests the bacterial translation apparatus [60]. AbiE, AbiI, AbiQ and AbiZ systems affect maturation of phage particles [59,65]. The AbiZ system, identified in 2007 by Durmaz and Klaenhammer, induces premature lysis of phage-infected cells, resulting in the release of the developing phage particles before completion of the maturation process. The timing of phage lysis is controlled by the phage holin protein; thus, AbiZ might interact cooperatively with the phage holin or with a holin inhibitor to make it active prematurely [59].

While the mechanism of cell death in the AbiZ system is self-explanatory, in case of other Abi systems is poorly elucidated. The most likely explanation for this phenomenon is that Abi proteins interfere with processes essential not only for phage, but also for bacterial development; therefore, death of individual bacterial cells is always observed following activation of the Abi systems [17,58-59]. As a consequence, release of progeny particles is limited and the bacterial population survives. Hence, the Abi systems constitute a barrier against bacteriophage proliferation, in which "altruistic suicide" of infected bacterial cells provides protection of the whole uninfected population [17,58].

### 6.5. CRISPR/*cas* systems

Another naturally-occurring distinct phage defense system recently described in Prokaryotes is CRISPR/*cas*. Besides RM mechanisms, this system is also directly engaged in protecting bacterial cells against invading genetic elements, such as phages or plasmids [66]. In brief, CRISPR-conferred phage resistance relies on incorporation of short phage-derived sequences within specific loci of the bacterial genome. In effect, the bacterial cell becomes immune to phages which carry homologous sequences.

CRISPR/*cas* systems are composed of two specific determinants: (i) clustered regularly interspaced short palindromic regions (CRISPR array) and (ii) regions encoding CRISPR-associated (Cas) proteins. The CRISPR arrays consist of non-coding sequences composed of unique phage-derived spacers (21-72 bp) separated by short direct repeated sequences (21-48 bp) of bacterial origin. The length of spacers and repeats within a single array is always the same, while their number may vary from 2-375, depending on the species. On the other hand, Cas proteins constitute a heterologous group of proteins, which contain various functional domains, e.g. typical for nucleases, helicases, nucleic acid binding proteins, etc. [66]. The specific role of individual Cas proteins vary as they were shown to be engaged at various stages of CRISPR-conferred resistance. Interestingly, *cas* genes were detected only in CRISPR-containing genomes, suggesting their tight association. The number of *cas* genes within a CRISPR locus varies from 4 to 20 [67]. Their position can be either upstream or downstream of repeat-spacer units, but always from the same side for a given CRISPR locus type. The CRISPR array and Cas-encoding genes are separated by an A-T rich leader region, suggested to be the promoter region of CRISPR transcription; yet, mechanisms regulating expression still remain to be elucidated [68]. Together these two elements, CRISPR spacer-repeat array and Cas proteins, provide "immunity" to the bacterial cell against invading foreign DNA molecules, including phages (for detailed review see: [67-69]). CRISPR arrays are widely distributed within the Prokaryotic world and are detected in the genomes of 40% of Bacteria and 90% of Archea [70]. Depending on the species, a single genome can carry up to 18 CRISPR loci, which are suggested to confer resistance to various phages [66].

The mechanism of CRISPR/*cas* conferred protection of bacterial cells against phage infection is rather complex and can be divided into three main stages: (i) adaptation, (ii) CRISPR expression and (iii) CRISPR-mediated interference. The first stage relies on incorporation into the bacterial genome within the CRISPR locus of short phage-derived fragments (proto-spacers). Despite the fact that the exact mechanism of spacer acquisition is not known, it is not accidental. Recognition of specific phage sequences for integration is suggested to be linked with sequences termed PAMs (proto-spacer adjacent motifs), located up- or downstream of the proto-spacer. Integration of new spacers occurs from the end of the leader region, between the palindromic repeats and involves certain Cas proteins. Stage 2 is CRISPR expression, which involves transcription of the whole CRISPR spacer-repeat array (pre-mRNA). The presence of palindromic repeat sequences within the transcript, leads to formation of secondary hair-pin like structures. These are subsequently processed into short CRISPR RNAs (crRNAs) by endonucleolytic digestion at a cleavage site located downstream from the last nucleotide forming the hairpin. Finally, the last stage of CRISPR/*cas* activity is based on interaction of mature crRNAs with invading foreign DNA elements (phages), which leads to silencing/degradation of the latter by a certain group of Cas proteins. By this activity, CRISPR/*cas*-carrying hosts are protected from invasion by phages carrying sequences homologous to the integrated spacers. Application of the CRISPR/*cas* system for developing novel phage resistant dairy starter strains may be an attractive alternative, which will be discussed in further parts of this chapter (see section: 12.4.).

## 6.6. Engineered defense systems

Besides the naturally-occurring defense mechanisms against recurrent phage infections (discussed above), new methods involving molecular techniques are designed to combat phages. The constantly growing knowledge on phage development and their genome sequences allows currently to develop engineered defense systems, which are otherwise not encountered in nature (for review see also: [71]). The idea of such systems relies on engineering bacterial strains in a way which impairs genes vital for phage development, e.g. phage replication proteins or other replication factors. Moreover, identification of homologues of these crucial genes within multiple phage genomes allows creating broad-range phage defense systems. As presented below, numerous studies deliver clear evidence that such engineered systems provide efficient protection against phage infections. The following parts of this chapter will delineate each of these systems in more details. Studies on developing engineered systems for lactic acid bacteria were performed in most part in *Streptococcus thermophilus* and *Lactococcus lactis* as strains from both species find wide applications in dairy fermentation processes.

### *6.6.1. Antisense RNA-based phage defense systems*

Bacterial-engineered expression of antisense RNA directed against phage transcripts has been described as one of the most efficient phage defense systems. The mode of action of such RNAs is hybridization to phage sense strand RNAs upon infection. By these means the system interferes with the phage life cycle, inhibiting translation of essential phage genes or degradation of their mRNAs [72].

An example are systems developed in *Streptococcus thermophilus,* which were shown to provide protection against Sfi21-type phages, including κ3 [73-74]. These systems are based on expression of antisense RNAs against genes from the replication module of the Sfi21-type phage κ3 genome, e.g. putative primase (*pri3.1*) or helicase (*hel3.1*) genes. Strong conservation of the whole replication module among the already sequenced Sfi21-type phages makes it a good target for inhibiting phage development [73]. Moreover, hybridization studies revealed that the Sfi21-type replication module is commonly encountered in majority of industrially isolated phages. This reinforced the choice to use it as a phage defense element [73,75].

To test the efficiency of the Sfi21-type module antisense RNA system, constructs expressing antisense RNA cassettes of different length were introduced into *S. thermophilus* strains, which were then challenged with phage infection. The most effective were constructs expressing antisense RNA covering the whole region of target (primase or helicase gene). Also shorter RNAs provided sufficient phage resistance, which was speculated to be due to the presence of specific structural or potential regulatory domains within these fragments. Furthermore, in case of constructs harboring antisense RNAs of similar length, more efficient were usually those comprising the RBS (ribosome-binding sequence) sites. Such effect was believed to be due to the fact that the antisense RBS sequences prevented gene translation by impeding efficient loading of ribosomes onto phage mRNAs [72]. Overall, expression of phage antisense RNAs in *S. thermophilus* was shown to interfere/delay the intracellular phage DNA replication, decrease phage plaque formation (EOP, efficiency of plating), lower the abundance of phage sense mRNA transcripts and reduce phage progeny particles released from infected cells [73,75].

Similar systems were also developed in *Lactococcus lactis* by expressing anti-sense RNAs directed against various phage genes (e.g. P335-type *gp18C* and *gp24C, gp15C* alone, or putative replication genes, 936-type phage F4-1 major coat protein (*mcp*) gene) [76-79]. In these cases, similarly as for *S. thermophilus* systems, the most efficient antisense RNAs in inhibiting phage development were those comprising the RBS site.

Current data allow to conclude that the most effective antiRNA-based phage defense systems, apart from some exceptions, are those which target: (i) genes vital for phage development (e.g. involved in synthesis of phage DNA), (ii) preferably early-expressed phage genes, (iii) genes expressed at low levels, (iv) genes which respective transcripts are unstable [73,79]. Sequencing of novel phage genomes and development of comparative genomics allows identification of other conserved phage genome regions that could serve as potential targets of antiRNAs.

### *6.6.2. Origin-derived phage-encoded resistance*

Defense systems that employ elements derived from lytic phage genomes are termed phage-encoded resistance (PER). One type of engineered PER systems is based on the origin (*ori*) of phage replication [71]. The principle of such systems relies on presenting *in trans* false targets (in this case, phage-derived *oris*), which titrate phage replication factors and make them inaccessible for the phage. In result, phage development is inhibited due to arrested replication of its DNA. These engineered systems resemble the naturally-occurring abortive infection mechanisms as they exploit the same principle (for details see: 6.4. Phage abortive infection systems).

One of the first phage origin-derived systems developed was for *Lactococcus lactis* and employed the *ori* of replication of an industrial phage Φ50 (*ori50*) [80]. Introduction of the Φ50 *ori* region on a high copy number plasmid into the *L. lactis* NCK203 strain provided resistance to not only to phage Φ50 itself, but also to other small isometric phage isolates from industrial environments [80-81]. It was suggested that all of these sensitive phages are part of the same family and most probably exhibit significant homologies within their *ori* regions. Additionally, replication of the *ori50*$^+$ plasmid was shown to be stimulated by Φ50 infection, implying that phage factors are engaged in the process [80]. Further studies determined that the system affects neither adsorption nor phage DNA injection, which suggested that this defense mechanism acts at a later stage of phage development, i.e. DNA replication. It was also clear from the study that the origin-derived phage-encoded resistance phenotype was strongly dependent on the plasmid copy number. Most probably, low copy number plasmids are insufficient in providing enough phage *ori* sites that could efficiently titrate and attract phage replication factors. Yet, on the other hand, when the copy number of *ori*$^+$ plasmids exceeded a certain level, resistant phage mutants were observed as a side-effect. Characterization of these mutants by DNA restriction analysis revealed mutations within the *ori* region, which enabled them to escape the phage defense system.

More recently, a similar origin-derived phage-encoded resistance system was developed for *S. thermophilus* strain Sfi1 based on the *ori* of phage Sfi21 [82]. The presence of this non-encoding phage DNA fragment rendered the Sfi1 host strain resistant to the concomitant phage infection by Sfi21 and 17 other *S. thermophilus* phages. Interestingly, all of them were found to exhibit homology within the *ori* region. However, resistant phages that could overcome this defense mechanism were also detected. They, on the other hand, exhibited differences in the *ori* sequence compared to the wild-type Sfi21-like *ori*. Examination of other *S. thermophilus* phage genomes (~ 30) allowed identifying other distinct replication *ori*s and to divide them into separate groups: replication group I, IIA and IIB [83]. Plasmid constructs harboring these three phage *ori* types increased phage resistance in certain host backgrounds. However, in some strains this origin-derived phage-encoded resistance was not observed. It is therefore speculated whether the efficiency of these systems could be also dependent on some still undetermined host factors.

Development of analogous systems for other lactic acid bacteria involves identification and functional characterization of *ori* regions of their respective phages. This approach can be especially useful for phage-sensitive strains for which other plasmid-encoded defense systems have not yet been determined.

### *6.6.3. Superinfection immunity and exclusion*

During the lysogenic life cycle of temperate phages, the lytic module is inactive due to the activity of the CI repressor. However, certain prophage genes - the superinfection-immunity (CI-like repressor) gene itself and the superinfection-exclusion gene, are actively expressed. Both functions were determined to provide protection to the lysogenic host against phage superinfection. Application of these genes to create engineered phage defense systems is yet another strategy of protecting bacterial cells from incoming infections. Multiple bacterial genomes carry prophage-derived sequences, which can count up to 10% of the total genomic content of the cell. Therefore, despite the fact that phage-related sequences are a burden for bacterial cells, they are also believed to provide some advantage to the host by increasing its fitness.

Genomic studies in *S. thermophilus* led to the identification of superinfection-immunity (*orf127*) and superinfection-exclusion (*orf203*) genes from the lysogeny module of the Sfi21 prophage [84-85]. Expression of *S. thermophilus* phage Sfi21 *orf127* gene from a plasmid vector conferred the phage resistance phenotype against homologous phage, but was ineffective against other heterologous phages [86]. Analysis of the respective ORF127 product revealed its structural homology with phage λ CI repressor and amino acid homology (15% identity) to a potential CI-like repressor of the lactococcal phage Tuc2009. Gel shift experiments allowed determining the ability of the Sfi21 CI-like repressor to bind to two operator sites identified in the genome of the superinfecting homologous phage Sfi21. Superinfection immunity genes (CI repressors) were also identified in phages of other lactic acid bacteria species (e.g. for *Lactococcus* phage TP901-1 and *Lactobacillus* phages A2 and Φadh) [87-89]. Their expression *in trans* was also reported to provide immunity against homologous phage infection.

In contrast, superinfection exclusion genes are not engaged in maintaining the lysogenic state, yet are also active during the lysogenic cycle. Experiments based on expression of the *S. thermophilus* phage Sfi21 *orf203* gene *in trans* in high copies determined that it confers resistance to superinfection of a range of heterologous lytic streptococcal phages [85]. Contrarily to the Sfi-21-derived superinfection immunity, in this case resistance to the Sfi21 phage itself was not observed. Moreover, the mechanistic background of the *orf203*-dependent resistance phenotype was shown to involve inhibition of phage DNA injection.

A superinfection exclusion system was also developed in *Lactococcus lactis* based on the *sie*$_{2009}$ gene from the temperate phage Tuc2009 [30]. When cloned *in trans*, *sie*$_{2009}$ provided resistance only to some 936-type phages used in the study. Moreover, neither c2- nor P335-type phages were affected. It was determined that Sie$_{2009}$ is a cell membrane-associated protein interfering at the stage of phage DNA injection. However, the exact mechanism by which Sie$_{2009}$ acts was not yet established. The ability of the designed system to confer resistance only to certain 936-type phages might indicate different mechanisms of DNA injection exhibited by various phages. A similar membrane protein was detected for *S. thermophilus* phage TP-J34, Ltp. Expression of the *ltp* gene provided protection against TP-J34 in *S. thermophilus* and, interestingly in *L. lactis* against a 936-type phage, P008 [32]. This, quite surprising observation of *sie*-encoded cross-resistance was argued to be due to a recent genetic transfer event between the two species. In both cases, it was noted that phage adsorption was not impaired, but there was significant inhibition of phage DNA accumulation within the host cell. Based on these observations it was proposed that the *ltp* gene product acts at the stage of phage DNA injection by either impairing insertion of the phage tail into the cytoplasmic membrane or by obscuring the host membrane protein responsible for inducing the release of phage DNA from the capsid.

Putative superinfection exclusion genes seem to be widespread among prophage-containing lactococcal and streptococcal strains and localized in the same genomic region limited by repressor and integrase gene from each side. Although *sie* genes lack significant homology, all currently identified Sie proteins are small with hydrophobic N′ tail and at least one transmembrane domain. Various studies of lactococcal Sie proteins allowed grouping them into several phylogenetic groups, depending on the subset of 936-type phages they target (sk1/jj50, bIL170/p008 or 712 group) [28]. At present, it is argued that all *sie* systems identified for lactococcal lysogenic strains interact or mask cell membrane associated factors engaged in phage DNA injection or come into direct contact with structural proteins of the infecting phage. The most probable theory is that Sie function is aimed against the tail tape measure protein (function implicated in phage DNA injection process), as its encoding region is among the few divergent genomic regions between the different subsets of 936-type phages.

Superinfection exclusion and immunity genes in natural conditions can also be provided by defective prophages. The nature of defective phages is that they cannot be efficiently induced by environmental factors; hence, cured from the host strain. Such lysogenic lactic acid bacterial strains (particularly *Lactobacillus* species), which exhibit no threat to the fermentation processes due to uncontrolled prophage induction, are of special interest to the dairy industry as naturally-resistant strains to superinfection events.

### *6.6.4. Phage-triggered suicide systems*

Phage-triggered suicide systems rely on expression of toxic elements under the strict control of phage-inducible promoters. Such specifically engineered systems most closely resemble the naturally-occurring abortive infection systems, which trap the phage within infected cells and lead to programmed cell death. Upon phage infection, host cells are lysed, disabling at the same time phage propagation and the concomitant spread of the phage. In effect, the uninfected bacterial population is saved (for details see: 6.4. Phage abortive infection systems). Suicide systems are based on three genetic components: (i) a lethal gene cassette, (ii) a phage promoter induced only after phage infection, and (iii) an appropriate vector, providing sufficient amount of the lethal gene product.

Such system, based on an inducible plasmid strategy, was created for *L. lactis* to control phage infections [90]. The system comprises a lethal three gene cassette, *llaIR*$^{+}$, encoding a restriction endonuclease of the *L. lactis* LlaI R/M system, cloned under the tight control of the phage Φ31 middle-expressed promoter (Φ31*p*) that is active at a significant level solely after Φ31 infection [91]. Expression of this plasmid-encoded suicide system was designed to restrict unmethylated host and phage DNA upon infection. During infection, induction of the *llaIR*+ cassette caused a significant drop of phage Φ31 EOP. Only a small fraction of infected cells produced progeny phage particles. The system provided also protection against other Φ31-like phages. Yet, despite the observed inhibition of phage development, some phages were found to escape this defense system. Phage mutants that emerged during the assays were all found to be altered in the sequence encoding the transcriptional activator of the Φ31*p* promoter [92]. Thus, these mutants escaped the system due to lack of efficient transcription of the *llaIR*+ cassette. The drawback of this system is the fact that it is only active against phages that can trigger the phage-derived promoter; in this case, against Φ31 and its closely related phages. Another disadvantage is the fact that the *llaIR*+ cassette is not expressed immediately, but after a time necessary for the infecting phage to synthesize transcription factors activating the middle-expressed promoter. This, in effect, allows for replication of a low number of phage particles that escape restriction. It would seem more appropriate to use early phage promoters; yet, these are usually host-controlled. Improvement of the efficiency of the already existing suicide systems should involve cautious selection of effectively controlled and adequately strong phage promoters and more proficient restriction endonucleases. This is best illustrated by the described earlier observation that application of a stronger promoter, despite a more efficient reduction of phage EOP, can have a negative effect on bacterial cell growth. As a method to enhance the efficacy of the suicide system, it was proposed to include within the suicide cassette another gene - *llaIC*, encoding a regulator protein [93]. The presence of this regulator protein was suggested to significantly increase the anti-phage restriction activity of LlaIR.

### *6.6.5. Subunit poisoning*

The subunit poisoning system is an engineered phage defense strategy that relies on expression *in trans* of truncated/mutated proteins which impair (poison) the function of their wild type variants. To achieve this, mutant proteins should be expressed at levels higher than their wild-type counterparts. Moreover, despite alterations in their amino acid sequences, they must have an intact structural form in order to titrate sites or substrates, or other protein components, away from the wild-type proteins.

An example of such system is based on the CI-like repressor of lytic *Lactococcus lactis* phage Φ31. The general idea of this strategy resembles very much the superinfection immunity approach, where *S. thermophilus* bacteria expressing the phage Sfi21 CI repressor were protected from closely-related phages (for details see: *6.6.3.* Superinfection immunity and exclusion). However, in this case, the exact mechanism is somewhat different [94]. Studies of the wild type CI repressor of phage Φ31 showed that it is non-functional and, when expressed in the *L. lactis* host, does not provide protection against superinfecting phages nor represses the transcription of phage lytic genes. Yet, when this wild-type Φ31 CI protein or its truncated variants were expressed *in trans*, they could efficiently inhibit infection of Φ31 and other lytic P335-type phages. Expression of Φ31-derived *c*I mutant genes from a high-copy number vector was shown to inhibit growth of Φ31 (to EOP $10^{-6}$ or lower, depending on the mutation) and of other lytic P335-type phages. The observed effect was determined to be due to the competitive binding of the non-functional Φ31 CI and phage-expressed CI repressors to two of the three wild-type operators identified within the genetic switch region. It was suggested that the truncated variants of the Φ31 CI repressor exhibit a higher affinity for these sites than the phage-encoded CI protein. In effect of Φ31 CI binding, expression of lytic phage functions was repressed, impairing phage DNA replication.

During the study resistant phages were also detected. Sequence analysis studies within the genetic switch regions revealed alterations in their operator sites, which impaired binding of the Φ31-derived CI repressor.

Another example of subunit poisoning phage defense is a system developed in *Streptococcus thermophilus*. The strategy was based on mutating the primase-encoding gene, an essential component of the replication module of the *S. thermophilus* Sfi-21-like κ3 phage [95]. Mutation of sequences within the highly conserved domains of the primase gene resulted in obtaining dysfunctional protein variants. Expression at high level of such primase derivatives *in trans* resulted in reduction of EOP of κ3 and several other Sfi-21-type phages due to inhibition of phage DNA replication. The mutated primase was implied to titrate replication factors and/or the origin of replication, making them inaccessible for the wild-type primase. This suggestion seems to be credible as introduction of the STOP codon upstream of the initial gene mutations restored the phage sensitivity phenotype. Such alteration lead to the synthesis of a truncated protein, which, most probably, could no longer mimic the structure of the native primase. A great advantage of this primase-based subunit poisoning system is lack of phage mutants resistant to the mutated primase proteins under study.

Overall, subunit poisoning is an approach that is believed to constitute a broad phage defense system, as it was shown to be effective against more than one lytic P335-type phage. In this aspect, it differs from the earlier described superinfection immunity systems, where expression of phage repressor genes from phages of various lactic acid bacterial species (e.g. *Streptococcus* phage Sfi21, *Lactococcus* phage TP901-1 or *Lactobacillus* A2 or Φadh phages) provided immunity against the respective single phage only [85,87-89].

### *6.6.6. Host-factor elimination*

Eliminating a genetic element from the genome of starter bacteria to obtain phage-resistant strains is yet another strategy of engineering a phage defense system. This approach can target different stages of the phage life cycle, which are often host-dependent, e.g. phage injection dependent on host membrane proteins, host factors necessary for phage DNA replication.

Among methods identifying such host-encoded factors is random mutagenesis using the pGhost::IS*S1* mutation vector. This approach allowed to identify genes necessary for phage development in the genome of the *S. thermophilus* Sfi strain [96]. The *orf394* gene, encoding a putative transmembrane protein, was one of the host loci determined to confer phage resistance to Sfi19 as well as more than 10 other heterologous *S. thermophilus* phages. After infection by phage Sfi19, no phage DNA synthesis was detected for such mutant strain. Based on this observation, it was suggested that this transmembrane protein may be implicated in the stage of phage DNA injection, analogously to the Pip (phage infection protein) of *Lactococcus lactis* [97]. Among advantages of the host-factor elimination system is the fact that it is food-grade and that the engineered phage resistant strains can be successfully used in production processes. Yet, it must be noted that before its application in the industry, the strain should be assayed for its phage resistance phenotype during several rotation rounds of culturing. There is also no data on the phage mutants that can evolve due to the continuous use of such Pip$^-$ mutant strains.

Other host factors that were suggested to be efficient targets for developing phage defense systems are auxotrophic genes. Pedersen et al. developed a strategy of impairing phage replication in an industrial *Lactococcus lactis* strain by deleting the thymidylate synthase (*thyA*) gene from its genome [98]. This patented strategy is based on the process of phage DNA replication [99]. Upon infection, phages take advantage of the DNA replication machinery of the host to amplify its own genetic material. However, when the host is lacking one of the main DNA building factors, formation of novel replicated DNA molecules is inhibited. The *thyA* gene is responsible for *de novo* synthesis of dTTP in the cell. Strains devoid of *thyA* cannot synthesize dTTP in the medium that lacks thymidine, such as milk. Under such conditions, the Δ*thyA* mutant was resistant to infection by selected P335- and 936-type phages, which efficiently infected the parental wild type strain, and what is important its acidifying properties remained undisturbed. Addition of thymidine to the milk medium restored phage sensitivity of the strain. Among the drawbacks of this system is the fact that the mutant strain lacking thymidylate synthase is impaired for growth. Therefore, in industrial conditions it must be inoculated into milk tanks at higher concentrations than the parental wild type strain in order to meet the technological criteria. A solution to this problem could be addition of limiting amounts of thymidine to promote growth of such starter strain. Yet, thymidine as an additive in the cheese industry is not allowed. Among various options proposed by Pedersen to obtain a phage-resistant *thy* mutant for industrial use could be construction of a thermosensitive mutant, in which expression of the *thy* gene is inhibited at temperatures at which technological processes are carried out [98]. The greatest concern when selecting for phage-resistant strains is their ability to prevail a broad range of phages over long periods of time. In this case, it seems that the host-factor elimination system is the most universal among the presented systems as it acts against all phage types. Moreover, at this point it seems that the probability of occurrence of phage mutants overcoming this resistance mechanism is low. However, it should also be noted that some phages are known to encode own *thy* genes or utilize nucleotides of the host by hydrolyzing its DNA, which can be a weak point [100-101].

# 7. Problem of phage contamination in dairy industry

There are no commercial LAB cultures available which would be completely insensitive to all phages. Even when a starter culture that is launched on the market appears to be phage resistant, phages are detected usually after a certain period of use.

Phage contaminations in dairy plants can cause 3 main serious drawbacks:

- problems in obtaining expected technological parameters and product quality consistence
- staff stress, decrease of motivation and engagement, irregular working hours, staff economical consequences, job resignation
- financial losses (failed production, non-standard product, lower unit price, delayed deliveries, customer losses).

# 8. Phage detection in dairy industry

## 8.1. Simple tools for phage detection at the dairy plant level

A simple test assessing acidification activity of currently used starter cultures on a daily basis can be used successfully to monitor phage contaminations in dairy plants. Briefly, a cheese whey sample from the last production vat of the shift is collected and, before use, sterilized by filtration (0.45 μm filter-pore size). In the case of dairy beverages, a sample of the final product, before its filtration, is clarified with addition of lactic acid and centrifuged. Processed pasteurized milk or sterilized milk reconstituted from powder is inoculated in duplicate with starter cultures (including a phage alternative culture) at a standard dosage. One sample of each culture is inoculated with a whey filtrate (usually 1-2%) and the second one - with a temperature sterilized whey filtrate. After incubation (the temperature and time depend on the culture and process), the pH of the milk is measured. When the pH of the milk containing the filtrate is 0.2 units higher in comparison to the sample containing the sterilized filtrate, it indicates that phage contamination is rather high and phage-unrelated culture rotation as well as disinfection with higher concentrations of active substances should be recommended.

To avoid direct measurements of pH, bromocresol purple (100 μg $ml^{-1}$) as a pH indicator may also be used. The test lasts around 6 h, for mesophilic starters, and 4 h, for thermophilic cultures. When pH of the milk drops below 5.4, the indicator turns from purple to yellow. If, at the same time, the color of the sample containing the non-sterilized filtrate becomes green or purple, it means, with high probability, that phages are present and may adversely influence the fermentation process [102].

Another approach of phage detection is continuous monitoring of pH during fermentation processes conducted in vats or tanks with short time intervals and plotting the data on a graph. Even in the case when delay of the fermentation process is not observed, but the graph shows an irregular shape not related to temperature deviation, phage contamination is suspected (Fig.1). However, in this method a delay in acidification can also result from other inhibitors than phages (e.g. antibiotics, detergents) present in the sample.

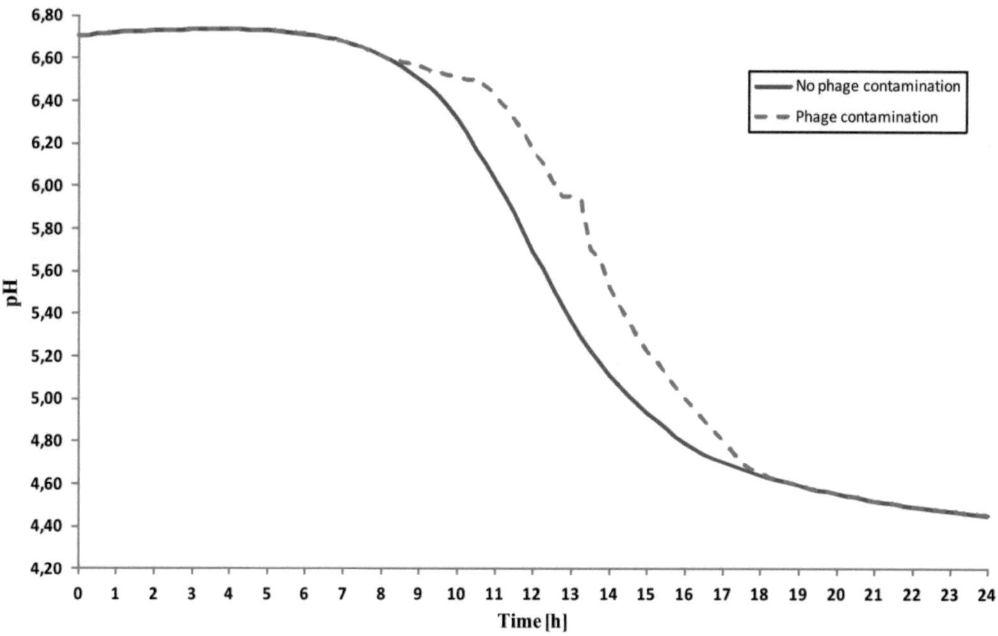


**Figure 1.** Example of pH curve during milk fermentation in the presence of virulent phages incubated with the multistrain and multispecies culture.

## 8.2. Routine service at culture supplier level

The most common and most useful method of phage enumeration is the plaque assay. The method is quite old and was first described by d'Herelle shortly after the discovery of bacteriophages. Currently it is used in many labs with some modifications, but its principle has not changed [103]. The most common, practical, cheap, without using large numbers of plates and sufficiently accurate method in the dairy industry is the semi-quantitative spot test method. Using this approach, results are available after 24-48 h. The method is well suitable for detection of phages of pure lactic acid bacterial strains at relatively low levels (< 100 phages $ml^{-1}$). Plague assays allow detecting the presence of phages as well as determining the number of phages in dairy samples against all individual strains present in the applied defined cultures. In case of phage contamination in a dairy plant, the method is a good tool for selecting the best phage-resistant alternative cultures. The method can also be used for hygiene monitoring by enumeration of phages in samples collected from critical places if the plant. For dairy culture producers, permanent phage monitoring can identify strains which are most sensitive in defined cultures. These strains can be systematically replaced with more phage resistant strains. Semi-solid medium supporting bacterial growth is used for multiplication of strains in form of a smooth opaque layer or lawn on the medium surface using standard Petri dishes. Serial dilutions of phage solution previously sterilized with a filter are placed (5-20 μl) on the surface of the opaque layer. When a single phage particle develops on a recipient bacterial lawn, it forms a plaque (clear spot, no bacterial lawn) visible to the naked eye. This plaque results from the destruction of bacterial cells by the phage progeny. Growth of the plaque is limited by slow diffusion of the phage in the semi-solid medium and bacterial cell growth stops, so phage growth is also inhibited due to the fact that host cells support phage growth. No visible plaques on the plate mean that the sample is not contaminated by phages. Large clear zones (no separate plaques) on the plate indicate with high probability that the level of phages is rather high and further dilutions of the sample are required to precisely determine the phage titer. The presence of a plaque means that: i) the tested sample contains phages; ii) the phage is virulent against the tested strain; iii) the strain is sensitive to the phage. Each phage particle that gives rise to a plaque is called a plaque-forming unit (PFU). One plaque corresponds to a single phage particle and phages can easily be counted. In result, the number of PFUs corresponds to the viable phage concentration in a given sample volume.

### 8.3. Sensitive methods (including ELISA and molecular DNA techniques) at the level of academic or innovation labs

Plaque assays and acidification tests are microbiological methods that are economically accessible and sensitive enough for detection of phages in the dairy industry. These techniques are time consuming, but provide many practical data for both dairy plants and starter producers. The polymerase chain reaction (PCR), ELISA and flow cytometry-based methods have been designed for detecting phages and are often used to complement microbiology tests. However, they have still many drawbacks to be applied for routine analyses in the dairy industry [104].

PCR-based methods detect virulent and non-virulent phages; thus, microbial methods should be used in parallel to precisely distinguish the virulent phages. PCR-based methods can also be too expensive and too specific (only phages targeted by specifically-designed primers are detected) for routine experiments. However, PCR is a fast method able to confirm the presence of bacteriophages within 30 minutes and can be applied to determine the potential utility and quality of big batches of milk. At the same time, the method could be handy in finding niches of phage accumulation, in order to reduce their impact in dairy fermentations [105-108].

ELISA techniques use for phage detection antibodies which are highly specific against structural proteins of phage capsids. Due to the wide phage diversity in the dairy environment, development of several antibodies detecting various groups of phages was required. ELISA is regarded as a highly useful method for monitoring specific phages in the dairy environment, but a single assay cannot be used to detect phages with different structural proteins. For this reason, the sensitivity of an ELISA method to detect phages in dairy a sample is rather low.

Flow-cytometry can also be used for detection of phages in dairy samples by discriminating the phage-infected cells from non-infected based on cell morphological changes leading to lysis. Running on the flow-cytometry of samples containing phages gives a broad distribution of cell mass (wide peak), which demonstrates the presence of both lysed and live cells, while non-infected samples give narrow peaks. Flow-cytometry allows detection of phages in real time, but expensive equipment and well-trained staff needed to perform the assays limits application of this technique in the dairy industry [104].

## 9. Sources of phage contamination

In dairy plants phages can originate from a variety of sources. The prime importance is to identify the potential sources of phage contamination and limit their entry to the fermentation process.

### 9.1. Raw milk

The most probable source of virulent phages is raw milk. LAB phages occur naturally in raw milk at low titers (between $10^1$-$10^3$ PFU $ml^{-1}$) and constitute a continuous supply of bacteriophages in dairy plants [109-110]. Phage concentrations in raw milk also depend on conditions of collecting, handling and storing of milk by the supplier (farm), on transport to the plant and, finally, handling of the milk in the plant itself. For example, reverse osmosis used to concentrate raw milk at a farm can impact the level of phages detected in milk. Almost 10% of 900 milk samples examined from various geographical areas in Spain contained *Lactococcus lactis* phages [110]. Using a multiplex PCR method *Streptococcus thermophilus* phages have been detected in more than one third of milk samples used for yoghurt production in Spain [106]. Phage biodiversity is increased by combining milk collected from different farms and these numbers can be even higher in processed milk.

### 9.2. Milk powder and whey protein concentrates

Reconstituted milk from powder is used in many countries for yoghurt, fresh cheese (tvarog and quark) and even maturated cheese production. Also whey proteins are used to standardize milk before the fermentation process or to improve the taste and texture as well as the nutrient value of the final product. Recently, the modern technology of milk powder and whey protein concentrate production applies often lower temperatures of treatment than during traditional technologies. Both milk powder and whey protein concentrates can be sources of high temperature-resistant phages and can influence the quality of the final product [111-112]. For separating whey proteins, ultrafiltration or/and microfiltration are more frequently employed. Applied separation processes result in higher concentrations of phages in the permeate or the retentate. Depending on which fraction is used in subsequent processes, different concentrations of phages in whey protein samples can be detected.

### 9.3. Starter cultures

The starter culture itself can be a source of phages, when strains contain temperate phages. Temperate phages are incorporated into the bacterial chromosome and their genome replicates in synchrony with the bacterial genome. Prophages are carried in many LAB strains. The analysis of bacterial genomes revealed that prophages are more widespread than previously considered [113-114]. Phages may be induced from lysogenic to lytic form by the manufacturing conditions. Serial subculturing of temperate phages in milk may result in their replacement by a virulent mutant. Prophage induction from multiple lysogenic starter culture strains has the potential to influence fermentation. Induction can occur under stress conditions, such as heat, salts, acidity, bacteriocins, starvation or UV [115-116], and can also occur naturally with a frequency of even up to 9% [117]. Starter culture producers make huge efforts to eliminate strains containing prophages using a screening assay for strain lysogeny. Usually, easily lysogenized strains are difficult to find in defined strain cultures. The main source of lysogenic strains are undefined cultures, which are still commonly used (for example, kefir grains). This is due to two main reasons: i) the exact strain composition of these starters is unknown; ii) elimination of lysogenic strains from undefined culture is very difficult.

### 9.4. Equipment/air

The one of the most probable sources of virulent phages is the dairy plant environment. Phages are commonly present on working surfaces. For propagation, phages need the presence of their bacterial hosts, in this case lactic acid bacteria. Due to this fact, they are usually found in places where conditions for LAB development are favorable. The most common sources of phage contaminations are valves, crevices and "dead ends" (difficult cleaning and disinfection places) of production lines. Also, the formation of biofilms on dairy equipment can lead to serious phage problems. Moreover, phages were detected at high levels on various equipments and objects found in cheese plants, such as walls, pipes, door handles, floors, office tables and even on cleaning materials [118]. Raw milk handling, cheese milk processed in open vats and whey handling can lead to spreading of phages in the air. Phage aerolization can occur during air displacements around contaminated places (fluids or surfaces) or by liquid splashes. Virulent phages can circulate through the air far away from their aerosolization source due to the ability to bind to small particles (< 2.1 μm) [118]. Taking into account high levels of phages detected in the air, it is hard to precisely determine whether phage propagation already took place or if it is likely to occur. Concentrations of up to $10^8$ PFU per $m^3$ of air have been detected in a cheese manufacturing plant in Germany; however, mainly in specific areas of the fermentation line [119-120].

## 10. Phage problem frequencies and consequences depend on product portfolio

Fermentation problems in the dairy plant can be related with: low starter activity, fermentation conditions (e.g. temperature fluctuations), milk composition (year, season, occurrence of mastitis, mineral levels, lactation period, microbial and enzymatic composition), presence of inhibitors in milk (antibiotics, detergents) and phage infections.

However, phages are the primary source of fermentation problems in the dairy industry. Bacteriophages can cause great economic losses due to fermentation failure in dairy plants. About one third of the annual world production of around 500 million tons is converted into fermented products. Two thirds of all processed milk is fermented by *Lactococcus lactis* and *Leuconostoc* spp. Thermophilic *Lactobacillus* and *Streptococcus thermophilus* spp. account for fermentation of the remaining major part of the milk. According to estimations, from 0.1% to 10% of all milk fermentations are negatively affected by virulent phages [102]. Phage contaminations can slow down or even halt the milk fermentation process. Consequences of the phage presence include: alteration of the product quality, such as taste, flavor, texture, and its microbiota composition. Phage contaminations due to the delay in lactic acid production can also lead to development of undesired microbiota during the fermentation process. In the worst cases, the inoculated milk must be discarded. The frequencies of phage contaminations and their consequences depend on the type of milk product produced. Phages can also sometimes turn a dairy staff life into a 'nightmare'.

## 10.1. Fermented milk beverages

Among dairy products, the least phage affected are fermented milk beverages (yoghurt, kefir, butter-milk, Actimel®-like products, etc.). There are many reasons behind this phenomenon. Milk for beverage production usually undergoes treatment at temperatures much higher than in cheese manufacturing. Moreover, some drinking yoghurts are produced from UHT milk. Beverages are made in relatively aseptic conditions, including more and more aseptic inoculation systems, where the fermented product is minimally exposed to the factory environment. In spite of that, phage contamination is sufficiently frequent and has become the primary source of fermentation problems in milk beverage production. Phage contaminations in this particular case lead to fermentation delays or inhibition, product alterations in taste and flavor as well as texture properties.

## 10.2. Ripened cheese

In cheese production the risk of phage infection is very high. A large cheese plant can process more than 500 tons of milk per day, very often in many vats, lasting more than one shift. Pasteurized milk (very often low temperature-treated milk or even raw milk) is used in cheese fermentation and many phages as well as microorganisms remain viable after pasteurization. Contamination, also by phages, increases during curd handling and whey separation in open vats. The consequences of phage infection in cheese production can be: delay or halt in milk acidification, cheese contamination with foreign microbiota, including pathogens, preferential growth of post-pasteurization microbiota, problems in whey separation (syneresis), higher water and lactose content in the final cheese product, abnormal or irregular holes (eyes), or no eyes, and alterations of flavor and texture [5]. To conclude, phage contamination may result in lower quality of cheese or cheese quality suitable only for processed cheese production and, in some extreme cases, complete loss of product.

### 10.3. Fresh cheese (cottage cheese, quark, tvarog)

Cottage cheese and traditional tvarog productions are the most sensitive processes to phages infection. Fermentation delays in production of cottage cheese lead often to complete loss of the final product. However, symptoms of phage contamination are most visible in production of traditional tvarog, where curd quality depends on the activity of lactic acid bacteria alone (rennet is not used). It is estimated that more than 70% of technological disruptions during tvarog manufacture is related to phage contaminations, which usually lead to the following consequences: delay or halt in milk acidification, curd lamination or its drop to the bottom of the tank or vat (which, in effect, causes problem with curd handling), prolonged process of whey separation due to the loss of the curd syneresis, low tvarog yield, contamination with foreign microbiota, including pathogens, intensive growth of post-pasteurization microbiota, off-flavor and texture alterations of the tvarog [121].

## 11. Phage control strategy

As previously stated, phages represent a constant threat of serious economic losses in the dairy industry. Dairy microbiologists have attempted for almost 80 years to eliminate or, at least, bring under better control, bacteriophages that interfere with the manufacture of fermented milk products. Phages rapidly disseminate in dairy environment and are difficult to eliminate. The important procedures for phage control are: adapted factory design, design of starters, cleaning and disinfection, and air control [102].

### 11.1. Cleaning and disinfection

The classical operations of cleaning and disinfection are an essential part of milk processing. Cleaning-in-place (CIP) procedures are usually applied in milk processing lines. The basic procedure consists of the following sequence operations: i) pre-rinse with cold water to remove gross residues; ii) circulation of alkali detergent to remove the remaining minor residues (from time to time acidic detergent is incorporated to remove precipitated minerals and milkstone deposits in the following sequence: alkali detergent, water rinse, acidic detergent); iii) rinse with cold water to flush out the detergent; iv) circulation of disinfectant to inactivate residual microorganisms and phages (still in many dairies this stage is not performed in each cleaning cycle); v) final rinse with cold water to flush out the detergent and cooling line [122]. The cleaning process can remove 90% or more of microorganisms associated with the surface, but cannot kill all of them. One of the drawbacks of the cleaning process is that residual live bacteria can redeposit and, in longer periods of time, can form a biofilm. The presence of LAB among the residual microorganisms increases phage risk contamination. The main role of disinfection is to kill microorganisms that survive the cleaning procedures.

| Disinfectant | Supplier/ Producer | Main active substances | Conditions recommended by supplier | | |
|---|---|---|---|---|---|
| | | | Concentration (%) | Temp. (°C) | Time* (min.) |
| Deptil PA 5 | Hypred | Hydrogen peroxide, Peracetic acid, Acetic acid | 0.1- antiseptic 2.5-fungicidal | < 30 | 20 |
| Divosan Hypochlorite VT3 | Johnson Diversey | Sodium hypochlorite | 0.1 – 3.0 | cool | 10 – 20 |
| Oxidan special 150 | Novadan | Hydrogen peroxide, Peracetic acid, Acetic acid | 0.1 – 0.35 | 5 – 40 | 5 – 60 |
| Hypochlor DES | Novadan | Sodium hypochlorite, Sodium hydride | 0.25 – 1.0 | 20 | 15 |
| Desinfect CL | Novadan | Sodium hypochlorite | 0.20 – 1.0 | 5 – 40 | 10 – 15 |
| P-3 Oxonia | Ecolab | Hydrogen peroxide | 0.5 – 1.0 | ca. 10 | 5 – 30 |
| P-3 Oxonia active 150 | Ecolab | Hydrogen peroxide, Peracetic acid, Acetic acid | 0.1 – 0.2 | ca. 10 | 5 – 30 |
| P3 – Oxysan ZS | Ecolab | Hydrogen peroxide, Peracetic acid, Acetic acid, Peroxyoctanoic acid | 0.10 | ca. 10 max. 40 | 5 – 30 |
| P-3 Hypochloran | Ecolab | Sodium hypochlorite, Sodium hydride | 0.2 – 0.5 | 20 – 60 | 15 |
| P-3 Horolith CD | Ecolab | Nitric acid, Phosphoric acid, Polyhexamethylene biguanide hydrochloride | 0.5 – 1.5 | 50 – 70 | 10 |
| Clarin spezial | Clarin | Peracetic acid, Hydrogen peroxide | 0.2 – 0.5 | 20 | 5 – 20 |

*exposure time

**Table 2.** Characteristics of CIP disinfectants used in the dairy industry.

Disinfection is becoming more and more important in the current strategies used by the dairy industry to limit bacteriophage infections. The virucidal efficacy of disinfectants against bacteria, yeasts, moulds, including pathogens, is well-documented in supplier specifications, but very seldom the information on the efficacy against phages is available. It is wrong to consider that disinfectants active against bacteria will also inactivate bacteriophages [123]. The virucidal activity of commercially available disinfectants is unknown or known only against lab reference phages proposed by the established in 1989

| Disinfectant | Supplier/ Producer | Main active substances | Conditions recommended by supplier | | |
|---|---|---|---|---|---|
| | | | Concentration (%) | Temp*. (°C) | Time** (min.) |
| Deptil Mycocide S | Hypred | Propan-2-ol<br>Didecyldimonium chloride | 0.3 – 2.5 | RT | 5 |
| Deptil HDS | Hypred | Ethanol<br>Sorbic acid | undiluted | RT | 5 |
| Deptil BFC | Hypred | Laurylamine dipropylenediamine | 1.0 | 20 - 90 | 5 - 15 |
| Tego 2000 VT 25 | Johnson Diversey | Amphoteric surfactants (amines, N-C10-C16- alkyl trimethylenedi, reaction products with chloroacetic acid) | 0.5 – 1.0 | TR max 50 | 15 - 60 |
| Divodes FG VT 29 | Johnson Diversey | Propan-1-ol<br>Propan-2-ol | 50 – 100 | RT | 5 - 15 |
| Divosan Extra VT 55 | Johnson Diversey | Benzalkonium chloride (CAS No 8001-54-5) | 0.4 – 0.8 | RT | 60 - 240 |
| Suredis VT1 | Johnson Diversey | Cationic surfactants (N-(3-aminopropyl)-N-dodecylpropane-1,3-diamine CAS: 2372-82-9<br>Sodium carbonate<br>Disodium tetraborate decahydrate | 0.5 – 2.0 | RT . max 50 | 5 - 30 |
| Tego Hygiene 2001 | Johnson Diversey | Trisodium nitrilotriacetate (CAS:5064-31-3)<br>N-Dodecylpropane-1,3-diamine (CAS: 5538-95-4)<br>2-methoxymethylethoxy propanol (CAS: 34590-94-8)<br>reaction product of alkylamino acetic acid and alkyl diazapentane (CAS: 139734-65-9) | 1.0 – 2.0 | RT max 50 | 15- 60 |
| Virocid | CID Lines | Benzalkonium chloride<br>Dimetylodidecyloammonium chloride<br>Glutaraldehyde<br>Propan-2-ol | 0.5 – 1.0 | RT | 60 |

| Disinfectant | Supplier/ Producer | Main active substances | Conditions recommended by supplier | | |
|---|---|---|---|---|---|
| | | | Concentration (%) | Temp*. (°C) | Time** (min.) |
| Eko Javel | PUT Ekoserwis | Sodium hypochlorite<br>Sodium hydride | 0.5 – 1.5 | RT | 15 |
| P-3 Topax 91 | Ecolab | Benzalkonium chloride (CAS No 8001-54-5) | 0.50 - 1 | RT | 10 – 20 |
| P-3 Topax 99 | Ecolab | Alkyl ammonium acetate<br>Acetic acid | 1.0 - static method<br>2.0 - foam method | RT | 10 – 20 |
| P-3 topactive DES | Ecolab | Hydrogen peroxide<br>Acetic acid<br>Amino-oxide | 1.0-3.0 | RT | 10-30 |
| P-3 Monodes | Ecolab | Benzyl alcohol<br>Propanol-2-ol<br>Ethanol | undiluted | RT | 0.5 |
| Anthium Dioxide 5% active chlorine | GSG | Chlorine dioxide<br>Activator – citric acid | 0.01 – 0.05 | RT | 10 |

* RT – room temperature, ** exposure time

**Table 3.** Characteristics of the disinfectants for surfaces, equipment, shoe baths and hands used in dairy industry.

CEN committee for harmonizing the method of evaluating the efficacy of disinfectants [124]. Factors influencing the efficiency of a given disinfectant are: concentration, temperature and exposure time. Among them, the most important is the concentration of active substances. Most of disinfectants are less effective against phages in the presence of interfering proteins (milk or whey) or hard water. The virucidal activity of most disinfectants is improved by increasing the temperature and is usually the lowest in cold water. Therefore, at low temperatures and/or in the presence of proteins, disinfectant concentration and/or contact time should be increased. It is always advisable to combine biocides and heat rather than use them separately at extreme conditions [125]. However, no disinfectant will be fully effective when sanitized surfaces are not cleaned and proteins or biofilm-living cells are present [126]. Under certain conditions phage particles may exist as aggregates, which may also impair complete inactivation. Peracetic acid and sodium hypochlorite are the most efficient biocides of the CIP system in the dairy industry; however, literature data indicate that some LAB phages may be resistant to sodium hypochlorite [125,127-130]. Nonetheless, the most recently available disinfectants are a combination of several biocides. Table 2 presents the chemical content of CIP disinfectants and conditions of their use in the dairy industry as recommended by the suppliers.

Disinfectants recommended mainly for surfaces, equipment, hands and shoe sanitization are listed in Table 3. Disinfectants are in liquid, foam or aerosol form, depending on their application. The efficacy of such disinfectants for phage inactivation, especially those based on alcohols, are lower in comparison to CIP disinfectants. Among biocides, particularly ineffective in phage inactivation is isopropanol [125]. However, taking into account a lower number of phages in an environment, it can be sufficient for their elimination.

### 11.2. Design of starter cultures rotation system based on phage contamination control

Starter cultures are a key factor influencing the diversity of phage population in a dairy plant. Application of undefined multispecies and multistrain cultures was the main strategy to overcome production problems related to phages in many factories (Flora Danica - Chr. Hansen, Probat 505 - Danisco) in the past. One complex culture (e.g. Flora Danica) allowed producing many products: maturated cheese, fresh cheese (tvarog and quark), butter, butter-milk and other mesophilic fermented beverages. Complex multispecies and multistrain cultures are relatively phage tolerant and even upon high phage contamination give products with small deviations that are accepted for marketing. In the past, when dairy plants produced a wide range of products, mainly for the local market, complex undefined cultures fulfilled the expectations of the dairy business.

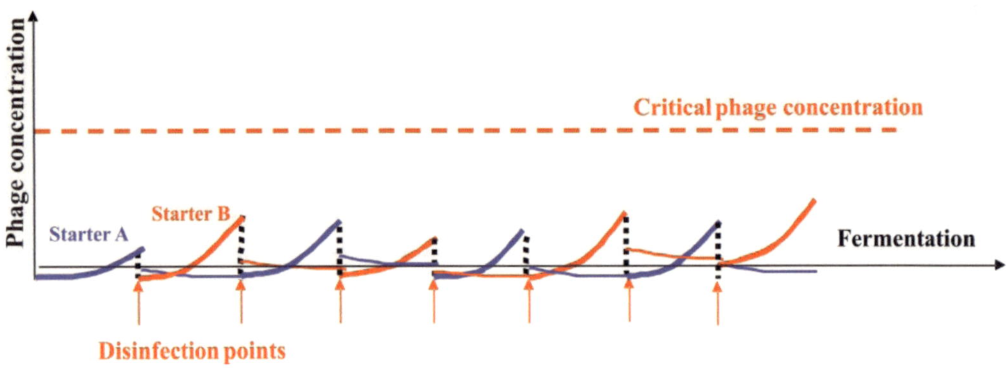


**Figure 2.** Example of well-designed culture rotation and disinfection frequency strategy for phage control in dairy plant.

Modern industrial fermentations increasingly rely on well-defined, direct vat inoculated (DVI), high concentrated (> $10^{10}$ cfu $g^{-1}$) and product-optimized starters, containing from two to five phage-unrelated strains [131-132]. Market share of bulk starters (semi-direct inoculation) diminished very fast in the last two decades and does not exceeded 20% for dairy beverages and 60% for cheese of the total global processed milk. The defined cultures have been widely adapted in large-scale production facilities due to the significant degree of control over fermentation processes and complementary fermentation properties, such as rapid acidification, gas formation, texturization, and development of flavor and aroma compounds. Each defined culture is designed in two or three phage-unrelated options, which can consistently enable producers to obtain high quality standard products. Rotation of defined phage-unrelated cultures is an efficient phage control method. Usually the rotation strategy in big dairy plants is elaborated in tight collaboration with culture suppliers based on individual phage monitoring programs. Ideally, sterilized products or whey samples are delivered on a routine basis at agreed intervals to the phage lab of the culture supplier. In longer perspective, successful cooperation of culture suppliers and users in monitoring different culture rotation strategies allows designing sequences of culture rotation and safe intervals between rotations as well as elaborate the cleaning and disinfection strategy adapted to specific dairy environments (Fig.2).

Rotation strategy of defined multiple strain cultures demands selection of strains resistant to a wide range of phages, which could replace infected strains. This aspect can be a drawback when considering continuous and effective use of this method. Moreover, continual rotation of multiple strains during fermentation processes has an effect on phage co-evolution and was shown to increase phage diversity and their abundance in the dairy environment [133]. It also requires constant selection of starter strains with specific fermentative properties. An alternative is the use of a single, highly specialized phage-resistant strain and its variants carrying phage resistance plasmids obtained from naturally resistant strains. This strategy was termed by Sing and Klaenhammer as the phage defense rotation strategy (PDRS) [134]. The success of designed rotations systems of phage-resistant single strain derivatives is assessed by the Heap-Lawrence starter culture activity test (SAT) performed usually in phage-contaminated milk or whey from earlier cycles [135]. Continuous rotation in repeated cycles of single starter lactococcal strain derivatives, where each carries a different type or a combination of various phage defense systems (e.g. R/M or Abi), has been recognized as an effective method of limiting phages during industrial processes [134,136]. Sing and Klaenhammer have shown that the rotation system of three *Lactococcus lactis* derivative strains encoding different phage defense mechanisms provided resistance to the culture during nine rotation cycles against $10^6$ PFU $ml^{-1}$ of whey composition containing as many as 160 phage isolates [134]. The strategy was then shown to demand precise determination of the type of defense systems to be used as well as the rotation order of the strains. Expression of several phage defense systems relying on different mechanisms conferred complementary defense against phage infection of single strain-derived cultures. Even if one defense system has been overcome, the phage can be inactivated by another. In the study of Durmaz and Klaenhammer (1995) three single starter *Lactococcus lactis* subsp. *lactis* derivatives, containing different plasmid-encoded phage defense mechanisms, were subjected to a 9-day rotation process challenged by two isometric phages (ul36 and Φ31) or a combination of 10 industrial phages at high titer [136]. Moreover, in most cases examined, an additive effect of different phage R/M and Abi defense systems was observed [136]. As assessed by SAT, the culture persisted incoming infections and only one Φ31-derived mutant phage was detected, but did not disturb culture growth during 17 rotation rounds. Based on these observations, it seems that continuous rotation of at least three derivatives of a single starter strain, where each carries a different phage defense system, is an attractive method to overcome phages as well as all types of resulting phage mutants. Moreover, the use of a limited number of strains, in this case one strain and its variants, limits the phage number as well as the occurrence of novel phages in fermentation plants [135,137]. A great advantage for the industry is also the use of only one indicator strain to monitor phage occurrence. Application of PDRS by construction of novel strains carrying newly identified phage-resistance mechanisms makes this strategy broad range with unlimited variants.

## 11.3. Production organization

An important element reducing the spread of phages in the dairy plant is the organization of production. The control of phage risk in dairy plants relies on development and implementation of a variety of procedures. To keep phages under control one should [5,102,123]:

- perform daily tests for phage detection
- avoid crossing paths for raw milk, pasteurized milk and whey
- reduce the diversity of products made on a given day in one production hall
- rotate manufacturing processes
- directly inoculate milk with high concentrated cultures
- rotate starter cultures
- use anti-phage media for bulk starter (BS) propagation
- perform aseptic inoculation where possible
- use air filtration (HEPA) and positive pressure in production facilities
- use positive pressure in fermentation tanks where possible
- use steam sterilization of production lines where possible, especially when phage contamination is high
- dispose stagnant zones of water, whey, milk and foam from production hall or other liquid pools containing live cultures
- clean and disinfect lines, floors, walls, bins and drains used immediately after the process completion
- redisinfect lines after longer production break (e.g. weekend, bank holiday, breakdown)
- disinfect of small equipment used in milk processing after each use (pH-electrode, temperature sensors, etc.)
- use footbaths with disinfecting agents at the entry of production facilities
- avoid using the same equipment for raw milk and whey transportation and treatment
- separate fermentation and packaging areas
- limit personnel path movements (staff in contact with raw milk has no admission to the production facilities)

Plant staff should be aware of the importance of phage control risk, well acquainted with procedures and follow them.

## 12. Selection of phage tolerant strains

### 12.1. Classical methods (isolation and selection of phage tolerant strains against the most aggressive phages from the dairy environment)

In order to isolate phage-resistant mutants, a secondary culture method can be used [138], in which sensitive strains undergo selective pressure of their specific phages. Sensitive strains are inoculated in liquid medium and subsequently infected with suspensions of a selected lytic phage at specific titer. Liquid cultures exhibiting complete lysis are incubated for 24-48 hours (secondary growth). After incubation, bacteria are streaked on adequate solid medium. The grown colonies are consecutively cultured in liquid medium with the same selected phage during at least three rounds. Resultant isolates that are able to grow normally in the presence of the specific phage are considered as true phage-resistant mutants [139].

Another means of natural selection of phage-resistant strains was developed by Viscardi and colleagues [140]. The approach is based on flow-cytometry technique that senses and selects bacterial cells to which phage particles that have been added to the medium did not adsorb. Two detection methods have been designed, which rely on recognition of either specifically labeled anti-phage antibodies or fluorochrome-stained phages. The presented method is an attractive alternative to other means of isolating phage-resistant strains (described earlier). In the study, several different *Streptococcus thermophilus* strains were analyzed for their potential to develop spontaneous phage resistance that could be detected by flow-cytometry technique. The designed selection methods proved quite sensitive, as phage-resistant cells could be detected after only one selection round. Nonetheless, a two-round selection based on selection with anti-phage antibodies or labeled phages and then with unlabeled phage alone was more efficient in obtaining stable and proper phage-resistant mutants. Phage adsorption assays determined that majority of the isolated mutants resisted phage infection at the level of phage adsorption. Moreover, several selection rounds using different labeled phages lead to isolating multi-phage resistant cells.

The great advantage of the method is its high sensitivity (detection of 2 out $10^7$ cells) and high analysis rates ($10^3$ cells per second). As the occurrence of spontaneous phage-resistant cells is rather low in nature, the method allows increasing the level of detection of such mutants. Furthermore, the selected *S. thermophilus* mutants were resistant to phage attack throughout multiple generations, indicating the stability of this property. The novelty of the method is the short amount (several days) of time necessary for obtaining phage-resistant mutants. This creates a possibility of fast selection of new resistant starter strains in the presence of novel phages, which constantly break away from the current defense systems.

### 12.2. BIM system - exposure of sensitive strains to lytic phages (spontaneous mutation in chromosomal or plasmidic genes)

Selection of BIMs (bacteriophage insensitive mutants) is a way to obtain phage-resistant strains without genetic manipulations. The idea of obtaining such cells is to infect a starter strain culture and select for mutants which have sustained phage attack.

This approach has its drawbacks, as it is based solely on the occurrence of random potential mutations in genes coding for receptor materials. The lack of a functional initial receptor for 936- and P335-type phages, such as a polysaccharide, is associated with mutations in genes involved in its synthesis or transport. It is well documented that phage insensitivity of *L. lactis* strains is correlated with loss of the galactose-associated receptor in the cell wall. This disturbs the synthesis of wall components and, as a consequence, insensitive strains often lose their industrial properties, such as the ability to produce acids, and reveal weaker growth in comparison to wild type strains.

Apart from altering cell growth, other two features, such as narrow phage specificity and spontaneous reversion to sensitive phenotype, limit exploitation of BIM mutants in industrial applications [17]. However, mutations in the *pip* gene, encoding a specific receptor for c2-type phages only (for further details on Pip function see: 4 and 6.1-2.), have no significant impact on vitality of lactococcal cells and resultant mutants are stably maintained [17,24]. Genetic engineering methods, which possess a huge potential for developing protection against phages, based on specific point mutations, and construction of stable mutants, might be the solution to this problem. However, at present methods utilizing recombinant DNA approaches restrict the industrial use of genetically modified strains. Mills and colleagues presented a simple 3-step approach, devoid of genetic engineering methods, for generating BIMs of *S. thermophilus* [141]. In the first step, sensitive bacteria were completely lysed in soft top agar plates by adding a selected industrial phage at a MOI > 1 (multiplicity of infection above 1). Subsequently, plates were incubated up to 48 hours after which appearance of resistant colonies was observed. In the next step, all colonies were collected and used to inoculate fresh liquid medium. Harvested bacteria from step 2 were used for conducting a continuous culture in milk with 20–25 passages in the presence of phage at a high concentration (MOI = 10). In order to obtain BIM colonies, the last passage was poured on solid agar from which phage-resistant BIMs were selected after overnight growth. Resistance to another phage could be generated by repeating the whole process on the resultant BIM strain. The insensitive phenotype was initially attributed to nonspecific mutations in receptor genes. However, further studies revealed that phage insensitivity is due to alteration of the CRISPR (clustered regularly interspaced short palindromic repeats) locus, not associated with the previously thought mutations [142] (for further details on CRISPRs see section 6.5 and 12.4).

## 12.3. Plasmid concept

Among the acknowledged and widely applied methods of obtaining starter strains resistant to phage infections is conjugational transfer of plasmids conferring phage resistance determinants [143-144]. In lactococci, there is a range of bacteriophage defense systems occurring naturally on plasmids (**natural, plasmid-encoded phage-resistance systems**). Among the plasmid-encoded phage resistance are such defense mechanisms as restriction/modification (R/M) or abortive infection ($Hsp^+$ or other $Abi^+$) (for more details see sections: 6.3. and 6.4.). First studies, which linked the presence of phage resistance mechanisms to plasmid molecules, were simple assays based on isolation of plasmids from resistant strains and their reintroduction into susceptible cells to obtain cells immune to attack by a particular phage. The later discovery of phage resistance determinants encoded on conjugational plasmids attracted great interest of the food production industry. Most of the data on conjugative plasmids conferring phage resistance comes from studies in *Lactococcus lactis*. In this species many various conjugal plasmids conferring phage-resistance have been identified, including: pTN20, pNP40 and pCI1750, carrying both conjugal transfer ($Tra^+$) and abortive infection ($Abi^+$) determinants, or pAJ1106, exhibiting $Tra^+$ and $Hsp^+$ phenotype [145-149]. Extensive studies of various research groups showed that indeed construction of phage-resistant strains via simple conjugational transfer is an effective means of generating phage resistant starter strains, some of which found application in the dairy industry [143,150].

Among the first conjugal plasmids discovered in *Lactococcus lactis* was pTR2030 isolated from strain ME2. It was characterized to encode heat-sensitive phage resistance ($Hsp^+$), restriction-modification (LlaIR/M) as well as conjugal transfer ($Tra^+$) genes [151]. Its introduction via conjugation into other lactococcal strains, including *Lactococcus lactis* subsp. *cremoris*, resulted in phage-resistance phenotypes [152]. Application of these genetic elements was hence proclaimed as an attractive and acceptable alternative for generating resistant strains, in contrast to strain construction using genetic engineering. The study of Sanders et al. (1986) described the successful attempt of introducing the pTR2030 plasmid via conjugation from a *L. lactis* donor into several industrial recipient strains, from both *lactis* and *cremoris* subspecies [143]. Resulting transconjugants proved resistant to homologous phage infection. Curing of pTR2030 from transconjugants restored phage-sensitive phenotypes, proving visibly that phage resistance is conferred by the plasmid. Noteworthy is the fact that selection of phage-resistant transconjugants was performed in an antibiotic-free background, which is most appropriate for manipulations with strains intended for food production. Another important advantage of this approach was the fact that transconjugant strains maintained their acid-producing properties. This aspect is quite important as it shows that conjugative plasmid manipulations do not alter the industrially attractive features of starter bacteria. The pTR2030 plasmid was maintained throughout multiple generations, indicating that phage resistance will be a stable feature during prolonged use of the transconjugant in industrial applications. Resistance mechanisms identified on conjugative plasmids were also applied in developing engineered bacterial phage defense systems, e.g. the LlaIR/M function encoded on the pTR2030 plasmid was used in constructing phage-triggered suicide systems (see section: *6.6.4.*).

The plasmid-concept of generating phage-resistant strains has also its limitations. First of all, it should be taken into account that many industrially-applied strains are hard to transform. Furthermore, there is a chance that introduction of new plasmids might destabilize industrially attractive strain properties that are also plasmid-encoded (issue of plasmid incompatibility). Introduction of plasmids transferring phage resistance into the bacterial chromosome could be a way of stabilizing this feature; yet, on the other hand, will demand approval of appropriate authorities. Furthermore, some industrially-exploited lactic acid bacteria species, e.g. *S. thermophilus*, carry few plasmids (including conjugal plasmids). This can be an obstacle in generating novel phage-resistant strains via conjugational events [153]. Yet, studies performed by Burrus et al. (2001) revealed the presence of an integrative conjugative element ICR*St1* in *S. thermophilus* strain CNRZ368, shown to encode a II-type R/M system that provided resistance to phage ϕST84 infection [154]. Identification of a phage defense system on an integrative element suggests that also such genetic elements as transposons can be responsible for the spread of phage-resistance mechanisms within bacterial populations.

## 12.4. CRISPR/*cas* defense in LAB

The CRISPR/*cas* defense system was first described in the 1980s for *E. coli,* but only recently recognized for lactic acid bacteria (2007), including such genera as *Lactobacillus, Bifidobacterium, Symbiobacterium, Enterococcus* and *Streptococcus*. Examination of more than 100 genomes of various LAB species allowed identifying over 60 different CRISPR loci, which were grouped into eight distinct families [155]. This indicates the highly diverse nature of LAB CRISPR loci. Additionally, it was observed that clustering of LAB CRISPRs was not in accordance with the classical phylogenetic correlations observed between the LAB phyla. This strongly implies that dissemination of CRISPR loci within the Prokaryotic world into separate lineages occurred by horizontal gene transfer events and their further evolution was imposed by the selective pressure due to phage infections. In general, CRISPR loci were determined to be located on the chromosome, except for one *E. faecium* strain found to carry the CRISPR array on a plasmid. Most LAB species harbor more than one CRISPR locus; yet, despite the common occurrence of CRISPR/*cas* systems, they have still not been identified for such species as *Lactococcus, Leuconostoc, Carnobacterium, Pediococcus,* and *Oenococcus*. This surprising absence of CRISPR loci was implied to be connected with an insufficient amount of sequencing data for these species in public databases. Examination of other strains of these species, involving genome sequencing, should be performed in order to fully resolve the issue on the existence of CRISPR/*cas* systems in these LABs. The identified various CRISPR arrays were determined to contain in total 100 different spacer sequences, including sequences of phage (26%) or prophage (47%) origin.

As CRISPR/*cas* systems confer phage resistance to host cells, they are quite of interest for the dairy industry where microbial production plays a significant role. Application of CRISPR/*cas* systems for construction of new LAB strain variants with differentiated resistance to phage infections is a novel alternative approach [67,142,156]. Moreover, such strains are regarded as safer for industrial applications, as the possibility for them to incorporate or disseminate foreign mobile genetic elements of unknown impact is low. Natural methods of selecting CRISPR-containing BIM cells (see section: 12.2.) of industrially applied bacteria could be an interesting solution for obtaining resistant strains, without deliberate genetic modifications. The first report on isolating CRISPR-containing lactic acid bacteria came from Barrangou et al. (2007) [67], who described the an approach of obtaining spontaneous *S. thermophilus* BIM cells by providing selection pressure due to phage infection. Protocols of isolating such strains have been later developed for dairy *S. thermophilus*, applied in the manufacturing of cheese and yoghurts [141]. The strategy is based on exposition of bacterial starter culture to high phage titers. Several rounds of growth in milk media under the constant selection pressure due to the phage presence resulted in obtaining phage-resistant mutants able to efficiently grow under industrial conditions. The great advantage of such approach is the fact that the presence of naturally acquired spacer sequences renders the strain resistant to phage infections, while preserving the industrially-attractive features of the initial starter cultures. Another strategy of constructing phage-resistant strains could be deliberate integration of synthetic spacers homologous to conserved sequences of industrial phage isolates into the CRISPR array of starter bacteria. However, this approach would involve certain molecular manipulations at the DNA level. Nonetheless, controlled modification of phage resistance of LAB strains using the CRISPR/*cas* regions is not considered by the food industry as a genetic modification method within the meaning of the existing rules in this area.

## Author details

A.K. Szczepankowska, R.K. Górecki and J.K. Bardowski
*Institute of Biochemistry and Biophysics of Polish Academy of Sciences, Warsaw, Poland*

P. Kołakowski *
*Danisco Biolacta, Innovation, Olsztyn, Poland*

## 13. References

[1] Whitehead HR, Cox GA. The occurrence of bacteriophage in lactic streptococci. N. Z. J. Dairy Sci. Technol. 1935;16 319–320.

[2] Daly C, Fitzgerald GF, Davis R. Biotechnology of lactic acid bacteria with special reference to bacteriophage resistance. Antonie Van Leeuwenhoek 1996;70(2-4) 99-110.

[3] Lawrence R C. Action of bacteriophages on lactic acid bacteria: consequences and protection. N. Z. J. Dairy Sci. Technol. 1978;13 129-136.

[4] Brüssow H, Hendrix RW. Phage genomics: small is beautiful. Cell 2002;108(1) 13-6.

[5] Kołakowski P, Rybka J. Causes of disorders of fermentation failures (in Polish). BIBIT (Information Bulletin Rhodia Food Biolacta) 2001;2(24) 7-14.

[6] Klaenhammer T, Altermann E, Arigoni F, Bolotin A, Breidt F, Broadbent J, Cano R, Chaillou S, Deutscher J, Gasson M, van de Guchte M, Guzzo J, Hartke A, Hawkins T, Hols P, Hutkins R, Kleerebezem M, Kok J, Kuipers O, Lubbers M, Maguin E, McKay L, Mills D, Nauta A, Overbeek R, Pel H, Pridmore D, Saier M, van Sinderen D, Sorokin A, Steele J, O'Sullivan D, de Vos W, Weimer B, Zagorec M, Siezen R. Discovering lactic acid bacteria by genomics. Antonie Van Leeuwenhoek 2002;82(1-4) 29-58.

* Corresponding Author

[7] Klaenhammer TR, Barrangou R, Buck BL, Azcarate-Peril MA, Altermann E. Genomic features of lactic acid bacteria effecting bioprocessing and health. FEMS Microbiol Rev. 2005;29(3) 393-409.

[8] Fox PF, McSweeney PLH. Cheese: An Overview. Cheese: Chemistry, Physics and Microbiology 2004;1 1-18.

[9] Ackermann HW, Kropinski AM. Curated list of prokaryote viruses with fully sequenced genomes. Res Microbiol. 2007;158(7) 555-566.

[10] Ackermann HW. Bacteriophage observations and evolution. Res Microbiol. 2003;154(4) 245-251.

[11] Ackermann HW. Frequency of morphological phage descriptions in the year 2000. Brief review. Arch Virol. 2001;146(5) 843-857.

[12] Hendrix RW. Bacteriophage genomics. Current Opinion in Microbiology 2003;6 506–511.

[13] Breitbart M, Salamon P, Andresen B, Mahaffy JM, Segall AM, Mead D, Azam F, Rohwer F. Genomic analysis of uncultured marine viral communities. Proc Natl Acad Sci USA 2002;99(22) 14250-5.

[14] Yu MX, Slater MR, Ackermann HW. Isolation and characterization of Thermus bacteriophages. Arch Virol. 2006;151(4) 663-79.

[15] Breitbart M, Hewson I, Felts B, Mahaffy JM, Nulton J, Salamon P, Rohwer F. Metagenomic analyses of an uncultured viral community from human feces. J Bacteriol. 2003;185(20) 6220-6223.

[16] Mc Grath S, Fitzgerald GF, van Sinderen D. Bacteriophages in dairy products: pros and cons. Biotechnology Journal 2007;2(4) 450-455.

[17] Forde A, Fitzgerald GF. Bacteriophage defence systems in lactic acid bacteria. Antonie Van Leeuwenhoek 1999;76(1-4) 89-113.

[18] Heller K, Braun V. Polymannose O-antigens of *Escherichia coli*, the binding sites for the reversible adsorption of bacteriophage T5+ via the L-shaped tail fibers. J. Virol. 1982;41(1) 222-7.

[19] Dupont K, Janzen T, Vogensen FK, Josephsen J, Stuer-Lauridsen B. Identification of *Lactococcus lactis* genes required for bacteriophage adsorption. Appl Environ Microbiol. 2004;70(10) 5825-32.

[20] Babu KS, Spence WS, Monteville MR, Geller BL. Characterization of a cloned gene (*pip*) from *Lactococcus lactis* required for phage infection. Dev Biol Stand. 1995;85 569-75.

[21] Mooney DT, Jann M, Geller BL. Subcellular location of phage infection protein (Pip) in *Lactococcus lactis*. Can J Microbiol. 2006;52(7) 664-72.

[22] Kraus J, Geller BL. Membrane receptor for prolate phages is not required for infection of *Lactococcus lactis* by small or large isometric phages. J Dairy Sci. 1998;81(9) 2329-2335.

[23] Hoier E, Janzen T, Rattray F, Sorensen K, Borsting MW, Brockmann E. The production, application and action of lactic cheese starter cultures. In: Law BA, Tamime AY. (eds.) Technology of Cheese Making. Oxford UK: Wiley-Blakwell; 2010. p166-192.

[24] Coffey A, Ross RP. Bacteriophage-resistance systems in dairy starter strains: molecular analysis to application. Antonie Van Leeuwenhoek 2002;82(1-4) 303-21.

[25] Tuncer Y, Akçelik M. A protein which masks galactose receptor mediated phage susceptibility in *Lactococcus lactis* subsp. *lactis* MPL56. International Journal of Food Science & Technology 2002;37(2) 139–144.

[26] Forde A, Fitzgerald GF. Molecular organization of exopolysaccharide (EPS) encoding genes on the lactococcal bacteriophage adsorption blocking plasmid, pCI658. Plasmid 2003;49(2) 130-42.

[27] Watanabe K, Ishibashi K, Nakashima Y, Sakurai T. A Phage-resistant mutant of *Lactobacillus casei* which permits phage adsorption but not genome injection. J Gen Virol. 1984;65 981 - 986.

[28] Mahony J, McGrath S, Fitzgerald GF, van Sinderen D. Identification and characterization of lactococcal-prophage-carried superinfection exclusion genes. Appl Environ Microbiol. 2008;74(20) 6206-15.

[29] Garvey P, Hill C, Fitzgerald GF. The lactococcal plasmid pNP40 encodes a third bacteriophage resistance mechanism, one which affects phage DNA penetration. Appl Environ Microbiol. 1996;62(2) 676-679.

[30] McGrath S, Fitzgerald GF, van Sinderen D. Identification and characterization of phage-resistance genes in temperate lactococcal bacteriophages. Mol Microbiol. 2002;43(2) 509-20.

[31] Ventura M, Canchaya C, Pridmore RD, Brüssow H. The prophages of *Lactobacillus johnsonii* NCC 533: comparative genomics and transcription analysis. Virology 2004;320(2) 229-242.

[32] Sun X, Göhler A, Heller KJ, Neve H. The ltp gene of temperate *Streptococcus thermophilus* phage TP-J34 confers superinfection exclusion to *Streptococcus thermophilus* and *Lactococcus lactis*. Virology 2006;350(1) 146-57.

[33] Blumenthal RM, Cheng X. Restriction-modification systems. In: Streips UN, Yasbin RE. (eds.) Modern microbial genetics, 2nd ed. New York: Wiley; 2002. p177–226

[34] Seegers JF, van Sinderen D, Fitzgerald GF. Molecular characterization of the lactococcal plasmid pCIS3: natural stacking of specificity subunits of a type I restriction/modification system in a single lactococcal strain. Microbiology 2000;146(2) 435-443.

[35] Smith MA, Read CM, Kneale GG. Domain structure and subunit interactions in the type I DNA methyltransferase M.EcoR124I. J. Mol. Biol. 2001;314(1) 41-50.

[36] Murray NE. Type I restriction systems: sophisticated molecular machines (a legacy of Bertani and Weigle). Microbiol. Mol. Biol. Rev. 2000;64(2) 412-434.

[37] Janscak P, Bickle TA. The DNA recognition subunit of the type IB restriction-modification enzyme EcoAI tolerates circular permutions of its polypeptide chain. J Mol Biol. 1998;284(4) 937-948.

[38] Janscak P, Dryden DT, Firman K. Analysis of the subunit assembly of the type IC restriction-modification enzyme EcoR124I. Nucleic Acids Res. 1998;26(19) 4439-45.

[39] O'Sullivan D, Twomey DP, Coffey A, Hill C, Fitzgerald GF, Ross RP. Novel type I restriction specificities through domain shuffling of HsdS subunits in *Lactococcus lactis*. *Mol. Microbiol.* 2000;36(4) 866-75.

[40] Gorecki RK, Koryszewska-Bagińska A, Gołębiewski M, Żylińska J, Grynberg M, Bardowski J. Adaptative potential of the *Lactococcus lactis* IL594 strain encoded in its 7 plasmids. PLoS ONE 2011;6(7) e22238.

[41] Adamczyk-Popławska M, Kondrzycka A, Urbanek K, Piekarowicz A. Tetra-amino-acid tandem repeats are involved in HsdS complementation in type IC restriction–modification systems. Microbiology 2003;149(11) 3311–3319.

[42] [42]Holubova I, Vejsadova S, Firman K, Weiserova M. Cellular localization of Type I restriction-modification enzymes is family dependent. Biochem Biophys Res Commun 2004;319(2) 375–380.

[43] Keatch SA, Su TJ, Dryden DT. Alleviation of restriction by DNA condensation and non-specific DNA binding ligands Nucleic Acids Res. 2004;32(19) 325841–5850.

[44] Pingoud A, Fuxreiter M, Pingoud V, Wende W. Type II restriction endonucleases: structure and mechanism. Cell Mol. Life Sci. 2005;62(6) 685-707.

[45] Pingoud A, Jeltsch A. Structure and function of type II restriction endonucleases. Nucleic Acids Res. 2001;29(18) 3705-27.

[46] Marshall JJ, Gowers DM, Halford SE. Restriction endonucleases that bridge and excise two recognition sites from DNA. J Mol Biol. 2007;367(2) 419-431.

[47] Gowers DM, Bellamy SRW, Halford SE. One recognition sequence, seven restriction enzymes, five reaction mechanisms. Nucl. Acids Res. 2004;32(11) 3469–3479.

[48] Marks P, McGeehan J, Wilson G, Errington N, Kneale G. Purification and characterisation of a novel DNA methyltransferase, M.AhdI. Nucleic Acids Res. 2003;31(11) 2803–2810.

[49] Szybalski W, Kim SC, Hasan N, Podhajska AJ. Class-IIS restriction enzymes–a review. Gene 1991;100 13–26.

[50] Christensen LL, Josephsen J. The methyltransferase from the LlaDII restriction modification system influences the level of expression of its own gene. J Bacteriol. 2004;186(2) 287-295.

[51] Su P, Im H, Hsich H, Kang AS, Dunn NW. LlaFI, a type III restriction and modification system in *Lactococcus lactis*. Appl Environ Microbiol. 1999;65(2) 686–693.

[52] Guinane CM, Kent RM, Norberg S, Hill C, Fitzgerald GF, Stanton C, Ross RP. Host specific diversity in *Lactobacillus johnsonii* as evidenced by a major chromosomal inversion and phage resistance mechanisms. Plos One 2011;6 e18740.

[53] Su P, Im H, Hsieh H, Kang AS, Dunn NW. LlaFI, a type III restriction and modification system in *Lactococcus lactis*. Appl. Environ. Microbiol 1999;65(2) 686-93

[54] Solow BT, Somkuti GA. Molecular properties of *Streptococcus thermophilus* plasmid pER35 encoding a restriction modification system. Curr Microbiol 2001;42(2) 122–128.

[55] Hao P, Zheng H, Yu Y, Ding G, Gu W, Chen S, Yu Z, Ren S, Oda M, Konno T, Wang S, Li X, Ji ZS, Zhao G. Complete sequencing and pan-genomic analysis of *Lactobacillus*

*delbrueckii* subsp. *bulgaricus* reveal its genetic basis for industrial yogurt production. PLoS ONE 2011;6(1) e15964.

[56] Janulaitis A, Petrusyte M, Maneliene Z, Klimasauskas S, Butkus V. Purification and properties of the *Eco*57I restriction endonuclease and methylase—prototypes of a new class (type IV). Nucleic Acids Res. 1992;20(22) 6043–6049.

[57] O'Sullivan O, O'Callaghan J, Sangrador-Vegas A, McAuliffe O, Slattery L, Kaleta P, Callanan M, Fitzgerald GF, Ross RP, Beresford T. Comparative genomics of lactic acid bacteria reveals a niche-specific gene set. BMC Microbiol. 2009;5 9–50.

[58] Chopin MC, Chopin A, Bidnenko E. Phage abortive infection in lactococci: variations on a theme. Curr. Opin. Microbiol. 2005;8(4) 473–479.

[59] Durmaz E, Klaenhammer TR. Abortive phage-resistance mechanism AbiZ speeds the lysis clock to cause premature lysis of phage-infected *Lactococcus lactis*. J Bacteriol. 2007;189(4) 1417–1425.

[60] Haaber J, Fortier LC, Moineau S, Hammer K. AbiV, a novel abortive phage infection mechanism on the chromosome of *Lactococcus lactis* subsp. *cremoris* MG1363. Appl. Environ. Microbiol. 2008;74(21) 6528–6537.

[61] Prevots F, Tolou S, Delpech B, Kaghad M, Daloyau M. Nucleotide sequence and analysis of the new chromosomal abortive infection gene *abiN* of *Lactococcus lactis* subsp. *cremoris* S114. FEMS Microbiol Lett. 1998;159(2) 331–336.

[62] O'Connor L, Tangney M, Fitzgerald GF. Expression, regulation, and mode of action of the AbiG abortive infection system of *Lactococcus lactis* subsp. *cremoris* UC653. Appl. Environ. Microbiol. 1999;65(1) 330–335.

[63] Twomey DP, De Urraza PJ, McKay LL, O'Sullivan DJ. Characterization of AbiR, a novel multicomponent abortive infection mechanism encoded by plasmid pKR223 of *Lactococcus lactis* subsp. *lactis* KR2. Appl Environ Microbiol. 2000;66(6) 2647–2651.

[64] Domingues S, McGovern S, Plochocka D, Santos MA, Ehrlich SD, Polard P, Chopin MC. The lactococcal abortive infection protein AbiP is membrane-anchored and binds nucleic acids. Virology 2008;373(1) 14–24.

[65] Emond E, Dion E, Walker SA, Vedamuthu ER, Kondo JK, Moineau S. AbiQ, an abortive infection mechanism from *Lactococcus lactis*. Appl Environ Microbiol. 1998;64(12) 4748–4756.

[66] Horvath P, Barrangou P. CRISPR/Cas, the immune system of Bacteria and Archea. Science 2010;327(5962) 167-170.

[67] Barrangou R, Fremaux C, Deveau H, Richards M, Boyaval P, Moineau S, Romero DA, Horvath P. CRISPR provides acquired resistance against viruses in prokaryotes. Science 2007;315(5819) 1709–1712.

[68] Lillestøl RK, Shah SA, Brügger K, Redder P, Phan H, Christiansen J, Garrett RA. CRISPR families of crenarcheal genus Sulfolobus: bidirectional transcription and dynamic properties. Mol Microbiol 2006;72(1) 259-272.

[69] Szczepankowska AK. Role of CRISPR/cas system in the development of bacteriophage resistance. In: Łobocka M, Szybalski WT. (eds.) Advances in Virus Research 82 (part A). USA: Academic Press (ELSEVIER); 2012. p289-338.

[70] Rousseau C, Nicolas J, Gonnet, M. CRISPI: A CRISPR Interactive database. Bioinformatics 2009;25(24) 3317–3318.

[71] Sturino JM, Klaenhammer TR. Engineered bacteriophage-defense systems in bioprocessing. Nat. Rev. Microbiol. 2006;4(5) 395-404.

[72] Inouye M. Antisense RNA: its functions and applications in gene regulation - a review. Gene 1988;72(1-2) 25–34.

[73] Sturino JM, Klaenhammer TR. Expression of antisense RNA targeted against *Streptococcus thermophilus* bacteriophages. Appl. Environ. Microbiol. 2002;68(2) 588–596.

[74] Sturino JM, Klaenhammer TR. Antisense RNA targeting of primase interferes with bacteriophage replication in *Streptococcus thermophilus.* Appl. Environ. Microbiol. 2004;70(3) 1735-1743.

[75] Brüssow H, Probst A, Frémont M, Sidoti J. Distinct *Streptococcus thermophilus* bacteriophages share an extremely conserved DNA fragment. Virology 1994;200(2) 854–857.

[76] Kim SG, Bor YC, Batt CA. Bacteriophage resistance in *Lactococcus lactis* subsp. *lactis* using antisense ribonucleic acid. J. Dairy Sci. 1992;75(7) 1761–1767.

[77] Chung DK, Chung SK, Batt CA. Antisense RNA directed against the major capsid protein of *Lactococcus lactis* subsp. *cremoris* bacteriophage F4-1 confers partial resistance to the host. Appl. Microbiol. Biotechnol. 1992;37(1) 79–83.

[78] Kim SG, Batt CA. Antisense mRNA-mediated bacteriophage resistance in *Lactococcus lactis* subsp. *lactis.* Appl. Environ. Microbiol. 1991;57(4) 1109–1113.

[79] McGrath S, Fitzgerald GF, van Sinderen D. Improvement and optimization of two engineered phage resistance mechanisms in *Lactococcus lactis.* Appl. Environ. Microbiol. 2001;67(2) 608–616.

[80] Hill C, Miller LA, Klaenhammer TR. Cloning, expression, and sequence determination of a bacteriophage fragment encoding bacteriophage resistance in *Lactococcus lactis.* J. Bacteriol. 1990;172(11) 6419–6426.

[81] O'Sullivan DJ, Hill C, Klaenhammer TR. Effect of increasing the copy number of bacteriophage origins of replication, *in trans*, on incoming-phage proliferation. Appl. Environ. Microbiol. 1993; 59(8) 2449–2456.

[82] Foley S, Lucchini S, Zwahlen MC, Brüssow H. A short noncoding viral DNA element showing characteristics of a replication origin confers bacteriophage resistance to *Streptococcus thermophilus*. Virology 1998;250(2) 377–387.

[83] Stanley E, Walsh L, van der Zwet A, Fitzgerald GF, van Sinderen D. Identification of four loci isolated from two *Streptococcus thermophilus* phage genomes responsible for mediating bacteriophage resistance. FEMS Microbiol. Lett. 2000;182(2) 271-277.

[84] Bruttin A, Desiere F, Lucchini S, Foley S, Brüssow H. Characterization of the lysogeny DNA module from the temperate *Streptococcus thermophilus* bacteriophage Sfi21. Virology 1997;233(1) 136–148.

[85] Bruttin A, Foley S, Brüssow H. DNA-binding activity of the *Streptococcus thermophilus* phage Sfi21 repressor. Virology 2002;303(1) 100–109.

[86] Bruttin A, Foley S, Brüssow, H. The site-specific integration system of the temperate *Streptococcus thermophilus* bacteriophage Sfi21. Virology 1997;237(1) 148-158.

[87] Engel G, Altermann E, Klein JR, Henrich B. Structure of a genome region of the *Lactobacillus gasseri* temperate phage phi adh covering a repressor gene and cognate promoters. Gene 1998;210(1) 61–70.

[88] Alvarez MA, Rodriguez A, Suárez JE. Stable expression of the *Lactobacillus casei* bacteriophage A2 repressor blocks phage propagation during milk fermentation. J. Appl. Microbiol. 1999;86(5) 812–816.

[89] Madsen PL, Johansen AH, Hammer K, Brøndsted L. The genetic switch regulating activity of early promoters of the temperate lactococcal bacteriophage TP901–1. J. Bacteriol. 1999;181(24) 7430–7438.

[90] Djordjevic GM, O'Sullivan DJ, Walker SA, Conkling MA, Klaenhammer TR. A triggered-suicide system designed as a defense against bacteriophages. J. Bacteriol. 1997;179(21) 6741–6748.

[91] O'Sullivan DJ, Walker SA, West SG, Klaenhammer TR. Development of an expression strategy using a lytic phage to trigger explosive plasmid amplification and gene expression. Biotechnology 1996;14(1) 82–87.

[92] Djordjevic GM, Klaenhammer TR. Bacteriophage-triggered defense systems: phage adaptation and design improvements. Appl. Environ. Microbiol. 1997;63(11) 4370-4376.

[93] O'Sullivan DJ, Klaenhammer TR. C *LlaI* is a bifunctional regulatory protein of the *llaI* restriction modification operon from *Lactococcus lactis*. Dev. Biol. Stand. 1995; 85: 591–595.

[94] Durmaz E, Madsen SM, Israelsen H, Klaenhammer TR. *Lactococcus lactis* lytic bacteriophages of the P335 group are inhibited by overexpression of a truncated CI repressor. J. Bacteriol. 2002;184(23) 6532–6544.

[95] Sturino JM, Klaenhammer TR. Inhibition of bacteriophage replication in *Streptococcus thermophilus* by subunit poisoning of primase. Microbiology 2007;153(10) 3295–3302.

[96] Lucchini S, Sidoti J, Brüssow H. Broad-range bacteriophage resistance in *Streptococcus thermophilus* by insertional mutagenesis. Virology 2000;275(2) 267-277.

[97] Garbutt KC, Kraus J, Geller BL. Bacteriophage resistance in *Lactococcus lactis* engineered by replacement of a gene for a bacteriophage receptor. J. Dairy Sci. 1997;80 1512–1519.

[98] Pedersen MB, Jensen PR, Janzen T, Nilsson D. Bacteriophage resistance of a Δ*thyA* mutant of *Lactococcus lactis* blocked in DNA replication. Appl. Environ. Microbiol. 2002;68(6) 3010–3023.

[99] Nilsson D, Janzen T. Method of preventing bacteriophage infection of bacterial cultures. International patent application no. PCT/DK99/00382 (1998).

[100] Kenny E, Atkinson T, Hartley BS. Nucleotide sequence of the thymidylate synthase gene (*thyP3*) from the *Bacillus subtilis* phage Φ3T. Gene 1985;34(2-3) 335–342.

[101] Powell IB, Tulloch DL, Hillier AJ, Davidson BE. Phage DNA synthesis and host DNA degradation in the life cycle of *Lactococcus lactis* bacteriophage c6A. J. Gen. Microbiol. 1992;138(5) 945–950.

[102] Moineau S, Levesque C. Control of bacteriophages in industrial fermentations. In: Kutter, Sulakvelidze A. (Eds.) Bacteriophages: Biology and Applications. Boca Raton, USA: CRC Press; 2005. p285-296.

[103] Carlson K. Appendix: Working with bacteriophages common techniques and methodological approaches. In: Kutter, Sulakvelidze A. (Eds.) Bacteriophages: Biology and Applications. Boca Raton, USA: CRC Press; 2005. p435-492.

[104] Garneau JE, Moineau S. Bacteriophages of lactic acid bacteria and their impact on milk fermentations. Microbial Cell Factories 2011;10(Suppl 1) S20, 1-10. http://www.microbialcellfactories.com/content/10/S1/S20.

[105] Labrie S, Moineau S. Multiplex PCR for detection and identification of lactococcal bacteriophages. Applied and Environmental Microbiology 2000;66(3) 987-994.

[106] Del Rio B, Binetti AG, Martin MC, Fernandez M, Magadan AH, Alvarez MA. Multiplex PCR for the detection and identification of dairy bacteriophages in milk. Food Microbiology 2007;24(1) 75-81.

[107] Martin MC, del Rio B, Martinez N, Magadan AH, Alvarez MA. Fast real-time polymerase chain reaction for quantitative detection of *Lactobacillus delbrueckii* bacteriophages in milk. Food Microbiology 2008;25(8) 978-982.

[108] Binetti AG, De Rio B, Martin MC, Alvarez MA. Detection and characterization of *Streptococcus thermophilus* bacteriophages by use of the antireceptor gene sequence. Appl. Environ. Microbiol. 2005;71(10) 6096-6103.

[109] Kleppen HP, Bang T, Nes IF, Holo H. Bacteriophages in milk fermentations: diversity fluctuations of normal and failed fermentations. International Dairy Journal 2011;21(9) 592-600.

[110] Madera C, Monjardin C, Suarez JE. Milk contamination and resistance to processing conditions determine the fate of *Lactococcus lactis* bacteriophages in dairies. Applied and Environmental Microbiology 2004;70(12) 7365-7371.

[111] Atamer Z, Ali Y, Neve H, Heller KJ, Hinrichs J. Thermal resistance of bacteriophages attacking flavor-producing dairy *Leuconostoc* starter cultures. International Dairy Journal, 2011;21(5) 327-334.

[112] Chopin MC. Resistance of 17 mesophilic lactic *Streptococcus* bacteriophages to pasteurization and spray-drying. Journal of Dairy Research 1980;47(1) 131-139.

[113] Canchaya C, Proux C, Fournous G, Bruttin A, Brüssov H. Prophage genomics. Microbiology and Molecular Biology Reviews 2003;67(2) 238-276.

[114] Mercanti DJ, Carminati D, Reinheimer JA, Quiberoni A. Widely distributed lysogeny in probiotic lactobacilli represents a potentially high risk for the fermentative dairy industry. International Journal of Food Microbiology 2011; 144(3) 503-510.

[115] Lunde M, Aastveit AH, Blatny JM, Nes IF. Effects of diverse environmental conditions on ϕLC3 prophage stability in *Lactococcus lactis*. Applied Environmental Microbiology 2005;71(2) 721-727.

[116] Madera C, Garcia P, Rodriguez A, Suarez JE, Martinez B. Prophage induction in *Lactococcus lactis* by the bacteriocin Lactococcin 972. International Journal of Food Microbiology 2009;129(1) 99-102.

[117] Lunde M, Blatny JM, Lillehaug D, Aastveit AH, Nes IF. Use of real-time quantitative PCR for the analysis of ϕLC3 prophage stability in lactococci. Applied Environmental Microbiology 2003;69(1) 41-48.

[118] Verreault D, Gendron L, Rousseau GM, Veillette M, Masse D, Lindsley WG, Moineau S, Duchaine C. Detection of airborne lactococcal bacteriophages in cheese manufacturing plants. Applied and Environmental Microbiology 2011;77(2) 491-497.

[119] Neve H, Kemper U, Geis A, Heller KJ. Monitoring and characterization of lactococcal bacteriophages in a dairy plant. Kieler Milchw. Forsh. 1994;46 167-178

[120] Neve H, Laborius A, Heller KJ. Testing of the applicability of battery-powered portable microbial air samplers for detection and enumeration of airborne *Lactococcus lactis* dairy bacteriophages. Kieler Milchw. Forsh. 2003;55(4) 301-315.

[121] Kołakowski P, Rybka J, Fetlinski A. Bacteriophages in milk industry (in Polish). BIBIT (Information Bulletin Rhodia Food Biolacta) 2001;3(25) 8-11.

[122] Simoes M, Simoes LC, Vieira MJ. A review of current and emergent biofilm control strategies. LWT-Food Science and Technology 2010;43(4) 573-583.

[123] Primrose SB. Controlling bacteriophage infections in industrial bioprocesses. Advances in Biochemical Engineering Biotechnology 1990;43 1-10.

[124] Deutsches Institut Fur Normung E.V. DIN EN 13610:2002. Chemical disinfectants-Quantitative suspension test for the evaluation of virucidal activity against bacteriophages of chemical disinfectants used in food and industrial areas- Test method and requirements (phase2, step1); 2003.

[125] Guglielmotti DM, Mercanti DJ, Reinheimer JA, Quiberoni AL. Rewiev: efficiency of physical and chemical treatments on the inactivation of dairy bacteriophages. Frontiers in Microbiology 2011;2(282) 282-297

[126] Simoes M, Simoes LC, Machado I, Pereira MO, Vieira MJ. Control of flow generated biofilms using surfactants – evidence of resistance and recovery. Food and Bioproducts Processing 2006;84(4) 338-345.

[127] Briggiler MM, De Antoni GL, Reinheimer JA, Quiberoni A. Thermal, chemical, and photocatalytic inactivation of *Lactobacillus plantarum* bacteriophages. Journal of Food Protection 2009;72(5) 1012-1019.

[128] Buzrul S, Ozturk P, Alpas H, Akcelik M. Thermal and chemical inactivation of lactococcal bacteriophages. LWT Food Science Technology 2007;40(10) 1671-1677.

[129] Capra ML, Quiberoni A, Reinheimer JA. Thermal and chemical resistance of *Lactobacillus casei* and *Lactobacillus paracasei* bacteriophages. Letters Applied Microbiology 204;38(6) 499-504.

[130] Suarez VB, Reinheimer JA. Effectiveness of thermal treatments and biocides in the inactivation of Argentinian *Lactococcus lactis* phages. Journal Food Protection 2002;65(11) 1756-1759.

[131] Thunell RK, Sandine WE. Types of starter cultures. In: Gilland SE. (ed.) Bacterial starter cultures for foods. Boca Raton USA: CRC Press; 1985. p127–144.

[132] Cogan TM, Peitersen N, Sellars RL. Starter systems. In: Bulletin of the International Dairy Federation, no. 263/1991. Practical phage control. International Dairy Federation, Brussels. 1991; p16–23.

[133] Heap HA, Lawrence RC. The contribution of starter strains to the level of phage infection in a commercial cheese factor. N. Z. J. Dairy Sci. Technol. 1977;12(4) 213.

[134] Sing WD, Klaenhammer TR. A strategy for rotation of different bacteriophage defenses in a lactococcal single-strain starter culture system. Appl. Environ. Microbiol. 1993;59(2) 365–372.

[135] Heap HA, Lawrence RC. The selection of starter strains for cheesemaking. N. Z. J. Dairy Sci. Technol. 1976;11(1) 16–20.

[136] Durmaz E, Klaenhammer TR. A starter culture rotation strategy incorporating paired restriction/modification and abortive infection bacteriophage defenses in a single *Lactococcus lactis* strain. Appl. Environ. Microbiol. 1995; 61(4) 1266-1273.

[137] Thunell RK, Sandine WE, Bodyfelt FW. Phage-insensitive, multiple-strain starter approach to cheddar cheesemaking. J. Dairy Sci. 1981;64(11) 2270-2277.

[138] Quiberoni A, Reinheimer JA, Tailliez P. Characterization of *Lactobacillus helveticus* spontaneous phage resistant mutants by RAPD–PCR fingerprints and phenotypic parameters. Food Research International 1998;31(8) 537–542.

[139] Guglielmotti DM, Reinheimer JA, Binetti AG, Giraffa G, Carminati D, Quiberoni A. Characterization of spontaneous phage-resistant derivatives of *Lactobacillus delbrueckii* commercial strains. Int J Food Microbiol. 2006;111(2) 126-33.

[140] Viscardi M, Capparelli R, Di Matteo R, Carminati D, Giraffa G, Iannelli D. Selection of bacteriophage-resistant mutants of *Streptococcus thermophilus*. J. Microbiol. Met. 2003;55(1) 109–119.

[141] Mills S, Coffey A, McAuliffe OE, Meijer WC, Hafkamp B, Ross RP. Efficient method for generation of bacteriophage insensitive mutants of *Streptococcus thermophilus* yoghurt and mozzarella strains. J Microbiol Methods 2007;70(1) 159-64.

[142] Mills S, Griffin C, Coffey A, Meijer WC, Hafkamp B, Ross RP. CRISPR analysis of bacteriophage-insensitive mutants (BIMs) of industrial *Streptococcus thermophilus*—implications for starter design. J Appl Microbiol. 2010;108(3) 945-55.

[143] Sanders ME, Leonhard PJ, Sing WD, Klaenhammer TR. Conjugal strategy for construction of fast acid-producing, bacteriophage-resistant lactic streptococci for use in dairy fermentations. Appl. Environ. Microbiol. 1986;52(2) 1001-1007.

[144] Pillidge CJ, Collins LJ, Ward LJH, Cantillon BM, Shaw BD, Timmins MJ, Heap HA, Polzin KM. Efficacy of four lactococcal phage resistance plasmids against phage in commercial *Lactococcus lactis* subsp. *cremoris* cheese starter strains. Int. Dairy J. 2000;10(9) 617– 625.

[145] McKay LL, Baldwin KA. Conjugative 40-megadalton plasmid in *Streptococcus lactis* subsp. *diacetylactis* DRC3 is associated with resistance to nisin and bacteriophage. Appl. Environ. Microbiol. 1983;47(1) 68-74.

[146] Higgins DL, Sanozky-Dawes RB, Klaenhammer TR. Restriction and modification activities from *Streptococcus lactis* ME2 are encoded by a self-transmissible plasmid, pTN20, that forms cointegrates during mobilization of lactose-fermenting ability. J. Bacteriol. 1988;170(8) 3435-3442.

[147] Coffey A, Fitzgerald GF, Daly C. Identification and characterization of a plasmid encoding phage abortive infection from *Lactococcus lactis* ssp. *lactis* UC811. Netherlands Milk and Dairy Journal 1989;43(3) 229-244.

[148] Jarvis AW, Heap HA, Limsowtin GKY. Resistance against industrial bacteriophages conferred on lactococci by plasmid pAJ1106 and related plasmids. Appl. Environ. Microbiol. 1989;55(6) 1537-1543.

[149] Harrington A, Hill C. Construction of a bacteriophage-resistant derivative of *Lactococcus lactis* subsp. *lactis* 425A by using the conjugal plasmid pNP40. Appl. Environ. Microbiol. 1991;57(12) 3405-3409.

[150] Alatossava T, Klaenhammer TR. Molecular characterization of three small isometric-headed bacteriophages which vary in their sensitivity to the lactococcal phage resistance plasmid pTR2030. Appl. Environ. Microbiol. 1991;57(5) 1346–1353.

[151] Klaenhammer TR, Sanozky RB. Conjugal transfer from *Streptococcus lactis* ME2 of plasmids encoding phage resistance, nisin resistance and lactose-fermenting ability: evidence for a high-frequency conjugative plasmid responsible for abortive infection of virulent bacteriophage. J. Gen. Microbiol. 1985;131(6) 1531-1541.

[152] Steenson LR, Klaenhammer TR. *Streptococcus cremoris* M12R transconjugants carrying the conjugal plasmid pTR2030 are insensitive to attack by lytic bacteriophages. Appl. Environ. Microbiol. 1985;50(4) 851-858.

[153] Mercenier A. Molecular genetics of *Streptococcus thermophilus.* FEMS Microbiol. Rev. 1990;87 61–78.

[154] Burrus V, Bontemps C, Decaris B, Guédon G. Characterization of a novel type II restriction-modification system, *Sth*368I, encoded by the integrative element ICE*St1* of *Streptococcus thermophilus* CNRZ368. Appl. Environ. Microbiol. 2001;67(4) 1522–1528.

[155] Horvath P, Coûté-Monvoisin AC, Romero DA, Boyaval P, Fremaux C, Barrangou R. Comparative analysis of CRISPR loci in lactic acid bacteria genomes. Int. J. Food Microbiol. 2009;131(1) 62–70.

[156] Horvath P, Romero DA, Coûté-Monvoisin AC, Richards M, Deveau H, Moineau S, Boyaval P, Fremaux C, Barrangou R. Diversity, activity, and evolution of CRISPR loci in *Streptococcus thermophilus*. J. Bacteriol. 2008;190(4) 1401–1412.

Chapter 3

# Redox Potential: Monitoring and Role in Development of Aroma Compounds, Rheological Properties and Survival of Oxygen Sensitive Strains During the Manufacture of Fermented Dairy Products

F. Martin, B. Ebel, C. Rojas, P. Gervais, N. Cayot and R. Cachon

Additional information is available at the end of the chapter

http://dx.doi.org/10.5772/51137

## 1. Introduction

Lactic acid bacteria can be found in a diversity of ecosystems, which is consistent with their ability to adapt to highly variable environments. Among the various parameters that characterize these environments (temperature, pH, water activity), redox is relatively recent. It has however already been addressed indirectly in studies relating to the impact of oxidative stress on lactic acid bacteria. Indeed, the concept of oxidation has often been associated with the presence of oxygen; however, oxidoreductive effects on microorganisms must not be limited to oxygen.

A broader vision could be proposed concerning the adaptation of lactic acid bacteria to extracellular redox. The metabolism of lactic acid bacteria, chemosynthetic organisms, involves a series of dehydrogenation (oxidation) and hydrogenation (reduction) reactions. This metabolism follows the principle of conservation of energy and matter, and therefore requires the availability of a terminal electron acceptor. In lactic acid bacteria, a carbon metabolic intermediate is reduced (mainly pyruvate). In homofermentative lactic acid bacteria, redox coenzymes ($NAD^+$/NADH) enable coupling between oxidation and reduction reactions. During anaerobic glycolysis, glucose is oxidized to 2 moles of pyruvate with the formation of 2 moles of NADH, which then further reduces pyruvate to form lactic acid. Consequently, the typical equation of homolactic fermentation is: 1 glucose $\rightarrow$2 lactate.

Such a perfect matching, theoretically consistent, must be qualified according to the environmental conditions, including the redox state of the extracellular medium. The adaptation of lactic acid bacteria to extracellular redox depends on their ability to positively or negatively interfere with oxidants (electron acceptors) or reducing molecules (electron donors). Carbon and electron flow management by the cell will thus be highly dependent on the ability of the microorganisms to interact with the redox environment.

Potentially, all biochemical reactions in the cell, and therefore the enzymatic activity, may be influenced by the redox state of the environment. Dissolved oxygen is an oxidizer and can reach concentrations of 8 $mg.L^{-1}$ of medium (equilibrium with air). Despite the strict anaerobic metabolism of some lactic acid bacteria, the majority are aerotolerant and can react with dissolved oxygen at varying levels. Lactic acid bacteria provided with NADH oxidase can reduce oxygen to water (reduction reaction coupled with the re-oxidation of NADH). This process influences both the intracellular and extracellular redox environment, and will result in a change in the metabolism, cellular physiology and physico-chemical environment surrounding the microorganism.

Changes in the extracellular environment can be monitored by measuring the redox potential ($E_h$). This parameter plays a key role in the quality of fermented dairy products, but is still rarely taken into consideration or is completely ignored during the manufacturing process. The reasons for this lack of interest can be attributed to difficulties associated with its measurement and control. Over the past ten years, several studies advocate the monitoring and control of $E_h$ in fermented products using lactic acid bacteria selected for their reducing ability, redox molecules, or heat treatment. In terms of food applications, the variation in $E_h$ must involve compounds that do not alter the product characteristics. So, modifying the $E_h$ using gas, which enables the product characteristics to be maintained, may be advantageously exploited in industry.

The aim of this chapter is to present the latest knowledge concerning the adaptation of lactic acid bacteria to their redox environment, and the interest of modifying $E_h$ using gas for lactic acid bacteria applications in the food industry.

## 2. Redox potential

$E_h$, like pH, is a parameter of the state of biological media which indicates the capacity to either gain or lose electrons. During oxidation, electrons are transferred from an electron donor to an electron acceptor, which is reduced. Electrochemical measurement of $E_h$ is not new but has attracted little attention as a parameter for controlling fermentation processes due to the sensitivity of its measurement. However, $E_h$ is already indirectly taken into account in industry through oxygen, of which the inhibitory effect on lactic acid bacteria is well-known. Indeed, oxygen modifies the growth capacity of microorganisms and the formation of end products, and so may contribute to the quality of fermented products [1, 2].

### 2.1. Definition of $E_h$

**Oxidation** is a reaction in which a molecule, atom or ion, loses electrons.

**Reduction** is a reaction in which a molecule, atom or ion, gains electrons.

An **oxidant** (also known as an oxidizing agent, oxidizer or oxidiser) can be defined as a substance that removes electrons from another reactant in a redox reaction.

A **reductant** (also known as a reducing agent or reducer) can be defined as a substance that donates an electron to another species in a redox reaction.

In the same way pH defines acid-base characteristics of a solution, $E_h$ defines the reducing and oxidizing characteristics.

Presented below is the reduction half-reaction of an oxidant (Ox) to its corresponding reduced species (Red):

$$Ox + ne^{-} \leftrightarrow Red \quad (1)$$

The Nernst equation gives the relationship between the redox potential and the activities of the oxidised and reduced species:

$$E_h = E_h^0 + 2.3 \times \left(\frac{RT}{nF}\right) \times \log\left(\frac{[Ox]}{[Red]}\right) \quad (2)$$

where:

$E_h$ = redox potential (mV) (in relation to a normal hydrogen electrode).
$E_h^0$ = standard redox potential (mV) (in relation to a normal hydrogen electrode) at pH 0
F = Faraday constant (96500 $C.mol^{-1}$)
n = number of electrons exchanged
R = gas constant (8.31 $J.mol^{-1}.K^{-1}$)
T = temperature in K
$2.3 \times \frac{RT}{F} = 59$ mV (at 25 °C)

However, chemical reactions in aqueous media involve protons, and the following half-reaction:

$$Ox + mH^{+} + ne^{-} \leftrightarrow Red + H_2O \quad (3)$$

From Equation (2) it can be written:

$$E_h = E_h^0 - 2.3 \times \left(\frac{mRT}{nF}\right) \times pH + 2.3 \times \left(\frac{RT}{nF}\right) \times \log\left(\frac{[Ox]}{[Red]}\right) \quad (4)$$

m = number of protons involved in the reaction

Equation (4) is used to determine $E_h^{0'}$ defined as the standard redox potential at pH 7, which is closer to biochemical and biological processes (Figure 1).

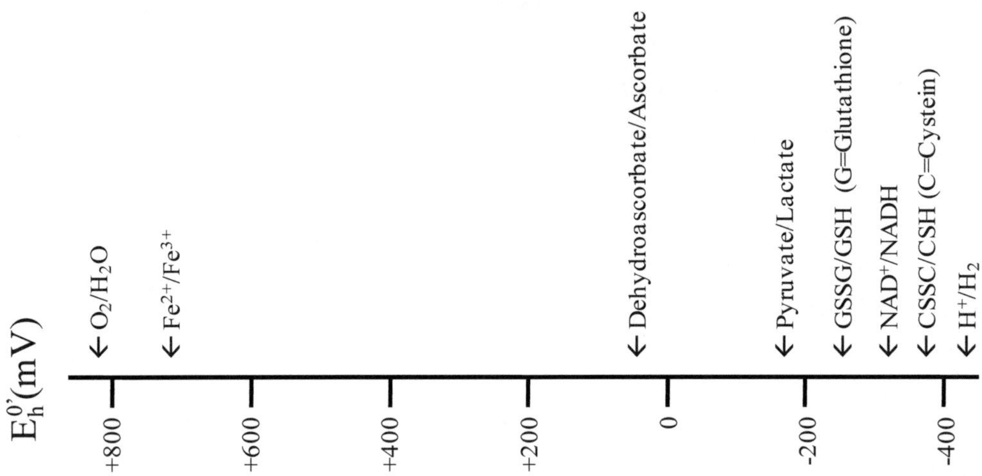


**Figure 1.** Standard reduction potential $E_h^{0'}$ (mV) of some important half-reactions involved in biological processes at 25 °C and pH 7.

## 2.2. Measurement of Eh

The first technique for measuring Eh is based on the use of coloured indicators (redox indicators), which are mostly indophenols or indigo derivatives with a reversible structure between oxidized (coloured) and reduced (colourless) state. However, the use of coloured indicators for measuring Eh, including biological media or food, is limited. Indeed, these molecules behave as electron donors and acceptors; they affect and can change the equilibrium. These compounds can also catalyse or inhibit biological reactions and may be toxic to microorganisms. Furthermore, in some cases it is difficult to appreciate a significant colour change and some Eh indicators also change colour with the pH of the medium. For these reasons, redox indicators are rarely used. They are more often used as indicators of redox thresholds, especially in the manufacture of strictly anaerobic culture media (resazurin) where maintaining a minimum level of reduction is essential for the growth of anaerobic microorganisms. Resazurin is also used to evaluate the reducing activity of starter cultures, for sterility testing and for the detection of microorganisms in dairy milk.

The second method commonly used in microbiology is a potentiometric technique which, contrary to redox indicators, is a direct method. The principle consists in measuring a potential difference determined between an inert electrode (usually made of platinum or gold) in contact with a redox couple in solution and a reference electrode. Electron exchange with the reduced and oxidised species takes place at the inert electrode. The inert electrode is made of stainless metals with a high enough standard potential to be electrochemically stable. These metals act as electron conductors between the measuring medium and the reference electrode. The reference system is the standard hydrogen electrode, but in practice two other references are used: the calomel electrode and the silver / silver chloride (Ag/AgCl) electrode. The redox potential is expressed in volts or millivolts. Redox values should always be expressed in relation to the hydrogen electrode. Consequently, potential measurements ($E_m$) using other references must be adjusted according to the reference potential of the hydrogen electrode ($E_r$):

$$E_h = E_m + E_r \quad (5)$$

For example, $E_r$ of the Ag/AgCl electrode is equal to 207 mV at 25 °C [3]. According to data from Galster [3], we propose the following equations linking $E_r$ and temperature for the two reference electrodes:

$$\text{Ag / AgCl (KCl 3M)} \qquad E_r = 207 + 0.8 \times (25 - T) \quad (6)$$

$$\text{Calomel (Saturated KCl)} \qquad E_r = 244 + 0.7 \times (25 - T) \quad (7)$$

Before use, the redox electrodes must be polished with fine aluminium powder to restore the platinum surface, and controlled in tap water. Three measurements in tap water should be compared and need to be within the confidence interval around their mean value (calculated at 20 mV, 95% confidence level) to ensure correct measurement [4].

Equation (4) shows the dependence of $E_h$ on pH. It is possible to overcome pH dependency by applying the Leistner and Mirna equation [5]:

$$E_{h7} = E_{h\beta} - \alpha \times (7 - \beta) \quad (8)$$

where:

$E_{h7}$ = redox potential (mV) at pH 7
$E_{h\beta}$ = redox potential (mV) at pH $\beta$
$\beta$ = pH of medium
$\alpha$ = Nernst $E_h$–pH correlation factor (mV/pH unit).

To calculate Eh7 in biological media, the Nernst factor ($\alpha$) must be determined experimentally by measuring Eh variation as a function of pH using an acid or a base. This value may vary according to the nature of the oxido-reducing molecules in the media. For example, the Nernst factor is 40 mV/pH unit in milk [6].

## 2.3. Use of gas to modify $E_h$

Gas applications in the food industry are numerous: modified atmosphere packaging (MAP), beverage distribution, cooling, freezing or carbonation. The advantage of using gases such as hydrogen ($H_2$), nitrogen ($N_2$) or carbon dioxide ($CO_2$) to modify $E_h$ is that they are not directly toxic to microorganisms. There are no safety issues for the product with these gases and they can be used sequentially. Finally, their use is authorized at European level. Of the gases used in the food industry, in this chapter we will focus more particularly on nitrogen and hydrogen.

Nitrogen ($N_2$) is odourless, colourless, tasteless, non-toxic, and non-flammable. It is used to extend the life of packaged products (authorized additive E941). It is used to expel oxygen from the packaging before it is closed, which prevents oxidative phenomena involving pigmentation, flavours and fatty acids. It is also used for rapid freezing and refrigeration of food during transport.

Hydrogen ($H_2$) has major potential in food as it is colourless, odourless and has no known toxic effects. It is already used in the food industry for the hydrogenation of liquid oils and their transformation into solid products such as margarine or peanut butter. Hydrogen is a powerful reducer in solution, even at very low concentrations. It has been used to demonstrate the effect of $E_h$ on the heat-resistance of bacteria [7]. Hydrogen is a special reducing agent: it imposes an $E_h$ value on the medium associated with the introduction of the $H^+/H_2$ couple ( $E_h^{0'}$ = -414 mV). This $E_h$ value is highly dependent on the concentration of this couple that mainly influences the stability of the $E_h$ imposed.

With the prospect of food use, hydrogen has the advantage over chemical reducing agents of not changing the product formulation, and therefore not altering the taste. Its industrial use has been rarely seen in this context because of its low flammability limit of 4% in air at 20 °C [8], this is why $N_2$-$H_2$ (96%-4%) is preferred to pure hydrogen. Its use in food technology is authorised at the European level (E949).

## 3. Effect of $E_h$ on a fermented dairy product: Yoghurt

### 3.1. Reminder regarding the manufacture of yoghurt

We chose to focus on the key steps in the manufacture of yoghurt, which are:

- Delivery of milk: The raw material can be either fresh milk, reconstituted milk (from skim milk powder), or a mixture. In all cases, it is generally accepted that a quality product can be made from an extremely high quality raw material. With this in mind, it is essential that when the milk and other raw materials are received methods are established to detect any potential defects as early as possible. Two parameters must therefore be analysed as soon as the milk is received:
    - Its microbiology: to ensure consumer health, prevent the degradation of milk components that persist in the finished product and eliminate any possible competition between the starter culture and the endogenous flora that may involve bacteriophages.
    - Its chemistry: a rapid analysis of the chemical composition of the milk is necessary in order to identify any problems such as colostrum and late-lactation milk. Furthermore, these data concerning the chemistry of the milk can be useful in the standardization of the mixture.
- Standardization of the mixture: each component in milk plays a role. Fat has an effect on the smoothness and the feeling of softness in the mouth, lactose is the raw material used by

lactic acid bacteria for acidification, proteins act on the texture and minerals help stabilize the gel. These components vary in cow's milk according to race, diet, stage of lactation of the animal and season, which is why, during yoghurt manufacture, it is necessary to standardize the milk fat and protein content to meet the nutritional and organoleptic characteristics of the product and obtain consistent quality throughout the year.

- Homogenization: homogenization has two main effects on milk fat and proteins. The fragmentation of the fat globules prevents the separation of the lipid phase and the rest of the mixture, thus preventing the cream rising to the top during fermentation. Homogenization also stabilizes the proteins.
- Heat treatment: This eliminates most of the microbial flora originally present in the milk, including pathogenic or spoilage flora. It denatures the whey proteins, improves the consistency and viscosity of fermented milks and prevents whey separation. The risk of syneresis is reduced.
  Heat treatment of milk also has a positive effect on enzyme activity by providing a supportive environment. The environment becomes reductive through the elimination of a high proportion of oxygen. This medium is more conducive to fermentation that takes place under anaerobic conditions.
- Cooling: After heat treatment, the mixture must be cooled to temperatures approaching 43 °C for inoculation and incubation of the starter culture.
- Fermentation: "Yoghurt" refers to a product fermented by *Streptococcus thermophilus* (*S. thermophilus*) and *Lactobacillus delbrueckii* ssp. *bulgaricus* (*Lb. bulgaricus*). In general, milk is fermented at 40-45 °C, the optimum growth temperature, with an incubation time of 2 and a half hours. However, a longer incubation period of 16-18 hours can be used at a temperature of 30 °C, or until the desired acidity is attained [9].

## 3.2. Yoghurt strains: Lactobacillus delbrueckii ssp. bulgaricus and Streptococcus thermophilus

The association of *S. thermophilus* and *Lb. bulgaricus* is called proto-cooperation. Each species produces one or more substances, initially absent from the culture medium, that stimulate the growth of the other species [10]. During the symbiosis observed in yoghurt, the growth phases of these two bacterial species are staggered. Initially, growth of *S. thermophilus* is observed which is then slowed by the inhibitory effect of the lactic acid produced; the growth rate of *Lb. bulgaricus* then increases [11].

*S. thermophilus* is a strain that often shows little proteolytic activity, due to general low activity or absence of a wall protease. Its growth is limited because the peptides and amino acids initially present in milk are insufficient to cover its needs. In contrast, *Lb. bulgaricus* membrane protease degrades milk caseins releasing small peptides and amino acids which can be used by *S. thermophilus* intracellular peptidases [12].

The cooperation between these two strains also involves the production by *S. thermophilus* of pyruvic acid, formic acid, and carbon dioxide ($CO_2$ obtained from the decarboxylation of milk urea by urease) which stimulates the growth of *Lb. bulgaricus* [9, 13]. However, formic acid is released late in fermentation and in small quantities. The two bacterial species also consume the formic acid resulting from the heat treatment of milk [14].

Some authors have also demonstrated that the association of *S. thermophilus* and *Lb. bulgaricus* affects the production of volatile compounds involved in flavour development in yoghurt [15]. *S. thermophilus* produces more acetaldehyde, acetoin and diacetyl than *Lb. bulgaricus*, contrary to the rest of the bibliography concerning acetaldehyde [9, 16, 17]. Quantities of these molecules and other carbonyl compounds are not crucial per se for yoghurt flavour, but there are relationships between them that give yoghurt its distinctive flavour.

Finally, Ebel *et al.* [18] showed that during the manufacture and storage of a fermented dairy product, the populations of *Lb. bulgaricus* and *S. thermophilus* are the same whatever the $E_h$ of the milk.

## 3.3. Texture

### 3.3.1. A look at yoghurt texture

The transformation of milk into yoghurt is called acid gelation. This gelation is a phenomenon that results in a remarkable change in the physical state of the system which changes from a liquid to a system with the characteristics of a solid. Several phases in the formation of a gel can be distinguished:

- The "solution" phase, where the polymer forms a solution: the macromolecules are not held together;
- The "gel" phase occurs when enough chains have joined together to form a network or gel, with dominant elastic rheological behaviour;
- Sometimes, additional aggregation of associated areas is observed; the gel becomes increasingly rigid and syneresis may occur with time: the gel shrinks and exudes some of the liquid phase.

The slow acidification of milk is due to bacteria that metabolize lactose and produce lactic acid. While casein micelles are stable at normal milk pH and room temperature, this supramolecular structure becomes unstable and leads to the formation of a gel with the slow progressive acidification of milk.

- From pH 6.7 to pH 5.8, the casein micelles seem to retain their integrity, shape and size.
- From pH 5.8 to pH 5.5, the micelles get closer together due to the decrease in the potential $\zeta$ and begin to form groups of micelles.
- From pH 5.5 to pH 5.0, significant changes in shape and size take place: micelle aggregates appear and these particles partially fuse. This is the phase transition between the solution and the acid gel.
- When the pH reaches 5, the solubilisation of micellar calcium, which occurs steadily from pH 6.8, is complete. From pH 5 to pH 4.8, rearrangements of the aggregates take place. At pH 4.9, gelation is complete. At pH 4.6, the acid gel is definitively formed. Aggregation of casein micelles at pH 4.6 is irreversible. Hydrophobic interactions are facilitated at this pH due to reduced electrostatic repulsion, leading to micelle aggregation.

Rheology is used to characterize the texture of yoghurt that specifically targets the mechanical properties. The rheological characterization of a product involves the application of a shear stress and measurement of the deformation, or application of a deformation (compression, stretching or shear) and measurement of its ability to withstand this distortion. Yoghurt can be defined as a viscoelastic fluid. It therefore has both the viscous properties of a liquid and the elastic properties of a solid.

### 3.3.2. *Effect of $E_h$ on a model acid skim milk gel*

It has been shown that dairy products are affected by $E_h$ [4, 19]. Delbeau *et al.* [19] showed that the use of gas to change the $E_h$ of milk can modify the sensory properties of a fermented dairy product. However, we do not know if these modifications are due to the impact of $E_h$ on physicochemical phenomena, lactic acid bacteria, or both. For this purpose, Martin *et al.* [20] wanted to determine to what extent chemical phenomena affect acid milk gelation under different $E_h$ conditions. Glucono-δ-lactone (GDL) was used to acidify milk to avoid variations caused by microorganisms sensitive to $E_h$.

Martin *et al.* [20] studied the effects of $E_h$ on model acidified skim milk gels obtained using GDL and prepared under different gaseous conditions. The milk prepared in air is an oxidizing medium; nitrogen, which is a neutral gas, can be used to remove oxygen from milk - even so the milk Eh remains oxidizing in these conditions - and hydrogen leads to a reducing $E_h$ (below 0). Martin *et al.* [20] focused on the effect of gas bubbling on gel structure through viscoelastic properties and measurement of whey separation (Table 1).

| Gaseous conditions applied to milk | pH | | $E_{h7}$ (mV) | | η (Pa.s) | WS (g/100g of GDL-gel) |
|---|---|---|---|---|---|---|
| | At t=0 | At t=3.5 hours | At t=0 | At t=3.5 hours | | |
| Air | 6.80 ± 0.03 | 4.6[a] ± 0.0 | 405 ± 22 | 414 ± 8 | 0.039[a] ± 0.000 | 4.74[a] ± 1.42 |
| Air bubbling | 6.70 ± 0.04 | 4.6[a] ± 0.0 | 433 ± 6 | 430 ± 5 | 0.032[c] ± 0.001 | 1.26[b] ± 0.26 |
| $N_2$ bubbling | 6.8 ± 0.06 | 4.6[a] ± 0.0 | 283 ± 13 | 288 ± 11 | 0.035[b] ± 0.001 | 1.93[b] ± 0.33 |
| $N_2 - H_2$ bubbling | 6.73 ± 0.04 | 4.6[a] ± 0.0 | - 349 ± 6 | - 83 ± 18 | 0.032[c] ± 0.001 | 0.59[c] ± 0.12 |

[a-c]: different letters indicate that groups were significantly different at an α risk of 5% (ANOVA test). Values in the same column should be compared.



**Table 1.** Characteristics of gel structure depending on the different $E_h$ conditions (milk acidified using GDL):
- Apparent viscosity η at 500 1/s of GDL-gel at pH 4.6 and 4 °C. Measurements were carried out 24 hours after addition of GDL.
- Evolution of average whey separation (WS) over 28 days in GDL-gels.

Values are means from triplicate experiments (mean value ± standard deviation).

The apparent viscosity of each gel was characterized at pH 4.6, 4 °C, 24 hours after addition of GDL under the different $E_h$ conditions (Table 1). For GDL-gels, apparent viscosity ranged from 0.032 to 0.039 Pa.s. GDL-gels produced in air had the highest apparent viscosity, whereas values obtained with air and $N_2 - H_2$ bubbling were similar and significantly lower than those obtained with $N_2$ bubbling. So, for GDL-gels, the viscosity was affected by bubbling. Martin *et al.* [20] showed that the type of gas used for bubbling has a significant influence but no clear trend can be deduced from these results in terms of the influence of an oxidizing or reducing environment.

The gel structure was then observed during storage for up to 28 days. The mean whey separation values of GDL-gels produced under different $E_h$ conditions are presented in Table 1. For each gaseous condition, the authors observed that whey separation occurred from the very first day of storage and the volume of whey separation was relatively constant during the 28 days of storage [20]. Whey separation ranged from 0.59 to 4.74 g / 100 g of GDL-gels. The highest whey separation was obtained with air but this value was lower than values reported in the literature: 18.48% of GDL-gels in the work by Lucey *et al.* [21] and 10% in a study by Fiszman *et al.* [22]. One explanation for is that in the study by Lucey *et al.* [21] the method used to measure whey separation was to remove the gels from their flasks and thus whey separation could have been over-estimated. Whey separation obtained with gas bubbling was lower (1.26 g / 100 g with air bubbling, 1.93 g / 100 g with N2 bubbling and 0.59 g / 100 g with $N_2 - H_2$ bubbling). The lowest whey separation was observed with GDL-gels made under $N_2 - H_2$. Adjusting the $E_h$ of milk to reducing conditions (under $N_2 - H_2$) could be a possible way of significantly decreasing the phenomenon of whey separation.

### *3.3.3. Effect of $E_h$ on a non-fat yoghurt*

In a second step, the authors proposed studying the extent to which lactic acid bacteria affect acid milk gelation under different $E_h$ conditions [23]. Indeed, oxygen modifies the growth capacity of bacteria and the formation of end products. So, $E_h$ may contribute to the quality of fermented products [2, 24, 25]. Martin *et al.* [23] wanted to determine the effects of $E_h$ on yoghurts made under various gaseous conditions. In this study they focused on exopolysaccharide production and gel structure (Table 2). The same gaseous conditions as in the study on the effect of $E_h$ on model acid skim milk gels were chosen.

*Lb. bulgaricus* and *S. thermophilus* produce exopolysaccharides (EPS) which can contribute to improving the texture and viscosity of fermented dairy products [26]. In standard yoghurts (produced in air) the concentration of EPS was 63.60 $mg.L^{-1}$, in accordance with the literature (50 to 350 $mg.L^{-1}$) [27, 28]. The concentration was lower in yoghurts produced with air bubbling (15.22 $mg.L^{-1}$) than in yoghurts produced with $N_2$ bubbling, which was lower than those made with $N_2 - H_2$ bubbling. The EPS concentration of yoghurts made in Air and with $N_2 - H_2$ bubbling were similar. In reducing $E_h$ conditions, lactic acid bacteria produced the same amount of EPS as in ambient air. This result has already been observed in the literature. Indeed, *Lactobacillus sake* 0-1 was reported to have optimal EPS production in anaerobic conditions [29], while higher EPS yields were correlated with a lower oxygen tension [30].

| Gaseous conditions applied to milk | pH | | $E_{h7}$ (mV) | | $C_{EPS}$ (mg/L) | η (Pa.s) | WS (g/100g of yoghurt) |
|---|---|---|---|---|---|---|---|
| | At t=0 | At t=3.5 hours | At t=0 | At t=3.5 hours | | | |
| Air | 6.80$^a$ ± 0.0 | 4.6$^a$ ± 0.0 | 425$^a$ ± 20 | 171$^a$ ± 2 | 63.60$^b$ ± 3.72 | 0.046$^a$ ± 0.00 | 1.98$^a$ ± 0.54 |
| Air bubbling | 6.80$^a$ ± 0.0 | 4.6$^a$ ± 0.0 | 435$^a$ ± 3 | 241$^a$ ± 8 | 15.22$^a$ ± 0.74 | 0.046$^a$ ± 0.00 | 1.76$^a$ ± 0.31 |
| $N_2$ bubbling | 6.81$^a$ ± 0.0 | 4.6$^a$ ± 0.0 | 285$^b$ ± 11 | 139$^b$ ± 5 | 25.29$^c$ ± 0.40 | 0.035$^b$ ± 0.00 | 1.03$^{ab}$ ± 0.27 |
| $N_2 - H_2$ bubbling | 6.81$^a$ ± 0.0 | 4.6$^a$ ± 0.0 | -345$^c$ ± 4 | -309$^c$ ± 10 | 62.70$^b$ ± 0.75 | 0.021$^c$ ± 0.01 | 0.59$^b$ ± 0.12 |

a, b, c: different letters indicate that groups were significantly different at an $\alpha$ risk of 5% (ANOVA test). Values in the same column should be compared.

Reprinted from Journal of Food Res. Int., Vol 431, Martin F, Cayot N, Vergoignan C, Journaux L, Gervais P, Cachon R, Impact of oxidoreduction potential and of gas bubbling on rheological properties of non-fat yoghurt, Pages No. 218-223, Copyright (2010), with permission from Elsevier.

**Table 2.** Characteristics of gel structure depending on the different $E_h$ conditions (milk acidified using lactic starters):

- Concentrations of exopolysaccharides ($C_{EPS}$) in yoghurts after one day of storage.
- Apparent viscosity η at 500 1/s of yoghurt at pH 4.6 and 4 °C. Measurements were made 24 hours after addition of starter culture.
- Evolution of average whey separation over 28 days (WS) in yoghurts.

Values are means from triplicate experiments.

The apparent viscosity of each yoghurt was characterized at pH 4.6 and 4 °C, 24 hours after addition of bacteria under the different Eh conditions (Table 2). The apparent viscosity ranged from 0.021 to 0.046 Pa.s. Yoghurts produced in air and with air bubbling had the highest apparent viscosity. The apparent viscosity of yoghurts made with $N_2$ bubbling was lower (0.035 Pa.s) than other oxidizing conditions (0.046 Pa.s), and values obtained with $N_2 - H_2$ bubbling were the lowest (0.021 Pa.s). Apparent viscosity is clearly affected by the gas type. A reducing environment reduces the apparent viscosity of yoghurt.

Apparent viscosity depends on the solid fraction in the gel as well as the relationships between the different solid elements. In yoghurt, solid particles include milk proteins, lactic acid bacteria and their EPS. Indeed, the gel of yoghurts produced under $N_2 - H_2$ conditions is weaker despite greater EPS production [23]. It is a common assumption that EPS produced by bacteria contribute to the rheological properties of yoghurt [31-33] but, as reported by Hassan *et al.* [34], van Marle [35] and Martin *et al.* [23], no correlation between the viscosity of yoghurt and EPS concentrations was found.

Whey separation of yoghurts produced under different $E_h$ conditions over 28 days of storage was then studied [23] (Table 2). Concerning GDL-gels, whey separation of yoghurts occurred from the very first day of storage and the volume of whey separation was relatively constant over the 28 days of storage. Whey separation ranged from 0.59 to

1.98 g/100 g of yoghurt. The highest whey separation was obtained with air and air bubbling and these values are in accordance with the literature [22]. Whey separation obtained with $N_2$ bubbling (1.03 g / 100 g) and $N_2$ - $H_2$ bubbling (0.59 g / 100 g) was lower. So, the more reducing the environment, the lower the whey separation. Adjusting the $E_h$ of milk to reducing conditions (under $N_2$ – $H_2$) could be a possible way of significantly decreasing the phenomenon of whey separation.

## 3.4. Aroma compounds

### *3.4.1. A look at yoghurt aroma compounds*

The typical flavours of fermented milk are mainly due to a blend of the following compounds: lactic acid, carbon compounds such as acetaldehyde, acetone, acetate and diacetyl, non-volatile acids such as pyruvic, oxalic and succinic acids, volatile acids such as acetic, propionic and formic acids and products from the thermal degradation of proteins, lipids or lactose.

Ott *et al.* [36] identified 91 aroma compounds (GC-olfactometry) in yoghurt among which 21 were detected more frequently and would thus have a major impact on flavour. Acetaldehyde is found in significant quantities and is responsible for the characteristic smell of yoghurt. Diacetyl, pentane-2,3-dione, and dimethyl sulphide also have a major impact on yoghurt flavour [36, 37].

Acetaldehyde was firstly reported by Pette *et al.* [38] as the main aromatic compound in yoghurt. During manufacture, production of this compound is only highlighted when a certain level of acidification is reached (pH 5.0). Concentrations found in the final product are 0.7 to 15.9 $mg.kg^{-1}$. The maximum amount is obtained at pH 4.2 and stabilizes at pH 4.0. The production of acetaldehyde and other flavour compounds by *S. thermophilus* and *Lb. bulgaricus* occurs during yoghurt fermentation and the final amount is dependent on specific enzymes which are able to catalyse the formation of carbon compounds from the various milk constituents.

Three metabolic pathways producing acetaldehyde were identified and some pathways may take place simultaneously [39]:

- From glucose in the glycolytic pathway,
- From the degradation of DNA,
- From L-threonine with threonine aldolase.

However, 90% of acetaldehyde produced by *Lb. bulgaricus* comes from glucose and 100% in the case of *S. thermophilus* [39].

Diacetyl and pentane-2,3-dione also have a significant impact on the final aroma of yoghurt: 1 mg of diacetyl and 0.1 mg of pentane-2, 3-dione per kg of yoghurt are produced by lactic acid bacteria during fermentation. These diketones are produced by decarboxylation of their precursors, 2-acetolactate and 2-aceto-hydroxybutyrate [39]. These compounds are thermally unstable and in the presence of oxygen are converted into their corresponding diketones [40, 41]. Moreover, during storage at 4 °C, the concentration of the two diketones increases slightly [41] due to the basal metabolic activity of the bacteria.

Agitating a mixed culture of *Lactococcus* and *Leuconostoc* promotes diacetyl production by allowing oxidative decarboxylation of 2-acetolactate [42, 43]. In unstirred cultures, the redox potential of the medium decreases rapidly at the start of fermentation. Only acetoin and 2-acetolactate are produced. The authors also showed that controlled oxygenation of the *Lactococcus lactis* ssp. *lactis* culture medium favoured diacetyl production by increasing the activity of diacetyl synthase [44].

Neijssel *et al.* [40] showed that the distribution of carbon flux from pyruvate depended on the NADH / $NAD^+$ ratio, intracellular redox potential or the concentration of metabolites and particularly that of pyruvate. Finally, the authors suggested adding air or oxygen to milk in order to increase the amount of diacetyl in cheese [45].

References [36] and [37] are the only articles that mention dimethyl sulphide as a compound having a significant impact on the flavour of yoghurt. The metabolic pathways involved in the synthesis of sulphur compounds are not well-known in yoghurt. However, the literature mentions these synthetic pathways in the development of cheese flavour.

In general, the majority of sulphur aromatic compounds come from methionine [46]. Methanethiol is easily oxidized to dimethyl disulphide and dimethyl trisulphide [47]. The appearance of these compounds is the direct result of the methanethiol content and is modulated by the low redox potential in Cheddar. Dimethyl sulphide is produced by a metabolic pathway that does not involve methanethiol, but that is different to that of dimethyl sulphide and trimethyl disulphide from methionine [48].

Studies have also shown that when the redox potential decreases, methanethiol and hydrogen sulphide concentrations increase [45]. Moreover, the cheeses to which reducing compounds (dithiothreitol or glutathione) were added contained higher amounts of sulphur compounds and had better qualitative and quantitative flavour performances [45]. It therefore seems that a reducing environment is essential for the production of aroma compounds by bacteria. If a cheese is exposed to air, the redox increases and this leads to the oxidation of sulphur compounds, resulting in lower quality aromatics.

### *3.4.2. Impact on aroma biosynthesis by lactic acid bacteria*

Studies on aroma biosynthesis by LAB usually take into account environmental factors such as pH and temperature. However, the $E_h$ of the medium has not yet been considered, although it is supposed to affect bacterial metabolism [49, 50]. Martin *et al.* [51] determined to what extent $E_h$ can affect the metabolic pathways involved in the production of aroma compounds in *Lb. bulgaricus* and *S. thermophilus.* Four aroma compounds (acetaldehyde, dimethyl sulphide, diacetyl and pentane-2,3-dione) were chosen as metabolic tracers of lactic acid bacteria metabolism. The same gaseous conditions as in the study of the effect of $E_h$ on model acid skim milk gels and non-fat yoghurt were chosen. The amounts of each of the four aroma compounds extracted using a headspace solid-phase micro-extraction technique (HS-SPME) and analysed using gas chromatography coupled with mass spectrometry (GC-MS) during 28 days of storage are reported in Table 3.

Firstly, the authors focused on the impact of these different $E_h$ conditions on the biosynthesis of these four aromas by bacteria after one day of storage [51]. In the standard yoghurt (made in ambient air), diacetyl was observed in the highest concentrations, and acetaldehyde the lowest. This result is contrary to the literature where the lowest concentrations were reported for dimethyl sulphide (0.013-0.070 $mg.kg^{-1}$; measured using dynamic and trapped headspace GC [37, 41]). In the same way, published concentrations were generally higher for acetaldehyde (0.7-15.9 $mg.kg^{-1}$) than in our standard yoghurt (0.18 $mg.kg^{-1}$). In the literature, the concentrations of diacetyl (0.31-17.3 $mg.kg^{-1}$) and 2,3-pentanedione (0.02-4.5 $mg.kg^{-1}$) were lower than in our standard yoghurt (162 $mg.kg^{-1}$ and 115 $mg.kg^{-1}$ respectively). An explanation for these differences can be put forward: the quantification technique used by Ott *et al.* [41] and Imhof *et al.* [37] was dynamic and trapped headspace GC. This technique requires Tenax® traps which may be saturated, as we showed in a preliminary experiment. Furthermore, in our study, to enable a more complete extraction of the aroma compounds, a saturated solution of NaCl was added to the yoghurt. Finally, we did not use the same species of LAB as Ott and Imhof, which may have resulted in different quantities of the various aroma compounds.

Yoghurts made with air bubbling had significantly higher concentrations of acetaldehyde and diacetyl compared to standard yoghurts. The concentration of dimethyl sulphide was significantly lower and that of pentane-2,3-dione was the same.

With $N_2$ bubbling, the concentration of acetaldehyde was similar to that in yoghurts made with air bubbling, whereas the concentration of dimethyl sulphide was lower. The concentration of diacetyl was the same as in standard yoghurts and the concentration of pentane-2,3-dione was not significantly different from that in yoghurts made in air (bubbling or not).

The authors also demonstrated that oxidative $E_h$ conditions clearly increased the production of aroma compounds [51]. These results are consistent with the bibliography. Oxidative conditions stimulated the production of volatile sulphur compounds such as dimethyl sulphide, and aldehydes such as acetaldehyde [49]. In the presence of oxygen, the oxidative decarboxylation of 2-acetolactate and 2-aceto-hydroxybutyrate to diacetyl and pentane-2,3-dione respectively was also favoured [40, 42, 44, 52]. For diacetyl, our result can be explained by the fact that in anaerobic conditions lactic acid bacteria dehydrogenate the NADH produced during glycolysis via lactate dehydrogenase (LDH) activity. Boumerdassi *et al.* [44] confirmed that oxygen increases NADH oxidase activity [53], which causes NADH re-oxidation at the expense of LDH, butanediol dehydrogenase and acetoin dehydrogenase activity [54]. Then, excess pyruvate is partially eliminated through acetolactate production, which increases diacetyl production [44].

Finally, bubbling with $N_2 – H_2$ (reducing conditions), the concentration of acetaldehyde and pentane-2,3-dione was the same as in standard yoghurts. The concentration of dimethyl sulphide was the same as in yoghurts made without oxygen and the concentration of diacetyl was significantly lower than under the other three $E_h$ conditions.

Then, [51] kept the yoghurts in Hungate tubes at 4 °C for 28 days in order to prevent exposure of the contents to oxygen, and the gaseous conditions applied to the milk are thus assumed to be constant during storage.

| Gaseous conditions and storage period (days) | Aroma compound (mg.kg) | | | |
|---|---|---|---|---|
| | ACH | DMS | DY | PTD |
| Ambient air | | | | |
| 1 | $0.18^{a} \pm 0.02$ | $10.16^{a} \pm 0.59$ | $162.08^{a} \pm 13.49$ | $115.25^{a} \pm 33.70$ |
| 7 | $0.13^{ab} \pm 0.02$ | $10.16^{a} \pm 0.99$ | $115.55^{cb} \pm 5.13$ | $84.33^{ab} \pm 1.86$ |
| 14 | $0.10^{b} \pm 0.00$ | $9.52^{a} \pm 0.42$ | $91.63^{c} \pm 8.02$ | $71.56^{ab} \pm 4.49$ |
| 21 | $0.18^{a} \pm 0.00$ | $13.06^{b} \pm 1.21$ | $141.29^{ab} \pm 11.77$ | $65.19^{b} \pm 5.49$ |
| 28 | $0.16^{ab} \pm 0.01$ | $11.82^{ab} \pm 0.48$ | $112.66^{c} \pm 7.56$ | $51.79^{b} \pm 2.95$ |
| Air bubbling | | | | |
| 1 | $0.28^{a} \pm 0.00$ | $5.27^{a} \pm 0.53$ | $299.90^{a} \pm 18.37$ | $123.47^{a} \pm 3.23$ |
| 7 | $0.14^{b} \pm 0.01$ | $6.34^{ab} \pm 0.30$ | $127.58^{bc} \pm 3.93$ | $83.98^{b} \pm 2.20$ |
| 14 | $0.11^{b} \pm 0.01$ | $6.85^{b} \pm 0.32$ | $104.23^{b} \pm 4.66$ | $78.10^{b} \pm 3.15$ |
| 21 | $0.24^{a} \pm 0.02$ | $9.33^{c} \pm 0.55$ | $143.84^{c} \pm 6.69$ | $67.89^{c} \pm 0.80$ |
| 28 | $0.13^{b} \pm 0.02$ | $6.78^{b} \pm 0.30$ | $101.91^{b} \pm 3.03$ | $54.36^{d} \pm 4.19$ |
| $N_2$ bubbling | | | | |
| 1 | $0.35^{a} \pm 0.04$ | $3.72^{a} \pm 0.22$ | $147.79^{a} \pm 9.91$ | $110.87^{a} \pm 4.49$ |
| 7 | $0.22^{bc} \pm 0.02$ | $5.68^{ab} \pm 0.34$ | $103.13^{bc} \pm 6.50$ | $75.67^{b} \pm 1.98$ |
| 14 | $0.18^{c} \pm 0.01$ | $6.16^{b} \pm 0.17$ | $78.80^{c} \pm 4.60$ | $58.16^{c} \pm 0.48$ |
| 21 | $0.28^{ab} \pm 0.04$ | $10.34^{c} \pm 1.37$ | $122.27^{ab} \pm 8.94$ | $51.95^{c} \pm 2.29$ |
| 28 | $0.18^{c} \pm 0.00$ | $7.16^{b} \pm 0.77$ | $84.76^{c} \pm 9.92$ | $38.72^{d} \pm 3.03$ |
| $N_2 - H_2$ bubbling | | | | |
| 1 | $0.17^{ab} \pm 0.00$ | $2.71^{a} \pm 0.57$ | $102.73^{a} \pm 9.10$ | $76.99^{a} \pm 5.90$ |
| 7 | $0.13^{bc} \pm 0.02$ | $5.49^{b} \pm 0.53$ | $89.61^{ab} \pm 7.70$ | $66.10^{ac} \pm 2.29$ |
| 14 | $0.11^{c} \pm 0.01$ | $6.07^{bc} \pm 0.43$ | $74.35^{b} \pm 5.42$ | $57.60^{bc} \pm 0.86$ |
| 21 | $0.19^{a} \pm 0.01$ | $7.48^{c} \pm 0.53$ | $112.09^{a} \pm 7.82$ | $52.10^{bd} \pm 0.59$ |
| 28 | $0.12^{c} \pm 0.01$ | $7.22^{c} \pm 0.63$ | $78.36^{b} \pm 9.30$ | $40.60^{d} \pm 7.93$ |

$^{a, b, c, d}$: different letters indicate that groups were significantly different at an $\alpha$ risk of 5% (ANOVA test).



**Table 3.** Evolution of average amounts of aroma compounds (mg.kg$^{-1}$) quantified in headspace of yoghurts made under different $E_h$ conditions (ambient air, bubbling with air, bubbling with $N_2$ and bubbling with $N_2 - H_2$) during 28 days of storage. ACH: Acetaldehyde (A); DMS: Dimethyl sulphide (B); DY: Diacetyl (C); PTD: pentane-2,3-dione (D). Values are means of experiments carried out in triplicate. Values in the same column should be compared.

During the 28 days of storage, for the standard yoghurt, the quantities of acetaldehyde and dimethyl sulphide produced were relatively stable, while diketone concentrations significantly decreased.

For yoghurts made with air bubbling, the aroma profiles remained almost constant. During storage, the concentration of acetaldehyde decreased slightly whereas that of dimethyl sulphide increased slightly. The diketone concentration significantly decreased.

For yoghurts made without oxygen (bubbling with $N_2$), the quantities of acetaldehyde, diacetyl and pentane-2,3-dione decreased during storage while that of dimethyl sulphide increased.

For yoghurts produced under reducing conditions (bubbling with $N_2 - H_2$), the aroma profiles during storage were the same as those made without oxygen. The concentration of acetaldehyde, diacetyl and pentane-2,3-dione decreased while that of dimethyl sulphide increased.

Furthermore, during storage, different profiles were observed for the four aromas depending on the $E_h$ conditions [51]. Under oxidizing conditions (+170 to +245 mV), the concentration of acetaldehyde was relatively stable during storage, which is in accordance with the literature [9, 55, 56] and the concentration of dimethyl sulphide was also stable. On the contrary, under reducing conditions (-300 to -349 mV), the concentration of acetaldehyde decreased and that of dimethyl sulphide increased. The metabolic pathways involved in the biosynthesis of sulphur compounds are still unclear. Under reducing conditions, it seems that another pathway promotes the production of dimethyl sulphide and that acetaldehyde may be reduced to ethanol. For diketones, whatever the $E_h$ conditions, the concentration decreased during storage. Diacetyl and pentane-2,3-dione can be reduced respectively to acetoin and pentane-2,3-diol [57].

## 4. Impact of $E_h$ on other dairy products

### 4.1. Probiotic dairy products

The use of gas to modify $E_h$ seems to be an interesting way of varying the organoleptic properties of dairy products as well as improving the survival of oxygen sensitive strains during storage in fermented dairy products containing probiotics. Indeed, these microorganisms are mainly anaerobes. Oxygen, which is a powerful oxidant, has a drastic effect on $E_h$ values and the viability of probiotic bacteria during manufacturing and storage [58-60]. So, many studies modify the redox potential to protect probiotics from oxygen toxicity in dairy products [1, 61-65]. However, these techniques sometimes have deleterious effects on the organoleptic properties of fermented milk. An alternative to these methods could be the use of gases. Indeed, Ebel *et al.* [18] showed that fermented dairy products made from milk gassed with $N_2$, and more particularly those made from milk gassed with $N_2 - H_2$, were characterized by a significant increase in *Bifidobacterium bifidum* survival during storage (Figure 2).

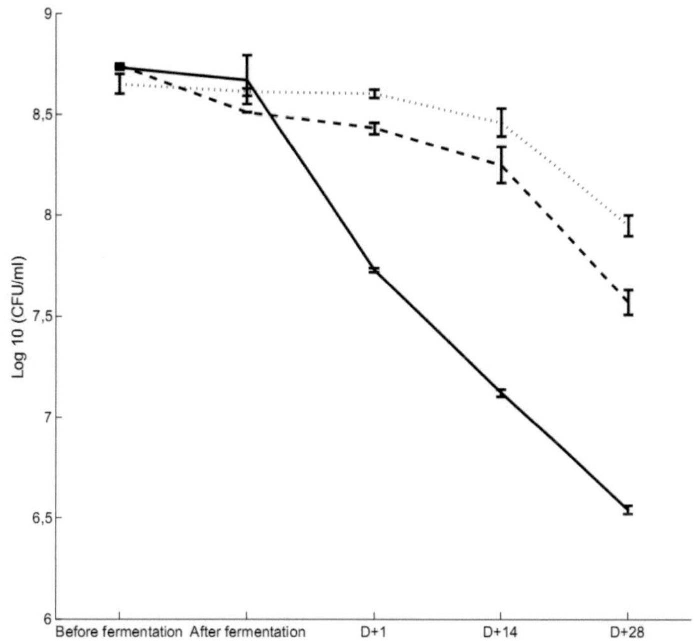


Reprinted from Journal of J. Dairy Sci., Vol 945, Ebel B, Martin F, Le LDT, Gervais P, Cachon R, Use of gases to improve survival of *Bifidobacterium bifidum* by modifying redox potential in fermented milk, Pages No. 2185-2191, Copyright (2011), with permission from Elsevier.

**Figure 2.** Evolution of a population of *Bifidobacterium bifidum* during fermentation and storage. Different gaseous conditions were applied to the milk: control (solid line), gassed with $N_2$ (dashed line), or gassed with $N_2 - H_2$ (dotted line).

After 28 days of storage, a difference in bacterial counts of 1.2 log and 1.5 log was observed between the control milk and after bubbling with $N_2$ or $N_2 - H_2$ respectively. No differences were highlighted during the fermentation process. It is interesting to note that this technique was set up without affecting the fermentation kinetics and survival of *S. thermophilus* and *Lb. bulgaricus*. The use of gas is a possible way of improving probiotic survival during storage without affecting acidification properties of yoghurt strains and consequently organoleptic properties.

## 4.2. Cheese

Controlling $E_h$ in cheese seems essential in governing aroma characteristics. Indeed, a reducing $E_h$ is necessary for the development of the characteristic flavour of certain fermented dairy products such as cheeses, notably through the production of thiol compounds [45, 66]. It has also been reported that Cheddar has a reducing $E_h$ and is an indicator of the establishment of the conditions required for the formation of aroma compounds [67]. As shown previously, $E_h$ can modify the metabolic pathways of aroma production by lactic bacteria [51]. Kieronczyk *et al.* [49] demonstrated that reducing $E_h$ conditions can stimulate carboxylic acid production in cheese, while oxidative $E_h$ conditions improve the production of volatile sulphur compounds and aldehydes. By ripening cheese under reducing $E_h$ conditions, the production of volatile fatty acids increased [68]. Adjusting the $E_h$ of the milk before cheese ripening could be a possible way of modifying the metabolism of lactic bacteria.

## 5. Conclusion

Pasteur defined fermentation as "life without air". In lactic acid bacteria, some exogenous electron acceptors may interfere significantly with the fermentative metabolism by acting on different cellular activities. A better understanding of the adaptive mechanisms to extracellular redox is still lacking, but the results in the literature show that lactic acid bacteria may use passive or active mechanisms. A remarkable feature in lactic acid bacteria is their ability to reduce the redox environment to low $E_h$ values.

With the prospect of food applications, changing $E_h$ using pure or a mixture of gases has the advantage of maintaining product safety as opposed to the use of oxidizing or reducing molecules. This chapter demonstrates the importance of $E_h$ both on the physico-chemistry of milk gels and bacterial metabolism and viability. The use of gas to modify $E_h$ seems to be an interesting way of varying the organoleptic properties of dairy products.

## Author details

F. Martin, B. Ebel, C. Rojas, P. Gervais, N. Cayot and R. Cachon
*Unité Procédés Alimentaires et Microbiologiques, UMR A 02.102, AgroSup Dijon/Université de Bourgogne, 1 esplanade Erasme, Dijon, France*

## 6. References

[1] Dave RI, Shah NP (1997) Effectiveness of ascorbic acid as an oxygen scavenger in improving viability of probiotic bacteria in yoghurts made with commercial starter cultures. Int. dairy j. 7(6-7): 435-443.

[2] Rödel W, Scheuer R (2000) Redox potential of meat and meat produits. IV. Recording criteria of quality in meat and meat products by measuring the redox potential. Fleischwirtschaft. 2: 46-48.

[3] Galster H (1991) pH Measurement: fundamentals, methods, applications, instrumentation. New York: VCH Publishers.

[4] Abraham S, Cachon R, Colas B, Feron G, De Coninck J (2007) $E_h$ and pH gradients in camembert cheese during ripening: Measurements using microelectrodes and correlations with texture. Int. dairy j. 178: 954-960.

[5] Leistner L, Mirna A (1959) Das redoxpotential von pökelladen. Fleischwirtschaft. 8: 659-666.

[6] Cachon R, Jeanson S, Aldarf M, Divies C (2002) Characterisation of lactic starters based on acidification and reduction activities. Lait. 823: 281-288.

[7] George SM, Richardson LCC, Pol IE, Peck MW (1998) Effect of oxygen concentration and redox potential on recovery of sublethally heat-damaged cells of *Escherichia coli* O157:H7, *Salmonella enteritidis* and *Listeria monocytogenes*. J. appl. microbiol. 84: 903-909.

[8] Medard (2002) Encyclopédie des gaz de l'air liquide 3rd ed. Amsterdam, The Netherlands: Lavoisier.

[9] Tamine AY, Robinson RK (1999) Yoghurt: science and technology. Washington DC: CRC Press.

[10] Fredrickon AG (1977) Behavior of mixed cultured of microorganisms. Annu. rev. microbiol. 31: 63-87.

[11] Rasic JL, Kurmann JA. Fermented fresh milk products (1978) In: Rasic JL, Kurmann JA, editors. Yoghurt scientific grounds, technology, manufacture and preparations. Copenhangen, Denmark: Technical Dairy Publishing House.

[12] Moreira M, Abraham A, De Antoni G (1999) Technological properties of milks fermented with thermophilic acid bacteria at suboptimal temperature. J. dairy sci. 83: 395-400.

[13] Zourari A, Accolas JP, Desmazeaud M (1992) Metabolism and biochemical characteristics of yoghurt bacteria: a review. Lait. 72: 1-34.

[14] Loones A. Laits fermentés par les bactéries lactiques (1994) In: De Roissart H, Luquet FM, editors. Bactéries lactiques. Paris, France: Lorica, Uriage. p. 133-154.

[15] Courtin P, Rul F (2003) Interactions between microorganisms in a simple ecosystem: yogurt bacteria as a study model. Lait. 84: 125-134.

[16] El-Soda MA, Abou-Donia SA, El-Shafy HK, Mashaly R, Ismail AA (1986) Metabolic activities and symbiosis in zabady isolated cultures. Egypt. j. dairy sci. 14: 1-10.

[17] Xanthopoulos V, Petridis D, Tzanetakis N (2001) Characterisation and classification of *Streptococcus thermophilus* and *Lactobacillus delbrueckii* subsp. *bulgaricus* strains isolated from traditional Greek yogurts. J. food sci. 66: 747-752.

[18] Ebel B, Martin F, Le LDT, Gervais P, Cachon R (2011) Use of gases to improve survival of *Bifidobacterium bifidum* by modifying redox potential in fermented milk. J. dairy sci. 945: 2185-2191.

[19] Cachon R, Feron G, Delbeau C, Ibarra D, Ledon H (2006) Process for modifying the sensory characteristics of a fermented milk product and its maturation during conservation of said product. EP1649755. L'Air Liquide, France.

[20] Martin F, Cayot N, Marin A, Journaux L, Cayot P, Gervais P, Cachon R (2009) Effect of oxidoreduction potential and of gas bubbling on rheological properties and microstructure of acid skim milk gels acidified with glucono-δ-lactone. J.dairy sci. 92: 5898-5906.

[21] Lucey JA, Tamehana M, Singh H, Munro PA (1998) A comparison of the formation, rheological properties and microstructure of acid skim milk gels made with a bacterial culture or glucono-δ-lactone. Food res. int. 312: 147-155.

[22] Fiszman SM, Lluch MA, Salvador A (1999) Effect of addition of gelatin on microstructure of acidic milk gels and yoghurt and on their rheological properties. Int. dairy j. 912: 895-901.

[23] Martin F, Cayot N, Vergoignan C, Journaux L, Gervais P, Cachon R (2010) Impact of oxidoreduction potential and of gas bubbling on rheological properties of non-fat yoghurt. Food res. int. 431: 218-223.

[24] Dave RI, Shah NP (1997) Effectiveness of ascorbic acid as an oxygen scavenger in improving viability of probiotic bacteria in yoghurts made with commercial starter cultures. Int. dairy j. 76-7: 435-443.

[25] van Dijk C, Ebbenhorst-Selles T, Ruisch H, Stolle-Smits T, Schijvens E, van Deelen W, Boeriu C (2000) Product and Redox Potential Analysis of Sauerkraut Fermentation. J. agr. food chem. 482: 132-139.

[26] Ruas-Madiedo P, Hugenholtz J, Zoon P (2002) An overview of the functionality of exopolysaccharides produced by lactic acid bacteria. Int. dairy j. 12(2-3): 163-171.

[27] Cerning J, Bouillanne C, Desmazeaud MJ, Landon M (1986) Isolation and characterization of exocellular polysaccharide produced *Lactobacillus bulgaricus*. Biotechnol. letters. 89: 625-628.

[28] Cerning J, Bouillanne C, Desmazeaud MJ, Landon M (1988) Exocellular polysaccharide production by *Streptococcus thermophilus*. Biotechnol. letters. 104: 255-260.

[29] van den Berg D, Robijn GW, Janssen AC, Giuseppin M, Vreeker R, Kamerling JP, Vliegenthart J, Ledeboer AM, Verrips CT (1995) Production of a novel extracellular polysaccharide by *Lactobacillus sake* 0-1 and characterization of the polysaccharide. Appl. environ. microb. 618: 2840-4.

[30] De Vuyst L, Vanderveken F, van de Ven S, Degeest B (1998) Production by and isolation of exopolysaccharides from *Streptococcus thermophilus* grown in a milk medium and evidence for their growth-associated biosynthesis. J. appl. microbiol. 846: 1059-1068.

[31] Hess SJ, Roberts RF, Ziegler GR (1997) Rheological properties of nonfat yogurt stabilized using *Lactobacillus delbrueckii* ssp. *bulgaricus* producing exopolysaccharide or using commercial stabilizer systems. J. dairy sci. 802: 252-263.

[32] van Marle ME, van den Ende D, de Kruif CG, Mellema J (1999) Steady shear viscosity of stirred yoghurts with varying ropiness. J. rheol. 43(6): 1643-1663.

[33] Rawson HL, Marshall VM (1997) Effect of 'ropy' strains of *Lactobacillus delbrueckii* ssp. *bulgaricus* and *Streptococcus thermophilus* on rheology of stirred yogurt. Int. j. food sci. technol. 323: 213-220.

[34] Hassan AN, Frank JF, Farmer MA, Schmidt KA, Shalabi SI (1995) Formation of yogurt microstructure and three-dimensional visualization as determined by confocal scanning laser microscopy. J. dairy sci. 7812: 2629-2636.

[35] van Marle ME. (1998) Structure and rheological properties of yoghurt gels and stirred yoghurts. Enschede, the Netherlands: Twente.

[36] Ott A, Fay LB, Chaintreau A (1997) Determination and origin of the aroma impact compounds of yogurt flavor. J. agric. food chem. 453: 850-858.

[37] Imhof R, Glättli H, Bosset JO (1994) Volatile organic aroma compounds produced by thermophilic and mesophilic mixed strain dairy starter cultures. LWT - Food sci. technol. 275: 442-449.

[38] Pette J, Lolkema H (1950) Yoghurt, III: Acid production and aroma formation in yoghurt. Neth. milk dairy j. 4: 261-273.

[39] Biavati B, Vescovo M, Torriani S, Bottazzi V (2000) Bifidobacteria: history, ecology, physiology and applications. Ann. microbiol. 502: 117-131.

[40] Neijssel OM, Snoep JL, Teixeira de Mattos MJ (1997) Regulation of energy source metabolism in streptococci. J. appl. microbiol. 83S1: 12S-19S.

[41] Ott A, Germond JE, Baumgartner M, Chaintreau A (1999) Aroma comparisons of traditional and mild yogurts: headspace gas chromatography quantification of volatiles and origin of α-diketones. J. agric. food chem. 476: 2379-2385.

[42] Monnet C, Schmitt P, Divies C (1994) Method for assaying volatile compounds by headspace gas chromatography and application to growing starter cultures. J. dairy sci. 777: 1809-1815.

[43] Starrenburg M, Hugenholtz J (1991) Citrate fermentation by *Lactococcus* and *Leuconostoc* spp. Appl. environ. microb. 57: 3535-3540.

[44] Boumerdassi H, Desmazeaud M, Monnet C, Boquien CY, Corrieu G (1996) Improvement of diacetyl production by *Lactococcus lactis* ssp. *lactis* CNRZ 483 through oxygen control. J. dairy sci. 795: 775-781.

[45] Green ML, Manning DJ (1982) Development of texture and flavour in cheese and other fermented products. J. dairy res. 4904: 737-748.

[46] Tachon S, Michelon D, Chambellon E, Cantonnet M, Mezange C, Henno L, Cachon R, Yvon M (2009) Experimental conditions affect the site of tetrazolium violet reduction in the electron transport chain of *Lactococcus lactis*. Microbiology+. 1559: 2941-2948.

[47] Singh H, Drake MA, Cadwallader KR (2003) Flavor of cheddar cheese: a chemical and sensory perspective. Compr. rev. food sci. f. 2: 139-162.

[48] Bonnarme P, Arfy K, Dury C, Helinck S, Yvon M, Spinnler HE (2001) Sulfur compound production by *Geotrichum candidum* from L-methionine: importance of the transamination step. FEMS microbiol. lett. 205: 247-252.

[49] Kieronczyk A, Cachon R, Feron G, Yvon M (2006) Addition of oxidizing or reducing agents to the reaction medium influences amino acid conversion to aroma compounds by *Lactococcus lactis*. J. appl. microbiol. 1015: 1114-1122.

[50] Riondet C, Cachon R, Waché Y, Alcaraz G, Diviès C (1999) Changes in the proton-motive force in *Escherichia coli* in response to external oxidoreduction potential. Eur. j. biochem. 2622: 595-599.

[51] Martin F, Cachon R, Pernin K, De Coninck J, Gervais P, Guichard E, Cayot N (2011) Effect of oxidoreduction potential on aroma biosynthesis by lactic acid bacteria in nonfat yogurt. J. dairy sci. 942: 614-622.

[52] Hugenholtz J, Kleerebezem M (1999) Metabolic engineering of lactic acid bacteria: overview of the approaches and results of pathway rerouting involved in food fermentations. Curr. opin. biotech. 105: 492-497.

[53] Condon S (1987) Responses of lactic acid bacteria to oxygen. FEMS microbiol. lett. 463: 269-280.

[54] Bassit N, Boquien CY, Picque D, Corrieu G (1993) Effect of Initial Oxygen Concentration on Diacetyl and Acetoin Production by *Lactococcus lactis* subsp. *lactis* biovar diacetylactis. Appl. environ. microb. 596: 1893-1897.

[55] Gardini F, Lanciotti R, Elisabetta Guerzoni M, Torriani S (1999) Evaluation of aroma production and survival of *Streptococcus thermophilus, Lactobacillus delbrueckii* subsp. *bulgaricus* and *Lactobacillus acidophilus* in fermented milks. Int. dairy j. 92: 125-134.

[56] Imhof R, Bosset JO (1994) Quantitative GC-MS analysis of volatile flavour compounds in pasteurized milk and fermented milk products applying a standard addition method. LWT - Food sci technol. 273: 265-269.

[57] Ott A, Germond JE, Chaintreau A (2000) Vicinal diketone formation in yogurt: 13C precursors and effect of branched-chain amino acids. J. agr food chem. 483: 724-731.

[58] Storz G, Imlay JA (1999) Oxidative stress. Curr. opin. microbiol. 22: 188-94.

[59] Talwalkar A, Kailasapathy K (2004) A review of oxygen toxicity in probiotic yogurts: Influence on the survival of probiotic bacteria and protective techniques. Compr. rev. food sci. f. 33: 117-124.

[60] Talwalkar A, Kailasapathy K (2004) The role of oxygen in the viability of probiotic bacteria with reference to *L. acidophilus* and *Bifidobacterium* spp. Curr. issues intest. microbiol. 51: 1-8.

[61] Bolduc MP, Bazinet L, Lessard J, Chapuzet JM, Vuillemard JC (2006) Electrochemical modification of the redox potential of pasteurized milk and its evolution during storage. J agric. food chem. 5413: 4651-4657.

[62] Bolduc MP, Raymond Y, Fustier P, Champagne CP, Vuillemard JC (2006) Sensitivity of bifidobacteria to oxygen and redox potential in non-fermented pasteurized milk. Int. dairy j. 169: 1038-1048.

[63] Dave RI, Shah NP (1997) Effect of cysteine on the viability of yoghurt and probiotic bacteria in yoghurts made with commercial starter cultures. Int. dairy j. 78-9: 537-545.

[64] Lourens-Hattingh A, Viljoen BC (2001) Yogurt as probiotic carrier food. Int. dairy j. 111-2: 1-17.

[65] Mortazavian AM, Ehsani MR, Mousavi SM, Reinheimer JA, Emamdjomeh Z, Sohrabvandi S, Rezaei K (2006) Preliminary investigation of the combined effect of heat treatment and incubation temperature on the viability of the probiotic micro-organisms in freshly made yogurt. Int. j. dairy technol. 59(1): 8-11.

[66] Kristoffersen T (1985) Development of flavor in cheese. Milchwissenschaft. 40(4): 197-199.

[67] Urbach G (1995) Contribution of lactic acid bacteria to flavour compound formation in dairy products. Int. dairy j. 58: 877-903.

[68] Ledon H, Ibarra D (2006) Method for modifying hygienic, physico-chemical and sensory properties of cheese by controlling the redox potential. US 2009/0214705 A1. L'Air Liquide, France.

Section 2

# Meat Products

Chapter 4

# Potential of Fermented Sausage-Associated Lactic Acid Bacteria to Degrade Biogenic Amines During Storage

Jirasak Kongkiattikajorn

Additional information is available at the end of the chapter

http://dx.doi.org/10.5772/51302

## 1. Introduction

Biogenic amines (BAs) are organic bases with aliphatic, aromatic or heterocyclic structures that can be found in several foods, in which they are mainly produced by microbial decarboxylation of amino acids, with the exception of physiological polyamines. BAs may be of endogenous origin at low concentrations in non-fermented food such as fruits, vegetables, meat, milk and fish. High concentrations have been found in fermented foods as a result of a contaminating microflora exhibiting amino acid decarboxylase activity (Silla-Santos, 1996). However, BAs can also trigger human health problems leading to palpitations, hypertension, vomiting, headaches and flushing if food containing high concentrations are ingested. In fermented foods, some lactic acid bacteria (LAB) are able to convert available amino acid precursors into BAs via decarboxylase or deiminase activities during or following ripening processes. For this reason, amino acid catabolism by LAB can affect both the quality and safety of fermented foods (Verges et al., 1999). The amount and type of BAs formed depends on the nature of food and particularly on the kind of microorganisms present. Enterobacteriaceae and certain LAB are particularly active in the production of BA (Beutling, 1996). These amine-producing microorganisms either may form part of the food associated population or may be introduced by contamination before, during or after processing of the food product. Therefore, microorganisms naturally present in raw materials, introduced throughout the processing or added as starter culture can critically influence BA production during the manufacture of fermented products (Bover-Cid et al., 2000).

Nham is a Thai-style fermented pork sausage. Nham ripening generally takes 3-5 days and relies mainly on adventitious microorganisms, which are normally found in raw materials.

LAB produce organic acids from carbohydrates and cause the pH drop, which contribute to Nham formation. *Micrococcus* and *Staphylococcus* are capable of reducing nitrate to nitrite, which is important in producing the characteristic pigmentation. Also, as a source of lipolytic and proteolytic enzymes, they may contribute to flavor production. Therefore, the acidification and the proteolytic process occurring during Nham ripening make the environment particularly favorable for BAs production.

During meat ripening, microbial growth, acidification and proteolysis provide favourable conditions for BA production. The species of lactobacilli most commonly found in meat and meat products are *Lactobacillus sake* and *Lactobacillus curvatus*, which together with *Lactobacillus bavaricus* and *Lactobacillus plantarum* constitute the main microbial flora isolated from fermented sausages. Other bacteria that can be found in relatively high numbers include enterococci (*E. faecalis* and *E. faecium*), which also contribute to the ripening process. However, the presence of enterococci might also reflect a given level of contamination or a poor curing process. Salt-tolerant, nitrate-reducing coagulase-negative staphylococci are also detected in relatively high numbers in ripened meat products. *Staphylococcus xylosus* is the main species found in Spanish fermented sausages, although *S. carnosus* can also be used as a starter culture. BAs can be degraded through oxidative deamination catalyzed by amines oxidase (AO) with the production of aldehyde, ammonia and hydrogen peroxide. Monoamine oxidases (MAOs) and diamine oxidases (DAOs) had been described from some genus of the family Enterobacteriaceae (Yamashita et al., 1993). The potential role of microorganisms with AO activity had become a particular interest in the last few years to prevent or reduce BA accumulation in food products, especially fermented foods. Mah and Hwang (2009) investigated the effect of *Staphylococcus xylosus* to inhibit BA formation in a salted and fermented anchovy. Reduction of tyramine during ripening of fermented sausages was achieved when *Micrococcus varians* was applied as starter culture (Leuschner and Hammes, 1998). Inoculation of *L. plantarum* in sauerkraut effectively suppressed the production of tyramine, putrescine and cadaverine (Kalac et al., 2000).

BAs are physiologically inactivated by AO, which are enzymes found in bacteria, fungi, plant and animal cells able to catalyse the oxidative deamination of amines with production of aldehydes, hydrogen peroxide and ammonia (Cooper, 1997). The sequential action (in the presence of an electron acceptor, such as $O_2$) of an AO and an aldehyde dehydrogenase leads to the production of an acid and ammonia, which can be used to support microbial growth (Parrot et al., 1987). MAO and DAO activity has been described in higher organisms as well as in bacteria (Murooka et al., 1976, 1979; Ishizuka et al., 1993). There are relevant differences between microbial AO in terms of substrate specificity and location, as stated by Cooper (1997). DAOs can oxidase several BA, such as putrescine and histamine, and their activity can be affected by substrate inhibition; aminoguanidine, antihistaminic drugs and foodborne inhibitors, such as ethanol, carnosine, thiamine, cadaverine and tyramine, reduce their activity (Lehane and Olley, 2000). The potential role of microorganisms involved in food ripenings with AO activity has been investigated with the aim to prevent or reduce the accumulation of BA in foods. Leuschner et al. (1998) tested in vitro the potential amine degradation by many bacteria isolated from foods and, in particular, in strains belonging to the genera *Lactobacillus*,

*Pediococcus, Micrococcus,* as well as to the species *S. carnosus* and *Brevibacterium linens.* They found that this enzymatic activity can be present at very different quantitative levels. Tyramine oxidase activity of several microbial strains was strictly dependent on pH (with an optimum at 7.0), temperature and NaCl, as well as glucose and hydralazine concentration. Moreover, this enzyme was characterised by a higher potential activity under aerobic conditions. Temperature has also an important effect on histamine degradation (Dapkevicius et al., 2000). The highest degradation rate of this amine was observed at 37 °C, but at 22°C and 15 °C, degradation was still considerable. The AO responsible for this degradation has its optimum temperature at 37°C and retains about 50% of its maximum activity at 20 °C (Schomburg and Stephan, 1993). Many *S. xylosus* strains isolated from artisanal fermented sausages in southern Italy showed the ability to degrade BA in vitro (Martuscelli et al., 2000). Among the strains tested, *S. xylosus* S81 completely oxidised histamine, but it degraded, under the adopted conditions, also a part of tyramine. Even if the AO activity in vitro of microorganisms is not quantitatively reproducible in vivo (due to the more severe conditions and, in particular, to the low $O_2$ tension, pH and salt concentration), reduction of histamine in dry sausages has been observed in the presence of AO-positive staphylococcal starter cultures (Leuschner and Hammes, 1998). In addition, important reduction of the concentration of tyramine and putrescine in the presence of AO positive *S. xylosus* starter cultures have been observed by Gardini et al. (2002). In other words, BA presence in foods is the consequence of a complex equilibrium between the composition of the food and the enzymatic activities of the microbial population. Together with the decarboxylating aptitude of the starter cultures, the presence and relative activity of AO should be considered as an important characteristic in the selection of starter cultures used in the production of fermented foods.

Since Nham is normally consumed without cooking, proper acid production is important to determine the quality and safety of Nham for consumption. Depending on the initial number of contamination, the occurrence of pathogens such as *Salmonella* spp., *Staphylococcus aureus*, and *Listeria monocytogenes* was found specially in Nham with pH higher than 4.6. Due to inconsistency of product quality and ambiguous product safety, improved process of Nham ripening has been developed by using a starter culture technology. Starter cultures are applied to improve and stabilize the quality of the final product and to shorten the ripening period of Nham production. Meanwhile, only little information is available on the effect of starter culture on BA reduction in Nham. Therefore, the objective of this study was to investigate the effectiveness of AOs activity of LAB in inhibiting BA accumulation during Nham ripening. In addition, the change of chemical and microbial properties of Nham during ripening and subsequently during 28 days stored at different temperature was investigated.

## 2. Materials and methods

### 2.1. Microbiological analysis

Nham sausages (25 g) were aseptically transferred into a stomacher bag, with 225 mL of peptone (0.85% of sodium chloride added) and then homogenized for two minutes. Further decimal dilutions were made and then 100 μL of each dilution was spread onto agar plates.

Aerobic plate count agar was used to determine total aerobic. BA producing bacteria were counted using differential media supplemented with amino acids as precursor of BAs (Joosten and Northolt, 1989). The media contained of tryptone (0.5%), yeast extract (0.5%), sodium chloride (0.5%), glucose (0.1%), Tween 80 (0.05%), $MgSO_4 \bullet 7\ H_2O$ (0.02%), $CaCO_3$ (0.01%), $MnSO_4 \bullet 4H_2O$ (0.005%), $FeSO_4 \bullet 7H_2O$ (0.004%), bromocresol purple (0.006%), amino acid (2%) and agar (2%). The medium contained the precursor amino acids (0.5% tyrosine di-sodium salt and 0.25% L-histidine monohydrochoride, L-ornithine monohydrochoride, L-lysine monohydrochoride, L- phenylalanine, and L-tryptophan), pyridoxal-5-phosphate as a codecarboxylase factor, growing factors and buffer compounds. All plates were then incubated for 48 h at 37 °C. Bacterial colonies which developed on each agar were then enumerated and expressed as log colony forming unit (CFU)/mL. Only bacterial colonies with purple halo in the differential media were counted as BAs producing bacteria.

## 2.2. Bacterial strains and growth conditions

Bacterial strains isolated from different fermented sausages were tested. LAB were grown in MRS broth.

## 2.3. Determination of amine degradation

An overnight culture was harvested, washed with 0.05 M phosphate buffer (pH 7) and the cell pellet resuspended in 0.05 M phosphate buffer supplemented with tyramine, histamine, tryptamine, phenylethylamine, putrescine, and cadaverine. The cell concentration was adjusted to $10^6$, $10^7$ and $10^8$ CFU/mL. The cell suspensions (20 mL) were incubated in a 100 ml flask and shaken at 200 rpm. Samples were taken and added to an equal amount of 1 M HCl. The mixture was boiled for 10 min and centrifugated at 9000 g. The supernatant was frozen at -15°C until HPLC analysis.

## 2.4. Preparation of starter culture

Starter cultures used in this study were *L. plantarum* + *L. sake*, which were isolated from sausage. A loop from a slant tryptic soy agar culture of each culture was inoculated in 10 mL of tryptic soy broth and incubated at 37 °C for 24 h. Five milliliters of the culture was then transferred to 100 mL of tryptic soy broth and incubated at 37 °C for another 24 h. The culture was centrifuged at 10,000 g for 10 min at 4°C and then washed with broth. Broth was prepared by homogenizing 1 part with 9 part of distilled water, filtered, adjusted to pH 7.0 and then autoclaved at 121°C for 15 min. After centrifugation, the cell pellet was resuspended in sterile fish broth, adjusted to approximately $10^7$ cell/g and used as starter culture in sauce ripening.

## 2.5. Nham preparation

Minced pork (56%), pieced cooked pork skin (37%), garlic (3.2%), cooked rice (2%), sodium polyphosphate (0.15%), sodium chloride (1.5%) and sodium erythrobate (0.15%) chili (1%)

were mixed thoroughly, packed into a plastic casing and sealed before incubation. Two separated batches of fermented sausage were prepared without starter culture and with different starter cultures (*L. plantarum* + *L. sake*) of approximately $10^7$ cell/g. After incubation the fermented sausages were homogenized for analysis.

## 2.6. Physical and chemical analyses

The pH was measured directly from samples using a microcomputerized pH meter, inserting the electrode into the middle of the sausage. Moisture was determined by drying the sample at 100–105°C until a constant weight was achieved. The color of Nham was determined by Minolta Model DP-301 colorimeter. Color values (L, a, and b) were measured. A white standard tile was used to calibrate the colorimeter (L= 100.01, a= -0.01, b= -0.02) before measurements. Therefore L measures lightness (luminosity) and varies from white to black. The chromatically (a and b values) gives designations of color as follows; a-value measures redness when positive, gray when zero, and greenness when negative, b-value measures yellowness when positive, gray when zero, and blueness when negative. The titratable acidity (TA) determined as total acid was estimated according to AOAC (2000) and expressed as g/100 g dry matter. TCA (trichloroacetic acid)-soluble peptide of the fermented sausages was measured by the method of Greene and Babbitt (1990). The oligopeptide content in the supernatant was determined according to by the method of Lowry et al (1951). Results were expressed as μmol/g (dry matter). Free α-amino acid was measured using TNBS according to Benjakul and Morrissey (1997) Results were expressed as μmol/g (dry matter).

## 2.7. Extraction of amino acids and BAs

10 ML of 10% (w/v) trichloroacetic acid (TCA) were added to 3 g-samples, and homogenization of the mixture was effected via shaking for 1 h. The extract was then filtered through Whatman No. 1 filter paper. To remove any fat, the samples were kept at -20 °C for 1 d, and then centrifuged at 7000 g for 15 min. The supernatants were collected and filtered through a 0.25 μm membrane filter.

## 2.8. Determination of BAs

Amines were determined by the high-performance liquid chromatography (HPLC) method described by Hernández-Jover et al. (1996). The method is based on the formation of ion pairs between amines extracted with 0.6 M perchloric acid from 5 to 10 g of sample, and octanesulphonic acid present in the mobile phase. Separation is preformed using a reversed phase column, then a postcolumn derivatization with *o*-phthalaldehyde (OPA) is followed by spectrofluorimetric detection. The method allows one to quantify, by an external standard procedure, 6 BAs, i.e., tyramine, histamine, tryptamine, phenylethylamine, putrescine, cadaverine. Samples for BA determination were stored at -15°C until required.

### 2.9. Determination of amino acids

Free amino acids (FAAs) in samples were determined using HPLC according to the method proposed by Rozan et al. (2000). A 20 μL aliquot of amino acid standard and digested sauce samples were transferred into vials and dried under vacuum. Then 20 μL of drying reagent containing methanol, water and triethylamine (ratio 2:2:1 v/v) was added. Then 20 μL of derivatizing reagent containing methanol, triethylamine, water and phenylisothiocyanate (PITC) (ratio 7:1:1:1 v/v) was added. The derivatized samples were then dissolved in 100 mL of buffer A that was used as mobile phase for HPLC. A Purospher® STAR RP-18e, 5 μm column was used with buffer A (0.1 M ammonium acetate, pH 6.5) and buffer B (0.1 M ammonium acetate containing acetonitrile and methanol, 44:46:10 v/v, pH 6.5) as mobile phase set for linier gradient at the flow rate of 1 mL/min. The injected sample volume was 20 μL and monitored at 254 nm of wavelength.

### 2.10. Statistical analysis

Data was analysed by one-way ANOVA and differences among treatment means were determined by Duncan's new multiple-range test.

## 3. Results and discussion

The effect of starter cultures of LAB on BAs and FAAs content was examined during the ripening process of Nham sausages. Microbial counts, pH and proteolysis-related parameters were also studied. The occurrence of amino acid-decarboxylase activity in 7 strains of LAB isolated from Nham sausages was investigated.

| Starter culture | Percent degradation (%) | | | | | |
|---|---|---|---|---|---|---|
| | Tryptamine | Phenylethylamine | Putrescine | Cadaverine | Histamine | Tyramine |
| *Lactobacillus curvatus* 1271 | 0 | 0 | 0 | 0 | 0 | 0 |
| *Lactobacillus farciminis* 1452 | 0 | 0 | 0 | 0 | 0 | 5.7 |
| *Lactobacillus kandleri* 2439 | 0 | 0 | 0 | 3.6 | 0 | 0 |
| *Lactobacillus kefir* 2045 | 0 | 0 | 0 | 0 | 0 | 0 |
| *Lactobacillus plantarum* 9825 | 0 | 0 | 12.6 | 9.3 | 0 | 19.4 |
| *Leuconostoc mali* 7412 | 0 | 0 | 0 | 0 | 0 | 0 |
| *Lactobacillus pentosus* 7054 | 0 | 0 | 5.2 | 4.7 | 0 | 0 |
| *Lactobacillus reuteri* 7498 | 0 | 0 | 4.8 | 0 | 0 | 0 |
| *Lactobacillus sake* 4127 | 0 | 0 | 17.3 | 8.2 | 0 | 14.5 |

**Table 1.** Strains exhibiting the potential to degrade BAs in a buffer system within 24 h at 30°C

The presence of BAs in a decarboxylase synthetic broth was determined by high performance liquid chromatography with OPA derivatization. Among the 9 LAB strains tested, 5 lactobacilli (in particular, *L. curvatus*) were amine producers and *L. plantarum* and *L.*

*sake*, were non-amine forming strains. The ability of AO exhibiting strains of LAB to degrade amine in vivo during sausage ripening was investigated.

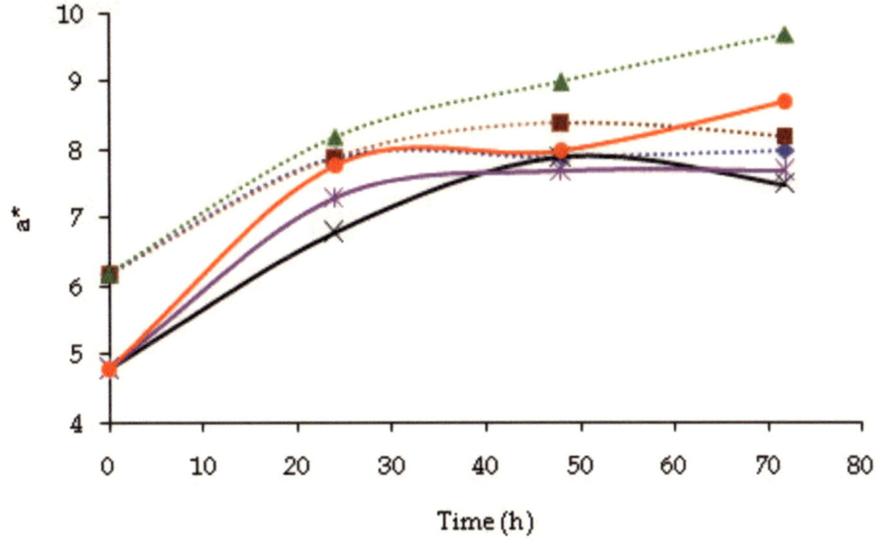


**Figure 1.** a* Value during ripening of Nham control at 25°C (◆), 30°C (■), 37°C (▲) and Nham with starters (*L. plantarum* + *L. sake*) at 25°C (×), 30°C (✳), 37°C (•).

Fig. 1 showed a* values represent red color of Nham during ripening time and temperature at 25°C, 30°C and 37°C, respectively. The results showed a value increased according to ripening and the a value of Nham control at 72 hours 37°C was higher than the other sample.

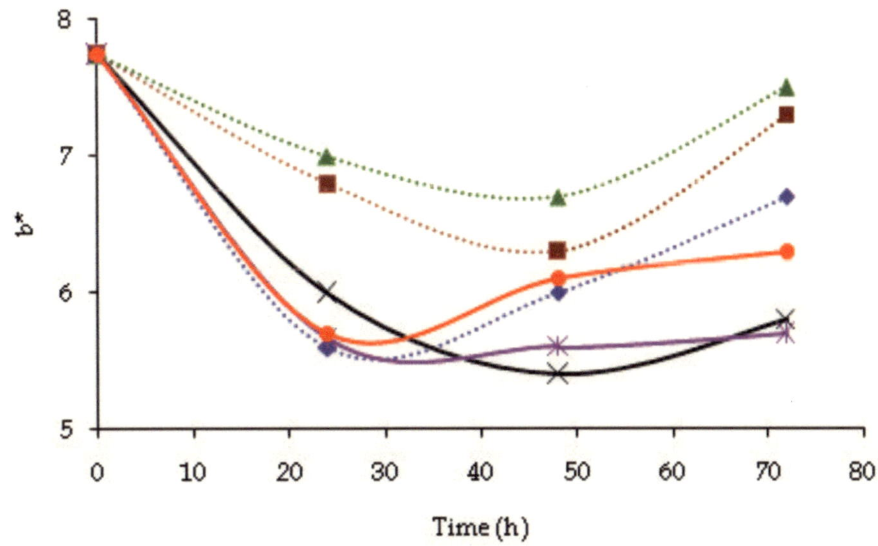


**Figure 2.** b* Value during ripening of Nham control at 25°C (◆), 30°C (■), 37°C (▲) and Nham with starters (*L. plantarum* + *L. sake*) at 25°C (×), 30°C (✳), 37°C (•).

Fig. 2 shows b* values represent yellow color of Nham during ripening time and temperature at 25°C, 30°C and 37°C, respectively. The results showed b value decreased according to storage and the b value of Nham with starters was lower than that of Nham control.

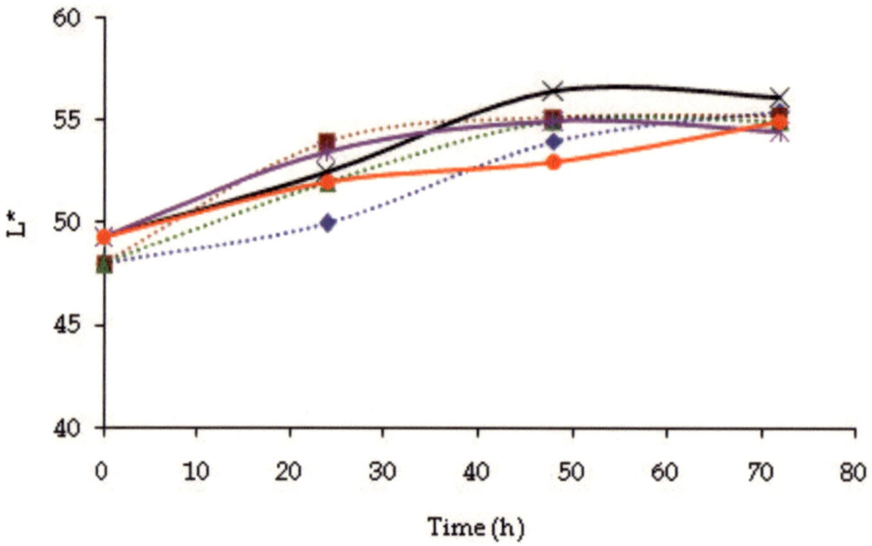


**Figure 3.** L* Value during ripening of Nham control at 25°C (◆), 30°C (■), 37°C (▲) and Nham with starters (*L. plantarum* + *L. sake*) at 25°C (×), 30°C (✳), 37°C (●).

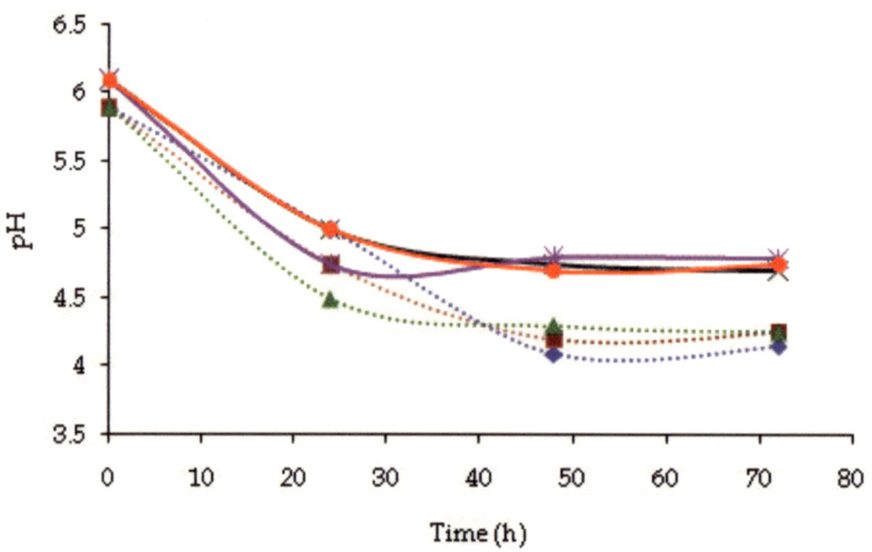


**Figure 4.** pH during ripening of Nham control at 25°C (◆), 30°C (■), 37°C (▲) and Nham with starters (*L. plantarum* + *L. sake*) at 25°C (×), 30°C (✳), 37°C (●).

Fig. 3 is represent L*values represent white color of Nham during ripening time and temperature at 25°C, 30°C and 37°C, respectively. The results showed L* value increased according to storage during 72 hour of ripening.

Fig. 4 shows that the initial pH of Nham samples ranged from 5.9 to 6.1. It then gradually decreased throughout the ripening process and there was significant difference at each time of sampling (*P* <0.05). The pH value reached 4.1 to 4.8 at the end of ripening (hour 72). However, there was significant difference (*P*<0.05) between the pH of Nham control and samples inoculated with starter cultures after 48 hour of ripening.

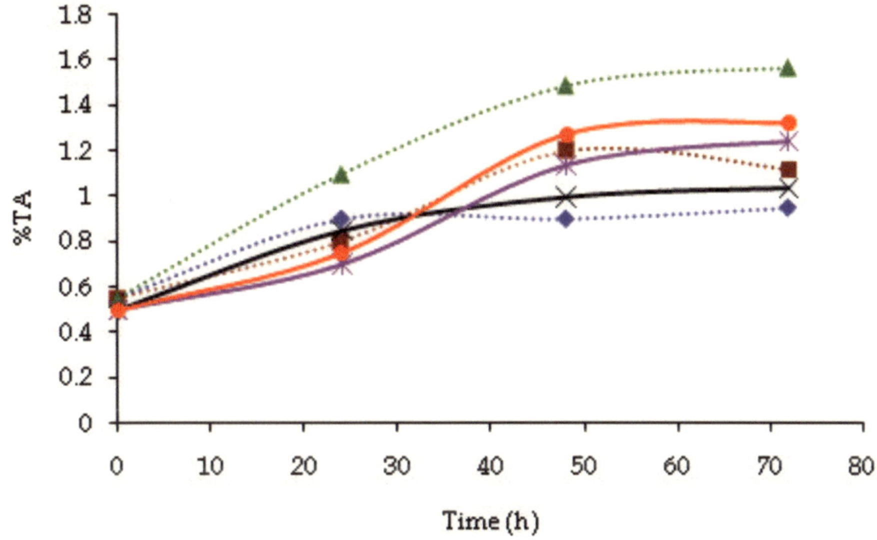


**Figure 5.** Total acid content during ripening of Nham control at 25°C (◆), 30°C (■), 37°C (▲) and Nham with starters (*L. plantarum* + *L. sake*) at 25°C (×), 30°C (✳), 37°C (●).

Fig. 5 shows that the initial total acid content of Nham samples ranged from 0.5 to 0.55. It then gradually increased throughout the ripening process and there was significant difference at each time of sampling (*P* <0.05). The total acid content of Nham control and Nham with starters reached 0.95% to 1.57% and 1.04 %to 1.32% at the end of ripening (hour 72). However, there was significant difference (*P*<0.05) between the total acid content of Nham control and samples inoculated with starter culture after 48 hour of ripening. The results was shown that Nham control fermented at 37°C contained total acid content higher than the other Nham samples.

Fig. 6 shows that TCA-soluble peptide content of Nham samples, the initial content was 9.02 µmol/g dry matter. It then gradually increased throughout the ripening process. The TCA-soluble peptide content of Nham control and Nham with starters reached 23.6 to 87.2 µmol/g dry matter and 24.1 %to 65.2 µmol/g dry matter, respectively, at the end of ripening (hour 72). However, there was not significant difference (*P*<0.05) between the TCA-soluble peptide content of Nham control and samples inoculated with starter culture after 48 hour

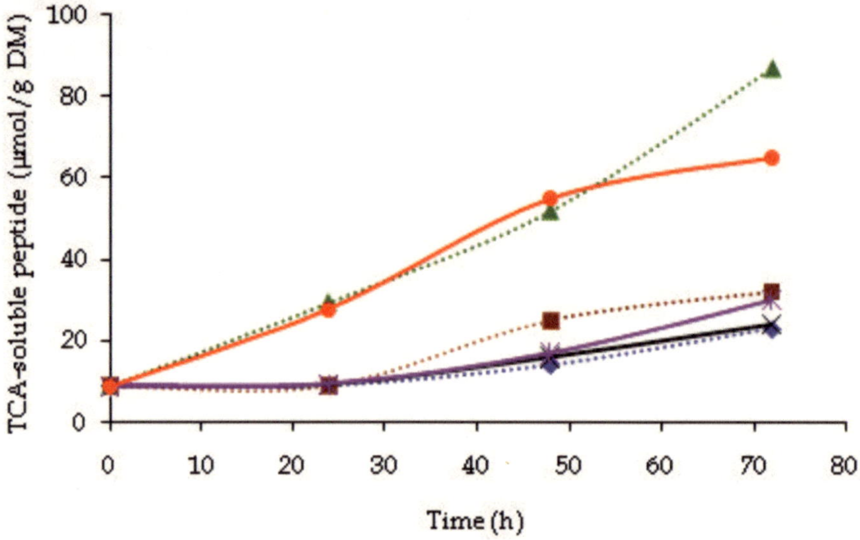


**Figure 6.** TCA-soluble peptide content during ripening of Nham control at 25°C (♦), 30°C (■), 37°C (▲) and Nham with starters (*L. plantarum* + *L. sake*) at 25°C (×), 30°C (✳), 37°C (•).

of ripening at each ripening temperature, and 72 hour of ripening at 25C and 30C. The results was shown that Nham control fermented at 37°C contained TCA-soluble peptide content higher than the other Nham samples after ripening for 72 hour.

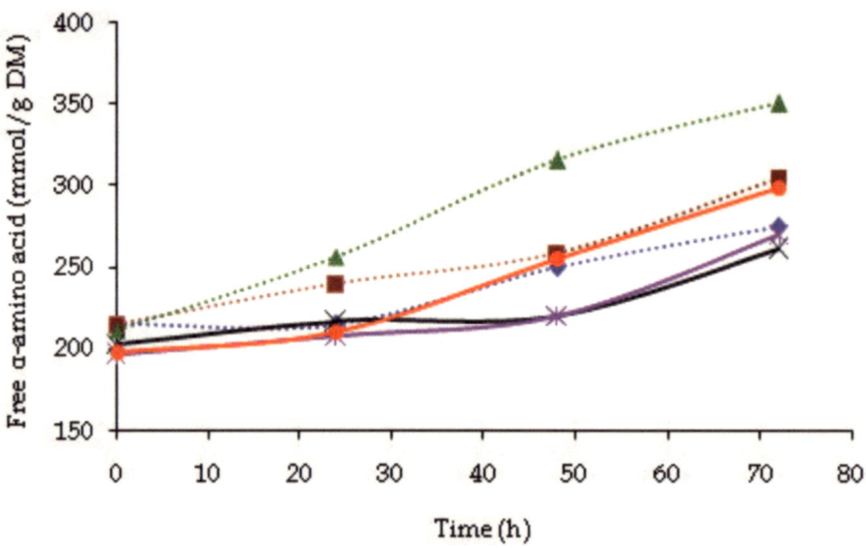


**Figure 7.** Free α-amino acid content during ripening of Nham control at 25°C (♦), 30°C (■), 37°C (▲) and Nham with starters (*L. plantarum* + *L. sake*) at 25°C (×), 30°C (✳), 37°C (•).

Fig. 7 shows that free α-amino acid content of Nham control samples and Nham with starters, the initial content were 216.2 mmol/g dry matter and 203.7 mmol/g dry matter, respectively. It then gradually increased throughout the ripening process. The free α-amino acid content of Nham control and Nham with starters reached 275.3 to 351.6 mmol/g dry matter and 262.4 to 302.2 mmol/g dry matter, respectively, at the end of ripening (hour 72). However, there was not significant difference ($P<0.05$) between the free α-amino acid content of Nham control and samples inoculated with starter culture during ripening at 25°C. The results was shown that Nham control fermented at 37°C contained free α-amino acid content higher than the other Nham samples throughout the ripening process.

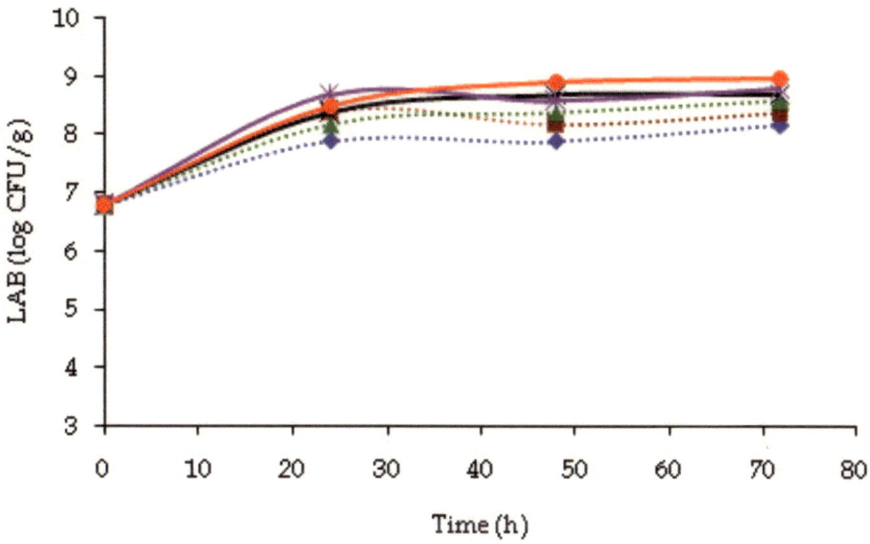


**Figure 8.** Total count of LAB during ripening of Nham control at 25°C (♦), 30°C (■), 37°C (▲) and Nham with starters (*L. plantarum* + *L. sake*) at 25°C (×), 30°C (✳), 37°C (●).

The differences between Nham in counts of LAB during ripening are shown in Fig. 8. LAB in Nham with starters was increase until the 72 h of ripening. Counts of LAB in Nham with starters (8.7 log CFU/g) were higher ($P < 0.05$) than in Nham control (7.7 log CFU/g).

Fig. 9 shows that cadaverine content of Nham samples, the initial content was 14.89 mg/kg dry matter. It then gradually increased throughout the ripening process. The cadaverine content of Nham control and Nham with starters reached 86.2 to 98.7 mg/kg dry matter and 42.4 to 51.6 mg/kg dry matter, respectively, at 72 hour of ripening. However, there was not significant difference ($P<0.05$) between the cadaverine content of Nham with starters during ripening at 25°C and 30°C. The results was shown that Nham control fermented at 37°C contained cadaverine content higher than the other Nham samples throughout the ripening process.

Fig. 10 shows that putrescine content of Nham samples, the initial content was 23.7 mg/kg dry matter. It then gradually increased throughout the ripening process and there was significant difference at each time of sampling ($P <0.05$). The putrescine content of Nham control and Nham with starters reached 115.4 to 242.6 mg/kg dry matter and 65.2 to 98.4 mg/kg dry matter, respectively, at 72 hour of ripening. However, there was not significant

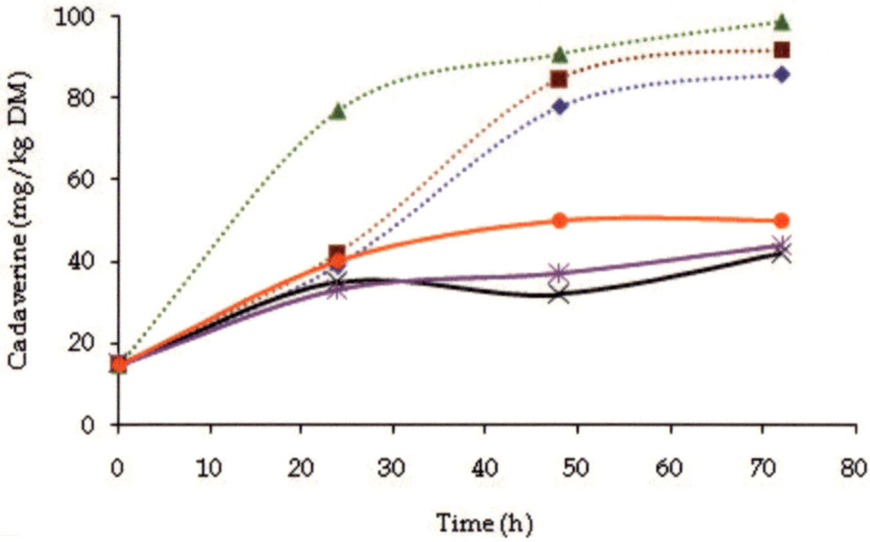


**Figure 9.** Cadaverine content during ripening of Nham control at 25°C (♦), 30°C (■), 37°C (▲) and Nham with starters (*L. plantarum* + *L. sake*) at 25°C (×), 30°C (✳), 37°C (•).

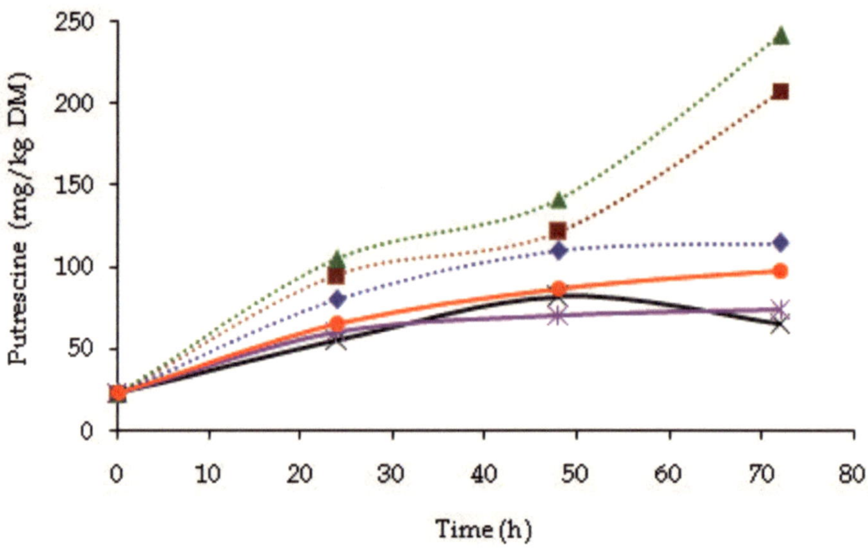


**Figure 10.** Putrescine content during ripening of Nham control at 25°C (♦), 30°C (■), 37°C (▲) and Nham with starters (*L. plantarum* + *L. sake*) at 25°C (×), 30°C (✳), 37°C (•).

difference ($P<0.05$) between the putrescine content of Nham with starters during ripening at 25°C and 30°C. The results was shown that Nham control fermented at 37°C contained putrescine content higher than the other Nham samples throughout the ripening process.

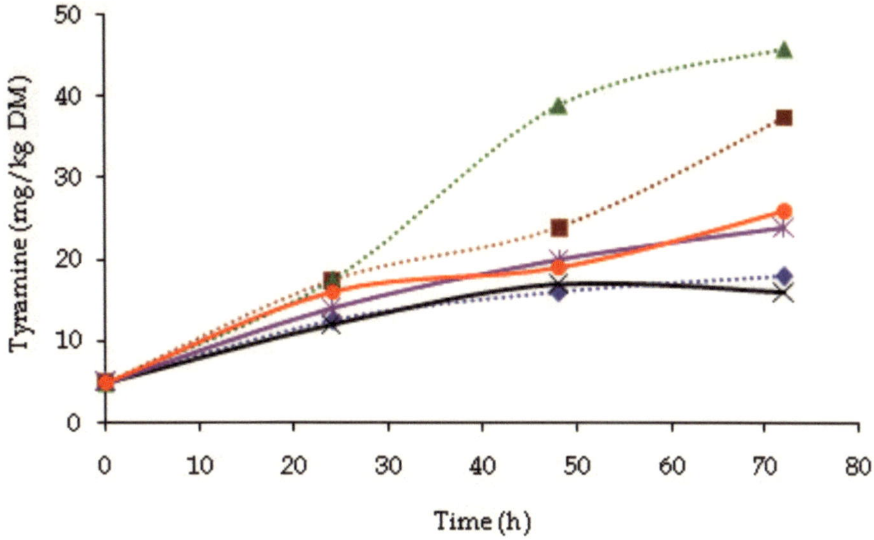


**Figure 11.** Tyramine content during ripening of Nham control at 25°C (◆), 30°C (■), 37°C (▲) and Nham with starters (*L. plantarum* + *L. sake*) at 25°C (×), 30°C (✳), 37°C (•).

Fig. 11 shows that tyramine content of Nham samples, the initial content was 5.63 mg/kg dry matter. It then gradually increased throughout the ripening process and there was significant difference at each time of sampling ($P$ <0.05). The tyramine content of Nham control and Nham with starters reached 17.6 to 46.4 mg/kg dry matter and 16.3 to 27.8 mg/kg dry matter, respectively, at 72 hour of ripening. However, there was not significant difference ($P$<0.05) between the tyramine content of Nham with starters during ripening at 30°C and 37°C and Nham control and Nham samples inoculated with starter culture during ripening at 25°C. The results was shown that Nham control fermented at 37°C contained tyramine content higher than the other Nham samples after 48 hour of the ripening process.

The effect of temperature on BA content was evaluated (Fig. 6-9). The storage temperature of Nham with starters at 30°C and 37°C were shown higher BA oxidation comparing Nham control, a low content was observed at 25°C. This suggested that at ripening temperature of 30 °C and 37°C, a strong oxidation of the AO activity of the starters was evident, whereas at 25°C activity was low for amino acid decarboxylase for lysine (precursor of putrescine) and tyrosine (precursor of tyramine) in Nham control.

Fig. 12 showed a* values represent red color of Nham during stored at 15°C, 4°C and 25°C. The initial a* values of Nham control ranged from 7.0 to 8.4. The results showed a* value increased according to 4 week storage for storage temperature at 15°C and 25°C. However, there was significant decrease ($P$<0.05) between the a* value of 4°C storage of the initial 1 week storage and after 4 week of storage. For Nham with starters, the initial a* values ranged from 7.5 to 8.5. The a* values of Nham with starter decreased according to 4 week storage for storage temperature at 4°C and 15°C. However, there was not significant difference ($P$<0.05) between the a* value of 25°C storage of the initial 1 week storage and after 4 week of storage.

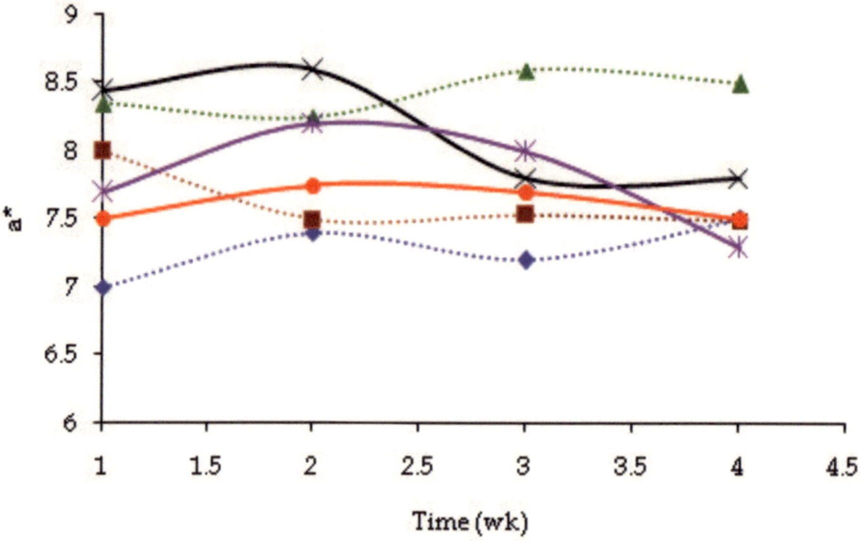


**Figure 12.** a* Value during storage of Nham control at 25°C (◆), 30°C (■), 37°C (▲) and Nham with starters (*L. plantarum* + *L. sake*) at 25°C (×), 30°C (✳), 37°C (●).

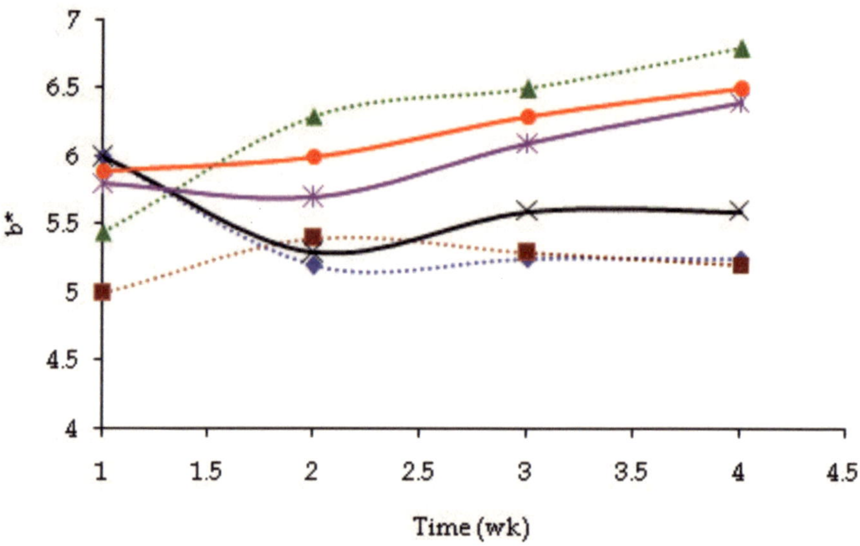


**Figure 13.** b* Value during storage of Nham control at 4°C (■), 15°C (◆), 25°C (▲) and Nham with starters (*L. plantarum* + *L. sake*) at 4°C (✳), 15°C (×), 25°C (●).

Fig. 13 showed b* values represent yellow color of Nham during stored at 15°C, 4°C and 25°C. The initial a* values of Nham control ranged from 5.0 to 6.1. The results showed a* value increased according to 4 week storage for storage temperature at 4°C and 25°C. However, there was significant decrease ($P<0.05$) between the b* value of 15°C storage of the initial 1 week storage and after 4 week of storage. For Nham with starters, the initial a* values ranged from 5.5 to 6.2. The b* values of Nham with starter increased according to 4 week storage for storage temperature at 4°C and 25°C. However, there was significant decrease ($P$<0.05) between the b* value of 15°C storage of the initial 1 week storage and after 4 week of storage.

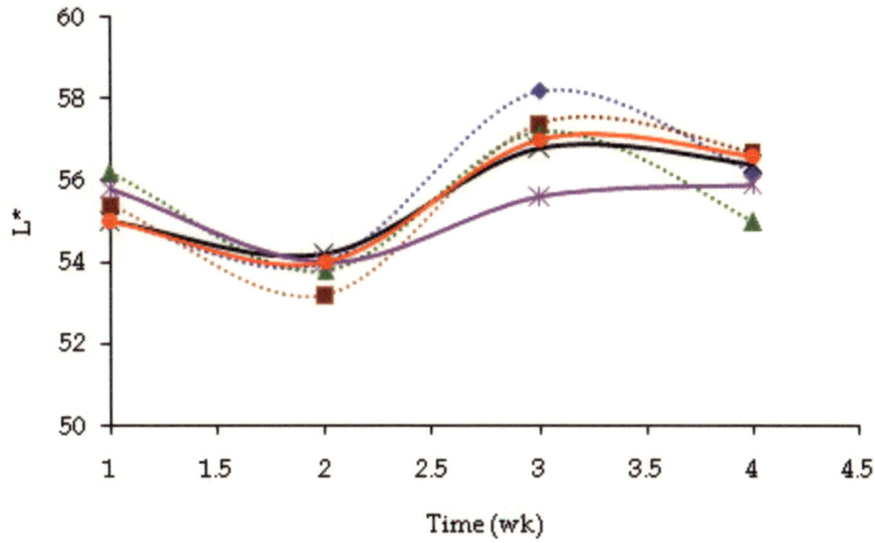


**Figure 14.** L* Value during storage of Nham control at 4°C (■), 15°C (♦), 25°C (▲) and Nham with starters (*L. plantarum* + *L. sake*) at 4°C (✳), 15°C (×), 25°C (•).

Fig. 14 showed L* values represent white color of Nham during stored at 15°C, 4°C and 25°C. The initial a* values of Nham control ranged from 55.1 to 56.4. The results showed L* value decreased after 2 week storage and then increased after 3 week storage for each storage temperature. However, there was not significant difference ($P$<0.05) between the L* value of 25°C storage of the initial 1 week storage and after 4 week of storage. For Nham with starters, the initial L* values ranged from 55.0 to 55.6. The L* values of Nham with starter decreased after 2 week storage then the L* values increased after 3 week storage for each storage temperature and after 4 week storage at 15°C and 25°C, the L* value was significant increased. However, there was no significant difference ($P$<0.05) between the L* value of 4°C storage of the initial 1 week storage and after 4 week of storage.

Fig. 15 shows that the initial pH of Nham samples ranged from 4.3 to 4.5. It then gradually decreased throughout the storage. The pH value reached 4.1 to 4.5 at 4 week of storage. The pH values at each storage temperature of Nham with starter were higher than Nham control at each time of sampling. The results was shown that pH value of Nham control stored at 25°C was lower than the other Nham samples throughout the storage process.

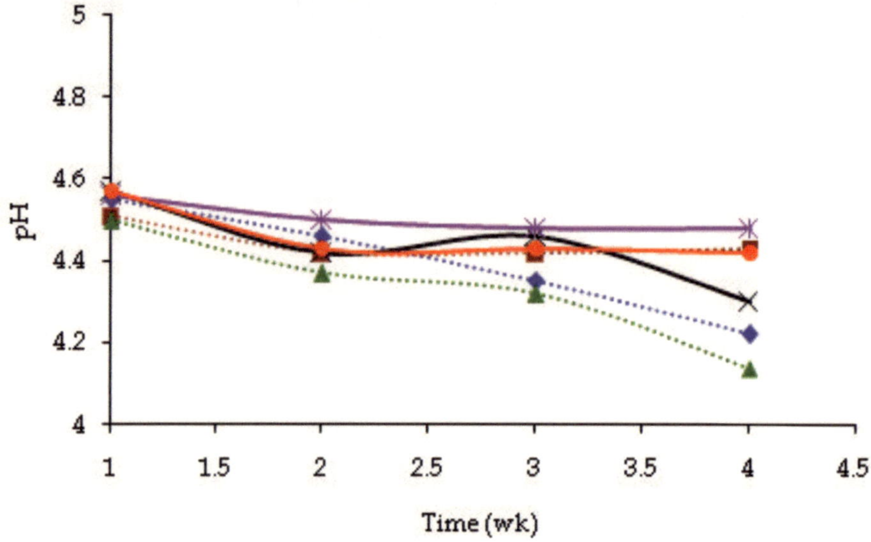


**Figure 15.** pH during storage of Nham control at 4°C (■), 15°C (◆), 25°C (▲) and Nham with starters (*L. plantarum* + *L. sake*) at 4°C (✳), 15°C (×), 25°C (•).

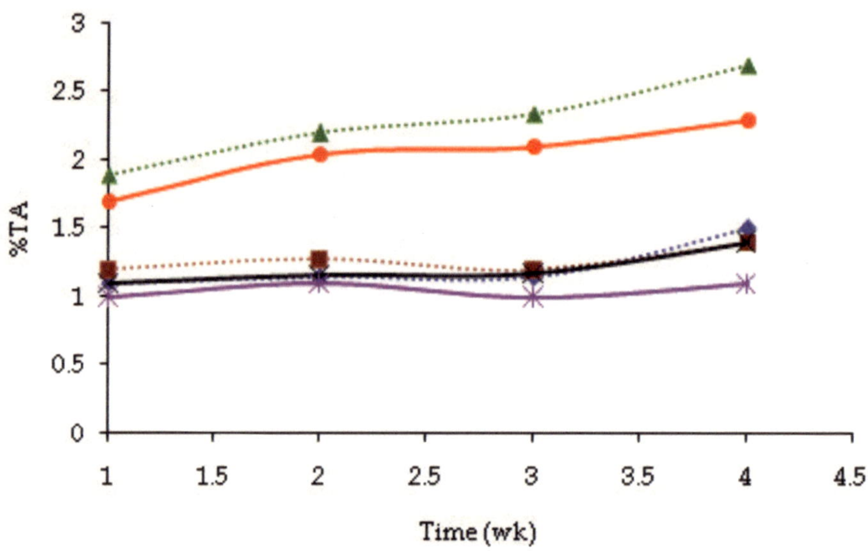


**Figure 16.** Total acid content during storage of Nham control at 4°C (■), 15°C (◆), 25°C (▲) and Nham with starters (*L. plantarum* + *L. sake*) at 4°C (✳), 15°C (×), 25°C (•).

Fig. 16 shows that the initial total acid content of Nham samples ranged from 1.1 to 1.7. It then gradually increased throughout the ripening process and there was significant difference at each time of sampling ($P < 0.05$). The total acid content of Nham control and Nham with starters reached 1.14% to 2.72% and 1.04 %to 2.32% at 4 week of storage, respectively. However, there was not significant difference ($P < 0.05$) between the total acid content of Nham control stored at 4°C and 15°C and Nham with starters stored at 15°C and from the results, the total acid content of Nham with starters stored at 4°C was not significant difference ($P$<0.05) during storage process. The total acid content of Nham control stored at 25°C was higher than the other Nham samples throughout the storage process.

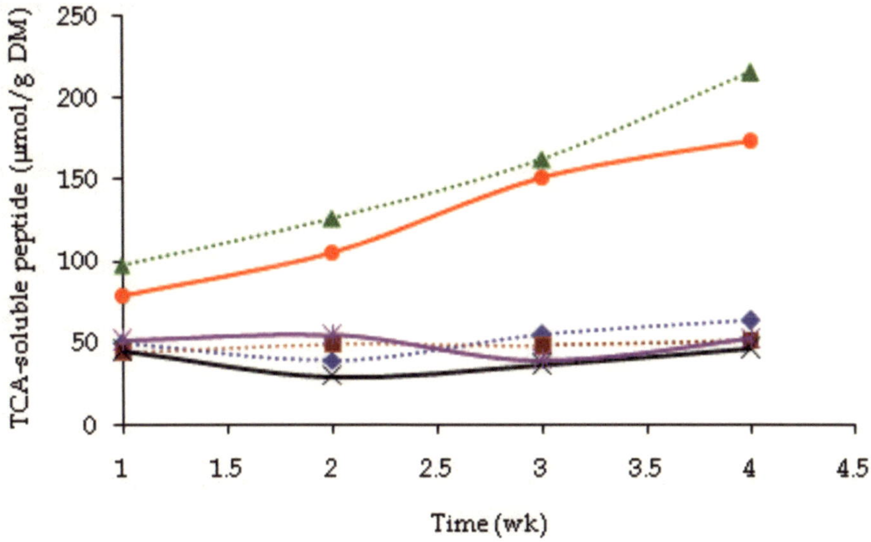


**Figure 17.** TCA-soluble peptide during storage of Nham control at 4°C (■), 15°C (◆), 25°C (▲) and Nham with starters (*L. plantarum* + *L. sake*) at 4°C (✳), 15°C (×), 25°C (●).

Fig. 17 shows that the TCA-soluble peptide of Nham control and Nham with starters ranged from 45.2 to 98.4 and 46.3 to 79.6 µmol/g dry matter, respectively. Nham control and Nham with starters stored at 25°C showed gradually increased throughout the storage process and there was significant difference at each time of sampling ($P$ <0.05). The TCA-soluble peptide of Nham control and Nham with starters stored at 4°C and 15°C. However, there was not significant difference ($P$<0.05) between the TCA-soluble peptide of Nham control and samples inoculated with starters culture throughout the storage process at 4°C and 15°C. From the results, the TCA-soluble peptide of Nham control stored at 25°C was higher than the other Nham samples throughout the storage process.

Fig. 18 shows that free α-amino acid content of Nham control samples and Nham with starters, the initial ranged from 342.3 to 603.4 and 346.6 to 507.2 mmol/g dry matter, respectively. It then gradually increased throughout the storage process at 15°C and 25°C and there was significant difference at each time of sampling ($P$ <0.05). The free α-amino acid content of Nham control and Nham with starters reached 375.2 to 1867.6 mmol/g dry matter and 359.4 to 1252.4 mmol/g dry matter, respectively, at 4 week of storage. However, there was not significant difference ($P$ <0.05) between the free α-amino acid content of Nham control and samples inoculated with starters during storage at 4°C. The results was shown that Nham control stored at 25°C contained free α-amino acid content higher than the other Nham samples after 3 week storage.

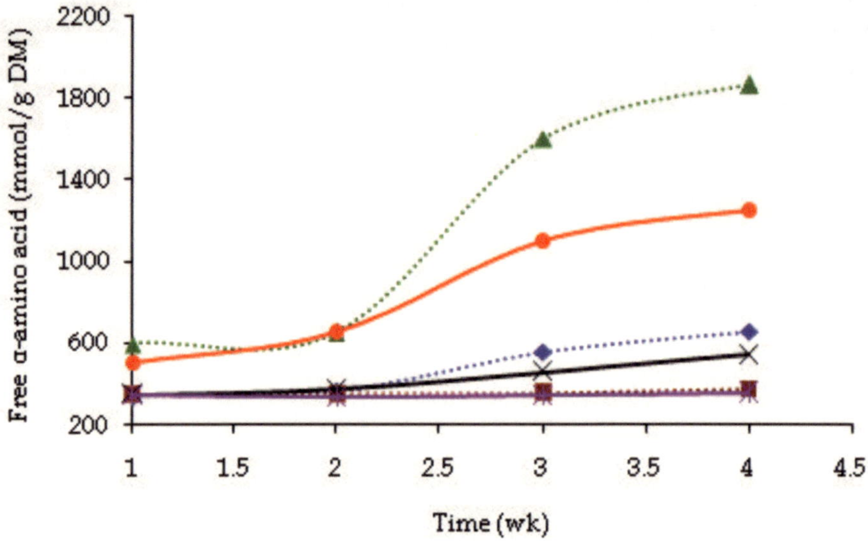


**Figure 18.** Free α-amino acid content during storage of Nham control at 4°C (■), 15°C (◆), 25°C (▲) and Nham with starters (*L. plantarum* + *L. sake*) at 4°C (✳), 15°C (×), 25°C (●).

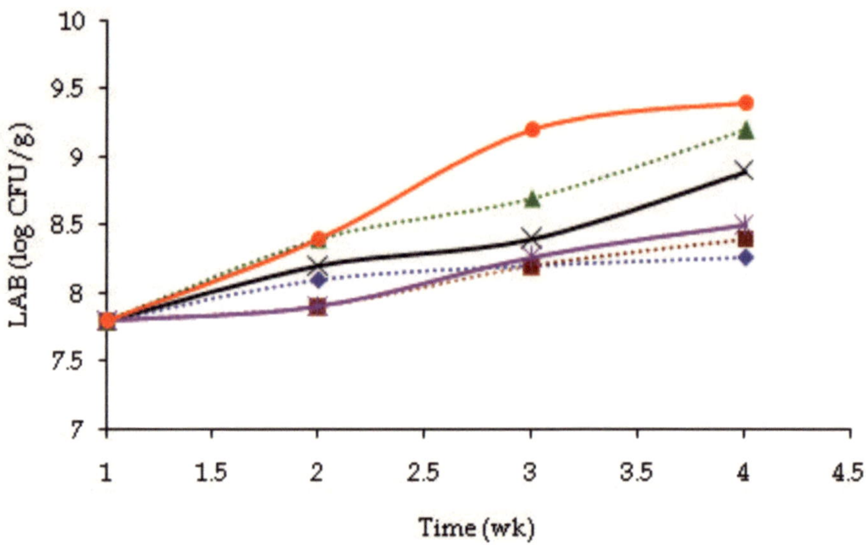


**Figure 19.** Total count of LAB during storage of Nham control at 4°C (■), 15°C (◆), 25°C (▲) and Nham with starters (*L. plantarum* + *L. sake*) at 4°C (✳), 15°C (×), 25°C (●).

The differences between Nham in counts of LAB during ripening are shown in Fig. 19. LAB in Nham with starters was increase until the 4 week of storage. Counts of LAB in Nham with starters stored at 25°C (9.4 log CFU/g) were higher ($P < 0.05$) than in Nham control stored at 25°C (9.1 log CFU/g). LAB counts in Nham increased steadily during storage, the dependence of the LAB counts of Nham control and Nham with starters on ripening at each storage temperature were significant differences. In the present study in Nham are concerned, total LAB counts in Nham with starters on 3 week of storage were higher ($P < 0.05$) in comparison with the Nham control produced at the same storage temperature. An increase of LAB in Nham with starters until 3 week of storage and consecutive increase till 4 week of storage was significant. LAB of Nham with starters produced increase steadily during ripening and stored at different temperatures, however, at 4°C storage, LAB counts storage was not different significant.

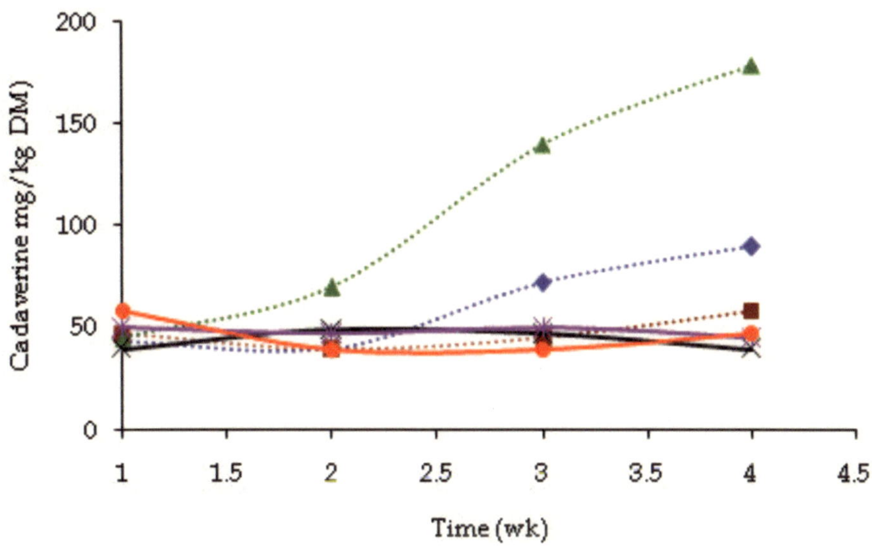


**Figure 20.** Cadaverine content during storage of Nham control at 4°C (■), 15°C (♦), 25°C (▲) and Nham with starters (*L. plantarum* + *L. sake*) at 4°C (*), 15°C (×), 25°C (•).

Fig. 20 shows that cadaverine content of Nham samples, the initial ranged from 43.7 to 58.2 mg/kg dry matter. In Nham control stored at 15°C and 25°C, it then gradually increased throughout the storage process. The cadaverine content of Nham control and Nham with starters reached 58.4 to 91.2 mg/kg dry matter and 41.6 to 47.3 mg/kg dry matter, respectively, at 72 hour of storage. However, there was not significant difference ($P<0.05$) between the cadaverine content of Nham with starters during stored at 4°C. The results was shown that Nham control stored at 25°C contained cadaverine content higher than the other Nham samples throughout the storage process and there was significant decreased ($P<0.05$) in the cadaverine content of Nham with starters stored at 25°C for 4 week.

Fig. 21 shows that putrescine content of Nham control and Nham with starters, the initial ranged from 124.6 to 176.3 mg/kg dry matter and 126.2 to 98.3 mg/kg dry matter. Nham control stored at 4°C, 15°C and 25°C gradually increased throughout the storage process. The putrescine content of Nham control and Nham with starters reached 175.3 to 339.4 mg/kg dry matter and 122.6 to 129.3 mg/kg dry matter, respectively, at 4 week of storage. However, there were significant increase ($P<0.05$) between the putrescine content of Nham with starters at each storage temperature for 2 week and then the putrescine content decreased and there was not significant difference ($P<0.05$) between the putrescine content of Nham with starters after 3 week storage. The results were shown that Nham control stored at 25°C contained putrescine higher than the other Nham samples throughout the storage process.

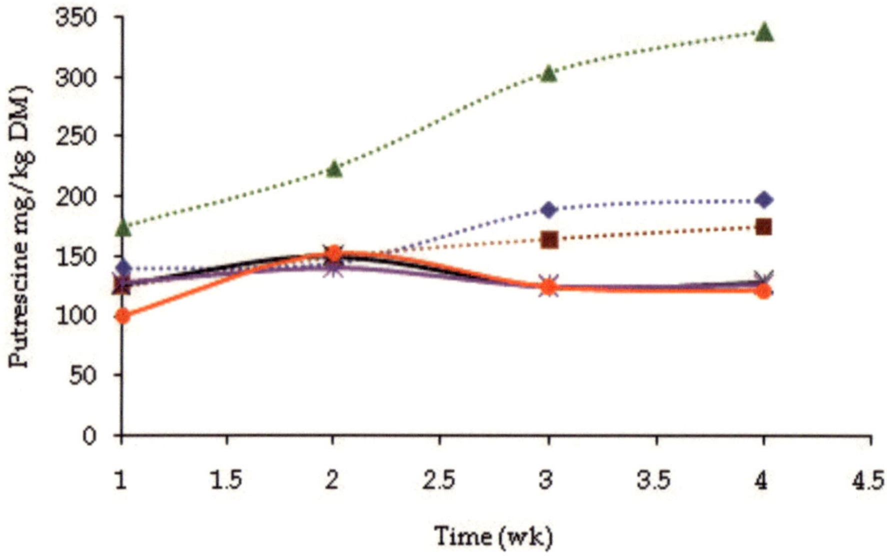


**Figure 21.** Putrescine content during storage of Nham control at 4°C (▪), 15°C (◆), 25°C (▲) and Nham with starters (*L. plantarum* + *L. sake*) at 4°C (✳), 15°C (×), 25°C (•).

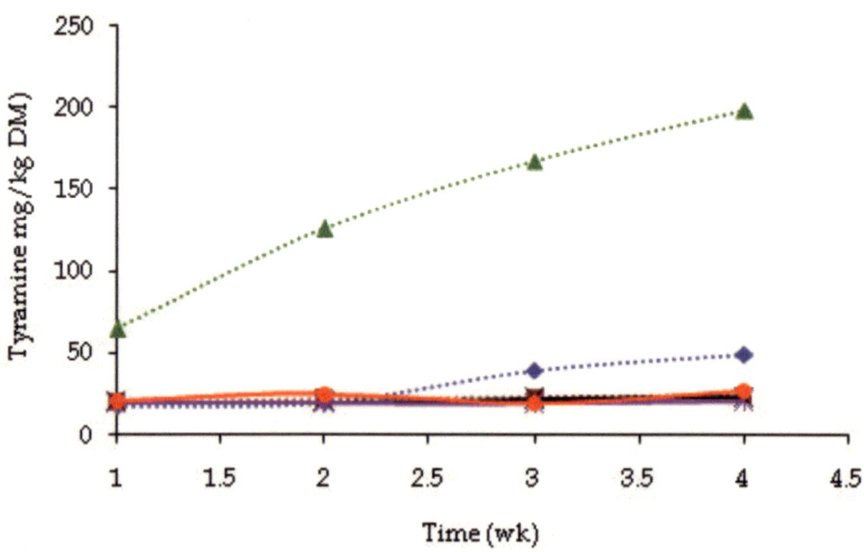


**Figure 22.** Tyramine content during storage of Nham control at 4°C (▪), 15°C (◆), 25°C (▲) and Nham with starters (*L. plantarum* + *L. sake*) at 4°C (✳), 15°C (×), 25°C (•).

Fig. 22 shows that tyramine content of Nham control and Nham with starters, the initial ranged from 18.2 to 65.3 mg/kg dry matter and 19.2 to 22.4 mg/kg dry matter. Nham control stored at 15°C and 25°C gradually increased during storage process. The tyramine content of Nham control and Nham with starters reached 25.2 to 198.6 mg/kg dry matter and 21.4 to 27.6 mg/kg dry matter, respectively, at 4 week of storage. There was not significant difference ($P<0.05$) between the tyramine content of Nham control stored at 4°C and Nham with starters at each storage temperature during storage time. The results was shown that Nham control stored at 25°C contained tyramine higher than the other Nham samples throughout the storage process.

One of the most important factors influencing BA formation in Nham is starter culture (Maijalaet al., 1995). Increase of LAB starters culture in Nham resulted in overgrowth more than the microflora and LAB producing BAs in Nham control and caused decrease in BA contents in Nham during ripening and storage. A higher amount of BAs was formed in the Nham control than in starters culture-ones. However, strains of the starters showed lower decarboxylase activity (lower total free amino acid content in Nham) in comparison with the Nham control. Moreover, from the fact that BA production increased in Nham control after the ripening was finished and Nham was stored at the 15°C and 25°C, which coincided with the temporary increase of total LAB, the presence of spontaneous decarboxylating microflora can be inferred, and the refrigerated storage should be recommended. Simultaneously, higher concentration of BAs was found in Nham fermented at high 30°C and 37°C as compared to 25°C at the end of ripening. However, as regards to the strongly hypothetical effects of some substances in the Nham spicing mixtures in connection with the BA formation, more research is needed. Decarboxylase activities present in microflora in Nham are influenced by pH, temperature (Gardini et al., 2001; Silla-Santos, 1996; Suzzi and Gardini, 2003). The decarboxylation of FAAs to BAs was found to be inhibited by low pH (Gardini et al., 2001). Though amino acid decarboxylase activities usually have acid pH optimum (Gale, 1946), the pH rise could favour the cell yield and growth (Maijala, 1994) of decarboxylase-positive microflora.

### 3.1. BAs contents of Nham

Occurrence of toxic compounds such as BAs is favoured by a high concentration of substrates (i.e., free amino acids) together with environmental and technological factors (e.g. NaCl content, chemico-physical variables, hygienic procedure adopted during production) promoting microbial growth and the decarboxylase activity of microorganisms (Silla-Santos, 1996). In this study, a high correlation among total BAs and total FAAs content was observed. Temperature markedly influences the formation of BAs, and at 15°C decarboxylases might be still active (Bover-Cid et al., 2001). During storage, the more temperature exceeds 14–15°C the more decarboxylase activities might release BAs from FAAs. In this respect, processing procedures for Nham based on low salt addition, high ripening temperatures (over 20°C), may favour proteolytic and decarboxylase activities. The high values of cadaverine, putrescine and tyramine detected in some Nham, may be ascribed to inadequate microflora and LAB producing BAs reduction occurring in some Nham control (Fig. 9- Fig. 11 and Fig. 20- Fig. 22).

The toxicological level of BAs depends on the individual characteristics and the presence of other amines (Brink et al., 1990; Halasz et al., 1994). Toxic doses of tyramine in foods were reported in the range 100–800 mg/kg, but average amounts of tyramine detected in analysed samples (Fig. 9- Fig. 11 and Fig. 20- Fig. 22) were below this range, even if in case of a few samples, the 100 mg/kg value was exceeded. Putrescine has been regarded as not toxic by itself, but as a potentiator for the toxic effect of tyramine and histamine if present (Hui and Taylor, 1985). However, it was probable to demonstrate significant relationship between the concentration of a specific FAA and its corresponding BA in meat products (Eerola et al., 1998). Fig. 9- Fig. 11 and Fig. 20- Fig. 22 shows the BAs content of Nham evidence the effect of starters on the decrease of the BAs occurrence in Nham after ripening and storage. Histamine was always below the minimum detectable, in spite of the abundance of their precursors (histidine) released during the process; phenylethylamine was also not detected.

The concentration of tyramine was high in Nham control while low concentration, of their precursors (tyrosine) released during the process. Moreover, tryptamine resulted absent in all the investigated samples. The sum of vasoactive BAs, VBA; (tyramine, phenylethylamine, histamine and tryptamine) lower than 200 mg/Kg has been suggested by Eerola et al. (1998) as a quality index (VBA index) for ripened meat products. It is interesting to note that the computed VBA index of Nham with starters with differently processed resulted appreciable samples (3.70 ± 2.46 mg/Kg). These results could be related to the specific characteristics of the product as well as to the process conditions adopted that could, in general, have limited the growth and activity of amino acid decarboxylase positive microorganisms (Suzzi and Gardini, 2003). Cadaverine, putrescine and tyramine were found in high amounts in Nham control. However, the occurrence of BAs in Nham control, and after the storage could be due to the microflora and LAB producing BAs that could have favoured their formation during ripening and storage. During ripening and storage of Nham control, putrescine and cadaverine show a marked increase with high amounts of their precursor, arginine and lysine, respectively, were detected. In fact, arginine may generate putrescine both via arginine deiminase pathway (ADI) leading to ornithine (Montel and Champomier, 1987) and their subsequent decarboxylation to putrescine, and via arginine decarboxylation to agmatine followed by deamination to putrescine and removal of urea (Moreno-Arribas et al., 2003). It seems reasonable to postulate that the large amounts of arginine could be the source of putrescine, which subsequently may be converted in spermine and spermidine by transamination reactions (Lehninger et al., 1999).

### 3.2. FAAs contents of Nham

FAAs were reported in Table 2- Table 5 as net amounts (mmol/g dry matter) in order to investigate the differences in contents due to starters in Nham during ripening and storage. FAAs were compared to evaluate if the extended storage times gave a similar increase in all of them or different patterns were detectable. Most single FAAs increased during ageing with particular reference to the lipophylic ones; a rise in lypophilic valine, phenylalanine and tryptophan processed following a traditional prolonged way (Ruiz et al., 1999). In the present study, stored Nham showed a FAA pattern enriched with glutamic acid, alanine, arginine, cysteine, serine, threonine and glycine, most FAAs displayed a rise during the extended storage. Arginine found in the most stored Nham was increase, due to changing of its content by proteolysis; and rise in arginine in stored Nham control was higher than stored Nham with starters. Arginine hydrolysis could be hydrolysed via the arginine deiminase pathway (ADI) leading to ammonia and ornithine. It seemed reasonable to postulate that ADI pathway enzymes (arginine deiminase and ornithine transcarbamylase) could be still active during storage times. Arginine catabolism, may be regarded as a source of the BA putrescine both via ADI ornithine generation (Montel and Champomier, 1987) and subsequent decarboxylation to putrescine, and via arginine decarboxylation to agmatine followed by deamidation to putrescine and removal of urea (Moreno-Arribas et al., 2003). The presence in Nham of environmental conditions suitable for decarboxylase activities together with large amounts of arginine may be consistent with the increase in putrescine.

The evolution during incubation/storage of the total free amino acid content, in both the Nham control and after inoculation with either of the two Lactobacillus strains selected, is shown in Table 2- Table 5, and encompassed 17 different amino acids. The control Nham showed the highest concentration of total amino acids at a 5% level of significance. The contents of total amino acids in Nham inoculated with *L. plantarum* and *L. sake,* increased throughout time, but at lower rates than the control. The contents of free amino acids and BAs in control and experimental Nham increase significantly throughout incubation/storage. However, specific lactic acid strains of the Lactobacillus genus can effectively prevent BAs from building-up excessively, putrescine (for quantitative reasons, owing to its level). This may lead to a favourable contribution to public health, especially in regions where Nham is frequently included in the diet. To have an overall evolution index of the proteases action in the Nham during processing the TCA-soluble peptide was evaluated (Fig. 6 and Fig. 17) (Toldr, 2005). More intense proteolytic activity occurred in the Nham control. The TCA-soluble peptide values of Nham control are quite high compared to those generally observed in other Nham with starters. This could be attributed to the microflora in Nham control slightly higher proteolytic activity during the process, in comparison with those Nham applied with starters. Proteolysis contributes to texture by breakdown of the muscle structure (Monin et al., 1997).

Table 2- Table 5 show the FAAs content of Nham during ripening and storage arginine and glutamic acid were the FFAs most representative; after ripening and storage a marked increase of alanine was observed. Table 6 shows the effect of the starters treatment on the evolution of the FAAs pattern of the Nham investigated during the ripening and storage: a significant increase in the concentration of all FAAs with respect to their initial occurred in Nham control and Nham with starters, resulting from the aminopeptidases activity of meat (Toldr, 2006) as well as microbial proteases (Dur et al., 2004; Molina and Toldrá, 1992; Rodrguez et al., 1998; Scannell et al., 2004). Moreover, starters in Nham seems to affect the production of some amino acids (Table 6). A lower concentration of lysine, threonine, glycine and proline was detected, after storage, in Nham processed. Arginine was the most abundant amino acid in all the final products, and its level was significantly higher in Nham control than in those subjected with starters to Nham. At the end of the ripening step, cysteine was also present in a relative higher concentration in Nham control, whereas significant larger amounts of proline, lysine, histidine, serine and threonine were reached in Nham control samples. The different profile of FAAs observed in Nham control and Nham with starters may be due to a different evolution of reactions and processes involving both production and consumption of amino acids that occur simultaneously during the various steps of the ripening process and storage and whose combined effects could give rise to an increase or, on the contrary, to a decrease of their concentration. The aminopeptidase activity is considered the main process implied in the FFA release in meat. Moreover, free amino acids concentration could be decreased either by chemical and enzymatic reactions where they act as substrates leading to the formation of secondary products (Ruiz et al., 1999; Ventanas et al., 1992) and/or by microbial amino acid decarboxylase activity with consequent BA production (Virgili et al., 2007).

In Nham control, an effect due to higher concentration of decarboxylase than that of Nham with starters, thus, their reaction with the free amino acids causing an increase of their BA concentration in these samples.

The ripened taste could be related to lysine and glutamic acid, while isoleucine and aspartic acid are implied in acid taste and unpleasant aroma (Buscailhon et al., 1994; Flores et al., 1998). In this study, the increase in concentration of lysine and glutamic acid was observed. The changes in the contents of free amino acids observed in fermented sausages during ripening are given in Table 2. The total free amino acid contents of the Nham control and Nham with starters constituted 212.7–216.4 mmol/g and 197.2–203.4 mmol/g dry matter, respectively (before ripening) on 0 day. An increase in the content of amino acids of Nham control and Nham with starters was observed and ranged between 275.2–349.8 mmol/g and 259.8–300.3 mmol/g dry matter during the ripening on day 3, and a further increase up to the range of 377.6–1851.7 mmol/g and 348.1 nmol/g–1256.0 mmol/g dry matter of total free amino acids was observed during storage at 4°C-25°C of Nham control and Nham with starters (4 weeks). The highest total free amino acid concentration of 1867.2 mmol/g was observed with Nham control stored at 25 °C for 4 week, whereas the lowest total free amino acid concentration of 359.6 mmol/g was observed with Nham with starters stored at 25 °C for 4 week. The hydrolysis of meat proteins generates polypeptides that can be further degraded to smaller peptides and free amino acids. This degradation can be produced by endogenous and microbial enzymes (De Masi et al., 1990; Hughes et al., 2002; Molly et al., 1997). The increase in the total free amino acid concentration was detected in all batches (Hierro et al., 1999, Bruna et al., 2000, Bolumar et al., 2001 and Hughes et al., 2002).

The main differences in the content of total free amino acids among batches were detected during 72 hour of ripening and during 4 week of storage. The amino acids in which differences, which were primarily responsible for the increase in total free amino acids during ripening, were observed were Glu (glutamic acid), Ala (alanine) and Arg (arginine) in Nham control and Nham with starters. Mateo et al. (1996) reported an increase in the total free amino acid content during the ripening. The change occurred during ripening and storage process indicating that the highest enzymatic activity took place during these stages

| Amino acid | 25°C | | | | | | 30°C | | | | | | 37°C | | | | | |
|---|---|---|---|---|---|---|---|---|---|---|---|---|---|---|---|---|---|---|
| | Control | | | *L. plantarum* + *L. sake* | | | Control | | | *L. plantarum* + *L. sake* | | | Control | | | *L. plantarum* + *L. sake* | | |
| | 24 (h) | 48 (h) | 72 (h) | 24 (h) | 48 (h) | 72 (h) | 24 (h) | 48 (h) | 72 (h) | 24 (h) | 48 (h) | 72 (h) | 24 (h) | 48 (h) | 72 (h) | 24 (h) | 48 (h) | 72 (h) |
| Ala | 12.5a | 14.5b | 16.3c | 13.9d | 13.6d | 15.4e | 14.8b | 15.7e | 17.9f | 13.4d | 14.2b | 16.6g | 16.6g | 20.4i | 22.1j | 12.9a | 15.7e | 18.6h |
| Arg | 90.9a | 105.9b | 119.5c | 94.2d | 95.6e | 108.0f | 101.0g | 107.2h | 126.6i | 90.3a | 95.5e | 115.3j | 109.8k | 134.0m | 149.8n | 90.7a | 109.4k | 127.2l |
| Asp | 1.9a | 2.3b | 2.5b | 1.9a | 1.9a | 2.2b | 2.1b | 2.3b | 2.7b | 1. 8 a | 1.9a | 2.3b | 2.2b | 2.7b | 3.0b | 1.9a | 2.2b | 2.7b |
| Cys | 10.9a | 13.3b | 14.6c | 11.2d | 11.2d | 13.2b | 12.0e | 13.3b | 15.5f | 10.1a | 11.1d | 13.1b | 13.1b | 15.7f | 17.4g | 10.4a | 12.8e | 15.5f |
| Glu | 54.5a | 65.8b | 72.9c | 55.1a | 54.9a | 64.2d | 61.6e | 67.7f | 80.7g | 51.4h | 56.8i | 69.1j | 64.2d | 80.4m | 88.3n | 52.5k | 65.0b | 77.0l |
| Gly | 6.6a | 7.9b | 8.7c | 6.7a | 6.7a | 8.0b | 7.3b | 8.1b | 9.4d | 6.3a | 6.9a | 8.4bc | 7.7b | 9.3d | 10.3e | 6.0a | 7.3b | 8.8c |
| His | 6.3a | 7.5b | 8.4c | 6.8a | 6.7a | 7.7b | 7.6b | 8.2c | 9.4d | 6.3a | 6.9a | 8.3c | 8.2c | 10.1de | 10.9e | 6.6a | 8.2c | 9.5d |
| Leu | 1.8a | 2.2ab | 2.4ab | 1.9a | 1.9a | 2.2ab | 2.1ab | 2.2ab | 2.7bc | 1.7a | 1.9a | 2.3ab | 2.1ab | 2.7bc | 3.0c | 1.8a | 2.2ab | 2.7bc |
| Lys | 4.3a | 5.1b | 5.6b | 4.5a | 4.6a | 5.2b | 5.0b | 5.5b | 6.5c | 4.3a | 4.7a | 5.7b | 5.3a | 6.5c | 7.4d | 4.4a | 5.3b | 6.2c |
| Ile | 1.2a | 1.4a | 1.5a | 1.2a | 1.2a | 1.4a | 1.3a | 1.4a | 1.6a | 1.1a | 1.2a | 1.4a | 1.4a | 1.7ab | 1.9b | 1.1a | 1.4a | 1.6a |
| Met | 0.4a | 0.4a | 0.5a | 0.4a | 0.4a | 0.4a | 0.4a | 0.5a | 0.5a | 0.4a | 0.4a | 0.5a | 0.4a | 0.5a | 0.6a | 0.4a | 0.4a | 0.5a |
| Phe | 0.9a | 1.0a | 1.2a | 0.9a | 0.9a | 1.0a | 0.9a | 1.0a | 1.2a | 0.8a | 0.9a | 1.0a | 1.1a | 1.3a | 1.4a | 0.9a | 1.0a | 1.3a |
| Pro | 2.6a | 3.1ab | 3.4b | 2.6a | 2.7a | 3.2b | 3.0b | 3.3b | 3.8b | 2.6a | 2.8a | 3.4b | 3.1ab | 3.8b | 4.3c | 2.5a | 3.0b | 3.6b |
| Ser | 7.6a | 9.2bc | 10.4c | 7.9a | 7.7a | 8.9b | 8.7b | 9.4c | 11.2d | 7.1e | 7.9f | 9.5c | 9.2bc | 11.6d | 12.8e | 7.7e | 9.1bc | 10.9c |
| Thr | | 8.4b | 9.3c | 7.7a | 7.5a | 8.7b | 8.3b | 9.2c | 10.5d | 7.2a | 7.8a | 9.3c | 8.6b | 10.3d | 11.3f | 6.6e | 8.1b | 9.4c |
| Tyr | 1.1a | 1.3a | 1.4a | 1.1a | 1.1a | 1.3a | 1.2a | 1.3a | 1.5a | 1.0a | 1.1a | 1.3a | 1.3a | 1.6a | 1.8a | 1.1a | 1.3a | 1.6a |
| Val | 2.2a | 2.6a | 2.9a | 2.3a | 2.2a | 2.6a | 2.5a | 2.7ab | 3.1b | 2.2a | 2.3a | 2.7ab | 2.7a | 3.3b | 3.6b | 2.2a | 2.6a | 3.2b |
| Total | 214.9a | 250.3b | 275.2c | 217.1a | 220.2a | 259.8b | 239.8c | 259.0b | 304.8d | 208.1a | 220.0a | 270.1c | 257.1b | 315.9f | 349.8g | 209.7a | 255.1b | 300.3e |

Control (without starter culture).
Results are expressed as means of three replicates in mmol/g dry matter.
Means with different letters along rows are significantly different ($P<0.05$).

**Table 2.** Amino acid content of Nham without and with starter cultures during ripening at different temperature.

(Verplaetse et al., 1989). A major release of free amino acids at the beginning of the process have been studied in coincidence with the ripening stage (Diaz et al., 1997). This increase has been attributed to the higher temperatures applied during ripening compared to the low temperature. The most significant increases occurred in the content of Arg (arginine) in the sample. The decrease in the content of amino acids may indicate their metabolism by bacteria (Bover-Cid et al., 2000; Ordonez et al., 1999; Sekikawa et al., 2003).

| Amino acid | Control | | | | *L. plantarum* + *L. sake* | | | |
|---|---|---|---|---|---|---|---|---|
| | Storage time (wk) | | | | Storage time (wk) | | | |
| | 1 | 2 | 3 | 4 | 1 | 2 | 3 | 4 |
| Ala | 21.4a | 22.3a | 23.0ab | 24.8b | 22.1a | 20.7a | 21.1a | 20.2a |
| Arg | 151.9a | 154.4a | 154.9a | 168.3b | 149.8a | 145.5a | 153.8a | 147.2a |
| Asp | 3.2a | 3.4a | 3.5a | 3.7a | 2.9a | 2.9a | 3.1a | 3.0a |
| Cys | 18.2a | 18.5a | 18.6a | 19.9b | 17.8ac | 17.0c | 17.7ac | 17.7ac |
| Glu | 96.0a | 98.4a | 99.6ab | 103.6b | 88.3c | 86.5c | 92.2d | 91.4d |
| Gly | 11.2ac | 11.4ac | 11.6ac | 12.4b | 10.8c | 10.4c | 10.8c | 10.6c |
| His | 10.5a | 10.5a | 10.6a | 11.3a | 10.8a | 10.2a | 10.7a | 10.2a |
| Leu | 2.9a | 3.0a | 3.1a | 3.3a | 2.9a | 2.8a | 2.9a | 2.9a |
| Lys | 7.7a | 7.8a | 7.9a | 8.5a | 7.4a | 7.0a | 7.5a | 7.3a |
| Ile | 2.0a | 2.0a | 2.1a | 2.2a | 1.9a | 1.8a | 1.9a | 1.9a |
| Met | 0.6a | 0.6a | 0.6a | 0.7a | 0.6a | 0.6a | 0.6a | 0.6a |
| Phe | 1.5a | 1.5a | 1.6a | 1.7a | 1.4a | 1.3a | 1.4a | 1.4a |
| Pro | 4.6a | 4.6a | 4.7a | 4.8a | 4.3a | 4.2a | 4.4a | 4.3a |
| Ser | 12.7a | 13.2a | 13.6ab | 14.4b | 12.3ac | 11.8c | 12.4ac | 12.1ac |
| Thr | 12.5ac | 12.8ac | 13.1ab | 14.0b | 12.2a | 11.9c | 12.3c | 12.3c |
| Tyr | 1.8a | 1.8a | 1.8a | 1.9a | 1.7a | 1.7a | 1.8a | 1.7a |
| Val | 3.7a | 3.8a | 3.9a | 4.2a | 3.6a | 3.4a | 3.5a | 3.5a |
| Total | 352.9a | 358.6a | 363.1ab | 377.6b | 344.9ac | 335.1c | 357.3a | 348.1ac |

Control (without starter culture).
Results are expressed as means of three replicates in mmol/g dry matter.
Means with different letters along rows are significantly different ($P<0.05$).

**Table 3.** Amino acid content of Nham without and with starter cultures during stored at 4°C.

| Amino acid | Control | | | | *L. plantarum* + *L. sake* | | | |
|---|---|---|---|---|---|---|---|---|
| | Storage time (wk) | | | | Storage time (wk) | | | |
| | 1 | 2 | 3 | 4 | 1 | 2 | 3 | 4 |
| Ala | 20.3a | 21.7ad | 31.2b | 37.9c | 23.0d | 23.1d | 27.3e | 31.8b |
| Arg | 147.5a | 159.4b | 237.5c | 288.0d | 155.9b | 162.4b | 191.7e | 231.4c |
| Asp | 3.2a | 3.3a | 5.0b | 6.3c | 3.1a | 3.2a | 3.9a | 4.8b |
| Cys | 18.5a | 19.5a | 28.5b | 36.1c | 18.6a | 19.0a | 23.4d | 27.8b |
| Glu | 91.6a | 97.3b | 146.3c | 183.7d | 91.1a | 93.2a | 114.0e | 138.8f |
| Gly | 11.0a | 11.5a | 16.9b | 21.1c | 11.1a | 11.5a | 14.1d | 16.8b |

| | | | | | | | | |
|---|---|---|---|---|---|---|---|---|
| His | 10.4a | 11.2a | 16.2b | 20.5c | 11.3a | 11.4a | 13.8d | 16.0b |
| Leu | 3.1a | 3.3a | 4.9b | 6.0c | 3.1a | 3.2a | 3.9a | 4.6b |
| Lys | 7.1a | 7.5a | 11.3b | 13.9c | 7.5a | 7.8a | 9.2d | 10.9e |
| Ile | 1.9a | 2.1a | 3.1ab | 3.8b | 1.9a | 2.0a | 2.5a | 3.0ab |
| Met | 0.6a | 0.6a | 0.9a | 1.2a | 0.6a | 0.6a | 0.8a | 0.9a |
| Phe | 1.4a | 1.5a | 2.3ab | 2.9b | 1.5a | 1.5a | 1.8a | 2.2ab |
| Pro | 4.3a | 4.5a | 6.7b | 8.3c | 4.3a | 4.6a | 5.6a | 6.7b |
| Ser | 12.8a | 13.9a | 20.7b | 26.2c | 13.0a | 13.1a | 15.8d | 19.3b |
| Thr | 11.7a | 12.5a | 18.7b | 23.1c | 12.7a | 12.8a | 15.4d | 18.4b |
| Tyr | 1.8a | 1.9a | 2.7b | 3.4c | 1.8a | 1.9a | 2.3ab | 2.7b |
| Val | 3.7a | 3.9a | 5.9b | 7.5c | 3.8a | 3.8a | 4.6a | 5.5b |
| Total | 348.8a | 367.2a | 542.4b | 657.7c | 359.1a | 374.0a | 461.2d | 547.1e |

Control (without starter culture).
Results are expressed as means of three replicates in mmol/g dry matter.
Means with different letters along rows are significantly different ($P<0.05$).

**Table 4.** Amino acid content of Nham without and with starter cultures during stored at 15°C.

| Amino acid | Control | | | | *L. plantarum + L. sake* | | | |
|---|---|---|---|---|---|---|---|---|
| | Storage time (wk) | | | | Storage time (wk) | | | |
| | 1 | 2 | 3 | 4 | 1 | 2 | 3 | 4 |
| Ala | 35.6a | 40.6b | 103.8c | 122.6d | 31.5e | 41.8b | 68.2f | 76.2g |
| Arg | 232.9a | 280.6b | 710.9c | 832.7d | 212.3e | 291.3f | 475.2g | 531.2h |
| Asp | 4.7a | 5.8ad | 14.7b | 17.1c | 4.1a | 5.6d | 9.5e | 11.0f |
| Cys | 28.7a | 34.6b | 86.9c | 101.8d | 24.0a | 32.3b | 55.1e | 64.4f |
| Glu | 140.9a | 174.2b | 433.8c | 512.6d | 118.7e | 160.1f | 270.5g | 313.2h |
| Gly | 16.9a | 20.3b | 50.6c | 59.2d | 14.2e | 19.3b | 32.9f | 37.8g |
| His | 17.8a | 21.0b | 52.8c | 62.4d | 15.1a | 20.4b | 34.1e | 39.5f |
| Leu | 4.8a | 5.9a | 14.6b | 17.0c | 4.1a | 5.6a | 9.4d | 10.9e |
| Lys | 10.9a | 13.5b | 34.6c | 39.8d | 9.9a | 13.3b | 21.8e | 25.0f |
| Ile | 3.0a | 3.6a | 8.9b | 10.1c | 2.6a | 3.4a | 5.9d | 6.7d |
| Met | 0.9a | 1.1ad | 2.8bc | 3.2c | 0.8a | 1.1ad | 1.9d | 2.1d |
| Phe | 2.3ac | 2.8a | 7.1b | 8.2b | 1.9c | 2.6a | 4.4d | 5.1d |
| Pro | 6.5a | 8.0b | 20.5c | 23.6d | 5.9a | 7.9b | 13.5e | 15.0f |
| Ser | 20.1a | 24.2b | 60.3c | 68.2d | 16.8e | 22.8e | 38.3f | 44.7g |
| Thr | 17.7a | 21.9b | 56.0c | 66.2d | 16.4a | 22.2b | 36.8e | 41.2f |
| Tyr | 2.8a | 3.4a | 8.4b | 9.8c | 2.4a | 3.2a | 5.5d | 6.4d |
| Val | 5.7a | 7.0b | 17.8c | 21.0d | 5.0a | 6.6b | 10.9e | 12.7f |
| Total | 541.0a | 651.8b | 1637.2c | 1851.7d | 472.0e | 647.8b | 1103.9f | 1256.0g |

Control (without starter culture).
Results are expressed as means of three replicates in mmol/g dry matter.
Means with different letters along rows are significantly different ($P<0.05$).

**Table 5.** Amino acid content of Nham without and with starter cultures during stored at 25°C.

Two types of fermented sausage differing in starter culture were produced in parallel with two different starter cultures (no starter and *L. plantarum* + *L. sake*). The sausages were ripened 3 days and subsequently stored 7, 14, 21 and 28 days at the 4 °C, 15 °C and 25 °C. Concentration of three most abundant amines, cadaverine, putrescine and tyramine increased significantly ($P < 0.05$) in Nham during ripening and and also during storage. The dominant BAs in the control were cadaverine – and tyramine and putrescine, to a lesser extent; the cadaverine, putrescine and tyramine content were lower if inoculation had added with *L. plantarum* + *L. sake*; whereas they ranked above 300 mg/kg in the control by 3 d. At the end of ripening, cadaverine (98.7 mg/kg dry matter), putrescine (242.6 mg/kg dry matter) and tyramine (46.4 mg/kg dry matter) content in the A-samples-sausage was higher ($P < 0.05$) than in Nham with starters (51.6, 98.4 and 27.8 mg/kg dry matter, respectively). Starter culture influenced significantly in decrease of ($P < 0.05$) cadaverine, putrescine and tyramine content in the sausage. Due to the significant ($P < 0.05$) increase of total aerobic counts in the Nham control between the end of ripening and during storage, followed by the significant ($P < 0.05$) increase of the sum of total BAs between the 72 hour of ripening (387.7 mg/kg dry matter) and the 4$^{th}$ week of storage at 25°C (629.2 mg/kg dry matter).

The main rate of BAs production was during the first two days, when a sharp pH decrease and the development of LAB occurred. Sausages fermented with starters had lower amounts of cadaverine, putrescine and tyramine than naturally fermented sausages (control) during storage at 15°C and 25°C. However, phenylethylamine, histamine and tryptamine were not detected.

Nham control showed proteolysis that was correlated with pH values higher than those with starters. However, no positive correlation was found between the proteolysis index and BAs production. Since proteolysis was stronger during the second half of the ripening process, the FAAs occurred later than the early amine production. No effect on pH development in the fermented sausage was observed when non-amine forming strain of *L. plantarum* + *L. sake* were present during 4 week of 4°C storage period. A study on the evolution of FAAs and BAs in Nham during 4 week at different temperatures of storage (4°C, 15°C and 25°C) was performed. FAAs and BAs were determined by RP-HPLC. Storage temperature of 15°C and 25°C promoted a significant increase of the contents of arginine, glutamic acid, cadaverine, putrescine and tyramine, expressed as g/kg of dry matter while storage temperature of 4°C decreased a significant of the contents of arginine, glutamic acid, cadaverine, putrescine and tyramine, expressed as g/kg of dry matter. These two amino acids and three BAs may serve as indicators of temperatures changes in stored fermented sausage.

## 4. Conclusions

The aim of this study was to investigate the effect of non-amine forming LAB as starter culture during ripening and storage time and temperature on the evolution of FAAs of Nham during processing. The correlation between FAAs and BAs content was also investigated. Larger increases of FAAs occurred in Nham without starter in the ripening and storage step. Total FAAs content was highly correlated with total BAs amount. Sausage ripening was further carried out with non-amine forming strain of *L. plantarum* + *L. sake* after ripening and stored at different temperature. The amount of amine in the product was significantly less than the control. The results obtained for BAs degradation by bacteria in a synthetic medium suggest that AO activity is strain dependent rather than being related to specific species. In all batchs, the total amino acid contents increased with time – and the predominant ones were arginine and glutamic acid. However, upon inoculation with non-amine forming strain, the total BAs contents remained considerably lower than those of the control. Hence, an efficient food-grade biological tool was made available that constrains buildup of BAs in fermented sausage during storage.

## Author details

Jirasak Kongkiattikajorn
*School of Bioresources and Technology, King Mongkut's University of Technology Thonburi, Thailand*

## Acknowledgement

This study was supported by Office of the Higher Education Commission (OHEC), Thailand.

## 5. References

AOAC. (2000). *Official methods of analysis* (17th ed.). Gaithersburg, Maryland: Association of Official Analytical Chemists.

Benjakul, S., & Morrissey, M. T. (1997). Protein hydrolysates from Pacific whiting solid wastes. *Journal of Agricultural and Food Chemistry*, Vol.45, pp. 3423–3430.

Beutling, D. (1996). Biogenic Amines in Nutrition (Biogene Amine in der Ernaehrung). Springer-Verlag, Berlin, Germany, pp. 59–67.

Bover-Cid, S., Hugas, M., Izquierdo-Pulido, M. & Vidal-Carou, M.C. (2000). Reduction of biogenic amine formation using a negative amino acid-decarboxylase starter culture for ripening of *fuet* sausage. *Journal of Food Protection*, Vol. 63, No.2, pp. 237–243.

Bover-Cid, S., Izquierdo-Pulido, M. & Vidal-Carou, M. C. (2000). Influence of hygienic quality of raw materials on biogenic amine production during ripening and storage of dry fermented sausage. *Journal of Food Protection*, Vol. 63, pp. 1544–1550.

Bover-Cid, S., Hugas, M., Izquierdo-Pulido, M. & Vidal-Carou, M. C. (2001). Amino acid-decarboxylase activity of bacteria isolated from fermented pork sausages. International *Journal of Food Microbiology*, Vol. 66, pp. 185–189.

Bolumar, T., Nieto, P. & Flores, J. (2001). Acidity, proteolysis and lipolysis changes in rapid cured fermented sausage dried at different temperatures. *Food Science and Technology International*, Vol.7, pp. 269–276.

Brink, B. ten., Damink, C., Joosten, H. M. L. J. & Huis in't Veld, J. H. J. (1990). Occurrence and formation of biologically active amines in foods. *International Journal of Food Microbiology*, Vol.11, pp. 73–84.

Bruna, J. M., Fernandez, M., Hierro, E. M., Ordonez, J. A. & De la Hoz, L. (2000). Combined use of pronase E and a fungal extract (*Penicillium aurantiogriseum*) to potentiate the sensory characteristics of dry fermented sausages. *Meat Science*, Vol.54, pp. 135–145.

Buscailhon, S., Berdagué, J. L., Bousset, J., Gandemer, G., Cornet, M. & Touraille, C. (1994). Relations between compositional traits and sensory qualities of French dry-cured ham. *Meat Science*, Vol.37, pp. 229–243.

Cooper, R.A., 1997. On the amine oxidases of *Klebsiella aerogenes* strain W70. *FEMS Microbiology Letters* 146, 85– 89.

Hernández-Jover, T., Izquierdo-Pulido, M.,Veciana-Nogués, M.T. & Vidal-Carou, M.C. (1996). Ion-pair high-performance liquid chromatographic determination of biogenic amines in meat and meat products. *Journal of Agricultural and Food Chemistry*, Vol. 44, No.9, pp. 2710–2715.

De Masi, T. W., Wardlaw, F. B., Dick, R. L. & Acton, J. C. (1990). Non protein nitrogen (NPN) and free amino acid contents of dry fermented and non fermented sausages. *Meat Science*, Vol.27, pp. 1–12.

Durà, M. A., Flores, M. & Tolrà, F. (2004). Effect of *Debaryomyces* spp. On the proteolysis of dry-fermented sausages. *Meat Science*, Vol.68, pp. 319–328.

Dapkevicius, M.L.N.E., Nout, M.J.R., Rombouts, F.M., Houben, J.H. & Wymenga, W., 2000. Biogenic amine formation and degradation by potential fish silage starter microorganisms. International *Journal of Food Microbiology*, Vol.57, pp. 107– 114.

Eerola, S., Roig-Saguèz, A. X. & Hirvi, T. K. (1998). Biogenic amines in Finnish dry sausages. *Journal of Food Safety*, Vol.18, pp. 127–138.

Flores, M., Spanier, A. M. & Toldrá, F. (1998). Flavour analysis of dry-cured ham. In Shahidi, F. (Ed.), Flavour of meat product and seafood pp. 320–341. London: Blackie.

Gale, E. F. (1946). The bacterial amino acid decarboxylases. In Nord, F. F. (Ed.), Advances in enzymology, Vol. 6, New York: Interscience Publisher.

Gardini, F., Martuscelli, M., Caruso, M. C., Galgano, F., Crudele, M. A. & Favati, F. (2001). Effects of pH, temperature and NaCl concentration on the growth kinetics, proteolytic activity and biogenic amine production of *Enterococcus faecali*. *International Journal of Food Microbiology*, Vol.64, pp. 105–117.

Gardini, F., Martuscelli, M., Crudele, M.A., Paparella, A. & Suzzi, G., 2002. Use of *Staphylococcus xylosus* as a starter culture in dried sausages: effect on the biogenic amine content. *Meat Science*, Vol. 61, pp. 275– 281.

Greene, D. H., & Babbitt, J. K. (1990). Control of muscle softening and protease–parasite interactions in arrowtooth flounder, Ateresthes stomias. *Journal of Food Science*, Vol.55, pp. 579–580.

Hierro, E., De la Hoz, L. & Ordonez, J. A. (1999). Contribution of the microbial and meat endogenous enzymes to the free amino acid and amine contents of dry fermented sausages. *Journal of Agriculture and Food Chemistry*, Vol.47, pp. 1156–1161.

Hughes, M. C., Kerry, J. P., Arendt, E. K., Kenneally, P. M., McSweeney, P. L. H. & O'Neill, E. E. (2002). Characterization of proteolysis during the ripening of semidry fermented sausages. *Meat Science*, Vol.62, pp. 205–216.

Ishizuka, H., Horinouchi, S. & Beppu, T., 1993. Putrescine oxidase of *Micrococcus rubens*: primary structure and *Escherichia coli*. *General Microbiology*, Vol.139, pp. 425–432.

Joosten, H.M.L.J. & Northolt, M.D. (1989). Detection, growth and amine producing capacity of lactobacilli in cheese. *Applied and Environmental Microbiology*, Vol. 55, No.9, pp. 2356–2359.

Kalac, P., Spicka, J., Krizek, M. and Pelikanova, T. (2000). The effect of lactic acid bacteria inoculants on biogenic amines formation in sauerkraut. *Food Chemistry*, Vol.70, pp. 355–359.

Lehane, L. & Olley, J., 2000. Histamine fish poisoning revisited. International *Journal of Food Microbiology*, Vol.58, pp. 1 –37.

Lehninger, A. L., Nelson, D. L. & Cox, M. M. (1999). Principi di biochimica, ed. Zanichelli, Bologna.

Leuschner, R.G.K. & Hammes, W.P., 1998. Tyramine degradation by micrococci during ripening of fermented sausages. *Meat Science*, Vol.49, pp. 289–296.

Lowry, Q. H., Rosebrough, N. J., Farr, L. A., & Randall, R. J. (1951). Protein measurement with the Folin phenol reagent. *Journal of Biological Chemistry*, Vol.193, pp. 256–275.

Mah, J.H. & Hwang, H.J. (2009). Inhibition of biogenic amine formation in a salted and fermented anchovy by *Staphylococcus xylosus* as a protective culture. *Food Control*, Vol. 20, pp. 796–801.

Maijala, R. (1994). Histamine and tyramine production by a Lactobacillus strain subjected to external pH decrease. *Journal of Food Protection*, Vol.57, pp. 259–262.

Maijala, R., Eerola, S., Lievonen, S., Hill, P. & Hirvi, T. (1995). Formation of biogenic amines during ripening of dry sausages as affected by starter culture and thawing time of raw materials. *Journal of Food Science*, Vol.60, pp. 1187–1190.

Martuscelli, M., Crudele, M.A., Gardini, F. & Suzzi, G., 2000. Biogenic amine formation and oxidation by *Staphylococcus xylosus* strains from artisanal fermented sausages. *Letters in Applied Microbiology*, Vol.31, pp. 228– 232.

Molina, I. & Toldrà, F. (1992). Detection of proteolytic activity in microorganisms isolated from dry-cured ham. *Journal of Food Science*, Vol. 57, pp. 1308–1310.

Montel, M. C. & Champomier, M. C. (1987). Arginine catabolism in *Lactobacillus sake* from meat. *Applied and Environmental Microbiology*, Vol.53, pp. 2683–2685.

Molly, K., Demeyer, D., Johansson, G., Raemaekers, M., Ghistelinck, M. & Geenen, I. (1997). The importance of meat enzymes in ripening and flavour generation in dry fermented sausages. First results of a European project. *Food Chemistry*, Vol.59, pp. 539–545.

Monin, G., Marinova, P., Talmant, A., Martin, J. F., Cornet, M. & Lanore, D. (1997). Chemical and structural changes in dry-cured hams (Bayonne hams) during processing and effects of the deharing technique. *Meat Science*, Vol.47, pp. 29–47.

Montel, M. C. & Champomier, M. C. (1987). Arginine catabolism in Lactobacillus sake from meat. *Applied and Environmental Microbiology*, Vol.53, pp. 2683–2685.

Moreno-Arribas, M. V., Polo, M. C., Jorganes, F., & Munoz, R. (2003). Screening of biogenic amine production by acid lactic bacteria isolated from grape must and wine. *International Journal of food Microbiology*, Vol.84, pp. 117–123.

Murooka, Y., Doi, N. & Harada, T., 1979. Distribution of membrane bound monoamine oxidase in bacteria. *Applied and Environmental Microbiology*, Vol.38, pp. 565– 569.

Murooka, Y., Higashiura, T. & Harada, T., 1976. Regulation of tyramine oxidase synthesis in *Klebsiella aerogenes*. *Journal of Bacteriology*, Vol.127, pp. 24– 31.

Ordonez, J. A., Hierro, E. M., Bruna, J. M. & De la Hoz, L. (1999). Changes in the components of dry-fermented sausages during ripening. *Critical Reviews in Food Science and Nutrition*, Vol.39, pp. 329–367.

Rodrìguez, M., Núñez, J. J., Cródoba, M. E., Bermúdez, M. E. & Asensio, M. A. (1998). Evolution of proteolytic activity of micro-organisms isolated from dry-cured ham. *Journal of Applied Microbiology*, Vol.85, pp. 905–912.

Rozan, P., Kuo, Y.H. & Lambein, F. (2000). Free amino acids present in commercially available seedlings sold for human consumption: a potential hazard for consumers. *Journal of Agricultural and Food Chemistry*, Vol. 48, No.3, pp. 716–723.

Ruiz, J., Garcia, C., Diaz, M. C., Cava, R., Tejeda, F. J. & Ventanas, J. (1999). Dry cured Iberian ham non-volatile components as affected by the length of the curing process. *Food Research International*, Vol.32, pp. 643–651.

Ruiz-Capillas, C. & Moral, A. (2001). Production of biogenic amines and their potential use as quality control index foer hake (*Merluccius merluccius* L.) stored in ice. *Journal of Food Science*, Vol.66, pp. 1030–1032.

Scannell, A. G. M., Kenneally, P. M. & Arendt, E. K. (2004). Contribution of starter cultures to the proteolytic process of a fermented non-dried whole muscle ham product. *International Journal of Food Microbiology*, Vol.93, pp. 219–230.

Schomburg, D. & Stephan, D., 1993. Enzyme Handbook. Springer-Verlag, Berlin, Germany.

Sekikawa, M., Kawamura, T., Fujii, H., Shimada, K., Fukushima, M. & Mikami, M.(2003). Effect of arginine on growth of lactic acid bacteria for fermented sausage under high concentration of salt. *Hokkaido Animal Science and Agriculture Society*, Vol.45, pp. 17–23.

Silla-Santos, M. H. (1996). Biogenic amines: Their importance in food. *International Journal of Food Microbiology*, 29, pp. 213–221.

Suzzi, G. & Gardini, F. (2003). Biogenic amines in dry fermented sausages: A review. *International Journal of Food Microbiology*, Vol.88, pp. 41–54.

Toldrá, F. (2005). Dry-cured ham. In Hui, Y. H., Culbertson, J. D., Duncan, S., Guerrero-Legarreta, I., Li-Chan, E. C. Y., Ma, C. Y., Manley, C. H., McMeekin, T. A., Nip, W. K., Nollet, L. M. L., Rahman, M. S., Toldrá, F., &. Xiong, Y. L (Eds.). Handbook of food science, technology and engineering (Vol. 1). Boca Raton, FL: USA, Marcel-Dekker Inc./CRC Press.

Toldrá, F. (2006). The role of muscle enzymes in dry-cured meat products with different drying conditions. *Trends in Food Science and Technology*, Vol.17, pp. 164–168

Ventanas, J., Cordoba, J. J., Antequera, T., Garcìa, C., López-Bote, C. & Asensio, M. A. (1992). Hydrolysis and Maillard reactions during ripening of Iberian ham. *Journal of Food Science*, Vol.57, No.4, pp. 813–815.

Verges, C.M.C., Zuniga, M., Morel-Deville, F., Perez-Martinez, G., Zagorec, M. & Ehrlich, S.D. (1999). Relationships between arginine degradation, pH and survival in *Lactobacillus sake*. *FEMS Microbiology Letter*, Vol. 180, pp. 297–304.

Verplaetse, A., De Bosschere, M. & Demeyer, D. (1989). Proteolysis during dry sausage ripening. In Proceedings of the 35$^{th}$ international congress on meat science and technology. Copenhaegen, Denmark.

Virgili, R., Saccani, G., Gabba, L., Tanzi, E. & Soresi Bordini, C. (2007). Changes of free amino acids and biogenic amines during estende ageing of Italian dry-cured ham. *LWT – Food Science and Technology*, Vol.40, No.5, pp. 871–878.

Yamashita, M., Sakaue, M., Iwata, M., Sugino, H. & Murooka, Y. (1993). Purification and characterization of monoamine oxidase from *Klebsiella aerogenes*. *Journal of Ripening and Bioengineering*, Vol. 76, pp. 289–295.

Chapter 5

# The Role of Lactic Acid Bacteria in Safety and Flavour Development of Meat and Meat Products

Lothar Kröckel

Additional information is available at the end of the chapter

http://dx.doi.org/10.5772/51117

## 1. Introduction

Lactic acid bacteria (LAB) are widespread in nature and commonly occur on all kind of plant materials, on mucous membranes, in saliva and, in feces. Consequently and unavoidably they are part of the contamination flora of fresh meats after slaughter. Under certain conditions, e.g. in packaged refrigerated meats or raw sausage meats, they are able to compete efficiently with accompanying microorganisms for nutrients and may reach substantial viable counts. Their metabolic activities may ultimately result in either a desired preservative effect due to the repression of pathogenic and spoilage microorganisms, a desired tasty meat product, such as raw fermented sausage, or in meat spoilage through undesired transformations of raw and cooked meats. Heterofermentative LAB of the *Carnobacterium*, *Leuconostoc* and *Weissella* genera are usually more involved in meat spoilage than the homofermentative *Lactobacillus* and *Pediococcus* genera. Therefore, commercially available meat starter cultures for dry-fermented sausage production exclusively belong to the latter two. Homofermentative LAB produce almost exclusively lactic acid from fermentable carbohydrates present in meats, which is relatively mild and palatable, while heterofermentative species produce significant amounts of less desirable fermentation end products, such as $CO_2$ gas, ethanol, acetic acid, butanoic acid and acetoin. However, under certain conditions *Lactobacillus* spp. may also produce significant amounts of acetic acid, ropy slime and, discolouration (greening) of meats [1,2].

In food industry starter and protective cultures are currently used in a number of products to safeguard the microbial and sensory quality. Lactic acid bacteria (LAB) are the main players in the natural transformation of agricultural primary products into safe, delicious and shelf stable foods for human consumption. In meat products there are three basic fields of application for the targeted use of such cultures: raw fermented sausages, raw cured hams, and pasteurised, sliced prepackaged meats (cold cuts) [3-8]. The use of protective cultures in prepackaged, refrigerated sliced Bologna-type sausage and cooked ham against pathogenic listeria is a much discussed, sustainable technology for improving the microbial safety and quality of these products. It helps to avoid chemical preservatives, such as sodium lactate/potassium acetate additives, or repasteurisation in package after slicing and packaging, which both have a negative impact on sensory product quality leaving a numb mouthfeel or warmed-over flavour, resp. [9,10].

## 2. Meat and meat products

### 2.1. Raw fermented sausages

The importance of starter and protective cultures for the manufacturing of safe and high-quality fermented sausages has been known for a long time and, lactobacilli play an important role in their production [11,5]. *Lb. sakei* and *Lb. curvatus* are quite often the predominant LAB in dry-fermented sausage while other lactobacilli, such as *Lb. versmoldensis, Lb. plantarum, Lb. brevis, Lb. farciminis, Lb. alimentarius, Weissella* species, pediococci, and leuconostocs, usually occur in significantly lower numbers [12]. This has recently been also shown for different traditional salamis from North Italy [13-15]. However, other recipes and ripening conditions may promote other LAB as well. LAB isolated from dry spontaneously fermented sausages from 15 different producers in Spain included mainly *Lb. sakei* (66%), *Lb. curvatus* (26%), and *Lb. plantarum* (8%) [16]. For dry fermented Spanish 'chorizo' sausage *Lb. sakei* (69%), *Lb. curvatus* (16%) and *Pediococcus* (9%) have been reported [17]. From naturally fermented Greek dry salami about 50% of the isolates belonged to *Lb. sakei/curvatus*, 30% to the *Weissella* genus, 10% to *Lb. plantarum* and 3% each to *Lb. farciminis* and *Enterococcus (Ec.) faecium* [18]. In "Alheira", a fermented sausage produced in Portugal, *Lb. plantarum* and *Ec. faecalis* prevailed while other LAB, such as *Lb. paraplantarum, Lb. brevis, Lb. rhamnosus, Lb. sakei, Lb. zeae, Lb. paracasei, Leuconostoc (Leuc.) mesenteroides, Pediococcus (Pc.) pentosaceus, Pc. acidilactici, Weissella (Ws.) cibaria, Ws. viridescens* and *Ec. faecium*, occurred in lower numbers [19].

The main role of LAB is to convert fermentable sugars in the sausage batter to lactic acid, thereby contributing to product safety by creating unfavourable conditions for pathogens and spoilage organisms. The production of lactic acid has also a direct impact on sensory product quality by providing a mild acidic taste, and by supporting the drying process which requires a sufficient decline in pH. Furthermore, LAB influence the sensory characteristics of the fermented sausages by the production of small amounts of acetic acid, ethanol, acetoin, pyruvic acid, carbon dioxide, and their ability to initiate the production of aromatic substances from proteinaceous precursors [20-22]. The selection criteria for lactic acid bacteria to be used in the production of fermented sausage include (i) fast production of lactic acid (ii) good growth at different temperatures, (iii) homofermentative metabolism, (iv) persistence over the whole fermentation and ripening process, (v) nitrate reduction, (vi) ability to express catalase, (vii) no fermentation of lactose, (viii) formation of flavour, (ix) no formation of peroxide, (x) no formation of biogenic amines, (xi) no formation of ropy slime, (xii) tolerance or even synergy to other microbial components of the starter, (xiii) antagonism against pathogens, (xiv) antagonism against technologically undesirable microorganisms, (xv) improvement of the nutritional value of the sausage and, (xvi) economic factors [23]. Many homofermentative LAB associated with cured meat products are quite resistant to nitrite up to 200 ppm [24]. A new starter culture for raw sausages, 'BITEC Advance LD-20' from Frutarom Savory Solutions, containing *Lb. sakei* and *S. carnosus* is marketed as consistently providing a 'pleasant mild taste' while rapidly deminishing the pH value of the sausage batter. Rapid acidification is important for product safety while a high competitiveness against the spontaneous lactic flora is important for product quality. The culture can be used for firm and fresh raw sausages as well as sausage spreads.

The use of homofermentative lactic acid bacteria is desirable because acetic acid has an unpleasant taste as compared with lactic acid [25].

It must be kept in mind, however, that, although lactic acid production and pH reduction by LAB provide quite unfavorable conditions for pathogenic bacteria thereby preventing them from growing and contributing to their reduction, several pathogenic microorganisms are able to survive in fermented sausages under certain conditions for extended periods, especially during refrigerated storage of sparsely dried sausages. Pathogenic strains of *Escherichia (E.) coli, Listeria (Li.) monocytogenes* and *Yersinia (Y.) enterocolitica* are inactivated better after the initial fermentation and ripening stage if stored at ambient rather than at refrigeration temperature. Inclusion of a maturation period above refrigeration temperatures before distribution may increase the safety of these products [26-29].

## 2.2. Dry-cured hams

Currently there are only a few publications which clearly substantiate the advantages of starter and protective cultures during raw cured ham production. On the other hand, starter cultures have been more and more implemented by meat industry into the production of dry-cured hams since the early 1980s [6,30]. These cultures are expected to be active under the harsh manufacturing conditions (low temperatures, high salt, lack of oxygen, presence of nitrite). LAB contribute to a moderate pH decrease which promotes the microbial stability as well as product texture, reduce stickiness and pH variations of the raw material. As an example, FSC-111 Bactoferm$^R$ from Chr.-Hansen A/S contains, besides a staphylococcal strain, also a strain of *Lb. sakei*.

The LAB induced acidification is usually more pronounced with injected or compound meats than with dry-salted ones. Modern turkey hams are produced by squeezing turkey breast over the screw of an extruder in the presence of (g/kg) nitrite curing salt (35), diphosphate (2,5), dextrose (2), water (100), starter culture and a spice compound, and subsequent tumbling until protein release. This mixture is then stuffed into fiber casings and left for 5 days at 2°C. This is followed by a fermentation step of around 16 hours at 22°C and 92-94% relative humidity until a pH below 5.4 is reached. Finally, the product is heated in a cabinet at 47 °C to a core temperature of 40°C. The desired result is a fresh looking product with a slightly hyaline appearance with an optimum safety against undesired and pathogenic microorganisms [30].

## 2.3. Fresh meats

In chilled vacuum-packaged beef, even close to the freezing point, psychrotrophic LAB are able to attain high population densities. At -1.5 °C LAB grew to 8-9 $log_{10}$ cfu $ml^{-1}$ drip in 16 weeks with maximum doubling times of around 2-4 days [31]. In this study, *Cb. divergens, Leuc. mesenteroides* and *Lb. delbruckii* dominated the LAB flora after 4, 8-12 and 16 weeks, respectively. At 2 °C other workers have reported *Lb. sakei, Lb. curvatus, Carnobacterium (Cb.) divergens, Cb. maltaromaticum, Leuconostoc spp.* and *Lactococcus raffinolactis* as relevant LAB with $t_d$ of around 19 hours and less [32]. After 25 days maximum LAB numbers of around 7-8 $log_{10}$ cfu $cm^{-2}$ were reached and after 8 weeks the the meat odour immediately after opening the bags was regarded "definitely off" ("slightly off" between 4-6 weeks).

LAB may be useful as protective cultures during the ripening of vacuum-packaged raw beef and, bioprotective cultures may also help to reduce *E. coli* O157:H7 in frozen ground-beef patties [33,34]. Peptides generated by LAB have been suggested as sensorial and hygienic biomarkers in meat conditioning and fermentation [35].

Today, meat industry is forced to produce meats with a shelf life long enough to fulfill logistic, retail sale and consumer demands. Besides general hygienic considerations, including appropriate temperature control modified-atmosphere packaging (MAP) with 30-40% $CO_2$ is used to prevent early spoilage. While Gram-negative spoilage bacteria are suppressed, psychrotrophic LAB are not [36-38].

## 2.4. Cooked meats

Cooked, sliced and prepackaged meat products are popular convenience foods. They are retailed under refrigeration with varying shelf lifes, e.g. at 5 to 7 °C for 14 to 28 days. During slicing and packaging the slices may be contaminated with microorganisms from the production environment. Especially certain psychrotrophic LAB may then attain high cell counts during cold storage and impair the sensory quality of the products [39-42]. More than 2/3 of the refrigerated sliced cooked meats from the German retail market contained LAB counts above 7 $log_{10}$ cfu $g^{-1}$ one week past the indicated shelf life (Figure 1) [43]. The LAB flora on Bologna-type sausage is mostly dominated by the *Lb. sakei/curvatus* cluster while *Leuc. carnosum* frequently dominates on cooked ham. Occasionally, also *Ws. viridescens, Cb. maltaromaticum* and *Leuc. mesenteroides* ssp. *mesenteroides* may occur in higher numbers. Independent from dominant occurrence, eight LAB species have been identified in German retail samples. The number of samples (n) out of 50 in which these species occurred were *Lb. sakei* (40), *Leuc. carnosum* (22), *Lb. curvatus* (18), *Ws. viridescens* (11), *Leuc. mesenteroides* ssp. *mesenteroides* (8), *Cb. maltaromaticum* (4), *Lactobacillus* sp. (4), *Lactococcus* sp. (4), *Cb. divergens* (2), *Leuc. gelidum* (1), *Leuconostoc* sp. (1) (Figure 2) [43].

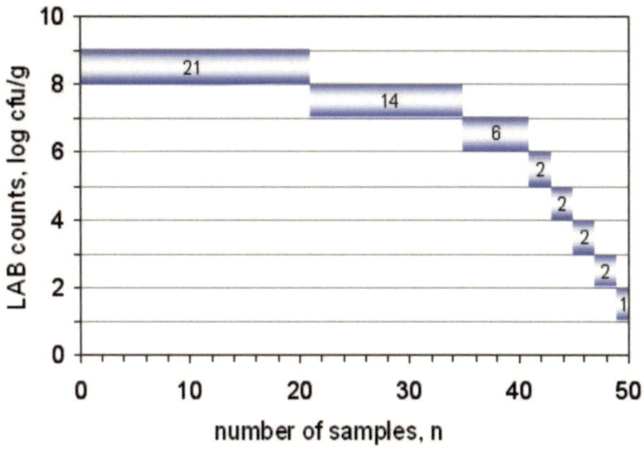


**Figure 1.** Distribution of samples of refrigerated sliced cooked meats from the German retail market with respect to different LAB counts one week past the indicated shelf life [43].

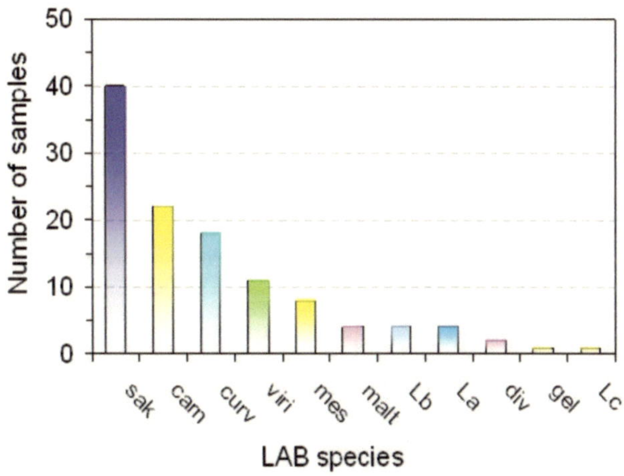


**Figure 2.** Abundancy of different LAB species in refrigerated sliced cooked meats from German retail (n=50). sak, *Lb. sakei*; carn, *Leuc. carnosum*; curv, *Lb. curvatus*; viri, *Ws. viridescens*; mes, *Leuc. mesenteroides*; malt, *Cb. maltaromaticum*; Lb, *Lactobacillus* sp.; La, *Lactococcus* sp.; div, *Cb. divergens*; gel, *Leuc. gelidum*; Lc, *Leuconostoc* sp. [43].

## 3. Biopreservation

Biopreservation of meats refers to the control of pathogenic and spoilage microorganisms by a competitive microflora of desired indigenous microorganisms or so-called starter and protective cultures. The development of starter cultures for meats is tightly coupled with the industrialisation of the traditional artisanal processes. The production of safe and tasty fermented sausages by traditional technologies requires expert knowledge and continous attention to guide the fermentation into the desired direction, i.e. to promote the development of the desired microorganisms and to suppress the development of undesired microorganisms. Mistakes are heavily paid for by dangerous and/or low quality outcomes.

Starter cultures, added at the beginning of fermentation, allow a standardization of the product quality and considerably reduce the risk of product defects. However, it should be kept in mind that starter cultures can not replace good manufacturing practice which besides the selection of the appropriate raw materials with acceptable hygienic parameters also includes the implementation and control of appropriate processing conditions. This is especially true with respect to the health risks associated with enterohaemorrhagic *E. coli*. Because of its increased acid tolerance and low infective dose for human infection, additional hurdles besides starter cultures have become very important for the production of safe raw fermented sausages. The hurdles principle for controlling undesired microorganisms in raw sausage fermentation has been illustrated by LEISTNER [27,44,45] and, in the meantime the implementation of HACCP (hazard analyis critical control point) concepts have become mandatory in food production [46].

Protective cultures may be distinquished from starter cultures by their lack of, or their reduced product transformation capabilities. Protective cultures my be used for a number of applications with the main focus on pathogen control, especially of *Li. monocytogenes*, but also of spoilage organisms such as LAB involved in the spoilage of deli meats, or of *Brochothrix thermosphacta* and *Clostridium estertheticum* in vacuum-packaged raw meats [47-50].

Of special interest are strains which excrete powerful anti-listerial bacteriocins *in situ* and, which at the same time have no or only a very weak spoilage potential [21, 51].

A strain of *Lactococcus (Lc.) lactis*, marketed as Bactoferm® Rubis by Chr.-Hansen A/S, is offered as a protective culture to be used instead of chemicals to preserve/stabilize the normal colour of vacuum packed or controlled atmosphere packaged, sliced, cured meat products [52].

The big retail chains and the official food control authorities look at high microbial counts in deli meats, regardless of the responsible microflora, usually with suspicion. The German Society for Hygiene and Microbiology (DGHM), e.g., recommends a maximum of $5x10^6$ cfu $g^{-1}$ [53]. In reality, however, many of the prepackaged sliced cold cuts display 10-100 times higher counts at the the end of their indicated shelf lives without being recognized as spoiled by sensory panels. On the other hand, unpleasant tastes and smells (not fresh, sour) are often associated with high LAB counts [54]. But a high count *per se* does not tell how long the product has been exposed to this high count already. Protective LAB cultures are looked at with suspicion because they have to be added in high numbers and, if metabolically too active, may reduce shelf life. Some authors generally view psychrotrophic LAB as spoilage organisms, regardless of their generally moderate role in spoilage [55]. There is no doubt that cold-cuts with protective cultures will differ from products without protective culture. But, as long as this difference is only manifested in a minor sour taste this kind of sensory deviation may be a reasonable price to pay for an increased food safety, especially with respect to *Li. monocytogenes*, without chemical preservatives and the control of more striking spoilage organisms, e.g. such as *Brochothrix thermosphacta*. Food preferences are changing, and presently many consumers tend to prefer products which are as much as possible free of chemical preservatives [8], processing aids and allergenic additives, and which are not overly treated by physical processes, such as heat, high pressure and irradiation. Nevertheless, many consumers also simply do not care, as long as the product is safe and affordable. Thus, protective cultures may be interesting for health and wellness-oriented consumers in countries with higher living standards. But less developed countries could also benefit, especially where cold-chain management is difficult and high-tech processing aids are not readily available. The challenge simply is to find the right LAB cultures for the particular product.

## 4. Sensory acceptance of bioprotective cultures on prepackaged cold cuts

As already mentioned, the application of bioprotective microbial cultures to prepackaged cold cuts is a much discussed innovative and sustainable technology for improving the microbiological safety and overall quality of these products. It could be an alternative to chemical preservatives or to a second pasteurisation step after packaging which both have a negative sensory impact. Although quite a number of lactic acid bacteria (LAB) have been suggested as protective cultures for sliced cooked meats, there is basically no information on consumer perception of products with added LAB. At the International Green Week Berlin 2010 the concept was introduced for the first time to a broader public and visitors were asked to participate in a sensory preference test [7].

Bologna-type sausages in 70 mm fiber casings were produced and stored at 2 °C until slicing. On the day of packaging the casings were removed and the sausages were briefely submersed in an aqueous suspension of a protective culture consisting of *Lb. sakei* strain Lb674 (sakacin P positive) and containing 8.5 $\log_{10}$ LAB $ml^{-1}$. Subsequently, the sausages were sliced, vacuum-packaged in polyethylene bags and kept refrigerated at 5°C until presentation to interested visitors. The consumers reacted predominantly positive on the possibility of safeguarding cold cuts with bioprotectants. Up to day 15 after packaging the inoculated samples reached a relative preference score (achieved points versus achievable points) of more than 45% (max. 60%) as compared to 60-70% for the freshly sliced samples without added LAB. Thereafter, the overall liking of the inoculated prepackaged sausage gradually decreased (Figure 3). The results indicate a potential market for more natural, microbiologically safe and sound cold cuts as a specialized segment of the convenience sector. As stated above, a mild acidic note may not be completely avoided when using protective cultures. But, this 'disadvantage' should be balanced against the risk of an uncontrolled growth of listeria on the one hand and the demand of many consumers for less chemical preservatives or thermal treatments on the other hand.

## 5. Probiotics

The steeply increasing business in the industrialised countries with health and wellness oriented foods in the 1990s, starting with probiotics in dairy products, has also raised interest in the development of probiotic meat products [56]. The concept of probiotics requires the intake of relevant amounts by the consumer of living probiotic microorgansims,

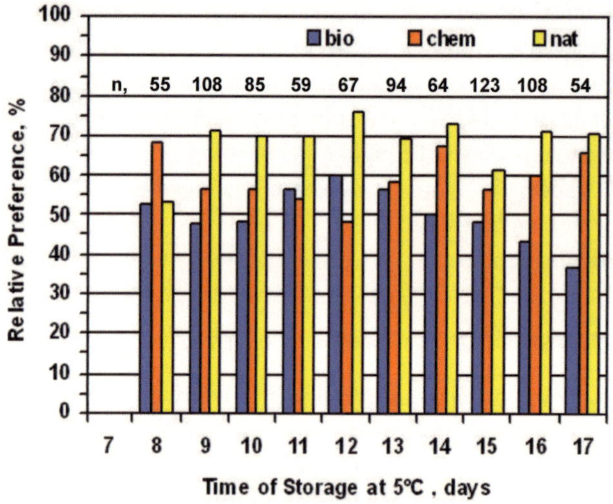


**Figure 3.** Consumer preference of vacuum-packaged Bologna-type sausage with *Lb. sakei* protective culture (bio) in comparison to non-packaged, sliced on-the-spot sausages without (nat) and with chemical (chem) preservatives presented at the International Green Week Berlin 2010. n, number of responses [7].

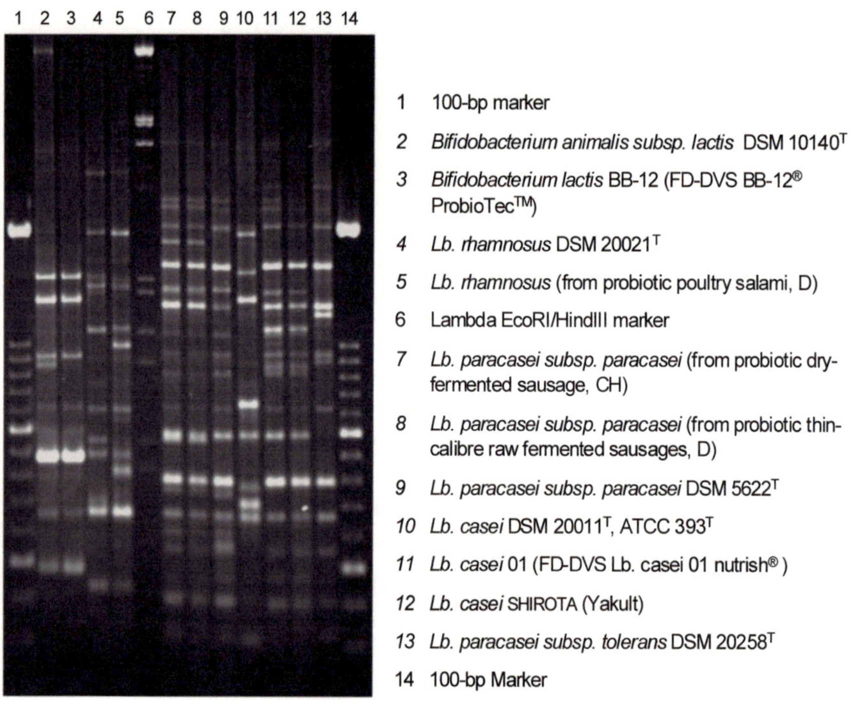


**Figure 4.** Genetic fingerprints of probiotic LAB and related reference strains using BOX-PCR [57].

and raw fermented sausages were considered as an appropriate vehicle for these probiotics. However, these environments are quite different from the human gastrointestinal (GI) tract, and the strains under consideration have to cope with and survive in the presence of nitrite, sodium chloride, reduced pH and water activity, various processing steps and, eventually, long-term storage. Due to the manufacturing process raw fermented sausages contain high numbers of lactic acid bacteria which, however, are not regarded as probiotics. On the other hand, most of the known probiotic bacteria are unable to establish themselves in the raw sausage environment. Exceptions thereof are microbial cultures belonging to the *Lactobacillus plantarum group* and to the *Lactobacillus casei* group [57-59]. The use of protective and probiotic cultures may be a useful and effective strategy to prevent or reduce pathogens in the food chain, improve food safety and consumer health.

Within a project investigating the possibilities for manufacturing high quality and microbiologically sound products from meat of mother sheep, salami-type raw fermented sausages were produced with added conventional (*Lb. sakei, Lb. plantarum*) and probiotic lactic starter cultures (*Lb. paracasei*). The products were subjected to microbiological and sensory evaluation for up to nine months. All sausage batches with added cultures resulted in microbiological safe and sensory appealing products. The *Lb. sakei* culture survived during the whole storage period on a high level (> $10^8$ cfu/g) while the two other cultures (*Lb. plantarum, Lb. paracasei* ) partly reached the threshold of $10^6$ cfu $g^{-1}$ already after 3 months and were replaced by indigenous lactic acid bacteria of the *Lb. sakei / curvatus* group. For some batches, however, an acceptable number of probiotic bacteria could still be detected after nine months. Overall, *Lb. paracasei* showed a better survival in the ripened sausage than *Lb. plantarum* [7].

One problem for official authorities involved in consumer protection is to verify the presence of the indicated probiotics at sufficiently high levels. In the absence of simple and relyable identification procedures this may be a challenging task. In such cases genetic fingerprinting of isolates recovered on suitable agar media at relevant dilutions is the method of choice (Figure 4) [57, 60]. In the past, *Lb. rhamnosus* and *Lb. paracasei* ssp. *paracasei* have been used in fermented sausages, and labelling was quite confusing (Table 1). As can be seen, *Lb. paracasei* survived in relatively high numbers even in very dry salami. More recently, also other LAB species have been suggested as probiotics, and microencapsulation of strains has been used to overcome survival problems in the sausage environment. Still, human verification studies for probiotic administration are quite rare [61].

## 6. Functional starter cultures

In fermented sausage production classical starter cultures are usually also protective cultures, especially with respect to the acid-sensitive microflora. Modern cultures may provide additional protective action, e.g. by producing bacteriocins inhibitory to listeria and/or undesired LAB, or they may possess an additional probiotic functionality. Strains combining these traits have been termed also 'functional starter cultures' [62].

## 7. Bacteriocin production

Strains from many LAB species excrete anti-listerial bacteriocins, of which nisin produced by *Lc. lactis* and pediocin produced by *Pc. acidilactici* are the most wellknown. Besides,

| Sausage no. | 1 | 2 | 3 | 4 |
|---|---|---|---|---|
| Type | soft, quickly ripened, smoked salami | very hard, air-dried salami | quickly ripened, thin-calibre, smoked sausage | smoked, dry-fermented salami with 30% weightloss |
| Characteristics | pH 4.7; $a_w$ 0.954 | pH 5.6 | pH 4.9 | n. d. |
| Origin | D | CH | D | D |
| Claims | 'probiotic poultry salami with three probiotic cultures (*Bifidus, Lb. casei, Lb. acidophilus*)' | ‚probiotic', with beef and pork | 'probiotic culture in high numbers ($5x10^8$ cfu/g)', with beef and pork | 'Probiotic !!!, naturally ripened', with beef and vegetable fat (no lard) |
| advertised culture detectable | no | not labeled | yes | not labeled |
| detected potentially probiotic culture | *Lb. rhamnosus* | *Lb. paracasei* subsp. *paracasei* | *Lb. paracasei* subsp. *paracasei* | *Lb. paracasei* subsp. *paracasei* |
| Viable counts (cfu/g) of probiotic culture | 1-6 x $10^7$ | 4-9 x $10^7$ | 3 x $10^7$ | 1 x $10^7$ |

n. d., not determined; D, Germany; CH, Switzerland.

**Table 1.** Detection of probiotic cultures in probiotic raw fermented sausages from retail [57].

bacteriocin-producing LAB with anti-listerial activity naturally occur on a wide range of ready-to-eat foods, including meats [63-65]. From a meat point of view the sakacins of *Lb. sakei* are the most interesting because of the high competitivity of this species in the meat environment [22,49]. The pediocin producer *Pc. acidilactici* is commonly used by the Spanish meat industry as a starter culture [66].

## 8. Hydrogen peroxide production

The demonstration of hydrogen peroxide formation by meat-borne lactic acid bacteria is of considerable importance for the characterization of individual strains, the selection of suitable starter and protective cultures for various applications for meat and meat products as well as for the search of potential microbiological causes for undesired sensory deviations (discolourations/'greening', rancidity). Many LAB are able to form hydrogen peroxide as a by-product of $O_2$-dependent metabolic pathways. Dependent on the environment, this trait may be desired or undesired [1,23,67,69,104].

In foods and feed it may contribute to the inhibition of an undesired accompanying microbiota [67]. The $H_2O_2$ formed by LAB acts bacteriostatic on GRAM-positive bacteria and bactericidal on Gram-negatives [12,68].

In a recent study a novel agar medium ('Prussian Blue' (PB) agar) was applied for the first time to lactic acid bacteria relevant to meat and meat products [69]. The PB agar detects $H_2O_2$ through the formation of Prussian Blue (Figure 5). It principally delivers similar results as the traditional manganese dioxide agar. However, it is more sensitive and, it is also more easily prepared and delivers results more quickly. A representative number of strains was used in the evaluation of the new medium (Table 2).

As to the production of $H_2O_2$, the study revealed large differences within the *Lb. sakei/curvatus* group. The bacteriocin producers frequently seemed to be relatively weak peroxide producers, while many commercial starter cultures were recognized as more or less strong peroxide producers. More recent field isolates of *Lb. sakei/curvatus* from prepackaged sliced Bologna-type sausage gave an essentially similar picture. In this case, however, only one of ten isolates of *Lb. curvatus* gave rise to a positive reaction.

| Species | strains[a)] | PB | $MnO_2$ |
|---|---|---|---|
| *Lb. sakei* (*Lb. bavaricus*) | DSM 20494 | - | nd |
| *Lb. sakei ssp. carnosus* (*Lb. curvatus ssp. melibiosus*) | DSM 15740 | + | ++ |
| *Lb. sakei ssp. carnosus* | Lb1047 | - | + |
| *Lb. sakei ssp. carnosus* | DSM 15831$^T$ | ++ | ++ |
| *Lb. sakei ssp. carnosus (ssp. sakei)* | 23K | +++ | ++ |
| *Lb. sakei ssp. carnosus* | Lb790 | +++ | ++ |
| *Lb. sakei ssp. sakei* | DSM 20017$^T$ | +++ | + |
| *Lb. brevis* | DSM 20054$^T$ | ++ | - |
| *Lb. farciminis* | DSM 20180$^T$ | - | nd |
| *Lb. hilgardii* | DSM 20176$^T$ | - | - |
| *Weissella paramesenteroides* | DSM 20288$^T$ | +++ | nd |
| *Weissella minor* | DSM 20014$^T$ | +++ | nd |
| *Leuconostoc carnosum* | Lb1259 | + | + |
| *Leuconostoc carnosum* | Lb1054 | ++ | + |
| *Leuconostoc carnosum* | Lb1045 | ++ | + |

a) strains have been obtained from the German Collection of Microorganisms (DSM) and the strain collections of MRI Location Kulmbach (Lb) and INRA at Jouy-en-Josas (23K)

**Table 2.** Reaction of different LAB species on PB agar with BHI or MRS base, and on $MnO_2$ agar. -, no production of $H_2O_2$; +/++/+++, moderate to strong production of $H_2O_2$; nd, not determined [69].

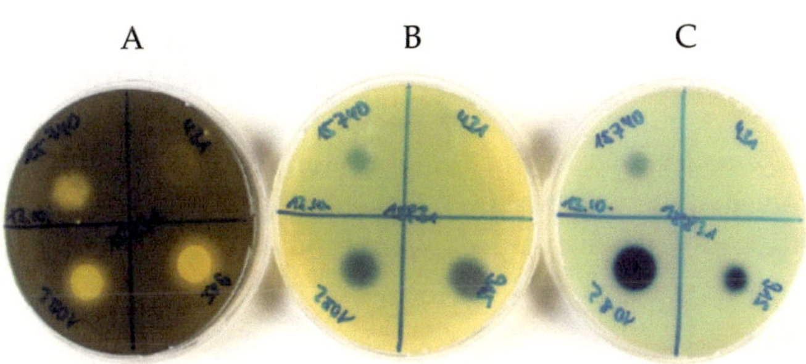


**Figure 5.** Reaction of different LAB species on $MnO_2$ agar (A) and on PB agar with MRS (B) or BHI (C) base. Production of $H_2O_2$ is indicated by bright and blue halos, resp. [69].

## 9. Formation of biogenic amines

Several LAB may produce biogenic amines by decarboxylation of amino acids, e.g. *Lb. buchneri, Lb. brevis, Lb. curvatus, Lb. hilgardii, Cb. maltaromaticum, Cb. divergens* [70]. Examples are such as tyramine and histamine during sausage fermentation. Strains of *Lb. plantarum, Lb. brevis* and *Lb. casei/paracasei,* and *Ec. faecium* and *Ec. faecalis* were identified as tyramine/histamine producers in the sausages [71]. Suitable starter cultures may contribute to reduction of biogenic amines in fermented sausages [72].

## 10. Identification of LAB

Identification of meat associated LAB is still wideley performed with phenotypic methods only, e.g API 50 CH [73]. These are, however, not always satisfying and may lead to misidentifications [74]. Nowadays, the application of PCR-DGGE and 16S rRNA gene sequencing allow the identification of a large number of strains in a quick and fast way [21,75]. Also various genomic fingerprinting methods are available. Nevertheless, conventional approaches remain important, especially when dealing with previously unknown species. Modern identification procedures rely on polyphasic approaches, integrating several lines of evidence to obtain a comprehensive description of a new species or of a microbiota [76].

## 11. Important LAB in meats

### 11.1. The Lb. sakei/curvatus cluster

In his 1983 review on lactic acid bacteria of meat and meat products EGAN mentions that according to recent findings of KANDLER and co-workers *Lb. sakei* (then *Lb. sake*) and *Lb. curvatus* were very common on German meat products [1]. Presently, two subspecies of *Lb. sakei* are known of which ssp. *carnosus* is the one characteristic for meats. It is common in fermented meat products, and is regularly found in vacuum-packaged meat and fermented plant material (sauerkraut). The subspecies *sakei* has been isolated from the Japanese sake starter and is regularly found in fermented meat products, vacuum-packaged meat, fermented plant material (sauerkraut), and human feces. The two subspecies can not be separated based on their physiological and biochemical characteristics [12]. The genomes of *Lb. sakei* 23K from a French dry-fermented sausage and *Lb. curvatus* CRL705 from an Argentinean artisanal fermented sausage have been sequenced [77,78]. Both genomes are highly similar. *Lb. curvatus* CRL705 lacks several genes present in *Lb. sakei* such as those related to fatty acid biosynthesis FASII, sucrose utilization, the arginine deiminase pathway, and citrate metabolism. The ones unique in *Lb. curvatus* CRL705 include genes for proteins and enzymes involved in the metabolism of carbohydrates, DNA, and fatty acids, as well as in the oxidative stress response and in bacteriocin production.

### 11.2. Lb. plantarum

The LAB species *Lb. plantarum* displays a high flexibility and versatility, and is able to colonize several ecological niches such as vegetables, meats, fish, milk substrates, and the human GI tract.

This is the basis of many applications in the food and health areas. As a starter culture for salamis *Lb. plantarum* is used since decades. More recently also probiotic strains have been described. With a size of 3.3 Mb its genome is one of the largest of LAB. A recent study on the phenetic and genetic diversity of the species revealed a high phenetic diversity which generally correlated with the origin of the isolates, e.g. from meat fermentations, kimchi, sourdough, egg plants and cheese. Four main clusters were determined: (i) meat, (ii) vegetable, (iii) sourdough, (iv) mixed sources with high meat content. On the genome level there were seven main clusters. The core genome contains more than 2000 genes, 121 genes being specific for *L. plantarum*. None of the strains could grow in milk, or at 4°C, or in the presence of 10% NaCl. A limited number grew at 17°C, or at 6% NaCl [79]. One of the earliest and most successful starter cultures for raw fermented sausages on the German market, "DuploFerment 66", contains a strain of *Lb. plantarum*. This is also the case for the "Saga II" starter from the US. In contrast to the first one, the latter strain does not grow at 10°C. Both strains are homofermentative for lactate and grow at 42°C but not at 8°C [25]. They provide rapid acidification of the raw sausage batter. On the other hand, *Lb. plantarum* is not very well adapted to meat and fails to maintain sufficiently high cell numbers to outcompete indigenous LAB. Sometimes it even does not grow in the meat batter [80,81]. In Italian natural fermented sausage the initial dominant populations of *Lb. plantarum* were accompanied by *Lb. sakei* and *Lb. curvatus* from the 10th day of fermentation and were finally competed out by the latter [21]. But, in certain traditional Greek fermented sausages *Lb. plantarum* and *Lb. plantarum/pentosus* may predominate [82,83].

### 11.3. Lb. brevis

In combination with *Pc. pentosaceus*, *Lb. brevis* has been used as an indigenous starter culture for a Vietnamese fermented meat product [84]. While *Lb. brevis* strongly acidifies the product, *Pc. pentosaceus* acts as a mild acidifier. The combination of both species resulted in a product with an intermediate taste (not too mild and not too sour) preferred by the sensory panel. Meat isolates of *Lb. brevis* may produce bacteriocins with antagonistic activity against *Li. monocytogenes* [85].

### 11.4. Lb. versmoldensis

This species was first reported in 2003 as the dominant LAB in some German raw fermented poultry salamis. The species was present in high numbers and frequently dominated the lactic acid bacteria (LAB) populations of the products [86]. Later, the species has been isolated also from Scandinavian fermented meats, Egyptian Domiati cheese and Japanese traditional fermented fish products [87-89]. There are no studies to date on the general behaviour of this species in meat ecosystems. The genome of strain KCTC 3814, an isolate from poultry salami, has been recently sequenced by the Korea Research Institute of Bioscience & Biotechnology [90].

### 11.5. Carnobacteria

Carnobacteria are non-aciduric and, therefore, are preferentially isolated from meats with elevated pH. *Cb. divergens* and *Cb. maltaromaticum* frequently constitute a major component of the microflora of packaged raw meats as well as of refrigerated, prepackaged, sliced cooked deli meats. Meat spoilage by *Cb. maltaromaticum* has been associated with "dairy", "spoiled-meat", and "mozarella cheese" perception [31,91,92]. The major volatiles on meat, acetoin, 1-octen-3-ol and butanoic acid, are volatile organic compounds with low sensory impacts. Butanoic acid in stored beef was also associated with *Cb. divergens*. It has a rancid cheese-like odor and can derive from leucine metabolism, microbial consumption of free amino acids via the Stickland reaction or from tributyrin hydrolysis.

The metabolites from leucine degradation are involved in dry fermented sausage aroma. The catabolism of leucine by a strain of *Cb. maltaromaticum* was studied directly in the growth medium with H-3-labelled leucine to investigate the effect of five parameters: phase of growth, pH, oxygen, glucose and alpha-ketoisocaproic acid. Leucine catabolism was most important during the exponential phase of growth. The addition of alpha-ketoisocaproic acid at 1%, glucose at levels of 0.5% to 2% and shaking of the growth medium increased leucine catabolism. At pH 5.4 and 7.2, the main metabolites detected were 3-methyl butanal, 3-methyl butanol and alpha-ketoisocaproic acid. At pH 6.5, the leucine catabolism was maximum and was characterised by a high production of 3-methyl butanoic acid [93].

Positive and negative effects of carnobacteria in the environment and in foods have recently been reviewed [94]. Because *Cb. divergens* and *Cb. maltaromaticum* show good growth in refrigerated meats and some of the strains produce potent anti-listerial bacteriocins, they may have some role as bioprotectants in meat environments. However, carnobacteria are associated with unpleasant spoilage metabolites in meats, such as acetic and butanoic acid as well as gas production in vacuum packed beef. An undesirable trait is also their ability to produce the biogenic amine tyramine from tyrosine. Carnobacteria are not regarded as human pathogens, but *Cb. maltaromaticum* is a well known fish pathogen and catagorised as a safety-level-2 microorganism. The genome of *Cb. maltaromaticum* ATCC 35586 carries putative virulence genes which probably play a role in fish pathogenesis [95]. Since carnobacteria are inhibited by acetate they do not grow well on routine LAB media such as MRS. A selective enumeration medium using a combination of three antibiotics (gentamicin, nalidixic acid, vancomycin) and an alkaline pH value (8.8) has recently been proposed for *Cb. maltaromaticum* from cheese [96].

## 11.6. Leuconostoc

*Leuc. gelidum* is a major spoilage organism in Finnish fresh meats [97]. Certain strains of *Leuc. gelidum* may produce yellow discolourations on prepackaged refrigerated German 'Weisswurst' and cold cuts (Figure 6, 7) [98]. Recently, the genome of a plant isolate of *Leuc. gelidum* has been sequenced [99].

The responsible pigment for the intensive 'neon-like' yellow discolouration is a bacterial carotenoid, the non-polar C30-carotenoid 4,4'-di-apo-7,8,11,12-tetra-hydro-lycopene. On fat-containing substrates this compound does not only stain the bacterial cells but also the substrate and, in the case of 'Weisswurst' does stain the natural casing (porc intestine) of the sausage as well as the sausage surface beneath. This triterpenoid is an intermediate in the microbial synthesis of 4,4'-diaponeurosporene which represents the main carotenoid in pigmented enterococci, Leuc. citreum and Lb. plantarum. Identification of the pigment was achieved by using UV-VIS spectroscopy in combination with available data from literature [100].

A report from Canada also described the yellow discolouration phenomenon on cooked sliced meats which had been stored for an extended time period under refrigeration [101]. These authors, employees of a big Canadian food company (then Canada Packers Inc.), tentatively identified an *Enterococcus* sp. as the causative agent.

**Figure 6.** Yellow discolourations on prepackaged refrigerated German 'Weisswurst' after targeted inoculation with *Leuc. gelidum* and incubation at 5°C for 14 days [98].

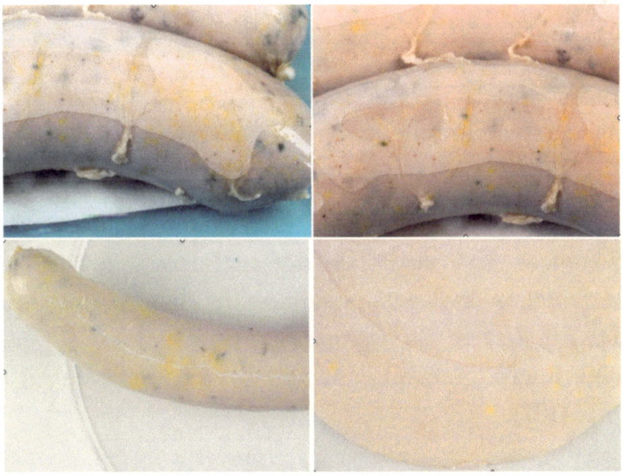

**Figure 7.** Yellow discolourations on pre-packaged meat products produced by *Leuc. gelidum*. A and B, 'Weisswurst' from organic production; C, grill sausage from conventional production; D, sliced cooked turkey breast from conventional production [98].

*Leuc. gasicomitatum* has been recognized as a specific spoilage organism in cold-stored Finnish MAP meats. It emerged as a spoilage problem of tomato-marinated, raw broiler meat strips. Due to $CO_2$ production the packages already showed clear bulging more than a week before the expected shelf life [102]. It is a psychrotrophic species and, because of its dominance in marinated meats and fish as well as in vegetable sausages, probably of plant origin. But, it was also detected in minced meat and high-oxygen modified-atmosphere packaged raw, beef steaks injected with sugar-salt solutions, so-called moisture-enhanced or value-added meats [97,103]. Recently, the genome of the type strain *Leuc. gasicomitatum* LMG 18811$^T$ has been sequenced [55].

### 11.7. Weissella

*Weissella* spp. are heterofermenters producing $CO_2$, ethanol and/or acetate from glucose. The species *Ws. viridescens*, *Ws. halotolerans* and *Ws. hellenica* have been associated with meat and meat products. *Ws. viridescens* is considered as heat resistant and may cause green discolouration in cured meats [104]. This species is frequently isolated from refrigerated sliced cooked meats [43] and was reported to produce cavities in the muscles of hams after cooking [105].

### 11.8. Pediococcus

The homofermentative pediococci are mostly applied for rapid and strong acidification at elevated temperature, especially in US summer sausage fermentation. Usually *Pc. acidilactici* and *Pc. pentosaceus* are the species involved. *Pediococcus* sp. are among the most common starter cultures in the US [11,21]. A pediocin producing *Pc. acidilactici* is also commonly used by the Spanish meat industry as a starter culture [66].

### 11.9. Enterococcus

In mediterranian traditional dry-fermented sausages enterococci are found in relevant numbers and are believed to contribute to the characterisic product flavor. *Ec. faecalis*, e.g., is common in Portuguese 'alheira' [19].

On the other hand, the presence of enterococci in foods is debatable, since some strains carry antibiotic resistances and virulence determinants relevant in human medicine [22,106]. Also, *Ec. faecium* and *Ec. faecalis* were identified as tyramine/histamine producers in the sausages [71]. The use of *Ec. faecium* strains has been suggested to control the growth of undesirable microorganisms such as listeria on material and environmental surfaces in meat plants [107].

## 12. Outlook

Meat and meat products provide a concentrated source of protein of high biological value and can make a valuable contribution to human diets. However, they are also highly perishable commodities which rapidly spoil and may even allow the growth of food-borne pathogenic microorganisms if no suitable preservative actions are taken. Meat fermentation involving beneficial LAB has become an important and sustainable preservation technology, and today a number of suitable species and strains are successfully applied as starter and protective cultures in various fermented meats all over the world. These cultures not only prevent the growth of common food pathogens but also of undesirable food spoilage bacteria, including heterofermentative LAB. The answer to the question which strains we should use for which products largely depends on consumer expectations and technological needs. Much has been learned over the years, however, we are still far from understanding the complex metabolic interactions of LAB in meats.

Systems biology has become an important approach in LAB microbiology and will become even stronger in the future [108]. It links quantitative microbial physiology with population dynamic modelling and ecological theories. In comparative systems biology of LAB, the so-called "omics"-techniques ("genomics", "proteomics", "transcriptomics", "metabolomics") and mathematical and statistical methods are of crucial importance [109, 110]. Comparative analyses between various species is expected to deliver understandable models of the metabolism of these species. Whole genome sequencing has made a quantum leap in the past few years and it is likely that very soon all genomes of meat associated LAB species and even of different strains will be available for comparative studies. Diversity and differences within each of the species at the strain level will have to be considered. The ripening, packaging and storage of meats could benefit from improved systems knowledge of the diverse meat microcosms with respect to microbial survival and growth, as well as desired and unwanted microbial transformations of meat components to ensure high-quality, healthy, safe and tasty products. The beneficial aspects of LAB in meat preservation could be explored using systems techniques and will decrease our dependence on chemical preservatives. Likewise, the impact of microbes on meat spoilage could be better managed with a systems understanding of the interplay of microbes, raw materials, additives and processing technologies.

In a global perspective, the role of starter and protective cultures for the safety and quality of meats is expected to increase. Although the chemical preservatives currently applied to prevent the growth of pathogens and spoilage bacteria in deli meats perfectly serve this purpose, there is an increasing consumer demand for more natural products. This is in part reflected by the so-called clean label strategies of the big manufacturers. Many chemical additives not only contribute to the sodium burden of the meats, but also leave an undesirable numb mouthfeel which negatively effects the sensory perception of the meat aroma. Innovations in fermented meat production will benefit from an improved knowledge of systems microbiology of LAB in the various meat environments on the one hand, and the gastrointestinal environment on the other. A future challenge will be to link intraspecies diversity to a specific sensory profile [21]. The application of probiotic starter microorganisms in dry-fermented sausages remains appealling for the wellness-oriented consumers even if immediate health claims should be difficult to establish. In this sense beneficial LAB will vitally contribute to a sustainable and diversified food production.

## Author details

Lothar Kröckel
*Max Rubner-Institute Location Kulmbach, Department of Safety and Quality of Meat, Kulmbach, Germany*

## 13. References

[1] Egan AF (1983) Lactic Acid Bacteria of Meat and Meat Products. Antonie van Leeuwenhoek 49: 327-336.

[2] Björkroth J, Ridell J, Korkeala H (1996) Characterization of *Lactobacillus sake* Strains Associated with Production of Ropy Slime by Randomly Amplified Polymorphic DNA (RAPD) and Pulsed-Field Gel Electrophoresis (PFGE) patterns. Int. j. food microbiol. 31: 59–68.

[3] Marshall VM (1987) Lactic Acid Bacteria: Starters for Flavour. FEMS microbiol. lett. 46: 327-336.

[4] Caplice E, Fitzgerald GF (1999) Food Fermentations: Role of Microorganisms in Food Production and Preservation. Int. j. food microbiol. 50: 131-149.

[5] Lücke FK (2000) Utilization of Microbes to Process and Preserve Meat. Meat sci. 56: 105-115.

[6] Schlafmann K, Meusburger AP, Hammes WP, Braun C, Fischer A, Hertel C (2002) Starter Cultures to Improve the Quality of Raw Ham. Fleischwirtschaft 82 (11): 108-114.

[7] Kröckel L, Dederer I, Troeger K (2011) Starter and Protective Cultures for Meat Products. Fleischwirtschaft 91 (3): 93-98.

[8] McIntyre L, Hudson JA, Billington C, Withers H (2012) Biocontrol of Foodborne Bacteria. In: McElhatton A, Sobral PJA, editors. Novel Technologies in Food Science – Integrating Food Science and Engineering Knowledge into the Food Chain 7. Springer Science+Business Media. pp. 183-204.

[9] Katla T, Moretro T, Sveen I, Aasen M, Axelsson L, Rorvik LM, Naterstad K (2002) Inhibition of *Listeria monocytogenes* in Chicken Cold Cuts by Addition of Sakacin P and Sakacin P-Producing *Lactobacillus sakei*. J. appl. microbiol. 93: 191-196.

[10] Vermeiren L, Devlieghere F, Vandekinderen I, Rajtak U, Debevere J (2006) The Sensory Acceptability of Cooked Meat Products Treated with a Protective Culture Depends on Glucose Content and Buffering Capacity: A Case Study with *Lactobacillus sakei* 10A. Meat sci. 74: 532-545.

[11] Hammes WP, Knauf HJ (1994) Starters in the Processing of Meat Products. Meat sci. 36: 155-168.

[12] Hammes WP, Hertel C (2009) Genus I. *Lactobacillus* Beijerinck 1901, 212AL. In: De Vos P, Garrity GM, Jones D, Krieg NR, Ludwig W, Rainey FA, Schleifer KH, Whitman WB, editors. The Firmicutes. Bergey's Manual of Systematic Bacteriology, 2nd Ed., Vol. 3. Springer, Dordrecht, Heidelberg London, New York; pp. 465- 511.

[13] Urso R, Comi G, Cocolin L (2006) Ecology of Lactic Acid Bacteria in Italian Fermented Sausages: Isolation, Identification and Molecular Characterization. Syst. appl. microbiol. 29: 671-680.

[14] Cocolin L, Dolci P, Rantsiou K, Urso R, Cantoni C, Comi G (2009) Lactic Acid Bacteria Ecology of Three Traditional Fermented Sausages Produced in the North of Italy as Determined by Molecular Methods. Meat sci. 82: 125-132.

[15] Cocolin L, Dolci P, Rantsiou K (2011) Biodiversity and Dynamics of Meat Fermentations: The Contribution of Molecular Methods for a Better Comprehension of a Complex Ecosystem. Meat sci. 89: 296-302.

[16] Hugas M, Garriga M, Aymerich T, Monfort JM (1993) Biochemical Characterization of Lactobacilli from Dry Fermented Sausages. Int. j. food microbiol. 18: 107-113.

[17] Santos EM, González-Fernández C, Jaime I, Rovira J (1998) Comparative Study of Lactic Acid Bacteria House Flora Isolated in Different Varieties of `Chorizo'. Int. j. food microbiol. 39: 123-128.

[18] Samelis J, Maurogenakis F, Metaxopoulos J (1994) Characterisation of Lactic Acid Bacteria Isolated from Naturally Fermented Greek Dry Salami. Int. j. food microbiol. 23: 179-196.

[19] Albano H, van Reenen CA, Todorov SD, Cruz D, Fraga L, Hogg T, Dicks LMT, Teixeira P (2009) Phenotypic and Genetic Heterogeneity of Lactic Acid Bacteria Isolated from "Alheira", a Traditional Fermented Sausage Produced in Portugal. Meat sci. 82: 389-398.

[20] Adams MR, Nicolaides L (1997) Review of the Sensitivity of Different Foodborne Pathogens to Fermentation. Food control 8: 227-239.

[21] Cocolin L, Rantsiou K (2012) Meat Fermentation. In: Hui YH, editor. Handbook of Meat and Meat Processing, Second Edition. CRC Press, pp. 557-572.

[22] Fontana C, Fadda S, Cocconcelli PS, Vignolo G (2012) Lactic Acid Bacteria in Meat Fermentations. In: Lahtinen S, Ouwehand AC, Salminen S, von Wright A, editors. Lactic Acid Bacteria – Microbiological and Functional Aspects, 4th Ed. CRC Press, Taylor & Francis, Boca Raton, London, New York. pp. 247-264.

[23] Buckenhüskes HJ (1993) Selection Criteria for Lactic Acid Bacteria to be Used as Starter Cultures for Various Food Commodities. FEMS microbiol. rev. 12: 253-271.

[24] Dodds KL, Collins-Thompson DL (1984) Nitrite Tolerance and Nitrite Reduction in Lactic Acid Bacteria Associated with Cured Meat Products. Int. j. food microbiol. 1: 163-170.

[25] Nordal J, Slinde E (1980) Characteristics of Some Lactic Acid Bacteria Used as Starter Cultures in Dry Sausage Production. Appl. environ. microbiol. 40: 472-475.

[26] Nissen H, Holck A (1998) Survival of *Escherichia coli* O157:H7, *Listeria monocytogenes* and *Salmonella kentucky* in Norwegian Fermented, Dry Sausage. Food microbiol. 15: 273–279.

[27] Leistner L (2000) Basic Aspects of Food Preservation by Hurdle Technology. Int. j. food microbiol. 55:181–186.

[28] Lindqvist R, Lindblad M (2009) Inactivation of *Escherichia coli*, *Listeria monocytogenes* and *Yersinia enterocolitica* in Fermented Sausages During Maturation/Storage. Int. j. food microbiol. 129: 59-67.

[29] Heir E, Holck AL, Omer MK, Alvseike O, Hoy M, Mage I, Axelsson L (2010) Reduction of Verotoxigenic *Escherichia coli* by Process and Recipe Optimisation in Dry-Fermented Sausages. Int. j. food microbiol. 141: 195–202.

[30] Erkes M (2011) Einsatz von Kulturen bei Rohpökelwaren. Fleischwirtschaft 91 (11): 39-43.

[31] Jones RJ (2004) Observations on the Succession dynamics of Lactic Acid Bacteria Populations in Chill-Stored Vacuum-Packaged Beef. Int. J. food microbiol. 90: 273– 282.

[32] Schillinger U, Lücke FK (1987) Lactic Acid Bacteria on Vacuum-Packaged Meat and their Influence on Shelf Life. Fleischwirtschaft 67: 1244-1248.

[33] Castellano P. Gonzalez C, Carduza F, Vignolo G (2010) Protective Action of *Lactobacillus curvatus* CRL705 on Vacuum-Packaged Raw Beef. Effect on Sensory and Structural Characteristics. Meat sci. 85: 394-401.

[34] Castellano P, Belfiore C, Vignolo G (2011) Combination of Bioprotective Cultures with EDTA to Reduce *Escherichia coli* O157:H7 in Frozen Ground-Beef Patties. Food control 22: 1461-1465.

[35] Fadda S, Lopez C, Vignolo G (2010) Role of Lactic Acid Bacteria During Meat Conditioning and Fermentation: Peptides Generated as Sensorial and Hygienic Biomarkers. Meat sci. 86: 66-79.

[36] Borch E, Kant-Muermans ML, Blixt Y (1996) Bacterial Spoilage of Meat and Cured Meat Products. Int. j. food microbiol. 33: 103–120.

[37] Korkeala HJ, Bjørkroth KJ (1997) Microbiological Spoilage and Contamination of Vacuum-Packaged Cooked Sausages. J. food prot. 60: 724–731.

[38] Nattress FM, Jeremiah LE (2000) Bacterial Mediated Off-Flavours in Retail-Ready Beef after Storage in Controlled Atmospheres. Food res. int. 33: 743-748

[39] Björkroth KJ, Vandamme P, Korkeala HJ (1998) Identification and Characterization of *Leuconostoc carnosum*, Associated with Production and Spoilage of Vacuum-Packaged, Sliced, Cooked Ham. Appl. environ. microbiol. 64: 3313-3319.

[40] Kröckel L (1998) Lactic Acid Bacteria as Protective Cultures in the Preservation of Meat. In: Adria-Normandie, editor. Les Bactérie Lactic – Quelles Souches? Pour quels Produits? Lactic Acid Bacteria – Which Strains for which Products? – Actes du colloque LACTIC 97, Caen, 10-12 Sept 1997 – Adria Normandie, Villers-Bocage, pp. 229-242.

[41] Laursen BG, Bay L, Cleenwerck I, Vancanneyt M, Swings J, Dalgaard P, Leisner JJ (2005) *Carnobacterium divergens* and *Carnobacterium maltaromaticum* as Spoilers or Protective Cultures in Meat and Seafood: Phenotypic and Genotypic Characterization. Syst. appl. microbiol. 28: 151-64.

[42] Lücke FK, Raabe C, Hampshire J (2007) Changes in Sensory Profile and Microbiological Quality During Chill Storage of Cured and Uncured Cooked Sliced Emulsion-Type Sausages. Arch. Lebensmittelhyg. 58: 57-63.

[43] Kröckel L (2008) Mikrobiologische Qualität Vorverpackter Aufschnittware - Aktuelle Untersuchungen von Erhitztem, Vorverpacktem Brühwurst- und Kochschinkenaufschnitt auf *Listeria monocytogenes*. Fleischwirtschaft 88 (11): 112-116.

[44] Leistner L (1995) Principles and Applications of Hurdle Technology. In: Gould GW, editor. New Methods of Food Preservation. Springer; pp. 1–21.

[45] Leistner L, Gorris LG (1995) Food Preservation by Hurdle Technology. Trends food sci technol 6: 41-46.

[46] Hui YH (2012) Hazard Analysis and Critical Control Point System. In: Hui YH, editor. Handbook of Meat and Meat Processing, 2nd Ed. CRC Press, pp. 741-767.

[47] Jacobsen T, Budde BB, Koch, AG (2003) Application of *Leuconostoc carnosum* for Biopreservation of Cooked Meat Products. J. appl. microbiol. 95: 242–249.

[48] Vermeiren L, Devlieghere F, Debevere J (2004) Evaluation of Meat Born Lactic Acid Bacteria as Protective Cultures for the Biopreservation of Cooked Meat Products. Int. j. food microbiol. 96: 149–164.

[49] Jones RJ, Wiklund E, Zagorec M, Tagg JR (2010) Evaluation of Stored Lamb Biopreserved Using a Three-Strain Cocktail of *Lactobacillus sakei*. Meat sci. 86: 955-959

[50] Jones RJ, Zagorec M, Brightwell G, Tagg JR (2009) Inhibition by *Lactobacillus sakei* of Other Species in the Flora of Vacuum Packaged Raw Meats During Prolonged Storage. Food microbiol. 26: 876-881.

[51] Ananou S, Maqueda M, Martínez-Bueno M, Valdivia E (2007) Biopreservation, an Ecological Approach to Improve the Safety and Shelf-Life of Foods. In: Méndez-Vilas A, editor. Communicating Current Research and Educational Topics and Trends in Applied Microbiology, pp. 475-486. Available: http://www.formatex.org/microbio/pdf/Pages475-486.pdf. Accessed 2012 Apr 3.

[52] www (2012) Ingredients with Potential: Chr. Hansen A/S - Bactoferm Rubis. Available: http://www.safoodcentre.com.au/__data/assets/pdf_file/0015/131451/innova_ingredients_with_potential.pdf. Accessed: 2012 Apr 4

[53] Anonymous (2011) Veröffentlichte Mikrobiologische Richt- und Warnwerte zur Beurteilung von Lebensmitteln (Stand: November 2011) Eine Empfehlung der Fachgruppe Lebensmittelmikrobiologie und -hygiene der Deutschen Gesellschaft für Hygiene und Mikrobiologie (DGHM); http://www.dghm.org/m_275. Accessed 2012 Apr 3.

[54] Kröckel L (2010) Mikrobiologisch-Genetische Ressourcen bei Fleisch - Biodiversität und Nachhaltige Nutzung bei der Herstellung von Fleisch-Erzeugnissen. Forschungsreport Ernährung, Landwirtschaft, Verbraucherschutz 1/2010: 24-26.

[55] Johansson P, Paulin L, Säde E, Salovuori N, Alatalo ER, Bjørkroth KJ, Auvinen P (2011) Genome Sequence of a Food Spoilage Lactic Acid Bacterium, *Leuconostoc gasicomitatum* LMG 18811(T), in Association with Specific Spoilage Reactions. Appl. environ. microbiol. 77: 4344-4351.

[56] Hugas M, Monfort JM (1997) Bacterial Starter Cultures for Meat Fermentation. Food chemistry 59: 547-554.

[57] Kröckel L (2006a) Use of Probiotic Bacteria in Meat Products. Fleischwirtschaft 86:109–13.

[58] Ammor MS, Mayo B (2007) Selection Criteria for Lactic Acid Bacteria to be used as Functional Starter Cultures in Dry Sausage Production: an Update. Meat sci. 76: 138-146.

[59] De Vuyst L, Falony G, Leroy F (2008) Probiotics in Fermented Sausages. Meat sci. 80: 75-78.

[60] Huang CH, Lee FL (2009) Development of Novel Species-Specific Primers for Species Identification of the *Lactobacillus casei* Group Based on RAPD Fingerprints. J. sci. food agric. 89: 1831-1837.

[61] Khan MI, Arshad MS, Anjum FM, Sameen A, Aneeq-ur-Rehman, Gill WT (2011) Meat as a Functional Food with Special Reference to Probiotic Sausages. Food res. int. 44: 3125-3133.

[62] Leroy F, Verluyten J, De Vuyst L (2006) Functional Meat Starter Cultures for Improved Sausage Fermentation. Int. j. food microbiol. 106: 270-285.

[63] Abee T, Kröckel L, Hill C (1995) Bacteriocins: Modes of Action and Potentials in Food Preservation and Control of Food Poisoning. Int. j. food microbiol. 28:169-185.

[64] Kelly WJ, Asmundson RV, Huang CM (1996) Isolation and Characterization of Bacteriocin-Producing Lactic Acid Bacteria from Ready-to-Eat Food Products. Int. j. food microbiol. 33: 209-218.

[65] Hugas M (1998) Bacteriocinogenic Lactic Acid Bacteria for the Biopreservation of Meat and Meat Products. Meat sci. 49: S139-S150.

[66] Nieto-Lozano JC, Reguera-Useros JI, Peláez-Martínez MC, Hardisson de la Torre A (2002) Bacteriocinogenic Activity from Starter Cultures Used in Spanish meat Industry. Meat sci. 62: 237-243.

[67] Lücke FK, Popp J, Kreutzer R (1986) Bildung von Wasserstoffperoxid durch Laktobazillen aus Rohwurst und Brühwurstaufschnitt. Chem. mikrobiol. technol. lebensm. 10: 78-81.

[68] Condon S. (1987) Responses of lactic acid bacteria to oxygen. FEMS Microbiol. rev. 46, 269-280.

[69] Kröckel L (2011) Evaluation of a Novel Agar Medium for the Detection of Hydrogen Peroxide Producing Lactic Acid Bacteria. Fleischwirtschaft 91 (10): 97-101.

[70] Silla-Santos MH (1996) Biogenic Amines: their Importance in Foods. Int. j. food microbiol. 29: 213–231.

[71] Komprda T, Sládková P, Petirová E, Dohnal V, Burdychová R (2010) Tyrosine- and Histidine-Decarboxylase Positive Lactic Acid Bacteria and Enterococci in Dry Fermented Sausages. Meat sci. 86: 870-877.

[72] Lu S, Xu X, Zhou G, Zhu Z, Meng Y, Sun Y (2010) Effect of Starter Cultures on Microbial Ecosystem and Biogenic Amines in Fermented Sausage. Food control 21: 444-449.

[73] Samappito W, Leenanon B, Levin RE (2011) Microbiological Characteristics of "Mhom", a Thai Traditional Meat Sausage. The Open Food Science Journal 5: 31-36.

[74] Sohier D, Coulon J, Lonvaud-Funel A (1999) Molecular Identification of *Lactobacillus hilgardii* and Genetic Relatedness with *Lactobacillus brevis*. Int. j. syst. bacteriol. 49: 1075-1081.

[75] Rantsiou K, Cocolin L (2006) New Developments in the Study of the Microbiota of Naturally Fermented Sausages as Determined by Molecular Methods: A Review. Int. j. food microbiol. 108: 255-267.

[76] Vandamme P, Pot B, Gillis M, de Vos P, Kersters K, Swings J (1996) Polyphasic taxonomy, a consensus approach to bacterial systematics. Microbiol rev. 60: 407-38.

[77] Chaillou S, Champomier-Verges MC, Cornet M, Crutz-Le Coq AM, Dudez AM, Martin V, Beaufils S, Darbon-Rongere E, Bossy R, Loux V, Zagorec M (2005) The Complete Genome Sequence of the Meat-Borne Lactic Acid Bacterium *Lactobacillus sakei* 23K. Nat. biotechnol. 23: 1527-1533.

[78] Hebert EM, Saavedra L, Taranto MP, Mozzi F, Magni C, Nader MEF, Font de Valdez G, Sesma F, Vignolo G, Rayaa RR (2012) Genome Sequence of the Bacteriocin-Producing *Lactobacillus curvatus* Strain CRL705. J. bacteriol. 194: 538-539.

[79] Siezen RJ, Tzeneva VA, Castioni A, Wels M, Phan HTK, Rademaker JLW, Starrenburg MJC, Kleerebezem M, Molenaar D, van Hylckama Vlieg JET (2010) Phenotypic and Genomic Diversity of *Lactobacillus plantarum* Strains Isolated from Various Environmental Niches. Environ. microbiol. 12: 758-773.

[80] Marchesini B, Bruttin A, Romailler N, Moreton RS, Stucchi C, Sozzi T (1992) Microbiological Events During Commercial Meat Fermentations. J. appl. bacteriol. 73: 203-209.

[81] Kröckel L (unpublished observations)

[82] Drosinos EH, Mataragas M, Xiraphi N, Moschonas G, Gaitis F, Metaxopoulos J (2005) Characterization of the Microbial Flora from a Traditional Greek Fermented Sausage. Meat sci. 69: 307-317.

[83] Drosinos EH, Paramithiotis S, Kolovos G, Tsikouras I, Metaxopoulos I (2007) Phenotypic and Technological Diversity of Lactic Acid Bacteria and Staphylococci Isolated from Traditionally Fermented Sausages in Southern Greece. Food microbiol. 24: 260-270.

[84] Ho TNT, Nguyen NT, Deschamps A, Hadj Sassi A, Urdaci M, Caubet R (2009) The Impact of *Lactobacillus brevis* and *Pediococcus pentosaceus* on the Sensorial Quality of "Nem Chua" – a Vietnamese Fermented Meat Product. Int. food res. j. 16: 71-81.

[85] Coventry MJ, Wan J, Gordon JB, Mawson RF, Hickey MW (1996) Production of Brevicin 286 by *Lactobacillus brevis* VB286 and Partial Characterization. J. appl. bacteriol. 80: 91-8.

[86] Kröckel L, Schillinger U, Franz CMAP, Bantleon A, Ludwig W (2003) *Lactobacillus versmoldensis* sp. nov., isolated from raw fermented sausage. Int. j. syst. env. microbiol. 53: 513-517.

[87] Klingberg TD, Axelsson L, Naterstad K, Elsser D, Budde BB (2005) Identification of Potential Probiotic Starter Cultures for Scandinavian-Type Fermented Sausages. Int. j. food microbiol. 105: 419-431.

[88] El-Baradei G, Delacroix-Buchet A, Ogier JC (2007) Biodiversity of Bacterial Ecosystems in Traditional Egyptian Domiati Cheese. Appl. environ. microbiol. 73: 1248-1255.

[89] An C, Takahashi H, Kimura B, Kuda T (2010) Comparison of PCR-DGGE and PCR-SSCP Analysis for Bacterial Flora of Japanese Traditional Fermented Fish Products, Aji-Narezushi and Iwashi-Nukazuke. J. sci. food agric. 90: 1796-1801.

[90] Kim DS, Choi SH, Kim DW, Kim RN, Nam SH, Kang A, Kim A, Park HS (2011a) Genome Sequence of *Lactobacillus versmoldensis* KCTC 3814. J. bacteriol. 193: 5589-5590.

[91] Casaburi A, Nasi A, Ferrocino I, Di Monaco R, Mauriello G, Villani F, Ercolini D (2011) Spoilage-Related Activity of *Carnobacterium maltaromaticum* Strains in Air-Stored and Vacuum-Packed Meat. Appl. environ. microbiol. 77: 7382-7393.

[92] Ercolini D, Ferrocino I, Nasi A, Ndagijimana M, Vernocchi P, La Storia A, Laghi L, Mauriello G, Guerzoni ME, Villani F (2011) Monitoring of Microbial Metabolites and Bacterial Diversity in Beef Stored Under Different Packaging Conditions. Appl. environ. microbiol. 77: 7372-7381.

[93] Larrouture-Thiveyrat C, Montel MC (2003) Effects of environmental factors on leucine catabolism by *Carnobacterium piscicola*. Int. j. food microbial. 81: 177–184.

[94] Leisner JJ, Groth Laursen B, Prevost H, Drider D, Dalgaard P (2007) *Carnobacterium*: Positive and Negative Effects in the Environment and in Foods. FEMS microbiol. rev. 31: 592–613.

[95] Leisner JJ, Hansen MA, Larsen MH, Hansen L, Ingmera H, Sørensen SJ (2012) The Genome Sequence of the Lactic Acid Bacterium *Carnobacterium maltaromaticum* ATCC 35586 Encodes Potential Virulence Factors. Int J. food microbiol. 152: 107–115.

[96] Edima HC, Cailliez-Grimal C, Revol-Junelles AM, Tonti L, Linder M, Millière JB (2007) A Selective Enumeration Medium for *Carnobacterium maltaromaticum*. J. microbiol. meth. 68: 516-21.

[97] Vihavainen E, Björkroth J (2007) Spoilage of Value-Added, High-Oxygen Modified-Atmosphere Packaged Raw, Beef Steaks by *Leuconostoc gasicomitatum* and *Leuconostoc gelidum*. Int. j. food microbiol. 119: 340–345.

[98] Kröckel L (2006b) Yellow Discolourations of Prepackaged Refrigerated German Weisswurst are Due to *Leuconostoc gelidum*. Fleischwirtschaft 86 (9): 129-133.

[99] Kim DS, Choi SH, Kim DW, Kim RN, Nam SH, Kang A, Kim A, Park HS (2011b) Genome Sequence of *Leuconostoc gelidum* KCTC 3527, Isolated from Kimchi. J. bacteriol. 193: 799–800.

[100] Kröckel L (2007) Ein Bakterielles Carotinoid Färbt Vorverpackte, Kühl Gelagerte Weißwurst Gelb. Mitteilungsblatt der Fleischforschung Kulmbach 46, Nr. 178: 223-230.

[101] Whiteley AM, D'Souza MD 1989. A Yellow Discoloration of Cooked Cured Meat Products - isolation and characterization of the causative organism. J. food protect. 52: 392– 395.

[102] Björkroth KJ, Geisen R, Schillinger U, Weiss N, De Vos P, Holzapfel WH, Korkeala HJ, Vandamme P (2000). Characterization of *Leuconostoc gasicomitatum* sp. nov., Associated with Spoiled Raw Tomato-Marinated Broiler Meat Strips Packaged Under Modified-Atmosphere Conditions. Appl. environ. microbiol. 66: 3764–3772.

[103] Nieminen TT, Vihavainen E, Paloranta A, Lehto J, Paulin L, Auvinen P, Solismaa M, Björkroth KJ (2011) Characterization of Psychrotrophic Bacterial Communities in Modified Atmosphere-Packed Meat with Terminal Restriction Fragment Length Polymorphism. Int. J. food microbiol. 144: 360–366.

[104] Björkroth J, Dicks LMT, Holzapfel WH (2009) Genus III. *Weissella*. In: De Vos P, Garrity GM, Jones D, Krieg NR, Ludwig W, Rainey FA, Schleifer KH, Whitman WB, editors. The Firmicutes. Bergey's Manual of Systematic Bacteriology, 2nd Ed., Vol. 3. Springer, Dordrecht, Heidelberg London, New York; pp. 643- 654.

[105] Comi G, Iacumin L (2012) Identification and Process Origin of Bacteria Responsible for Cavities and Volatile Off-Flavour Compounds in Artisan Cooked Ham. Int. j. food sci. technol. 47, 114-121.

[106] Mathur S, Singh R (2005) Antibiotic Resistance in Food Lactic Acid Bacteria - A Review. Int. j. food microbiol. 105: 281-295.

[107] Ammor S, Tauveron G, Dufour E, Chevallier I (2006) Antibacterial Activity of Lactic Acid Bacteria Against Spoilage and Pathogenic Bacteria Isolated from the Same Meat Small-Scale Facility: 2 - Behaviour of Pathogenic and Spoilage Bacteria in Dual Species Biofilms Including a Bacteriocin-Like-Producing Lactic Acid Bacteria. Food control 17: 462-468.

[108] Teusink B, Bachmann H, Molenaar D (2011) Systems Biology of Lactic Acid Bacteria: a Critical Review. Microb. Cell. Fact. 10 (suppl 1): S11.

[109] McLeod A, Zagorec M, Champomier-Vergès MC, Naterstad K, Axelsson L (2010) Primary Metabolism in *Lactobacillus sakei* Food isolates by Proteomic Analysis. BMC microbiol. 10: 120.

[110] Nyquist OL, McLeod A, Brede DA, Snipen L, Aakra Å., Nes IF (2011) Comparative Genomics of *Lactobacillus sakei* with Emphasis on Strains from Meat. Molec. genetics and genomics 285: 297-311.

Section 3

# Vegetable & Cereal Products

Chapter 6

# Lactic Acid Bacteria in Biopreservation and the Enhancement of the Functional Quality of Bread

Belal J. Muhialdin, Zaiton Hassan and Nazamid Saari

Additional information is available at the end of the chapter

http://dx.doi.org/10.5772/51026

## 1. Introduction

LAB have a long history in preserving foods from spoilage microorganisms - they are commonly used in food fermentation, may produce several metabolites with beneficial health effects and, thus, are generally recognized as safe (GRAS). The increasing resistance of food spoilage microorganisms to current preservatives, the consumer's high demand for safe, minimally processed foods and the hazards associated with the use of high doses of chemical preservatives has led to the need for finding safer alternatives in food preservation. The application of LAB with the simultaneous control of factors that affect fungal growth can help to minimize food spoilage. The selection and addition of novel isolates of LAB may be the key to reducing the use of chemicals, enhancing nutrients and extend the shelf life of bakery products. In this chapter, the focus will be on the use of LAB as biopreservative agents to extend the shelf life of bakery products and the inhibition of the common spoilage fungi of bread.

## 2. Sources of LAB

LAB are found in many habitats and occur naturally in a variety of food products, such as dairy, vegetables and meat products (Carr et al., 2002), all of which are rich in the nutrients required for the fastidious metabolism of LAB (Björkroth & Holzapfel, 2003; Hammes & Hertel, 2003). Some LAB are associated with the mouth flora, intestine and vagina of mammals (Whittenbury, 1964), while others are present in fermented seafood, such as *Lactobacillus plantarum* (IFRPD P15) and *L. reuteri* (IFRPD P17), which are reported to be associated with plaa-som fermented Thai fish (Saithong et al., 2010). LAB are the most important bacteria used in the fermentation industry of dairy products, such as yogurt, cheese, sour milk and butter, and in combination with yeast are commonly used to ferment cereal products such as dough (Lavermicocca et al., 2000; Muhialdin et al., 2011a; Ryan et al., 2008).

## 3. Spoilage fungi in food

The economic losses and the health hazards of the mycotoxins produced by spoilage fungi are the main concerns of the food industry (Gray & Bemiller, 2003). According to Gerez et al., (2009) the spoilage of bakery products by fungi is more common in countries with a high humidity and temperature. Pitt and Hocking (1999) estimated that about 5-10% of food production is spoiled by the growth of yeast and fungi in food materials. Similarly, in Western Europe, the growth of the spoilage fungi of bread is estimated to reach more than 200 million Euros per year (Legan, 1993; Schnürer & Magnusson, 2005). The history conditions of the food can be a major factor in determining any fungal spoilage - for example, stored and processed foods are more sensitive to spoilage when compared with fresh and prepared foods. *Aspergillus* and *Penicillium* species are the most common spoilage fungi for many foods and feeds while *Fusarium* species are reported to attack the cereal grains in the field (Samson et al., 2000).

The most widespread species of fungi that contaminate bakery products belong to the genera *Aspergillus, Penicillium, Eurotium* (Abellana et al., 1997; Guynot et al., 2005), *Monilia, Mucor, Endomyces, Cladosporium, Fusarium* and *Rhizopus* (Lavermicocca et al., 2000, 2003). In addition, fungi may be responsible for off-flavours, the production of mycotoxins and allergenic compounds. There are more than 400 known mycotoxins produced by different fungi (Filtenborg et al., 1996). Mycotoxigenic fungi such as *Aspergillus, Fusarium* and *Penicillium* are serious hazards for human health. The six classes of mycotoxins frequently encountered in different food systems are: aflatoxins, fumonisins, ochratoxins, patulin, trichothecenes and zearalenone (Dalié et al., 2009).

## 4. Common techniques to control spoilage fungi in bakery products

Two types of techniques/factors are commonly used to control spoilage fungi: physical ones such as drying, freeze drying, cold storage, modified atmosphere storage, irradiation, the pasteurization of packaged bread and heat treatment; and chemical ones, in general based on the use of organic acids such as propionic acid and its salts (Farkas, 2001; Legan, 1993). Heat treatment is one of the most important physical factors in controlling fungi growth and mycotoxin production, as mycotoxins are destroyed by heat, although the effectiveness of destruction is affected by the food matrix and the composition of the mycotoxin (Scott, 1984). Mycotoxins have different heat stability - for example, ochratoxin A is highly stable even at 200 ºC (Trivedi et al., 1992), aflatoxins are destroyed only at temperatures of approximately 250 ºC (Levi, 1980), while zearalenone and fumonisin require high temperatures between 150-200 ºC to be efficiently destroyed (Bennett et al., 1980). Microwaves are effective in destroying mycotoxins - the aflatoxin in peanuts is reported to be destroyed using microwaves at a power level of 1.6 kW for 16 min and at 3.2 kW for 5 min (Luter et al., 1982). Among the physical methods, a modified atmosphere and gamma irradiation are preferred to the chemical methods and they have been used successfully in grain storage (Shapira & Paster, 2000).

Chemical methods that use weak acids and salts such as propionic, sorbic and benzoic acids, are usually applied only to inhibit the growth of spoilage microorganisms. The allowable concentrations of sorbate, propionate and ethanol have a limit up to 0.2% (wt/wt), 0.3% (wt/wt) and 2% (wt/wt) respectively. The use of such low concentration may not be sufficient to prevent the growth of spoilage fungi (Dantigny et al., 2005; European Union, 1995). Propionic acid is inhibitory to fungi and *Bacillus* spores and has commonly been used to preserve bakery products. Its activity relies on the un-dissociated form which, at low pH, has optimum activity (Coda et al., 2008; Pattison et al., 2004). The use of propionic acid at a concentration of 4% led to the appearance of cancer-like tumours in rats and eventually led to the prohibition of the use of calcium propionate in some European countries (Pattison et al., 2004). There is a major concern with microorganisms that can develop resistance to chemical preservatives, namely food spoilage and human pathogen fungi resistant to antibiotics and chemicals additives, such as sorbic and benzoic acids (Brul & Coote, 1999; Lourens-Hattingh & Viljoen, 2001). Calcium propionate has been reported to inhibit the growth of many fungi but, after a lag phase, it stimulated the growth of resistant strains of *Penicillium roqueforti* (Suhr & Nielsen, 2004). Interest in natural bio-preservation from LAB has been on the rise as an alternative to chemical preservatives.

## 5. Significance of the metabolites of LAB

LAB are well known for their antifungal activity, which is related to the production of a variety of compounds including acids, alcohols, carbon dioxide, diacetyl, hydrogen peroxide, phenyllactic acid, bacteriocins and cycle peptides (Gerez et al., 2009; Lavermicocca et al., 2000; Magnusson et al., 2003; Prema et al., 2008). These compounds were added to several foods in order to conserve them from food-borne and spoilage microorganisms. Organic acids are the main product of LAB in the fermentation systems of the raw materials. The main acids produced by LAB are lactic acid and acetic acid, besides certain other acids depending upon the strain of LAB (El-Ziney, 1998). These acids will be diffused through the membrane of the target organisms in their hydrophobic un-dissociated form and then used to reduce the cytoplasmic pH and stop metabolic activities (Piard & Desmazeaud, 1991). Other factors that contribute to the preservative action of the acids are the sole effect of pH, the extent of the dissociation of the acid and the specific effect of the molecule itself on the microorganisms (Axelsson, 1998).

Bacteriocins exhibit good potential for use in the food industry and as bio-preservation agents (Ennahar et al., 1999). Bacteriocins are small, ribosomally synthesized, antimicrobial peptides or proteins that display inhibition activity toward related species, with no reports about fungal inhibition (Cotter & Ross, 2005). The notable property of LAB supernatant is the heat stability of the antifungal compounds present in it. This will promote the use of LAB supernatant and/or antifungal compounds in heat-treated foods. The supernatant of certain LAB observed to be active within a wide range of pH, starting from as low as 3 and up to 9 depending upon the strain (Muhialdin et al., 2011b). This could be considered as a major factor whereby LAB are used in food preservation when compared with the chemical preservative which are usually active at low pH between 3 and 4.5. Additionally, LAB have a broad spectrum of antifungal activity against several food spoilage and mycotoxin-producing fungi while commercial preservatives are usually used to control only one or few fungi.

## 6. Bioactive compounds as antifungal agents

Several lactobacilli species are reported to have antifungal activity (Gerez et al., 2009; Muhialdin et al., 2011b; Plockova et al., 2001; Stiles et al., 1999). The antifungal compounds consist of organic acids, reuterin, hydrogen peroxide and other peptides (Table 1). The organic acids are active at low pH and the activity relies on the un-dissociated form of the acids. Recently, interest has dramatically increased in the use of bioactive peptides produced by LAB as an antifungal agent. The use of protein-like compounds are preferred over the use of acids because their activity is present over a wide range of pH and they are heat stable compounds which are ideal for use in heat processed foods (Muhialdin et al., 2011a). Cyclic dipeptides cyclo (Phe-Pro) and cyclo (Phe-OH-Pro) were produced by the *L. coryniformis* subsp. *coryniformis* Si3 strain and were inhibitory to *Aspergillus* sp. (Magnusson, 2003; Ström et al., 2002). Ryan et al. (2011) observed that sourdough made with *L. amylovorus* DSM 19280 had a longer shelf life compared with bread produced with calcium propionate. The selected strain inhibited the growth of *Fusarium culmorum* FST4.05, *Aspergillus niger* FST4.21, *Penicillium expansum* FST4.22, *Penicillium roqueforti* FST4.11 and *L. amylovorus* DSM 19280 and produced seventeen antifungal compounds.

| Compound | Producer | Inhibited fungi | References |
|---|---|---|---|
| Possibly proteinaceous | *Pediococcus acidilactici* | *Saccharomyces cerevisiae* | Vandenbergh & Kanka (1989) |
| Possibly proteinaceous | *L. lactis* subsp. *Lactis* CHD 28.3 | *A. flavus, A. parasiticus, Fusarium* spp. | Roy et al. (1996) |
| Caproic acid, propionic acid, butyric acid, valeric acid | *L. sanfranciscencis* CB1 | *Fusarium* spp., *Penicillium* spp., *Aspergillus* spp., *Monilia* spp. | Corsetti et al. (1998) |
| Benzoic acid, methylhydantoin, mevalonolactone, | *L. plantarum* VTT E78076 | *F. avenaceum* | Niku-Paavola et al. (1999) |
| phenyllactic and 4-hydroxy-phenyllactic acids | *L. plantarum* 21B | Broad spectrum against bakery spoilage fungi | Lavermicocca et al. (200) |
| 3-Phenyllactic acid, cyclo (Phe-OH-Pro), cyclo (Phe-Pro). | *L. plantarum* MiLAB 393 | *F. sporotrichioides* and *A. fumigatus* | Ström et al. (2002) |
| Hydroxy fatty acids, phenyllactic acid, cyclo(Phe-Pro), cyclo(Phe-OH-Pro), | *L. plantarum* MiLAB14 | Broad spectrum | Magnusson et al. (2003) |

| Compound | Producer | Inhibited fungi | References |
|---|---|---|---|
| Possibly cyclic dipeptide | *P. pentosaceus* | *P. expansum* | Rouse et al. (2008) |
| diacetyl and hydrogen peroxide | *L. fermentum* and *Leuconostoc mesenteroides* | *Rhizopus oryzae, A. niger, A. flavus, Penicillium* sp and *F. oxysporum* | Ogunbanwo et al. (2008) |
| Acetic acid, phenyllactic acid | *L. reuteri* 1100 | *F. graminearum* | Gerez et al. (2009) |
| (cyclo(Leu–Leu)) | *L. plantarum* AF1 | *Aspergillus flavus* ATCC 22546 | Yang & Chang (2010) |
| Four peptides and organic acid mixture | *L. plantarum* LB1 and *L. rossiae* LB5 | *Penicillium roqueforti* DPPMAF1 | Rizzello et al. (2011) |
| Mixture of peptides | *L. plantarum* 1A7 (S1A7) | Broad spectrum | Coda et al. (2011) |
| Possibly protein-like | *L. fermentum* Te007, *P. pentosaceus* Te010, *L. pentosus* G004, and *L. paracasi* D5 | *A. niger* and *A. oryzae* | Muhialdin et al. (2011a) |
| nine carboxylic acids, two nucleosides, sodium decanoate and five cyclic dipeptides | *L. amylovorus* DSM 19280 | *A. niger* FST 4.21, *A. fumigatus* J9, *F. culmorum* TMW 4.0754 *P. expansum* FST 4.22 and *P. roqueforti* FST 4.11 | Ryan et al. (2011) |
| 3-phenyllactic acid and Benzene acetic acid, 2- propenyl ester | *L. plantarum* IMAU10014 | *Botrytis cinerea, Glomerella cingulate, Phytophthora drechsleri Tucker, P. citrinum, P. digitatum* and *F. oxysporum* | Wang et al. (2012) |

**Table 1.** Antifungal compounds produced by lactic acid bacteria and their target fungi

## 7. Method for determining antifungal activity

Rapid, reliable and sensitive methods for the detection of the antifungal activity of LAB becomes essential in the search for new replacements for chemical preservatives with potential industrial applications.

## 7.1. Dual agar overlay method

This method has been described by several authors (Magnusson & Schnürer, 2001; Ström et al., 2002; Hassan & Bullerman, 2008) and it is accurate and simple for determining the antifungal activity of LAB isolates. The method consists of inoculating the LAB cells in two 2-cm-long lines and/or small circle spots on a MRS agar surface then incubating the plates at 30 °C for 24-48 h in anaerobic jars. The plates are overlaid with 10 ml of malt extract soft agar (2% malt extract, 0.7% agar; Oxoid) containing different concentrations of the spore inoculant of $10^4$ and $10^5$ spore/ml. The plates are then incubated aerobically at 30 °C for 48-72 h. The inhibition activity is indicated by the clear zones around the bacterial streaks. The scale for measuring the activity can be recorded as follows: -, no activity; +, no fungal growth on 0.1 to 3% of the plate area; ++, no fungal growth on 3 to 8% of the plate area; and+++, no fungal growth on 8% of plate area. Another way to measure the activity is by recording the clear zone diameter around the isolates streak, which refers to the inhibition of the fungi growth. The dual agar overlay method is also a good method for the screening of the antifungal activity of the supernatant of LAB isolates. The supernatant can be mixed with the de Man, Rogosa and Sharpe (MRS) agar or potato dextrose agar (PDA) and poured into Petri dishes followed by a similar step, mentioned previously. The supernatant can be added to the agar before it is autoclaved in order to determine the heat stability of the antifungal compounds present in the supernatants, which is a good indicator of whether the supernatant is used in heat processed foods.

## 7.2. Agar well diffusion method

The well diffusion method is another approach for determining the antifungal activity of LAB, described as a simple, accurate and flexible method. It is suitable to determine the inhibition activity of LAB supernatant. A fungi numbering $10^4$ -$10^5$ spore/ml are mixed with the selected agar and allowed to solidify. The wells can be made on a variety of agar surfaces - for example, wells are made on potato dextrose agar if the target is a fungi or on a nutrient agar if the target is a bacteria; the wells are made by using a sterilized cork borer with a diameter of 3 or 5 mm. 50 µl of the same agar is added to each well in order to seal the base so as to avoid leakage. The cell-free supernatants are then added to wells in amounts of 30-80 µl and incubated at room temperature for 3-6 h in order to allow the supernatant to be diffused through the agar. The antifungal activity is recorded by measuring the clear zones' diameters around the wells.

## 7.3. Dry weight of biomass

The reduction of the biomass of the fungi can be a tool for determining the growth inhibition activity of the supernatant. 50 ml of the supernatant is inoculated into a 250 ml flask containing the growth medium for the target fungi and then the suspension of the fungi spores is added at a concentration of $10^5$. The fungal mass is harvested on filter paper and dried in an oven at 50 °C for 2 days. The average of the fungal biomass inhibition can be calculated by comparing the weight of treated fungi with the positive control which contains the fungi and the growth medium with no supernatant.

### 7.4. Micro-titter 96 well plate

The method is simple, inexpensive and practical for determining antibacterial and antifungal activity. The supernatant of LAB is placed into the wells of 190 µl and inoculated with 10 µl of a conidial suspension containing about $10^4$-$10^5$ spore/ml. The plates are then incubated at 25-30 °C. The control is a conidial suspension placed in the wells in equal amounts without the addition of the LAB supernatant. Fungal growth is observed by the naked eye and determined by measuring the optical density at 560-580 nm, starting from 0 h and repeated every 24 h with a spectrophotometer. The result can be obtained by comparing the OD readings of the control with the treated wells. The method is appropriate for evaluating the MIC, heat stability, enzyme activity and effects of pH for the LAB supernatant.

## 8. Effect of the addition of LAB on bread quality

### 8.1. Shelf life

Traditionally, chemical preservatives and fungicides are used to inhibit fungal growth but concerns about environmental pollution and consumer health, along with problems of microbial resistance, favour the demand for alternative methods in controlling the growth of fungi (Druvefors et al., 2005). The shelf life of bread has been reported to be extended when certain LAB strains were added to bread formulations (Muhialdin et al., 2011a; Ogunbanwo et al., 2008; Rizzello et al., 2010; Ryan et al., 2011) (Table 2). The use of safe microbes in bread to extend the shelf life of the product is a great research area. Since LAB isolates are safe for use in foods, they are a significant alternative to chemical preservatives. Several researchers in the area of the bakery industry have successfully added LAB to dough and these strains grew well, producing the desired antifungal compounds in the dough.

Various fungi isolated from bakeries were inhibited by *L. plantarum* (LB1) and *L. rossiae* (LB5) isolated from raw wheat germ. Organic acids and peptides synthesized during fermentation were responsible for the antifungal activity; formic acid had the highest inhibition activity (Rizzello et al., 2011). However, the inhibitory compounds characterized were different, depending upon the LAB strains and flour type used. Dal Bello et al., (2007) characterized lactic acid, phenyllactic acid (PLA), cyclic dipeptides cyclo (L-Leu–L-Pro) and cyclo (L-Phe–L-Pro) produced by *L. plantarum* FST 1.7 and found them to inhibit the growth of *Fusarium* spp. in wheat bread. Ryan et al., (2008) reduced the use of calcium propionate from 3000 ppm to 1000 ppm when using sourdough fermented with *L. plantarum* FST 1.7 (LP 1.7) and *L. plantarum* FST 1.9 (LP 1.9), in which the growth of *A. niger*, *F. culmorum* and *P. expansum* was delayed for over six days while the growth of *P. roqueforti* appeared after three days of incubation at 30 °C. *L. plantarum* VTT E-78076. *Pediococcus pentosaceus* VTT E-90390 was reported to inhibit the growth of rope-forming *Bacillus subtilis* and *Bacillus licheniformis* in laboratory conditions and in the bread when the selected strains were inoculated to sourdough and subsequently 20-30 g of the inoculated sourdough was added to 100 g of wheat dough (Katina et al., 2002). Lavermicocca et al. (2000) found that *L.*

| Strains | No. of days | Target fungi | Storage temperature °C | Reference |
|---|---|---|---|---|
| *L. plantarum* 21B | 7 | Broad spectrum | 20 | Lavermicocca et al., (2000) |
| *L. plantarum* | 12 | *Rhizopus oryzae A. niger A. flavus Penicillium sp. F. oxysporum* | 27 | Ogunbanwo et al., (2008) |
| *L. brevis* AM7 | 21 | *P. roqueforti* DPPMAF1 | 25 | Coda et al., (2008) |
| *L. plantarum* | 10 | *A. niger, F. culmorum,* and *P. expansum* | 25 | Ryan et al., (2008) |
| *L. plantarum* CRL 778, *L. reuteri* CRL 1100, and *L. brevis* CRL 772 and CRL 796 | 8 | *Aspergillus, Fusarium,* and *Penicillium* | 30 | Gerez et al., (2009) |
| *L. plantarum* 1A7 (S1A7) | 28 | *P. roqueforti* DPPMAF1 | 25 | Coda et al., (2011) |
| *L. amylovorus* DSM 19280 | 14 | *F. culmorum* FST 4.05, *A. niger* FST4.21, *P. expansum* FST 4.22, *P. roqueforti* FST 4.11 | 25 | Ryan et al., (2011) |
| *L. fermentum* Te007, *P. pentosaceus* Te010, *L. pentosus* G004, and *L. paracasi* D5 | 9-12 | *A. niger* and *A. oryzae* | 30 | Muhialdin et al., (2011) |

**Table 2.** Delay of the appearance of fungal growth on bread with added lactic acid bacteria cells

*plantarum* 21B inhibited the bread spoilage fungi *Aspergillus, Fusarium, Penicillium* and *Eurotium*; the active compounds were phenyllactic and 4-hydroxyphenyllactic acids. The growth of *Aspergillus niger* appeared after two days in the control sample while *L. plantarum* 21B delayed the growth of the stated fungi for seven days at 20 °C.

## 8.2. Flavour

Flavour is one of the most valued sensory attributes in bread - volatile and non-volatile compounds produced during the fermentation of dough contribute to bread's flavour. Reports show that the fermentation of dough with LAB can enhance the aroma and flavour

(Ryan et al., 2011; Muhialdin et al., 2011a). The growth of fungi is responsible for the formation of off-flavours and the production of mycotoxins; adding LAB to dough can prevent the growth of fungi and enhance the flavour of bread. The produced compound plays an important role for any technological application to enhance the flavour, such as diacetyl which gives a buttery flavour. Sourness in white bread indicates spoilage in contrast to the sourness of sourdough bread; for this reason, the search for new LAB for application in white bread becomes essential. Finding a new LAB strain that produces less acid and does not drop the pH below 4 will mark a good strategy for resolving such an issue. The addition of *L. paracasi* D5 and *L. fermentum* Te007 in the production of white bread resulted in an improved aroma and a pleasant caramel-like flavour in the baked bread itself (Muhialdin et al., 2011a).

## 8.3. Quality and acceptability

The quality of bread produced with LAB as a starter culture was reported to improve the texture and the quality of bread by increasing the air cells (Coda et al., 2008; Katina et al., 2002; Lavermicocca et al., 2000). Baker's yeast - also referred to as 'baking yeast' (*Saccharomyces cerevisiae*) - has the ability to ferment different carbohydrates and produce $CO_2$; the most important factor involving baking yeast in bread manufacturing is to leaven the dough during the bread's preparation. The presence of antimicrobials in the dough is used to inhibit the growth of spoilage microorganisms that can affect the growth of the baker's yeast and delay the fermentation of dough, thereby resulting in economic losses to the bakery industry (Pattison & von Holy, 2001). Baking yeast is a excellent producer of the necessary flavour and aroma compounds from the products of secondary metabolism (Evans 1990).

Pattison & von Holy (2001) found that the presence of propionic salts reduced the baking yeast activity by up to 34.4% in an *in vitro* study carried out using several natural antimicrobials with positive control calcium propionate. In comparison, lactic acid and acetic acid displayed slight effects on the activity reduction of the yeast compared with the positive control. Baking yeast and lactic acid bacteria commonly have live symbiotically in the natural ecosystem of fermenting food and beverages (Kenns et al., 1991). The volume of the dough was increased by adding sourdough containing *L. amylovorus* DSM 19280 when compared with chemical acidification (Ryan et al., 2011). Rizzello et al. (2010) reported the improvement of bread texture properties and the delaying of the staling of the bread because of the anti-staling effect produced by LAB and the synthesis of antifungal compounds. As mentioned previously, *S. cerevisiae* is responsible of leaving the dough and giving the most desirable texture to the bread.

The key role in achieving the optimum growth and activity of the bakery yeast is played by selecting a LAB that does not exhibit inhibition activity against the bakery yeast. Before choosing the LAB to be added to the dough as a co-starter, a simple experiment can be conducted in order to examine the tolerance of the bread yeast to the selected LAB strain. In a test tube mix of 10 ml water, 5 g of white flour, the LAB strain and baking yeast, we incubate and observe the production of gas at the top of the tube, which is a good indicator of the yeast activity. Ogunbanwo et al. (2008) isolated LAB from retted cassava and studied the effects of lactic acid bacteria as a starter co-culture in combination with *S. cerevisiae* in order to produce cassava-wheat bread. The improvement in the nutritional contents, physical properties and the extension of the shelf life were reported. Bread produced using *L. acidophilus* and *L._brevis* had the highest acceptability on average in relation to the bread produced with other strains of LAB. The use of LAB in bread in terms of improving the quality of wheat bread, bread volume and crumb structure has been reported (Clarke et al., 2002; Zannini et al., 2009).

### 8.4. Enhancement of a specific nutrient

LAB fermentation in dough has been approved for enhancing the nutritional value and digestibility of bread. Vitamin B, organic acids and the free amino acids produced through the fermentation of LAB can enhance the nutrients' presence in bread. The human body cannot synthesize B-group vitamins and this is why the body needs an external source of the vitamins. Certain LAB has been proven to synthesize B-group vitamins during the fermentation of foods; at the same time, LAB are considered to be the perfect vehicle for delivering the vitamins to the human body.

There are reports about the production of B-group vitamins by LAB isolates. Keuth and Bisping (1993) described the production of Riboflavin (Vitamin B 2) by *Streptococcus* and *Enterococcus* isolated from tempeh (Indonesian fermented food). Folates were observed to be produced by *L. plantarum* in low amounts (Sybesma et al., 2003). Vitamin B 12 (Cobalamin) was also produced by *L. reuteri* as well as the other groups of vitamin B (Santos et al., 2008). LAB enzymatic activity by proteases that take place during dough fermentation will release small peptides and free amino acids, which are considered to be important nutrients that should be present in bread in high quantities (Thiele et al., 2002). Essential amino acids, including lysine, threonine, phenylalanine and valine were reported to be produced by LAB (Gerez et al., 2006). The enzymes produced by LAB including amylases, proteases, phytases and lipases improve the food quality through the hydrolysis of polysaccharides, proteins, phytates and lipids. Anti-nutrients such as phytic acid and tannins can be reduced by LAB fermentation in food, leading to increased sensory properties of the bread (Chelule et al., 2010). The growth of fungi in food materials can cause the synthesis of allergenic spores and hazardous mycotoxins, which will lead to the reduction of the nutritional value of food stuffs. Adding 4% of fermented sourdough to the white wheat flour improved the texture and physical sensation of the bread. Furthermore, it enhanced the free amino acids, protein digestibility, phytase and antioxidant activities (Rizzello et al., 2010).

## 9. Starter cultures for the bread industry

Lactic acid bacteria were reported as being used as a starter culture or co-culture in the bread industry with success in terms of survivability in dough (Lavermicocca et al., 2000;

Rezzillo et al., 2011). The use of lactic acid bacteria as an antifungal agent or as a starter culture for bakery and processed foods can solve two global issues; firstly, it can extend the shelf life of the food products, which will reduce their cost and the need for low temperatures, secondly, it will satisfy the high demand of modern consumers for high quality food that is free of chemicals. Above all, the product must be safe with an extended shelf life and good sensory properties.

## 10. Production of LAB cells and inhibitory compounds

### 10.1. Growth medium

The growth of LAB and the production of antifungal compounds are largely affected by the food matrix itself (Helander, 1997). Most of the studies regarding the antifungal activity of LAB were done using the universal MRS agar. As demonstrated earlier, there are few studies that evaluate the ability of LAB isolates to produce the active compounds in non-defined media as well as few *in situ* studies. The challenge for the food industry is the need for the high production of biomass and the bioactive compounds using an inexpensive fermentation growth medium. A defined medium is all well and necessary for laboratory screening purposes but it is not suitable for heavy industrial plant. The question here is whether the selected LAB can produce the biomass and maintain the antifungal activity. In our laboratory, *L. fermentum* Te007, *Pediococcus pentosaceus* Te010, *L. pentosus* G004 and *L. paracasi* D5 were used to ferment white bread dough and they maintained the antifungal activity, as detected using MRS agar, indicating that these isolates produced the antifungal compounds in the bread dough (Muhialdin et al., 2011a). *Pediococcus pentosaceus* Te010 was further investigated for its ability to grow in formulated media from plant extracts supplemented with the basic growth needs of LAB, such as vitamins, carbohydrates, nitrogen sources and salts. The results indicated that the selected isolate was able to grow in the formulated media and maintain the production of the antifungal activity but, unfortunately, the compounds have not yet been characterized (unpublished data).

### 10.2. Growth conditions

The growth conditions of any microbe are the key to success during the fermentation process. As for LAB, the generally optimum temperature for growth is 37 °C for 48 h in anaerobic conditions. This is not exactly what can be applied for the production of antagonistic fungal inhibitor compounds. Some of the LAB are psychrophilic and prefer low temperatures for their growth while others are thermophilic and prefer high temperatures for their growth. This should be considered as a significant factor because the optimum growth temperature has a significant impact on the production of antifungal compounds. As well as temperature, the incubation time has a significant effect on the production of antifungal compounds with respect to the availability of nutrients in the growth medium and the production of primary or secondary metabolites.

## 11. Future research

The high demand by consumers for foods free of chemical preservatives has led to increasing amounts of research to provide alternatives for these chemicals. LAB provides technologically practicable alternatives for the replacement of chemical preservatives. The achievement of selecting LAB as starter cultures or co-cultures in fermentation processes can improve the desired properties of bread, at the same time providing consumers with new chemical-free foods. There is a need to study the interaction between the food matrix and the kinetics of the starter culture of LAB in bread; such studies will contribute to the bread industry by increasing the yield of the antifungal and nutritional compounds produced by LAB. Besides using the LAB cells in bread formulations, the use of the supernatant of LAB should be considered, especially the supernatant of LAB that are grown in non-conventional media such as plant extract and other cheap materials. Additional studies on the contribution of bioactive molecules to the quality and shelf life of foods will surely widen the use of LAB strains as a novel bio-control strategy in bakery products.

## 12. Conclusion

LAB can be used as a starter culture or a co-culture in the bread industry to enhance the sensory properties of bread and extend the shelf life. The nutritional value of the bread is enhanced due to the production of free amino acids, organic acids and a variety of Group-B vitamins. The antifungal compounds produced by LAB are important for the food industry for replacing or reducing the use of chemical preservatives. Several methods have been developed to determine the antifungal activity of the cells and the free cell supernatant. Natural sources of food preservatives - especially LAB - are important and reflect one possibility for fulfilling the needs of modern consumers of bakery products that are free of chemicals. Challenges are evident in finding new and novel isolates of LAB that can be applied in bread and which do not affect the activity of the yeast or inhibit their growth. Future works should consider the use of the LAB supernatant as well as the cells because the active compounds can be present in the supernatant. Inexpensive media are also important for high-scale industry, especially the use of plant extracts that are rich in carbohydrates and which can be supplied in bulk over the course of the year.

## Author details

Belal J. Muhialdin and Nazamid Saari
*Universiti Putra Malaysia (UPM),*
*Malaysia*

Zaiton Hassan
*Universiti Sains Islam Malaysia (USIM),*
*Malaysia*

## 13. References

Abellana, M. L., Torres, V. S. and A. J. Ramos. 1997. Caracterizacio´n de diferentesproductos de bollería industrial. II. *Estudio de la Microflora. Alimentaria,* 287: 51-56.

Axelsson, L. 1998. Lactic acid bacteria: Classification and physiology. In *Lactic Acid Bacteria: Microbiology and functional aspects,* 2nd Edition, Revised and Expanded. Edited by S. Salminen & A. von Wright. pp. 1-72. Marcel Dekker, Inc. New York.

Bennett, G. A., Shotwell, O. L. and Hesseltine, C. W. 1980. Destruction of zearalenone in contaminated corn. *Journal of the American Oil Chemists' Society,* 57: 245-7.

Björkroth, J. and Holzapfel, W. 2003. Genera *Leuconostoc, Oenococcus* and *Weissella,* In M. Dworkin et al., eds., *The Prokaryotes: An Evolving Electronic Resource for the Microbiological Community,* 3rd edition, Springer-Verlag, New York,
http://link.springer-ny.com/link/service/books/10125/

Brul, S. and Coote, P. 1999. Preservative agents in foods-mode of action and microbial resistance mechanisms. International *Journal of Food Microbiology,* 50: 1-17.

Carr, F. J., Chill, D. and Maida, N. 2002. The lactic acid bacteria: a literature survey. *Critical Reviews in Microbiology,* 28: 281-370.

Chelule, P. K., Mokoena, M. P. and Gqaleni, N. 2010. Advantages of traditional lactic acid bacteria fermentation of food in Africa. *RORMATEX,* pp. 1164-1167,
http://www.formatex.info/microbiology2/1160-1167.pdf

Clarke, C., Schober, T. J. and Arendt, E. K. 2002. Effect of single strain and traditional mixed strain starter cultures in rheological properties of wheat dough and bread quality. *Cereal Chemistry,* 79: 640-647.

Coda, R., Rizzello, C. G., Nigro, F., De Angelis, M., Arnault, P., and Gobbetti, M. 2008. Long-term fungi inhibitory activity of water-soluble extract from Phaseolus vulgaris cv Pinto and sourdough lactic acid bacteria during bread storage. *Applied and Environmental Microbiology,* 74: 7391-7398.

Coda, R., Cassone, A., Rizzello, C. G., Nionelli, L., Cardinali, G. and Gobbetti, M. 2011. Antifungal Activity of *Wickerhamomyces anomalus* and *Lactobacillus plantarum* during Sourdough Fermentation: Identification of Novel Compounds and Long-Term Effect during Storage of Wheat Bread. *Applied and Environmental Microbiology,* 77: 3484-3492.

Corsetti, A., Gobetti, M., Rossi, J. and Damiani, P. 1998. Antimould activity of sourdough lactic acid bacteria: identification of a mixture of organic acids produced by *Lactobacillus sanfrancisco* CB1. *Applied Microbiology and Biotechnology,* 50: 253-256.

Cotter, P. D., Hill, C. and Ross, R. P. 2005.Bacteriocins: developing innate immunity for food. *Nature Reviews Microbiology,* 3: 777-788.

Dal Bello, F., Clarke, C. I., Ryan, L. A. M., Ulmer, H., Schober, T. J., Strom, K., Sjorgren, J., Van Sinderen, D., Schnurer, J. and Arendt, E. K. 2007. Improvement of the quality and shelf life of wheat bread by fermentation with the antifungal strain *Lactobacillus plantarum* FST 1.7. *Journal of Cereal Science,* 45: 309-318.

Dalié, D. K. D., Deschamps, A. M. and Richard-Forget, F. 2009. Lactic acid bacteria - Potential for control of mould growth and mycotoxins.A review. *Food Control,* 21: 370-380.

Dantigny, P., Guilmart, A., Radoi, F., Bensoussan, M., and Zwietering, M. 2005. Modelling the effect of ethanol on growth rate of food spoilage moulds. *International Journal of Food Microbiology*, 98: 261-269.

Druvefors, U. A., Passoth, V. and Schnurer, J. 2005. Nutrient effects on biocontrol of *Penicillium roqueforti* by *Pichia anomala* J121 during airtight storage of wheat. *Applied Environmental Microbiology*, 71: 1865-1869.

El-Ziney, M. G. and Debevere, J. M. 1998. The effect of reuterin on *Listeria monocytogenes* and *Escherichia coli* O157:H7 in milk and cottage cheese. *Journal of Food Protection*, 61: 1275-1280.

Ennahar, S., Sonomoto, K. and Ishizaki, A. 1999. Class IIa bacteriocins from lactic acid bacteria: antibacterial activity and food preservation. *Journal of Bioscience and Bioengineering*, 87: 705-716.

European Union. 1995. European Parliament and Council directive no. 95/2/EC of 20 February 1995 on food additives other than colours and sweeteners, 53. Office for Official Publications of the European Communities, Luxembourg. http://europa.eu.int/eur-lex/en/consleg/pdf/1995 /en_1995L0002_do_001.pdf.

Evans, I. H. (1990). Yeast strains for baking. In: *Yeast Technology*, eds., Spencer, J. F. T. and Spencer, D. M. pp. 13±45. Berlin, Germany: Springer-Verlag.

Farkas, J. 2001. Physical methods for food preservation. In *Food microbiology: Fundamentals and frontiers*. Edited by M. P. Doyle, L. R. Beuchat & T. J. Montville. pp. 567-592. ASM press. Washington, USA.

Filtenborg, O., Frisvad, J. C. and Thrane, U. 1996. Moulds in food spoilage. International *Journal of Food Microbiology*, 33: 85-102.

Hammes, W. P. and Hertel, C. 2003. The Genera *Lactobacillus* and *Carnobacterium*. In *The Prokaryotes: An Evolving Electronic Resource for the Microbiological Community*, Edited by M. Dworkin. Springer-Verlag, New York. http://link.springer-ny.com/link/service/books/10125/.

Hassan, Y. I. and Bullerman, L. B. 2008. Antifungal activity of *Lactobacillus paracasei* ssp. *tolerans* isolated from a sourdough bread culture. *International Journal of Food Microbiology*, 121: 112-115.

Helander, I. M., von Wright, A. and Mattila-Sandholm, T-M. 1997. Potential of lactic acid bacteria and novel antimicrobials against Gram-negative bacteria. *Trends in food Science and Technology*, 8: 146-150.

Gerez, C. L., Rollán, G. and Font de Valdez, G. 2006. Gluten breakdown by lactobacilli and pediococci strains isolated from sourdough. *Letters in Applied Microbiology*, 42: 459-464.

Gerez, C. L., Torino, M. I., Rollan, G. and Font de Valdez, G. 2009. Prevention of bread mould spoilage by using lactic acid bacteria with antifungal properties. *Food Control*, 20: 144-148.

Gray, J. and Bemiller, J. 2003. Bread staling: Molecular basis and control. *Comprehensive Reviews in Food Science and Safety*, 2: 1-21.

Guynot, M. E., MarÍn, S., Setu, L., Sanchis, V. and Ramos, A. J. 2005. Screening for antifungal activity of some essential oils against common spoilage fungi of bakery products. *Food Science and Technology International*, 11: 25-32.

James, S. and Stratford, M. 2003. Spoilage yeast with emphasis on the genus *Zygosaccharomyces*. In *Yeasts in food*. Edited by B. T & R. V. Behrs, Verlag, pp. 171-191.

Katina, K., Sauri, M., Alakomi, H. L. and Mattila-Sandholm T. 2002. Potential of Lactic Acid Bacteria to Inhibit Rope Spoilage in Wheat Sourdough Bread. *Lebensm.-Wiss. u.-Technology*, 35: 38-45.

Kenns C, Veiga, M. C., Dubourgular, H. C., Touzel, J. P., Albengae, G., Navean, H. and Nyns, E. J. 1991. Tropic relationship between *Saccharomyces cerevisiae* and *Lactobacillus plantarum* and their metabolism of glucose and citrate. *Journal of Applied Environmental Microbiology*, 57:1047-1051.

Keuth, S., and Bisping, B. 1993. Formation of vitamins by pure cultures of tempe moulds and bacteria during the tempe solid substrate fermentation. *Journal of Applied Bacteriology*, 75: 427-434.

Lavermicocca, P., Valerio, F., Evidente, A., Lazzaroni, S., Corsetti, A. and Gobbetti, M. 2000. Purification and characterization of novel antifungal compounds by sourdough *Lactobacillus plantarum* 21B. *Applied and Environmental Microbiology*, 66: 4084-4090.

Lavermicocca, P., Valerio, F. and Visconti, A. 2003. Antifungal activity of phenyllactic acid against molds isolated from bakery products. *Applied Environmental Microbiology*, 69: 634-640.

Legan, J. D. 1993. Mould spoilage of bread: the problem and some solutions. *International Biodeterioration and Biodegradation*, 32: 33-53.

Levi, C. 1980. Mycotoxins in coffee. *Journal - Association of Official Analytical Chemists*, 63: 1282-1285.

Lourens-Hattingh, A. and Viljoen, B. C. 2001.Yoghurt as probiotic carrier food. *International Dairy Journal*, 11: 1-17.

Luter, L., Wyslouzil, W. and Kashyap, S. C. 1982. The destruction of aflatoxin in peanuts by microwave roasting. *Canadian Institute of Food Science and Technology Journal*, 15: 236-238.

Magnusson, J. and Schnurer, J. 2001. *Lactobacillus coryniformis* subsp. *coryniformis* strain Si3 produces a broad-spectrum proteinaceous antifungal compound. *Applied and Environmental Microbiology*, 67: 1-5.

Magnusson, J. 2003. Antifungal activity of lactic acid bacteria. Ph.D. Thesis, Agraria 397, Swedish University of Agricultural Sciences, Uppsala, Sweden.

Magnusson, J., Ström, K., Roos, S., Sjögren, J. and Schnürer, J. 2003. Broad and complex antifungal activity among environmental isolates of lactic acid bacteria. *FEMS Microbiology Letters*, 219: 129-135.

Muhialdin, B. J., Hassan, Z. and Sadon, S. K. 2011a. Antifungal Activity of *Lactobacillus fermentum* Te007, *Pediococcus pentosaceus* Te010, *Lactobacillus pentosus* G004 and *L. paracasi* D5 on Selected Foods. *Journal of Food Science*, 76: 493-499.

Muhialdin, B. J., Hassan, Z. Sadon, S. K. NurAqilah, Z. and AZFAR, A. A. 2011b. Effect of pH and Heat Treatment on Antifungal Activity of *Lactobacillus fermentum* Te007, *Lactobacillus pentosus* G004 and *Pediococcuspentosaceus* Te010. *Innovative Romanian Food Biotechnology*, 8: 41-53.

Muhialdin, B. J. and Hassan, Z. 2011. Screening of Lactic Acid Bacteria for Antifungal Activity against *Aspergillus oryzae*. *American Journal of Applied Science*, 8: 447-451.

Niku-Paavola, ML., Laitila, A. Mattila-Sandholm, T. and Haikara, A. 1999. New types of antimicrobial compounds produced by *Lactobacillus plantarum*. *Journal of Applied Microbiology*, 86: 29-35.

Ogunbanwo, S. T., Adebayo, A. A., Ayodele, M. A., Okanlawon, B. M. and Edema, M. O. 2008. Effects of lactic acid bacteria and *Saccharomyces cerevisiae* co-cultures used as starters on the nutritional contents and shelf life of cassava-wheat bread. *Journal of Applied Biosciences*, 12: 612-622.

Pattison, T. L. and von Holy, A. 2001. Effect of selected natural antimicrobials on Baker's yeast activity. *Letters in Applied Microbiology*, 33: 211-215.

Pattison, T. L., Lindsay, D. and von Holy, A. 2004. Natural antimicrobial as potential replacements for calcium propionate in bread. *South Africa Journal of Science*, 100: 339-342.

Piard, J. C. and Desmazeaud, M. 1991. Inhibiting factors produced by lactic acid bacteria. 1. Oxygen metabolites and catabolism end-products. *Le Lait*, 71: 525-541.

Pitt, J. J. & Hocking, A. D. 1999. Fungi and food spoilage Second ed. Aspen Publications. pp.

Plockova, M., Stiles, J. and Chumchalova, J. 2001 Control of mould growth by Lactobacillus rhamnosus VT1 and Lactobacillus reuteri CCM 3625 on milk agar plates. *Czech Journal of Food Science*, 19: 46-50.

Prema, P., Smila, D., Palavesam, A. and Immanuel, G. 2008. Production and Characterization of an Antifungal Compound (3-Phenyllactic Acid) Produced by *Lactobacillus plantarum* Strain. *Food Bioprocess Technology*, 3: 379-386.

Rizzello, C. G., Nionelli, L., Coda, R., Di Cagno, R. and Gobbetti, M. 2010. Use of sourdough fermented wheat germ for enhancing the nutritional, texture and sensory characteristics of the white bread. *European Food Research and Technology*, 230: 645-654.

Rizzello, C. G., Cassone, A., Coda, R. and Gobbetti, M. 2011. Antifungal activity of sourdough fermented wheat germ used as an ingredient for bread making. *Food Chemistry*, 127: 952-959.

Rouse, S., Harnett, D., Vaughan, A. and van Sinderen, D. 2008. Lactic acid bacteria with potential to eliminate fungal spoilage in foods. *Journal of Applied Microbiology*, 104: 915-923.

Roy, U., Batish, V. K., Grover, S. and Neelakantan, S. 1996. Production of antifungal substance by *Lactococcus lactis* subsp. *lactis* CHD-28.3.*International Journal Food Microbiology*, 32: 27-34.

Ryan, L. A. M., Dal Bello, F. and Arendt, E. K. 2008. The use of sourdough fermented by antifungal LAB to reduce the amount of calcium propionate in bread. *International Journal of Food Microbiology*, 125: 274-278.

Ryan, L. A. M., Zannini, E., Dal Bello, F., Pawlowska, A., Koehler, P. and Arendt, E. K. 2011. *Lactobacillus amylovorus* DSM 19280 as a novel food-grade antifungal agent for bakery products. *International Journal of Food Microbiology*, 146: 276-283.

Saithong, P., Panthavee, W., Boonyaratanakornkit, M. and Sikkhamondhol, C. 2010. Use of a starter culture of lactic acid bacteria in plaa-som, a Thai fermented fish. *Journal of Bioscience and Bioengineering*, 110: 553-557.

Samson, R. A., Seifert, K. A., Kuijpers, A. F. A., Houbraken, J. A. M. P. and Frisvad, J. C. 2004. Phylogenetic analysis of *Penicillium* subgenus *Penicillium* using partial b-tubulin sequences. *Studies in Mycology*, 49: 175-200.

Santos, F., Vera, J. L., van der Heijden, R., Valdez, G., de Vos, W. M., Sesma, F. and Hugenholtz, J. 2008. The complete coenzyme B 12 biosynthesis gene cluster of *Lactobacillus reuteri* CRL1098. *Microbiology*, 154: 81-93.

Scott, P. M. 1984. Effect of food processing on mycotoxins. Journal of Food Protection, 47: 489-499.

Schnürer, J. and Magnusson, J. 2005. Antifungal lactic acid bacteria as biopreservatives. *Trends in Food Science and Technology*, 16: 70-78.

Shapira, R. and Paster, N. 2004.Control of mycotoxins in storage and techniques for their decontamination. Mycotoxins in food. Woodhead Publishing Limited, Abington Hall, Abington, Cambridge CB1 6AH, England. pp. 190-223.

Stiles, J., Plockova, M., Toth, V. and Chumchalova, J. 1999. Inhibition of *Fusarium sp.* DMF 0101 by *Lactobacillus* strains grown in MRS and Elliker broths. *Advances in Food Science*, 21: 117-121.

Ström, K., Sjörgen, J., Broberg, A. and Schnürer, J. 2002. *Lactobacillus plantarum* MiLAB 393 produces the antifungal cyclic dipeptides cyclo(L-Phe-L-Pro) and cyclo(L-Phe-trans-4-OH-L-Pro) and 3 phenyllactic acid. *Applied and Environment Microbiology*, 68: 4322-4327.

Suhr, K. I. and Nielsen, P. V. 2004. Effect of weak acid preservatives on growth of bakery product spoilage fungi at different water activities and pH values. *International Journal of Food Microbiology*, 95: 67-78.

Sybesma, W., Starrenburg, M., Tijsseling, L., Hoefnagel, M. H. and Hugenholtz, J. 2003. Effects of cultivation conditions on folate production by lactic acid bacteria. *Applied and Environmental Microbiology*, 69: 4542-4548.

Thiele, C., Gänzle, M. G. and Vogel, R. F. 2002. Contribution of sourdough lactobacilli, yeast, and cereal enzymes to the generation of amino acids in dough relevant for bread flavor. *Cereal Chemistry*, 79: 45-51.

Trivedi, A. B., Doi, E. and Kitabatake, N. 1992. Detoxification of ochratoxin A on heating under acidic and alkaline conditions. *Bioscience, Biotechnology and Biochemistry*, 56: 741-755.

Vandenbergh, P. A. and Kanka, B. S. 1989. Antifungal product. *United States Patent*. 4,877,615.

Wang H., Yan Y., Wang J., Zhang H., Qi W. (2012). Production and Characterization of Antifungal Compounds Produced by *Lactobacillus plantarum* IMAU10014. *PLoS ONE*, 7(1): e29452. doi:10.1371/journal.pone.0029452

Whittenbury, R. 1964. Hydrogen peroxide formation and catalase activity in the lactic acid bacteria. *Journal of General Microbiology*, 35: 13-26.

Yang, E. J. and Chang, H. C. 2010. Purification of a new antifungal compound produced by *Lactobacillus plantarum* AF1 isolated from kimchi. *International Journal of Food Microbiology*, 139: 56-63.

Zannini, E., Garofalo, C., Aquilanti, L., Santarelli, S., Silvestri, G. and Clementi, F. 2009. Microbiological and technological characterization of sourdoughs destined for bread-making with barley flour. *Food Microbiology*, 26: 744-753.

Chapter 7

# Fermentation of Vegetable Juices by *Lactobacillus Acidophilus* LA-5

Lavinia Claudia Buruleanu, Magda Gabriela Bratu,
Iuliana Manea, Daniela Avram and Carmen Leane Nicolescu

Additional information is available at the end of the chapter

http://dx.doi.org/10.5772/51309

## 1. Introduction

Probiotics foods represent one of the largest sectors in functional food markets. Most of the available probiotic products are some form of dairy, despite the continuous growth of the non-dairy probiotic sector, with products like soy-based drinks, fruit-based foods, and other cereal-based products. Both non-dairy (in general) and soy-based probiotic products represent a huge growth potential for the food industry, and may be widely explored through the development of new ingredients, processes, and products. For this purpose, new studies must be carried out to: test ingredients, explore more options of media that have not yet been industrially utilized, reengineer products and processes, towards potentially meet the demands of lactose-intolerant and vegetarian consumers for new nourishing and palatable probiotic products [1].

Lactic acid bacteria are among the most important probiotic microorganisms typically associated with the human gastrointestinal tract. Traditionally, lactic acid bacteria have been classified on the basis of phenotypic properties, e.g. morphology, mode of glucose fermentation, growth at different temperatures, lactic acid configuration, and fermentation of various carbohydrates. However some species, like the so-called *Lactobacillus acidophilus* group and some bifidobacteria, are not readily distinguishable by phenotypic characteristics [2]. From the physiological point of view, *Lactobacillus acidophilus* strains were characterized as lactic acid bacteria with strictly homofermentative metabolism (> 85% lactic acid). The hexoses are preferential fermented via Embden – Meyerhof – Parnas (EMP), (as the strains produce aldolase and phosphoketolase), and only then the pentoses and gluconate are fermented. LAB of the *Lactobacillus acidophilus* group as well as of the *Bifidobacterium* group isolated from the human faeces or intestine are thought to have beneficial effects on health being thus considered to be probiotic bacteria [3].

For use in food, important criteria for probiotics must be met, in particular that they should not only be capable of surviving passage through the digestive tract, by exhibiting acid and bile tolerance, but also have the capability to proliferate in the gut.

Probiotics must be able to exert their benefits on the host through growth and/or activity in the human body. Although generally recognised as safe a probiotic strains must be characterized by a set of tests that assure its safety to consumer (1, 2, 3, 5, 6).

Inclusion of probiotic bacteria in fermented dairy products enhances their value as better therapeutic functional foods. However, insufficient viability and survival of these bacteria remain a problem in commercial food products. By selecting better functional probiotic strains and adopting improved methods to enhance survival, including the use of appropriate prebiotics and the optimal combination of probiotics and prebiotics (synbiotics), an increased delivery of viable bacteria in fermented products to the consumers can be achieved [5].

The fermentation of vegetable products, applied as a preservation method for the production of finished and half-finished food products, is considered as an important technology, though requiring more research, as a growing number of raw materials are being processed in this way by the food industry. The main reasons for this interest are nutritional, physiological and hygienic aspects of the process [6]. Thus, according to Kelwicka, (2010) [7], the fermentation of beetroot juice requires selected starter cultures made of LAB, naturally present in this vegetable although their number is usually very small. This makes them un-appropriate to, alone, conducting a fermentation that ensures satisfying sensory properties of the fermented juice, with improved health promoting activity.

Thus, probiotic juices represent an alternative to dairy products that suits consumers who don't want to eat dairy foods or are lactose intolerant. Adding probiotics to juices is more complex than formulating in dairy products where the bacteria can be easily added to other cultures.

Despite its potential for healthy products development, there is very little research activity addressing the fermentation of vegetable juices using probiotic bacteria.

## 2. Materials and methods

### 2.1. Vegetables treatments

Fresh vegetables (carrots, cucumbers, beetroot, white cabbage, red cabbage) were purchased from a retail market and specifically processed by removing the non-edible pieces. The raw material processing was made faster, because the possibility of contamination and proliferation of microorganisms in the products is very high in comparison with their intact counterparts (Lee, 2011). Using a domestic extractor the vegetables were turned into juice. The heating treatment of the juice, applied at $80^0$C with a view to destroy the undesirable microorganisms under the limit of detection, was followed by cooling at $40^0$C.

## 2.2. Microorganisms and fermentation conditions

The strain *Lactobacillus acidophilus* LA-5 from Christian Hansen (Romania) was used in this study.

The lyophilized culture was aseptically inoculated into the vegetable juices and vigorously homogenized for 15 min, according to the producer's specification. The fermentation experiments were carried out using Erlenmeyer flasks containing 50ml of juice, without pH adjustment. The flasks were incubated statically in an incubator chamber at 37±0.2$^0$C. Sampling was taken at regular interval of times for physico-chemical and microbiological analysis.

The tested supplements were: L-cysteine hydrochloride monohydrate (Merck, Darmstadt, Germany), L-lysine hydrochloride (Merck), L-valine (Merck), L-leucine (Calbiochem, San Diego, CA, USA) and yeast extract (Merck). Cysteine, lysine, valine and leucine were separately added in quantity by 0.1% (w/v) into carrot juice, while amounts by 0.2% (w/v) were tested, also individual, in the case of the yeast extract and cysteine. A control sample without supplements was carried out for each experiment.

## 2.3. Physico – Chemical analysis

Metabolic activity of the strain LA-5 in the conditions mentioned above was evaluated based on the dynamics of pH, respectively end products of fermentation. The pH values were measured with a HACH pH-meter. Lactic acid was determined using commercial kits (K-DLATE from Megazyme International). The calculations were made with Megazyme Mega-Calc™ and expressed as g lactic acid/l. Reducing sugars were analyzed applying the spectrophotometric method with 3.5-dinitrosalicilic acid (DNS) after the removing of other substances with reducing character using basic lead acetate and expressed as g glucose/l. Ascorbic acid was determined applying the 2,6-dichloroindophenol titrimetic method, based on the reduction of the sodium salt of the dye by ascorbic acid (AOAC method). It was expressed as mg/100ml. The amino acids content, expressed as g glycine/100ml, was determined through the Sörensen method.

## 2.4. Microbiological analysis

The amount of viable cells of *Lactobacillus* sp. was determined by serial tenfold dilution with sterile peptone water. Aliquots of 1ml were plated, in duplicate, in plates with Man-Rogosa-Sharpe agar, enriched with L-cysteine HCl. The Petri plates were incubated for 48-72h at 37°C and the results were expressed as log colony forming units (CFU)/ml juice.

The optical density of biomass was measured with the UV-Visible spectrophotometer at 610nm. In the preparation of the calibration curve for optical density vs. dry cell weight several dilutions of the juices were made. According Altiok [8], for each dilution 2 ml of sample was used to obtain optical densities at 610 nm wavelength and 15 ml of sample was filtered with a pre-weighed cellulose acetate membrane filter having a pore size of 0.45 μm using a vacuum pump. The biomass collected on the filters was washed with 15 ml of water and the filters were dried at 100ºC for approximately 24 h until constant weight was observed. The results were expressed as g.

### 2.5. Statistical analysis

Statistical analysis was carried out using the software SPSS (Statistical Package for the Social Science 17.0 trial version).

## 3. Results and discussions

### 3.1. Effect of inoculum size on the lactic acid accumulation and biomass growth

A comparative study of the dynamics of lactic acid fermentation of carrot juice using three different concentrations of lyophilized pure culture was realized (Figure 1).

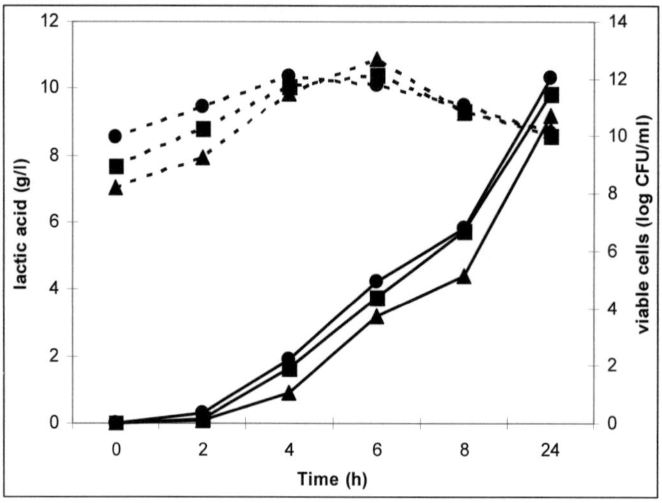


**Figure 1.** Correlation between lactic acid production by *Lactobacillus acidophilus* LA-5 and number of viable cells during fermentation of carrot juice with different inoculum size ▲ 0.2g/l; ■ 0.3g/l; ● 0.4 g/l (smooth lines - lactate, dashed lines - viable cells count)

Relative higher differences concerning the lactate increasing were observed between the variant with 0.2g/l pure culture initial added and the other two within 24 hours of fermentation. Thus, at the end of this interval, the excess was by 7.06% in the juice with 0.3g/l inoculum and 12.06% in the juice with 0.4g/l inoculum respectively. However, in all the batches the lactic acid accumulation, higher than 9g/l, could be considered satisfactory for the shelf life of the final products. From the other part, the number of viable cells is decisive for the probiotic feature of these ones. A direct proportionality between the amount of the lyophilised culture initial added and the viable cells was observed only in the first 4h of the fermentation. As a general characteristic, in the interval 6 - 24h pH values less than 4.5 have become inhibitory for the useful microbiota in all the experimental samples.

The initial concentration of reducing sugars of the carrot juices, by 25.2g/l, was favourable for the growth of *Lactobacillus acidophilus* LA-5. Testing two strains of Lactobacillus (one genetically selected Mont4+ and the other genetically altered, Mont4+pxyAB-mod). Kiouss [9] established that the Mont4+ had the highest yield of lactic acid fermenting with six percent concentration of glucose, whereas the L strain utilized the sugar best at the four percent concentration. In the same time temperature and pH seemed to play the largest role in the organisms ability to grow and thus affecting its production of lactic acid.

Concluding, higher inoculum densities of *Lactobacillus acidophilus* LA-5 were not significantly influenced the survival yield of the useful microbiota in the lactic acid fermented juices after 24h. In the same time, no parallel relationships between lactic acid concentration and the inoculum size were determined. The result agrees to those obtained by Agarwal, Dutt, Meghwanshi and Saxena [10] using *Enterococcus flavescens* for production of lactic acid. In their opinion, beyond a certain concentration lactic acid yield dropped due to high cell density resulting in fast depletion of essential nutrients, limiting further growth and reducing the yield. Referring to bifidobacteria, Dave and Shah [11] reported also that a higher inoculum did not always improve their viability to a satisfactory level. No data referring to *Lactobacillus acidophilus* were found in the literature.

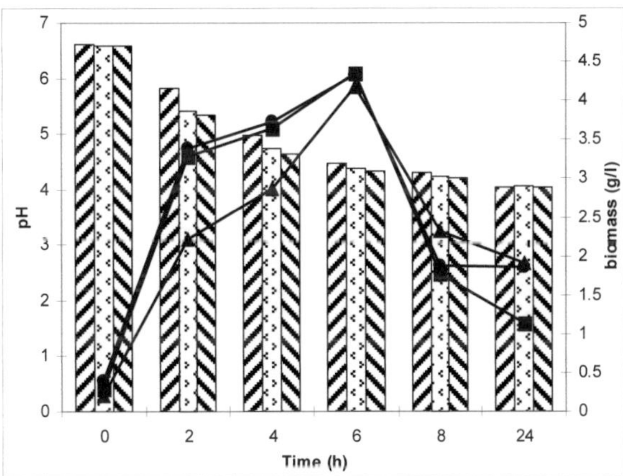


**Figure 2.** pH and biomass evolution during lactic acid fermentation of carrot juice with different inoculum of *Lactobacillus acidophilus* LA-5: 0.2g/l ( and ▲); 0.3g/l ( and ■); 0.4 g/l ( and •); columns - pH values, lines - biomass

Although the pH dynamics was quite different in the first 6h of the process, the initial amount of the pure culture did not affect the subsequent evolution or the final value of this parameter (Figure 2).

The sharp decrease in biomass from 6 to 8h has been correlated with the viable cells tendency, as result of reaching pH values by 4.34 to 4.47. Being known that *Lb. acidophilus* is more sensitive in acidic environment, this result underlines the necessity to manage the size of inoculum in order to obtain a balance between the lactic acid accumulation and the survival of the probiotic microorganisms.

The maximum rate of acidification $v_{max}$ was calculated as the time variation of pH (dpH/dt) and expressed as pH units/min (Table 1). Other kinetic parameters were also calculated: time to reach $v_{max}$ ($t_{max}$, hours), time to reach pH 5.0 ($t_{pH\ 5.0}$, hours), time to complete the fermentation ($t_{pH\ 4.2}$, hours).

| Inoculum, g/l | $v_{max} \cdot 10^{-3}$ (units/min.) | $t_{max}$ (h) | $t_{pH\ 5.0}$ (h) | $t_{pH\ 4.2}$ (h) |
|---|---|---|---|---|
| 0.2 | 7.08 | 4 | 2.95 | 8.4 |
| 0.3 | 9.83 | 2 | 2.88 | 8.2 |
| 0.4 | 10.41 | 2 | 2.67 | 8.05 |

**Table 1.** 1. Acidification kinetic parameters of fermentation of carrot juices by *Lactobacillus acidophilus* LA-5

A double amount of inoculum had an insignificant influence on the time to reach pH 5.0, important parameter from the shelf life of the fermented juices. Thus, $t_{pH\ 5.0}$ (h) was 1.1-fold higher in the case of the batch with 0.2g/l lyophilized pure culture initial added to juice than that one with 0.4g/l. A different situation was registered concerning the maximum rate of acidification ($v_{max}$) and the time to reach this rate ($t_{max}$). Thus, a polynomial equation of the form $y = -108.5x^2 + 81.75x - 4.93$ correlated the size of inoculum with the corresponding values of $v_{max}$ at R squared = 1. Although at the initial moment of fermentation seems to be advantageous to use a higher amount of pure culture, this aspect lessen in time, from the economic point of view being important to obtain a balance between the quantity of inoculum and the targeted parameters which ensure the preservation of the final product.

The values of the biomass content became close after about 6h of fermentation. No parallel relationship between lactic acid concentration and biomass was observed, result that agrees to those obtained by Amrane [12] and Kotzamanidis [13].

However, taking into account the lactic acid accumulation and the dynamics of the number of viable cells, it was obvious that the utilization of higher amount of inoculum is not justified.

### 3.2. Effect of temperature on the dynamics of fermentation

According to the information provided by the producer of the lactic culture, respectively to the data found in literature, two different incubation temperatures were tested: 37⁰C and 41⁰C respectively.

The dynamics of both pH and lactic acid (Figure 3) emphasizes the influence of the higher temperature on the rate of acidification. After 24h no significant differences between the pH values were determined, while the lactic acid content of the samples fermented at 41⁰C was 1.24-fold higher comparatively with those fermented at 37⁰C. This situation may be due to the higher amino acids content in the samples fermented at 41⁰C, that act as buffer. Thus, expressed as glycin, the total amount was by 0.165g/100ml at the end of the analyzed interval, which represented an increase by 10% comparatively with the batch fermented at lower temperature.

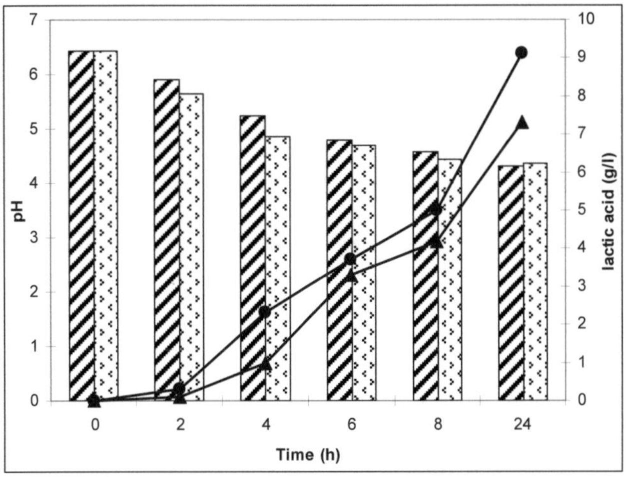


**Figure 3.** pH and lactic acid dynamics during the lactic acid fermentation of carrot juice at different temperatures: 37⁰C ( and ▲) and 41⁰C ( and •); columns - pH values, lines - lactic acid content

The rate of acidification has been correlated with the glucose consumption: 38.9% in the case of the juice fermented at 37⁰C, respectively 53.89% in the case of the juice fermented at 41⁰C. The different tendency of this parameter became obviously after 4h of fermentation (Figure 4), being the consequence of the different rate of growth of *Lactobacillus acidophilus*, expressed as optical density at 610nm.

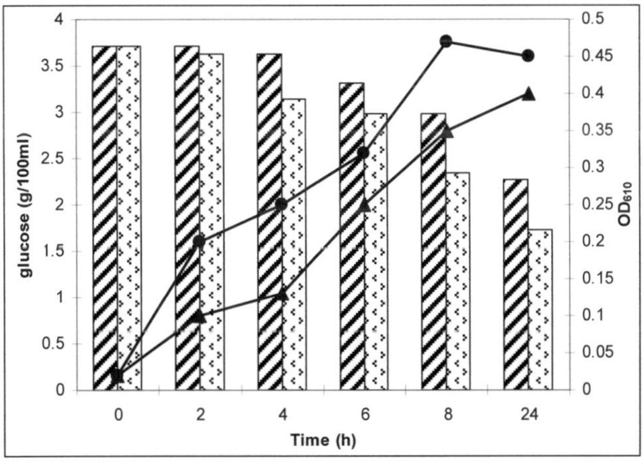


**Figure 4.** Glucose consumption and microbial evolution during the lactic acid fermentation of the carrot juices at different temperatures: 37⁰C ( and ▲) and 41⁰C ( and •); columns - glucose, lines - optical density at wavelength by 610nm

Although close, the yields of glucose conversion to lactic acid have inclined the balance in favour of the juices fermented at 37⁰C, the corresponded value being by 0.5, unlike 0.45 in the case of the juices incubated at 41⁰C.

The faster consumption of the carbon source, correlated with the growth of the useful microbiota at higher temperature, respectively with the increase of the lactic acid content until the value by 9.1g/l, was followed at 24h by the decline of the viability of *Lactobacillus acidophilus*. Taking into account the dynamics of all the above mentioned parameters, the incubation temperature applied in the further studies was by 37±0.1$^0$C.

## 3.3. The behaviour of different raw materials during the lactic acid fermentation by Lactobacillus acidophilus LA-5

Fresh white cabbage (*Brassica oleracea* L.), red cabbage (*Brassica oleracea* var. *capitata* f. *rubra*), red beet (*Beta vulgaris* var. *vulgaris*), cucumbers (*Cucumis sativus*) and red onion (*Allium cepa* var. *ascalonicum*) were chosen in order to perform different experimental batches, as follows: Cb - cabbage juice, RCb - red cabbage juice, Rb - red beet juice, Cc - cucumber juice, CcO - cucumber juice with 0.1% (v/v) onion juice added after the heating and cooling of the batches.

pH and lactic acid dynamics during the lactic acid fermentation of vegetable juices with *Lb. acidophilus* are shown in Figure 5 and Figure 6 respectively. The pH values ranged from 6.29 to 3.74, no significant differences between the analyzed batches being observed, excepting the red beet juice. Thus, after one day a higher value by 4.28 was determined, the prolongation of the time of fermentation with other 24h hadn't a positive influence on this parameter.

After 24h, the highest decrease of pH was determined in the case of the cucumber juice (2.51 units), correlated with the increase of the lactic acid amount until 9.36g/l. Although the pH values of the samples Cc and Cb were close during the process development, the maximum rate of acidification $v_{max}$ registered a better value of $9.33 \cdot 10^{-3}$ units/min. in the case of the cucumber juice. This could explain the fermentation slowdown in the batch Cb the interval 6 - 8 hours. Correlated with the results of the microbiological analysis, it seems that this time the process was directed towards the growth of the useful microbiota. A minimum value of the maximum rate of acidification, by $6.66 \cdot 10^{-3}$ units/min., was determined in the case of CcO, while the time to reach pH 5.0 ($t_{pH\ 5.0}$, hours) ranged between 1.9 (Cb) to 3.5 (CcO).

A relative distinct behaviour was observed in the case of red cabbage juice, red beet juice and cucumber juice with onion juice added, in the sense of the slowdown of the metabolism objectified in the dynamics of the parameters that describe the process unfolding. The differences could be explained through the presence of some chemical constituents which can act as inhibitors on useful bacteria, like anthocyanins in the red cabbage, betacyanins in red beet, respectively constituent sulfides in the onion juice. According [14], sulfides, especially those with three or more sulfur atoms, apparently possess potent antimicrobial activity. However, concerning the batch with onion juice added the initial trend was attenuated after 6 hours of fermentation, the oils and their sulfides constituent showing weak antimicrobial activity ([15]).

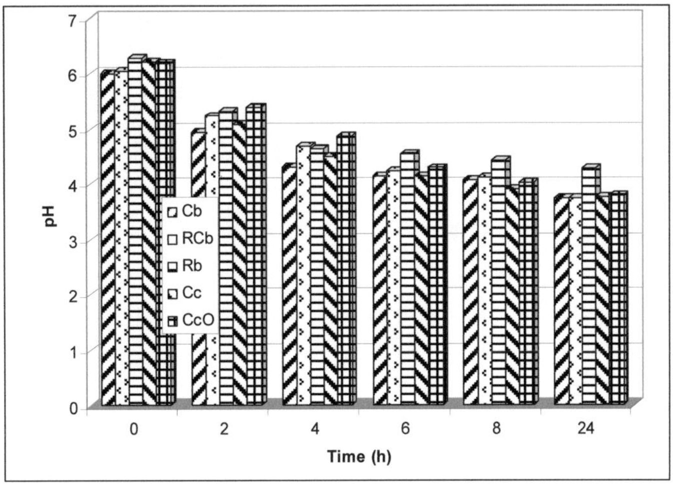


**Figure 5.** pH dynamics in vegetable juices obtained from different raw materials, during fermentation with *Lactobacillus acidophilus* LA-5

Referring to the red cabbage juice, although after 24 hours of fermentation the pH values were similar, the lactic acid content was lesser with about 1.5g/l compared with the white cabbage juice. This can be due to the amphoteric nature of the anthocyanins.

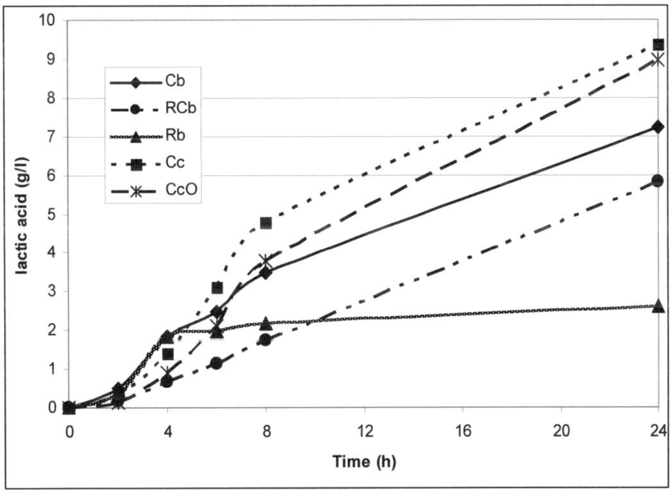


**Figure 6.** Lactic acid accumulation in vegetable juices obtained from different raw materials, during fermentation with *Lactobacillus acidophilus* LA-5

[16] studied the fermentation of cucumber juices with a 0.5%, 1% and 2% additions of the onion juices by *Lb. plantarum* CCM 7039. It was found that in the initial stages of fermentation, the presence of onion in the juices positively influenced lactic and acetic acid production. However, in further course of fermentation, slight inhibition effects of onion in the fermented juices were observed, especially at elevated onion/cucumber ratio.

The correlation between the biomass amount and the production of lactic acid (Figure 7) in the case of lactic acid fermentation of red beet juices with *Lactobacillus acidophilus* in the first 24 hours, was described using the Luedeking & Piret model [17]. According to this model, the instantaneous rate of lactic acid formation (dP/dt) can be related to the instantaneous rate of bacterial growth (dN/dt), and to the bacterial density (N), throughout fermentation at a given pH, by the expression:

$$dP/dt = \alpha \, dN/dt + \beta N$$

where the constants $\alpha$ and $\beta$ are determined by the pH of the fermentation.

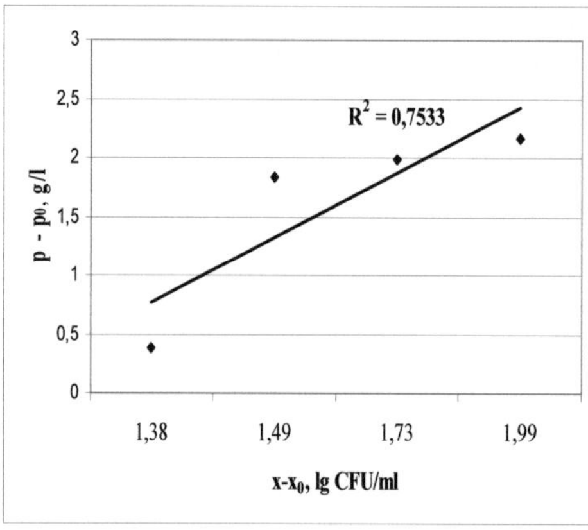


**Figure 7.** The correlation between the lactic acid production and viable cells count of *Lactobacillus acidophilus* LA-5 growing on red beet juices

A simplified presentation of the above model relates to the linear part of the equation which is presented as:

$$(p - p_0) = \alpha \, (x - x_0)$$

where $p_0$ and $p$ are the concentrations of lactic acid (g/l) initially and at time $t$, respectively, and $x_0$ and $x$ are the increases of the biomass (log CFU/mL) initially and at time $t$, respectively.

The R squared coefficient closed by the ideal value "1" ($R^2 = 0.9989$) in the case of the carrot juices fermented with *Lactobacillus acidophilus* LA-5 (data not shown) highlights a better linear correlation, respectively a strong connection between the lactic acid production and the lactic acid bacteria growth. Not the same situation has registered in the lactic acid fermentation of the red beet juices with the same strain. The highest value of the coefficient ($1 - R^2$) it is caused by the increase of the lactic acid amount in the first 4 hours, followed by a steady interval of evolution of this parameter. From the other hand, according [18], the deviations from the linear dependence are mostly caused by nutritive limitations of the substrates, and are related to the specific bacterial species. Not at least, the initial content of reducing sugars of the red beet, by 21.2g/l, could be limiting. However, taking into account the fact that the cucumber juice underwent a tumultuous fermentation although its content was only with 15.09% higher, it seems that other chemical constituents of the raw materials are responsible for the above mentioned differences.

The initial content of sugars in cucumber juice was situated at the maximum limit determined by [19], while in the case of the white cabbage juice was close to that one determined by [20].

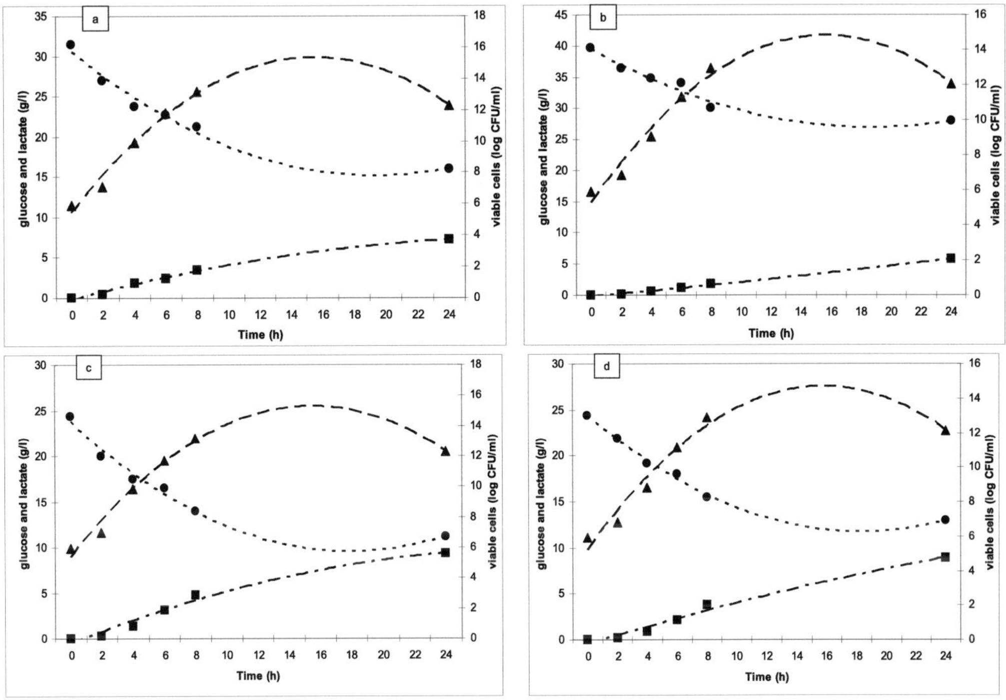


**Figure 8.** Correlation between the substrate consumption, lactate production and viable cells Cb (a), RCb (b), Cc (c) and CcO (d)

• - glucose, ■ - lactate, ▲ - viable cells (points - experimental data, smooth lines - predicted values)

The metabolization of the reducing sugars after 24h of lactic acid fermentation of vegetable juices with *Lb. acidophilus* LA-5 ranged between 26.66% (Rb) to 54.09% (Cc). Relative close values were obtained by other authors in lactic acid fermentation of vegetable juices. Thus, the utilization of sugar during fermentation in a mixture of beetroot juice and carrot juice and different content of brewer's yeast autolysate with *Lb. plantarum* A112 and with *Lb. acidophilus* NCDO 1748 varied from 19.4 to 24.1% ([21]).

The tested pure culture, routinely used for dairy products, was found to be capable of growing on pure vegetable juices without nutrients added. In the batches obtained from cabbage, respectively cucumber, the maximum volumetric productivity was determined after 8 hours as follows: 19.25x10$^{14}$ CFU/(l·h) for Cb, 11.9x10$^{14}$ CFU/(l·h) for RCb, 18.6x10$^{14}$ CFU/(l·h) for Cc and 10.25x10$^{14}$ CFU/(l·h) for CcO respectively.

The relationship between the growth of *Lactobacillus acidophilus*, the substrate metabolization and the lactic acid accumulation is shown in Figure 8. The prediction functions of the values of the analyzed parameters in all the samples were defined as polynomial, the R squared being very close to unit.

Correlating the number of viable cells with the dynamics of the lactic acid, the values were lower until 6 hours in the red cabbage juice and cucumber juice with onion juice added respectively. The differences were lessened in the next period of the process. However, the final yield of the lactic acid production was better in the sample CcO, by 0.78, comparatively with 0.7 in the sample Cc.

## 3.4. Effect of growth factors on the dynamics of the lactic acid fermentation of the carrot juices by Lactobacillus acidophilus LA-5

Kinetic parameters such as the time to reach pH 5.0 and the maximum rate of acidification are important in terms of the shelf life of the fermented vegetable juices. These ones were differently modified by the presence of the amino acids or of the yeast extract at the initial moment of fermentation. From Table 2 we deduced that a highest influence on both $t_{pH\ 5.0}$ and $v_{max}$ was exerted by cysteine, added to the juice in amount by 0.2% (w/v). Compared with the other supplements, the yeast extract had a relative good effect on the analyzed parameters. At the used concentrations, the behavior of valine and lysine seems to be unobservable from this point of view, excepting the poor effect of lysine on the maximum rate of acidification. Time to complete the fermentation ($t_{pH\ 4.2}$, h) ranged between 7.4 (YE) and 10.42 (Leu), trend that underline the statement that in the above mentioned experimental conditions *Lactobacillus acidophilus* growing faster.

| Kinetic parameter | Supplements[1)] | | | | | |
|---|---|---|---|---|---|---|
| | Cys_1 | Leu | Val | Lys | Cys_2 | YE |
| Time-decreasing of $t_{pH\ 5.0}$[2)] | 1.28 | 0.85 | 0 | 0 | 1.69 | 1.1 |
| Time-increasing of $v_{max}$[3)] | 0.82 | 0.84 | 0.98 | 1.05 | 1.1 | 1.05 |

[1)]The notations used for the samples are in agreement with the nutrients added, as follows: L-Cysteine (Cys_1 sample with 0.1% cysteine and Cys_2 sample with 0.2% cysteine), L-Leucine (Leu), L-Valine (Val), L-Lysine (Lys) and yeast extract (YE) respectively

[2)]The data were obtained by dividing the kinetic parameters of the control to the corresponding values of the samples

[3)]The data were obtained by dividing the kinetic parameters of the samples to the corresponding values of the control

Subunit or null values should be considered as lack of effect on the analyzed parameters.

**Table 2.** Effect of supplements on the kinetic parameters

MRS broth used for lactobacilli enumeration often incorporates L-cysteine to improve the recovery of these ones, especially due to the fact that *Lactobacillus acidophilus* LA-5 is micro-aerophilic. Cysteine, a sulfur containing amino acid, could provide amino nitrogen as a growth factor while reducing the redox potential. [22] reported that the incubation time to reach a pH of 4.5 was greatly affected by the addition of cysteine in yogurts made with different commercial cultures, although their viability was adversely affected in function of the amount of supplement and the type of the starter culture.

Lactic acid is the major metabolite of *Lactobacillus acidophilus,* influencing both the preservation of the fermented products and the sensorial characteristics of these ones. The effect of the amino acids and of the yeast extract on the dynamics of the lactic acid, assessed against the control, is underlined through the data from Table 3. The buffering capacity of the amino acids prevented a direct proportionality between the pH values and the lactic acid content.

| Time, h | Cys_1 | Leu | Val | Lys | Cys_2 | YE |
|---|---|---|---|---|---|---|
| 2 | 8.737864 | -12.6214 | 21.52778 | -29.8611 | 107.6923 | 15.38462 |
| 4 | 17.66784 | -23.6749 | -5.55556 | -2.77778 | 28.125 | 12.5 |
| 6 | 16.98113 | -1.50943 | 11.71717 | 3.636364 | 1.818182 | 5.454545 |
| 8 | 20.63492 | -1.5873 | 8.571429 | 1.428571 | -11.1111 | 15.87302 |
| 24 | 0.925926 | -0.92593 | 5.076142 | 4.568528 | -14.433 | 11.34021 |

**Table 3.** Time-increasing of lactic acid during 24h of lactic acid fermentation of carrot juices by *Lactobacillus acidophilus* LA-5

The values were expressed in percents by reporting the difference between sample and control to the control, at the same moment of time

Negative values shows that for the corresponding interval of time the supplements had not influence on the lactic acid production at the used levels.

Analyzing the whole process, only the samples with a minimum amount of cysteine added and those with yeast extract have been a great effect on the time-increasing of lactic acid. At the other opposite were found the samples with leucine added, this amino acid with non-polar hydrophobic chains clumsying the fermentation. From the viewpoint of increase the lactic acid content in the final stages of the process, the supplementation of the carrot juices with 0.2% (w/v) cysteine seems to be undesirable.

The beneficial effect of cysteine on the lactic acid accumulation in vegetable juices can occur due to its buffering capacity, which may diminish the toxic effects of organic acids on lactobacilli. Referring to the yeast extract, which contains more cell growth factors, being used generally as a source of assimilable nitrogen, vitamins and minerals, its influence at the level of 0.2%(w/v) on the time-increasing of lactic acid could be characterized as moderate. If some authors reported different maximum lactic acid concentration in media supplemented with yeast extract, several possible explanations include the strain of microorganism, the chemical composition of the substrate, the fermentation system, and generally the conditions employed during fermentation ([12]).

Effect of supplements on the performance of lactic acid production was evaluated based on lactic acid productivity and lactic acid yield, respectively on glucose ratio (Table 4).

The previous conclusion referring to the positive influence of the yeast extract and cysteine (in minimum amount) on the development of the lactic acid fermentation of vegetable juices is confirmed by the data from Table 4. Good values of lactic acid productivity were obtained after 24 h of fermentation in the samples with valine and lysine added, although in these ones the substrate consumption seems to be directed to the increasing of biomass, aspect emphasized by the average values of the lactic acid yield.

| Parameter | Cys_1 | Leu | Val | Lys | Cys_2 | YE |
|---|---|---|---|---|---|---|
| Lactic acid yield[2)] | 1.1 | 0.85 | 0.88 | 0.79 | 0.85 | 1.15 |
| Lactic acid productivity[3)] | 1.01 | 0.99 | 1.06 | 1.05 | 0.7 | 1.13 |
| Glucose conversion ratio[4)] | 1.1 | 0.9 | 1.2 | 1.05 | 0.92 | 1.25 |

[1)]The data from the table were obtained by dividing the corresponding values for the samples to those of the control
[2)]Lactic acid yield was calculated by dividing the amount of lactic acid produced to the amount of glucose consumed
[3)]Lactic acid productivity was defined as the amount of lactic acid produced per hour per liter
[4)]Glucose conversion ratio was calculated by dividing the amount of glucose consumed to the initial amount of glucose.

**Table 4.** Effect of supplements on lactic/acetic acid production after 48 h of fermentation[1)]

The effect of supplements (amino acids and yeast extract) on the ascorbic acid dynamics is shown in Figure 9. L-Ascorbic acid (AA), also known as vitamin C, is a representative water-soluble vitamin possessing a variety of biological, pharmaceutical, and dermatological functions; it promotes collagen biosynthesis, provides photoprotection, causes melanin reduction, scavenges free radicals, and enhances immunity ([23]).

Due to the heat treatment applied with a view to destroy the epiphytic microbiota of the fresh vegetable juices, the losses occurred in the ascorbic acid content represented about 65%.

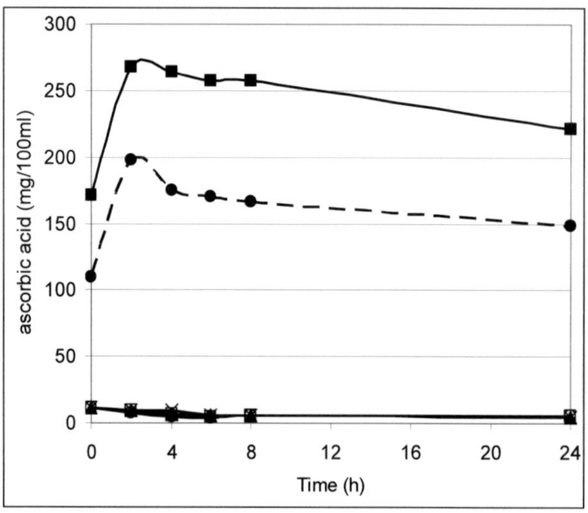


**Figure 9.** Time-course (0-24h) of the relative levels of ascorbic acid
(•Cys_1, ▪Cys_2, ○YE, □Leu, ▲Val, x Lis). The data shown are average values of two independent replicate experiments

The presence of ascorbic acid into vegetable juices submitted to fermentation by probiotic bacteria, especially by *Lactobacillus acidophilus* strains, is desired not only from the nutritional point of view, but also due to the fact that it could promote anaerobic conditions, acting as an oxygen scavenger. [24] have shown also that the fruit juices may be an alternative vehicle for the incorporation of probiotics because they are rich in nutrients and do not contain starter cultures that compete for nutrients with probiotics. Furthermore, fruit juices are often supplemented with oxygen scavenging ingredients such as ascorbic acid, thus promoting anaerobic conditions.

L-cysteine, a sulfur-containing amino acid known as a powerful reducing agent, caused the reduction of dehydroascorbic acid to ascorbic acid, which led a different behavior of the samples Cys_1 and Cys_2 by the others. The increase of this parameter was by 80% and 56.4% respectively, after 2h from the initial moment of fermentation. Subsequently, the analyzed parameter had the same diminishing tendency as in the other batches.

The losses occurred after 24h of lactic acid fermentation of carrot juices with *Lactobacillus acidophilus* LA-5 ranged from 48.39% (YE) to 61.9% (control). The possible reason could be the oxygen traces that cause the chemical oxidation of the vitamin C.

In order to evaluate the probiotic feature of the vegetable juices, the study of the effect of supplements on *Lactobacillus acidophilus* growth is from overwhelming importance, both during the lactic acid fermentation and during the storage of the final products.

Between the analyzed samples, those with yeast extract and 0.1% (w/v) cysteine added registered a higher increase of the number of viable cells till 14.4 - 14.5 log CFU/ml in the first 8h of the process. Concerning the yeast extract, the most possible explanation is due to an enhanced availability of minerals, which are growth promoters for *L. acidophilus* ([25]), while discussing the factors that affect the activity of endogenous probiotics, (26) mentioned that some of the growth promoters in cow milk were apparently cysteine-containing peptides.

Referring to the juices with leucine, lower values were determined comparative with the control during 24h, while in the samples with 0.2% (w/v) cysteine added the trend of the survival of lactobacilli was slow down in the period 6 - 8h, the level being by 13.5 and 13.6 log CFU/ml respectively. The last observation agrees with this one of [27], which have shown that the increasing of cysteine concentration improved the viability of *B. bifidum* in bio-yogurt, although it had no important effect on the viability of *Lactobacillus acidophilus.*

The batches supplemented with valine and lysine had occupied an intermediate position, the growth until 14.2 log CFU/ml after 8h of fermentation making from the utilization of these amino acids a promising variant in the future, with a view to optimize the conditions of the process unfolding. In the period 8 - 24h the number of viable cells decreased, as result of the lack of tolerance at lower pH of the analyzed strain.

The correlation between the most important parameters of the lactic acid fermentation of the carrot juices with *Lactobacillus acidophilus* LA-5 were evaluated using Pearson correlation analysis (significance level $p < 0.01$; confidence level of 99%).

| Analytical variables | pH | lactic acid | glucose | viable cells | glycine | ascorbic acid |
|---|---|---|---|---|---|---|
| pH | 1 | -0.889** | 0.829** | -0.940** | 0.099* | -0.184* |
| lactic acid | | 1 | -0.891** | 0.843** | -0.201* | 0.016* |
| glucose | | | 1 | -0.789** | 0.093* | 0.084* |
| viable cells | | | | 1 | -0.061* | 0.066* |
| glycine | | | | | 1 | -0.103* |
| ascorbic acid | | | | | | 1 |

** Correlation is significant at the 0.01 level (2-tailed)
* Not significant

**Table 5.** The Pearson coefficients for the experimental batches

The correlations are strong between pH and lactic acid, respectively pH and glucose, while a very strong relationship pH - viable cells could be considered (Table 5). A non-existent relationship between ascorbic acid / amino acids content (expressed as glycine) and the other analyzed parameters was determined.

A firm correlation between glucose and lactic acid was expected, but on the one hand it is known that the practical yield of sugars conversion to lactic acid of the strains of the group *Lb. acidophilus* is about 85%, while on the other hand the analysis does not include supplementary data referring to other factors that might be involved in the dynamics of the lactic acid fermentation of vegetable juices.

Factor Analysis (FA) is a multidimensional statistic method whose purpose is the analysis of the structure of mutual dependences of variables. The method is similar to the Principal Component Analysis (PCA) with the exception of the factor weights that are scaled ([28]).

Applying FA to the experimental data, the analytical variables were reduced to two principal components, which accounted for 59.72% (PC1) and respectively 18.95% (PC2) from the total variance. According to the component matrix, respectively to the values of the component loadings expressed by the first second principal components (rotation method: Varimax with Kaiser normalization), the most notable variables were pH and lactic acid (equal loading values by 0.954). Higher values were obtained also for viable cells (loading 0.939) and glucose (loading 0.933).

The combination of PC1 and PC2 (Figure 10) underlined the lack of correlation between amino acids content / ascorbic acid and all the other parameters taking into account both control and supplemented samples. While PC1 affected the dependent and independent variables involved in the progress of the lactic acid fermentation of vegetable juices, respectively in their probiotic feature, PC2 separated the variables which contribute to the nutritional characteristics of the final products.

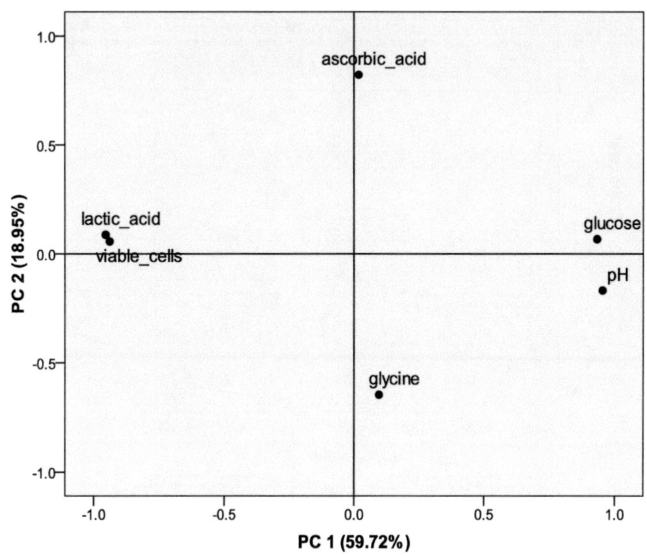


**Figure 10.** Component plot in rotated space

Applying PCA to the lactic acid fermentation of cabbage juices with various microorganisms, [29] established that the original 7 analytical variables were reduced also to 2 independent components that explained 88.2% from total variance of input data (PC1 66.9% and PC2 21.3%).

Cluster Analysis (CA) is a statistic method whose purpose is to join data into clusters with a view to increase their withingroup homogeneity. Usually, the FA is considered the first step of CA, with a view to reduce the data dimensionality. In order to better distinguish among experimental samples, the cluster method of the nearest neighbour was used. The distances between objects were measured as squared Euclidean distance. K-Means Cluster Analysis divided the experimental data into three groups, characterized by similar analytical properties, as follows:

- cluster 1: all the carrot juices (control samples and the batches with amino acids and yeast extract added) at the initial moment of fermentation, respectively at $2^{th}$ h of fermentation. Supplementary, this cluster included the control and the sample with leucine at $4^{th}$ h of fermentation (C_4 and Leu_4);
- cluster 2: all the carrot juices at $24^{th}$ h of fermentation and the sample with lysine added at $8^{th}$ h of fermentation ;
- cluster 3: the carrot juices with leucine and lysine added, respectively the control, at $6^{th}$ and $8^{th}$ h of fermentation (Leu_6, Leu_8, Lys_6, Lys_8, C_6, C_8); the carrot juices with cysteine, valine, respectively yeast extract from $4^{th}$ to $8^{th}$ h of fermentation (Cys_4 - Cys_8, (Val_4 - Val_8), (YE_4 - YE_8).

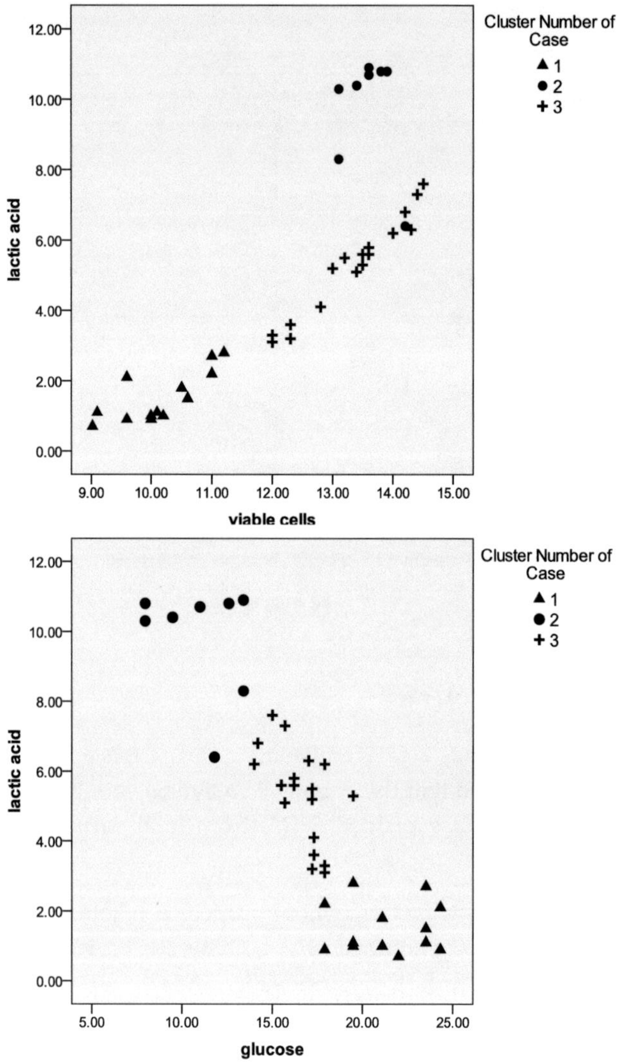


**Figure 11.** Clusters plotting in coordinate of two selected variables: lactic acid - viable cells and lactic acid - glucose

The clusters in axes of two selected variables (Figure 11) denote that the samples from the first cluster were marked with a higher content of substrate, null or very lower lactate amount and pH values more than 5. The corresponding time was both 0 and 2h (C, Val, Leu, and YE) or the entire interval 0 - 4h (Leu and control).

The samples at the final moment of fermentation and those with lysine after 8h of the process were included in the second cluster, characterized through lower or average values of glucose content, higher lactic acid amount and pH values close to 4.2. This cluster marks the achievement of the optimum characteristics of the lactic acid fermented products.

The samples included in the third cluster best describes a vigorous process, being characterized through average values of the main parameters involved in the dynamics of the lactic acid fermentation of vegetable juices.

The usefulness of the methods of statistical analysis is underlined by a lot of applications of CA that could be reported: in evaluation of analytical and sensory characteristics of vegetable juices ([28], [29]), in distinguishing between wines aged a different number of months ([30]).

## 4. Further research

The importance of consuming probiotic foods for the improvement of the quality of life increasingly more in the last years, being underlined by the scientific literature. The diversification of the market from this point of view could be strong correlated with the increasing of the life expectancy worldwide.

Our further researches are needed in order to optimize the level of nutrients (individually and in combination) and in the same time their influence on growth and viability of probiotics (in particular of *Lactobacillus acidophilus*, single strain or in combination with other probiotics), not only during fermentation but especially during the storage of the final products.

## 5. Conclusions

Different vegetable juices are suitable and alternative food matrices for the production of functional foods with *Lactobacillus acidophilus* LA-5, a probiotic strain which is not present in the epiphytic microbiota. Although some differences between the growths trends were determined, all the analyzed vegetables could be considered proper in order to obtain lactic acid fermented juices with a higher self-life. Application of Principal Component Analysis selected the most important parameters from analytical point of view: pH, lactic acid, biomass and viable cells, while the Cluster Analysis divided the experimental variables into three groups.

## Author details

Lavinia Claudia Buruleanu, Magda Gabriela Bratu,
Iuliana Manea, Daniela Avram and Carmen Leane Nicolescu
*Department of Food Engineering, Faculty of Environmental Engineering and Biotechnology,*
*Valahia University of Targoviste, Romania*

## Acknowledgement

The research was funded by Executive Unit for Financing Higher Education, Research, Development and Innovation (UEFISCDI) in the frame of the Project PN-II-ID-PCE-2008-2 (ID_1359). The microorganisms were kindly provided by Chr. Hansen, Romania.

## 6. References


[1] Granato D, Branco GF, Nazzaro F, Cruz AG, Faria JAF (2010) Functional Foods and Nondairy Probiotic Food Development: Trends, Concepts and Products. Compr. Rev. Food Sci. Food Saf. 9: 292-302.

[2] Holzapfel WH, Haberer P, Geisen R, Björkroth J, Schillinger U (2001)Taxonomy and Important Features of Probiotic Microorganisms in Food and Nutrition. Am. J. Clin. Nutr. 73(suppl): 365S-73S.

[3] Saito T (2004) Selection of Useful Probiotic Lactic Acid Bacteria from the *Lactobacillus acidophilus* Group and their Applications to Functional Foods. Animal Sci. J. 75:1-13.

[4] Pineiro M, Stanton C (2007) Probiotic Bacteria: Legislative Framework - Requirements to Evidence Basis. J. Nutr. 850S-853S.

[5] Kailasapathy K, Chin J (2000) Survival and Therapeutic Potential of Probiotic Organisms with Reference to *Lactobacillus acidophilus* and *Bifidobacterium* spp. Immun. Cell Biol. 78: 80–88.

[6] Karovičová J, Kohajdová Z (2003) Lactic Acid Fermented Vegetable Juices. Hort Sci.(Prague), 30 (4): 152-158.

[7] Klewicka E (2010) Antimutational Activity of Beetroot Juice. Food Technol. Biotechnol. 48(2): 229–233

[8] Altiok D (2004) Kinetic Modelling of Lactic Acid Production from Whey, Ph D Thesis, Izmir Institute of Technology, Turkey

[9] Kious, Jessica J. Lactobacillus and Lactic Acid Production [online]. Energy Research Undergraduate Laboratory Fellowship Program (ERULF), National Renewable Energy Laboratory, Colorado, 2000. Portable Document Format. Available from Internet: http://www.nrel.gov/docs/gen/fy01/NN0017.pdf.

[10] Agarwal L, Dutt K, Meghwanshi GK, Saxena RK (2008) Anaerobic Fermentative Production of Lactic Acid using Cheese Whey and Corn Steep Liquor. Biotechnol. Lett. 30: 631–635.

[11] Dave RI, Shah NP (1997) Viability of Yoghurt and Probiotic Bacteria in Yoghurts Made from Commercial Starter Cultures. Int. Dairy J. 7: 31-41.

[12] Amrane A (2005) Analysis of The Kinetics of Growth and Lactic Acid Production for *Lactobacillus Helveticus* Growing on Supplemented Whey Permeate. J. chem. technol. biotechnol. 80: 345-352.

[13] Kotzamanidis C, Roukas T, Skaracis G (2002) Optimization of Lactic Acid Production from Beet Molasses by Lactobacillus delbrueckii NCIMB 8130. World j. microbiol. biotechnol. 18: 441-448.

[14] Kim JW, Kim YS, Kyung KH (2004) Inhibitory Activity of Essential Oils of Garlic and Onion against Bacteria and Yeasts. J. food prot. 67(3): 499-504.

[15] Kim JW, Huhi JE, Kyung SH, Kyung KH (2004) Antimicrobial Activity of Alk(En)Yl Sulfides Found in Essential Oils of Garlic and Onion. Food sci. biotechnol. 13(2): 235-239.

[16] Kohajdová Y, Karovičová J, Greifová M (2007) Analytical and Organoleptic Profiles of Lactic Acid-Fermented Cucumber Juice with Addition of Onion Juice. J. food nutr. res. 46(3): 105-111.

[17] Rakin M, Baras J, Vukašinović M, Maksimović M (2004) The Examination of Parameters for Lactic Acid Fermentation and Nutritive Value of Fermented Juice of Beetroot, Carrot an Brewer's Yeast Autolysate, J. serbian chem. soc. 69 (8-9): 625-634.

[18] Luedeking R, Piret EL (1959) A Kinetic Study of the Lactic Acid Fermentation. Batch Process at Controlled pH, J. biochem. microbiol. technol. eng., 1: 393-412.

[19] Amrane A, Prigent Y (1999) Analysis of Growth and Production Coupling for Batch Cultures of Lactobacillus Helveticus with Help of Anunstructurated Model. Process biochemistry, 34: 1-10.

[20] Lu Z, Fleming HP, McFeeters RF (2002) Effects of Fruit Size on Fresh Cucumber Composition and the Chemical and Physical Consequences of Fermentation. J. food sci. 67(8): 2934-2939.

[21] Yoon KY, Woodams EE, Hang YD (2005) Production of Probiotic Cabbage Juice by Lactic Acid Bacteria. Bioresour. technol. 97(12): 1427-1430.

[22] Rakin MB, Baras JK, Vukašinović MS (2005) Lactic Acid Fermentation in Vegetable Juices Supplemented with Different Content of Brewer's Yeast Autolysate. Acta periodica technol. 36: 71-80.

[23] Dave RI (1998) Factors Affecting Viability of Yoghurt and Prbiotic Bacteria in Commercial Starter Cultures. PhD thesis, Victoria University of Technology, Werribee Campus, Victoria, Australia.

[24] Jang KI, Lee HG (2008) Influence of Acetic Acid Solution on Heat Stability of L-Ascorbic Acid. Food sci. biotechnol. 17: 637-641.

[25] Ding WK, Shah NP (2008) Survival of Free and Microencapsulated Probiotic Bacteria in Orange and Apple Juices. Int. food res. J. 15: 219-232.

[26] Lourens-Hattingh A, Viljoen BC (2001) Yogurt as Probiotic Carrier Food. Int. dairy j. 11: 1-17.

[27] Bezkorovainy A. (2001) Probiotics: Determinant of Survival and Growth in the Gut. Am. j. clin. nutr. 73(suppl): 399S-405S.

[28] Bari MR, Ashrafi R, Alizade M, Rofegarineghad L. (2009) Effects of Different Contents of Yogurt Starter/Probiotic Bacteria, Storage Time and Different Concentration of Cysteine on the Microflora Characteristics of Bio-yogurt. Res. j. biol. sci. 4: 137-142.

[29] Karovičová J, Kohajdová Z (2002) The Use of PCA, FA, CA for Evaluation of Vegetable Juices Processed by Lactic Acid Fermentation. Czech j. food sci. 20: 135-143.

[30] Kohajdová Z, Karovičová J. (2004) Optimisation of Method of Fermentation of Cabbage Juice. Czech j. food sci. 22: 39-50.

[31] Li Z, Pan QH, Jin ZM, He JJ, Liang NN, Duan CQ (2009) Evolution of 49 Phenolic Compounds in Shortly-Aged Red Wines Made from Cabernet Gernischt (*Vitis vinifera* L. cv.). Food sci. biotechnol. 18: 1001-1012.

Section 4

# Health Applications Purposes

Chapter 8

# Lactic Acid Bacteria as Probiotics: Characteristics, Selection Criteria and Role in Immunomodulation of Human GI Muccosal Barrier

Daoud Harzallah and Hani Belhadj

Additional information is available at the end of the chapter

http://dx.doi.org/10.5772/50732

## 1. Introduction

As it was reported by Chow (2002), the notion that food could serve as medicine was first conceived thousands of years ago by the Greek philosopher and father of medicine, Hippocrates, who once wrote: 'Let food be thy medicine, and let medicine be thy food'. However, during recent times, the concept of food having medicinal value has been reborn as 'functional foods'. The list of health benefits accredited to functional food continues to increase, and the gut is an obvious target for the development of functional foods, because it acts as an interface between the diet and all other body functions. One of the most promising areas for the development of functional food components lies in the use of probiotics and prebiotics which scientific researches have demonstrated therapeutic evidence. Nowadays, consumers are aware of the link among lifestyle, diet and good health, which explains the emerging demand for products that are able to enhance health beyond providing basic nutrition. Besides the nutritional valaes, ingestion of lactic acid bacteria (LAB) and their fermented foods has been suggested to confer a range of health benefits including immune system modulation, increased resistance to malignancy, and infectious illness (Soccol, et al., 2010). LAB were first isolated from milk. They can be found in fermented products as meat, milk products, vegetables, beverages and bakery products. LAB occur naturally in soil, water, manure, sewage, silage and plants. They are part of the microbiota on mucous membranes, such as the intestines, mouth, skin, urinary and genital organs of both humans and animals, and may have a beneficial influence on these ecosystems. LAB that grow as the adventitious microflora of foods or that are added to foods as cultures are generally considered to be harmless or even an advantage for human health. Since their discovery, LAB has been gained mush interest in various applications, as starter cultures in food and feed fermentations, pharmaceuticals, probiotics and as biological control agents. In food industry, LAB are widely used as starters to achieve favorable changes in texture, aroma, flavor and acidity (Leory and De Vuyst, 2004). However, there has been an important interest in using bacteriocin and/or other inhibitory substance producing LAB for non-fermentative biopreservation applications. Du to their antimicrobial and antioxidant activities some LAB strains are used in food biopreservation. However, LAB are generally regarded as safe (GRAS) to the consumer and during storage, they naturally dominate the microflora of many foods (Osmanağaoğlu and Beyatli, 1999; Parada et al., 2007). Many of the indications for probiotic activity have been obtained from effects observed in various clinical situations. Even, there are few strains that have officially gained the status of pharmaceutical preparation; each of these effects is gradually being supported by a number of clinical studies or human intervention trials, performed in a way that resembles the traditional pharmacological approach (placebo-controlled, double blind, randomized trials) and the strains used in these studies belong to different microbial species, but are mostly lactic acid bacteria (Mercenier et al, 2003).

## 2. LAB as probiotic agents

### 2.1. Overview of probiotics

The most tried and tested manner in which the gut microbiota composition may be influenced is through the use of live microbial dietary additions, as probiotics. In fact, the concept dates back as far as prebiblical ages. The first records of ingestion of live bacteria by humans are over 2,000 years old. However, at the beginning of this century probiotics were first put onto a scientific basis by the work of Metchnikoff (1908). He hypothesised that the normal gut microflora could exert adverse effects on the host and that consumption of 'soured milks' reversed this effect. The word ''probiotics'' was initially used as an anonym of the word ''antibiotic''. It is derived from Greek words pro and biotos and translated as ''for life''. The origin of the first use can be traced back to Kollath (1953), who used it to describe the restoration of the health of malnourished patients by different organic and inorganic supplements. Later, Vergin (1954) proposed that the microbial imbalance in the body caused by antibiotic treatment could have been restored by a probiotic rich diet; a suggestion cited by many as the first reference to probiotics as they are defined nowadays. Similarly, Kolb recognized detrimental effects of antibiotic therapy and proposed the prevention by probiotics (Vasiljevic and Shah, 2008) Later on, Lilly and Stillwell (1965) defined probiotics as "…microorganisms promoting the growth of other microorganisms". Following recommendations of a FAO/WHO (2002) working group on the evaluation of probiotics in food, probiotics, are live microorganisms that, when administered in adequate amounts, confer a health benefit on the host (Sanders, 2008; Schrezenmeir and De Vrese, **2001)**. The idea of health-promoting effects of LAB is by no means new, as Metchnikoff proposed that lactobacilli may fight against intestinal putrefaction and contribute to long life. Such microorganisms may not necessarily be constant inhabitants of the gut, but they should have a "…beneficial effect on the general and health status of man and animal"

| Probiotic | Human disease in which benefit is shown | Animal model in which benefit is shown |
|---|---|---|
| **Yeast** | | |
| *Saccharomyces boulardii* | *Clostridium difficile* infection | *Citrobacter rodentium*-induced colitis |
| **Gram-negative bacteria** | | |
| *Escherichia coli* Nissle 1917 | NA | DSS-induced colitis |
| **Gram-positive bacteria** | | |
| *Bifidobacteria bifidum* | NA | Rat model of necrotizing enterocolitis |
| *Bifidobacteria infantis* | IBS29 | NA |
| *Lactobacillus rhamnosus GG* (used with lactoferrin) | Sepsis in very low birth weight infants | NA |
| *Lactococcus lactis* (engineered to produce IL-10 or trefoil factors) | Crohn's disease | DSS-induced colitis and IL-$10^{-/-}$ mice (spontaneous IBD) |
| *Lactobacillus plantarum* 299v | Antibiotic-associated diarrhea | IL-$10^{-/-}$ mice (spontaneous IBD) |
| *Lactobacillus acidophilus* | NA | Visceral hyperalgesia 40 and *C. rodentium*-induced colitis |
| *Lactobacillus rhamnosus* | Pediatric antibiotic-associated diarrhea | – |
| *Lactobacillus casei* | NA | DNBS-induced colitis |
| *Bacillus polyfermenticus* | NA | DSS-induced colitis and TNBS-induced colitis |
| **Combination regimens** | | |
| *Lactobacillus rhamnosus GG* combined with *Bifidobacterium lactis* | Bacterial infections | NA |
| *Lactobacillus rhamnosus* combined with *Lactobacillus helveticus* | NA | *C. rodentium*-induced colitis, chronic stress, and early life stress |
| VSL#3 (*Lactobacillus casei, Lactobacillus plantarum, Lactobacillus acidophilus, Lactobacillus bulgaricus, Bifidobacterium longum, Bifidobacterium breve, Bifidocacterium infantis and Streptococcus thermophilus*) | Pouchitis and pediatric ulcerative colitis | DSS-induced colitis, IL-$10^{-/-}$ mice (spontaneous IBD; DNA only), and SAMP mouse model of spontaneous IBD |

Abbreviations: DNBS, dinitrobenzene sulfonic acid; DSS, dextran sodium sulfate; IL-10, interleukin 10; NA, not available; TNBS, trinitrobenzene sulfonic acid.

**Table 1.** Selected organisms that are used as probiotic agents (Gareau et al., 2010).

(Holzapfel et al., 2001; Belhadj et al., 2010). Other definitions advanced through the years have been restrictive by specification of mechanisms, site of action, delivery format, method, or host. Probiotics have been shown to exert a wide range of effects. The mechanism of action of probiotics (e.g, having an impact on the intestinal microbiota or enhancing immune function) was dropped from the definition to encompass health effects due to novel mechanisms and to allow application of the term before the mechanism is confirmed. Physiologic benefits have been attributed to dead microorganisms. Furthermore, certain mechanisms of action (such as delivery of certain enzymes to the intestine) may not require live cells. However, regardless of functionality, dead microbes are not probiotics (Sanders, 2008). In relation to food, probiotics are considered as "viable preparations in foods or dietary supplements to improve the health of humans and animals". According to these definitions, an impressive number of microbial species are considered as probiotics. (Holzapfel et al., 2001). For gastrointestinal ecosysteme, however, the most important microbial species that are used as probiotics are lactic acid bacteria (LAB) (Table 1).

## 2.2. Selection of probiotics

Many in vitro tests are performed when screening for potential probiotic strains. The first step in the selection of a probiotic LAB strain is the determination of its taxonomic classification, which may give an indication of the origin, habitat and physiology of the strain. All these characteristics have important consequences on the selection of the novel strains (Morelli, 2007). An FAO/WHO (2002) expert panel suggested that the specificity of probiotic action is more important than the source of microorganism. This conclusion was brought forward due to uncertainty of the origin of the human intestinal microflora since the infants are borne with virtually sterile intestine. However, the panel also underlined a need for improvement of in vitro tests to predict the performance of probiotics in humans. While many probiotics meet criteria such as acid and bile resistance and survival during gastrointestinal transit, an ideal probiotic strain remains to be identified for any given indication. Furthermore, it seems unlikely that a single probiotic will be equally suited to all indications; selection of strains for disease-specific indications will be required (Shanahan, 2003).

The initial screening and selection of probiotics includes testing of the following important criteria: phenotype and genotype stability, including plasmid stability; carbohydrate and protein utilization patterns; acid and bile tolerance and survival and growth; intestinal epithelial adhesion properties; production of antimicrobial substances; antibiotic resistance patterns; ability to inhibit known pathogens, spoilage organisms, or both; and immunogenicity. The ability to adhere to the intestinal mucosa is one of the more important selection criteria for probiotics because adhesion to the intestinal mucosa is considered to be a prerequisite for colonization (Tuomola et al., 2001). The table below (Table 2) indicates key creteria for sellecting probiotic candidat for commercial application, and figure 1 presents major and cardinal steps for sellecting probiotic candidats.

It is of high importance that the probiotic strain can survive the location where it is presumed to be active. For a longer and perhaps higher activity, it is necessary that the strain can proliferate and colonise at this specific location. Probably only host-specific microbial strains are able to compete with the indigenous microflora and to colonise the niches. Besides, the probiotic strain must be tolerated by the immune system and not provoke the formation of antibodies against the probiotic strain. So, the host must be immuno-tolerant to the probiotic. On the other hand, the probiotic strain can act as an adjuvant and stimulate the immune system against pathogenic microorganisms. It goes without saying that a probiotic has to be harmless to the host: there must be no local or general pathogenic, allergic or mutagenic/carcinogenic reactions provoked by the microorganism itself, its fermentation products or its cell components after decrease of the bacteria (Desai, 2008).

| General | Property |
|---|---|
| Safety criteria | Origin |
| | Pathogenicity and infectivity |
| | Virulence factors—toxicity, metabolic activity and intrinsic properties, i.e., antibiotic resistance |
| Technological criteria | Genetically stable strains |
| | Desired viability during processing and storage |
| | Good sensory properties |
| | Phage resistance |
| | Large-scale production |
| Functional criteria | Tolerance to gastric acid and juices |
| | Bile tolerance |
| | Adhesion to mucosal surface |
| | Validated and documented health effects |
| Desirable physiological criteria | Immunomodulation |
| | Antagonistic activity towards gastrointestinal pathogens, i.e., Helicobacter pylori, Candida albicans |
| | Cholesterol metabolism |
| | Lactose metabolism |
| | Antimutagenic and anticarcinogenic properties |

**Table 2.** Key and desirable criteria for the selection of probiotics in commercial applications (Vasiljevic and Shah, 2008).

When probiotic strains are selected, attributes important for efficacy and technological function must be assessed and a list of characteristics required for all probiotic functions is required. Basic initial characterization of strain identity and taxonomy should be conducted, followed by evaluation with validated assays both in studies of animal models and in controlled studies in the target host. In vitro assays are frequently conducted that have not been proved to be predictive of in vivo function. Technological robustness must also be determined, such as the strain's ability to be grown to high numbers, concentrated, stabilized, and incorporated into a final product with good sensory properties, if applicable, and to be stable, both physiologically and genetically, through the end of the shelf life of the product and at the active site in the host. Assessment of stability can also be a challenge, since factors such as chain length and injury may challenge the typical assessment of colony-forming units, as well as in vivo function (Sanders, 2008). Dose levels of probiotics should be based on levels found to be efficacious in human studies. One dose level cannot be assumed to be effective for all strains. Furthermore, the impact of product format on

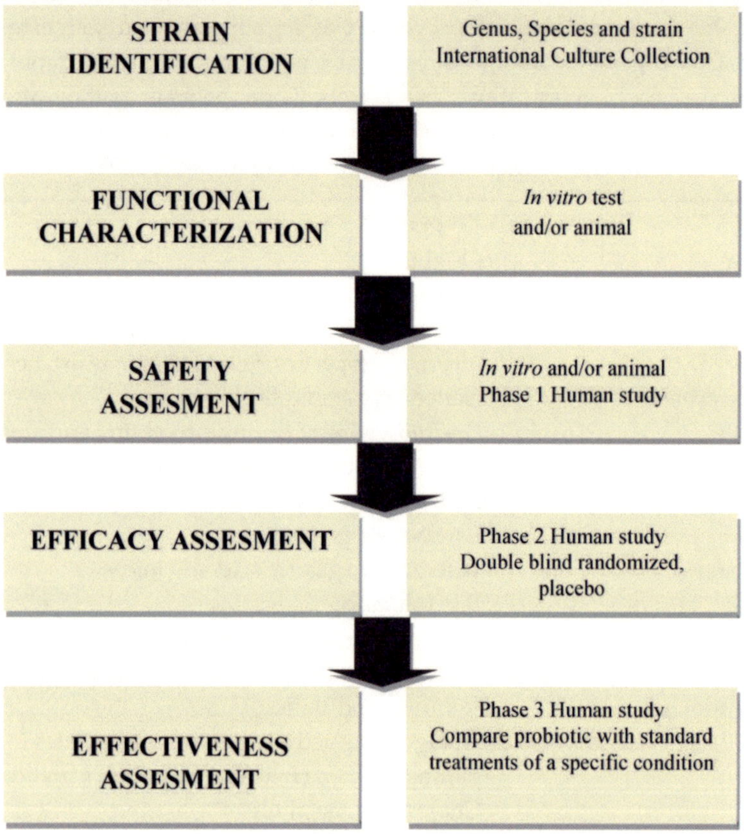


**Figure 1.** Scheme of the Guidelines for the Evaluation of Probiotics for Food Use. (Adapted from, Collado et al., 2009).

probiotic function has yet to be explored in depth. The common quality-control parameter of colony-forming units per gram may not be the only parameter indicative of the efficacy of the final product. Other factors, such as probiotic growth during product manufacture, coating, preservation technology, metabolic state of the probiotic, and the presence of other functional ingredients in the final product, may play a role in the effectiveness of a product. More research is needed to understand how much influence such factors have on in vivo efficacy **(Sanders, 2008).**

## 2.3. Potential mechanisms of action of probiotics

A wide variety of potential beneficial health effects have been attributed to probiotics (Table 3). Claimed effects range from the alleviation of constipation to the prevention of major life-threatening diseases such as inflammatory bowel disease, cancer, and cardiovascular incidents. Some of these claims, such as the effects of probiotics on the shortening of intestinal transit time or the relief from lactose maldigestion, are considered well-established, while others, such as cancer prevention or the effect on blood cholesterol levels, need further scientific backup (Leroy et al., 2008). The mechanisms of action may vary from one probiotic strain to another and are, in most cases, probably a combination of activities, thus making the investigation of the responsible mechanisms a very difficult and complex task. In general, three levels of action can be distinguished: probiotics can influence human

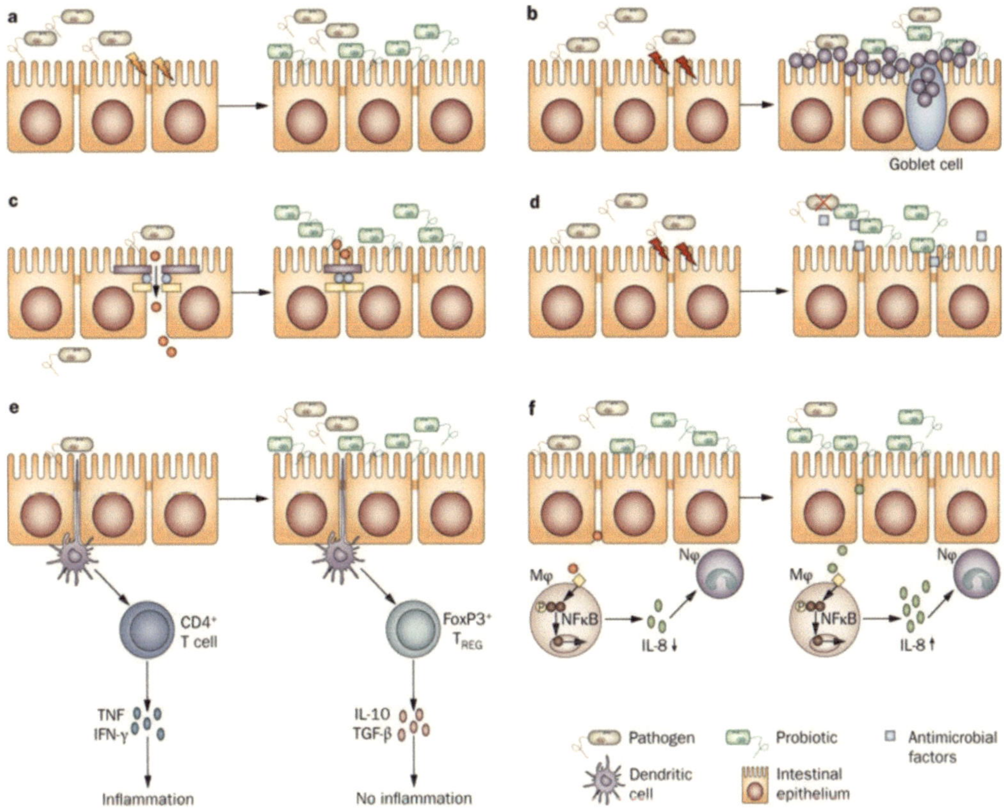


Probiotic organisms can provide a beneficial effect on intestinal epithelial cells in numerous ways. **a**: Some strains can block pathogen entry into the epithelial cell by providing a physical barrier, referred to as colonization resistance or **b**: create a mucus barrier by causing the release of mucus from goblet cells. **c**: Other probiotics maintain intestinal permeability by increasing the intercellular integrity of apical tight junctions, for example, by upregulating the expression of zona-occludens 1 (a tight junction protein), or by preventing tight junction protein redistribution thereby stopping the passage of molecules into the lamina propria. **d**: Some probiotic strains have been shown to produce antimicrobial factors. **e**: Still other strains stimulate the innate immune system by signaling dendritic cells, which then travel to mesenteric lymph nodes and lead to the induction of TREG cells and the production of anti-inflammatory cytokines, including IL-10 and TGF-β. f: Some probiotics (or their products) may also prevent (left-hand side) or trigger (right-hand side) an innate immune response by initiating TNF production by epithelial cells and inhibiting (or activitating) NFκB in Mϕ and dampening (or priming) the host immune response by influencing the production of IL-8 and subsequent recruitment of Nϕ to sites of intestinal injury. Abbreviations: Mϕ, macrophage; Nϕ, neutrophil; TREG cell, regulatory T cell. Reproduced from, *Gareau M. G., P. M. Sherman & W. A. Walker (2010) Nature Reviews Gastroenterology and Hepatology 7, 503-514.*

**Figure 2.** Potential mechanisms of action of probiotics.

| Health benefit | Proposed mechanism(s) |
|---|---|
| Cancer prevention | Inhibition of the transformation of pro-carcinogens into active carcinogens, binding/inactivation of mutagenic compounds, production of anti-mutagenic compounds, suppression of growth of pro-carcinogenic bacteria, reduction of the absorption of carcinogens, enhancment of immune function, influence on bile salt concentrations |
| Control of irritable bowel syndrome | Modulation of gut microbiota, reduction of intestinal gas production |
| Management and prevention of atopic diseases | Modulation of immune response |
| Management of inflammatory bowel diseases (Crohn's disease, ulcerative colitis, pouchitis) | Modulation of immune response, modulation of gut microbiota |
| Prevention of heart diseases/influence on blood cholesterol levels | Assimilation of cholesterol by bacterial cells, deconjugation of bile acids by bacterial acid hydrolases,<br>cholesterol-binding to bacterial cell walls, reduction of hepatic cholesterol synthesis and/or redistribution of cholesterol from plasma to liver through influence of the bacterial production of short-chain fatty acids |
| Prevention of urogenital tract disorders | Production of antimicrobial substances, competition for adhesion sites, competitive exclusion of pathogens |
| Prevention/alleviation of diarrhoea<br>caused by bacteria/viruses | Modulation of gut microbiota, production of antimicrobial substances, competition for adhesion sites, stimulation of mucus secretion, modulation of immune response |
| Prevention/treatment of *Helicobacter pylori* infections | Production of antimicrobial substances, stimulation of the mucus secretion, competition for adhesion sites, stimulation of specific and non-specific immune responses |
| Relief of lactose indigestion | Action of bacterial β-galactosidase(s) on lactose |
| Shortening of colonic transit time | Influence on peristalsis through bacterial metabolite production |

**Table 3.** Potential and established health benefits associated with the usage of probiotics (Leroy et al., 2008).

health by interacting with other microorganisms present on the site of action, by strengthening mucosal barriers, and by affecting the immune system of the host (Leroy et al., 2008), and the figure 2 shows the most important mechanisms by whiche probiotics exerce their action inside the gut.

# 3. Probiotics and gut health

## 3.1. Gut microbiota

The human gastrointestinal tract is inhabited by a complex and dynamic population of around 500-1000 of different microbial species which remain in a complex equilibrium. It has been estimated that bacteria account for 35–50% of the volume content of the human colon. These include *Bacteroides, Lactobacillus, Clostridium, Fusobacterium, Bifidobacterium, Eubacterium, Peptococcus, Peptostreptococcus, Escherichia and Veillonella*. The bacterial strains with identified beneficial properties include mainly *Bifidobacterium* and *Lactobacillus* species. The dominant microbial composition of the intestine have been shown to be stable over time during adulthood, and the microbial patterns are unique for each individual. However, there are numerous external factors that have potential to influence the microbial composition in the gut as host genetics, birth delivery mode, diet, age, antibiotic treatments and also, other microorganisms as probiotics. (Collado et al., 2009). The intestine is one of the main surfaces of contact with exogenous agents (viruses, bacteria, allergens) in the human body. It has a primary role in the host defense against external aggressions by means of the intestinal mucosa, the local immune system, and the interactions with the intestinal microbiota (resident and in transitbacteria). Gut microbiota influences human health through an impact on the gut defense barrier, immune function, nutrient utilization and potentially by direct signaling with the gastrointestinal epithelium (Collado et al., 2009). Only a limited fraction of bacterial phyla compose the major intestinal microbiota. In healthy adults, 80% of phylotypes belong to four major phylogenetic groups, which are the *Clostiridium leptum, Clostridium coccoides, Bacteroides* and *Bifidobacteria* groups. However, a large fraction of dominant phylotypes is subject specific. Also, studies have found that mucosal microbiota is stable along the distal gastrointestinal tract from ileum to rectum, but mucosa-associated microbiota is different from fecal microbiota. The difference has been estimated to be between 50–90%.

The intestinal microbiota is not homogeneous. The number of bacterial cells present in the mammalian gut shows a continuum that goes from $10^1$ to $10^3$ bacteria per gram of contents in the stomach and duodenum, progressing to $10^4$ to $10^7$ bacteria per gram in the jejunum and ileum and culminating in $10^{11}$ to $10^{12}$ cells per gram in the colon (Figure 3a). Additionally, the microbial composition varies between these sites. In addition to the longitudinal heterogeneity displayed by the intestinal microbiota, there is also a great deal of latitudinal variation in the microbiota composition (Figure 3b). The intestinal epithelium is separated from the lumen by a thick and physicochemically complex mucus layer. The microbiota present in the intestinal lumen differs significantly from the microbiota attached and embedded in this mucus layer as well as the microbiota present in the immediate

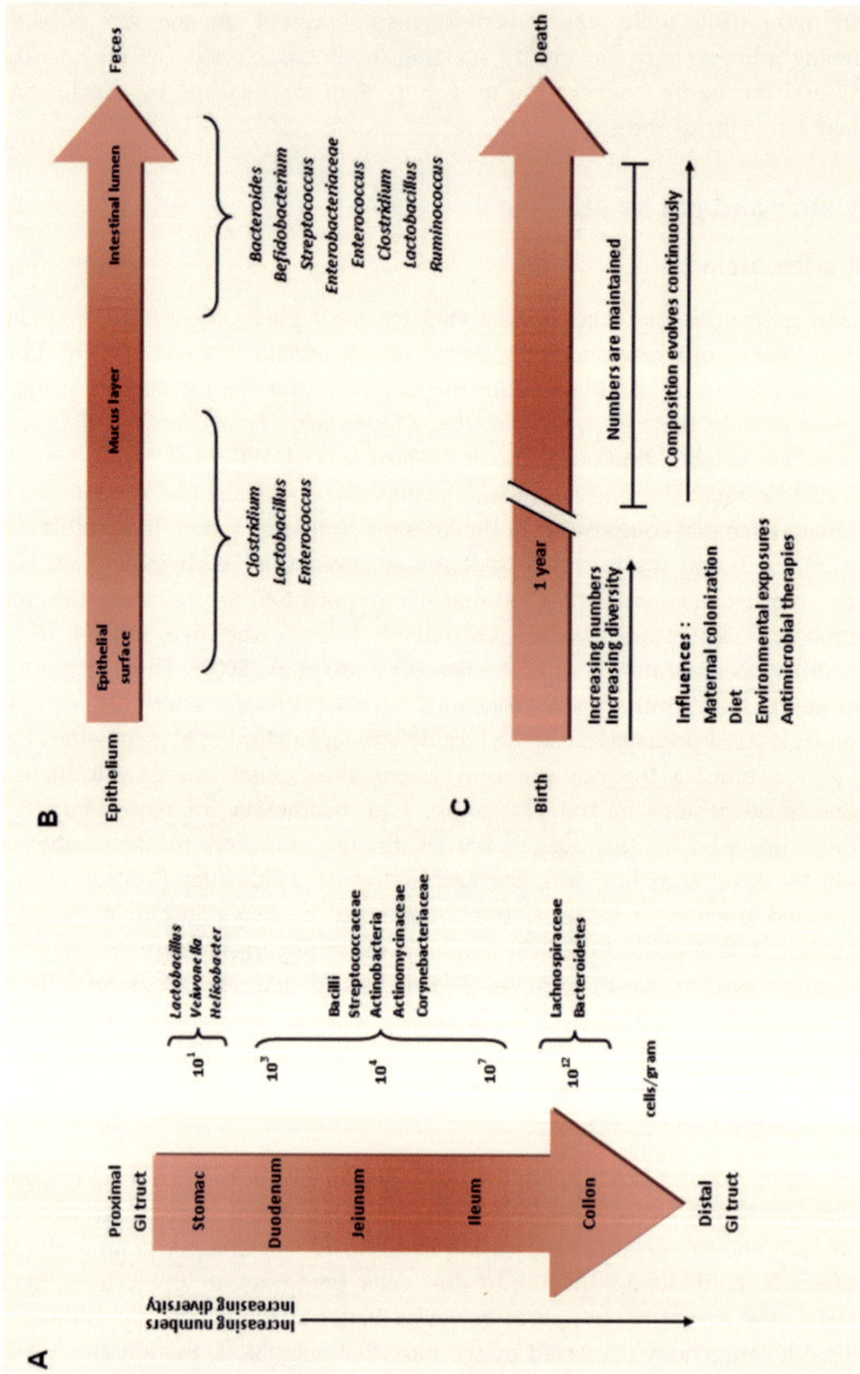


*a*: variations in microbial numbers and composition across the length of the gastrointestinal tract. *b*: longitudinal variations in microbial composition in the intestine. *c*: temporal aspects of microbiota establishment and maintenance and factors influencing microbial composition. (Sekirov et al., 2010).

**Figure 3.** Spatial and temporal aspects of intestinal microbiota composition.

proximity of the epithelium. For instance, *Bacteroides, Bifidobacterium, Streptococcus,* members of *Enterobacteriacea, Enterococcus, Clostridium, Lactobacillus,* and *Ruminococcus* were all found in feces, whereas only *Clostridium, Lactobacillus,* and *Enterococcus* were detected in the mucus layer and epithelial crypts of the small intestine (Sekirov et al., 2010). Colonization of the human gut with microbes begins immediately at birth (Figure 3c). Upon passage through the birth canal, infants are exposed to a complex microbial population. After the initial establishment of the intestinal microbiota and during the first year of life, the microbial composition of the mammalian intestine is relatively simple and varies widely between different individuals and also with time. However, after one year of age, the intestinal microbiota of children starts to resemble that of a young adult and stabilizes (Figure 3c) (Sekirov et al., 2010).

## 3.2. Survival and antagonism effects of probiotics in the gut

The intestinal epithelium is the largest mucosal surface in the human body, provides an interface between the external environment and the host. The gut epithelium is constantly exposed to foreign microbes and antigens derived from digested foods. Thus, the gut epithelium acts as a physical barrier against microbial invaders and is equipped with various elements of the innate defense system. In the gut, two key elements govern the interplay between environmental triggers and the host: intestinal permeability and intestinal mucosal defense. Resident bacteria can interact with pathogenic microorganisms and external antigens to protect the gut using various strategies.

According to the generally accepted definition of a probiotic, the probiotic microorganism should be viable at the time of ingestion to confer a health benefit. Although not explicitly stated, this definition implies that a probiotic should survive GI tract passage and, colonize the host epithelium. A variety of traits are believed to be relevant for surviving GI tract passage, the most important of which is tolerance both to the highly acidic conditions present in the stomach and to concentrations of bile salts found in the small intestine. These properties have consequently become important selection criteria for new probiotic functionality. In addition to tolerating the harsh physical-chemical environment of the GI tract, adherence to intestinal mucosal cells would be necessary for colonization and any direct interactions between the probiotic and host cells leading to the competitive exclusion of pathogens and/or modulation of host cell responses. Moreover, As enteropathogenic Escherichia coli are known to bind to epithelial cells via mannose receptors, probiotic strains with similar adherence capabilities could inhibit pathogen attachment and colonization at these binding sites and thereby protect the host against infection (Marco et al., 2006).

Probiotic bacteria can antagonize pathogenic bacteria by reducing luminal pH, inhibiting bacterial adherence and translocation, or producing antibacterial substances and defensins. One of the mechanisms by which the gut flora resists colonization by pathogenic bacteria is by the production of a physiologically restrictive environment, with respect to pH, redox potential, and hydrogen sulfide production. Probiotic bacteria decrease the luminal pH, as has been demonstrated in patients with ulcerative colitis (UC) following ingestion of the probiotic preparation VSL#3. In a fatal mouse Shiga toxin-producing *E. coli O157:H7* infection model, the probiotic Befidobacterium breve produced a high concentration of acetic acid, consequently lowering the luminal pH. This pH reduction was associated with increased animal survival (Ng et al., 2009).

Production of antimicrobial compounds, termed bacteriocins, by probiotic bacteria is also likely to contribute to their beneficial activity. Several bacteriocins produced by different species from the genus *Lactobacillus* have been described. The inhibitory activity of these bacteriocins varies; some inhibit taxonomically related Gram-positive bacteria, and some are active against a much wider range of Gram-positive and Gram-negative bacteria as well as yeasts and molds. For example, the probiotic *L. salivarius subsp. salivarius* UCC118 produces a peptide that inhibits a broad range of pathogens such as *Bacillus, Staphylococcus, Enterococcus, Listeria,* and *Salmonella* species. Lacticin 3147, a broad-spectrum bacteriocin produced by *Lactococcus lactis,* inhibits a range of genetically distinct *Clostridium difficile* isolates from healthy subjects and patients with IBD. A further example is the antimicrobial effect of *Lactobacillus* species on *Helicobacter pylori* infection of gastric mucosa, achieved by the release of bacteriocins and the ability to decrease adherence of this pathogen to epithelial cells (Gotteland et al., 2006). Probiotics can reduce the epithelial injury that follows exposure to *E. coli O157:H7* and *E. coli O127:H6*. The pretreatment of intestinal (T84) cells with lactic acid-producing bacteria reduced the ability of pathogenic *E. coli* to inject virulence factors into the cells or to breach the intracellular tight junctions. Adhesion and invasion of an intestinal epithelial cell line (Intestine 407) by adherent invasive *E. coli* isolated from patients with Crohn's disease (CD) was substantially diminished by co- or preincubation with the probiotic strain *E. coli* Nissle 1917 (Wehkamp et al., 2004 ; Schlee et al., 2007). These findings demonstrate that probiotics prevent epithelial injury induced by attaching-effacing bacteria and contributes to an improved mucosal barrier and provide a means of limiting access of enteric pathogens (Sherman et al., 2005).

## 4. Probiotics and the mucous layer

Most mucosal surfaces are covered by a hydrated gel formed by mucins. Mucins are secreted by specialized epithelial cells, such as gastric foveolar mucous cells and intestinal goblet cells, Goblet cells are found along the entire length of the intestinal tract, as well as other mucosal surfaces. Mucins, are abundantly core glycosylated (up to 80% wt/wt) and either localized to the cell membrane or secreted into the lumen to form the mucous layer (Turner, 2009). Of the 18 mucin-type glycoproteins expressed by humans, MUC2 is the predominant glycoprotein found in the small and large bowel mucus. The NH2- and COOH-termini are not glycosylated to the same extent, but are rich in cysteine residues that form intra- and inter-molecular disulfide bonds. These glycan groups confer proteolytic resistance and hydrophilicity to the mucins, whereas the disulfide linkages form a matrix of glycoproteins that is the backbone of the mucous layer (Ohland and MacNaughton, 2010). Although small molecules pass through the heavily glycosylated mucus layer with relative ease, bulk fluid flow is limited and thereby contributes to the development of an unstirred layer of fluid at the epithelial cell surface. As the unstirred layer is protected from convective mixing forces, the diffusion of ions and small solutes is slowed (Turner, 2009). This gel layer provides protection by shielding the epithelium from potentially harmful antigens and molecules including bacteria from directly contacting the epithelial cell layer, while acting as a lubricant for intestinal motility. Mucins can also bind the epithelial cell surface carbohydrates and form the bottom layer, which is firmly attached to the mucosa, whereas the upper layer is loosely adherent. The mucus is the first barrier that intestinal bacteria meet, and pathogens must penetrate it to reach the epithelial cells during infection (Ohland and MacNaughton, 2010).

Probiotics may promote mucus secretion as one mechanism to improve barrier function and exclusion of pathogens. In support of this concept, probiotics have been shown to increase mucin expression in vitro, contributing to barrier function and exclusion of pathogens. Several studies showed that increased mucin expression in the human intestinal cell lines Caco-2 (MUC2) and HT29 (MUC2 and 3), thus blocking pathogenic *E. coli* invasion and adherence. However, this protective effect was dependent on probiotic adhesion to the cell monolayers, which likely does not occur in vivo (Mack et al., 2003; Mattar et al., 2002). Conversely, another study showed that *L. acidophilus A4* cell extract was sufficient to increase MUC2 expression in HT29 cells, independent of attachment (Kim et al., 2008). Additionally, intestinal trefoil factor 3 (TFF3) is coexpressed with MUC2 by colonic goblet cells and is suggested to promote wound repair (Gaudier et al., 2005 ; Kalabis et al., 2006). However, healthy rats did not display increased colonic TFF3 expression after stimulation by VSL#3 probiotics (Caballero-Franco et al., 2007). Furthermore, mice treated with 1% dextran sodium sulfate (DSS) to induce chronic colitis did not exhibit increased TFF3 expression or wound healing when subsequently treated with VSL#3. This observation indicates that probiotics do not enhance barrier function by up-regulation of TFF3, nor are they effective at healing established inflammation. Therefore, use of current probiotics is likely to be effective only in preventing inflammation as shown by studies in animal models (Ohland and MacNaughton, 2010).

## 5. Interaction of probiotic bacteria with gut epithelium

The composition of the commensal gut microbiota is probably influenced by the combination of food practices and other factors like the geographical localization, various levels of hygiene or various climates. The host-microbe interaction is of primary importance during neonatal period. The establishment of a normal microbiota provides the most substantial antigenic challenge to the immune system, thus helping the gut associated lymphoid tissus (GALT) maturation. The intestinal microbiota contributes to the anti-inflammatory character of the intestinal immune system. Several immunoregulatory mechanisms, including regulatory cells, cytokines, apoptosis among others, participate in the control of immune responses by preventing the pathological processes associated with excessive reactivity. An interesting premise for probiotic physiological action is their capacity to modulate the immune system. Consequently, many studies have focused on the effects of probiotics on diverse aspects of the immune response. Following consumption of probiotic products, the interaction of these bacteria with intestinal enterocytes initiates a host response, since intestinal cells produce various immunomodulatory molecules when stimulated by bacteria (Delcenseri et al., 2009). Furthermore, the indigenous microbiota is a natural resistance factor against potential pathogenic microorganisms and provides colonization resistance, also known as gut barrier, by controlling the growth of opportunistic microorganisms. It has been suggested that commensal bacteria protect their host against microbial pathogens by interfering with their adhesion and toxic effects (Myllyluoma, 2007).

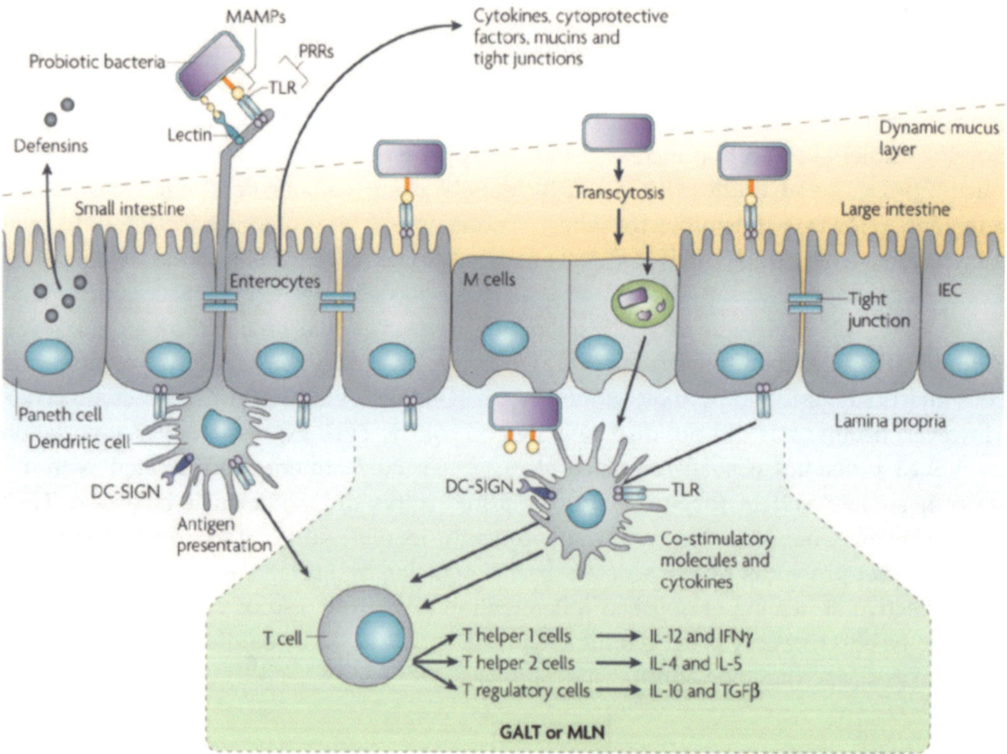


A fraction of ingested probiotics are able to interact with intestinal epithelial cells (IECs) and dendritic cells (DCs), depending on the presence of a dynamic mucus layer. Probiotics can occasionally encounter DCs through two routes: DCs residing in the lamina propria sample luminal bacterial antigens by passing their dendrites between IECs into the gut lumen, and DCs can also interact directly with bacteria that have gained access to the dome region of the gut-associated lymphoid tissue (GALT) through specialized epithelial cells, termed microfold or M cells. The interaction of the host cells with microorganism-associated molecular patterns (MAMPs) that are present on the surface macromolecules of probiotic bacteria will induce a certain molecular response. The host pattern recognition receptors (PRRs) that can perceive probiotic signals include Toll-like receptors (TLRs) and the C type lectin DC-specific intercellular adhesion molecule 3-grabbing non-integrin (DC-SIGN). Some molecular responses of IECs depend on the subtype of cell, for example, Paneth cells produce defensins and goblet cells produce mucus. Important responses of DCs against probiotics include the production of cytokines, major histocompatibility complex molecules for antigen presentation, and co-stimulatory molecules that polarize T cells into T helper or CD4+CD25+ regulatory T cells in the mesenteric lymph nodes (MLNs) or subepithelial dome of the GALT. IFNγ, interferon-γ; IL, interleukin; TGFb; transforming growth factor-β. Reproduced from: *S. Lebeer, J. Vanderleyden & S. C. J. De Keersmaecker (2010). Host interactions of probiotic bacterial surface molecules: comparison with commensals and pathogens. Nature Reviews Microbiology 8, 171-184.*

**Figure 4.** Interaction of probiotic bacteria with IECs and DCs from the GALT.

The tight epithelial cell barrier forms the another line of defence between the gut luminal contents and the host. Epithelial cells lining the gastrointestinal tract are able to respond to infection by initiating either nonspecific or specific host-defence response (Kagnoff and Eckmann 1997, Strober 1998). Bacterial adhesion to the host cell or recognition by the host cell is often an essential first stage in the disease process. A wide range of gastrointestinal cell surface constituents, such as several glygoconjucates, can serve as receptors for bacterial adherence (Servin and Coconnier 2003, Pretzer et al., 2005). Furthermore, epithelial cells express constitutively host pattern recognition receptors (PRRS), such as Toll-like receptors (TLR). These are a family of transmembrane receptors that recognize repetitive patterns, i.e. the pathogen-associated molecular patterns present in diverse microbes, including gram-positive and gram-negative bacteria (Bäckhed and Hornef 2003, Takeda et al., 2003). TLRs are also found on innate immune cells, such as dendritic cells and macrophages (Vinderola et al., 2005). TLR4 recognizes lipopolysaccharide and gram-negative bacteria, while TLR2 recognizes a variety of microbial components, such as peptidoglycan and lipoteichoic acids, from gram-positive bacteria (Abreu 2003, Matsuguchi et al., 2003, Takeda et al., 2003). Also, several other TLRs with specific actions are known, such as TLR5, which responds to the bacterial flagella (Rhee et al., 2005), and TLR9, which is activated by bacterially derived short DNA fragments containing CpG sequences (Pedersen et al., 2005). Other known recognition receptors are nucleotide-binding oligomerization domain proteins, which recognize both gram-positive and gram-negative bacteria. They are located in cell cytoplasm and are implicated in the induction of defensins. Increased epithelial barrier permeability is frequently associated with gastrointestinal disorders contributing to both disease onset and persistence (Lu and Walker 2001, Berkes 2003). The gatekeeper of the paracellular pathway is the tight junction, which is an apically located cell-cell junction between epithelial cells. The tight junction permits the passage of small molecules, such as ions, while restricting the movement of large molecules, such as antigens and microorganisms, which can cause inflammation. The integral membrane protein family, which are mainly claudins, occluding and zonula occludens 1, are implicated in the formation of the paracellular channels (Berkes et al., 2003).

## 6. Origine and safety of probiotics

An old dogma of probiotic selection has been that the probiotic strains should be of "human origin". One may argue that from evolutionary point of view, describing bacteria to be of human origin does not make much sense at all. The requirement for probiotics to be of human origin relates actually to the isolation of the strain rather than the "origin" itself. Usually, the strains claimed to be "of human origin" have been isolated from faecal samples of healthy human subjects, and have therefore been considered to be "part of normal healthy human gut microbiota". In reality the recovery of a strain from a faecal sample does not necessarily mean that this strain is part of the normal microbiota of this individual, since microbes passing the GI tract transiently can also be recovered from the faecal samples (Forssten et al., 2011). In practice it is impossible to know the actual origin of the probiotic strains, regardless of whether they have been isolated from faecal samples, fermented dairy products or any other source for that matter. Isolation of a strain from faeces of a healthy individual is also not a guarantee of the safety of the strain—such a sample will also always contain commensal microbes which can act as opportunistic pathogens, or even low levels of true pathogens, which are present in the individual at sub-clinical levels. Therefore, it has been recommend that instead of concentrating on the first point of isolation, the selection processes for new potential probiotic strains should mainly focus on the functional properties of the probiotic strains rather than the "origin" (Forssten et al., 2011; Ouwehand and Lahtinen 2008).

As viable, probiotic bacteria have to be consumed in large quantities, over an extended period of time, to exert beneficial effects; the issue of the safety of these microorganisms is of primary concern (Leroy et al., 2008). Until now, reports of a harmful effect of these microbes to the host are rare. However, many species of the genera Lactobacillus, Leuconostoc, Pediococcus, Enterococcus, and Bifidobacterium were isolated frequently from various types of infective lesions. According to Gasser (1994), L. rhamnosus, L. acidophilus, L. plantarum, L. casei, Lactobacillus paracasei, Lactobacillus salivarius, Lactobacillus lactis, and Leuconostoc mesenteroides are some examples of probiotic bacteria isolated from bacterial endocarditis; L. rhamnosus, L. plantarum, Leuconostoc. mesenteroides, Pediococcus acidilactici, Bifidobacterium eriksonii, and Bifidobacterium adolescentis have been isolated from bloodstream infections and many have been isolated from local infections. Although minor side effects of the use of probiotics have been reported, infections with probiotic bacteria occur and invariably only in immunocompromised patients or those with intestinal bleeding (Leroy et al., 2008).

An issue of concern regarding the use of probiotics is the presence of chromosomal, transposon, or plasmid-located antibiotic resistance genes amongst the probiotic microorganisms. At this moment, insufficient information is available on situations in which these genetic elements could be mobilised, and it is not known if situations could arise where this would become a clinical problem (Leroy et al., 2008). When dealing with the selection of probiotic strains, the FAO/WHO Consultancy recommends that probiotic microorganisms should not harbor transmissible drug resistance genes encoding resistance to clinically used drugs (FAO/WHO, 2002). For the assessment of the safety of probiotic microorganisms and products, FAO/WHO has formulated guidelines, recommending that probiotic strains should be evaluated for a number of parameters, including antibiotic susceptibility patterns, toxin production, metabolic and haemolytic activities, and infectivity in immunocompromised animals (FAO/WHO, 2002). In vitro safety screenings of probiotics may include, among others, antibiotic resistance assays, screenings for virulence factors, resistance to host defence mechanisms and induction of haemolysis. Several different animal models have been utilized in the safety assessment of probiotics. These include models of immunodeficiency, endocarditis, colitis and liver injury. In some cases even acute toxicity of probiotics has been assessed. Last but not least, also clinical intervention trials have yielded evidence on the safety of probiotics for human consumption (Forssten et al., 2011).

## 7. Conclusion

The individual diversity of the intestinal microflora underscores the difficulty of identifying the entire human microbiota and poses barriers to this field of research. In addition, it is apparent that the actions of probiotics are species and strain specific. It is also apparent that even a single strain of probiotic may exert its actions via multiple, concomitant pathways. Probiotics have long been used as an alternative to traditional medicine with the goal of maintaining enteric homeostasis and preventing disease. However, the actual efficacy of this treatment in still debated. Clinical trials have shown that probiotic treatment can reduce the risk of some diseases, especially antibiotic-associated diarrhea, but conclusive evidence is impeded owing to the wide range of doses and strains of bacteria used. The mechanism of action is also an area of interest (Ohland and MacNaughton, 2010). Many studies, as discussed above, have shown that probiotics increase barrier function in terms of increased mucus, antimicrobial peptides, and sIgA production, competitive adherence for pathogens, and increased TJ integrity of epithelial cells. Current investigation into the mechanism of action of specific probiotics has focused on probiotic-induced changes in the innate immune functions involving TLRs and its downstream systems Like NF-κB, and other pathways (Yoon and Sun, 2011). Although the immunomodulatory effects of probiotics have been demonstrated in experimental animal models of allergy, autoimmunity, and IBD, information from clinical trials in humans is scarce. Furthermore, some studies suggest that probiotics could induce detrimental effects. Therefore, more research, especially in the form of well-designed clinical trials, is needed to evaluate the efficacy and safety of probiotics (Ezendam and Van Loveren, 2008). With evolving knowledge, efective probiotic therapy will be possible in the future.

## Author details

Daoud Harzallah and Hani Belhadj
*Laboratory of Applied Microbiology, Faculty of Natural and Life Sciences, University Ferhat Abbas, 19000, Sétif, Algeria*

## 8. References

Abreu M.T. (2003). Immunologic regulation of toll-like receptors in gut epithelium. Curr. Opin. Gastroenterol; 19:559-64.

Bäckhed F, and M. Hornef (2003). Toll-like receptor 4-mediated signaling by epithelial surfaces: necessity or threat? Microbes Infect.; 5:951-9.

Belhadj, H., Harzallah, D., Khennouf, S., Dahamna, S., Bouharati, S., and Baghiani, A. (2010). Isolation, identification and antimicrobial activity of lactic acid bacteria from Algerian honeybee collected pollen. Acta Hort. (ISHS) 854: 51 – 58

Berkes J, V.K. Viswanathan, S.D. Savkovic, G. Hecht (2003). Intestinal epithelial responses to enteric pathogens: effects on the tight junction barrier, ion transport, and inflammation. Gut;52:439-51.

Caballero-Franco C., K. Keller, C. De Simone, K. Chadee (2007). The VSL#3 probiotic formula induces mucin gene expression and secretion in colonic epithelial cells. Am J Physiol Gastrointest Liver Physiol 292: G315–G322.

Chow J., (2002). Probiotics and prebiotics: A brief overview, J. Ren. Nutr. 12 76–86.

Collado M. C., E. Isolauri, S. Salminen and Y. Sanz (2009). The Impact of Probiotic on Gut Health. Current Drug Metabolism, , 10, 68-78.

Desai A. (2008) Strain Identification, Viability And Probiotics Properties Of Lactobacillus Casei. PhD Thesis Victoria University, Victoria, Australia 228 pp.

Ezendam J., and H. Van Loveren (2008). Probiotics: Immunomodulation and Evaluation of Safety and Efficacy. Nutrition Reviews, 64, (1) : 1-14.

FAO/WHO. (2002). Guidelines for the evaluation of probiotics in food – Joint Food and Agricultural Organization of the United Nations and World Health Organization Working Group Meeting Report, London Ontario, Canada.

Forssten S. D., S. J. Lahtinen A. C. Ouwehand, and J. J. Malago et al. (eds.), (2011) Chapter 2 :The Intestinal Microbiota and Probiotics 24 pp. In : Probiotic Bacteria and Enteric Infections. Springer Science Business Media B.V.

Gareau M. G., P. M. Sherman and W. A.Walker (2010). Probiotics and the Gut Microbiota in Intestinal Health and Disease. Nature Reviews Gastroenterology and Hepatology 7, 503-514.

Gasser F. (1994) Safety of lactic acid bacteria and their occurrence in human clinical infections. Bull Inst Pasteur; 92:45–67.

Gaudier E., C. Michel, JP. Segain, C. Cherbut, and C. Hoebler (2005). The VSL# 3 probiotic mixture odifies microflora but does not heal chronic dextran- sodium sulfate-induced colitis or reinforce the mucus barrier in mice. J Nutr. 135: 2753–2761.

Gotteland M., O. Brunser, S. Cruchet (2006) Systematic review: are probiotics useful in controlling gastric colonization by Helicobacter pylori? Aliment. Pharmacol. Ther.23:1077–1086.

Holzapfel, W.H., P. Haberer, R. Geisen, J. Björkroth, U. Schillinger (2001) Taxonomy and important features of probiotic microorganisms in food nutrition. Am. J. Clin. Nutr. 73, 365S-373S.

Kagnoff M.F., and L. Eckmann (1997) Epithelial cells as sensors for microbial infection. J Clin Invest;100:6-10.

Kalabis J, I. Rosenberg, and DK. Podolsky (2006). Vangl1 protein acts as a downstream effector of intestinal trefoil factor (ITF)/TFF3 signaling and regulates wound healing of intestinal epithelium. J Biol Chem 281:6434–6441.

Kim Y, SH. Kim, KY. Whang, YJ. Kim, and S. Oh (2008). Inhibition of Escherichia coli O157:H7 attachment by interactions between lactic acid bacteria and intestinal epithelial cells. J Microbiol Biotechnol 18: 1278–1285.

Kollath, W. (1953) Ernährung und Zahnsystem. Deutsche Zahnärztliche Zeitschrift 8: 7-16.

Lebeer S., J. Vanderleyden and S. C. J. De Keersmaecker (2010). Host interactions of probiotic bacterial surface molecules: comparison with commensals and pathogens. *Nature Reviews Microbiology* 8, 171-184.

Leroy, F. and De Vuyst, L. (2004). Functional lactic acid bacteria starter cultures for the food fermentation industry. Trends in Food Science and Technology 15, 67-78.

Leroy F., G. Falony, and L. de Vuyst (2008) Latest Developments in Probiotics. In Toldrà F. (ed.), Meat Biotechnology, Springer Science+Business Media, LLC. 217-229 pp.

Lilly D.M, and R.H. Stillwell (1965) Probiotics: growth promoting factors produced by microorganisms. Science. 147: 747–8.

Lu L, and W.A. Walker (2001). Pathologic and physiologic interactions of bacteria with the gastrointestinal epithelium. Am. J. Clin. Nutr.;73:1124S-30S.

Mack DR, S. Ahrne, L. Hyde, S. Wei, and MA. Hollingsworth (2003). Extracellular MUC3 mucin secretion follows adherence of Lactobacillus strains to intestinal epithelial cells in vitro. Gut 52: 827–833.

Marco M. L, S. Pavan and M. Kleerebezem (2006). Towards understanding molecular modes of probiotic action . Curr Opin Biotechnol. 17:204–210.

Mattar AF., DH. Teitelbaum, RA. Drongowski, F. Yongyi, CM. Harmon, and AG. Coran (2002). Probiotics up-regulate MUC-2 mucin gene expression in a Caco-2 cell-culture model. Pediatr Surg. Int. 18: 586–590.

Matsuguchi T, A. Takagi, T. Matsuzaki, M. Nagaoka, K. Ishikawa, T. Yokokura, and Y. Yoshikai (2003). Lipoteichoic acids from Lactobacillus strains elicit strong tumor necrosis factor alpha-inducing activities in macrophages through Toll-like receptor 2. Clin Diagn Lab Immunol; 10:259-66.

Mercenier A., S. Pavan and B. Pot (2003) Probiotics as Biotherapeutic Agents: Present Knowledge and Future Prospects. Current Pharmaceutical Design, 8 : 99-110.

Metchnikoff, E. (1908). The Prolongation Of Life, New York, G.P. Putnam's Sons, 376 p.

Morelli, L. (2007). In vitro assessment of probiotic bacteria: From survival to functionality. International Dairy Journal, 17, 1278–1283.

Myllyluoma E. (2007). The Role Of Probiotics In Helicobacter pylori Infection. Dessertation, 86p. Nstitute Of Biomedicine Pharmacology University Of Helsinki, Helsinki, Finland.

Ng S.C., A.L. Hart, M.A. Kamm, A.J. Stagg, and S.C. Knight (2009) Mechanisms of Action of Probiotics: Recent Advances Inflamm Bowel Dis.15:300 –310.

Ohland C. L. and W. K. MacNaughton (2010). Probiotic bacteria and intestinal epithelial barrier function.Am.J. Physiol. Gastrointest. Liver Physiol. 298: G807–G819.

Osmanagaoglu O., and Y. Beyatli (1999) The Use of Bacteriocins Produced by Lactic Acid Bacteria in Food Biopreservation. Türk. Mikrobiyol. Cem. Derg. 32: 295-306.

Ouwehand A.C., Lahtinen S.J. (2008). The (non-) sense of human origin of probiotics. NutraCos, 7:8–10.

Parada J. L., C. R.Caron, A. B. P. Medeiros and C. R. Soccol (2007) Bacteriocins from Lactic Acid Bacteria: Purification, Properties and use as Biopreservatives. Brazilian Archives of Biology And Technology. Vol.50 (3):521-542.

Pedersen G, L. Andresen, M.W. Matthiessen, J. Rask-Madsen, J. Brynskov (2005). Expression of Toll-like receptor 9 and response to bacterial CpG oligodeoxynucleotides in human intestinal epithelium. Clin Exp Immunol;141:298-306.

Pretzer G, J. Snel, D. Molenaar, A. Wiersma, P.A. Bron, J. Lambert, WM de Vos, R. Ven Der Meer, M.A. Smits, and M. Kleerebezem (2005). Biodiversity-based identification and functional characterization of the mannose-specific adhesin of Lactobacillus plantarum. J Bacteriol. 187:6128-36.

Rhee SH, E. Im, M. Riegler, E. Kokkotou, M. O'Brien, C. Pothoulakis. (2005). Pathophysiological role of Toll-like receptor 5 engagement by bacterial flagellin in colonic inflammation. Proc. Natl. Acad .Sci.; 102:13610-5.

Sanders M. E. (2008) Probiotics: Definition, Sources, Selection, and Uses. Clinical Infectious Diseases. 46:S58–61.

Shanahan F. (2003) Probiotics: a perspective on problems and pitfalls, Scand. J. Gastroenterol. 38 (S) :34– 36.

Schlee M., J. Wehkamp, A. Altenhoefer, et al. (2007) Induction of human beta-defensin 2 by the probiotic Escherichia coli Nissle 1917 is mediated through flagellin. Infect Immun.75:2399–2407.

Schrezenmeir J., and M. De Vrese (2001) probiotics, prebiotics, and symbiotics – approaching a definition. Am. J. Clin. Nutr. 73:361S-364S.

Sherman PM, K.C. Johnson-Henry, H.P. Yeung, et al. (2005) Probiotics reduce enterohemorrhagic Escherichia coli O157:H7- and enteropathogenic E. coli O127:H6-induced changes in polarized T84 epithelial cell monolayers by reducing bacterial adhesion and cytoskeletal rearrangements. Infect Immun.73:5183–5188.

Sekirov I., S. L. Russell, L. C. M. Antunes, And B. B. Finlay (2010). Gut Microbiota in Health and Disease. Physiol Rev 90: 859–904.

Servin A.L., M.H. Coconnier (2003). Adhesion of probiotic strains to the intestinal mucosa and interaction with pathogens. Best Pract. Res. Clin. Gastroenterol;17:741-54.

Soccol C. R., L. P. de Souza Vandenberghe, M. R. Spier, A. B. Pedroni Medeiros, C. T. Yamaguishi, J. De D. Lindner, A. Pandey and V. Thomaz-Soccol1(2010). The Potential of Probiotics: A Review. Food Technol. Biotechnol. 48 (4): 413–434.

Strober W. (1998). Interactions between epithelial cells and immune cells in the intestine. Ann N.Y Acad. Sci.; 859:37-45.

Takeda K, T. Kaisho, and S. Akira (2003). Toll-like receptors. Annu. Rev. Immunol.; 21:335-76.

Tuomola, E., R. Crittenden, M. Playne, E. Isolauri, and S. Salminen. (2001). Quality assurance criteria for probiotic bacteria. Am J Clin Nutr.; 73:393S -398S.

Turner J. R. (2009). Intestinal mucosal barrier function in health and disease. Nature Reviews Immunology. 9: 799-809.

Vasiljevic and N.P. Shah, (2008) Review: probiotics — from Metchnikoff to bioactives, International Dairy Journal 18, pp. 714–728.

Vinderola G, C. Matar, and G. Perdigon. (2005). Role of intestinal epithelial cells in immune effects mediated by gram-positive probiotic bacteria: involvement of toll-like receptors. Clin. Diagn. Lab Immunol; 12:1075-84.

Vergin F. (1954) Anti- and probiotika. Hippokrates 25: 16–19.

Yoon S. S., and J. Sun (2011). Probiotics, Nuclear Receptor Signaling, and Anti-Inflammatory Pathways. Gastroenterology Research and Practice Volume 111: 1-16.

Wehkamp J., J. Harder, K. Wehkamp, et al. (2004). NF-kappaB- and AP-1-mediated induction of human beta defensin-2 in intestinal epithelial cells by Escherichia coli Nissle 1917: a novel effect of a probiotic bacterium. Infect Immun.72:5750–5758.

Chapter 9

# Probiotics and Intestinal Microbiota: Implications in Colon Cancer Prevention

Katia Sivieri, Raquel Bedani, Daniela Cardoso Umbelino Cavallini and Elizeu A. Rossi

Additional information is available at the end of the chapter

http://dx.doi.org/10.5772/51696

## 1. Introduction

Colon cancer (CC) is one of the commonest causes of death among all types of cancers [1]. The development of cancer is a multifactorial process influenced by genetic, physiological, and environmental factors [2,3]. Regarding environmental factors, the lifestyle, particularly dietary intake, may affect the risk of CC developing [1,4]. Western diet, rich in animal fat and poor in fiber, is generally associated with an increased risk of colon cancer [5,6,7]. Thus, it has been hypothesized that the connection between the diet and CC, may be the influence that the diet has on the colon microbiota and bacterial metabolism, making both relevant factors in the etiology of the disease [8,9]. Additionally, it has been clearly demonstrated that the gut microbiota may be modulated by many factors including diet [10].

Several studies have indicated that the intestinal microbiota is an important determinant for general health of the human body [1]. Therefore, a beneficial modulation of the composition and metabolic activity of the gut microbiota might represent an interesting approach to improve health, reducing the risk of CC development. This modulation may be though about probiotic consumption.

Probiotics are defined as live microorganisms which when administered in adequate amounts confer a health benefit on the host [11]. Among the best known probiotic microorganisms are strains belonging to the *Lactobacillus* and *Bifidobacterium* genera. However, other microorganisms, such as *Enterococcus* spp., *Streptococcus* spp., *Escherichia coli* Nissle 1917, some bacilli, and *Saccharomyces cerevisiae* subsp. *boulardii* have also been considered for use as probiotics [12].

Even though the mechanisms by which probiotics may inhibit colon cancer are not fully elucidated, certain potential mechanisms have been disclosed, such as the alteration of the composition and the metabolic activities of the intestinal microbiota, the changing physicochemical conditions in the colon, the binding of dietary carcinogens, the production of short chain fatty acids (SCFA), the protection of the colonic mucosa and enhancement the immune system [1,3].

The anticarcinogenic effects of probiotic microorganisms *in vitro* and in animal studies are well documented [3]. In clinical trials, the probiotics are thought to play a protective role in the initial process of carcinogenesis. Nevertheless, it is important to determine whether the long-term administration of these microorganisms might result in changes in the incidence of CC in humans [13]. Additionally, there are several challenges for the development of probiotics, including the selection of the appropriate microorganisms, control of dietary intake, time and frequency of probiotic dosing and the use of accepted biomarkers for raised cancer risk that might be monitored during clinical trials [4,13]. Further experimental models are needed to understand the exact mechanisms involved in the influence of probiotics on colon cancer development.

Therefore, this chapter will discuss the effects of probiotics in colon cancer prevention and the possible mechanisms of action these microorganisms. Additionally, this chapter will also show the results of original work, carried out by our research group, about the effects of probiotic *Enterococcus faecium* CRL 183 (strain isolated from Tafí cheese, a homemade traditional highlands cheese the province of Tucumán, Argentina) on intestinal microbiota and colon cancer prevention.

## 2. Colon cancer

Social and economic transformations related to urbanization and industrialization in Brazil resulted in changes in the morbimortality profile of the population. While, in the first half of the 20th century infectious disease event were the most frequent, from the 1960 metabolic diseases and noncommunicable grievances occupied the first place, contributing to the process of demographic transition, which favours the spread of cardiovascular and respirator disease, cancer and diabetes, as does the nutritional transition, with a marked reduction of malnutrition and large growth in the number of overweight people [14].

Known for many centuries, cancer was widely regarded as a disease of developed countries with large financial resources. However, for approximately four decades, this situation has undergone transformation, and most of the global burden of cancer can be observed in developing countries, especially those with low to medium resources [15].

Cancer has become a global public health problem of course, since the World Health Organization (WHO) estimates that in the year 2030 there will be 27 mil new cases of cancer, 17 million deaths and 75 million people living with the disease [15].

Cancer of the colon and rectum is the third commonest type of cancer among men and the commonest in women. It is estimated that in 2011, in Brazil, 14,180 new cases of colon cancer and rectum, occurred in men and women. These values correspond to a perceived risk 15 new cases per 100 thousand men and 16 per 100 thousand women [15].

This neoplasia is considered to have a good prognosis when diagnosed in the early stages. Colon cancer like others forms of cancer develops as a result of interaction between endogenous and environmental factors. Among the factors that may affect the risk of developing this disease are age, eating habits, physical activity, alcohol consumption, smoking, nutritional status, presence of polyps, cancer history of self and family, cases of ulcerative enterocolitis and chronic constipation [15,16].

Most cases of CC occur sporadically, being the most common type of adenocarcinoma, which develops from glandular cells that cover the wall of the intestine [17]. Adenocarcinomas grow from normal epithelium through an accumulation of mutations that result in malignant transformation [19].

Genomic instability is fundamental to this process and is related to the rearrangement of genes, or loss of DNA fragments, aneuploidy and loss of heterozygosis [19]. In addition, inactivation of tumor suppressor genes, such as APC, DCC, DPC4 and p53, along with the activation of oncogenes, of which the family of *ras* genes are the best well described, play important parts in the appearance of malignancy [17].

Generally, the colon tumor is detected for the first time as a polyp (mass of cells growing out of the wall of the colon), although nowadays it is possible to detect small lesions affecting the crypts, called aberrant crypts foci (ACF) [18]. ACF are not only morphologically but also genetically distinct lesions and are precursors of adenoma and cancer. Tumors can appear anywhere in the colon, although most sporadic rectal colon cancers are located on the left side of the distal colon (including the rectum and sigmoid colon) [19].

Epidemiological studies have pointed to the high consumption of red meat, fat and low fiber intake, typical of the Western diet as risk factors in the etiology of this type of cancer [20].

One of the possible effects of a Western diet on colon cancer is related to increased excretion of bile acids [21]. In addition, the increased ammonia production in rats consuming a diet rich in protein has also been linked to an increased risk of cancer [22]. However, high consumption of fruits, cereals, fish and calcium may reduce the risk of developing colon cancer [23].

The effect of diet on carcinogenesis can be modulated by changes in metabolic activity and composition of the intestinal microbiota [23]. Several studies have trial to establish relationships between bacteria and colon cancer. We know that various bacterial metabolites are carcinogenic, examples being, the nitrosamines, phenol, indole, ammonia and amines [13].

There is multiple evidence that bacteria play a key role in the emergence of chronic inflammatory bowel diseases. Experimental studies demonstrate the impossibility of developing this inflammation in the absence of bacteria and researches have tried for many years trying to identify a possible causative agent of inflammatory bowel diseases. Studies suggest that chronic inflammatory intestinal activity seems, paradoxically to be triggered by bacteria belonging to the normal commensal which take on microbiota in situations as yet unknown, a pathological role that can activate the local immune apparatus [24].

There are many types of intestinal bacteria that produce a variety of metabolites that modulate the normal development and functioning of the host. On the other hand, the metabolic activity of intestinal microbiota can generate compounds that are harmful such as reactive oxygen intermediates. These molecules, which include superoxide, hydrogen peroxide, hypochlorous acid, singlet oxygen and hydroxyl radical, can cause oxidative damage to cellular DNA and increase the risk of colon cancer [25]. Studies have shown that *Enterococcus faecalis* can produce superoxide and hydrogen peroxide, causing damage to DNA in skin cells, in both *in vitro* and *in vivo* tests [26].

Given the role of intestinal microbiota in colon carcinogenesis, it is suggested that factors that modulate beneficially the composition and/or activity of the microbiota could inhibit the development of CC.

## 3. Evidences for relationship among intestinal microbiota, probiotics and colon cancer

### 3.1. The intestinal microbiota

The gastrontestinal (GI) microbiota undergoes changes in quantity and quality, depending on the location of colonization in the GI. Traditional culture-based characterization may take into account no more than 30% or so of the microorganisms that can be seen and enumerated by microscopic observation. The worldwide species diversity of commensal intestinal bacteria is immense. In that respect, the use of molecular tools has indicated that the majority of the dominant bacterial species observed in the faecal microbiota of an individual (approximately 80%) are specific to this individual [27]. Also, these species are not distributed homogeneously along the length of the GI, so the bacterial activities are considerably variable in different parts of the intestine [28].

The stomach and the small intestine contain few species, whereas the colon contains a complex and dynamic microbial ecosystem, with a great concentration of bacteria. Among these are the bifidobacteria and lactobacilli, considered non-pathogenic or beneficial bacteria [29]. The bacterial population in the large intestine is very large and reaches a maximum count of $10^{12}$ CFU.g$^{-1}$. In the small intestine, bacterial contents are considerably smaller from $10^4$ to $10^7$ CFU.g$^{-1}$, while in the stomach only $10^1$ at $10^2$ CFU.g$^{-1}$ are found in function of low pH on this site. In total, the number of intestinal bacteria is approximately ten times the number of cells that make up the human body [30].

On the basis of rRNA sequencing 40,000 strains of intestinal bacteria can be indentified, including non-cultivable bacteria [31]. It was noted that 99% of intestinal bacteria consist of four phyla, Proteobacteria, Actinobacteria, and two main phyla Bacteroidetes and Firmicutes [32]. While the species in the phylum Bacteroidetes show a great variety between individuals, a large number of species in the phylum Firmicutes belong to clusters of clostridial butyrate producers [33].

With advances in molecular biology, it is known that the intestinal microbiome, contains 100 times more genes than the whole human genome [34]. Thus, a close relationship is evolving between the human gut microbiota. The human intestine exhibits to a symbiotic relationship that plays a key role in human homeostasis, including metabolism, growth and immunity [35].

One of the primary functions of the intestinal microbiota is the harnessing of energy from elements of the diet that could be lost through excretion [36]. The polysaccharides are not absorbed in the colon, but metabolized by resident microorganisms to short chain fatty acids (SCFA), such as propionate and butyrate, which are absorbed by passive diffusion [37]. SCFA production is dependent on the available fermentation of substrate, such as, starch or other polysaccharides, results butyrate, acetate and propionate [37]. SCFA concentrations are higher on the right side of the colon than on the left and this is probably due to the greater availability of carbohydrates [29]. The SCFA have an important role in the maintenance of the epithelial layer. Studies show that epithelial cells acquire about 70% of their butyrate oxidation [29]. The butyrate also acts as a trophic factor for cells in intact tissues [38]. In addition, it has been proposed that butyrate lowers the risk of colon cancer by its ability to inhibit the genotoxic activity of nitrosamines and hydrogen peroxide, as well as to induce various levels of apoptosis, differentiation and the cell cycle stop colon cancer in animal models [39].

Other researchers also cite the effect of butyrate on mediators of inflammation, it has been proved that this SCFA is able to inhibit the expression of some cytokines (TNF, IL-6, IL-1) and to inhibit the activation of nuclear factor κB (NF- κB) [40]. Other functions of the gastrointestinal microbiota include digestion of poorly digested nutrients, modification of bile acids, and nutritional supplementation by auxotrophic of mutants additional compounds that cannot be acquired by food consumption, such as folic acid and biotin [41].

The non-pathogenic commensal microbiota has a profound impact on the normal physiology of the GI tract. It ensures the efficiency of bowel motility, intestinal growth and immunity, as well as digestion, nutrient absorption and fortification of the mucus barrier [42].

Researchers have made advances in the characterization of GI microbiota defining the responses that may contribute to the development of inflammatory bowel diseases, such as, colon cancer [43]. Given the importance of a better understanding of intestinal microbiota, the TGI has been often studied. In recent decades, various intestinal simulators have been and are being developed, to facilitate the study of the intestinal microbial ecosystem and its interactions [44, 45].

### 3.2. Methods for *in vitro* evaluation of effects of probiotics on intestinal microbiota

The FAO/WHO refers to probiotics as live microorganisms that administered in adequate doses, benefit the health of the host [11]. The beneficial effects of ingesting probiotics enhanced relief of the symptoms of lactose intolerance, treatment for diarrhea, reduction of serum cholesterol, enhanced immune response and anticarcinogenic effects [46].

The rising consumption of probiotic products by Europeans is mainly is in the form of dairy products containing generally *Lactobacillus* spp. and *Bifidobacterium* spp. However there are products in which the microorganisms used are strains of *Enterococcus* spp. or yeasts such as *Saccharomyces boulardii* [47]. Foods for human consumption containing lactic acid bacteria (LAB) include fermented milk, fruit juices, wine and sausages. Simple cultures or mixed microorganisms are used in probiotic preparations [48].

Several experimental observations have pointed to the potential protective effect of LAB against the development of tumors in the colon [49]. Within the intestinal microbiota, the LAB complex constitutes part of those bacteria able to promote a beneficial effect. They have an important role in retarding colon carcinogenesis by possibly of influencing metabolic, protective and immunological functions in the intestine [39]. The effect of intake of probiotics on intestinal native microbiota can be assessed through *in vivo* or *in vitro* models. *In vivo* models may involve healthy human volunteers, hospitalized patients or an animal model, but these models have some limitations such as high cost, delay in obtaining results and the type of food or drugs administered [50], whereas, *in vitro* models enable you to simplify the system and study separately the metabolism of native and added microbiota, in the presence of specific substrates [50].

*In vitro* fermentation models range from a simple batch system to more complex systems of continuous flow and multi-stage. *In vitro* gut fermentation models enable the stable cultivation of a complete intestinal microbiota for a defined and model-specific period of time. Selection of the appropriate model requires careful evaluation of the study objectives given the advantages and limitations exhibited by each type of system. Some existing systems are included in the batch, continuous culture, multi-stage continuous culture, continuous artificial digestive system and stationary systems [51].

Batch fermentation is the growth of a pure or mixed bacterial suspension in a carefully selected medium without the further addition of nutrients. These models are generally closed systems is sealed bottles or reactors containing suspensions of fecal material which are maintained under anaerobic conditions. Several studies have already been carried out, using this type of model in research on the prebiotic potential of fructans. This template is particularly useful to investigate metabolic profiles of SCFAs arising from active metabolism of dietary compounds by intestinal microbiota [50].

Continuous culture fermentation models exist as either single- or multistage systems and are necessary to perform long-term studies, as substrate replenishment and toxic product removal are facilitated. Single-stage continuous fermentation models are often used to elucidate proximal colon function and metabolic activity as the mixing of digest from both the caecum and ascending colon is well simulated in these models [52].

These models have several advantages, such as: the ease of use of the system, the possibility of using radioactive substances and the low operation cost [28].

A major advance for *in vitro* fermentation systems was the development of continuous multi-stage models, which allow the simulation of horizontal processes. This type of system makes it easy to study the nutritional and physicochemical properties of intestinal microbiota, through the combination of three reactors connected in series, simulating the proximal, distal and transversal colon (see Figure 1). Later, Molly et al. [44] developed the human microbial ecosystem simulator (SHIME ®), which consists of a succession of five connected reactors, which represent the different parts of the human gastrointestinal tract with their respective values of pH, residence time and volumetric capacity (Figure 1). The five reactors are continually agitated and kept at a temperature of 37 ºC by means of a thermostat. The medium is kept in the anaerobic state, by daily injection of $N_2$. The appropriate pH for each portion of the GI tract is controlled automatically by adding 1N NaOH or concentrated HCl [44, 45].

**Figure 1.** Computer controlled simulation of human microbial ecosystem (SHIME ®) housed in the Probiotics Research Laboratory of FCF/UNESP-Brazil. Sivieri et al.[53]

The adaptation, survival and proliferation of a human intestinal microbiota in continuous fermentation *in vitro* models are depended on environmental parameters such as pH, retention time, temperature, flow rate and oxygen depletion. The rigorous control of these factors allows steady established state in conditions the microbial composition and metabolic activity, creating a reproducible system.

The continuous cultivation model has been used in research on the metabolism and ecology of intestinal microbiota, with an emphasis on the use of probiotics [51, 54], prebiotics [55, 56] and the formation of fermentation products [57]. The *in vitro* modeling of host digestive functions in vitro coupled with multistage continuous fermentation, represents the most advanced attempt thus far at simulating interdependent physiological functions within the human gut, stomach lumen and small intestine. Human digestive functions that are reproduced in the TIM-1 small intestine model include bile secretion, motility, pH and absorption capacity of the upper intestine. Proximal colon simulator models such as TIM-2 include other host functions such as peristaltic mixing and water and metabolite absorption. The combination of TIM-1 and TIM-2 models led to the creation of an artificial digestive system which has been used to investigate pharmaceutical drug delivery and advanced nutritional studies [58, 59].

The use of a multidisciplinary biological systems approach, in combination with '-omics' platforms as outlined will facilitate the most advanced system for unraveling the complex microbial and host factors governing human gut microbiota functionality [60].

In vitro fermentation models are an innovative technological platform where the greatest advantages are exhibited by the virtually limitless experimental capacity as experimentation is not restricted by ethical concerns. Host intestinal function is only partially simulated in some model designs (e.g. TIM-1 and TIM-2) and together with microbial population balancing remains a major challenge of in vitro gut fermentation modeling.

### 3.3. Inhibition on colon cancer by probiotics and the possible action mechanisms of these microorganisms

The evidence pointing to the beneficial effects of probiotics on colon cancer comes from *in vitro* tests, experiments with animals and clinical trials. Additionally, these has been much discussed on which step in the process of carcinogenesis might the effect by probiotics. It is likely that different probiotic strains act on different stages of carcinogenesis [20].

In general, the probiotics do not colonize the human gut, but some strains are can permanently colonize the indigenous microbiota [61].

The mechanisms by which probiotics may inhibit colon cancer are not yet fully characterized. However, several explanations have been suggested including: alteration of the metabolic activities of the intestinal microbiota; quantitative and qualitative changes in the intestinal microbial compositin; alteration of physicochemical conditions in the colon; binding and/or degradation of potential carcinogens; SCFA production; production of anti-tumorigenic or anti-mutagenic compounds; modulation of hosts's immune response, and/or physiology [3,62, 63].

Probiotics may modulate the metabolic activities of the intestinal microbiota by three possible mechanisms: competing with and displacing other components of the microbiota; producing antibacterial substances, including bateriocins, to control the growth of other members of the microbiota; producing lactic and other organic acids, which might lower the luminal pH and thus modulate enzyme activity [20,64].

Several investigations have shown that probiotics can influence bacterial enzymes activity related to the production of carcinogenic compounds, such as beta-glucuronidase, nitroreductase and azoreductase [65, 66, 67].

Bacterial glucuronidase appears to have an important role in the initiation of colon cancer, due to its ability to hydrolyze several glucuronides and carcinogenic aglycones in the intestinal lumen [65,68]. The nitroreductase and azoreductase take past in to the formation of aromatic amines harmful to the body [69].

Both harmful and beneficial bacteria are commonly found in the intestines and differ in their enzymatic activity [70]. In general, bacteria from the genera *Bifidobacterium* and *Lactobacillus* produced a very little activity of enzymes that convert pro-carcinogens into carcinogens, compared with bacteria from the genera *Bacteroides* and *Clostridium* [71]. Therefore, the activities of these enzymes in the lumen might be correlated with the number of lactic acid bacteria (LAB) in the intestine [72]. This suggests that increasing the proportion of LAB in the gut could diminish the levels of xenobiotic metabolizing enzymes [71]. Thus, the effect of probiotic microorganisms on fecal enzyme activities might be explained by this mechanism.

In a preliminary study, on feces of small animal, the animal supplementation of a high cholesterol diet with a mixture of probiotic strains of *L. johnsonii* and *L. reuteri* for 5 weeks significantly decreased the activity of fecal-glucuronidase and azoreductase [67].

Gorbach and Goldin [65] studied, in humans, the effect of ingestion of *L. acidophilus* NCFM strains about the activity of-glucuronidase, nitroreductase and azoreductase. Both strains had a similar effect and caused a significant decline in the activity of these three enzymes. A reverse effect was found 10 to 30 days after the end of the intake of these bacteria, suggesting that continuous consumption of *L. acidophilus* is necessary for maintaining.

Benno and Mitsuoka [73] and Spanhaak et al. [66] also found in humans, a significant reduction in the activity glucuronidase after intake of *Bifidobacterium longum* and *L casei* Shirota, respectively. On the other hand, Marteau et al. [74] verified in healthy volunteers that the regular consumption of a fermented dairy product (100 g three times per day) containing *L. acidophilus, B. bifidum, Streptococcus thermophilus* and *S. cremoris* for 3 weeks decreased the feces nitroreductase activity from baseline but not that of β-glucuronidase or azoreductase.

Feces metabolites are also indicators of bacterial activity. Changes in enzyme activities and the concentration of ammonia, phenol and cresol have been detected in volunteers who consumed Lactobacilli [65]. Other metabolites with possible adverse effects are N-nitroso compounds, diacylglycerol and secondary bile acids [49].

A wide variety of microrganisms can produce ammonia, for example, enterobacteria, bacteroides and clostridia. Ammonia is considered a potential promoter of tumor in the colon and it can increase the rate of neoplastic transformation in the intestine. According to Benno and Mitsuoka [73], reducing the proportion of clostridia and bacteroides could explain the decrease in the concentration of ammonia in individuals who consumed fecal *B. longum.*

Epidemiological studies indicate an association between the risk of developing colon cancer and the consumption of high fat diets [7,75, 76]. For the digestion of fats, bile acids conjugated to glycine or taurine molecules are released into the small intestine and reabsorbed in the same location. It is believed that the deoxycholic acids may be cytotoxic to the epithelial cells, which could lead to the development of colon cancer [71]. Probiotic modulation of the intestinal microbiota may affect the activity of one of the enzymes (7a-dehydroxylase) forming these toxic products, but probiotics may also reduce the toxicity of bile salts that bind to them [77]. Lidbeck et al. [68] found that administering *L. acidophilus* to colon cancer patients for 6 weeks resulted in reduction in the concentration of soluble bile acids in the stool.

The consumption of fermented milk containing *L. acidophilus* may reduce the population of harmful bacteria, such as coliforms, and increased levels of lactobacilli in the intestine [78], suggesting that supplementation with this microorganism can have a beneficial effect since it inhibits the growth of bacteria that harmful are possibly involved in the production of tumor promoters and pro-carcinogens. Savard et al. [79] assessed the impact of four week's consumption of commercial yoghurt with *Bifidobacterium animalis* subsp. lactis (BB-12) and *Lactobacillus acidophilus* (LA-5) on fecal bacterial counts in healthy adults. The yoghurt had a positive effect on the bacterial population in that a the increase in beneficial bacteria and the reduction of potentially pathogenic bacteria was observed.

Not all studies show a correlation between the administration of probiotics and the activity of intestinal microbiota. Bartram et al. [80] argued that the fecal microbiota is relatively stable and generally unaffected by the administration of probiotics. In an intervention study, 12 individuals consumed yogurt (500 mL) enriched with *B. longum*. No significant difference was found in fecal weight, pH, concentration of fecal short chain fatty acids, bile acids and neutral sterols after 3 weeks of intervention. Despite the rise in the fecal concentration of *B. longum*, the results suggested litlle or no modulation of resident microbiota.

Some researchers have suggested that a high intestinal pH may be related to increased risk of colon cancer, whereas acidification of the colon could prevent the formation of carcinogens. Benno and Mitsuoka [73] found a significant reduction of faecal pH in health men who ingested *B. longum* for 5 weeks.

Evidence indicates that a high concentration of short chain fatty acids (acetate, propionate and butyrate) can assist in maintaining an appropriate pH in the lumen of the colon for the expression of many bacterial enzymes that probably metabolize carcinogens in the gut [81]. The activity of some dietary carcinogens, such as nitrosamines (resulting from commensal bacterial metabolic activity in individuals who consume a diet rich in proteins) can be neutralized by butyric acid produced by some probiotics [82]. Furthermore, production of ammonia, nitrosamines and secondary bile acids in the intestinal environment can be reduced by lowering the pH [83].

Butyrate, particularly, has received much attention as a potential chemopreventive agent [1,84]. While acting as an energy source for untransformed cells, butyrate possibly reduces survival of tumor cells by inducing apoptosis and differentiation, as well as by inhibiting proliferation. These mechanisms may play an important role in the reduction and/or inhibition of promotion and progression of cancer [1, 85].

Studies show that the LAB may be involved in the detoxification of various carcinogens such as polycyclic aromatic hydrocarbons and heterocyclic aromatic amines [86]. The mechanisms of action of these bacteria are poorly known, but it is possible that the LAB bind directly to the carcinogen and catalyze detoxification reactions [62]. It is worth noting that the protective effects conferred by LAB only appear when these are at a high density and when there is a regular intake [87].

Evidence is accumulating that heterocyclic aromatic amines (HCAs), which are derived from amino acids in meat during cooking, might be involved in the etiology of human cancer [88]. Zsivkovits et al. [89] showed that *L. bulgaricus* 291, *S. thermophilus* F4, *S. thermophilus* V3 and *B. longum* BB536 are highly protective against the genotoxic effects of HCAs in rats. Additionally, the inhibition of HCAs induced DNA damage was dose dependent and significant when 1 x $10^7$ cells/animal were administered. Other authors showed that *L. casei* DN 114001 may metabolize or adsorb HCAs and reduce their genotoxicity *in vitro* [89].

*In vivo* evidence that probiotics bond the carcinogens are still not conclusive. Hayatsu Hayatsu (1993) demonstrated the marked suppressive effect of orally administered *L. casei Shirota* (LcS) on the urinary mutagenicity arising from ingestion of fried ground beef by humans. In another clinical trial, the consumption of *L. acidophilus* decreased the urinary and fecal excretion of mutagens [68]. In view of the *in vitro* results, it is possible that the LAB supplements are influencing excretion of mutagens by simply binding them in the intestine [62]. Even though the binding of carcinogens is a possible mechanism for the inhibition of genotoxicity and mutagenicity by LAB in vitro, some researchers have reported that it does not appear to have any influence *in vivo* [90]. Additionally, the extent of the binding depends on the mutagen and bacterial strain used [71].

Several studies have also reported the effect of probiotics on the promotion phase of carcinogenesis. Rowland et al. [91] found that administration of *B. longum* (6 x $10^9$ CFU/day) inhibited the formation of aberrant crypt foci (ACF) in rats that received an induced of carcinogenesis (azomethane). As the probiotic treatment began 1 week after exposure to the carcinogen, these results indicate an effect on the early promotional phase of carcinogenesis [71].

Goldin et al. [92] observed a lower incidence of colonic tumors in rats who consumed *Lactobacillus* GG before, during and after chemical induction with dimethylhydrazine (DMH) than in animals that were fed the probiotic after receiving carcinogen. The researchers concluded that probiotics acted by inhibiting the initiation stage of carcinogenesis.

Kumar et al. [93] tested the efficacy of *L. plantarum* AS1 in the suppression of colorectal cancer induced by DMH in rats and formed that AS1 was capable of diminishing colon tumor through its antioxidant activity. However, long-term administration of this strain was necessary to achieve the maximum inhibitory effect.

On the other hand, not all studies have shown significant effects of probiotic on carcinogen-induced ACF. Gallanger et al. [94] using na ACF promotion protocol together with *B. longum* and *L. acidophilus*, obtained inconsistent results, which they attributed to differences in the ages of rats when DMH was administered.

Several studies have correlated the effect of probiotic on colon cancer with the modulation of the immune system. There is evidences that probiotics may contribute to the development of the mucosal immune system by influencing the innate inflammatory response and reducing mucosal inflammation. Additionally, probiotics also act on dendritic and epithelial cells and native T cells in the lamina propria of the gut and can thus influence adaptive immunity [13, 95].

Probiotics may influence the immune system by the action of products, such as metabolites, cell-wall components and DNA. Thus, immune modulatory effects might even be achieved by dead probiotic microorganisms or just probiotic derived components such as peptidoglycan fragments or DNA. Probiotic products are recognized by host cells sensitive to them these because they are equipped with recognition receptors adhesion. The main target cells in this context are therefore gut epithelial and gut-associated immune cells. The adhesion of probiotics to epithelial cells might itself might already trigger a signaling cascade leading to immune modulation [96].

Recent advances in the understanding of the immunomodulatory activity of probiotics have resulted from the discovery of Toll-like pattern recognition receptors (TLRs). These are transmembrane proteins present on the surface of cells such as macrophages, monocytes, dendritic cells and epithelial cells [97].

The innate immune system recognizes a large number of molecular structures from bacteria, such as, lipopolysaccharides and lipoteichoic acid, and is able to distinguish whether a particular microorganism is part of its microbiota or not. Different structures can activate different TLRs [98]. For example, TLR-2 recognizes the peptidoglycan, lipoteichoic acid, which is a component of the wall of Gram-positive bacteria such as lactobacilli and bifidobacteria [99], whereas TLR-4 is the most important receptor for lipopolysaccharide, the main component of the wall of Gram-negative bacteria [100].

Rachmilewitz et al. [101] using a probiotic mixture of 8 strains of freeze-dried lactic acid bacteria (*Bifidobacterium longum, B. infantis, B. breve, Lactobacillus acidophilus, L. casei, L. delbrueckii subsp. bulgaricus, L. plantarum, Streptococcus salivaris subsp. thermophilus*), reported that the chromosomal DNA of this mixture was responsible, via TLR-9 receptors for an anti-inflammatory effect observed in mice with colitis.

The connection of components of microorganisms to these receptors can lead to a cascade of inflammatory reactions via the activation of nuclear factor-kB (NF-kB), with subsequent release of cytokines, epitope chemokines and lipid mediators of reactive oxygen and nitrogen species [102]. Studies have shown that probiotics can activate elements responsible for the formation of cytokines and epitope chemokine's, although that response was weaker for *L. rhamnosus* if than for a Gram-positive pathogen (*Streptococcus pyogenes*) [103]. Some authors have suggested that a possible mechanism of action of probiotics would be the inhibition of NF-kB activation by reducing intestinal inflammation [104]. However, the possible mechanisms of probiotics against carcinogenesis, regarding the modulation of the immune system, are complex and still need to be better further elucidated.

An inflammatory immune response produces monocytes and macrophages, activated by cytokines that release cytotoxic molecules capable of the lyzing tumor cells *in vitro* [105]. The cytokines IL-1 and inflammatory TNF (tumor necrosis factor) exert cytotoxic and cytostatic effects on neoplastic cells *in-vitro* [106]. Natural-killer cells (NK) are effective against tumor cells and low activity of this cell type has been linked to a risk of cancer [107]. Matsuzaki and Chin [108] found that in mice, NK cell activity and inflammatory responses increased with the administration of probiotic strains.

Several studies in humans have shown an increase of NK cells in response to the consumption of probiotics [109, 110], and the same has been in animal models. When Takagi et al. [111] administered the strain *L. casei* Shirota, in order to inhibit tumor development induced by methylcholantracene in mice, there were high levels of NK cells in the group treated with the probiotic, which slowed the early development of the tumor, compared to the control group.

On the other hand, Berman et al. [112] did not observed any increasing in NK cells in healthy subjects who consumed during 8 weeks a formulation containing 4 species of probiotics (*L. rhamnosus, L. plantarum, L. salivarus* and *B. bifidum*). However, the researchers did note an increase in phagocytosis by neutrophils and monocytes.

Evidence has shown that the probiotic *Lactobacillus casei* Shirota has anti-tumor effects and antineoplastic action in rodents (biologically or chemically induced). Intrapleural administration of the strain in mice with tumor induced the production of various cytokines, such as interferon IL-1 and TNF in the thoracic cavity, which resulted in tumor inhibition and increased survival [113]. A study on *B. longum* and *B. animalis* showed that these bacteria induce the production of inflammatory cytokines (IL-6 and TNF-) [114].

In a clinical trial, the effect of *L. casei* Shirota on NK cell activity in humans was investigated. The activity of NK was increased as a likely consequence of *L. casei* Shirota-induced IL-12 production which was detected in *in vitro* assays [115].

According to the results of the various studies mention here, the probiotic microorganisms are capable of modulating the immune system in a strain-specific manner [116]. Therefore, different strains may induce different immune responses that might lead to the inhibition of carcinogenesis.

## 4. Effects of *Enterococcus faecium* CRL 183 on intestinal microbiota and colon cancer

*Enterococcus* spp. are Gram-positive, non-sporulating, catalase and oxidase negative facultative anaerobes [72]. Species of this genus are natural constituents of the intestinal microbiota of humans and comprise the third-largest genus of lactic acid bacteria (LAB), after *Lactobacillus* spp. and *Streptococcus* spp. [117].

It is hard to determine the exact number of enterococci species, but from a microbiological and functional point of view, *Enterococcus faecalis* and *Enterococcus faecium* are considered the most important [117, 118].

Some strains of *Enterococcus* spp. exhibit antibiotic resistance, possess virulence factors (adhesions, invasins, pili and haemolysin) and may cause bacteremia, endocarditis and other infections [117]. However, commercial pharmaceutical preparations of enterococci include *Enterococcus faecium* SF68® (NCIMB 10415, produced by Cerbios-Pharma SA, Barbengo, Switzerland) and *Enterococcus faecalis* Symbioflor 1 (SymbioPharm, Herborn, Germany), are on the market without reported health problems. Since 2008, *Enterococcus faecium* has been authorized for use in food and recognized as a probiotic microorganism in Brazil [119].

Currently, several strains of *Enterococcus faecium* are considered safe for human consumption, being used as starter cultures in cheese making and other fermented products and recognized as probiotic microorganisms [120]. The use of *Enterococcus faecium* as a starter culture in various fermented foods can be explained by its resistance to high concentrations of NaCl and low pH, and its ability to produce different aromas.

The strain of *Enterococcus faecium* CRL 183 was isolated by researchers from at the Reference Center for Lactobacillus (Cerela-Argentina), from cheese samples of Tafí – a traditional homemade cheese from the highland province of Tucuman, Argentina [121]. *In vitro* and *in vivo* studies showed that *Enterococcus faecium* CRL 183 is able to adhere to the intestinal cells, resists the gastrointestinal environment and colonizes the large intestine of rats, thus satisfying the requirement for a probiotic microorganism [122, 123]. Furthermore, this strain has no antibiotic resistance and no virulence factors, ensuring its safe use as a starter culture [121].

*Enterococcus faecium* CRL 183 has been investigated by our research group for about 20 years, with the objective of defining its functional properties in the free form or associated with food products [122,123,124,125,126, 127, 128, 129, 130].

The best functional effects of *Enterococcus faecium* CRL 183 were obtained when this microorganism was used as a starter culture of a yogurt-like fermented soy product (soy yogurt) [129]. This product has sensorial and technological properties similar to fermented-milk yogurt drinks and has exhibited functional properties in animal tests and clinical trials. Among the beneficial effects of the soy product fermented with *Enterococcus faecium* CRL 183, the following deserve special attention: improved of lipid profile, modulation of the immune system, positive changes in the intestinal microbiota, and reduction of colon cancer development [123, 125,128,130] .

Sivieri et al [123] studied the effect of daily ingestion of *Enterococcus faecium* CRL 183 (8 log CFU/mL) on the incidence of colorectal tumors induced by 1,2 dimethylhydrazine (DMH) in rats (20 mg/kg body weight, in a weekly dose, for 14 weeks). The experiment was conducted over 42 weeks and the rats were allocated to three groups: G1 - Control (not induced); G2 – Induced with DMH; G3 – Induced with DMH + *E. faecium* CRL 183. Thioglycollate-elicited peritoneal exudate cells (PECs) were harvested from animals in PBS and the adherent cells were obtained after incubation with LPS or RPMI-1640 (CO2 - 95:5, v/v). The cytokine levels (IL-4, IFN- $\gamma$ and TNF- $\alpha$) were determined in the supernatant of of the cell culture by ELISA. After euthanasia, colons were removed for histological analysis. The animals with induced colorectal cancer and that received the suspension of *Enterococcus faecium* CRL 183 (G3) showed a 50% reduction in average number of tumors compared to G2 ($P < 0.001$) (Figures 2 and 3). The total number of aberrant crypt foci (ACF), the total ACF/mm$^2$, the number of crypts per ACF and the adenocarcinoma were also reduced in G3. In addition, G3 exhibited increased production of IL-4, IFN- $\gamma$ and TNF- $\alpha$ by PECs compared to G2.

Anti-tumor activities of probiotic acid lactic bacteria have been attributed to an enhanced immune response [132]. The induction of TNF- $\alpha$ by probiotic bacteria would be necessary to initiate cross-talk between the immune cells associated with the lamina propria and the intestinal epithelial cells. IFN- $\gamma$ is involved in the maturation of immune cells (dendritic cells), controls their cellular proliferation at the intestinal level and induce other cytokines, especially IL-4, IL-5 and IL-10. Because of its role in mediating macrophage and NK cell activation, IFN- $\gamma$ is important in the host defense against intracellular pathogens, viruses and tumors [133]. According to Perdigón et al. [134] IL-4 exerts control over the inflammatory response induced by the carcinogen. In that study, the antitumor activity of *Enterococcus faecium* CRL 183 was attributed to its ability to modulate the immune response.

It has been suggested that increasing the consumption of red meat and animal fat lead to an increased risk of developing cancer colon, in comparison with a vegetarian diet [23]. Several studies have demonstrated that the microbiota of the colon is involved in the etiology of the colon cancer and that the some strains of probiotic microorganism can have beneficial effects on the composition of the intestinal microbiota, stimulates the production of short chain fatty acids (SCFA) and inhibit the activity of enzymes that convert pro-carcinogens into carcinogens [39,49, 135, 136].

Based in these evidence, a study was carried out to determine if consumption of a soy product fermented with *E. faecium* CRL 183 was able to modify the fecal microbiota of rats fed a diet containing red meat [122]. The experiment was conducted during over days and the animals were randomly divided into six groups: GI - standard casein-based rodent feed; GII to GVI - beef-based feed. From the 30th day, G III–VI also ingested the following products: G III, *E. faecium*-fermented soy product; G IV, pure suspension of *E. faecium*; G V, sterilized fermented soy product; and G VI, unfermented soy product (3 mL kg$^{-1}$ BW day$^{-1}$). The feces of each animal were collected at the start (T0) and on the 30th (T30) and 60th (T60) days of the experiment, to determine the viable cell counts of total aerobic and anaerobic bacteria, *Enterococcus* spp., Enterobacteria, *Lactobacillus* spp., *Clostridium* spp., *Bacteroides* spp. and *Bifidobacterium* spp.

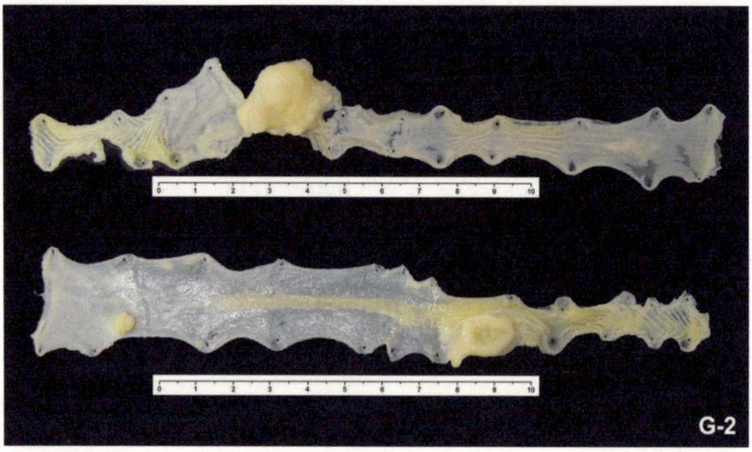


**Figure 2.** Topographic view of macroscopic growths by G2 – Induced with DMH. Sivieri et al [123]

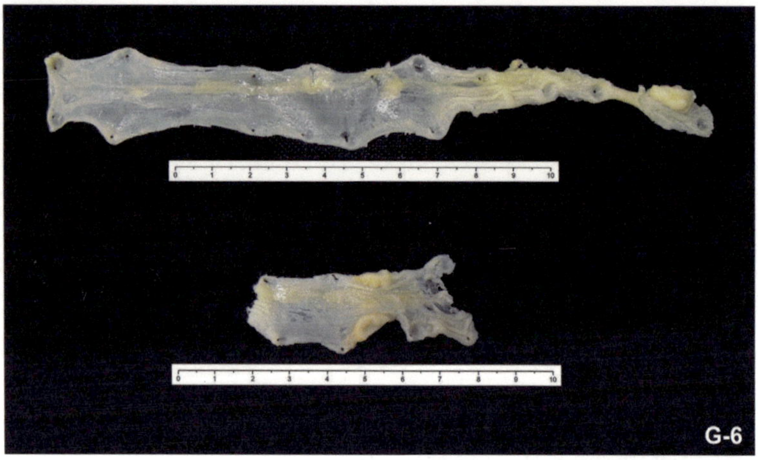


**Figure 3.** Topographic view of macroscopic growths by G3 – Induced with DMH + *E. faecium* CRL 183. Sivieri et al [123]

By day T30 days of experiment, rats on a red meat-based diet exhibited an increase in the population of total anaerobes, enterobacteria and enterococci and a decrease in the numbers of lactobacilli and bifidobacteria. From T30 to T60, the obtained results showed that fermented soy product and pure *Enterococcus faecium* CRL 183 suspension promoted an increase in the numbers of lactobacilli (0.45 log CFU $g^{-1}$ and 1.83 log CFU $g^{-1}$, respectively). During the same period, only the animals treated with pure *Enterococcus faecium* CRL 183 suspension showed a rise in the fecal bifidobacterium population. The fermented soy product promoted a slight fall in the *Bacteroides* spp. population (2.80 ± 0.20 to 2.34 ± 0.07 log CFU g–1), but the counts of *clostridia*. and enterobacteria were unchanged.

Another study, using New Zealand rabbits with induced hypercholesterolemia as an animal model, was conducted to investigate the possible correlation between fecal microbiota, serum lipid parameters and atherosclerotic lesion development. It was show that, after 60 days of the experiment, intake of the probiotic soy product (with or without isoflavones) was correlated with significant increases ($P<0.05$) in *Lactobacillus* spp., *Bifidobacterium* spp. and *Enterococcus* spp. and a decrease in the enterobacteria population (Cavallini et al., 2011).

The studies conducted by Bedani et al (2010) and Cavallini et al. (2011) suggest that daily ingestion of the soy product fermented with *Enterococcus faecium* CRL 183, or the pure culture of this probiotic microorganism, may contribute to a beneficial balance of the fecal microbiota.

Currently, other studies, using animal models and an *in vitro* simulator of human intestinal microbial ecosystem (SHIME), are being conducted by our research group in order to elucidate the possible mechanisms involved in the protective effect of *Enterococcus faecium* CRL 183 against colon cancer and the importance of the modulation of fecal microbiota and stimulation of the immune system in the disease pathogenesis [53].

## 5. Conclusions

From the above discussion, it is evident that probiotics have the capacity to modulate the intestinal microbiota and the immune system, to the benefit of the host organism, reducing the risk of many chronic degenerative diseases, among them colon cancer. It appears also that the actions performed by probiotics are species-strain-dependent, so that several effects or actions can occur with the same bacterial genus. However, the results of several experiments reported in the literature, highlight a degree of controversy concerning the effects observed, especially regarding the various types of cancers and it is difficult to compare these studies. Such controversies are due mainly to large variations in the time of the experiment - usually prevailing those of short duration the experimental models, bacterial strains and the doses and frequencies of administration of probiotics. In this sense, it is important that further studies be done to define and standardize these variables mentioned, and especially to elucidate the mechanisms involved in each of the observed effects.

It showed also be mentioned that, according there is also in the literature, that probiotics studied are taken almost exclusively in milk as can be observed in the products available on the market. This condition often makes them inappropriate for certain lactose intolerant population groups on those and allergic to milk proteins. Thus, alternative vehicles for probiotics, free of lactose and of β-lactoglobulin, such as the aqueous extract of soybeans, for example, deserve special attention from researchers seeking to develop products with a good nutritional profile and suitable to transport the probiotic specified for the purpose desire. It is expected that in the near future, as results of the interaction of various fields of study such as food science and technology, nutrition, microbiology, genetic engineering and molecular biology the market can offer consumers products that are more accessible and effective, reducing the risk of certain diseases, particularly certain types of cancer, and acting as adjuvants in specific treatments for existing diseases.

Finally, from the results obtained by our research group in the studies of probiotics in relation to colon cancer, and even other diseases, it appears that these was always variability between individuals, either in clinical trials or in studies with animal models, suggesting a possible specificity of these individuals in relation to consumption of given probiotics. This leads us to wonder, if today nutrigenomics is already a reality, is it not the moment to propose studies on something like "probiogenomics" or even about self-probiotics? Certainly the future will provide an answer to that question

## Author details

Katia Sivieri,
*Department of Food & Nutrition, Faculty of Pharmaceutical Sciences, São Paulo State University, Araraquara, SP, Brazil*

Raquel Bedani
*University of São Paulo, Faculty of Pharmaceutical Sciences, São Paulo, SP, Brazil.*

Daniela Cardoso Umbelino Cavallini
*Department of Food & Nutrition, Faculty of Pharmaceutical Sciences, São Paulo State University, Araraquara, SP, Brazil*

Elizeu A. Rossi
*Department of Food & Nutrition, Faculty of Pharmaceutical Sciences, São Paulo State University, Araraquara, SP, Brazil*

## 6. References

[1] Stein K, Borowicki A, Scharlau D, Schettler A, Scheu K, Obst U, Glei M (2011). Effects of synbiotic fermentation products on primary chemoprevention in human colon cells. J Nutr Biochem. (in press).

[2] Turpin W, Humblot C, Thomas M, Guyot JP (2010).Lactobacilli as multifaceted probiotics with poorly disclosed molecular mechanisms. Int J Food Microbiol. 143:87–102.

[3] Lyra A, Lahtinen S, Ouwehand AC. Gastrointestinal benefits of probiotics – clinical evidence. In Salminen, S.; von Wright, A.; Lahtinen, S.; Ouwehand, A., eds. Lactic acid bacteria: microbiological and functional aspects. 4th ed. Boca Raton: CRC Press, 2012. p. 509-523.

[4] Pearson JR, Gill CIR. Rowland, IR (2009). Diet, fecal water, and colon cancer – development of biomarker. Nutr Rev. 67:509-526.

[5] Reddy BS (2000). Novel approaches to the prevention of colon cancer by nutritional manipulation and chemoprevention. Cancer Epidem Biomar. 9:239–247.

[6] Norat T, Lukanova A, Ferrari P, Riboli E (2002). Meat consumption and colorectal cancer risk: dose-response meta-analysis of epidemiological studies. Int J Cancer. 98:241–256.

[7] O'Keefe SJD, Ou J, Aufreiter S, O'Connor D, Sharma S, Sepulveda J, Fukuwatari T, Shibata K, Mawhinney T (2009). Products of the colonic microbiota mediate the effects of diet on colon cancer risk. Journal of Nutrition, v. 139, p. 2044–2048, 2009.
[8] McGarr SE, Ridlon JM, Hylemon PB (2005). Diet, anaerobic bacterial metabolism, and colon cancer. J Clin Gastroenterol. 39:98–109.
[9] Hatakka K, Holma R, El-Nezami H, Suomalainen T, Kuisma M, Saxelin M, Poussa T, Mykkänen H, Korpela R (2008). The influence of Lactobacillus rhamnosus LC705 together with Propionibacterium freudenreichii ssp. shermanii JS on potentially carcinogenic bacterial activity in human colon. Int J Food Microbiol. 128:406–410.
[10] Gong J, Yang C (2012). Advances in the methods for studing gut microbiota and their relevance to the research of dietary fiber functions. Food Res Int. (in press).
[11] FAO/WHO. Working Group Report on Drafting Guidelines for the Evaluation of Probiotics in Food. London, Ontario, Canada, april 30 and may 1, 2002.
[12] Rauch M, Lynch SV (2011). The potential for probiotic manipulation of the gastrointestinal microbiome. Curr Opin Biotech.23:191-201.
[13] Zhu Y, Luo TM, Jobin C, Young, H.A (2011). Gut microbiota and probiotics in colon tumorigenesis. Cancer Letters. 309: 119–127.
[14] CVE. CENTRO DE VIGILÂNCIA EPIDEMIOLÓGICA. Doenças crônicas não transmissíveis. São Paulo, 2012. Disponível em: www.cve.saude.sp.gov.br/htm/cve_dcnt.html Acesso em: 18 jan. 2012.
[15] INCA – INSTITUTO NACIONAL DE CÂNCER. Ações de prevenção primária e secundária no controle do câncer. Rio de Janeiro: Inca, 2008. 628p. Disponível em: www1. inca.gov.br/enfermagem/docs/cap5.pdf Acesso em: 08 jan. 2012.
[16] Neves FJ, Mattos IE, Koifman RJ (2005). Mortalidade por câncer de cólon e reto nas capitais brasileiras no período 1980-1997. Arquivos de Gastroenterologia, 42:63 – 70.
[17] Hope ME, Hold GL, Kain R (2005). Sporadic colorectal cancer – role of the commensal microbiota. FEMS Microbiol. Lett. 244:1-7.
[18] Grady WM (2004). Genomic instability and colon cancer. Genomic Metastasis Reviews, 23:11 – 27.
[19] Rabeneck L, Davila JA, El-Serag HB (2003). Is there a true "shift" to the right colon the incidence of colorectal cancer? Am. J. Gastroenterol. 98:1400-1409.
[20] Commane D, Hughes R, Shortt C (2005). The potential mechanisms involved in the anti-carcinogenic action of probiotics. Mutat. Res.591:276-289.
[21] Bartram HP, Scheppach W, Schmid H (1993). Proliferation of human colonic mucosa as an intermediate biomarker of carcinogenesis: effects of butyrate, deoxycholate, calcium, ammonia, and pH. Cancer Res. 53:3283-3288.
[22] Macbain AJ, Macfarlane GT (2007). Ecological and physiological studies on large intestinal bacteria in relation to production of hydrolytic and reductive enzymes involved in formation of genotoxic metabolites. J. Med. Microbiol. 47: 407-416.
[23] Guarner F, Malagelada JR (2003). Gut flora in health and disease. Lancet 361:512–519.
[24] Pinho MA (2008). biologia molecular das doenças inflamatórias intestinais. Rev bras colo-proctol. 28:119 – 123.

[25] Owen RW, Spiegelhalder B, Bartsch H (2000). Generation of reactive oxygen species by the faecal matrix. Gut. 46:225-232.

[26] Huycke MM, Abrams V, Moore DR (2002). Enterococcus faecalis produces extracellular superoxide and hydrogen peroxide that damages colonic epithelial cell DNA. Carcinogenesis. 23:529-536.

[27] Montalto M, D'Onofrio F, Gallo A, Cazzato A, Gasbarrini G. Intestinal microbiota and its functions (2009). Digestive and Liver Disease Supplements, 3:30-34.

[28] Gibson GR, Fuller R (2000). Aspects of in vitro and in vivo research approaches directed toward identifying probiotics and prebiotics for human. J. Nutr.130:S391-S395.

[29] Cummings JH, Macfarlane GT (1997). Colonic microflora: nutrition and health. Nutrition.13:476-478.

[30] Whitman W.B., Coleman D.C., Wiebe WJ (1998). Prokaryotes: the unseen majority. Proc. Natl. Acad. Sci. USA. 95:6578–6583

[31] Frank DN, St Amand AL, Feldman RA, Boedeker EC, Harpaz N, Pace NR (2007). Molecular-phylogenetic characterization of microbial commu- nity imbalances in human inflammatory bowel diseases. Proc. Natl. Acad. Sci. USA. 104: 13780 –13785.

[32] Mason KL, Huffnagle GB, Noverr MC, Kao JY (2008). Overview of gut immunology. In: GI Microbiota and Regulation of the Immune System, edited by GB Huffnagle and MC Noverr. Austin, TX: Landes Bioscience and Springer Science®Business Media, 2008, p. 1 - 14.

[33] Eckburg PB (2005). Diversity of the human intestinal microbial flora. Science. 308:1635–1638

[34] Tsai F, Coyle WJ. (2009) The microbiome and obesity: is obesity linked to our gut flora? Curr Gastroenterol Rep 11:307–313.

[35] Metges CC (2000). Contribution of microbial amino acids to amino acid homeostasis of the host. J. Nutr. 130:1857S–1864S

[36] Cummings JH., Macfarlane GT, Englyst HN (2001). Prebiotic digestion and fermentation. Am. J. Clin. Nutr. 73:415S–20S.

[37] Pryde SE, Duncan SH, Hold G.L (2002). The microbiology of butyrate formation in the human colon. FEMS Microbiol. Lett. 17:133-139.

[38] Csordas A (1996). Butyrate, aspirin and colorectal cancer. Eur. J. Cancer. Prev. 5:221-231.

[39] Wollowski I, Rechkemmer G, Pool-Zobel B(2001). Protective role of probiotics and prebiotics in colon cancer. Am. J. Clin. Nutr. 73:451S–455S

[40] Segain JP, Raingeard de la Bletiere D, Bourreille A (2000). Butyrate inhibits inflammatory responses through NFkappaB inhibition: implications for Crohn's disease. Gut: 47:397-403.

[41] Jones BV, Begley M, Hill C, Gahan CG, Marchesi JR (2008). Functional and comparative metagenomic analysis of bile salt hydrolase activity in the human gut microbiome. Proc Natl Acad Sci USA 105: 13580–13585.

[42] Round JL, Mazmanian SK (2009). The gut microbiota shapes intestinal immune responses during health and disease. Nat Rev Immunol 9: 313–323.

[43] Azcárate-Peril MA, Sikes M, Bruno-Bárcena JM (2011). The intestinal microbiota, gastrointestinal environment and colorectal cancer: a putative role for probiotics in

prevention of colorectal cancer? Am. J. Physiol. Gastrointest. Liver Physiol. 301: G401–G424.

[44] Molly K, Woestyne MV, Verstraete (1993). Development of a 5-step multichamber reactor as a Simulation of the Human Intestinal Microbial Ecosystem. Appl. Microbiol. Biotechnol, 39:254–2583.

[45] Possemiers S, Verthé K, Uyttendaele S, Verstraete W (2004). PCR-DGGE-based quantification of stability of the microbial community in a simulator of the human intestinal microbial ecosystem. Fems Microbiol Ecol. 49: 495–507.

[46] Saavedra L, Taranto MP, Sesma F, Valdez GF (2003). Homemade traditional cheeses for the isolation of probiotic Enterococcus faecium strains. Int J Food Microbiol. 88:241-245.

[47] Mercenier A, Pavan S, Pot B (2002). Probiotic as biotherapeutic agents: present knowledge and future prospects. Curr. Pharm. Design.8:99-110.

[48] Berg RD (1998). Probiotic, probiotics or conbiotics? Trends Microbiol. 6:89-92.

[49] Salminen S, Bouley C, Boutron-Ruault MC, Cummings JH, Franck A, Gibson GR (1998). Functional food science of gastrointestinal physiology and function. Br. J. Nutr. 80:147S–171S

[50] MacFarlane GT, MacFarlane S (2007). Models for intestinal fermentation: association between food components, delivery systems, bioavailability and functional interactions in the gut. Curr Opin Biotechnol. 18:156-162.

[51] Payne S, Gibson G, Wynne A, Hudspith B, Brostoff J, Tuohy K (2003). In vitro studies on colonization resistance of the human gut microbiota to Candida albicans and the effects of tetracycline and Lactobacillus plantarum LPK. Curr. Issues Intest. Microbiol. 4:1-8.

[52] Gibson GR, Roberfroid MB (1995). Dietary modulation of the human colonic microbiota: introducing the concept of prebiotics. J Nutr. 125:.1401 1412.

[53] Sivieri K, Bianchi F, Rossi EA (2011). Fermentation by gut microbiota cultured in a simulator of the human intestinal microbial ecosystem is improved probiotic Enterococcus faecium CRL 183. Functional foods in health and disease, 10:389-402.

[54] Pereira DIA, MCCartney, AL, Gibson G (2000). An in vitro study of the probiotic potential of a bile salt hydrolyzing Lactobacillus fermentum strain, and determination of its cholesterol-lowering properties. Appl Environ Microbiol. 69:4743-4752

[55] Rycroft CE, Jones MR, Gibson GR, Rastall RA (2001). Acomparative "in vitro" evaluation of the fermentation properties of prebiotic oligosaccharides. J. Applied Microbial. 91:878-887.

[56] Tzortzis G, Goulas AK, Gee GM, Gibson GR (2005). A Novel Galactooligosaccharide Mixture Increases the Bifidobacterial Population Numbers in a Continuous In Vitro Fermentation System and in the Proximal Colonic Contents of Pigs In Vivo. J. Nutr. 135:1726-1731.

[57] McFarlane S, Furrie E, Cummings JH, McFarlane, GT (2004). Chemotaxonomic analysis of bacterial populations colonizing the rectal mucosa in patients with ulcerative colitis. Clin Infect Dis. 38:1690-1699.

[58] Minekus, M. (1995). A multi compartmental dynamic computer-controlled model simulating the stomach and small intestine. Altern. Lab. Anim. (ALTA) 23:197–209

[59] Souliman, S. (2007) Investigation of the biopharmaceutical behavior of theophylline hydrophilic matrix tablets using USP methods and an artificial digestive system. Drug Dev. Ind. Pharm. 33, 475–483.

[60] Payne A N , Zihler A, Chassard C, Lacroix C (2012). Advances and perspectives in in vitro human gut fermentation modeling. Trends Biotechnol. 30:17–25

[61] Guiemonde M, Salminen S (2006). New methods for selecting and evaluating probiotics. Dig. Liver. Dis. 38: 242 S-247 S.

[62] Rafter J (2003). Probiotics and colon cancer. Best. Pract. Res. Clin. Gastroenterol. 17:849-859.

[63] Fotiadis C, Stoidis CN, Spyropoulos BG, Zografos ED (2008). Role of Probiotics, Prebiotics and Synbiotics in Chemoprevention for Colorectal Cancer. World J. Gastroenterol. 14:6453-6457.

[64] Oelschlaeger T (2010). Mechanisms of probiotic actions. International Journal of Medical Microbiology 300:57-62.

[65] Goldin BR, Gorbach SL (1984). Alterations of the intestinal microflora by diet, oral antibiotics and Lactobacillus: decreased production of free amines from aromatic nitro compounds, azo dyes and glucuronides. J. Natl. Cancer. Inst. 73:689-695.

[66] Spanhaak S, Havenaar R, Schaafsma G (1998). The effect of consumption of milk fermented by Lactobacillus casei strain Shirota on the intestinal microflora and immune parameters in humans. Eur. J. Clin. Nutr. 52:899-907.

[67] Haberer P, Toit M, Dicks LMT (2003). Effect of potentially probiotic lactobacilli on faecal enzyme activity in minipigs on high-fat, high-cholesterol diet – a preliminary in vivo trial. Int. J. Food Microbiol. 87:287-291.

[68] Lidbeck A, Nord CE, Gustafsson JA, Rafter J (1992). Lactobacilli, anticarcinogenic activities and human intestinal microflora. Eur. J. Cancer. Prev. 1:341-353.

[69] Ling WH, Hänninen O, Mykkänen H, Heikure M (1992). Colonization and fecal enzyme activities after oral Lactobacillus GG administration in eldery nursing home residents. Ann. Nutr. Metab. 36:162-166.

[70] Mital BK, Garg SK (1995). Anticarcinogenic, hypocholesterolemic, and antagonistic activities of Lactobacillus acidophilus. Crit. Rev. Microbiol. 21:175-214.

[71] Burns AJ, Rowland I (2000). Anti-Carcinogenicity of Probiotics and Prebiotics. Curr. Issues Intest. Microbiol. 1: 13-24.

[72] Foulquié Moreno MR, Sarantinopoulos P, Tsakalidou E, De Vuyst L (2006). The role and application of enterococci in food and health. Intern J Food Microbiol. 106:1-24.

[73] Benno Y, Mitsuoka T (1992). Impact of Bifidobacterium on human fecal microflora. Microbiol. Immunol. 36:683-694.

[74] Marteau P, Pochart P, Flourié B, Pellier P, Santos L, Desjeux JF, Rambaud JC (1990). Effect of Chronic Ingestion of a Fermented Dairy Product Containing Lactobacillus acidophilus and Bifidobacterium bifidum on Metabolic Activities of the Colonic Flora in Humans. Am J. Clin. Nutr. 52:685-688.

[75] Reddy BS (2000). Novel approaches to the prevention of colon cancer by nutritional manipulation and chemoprevention. Cancer Epidemiol Biomarkers Prev. 9:239–247.

[76] Norat T, Lukanova A, Ferrari P, Riboli E (2002). Meat consumption and colorectal cancer risk: dose-response meta-analysis of epidemiological studies. Int. J. Cancer. 98:241–256.

[77] Iannitti T, Palmieri B (2010). Therapeutical Use of Probiotic Formulations in Clinical Practice. Clin. nutr. 29:701-725.

[78] Shahani KM, Ayebo AD (1980). Role of dietary lactobacilli in gastrointestinal microecology.Am J Clin Nutr. 33:2448-57.

[79] Savard P, Lamarche B, Paradis M.E, Thiboutot H, Laurin E, Roy D (2011). Impact of Bifidobacterium animalis subsp. lactis BB-12 and, Lactobacillus acidophilus LA-5-containing yoghurt, on fecal bacterial counts of healthy adults. Int. j. food microbiol. 149:50-57.

[80] Bartram HP, Scheppach W, Gerlach S. (1994). Does yoghurt enriched with Bifidobacterium longum affect colonic microbiology and fecal metabolites. Am. J. Clin. Nutr. 59:428-432.

[81] Kopp-Hoolihan L (2001). Prophylactic and therapeutic uses of probiotics: a review. J. Am. Diet. Assoc.101:229-241.

[82] Wollowski I, Rechkemmer G, Pool-Zobel B(2002). Protective role of probiotics and prebiotics in colon cancer. Am. J. Clin. Nutr. 73:451S–455S

[83] Zampa A, Silvi S, Fabiani R, (2004). Effects of different digestible carbohydrates on bile acid metabolism and SCFA production by human gut micro-flora grown in an in vitro semi-continuous culture. Anaerobe. 10:19-26.

[84] Scharlau D, Borowicki A, Habermann N, Hofmann T, Klenow S, Miene C (2009). Mechanisms of primary cancer prevention by butyrate and other products formed during gut flora-mediated fermentation of dietary fibre. Mutat. Res. 682:39-53.

[85] Sengupta S, Muir JG, Gibson PR (2006). Does butyrate protect from colorectal cancer? J Gastroenterol Hepatol; 21:209–18.

[86] Knasmuller S, Steinkellner H, Hirschl AM. Impact of bacteria in dairy products and of the intestinal microflora on the genotoxic and carcinogenic effects of heterocyclic aromatic amines. Mutat Res 2001;480:129-138.

[87] Tannock GW, Munro K, Harmsen HJM (2000). Analysis of the fecal microflora of human subjects consuming a probiotic product containing Lactobacillus rhamnosus DR20. Appl Environ Microbiol. 66:2578-2588.

[88] Nowak A, Libudzisz Z (2009). Ability of Probiotic Lactobacillus casei DN 114001 to Bind or/and Metabolise Heterocyclic Aromatic Amines In Vitro. Eur. J. nutr. 48:419–427.

[89] Zsivkovits M, Fekadu K, Sontag G, Nabinger U, Huber W.W, Kundi M, Chakraborty A, Foissy H, Kasmuller S (2003). Prevention of heterocyclic amine-induced DNA damage in colon and liver of rats by different Lactobacillus strains. Carcinogenesis. 24:1913-1918.

[90] Bolognani F, Rumney C.J, Rowland I.R (1997). Influence of carcinogen binding by lactic acid producing bacteria on tissue distribution and in vivo mutagenicity of dietary carcinogens. Food Chem. Toxicol. 35: 535-545.

[91] Rowland I, Rumney C, Counts J, (1998). Effects of Bifidobacterium longum and inulin on gut bacterial metabolism and carcinogen induced aberrant crypt foci in rats. Carcinogenesis.19:281-285.

[92] Goldin B, Gualtieri L, Moore R (1996). The effect of Lactobacillus GG on the initiation and promotion of DMH induced intestinal tumours in the rat. Nut. Cancer 25:197-204.

[93] Kumar RS, Kanmani P, Yuvaraj N, Paari KA, Pattukumar V, Thirunavukkarasu C, Arul V (2012). Lactobacillus plantarum AS1 Isolated from South Indian Fermented Food Kallappam Suppress 1,2-Dimethyl Hydrazine (DMH)-Induced Colorectal Cancer in Male Wistar Rats. Appl. biochem. biotechnol 166:620–631.

[94] Gallaher DD, Stallings WH, Blessing LL, Busta FF, Brady LJ (1996). Probiotics cecal microflora, and aberrant crypts in the rat colon. J. Nutr. 126: 1362-1371.

[95] Saad SMI, Bedani R, Mamizuka EM. Benefícios à saúde dos probióticos e prebióticos. In: Saad, S.M.I.; Cruz, A.G.; Faria, J.A.F. eds. Probióticos e prebióticos em alimentos: fundamentos e aplicações tecnológicas. São Paulo: Livraria Varela, 2011.p.51-84.

[96] Oelschlaeger T (2010) Mechanisms of probiotic actions. International Journal of Medical Microbiology 300:57-62.

[97] Corthésy B, Gaskins HR, Mercenier A (2007). Cross-talk between probiotic bacteria and the host immune system. J Nutr. 137:781S-790S.

[98] Britti MS, Roselli M, Finamore A (2006). Regulation of immune response at intestinal and peripheral sites by probiotics. Biologia 61:735-740.

[99] Neuhaus FC, Baddiley J (2003). A continuum of anionic charge: structures and functions of d-alanyl-teichoic acids in gram-positive bacteria. Microbiol. Mol. Biol. Rev. 67:686-723.

[100] Chow JC, Young DW, Golenbock DT (1999). Toll-like Receptor-4 Mediates Lipopolysaccharide-induced Signal Transduction. J. Biol. Chem. 274:10689-10692.

[101] Rachmilwitz D, Katakura K, Karmeli F, Hayashi T, Reinus C, Rudensky B, Akira S, Takeda K, Lee J, Takabayashi K, Raz E (2004). Toll like receptor 9 signaling mediates the antiinflammatory effects probiotics in murine experimental colitis. Gastroenterology, 126:520-528.

[102] Watson JL, Mckay DM. The immunophysiological impact of bacterial CpG DNA on the gut Clin Chim Acta 2006;364:1-11.

[103] Veckman, V, Miettinen, M, Siren, J, (2004). Streptococcus pyogenes and Lactobacillus rhamnosus differentially induce maturation and production of Th1-type cytokins and chemokines in human monocyte-derived dendritic cells. J. Leukoc. Biol.75:764-771.

[104] Petroff EO, Kojima K, Ropeleski MJ (2004). Probiotics inhibit nuclear factor-κB and induce heat shock proteins in colonic epithelial cells through proteasome inhibition.Gatroenterology.127:1474-1487.

[105] Philip R, Epstein I(1986). Tumour necrosis factor as immunomodulator and mediator of monocyte citotoxicity induced by itself, · -interferon and interleukin-1. Nature. 323:86-89.

[106] Raitano A, Kore M (1993). Growth inhibition of a human colorectal carcinoma cell line by interleukin-1 associated with enhanced expression of · -interferon receptors. Cancer Res. 53:636-640.

[107] Takeuchi H, Maehara Y, Tokunaga E (2001). Prognostic significance of natural killer cell activity in patients with gastric carcinoma: a multivariate analysis. Am. J. Gastroenterol. 96:574-578.

[108] Matsuazaki T, Chin J (2000). Modulating immune responses with probiotic bacteria. Immunol. Cell Biol. 78:67-73.

[109] Nagao F, Nakayama M, Muto T (2000). Effects of a fermented milk drink containing Lactobacillus casei strain Shirota on the immune system in healthy human subjects. Biosci. Biotechnol. Biochem. 64:2706-2708.

[110] Gill HS, Rutherford KJ, Cross ML (2001). Enhancement of immunity in the elderly by dietary supplementation with the probiotic Bifidobacterium lactis HN019. Am J Clin Nutr 74:833-839.

[111] Takagi A, Matsuzaki T, Sato M (2001). Enhancement of natural killer cell cytotoxicity delayed murine caecinogenesis by a probiotic microorganism. Carcinogenesis;22:599-605.

[112] Berman SH, Eichelsdoerfer P, Yim D (2006). Daily ingestion of a nutritional probiotic supplement enhances innate immune function in healthy adults. Nutr. Res.26:454-459.

[113] Matsuzaki, T (1998). Immunomodulation by treatment with Lactobacillus casei strain Shirota. Int. J. Food Microbiol. 16:133–140.

[114] Sekine K, Kawashima T, Hashimoto Y (1994). Comparison of the TNF-$\alpha$ levels induced by human-derived Bifidobacterium longum and rat-derived Bifidobacterium animalis in mouse peritoneal cells. Microflora;39:79-89.

[115] Takeda K, Suzuki T, Shimada SI, Shida K, Nanno M, Okumura K (2006). Interleukin-12 is involved in the enhancement of human natural killer cell activity by Lactobacillus casei Shirota. Clin. exp.immunol. 146:109–115.

[116] Oelschlaeger T (2010) Mechanisms of probiotic actions. International Journal of Medical Microbiology 300:57-62.

[117] Franz CMAP, Huch M, Abriouel H, Holzapfel W, Gálvez A (2011). Enterococci as probiotics and their implications in food safety. Intern. J. Food. Microbiol. 151:125-140.

[118] Foulquié-Moreno MR, Sarantinopoulos P, Tsakalidou E, De Vuyst L (2006). The role and application of enterococci in food and health. Int. J. Food Microbiol.106:1–24.

[119] AGÊNCIA NACIONAL DE VIGILÂNCIA SANITÁRIA. Alimentos com alegações de propriedades funcionais e ou de saúde, novos alimentos/ingredientes, substâncias bioativas e probióticos. Disponível em: http://www.anvisa.gov. br/alimentos/comissoes/tecno_lista_alega.htm. Acesso em: 01 abr. 2012.

[120] Gardiner GE, Ross RP, Wallace JM, Scanlan FD, Jägers PPJM, Fitzgerald GF, Collins K, Stanton C (1999). Influence of a Probiotic Adjunct Culture of Enterococcus faecium on the Quality of Cheddar Cheese. J. Agric. Food Chem. 47:4907-4916.

[121] Saavedra JM (2001). Clinical application of probiotic agents. Am. J Clin. Nutr. 73:1147S-1151S.

[122] Bedani R, Pauly-Silveira, ND, Roselino, MN, Valdez GF, Rossi EA. Effect of fermented soy product on the fecal microbiota of rats fed on a beef-based animal diet (2010). J Sci Food Agric. 90:233–238.

[123] Sivieri K, SpinardI-Barbisan ALT, Barbisan LF, Bedani R, Pauly ND, Carlos IZ, Benzatti F, Vendramini RC, Rossi EA (2008). Probiotic Enterococcus faecium CRL 183 inhibit chemically induced colon cancer in male Wistar rats. Eur Food Res Technol. 228:231-237.

[124] Cavallini DCU, Bedani R, Pauly ND, Bondespacho LQ, Vendramini RC, Rossi EA (2009). Effects of probiotic bacteria, isoflavones and simvastatin on lipid profile and atherosclerosis in cholesterol-fed rabbits. Lipids Health Dis.8:1-10.

[125] Manzoni MSJ (2008). Fermented soy product supplemented with isofl avones affects adipose tissue in a regional-speccifi c manner and improves HDL cholesterol in rats fed on a cholesterol-enriched diet. Eur. Food Res. Technol. 227:1591-1597.

[126] Rossi EA (1994). In vitro effect of Enterococcus faecium and Lactobacillus acidophilus on cholesterol. Microbiol. Alim. Nutr. 12: 267-270.

[127] Rossi EA, Cavallini DCU, Carlos IZ, Vendramini RC, Dâmaso AR, Valdez GF (2008). Intake of isoflavone-supplemented soy yogurt fermented with Enterococcus faecium lowers serum total cholesterol and non-HDL cholesterol of hypercholesterolemic rats. European Food Research and Technology. 228:275-282.

[128] Rossi EA, Cavallini DCU, Carlos IZ, Oliveira MG, Valdez GF (2003). Efeito de um novo produto fermentado de soja sobre lípides séricos de homens adultos normocolesterolêmicos. Arch. Lat. Nutr. 53:47-51.

[129] Rossi EA, Vendramini RC, Carlos IZ, Pei, YC, Valdez GF (1999). Development of a novel fermented soymilk product with potential probiotic properties. Eur Food Res Technol , 209:305-307.

[130] Rossi EA, Vendramini RC, Carlos IZ, Ueiji IS, Squinzari MM, Silva Júnior SI, Valdez GF (2000). Effects of a Novel Fermented Soy Producton the Serum Lipids of Hypercholesterolemic Rabbits. Arq Bras Cardiol. 74:213-216.

[131] Cavallini DCU, Suzuki JY, Abdalla DSP, Vendramini RC, Pauly-Silveira, ND, Pinto R, Rossi, EA (2011). Influence of a probiotic soy product on fecal microbiota and its association with cardiovascular risk factors in an animal model. Lipids Health Dis.10:1-9.

[132] Vinderola G, Matar C, Perdigon G (2007). Milk fermentation products of L. helveticus R389 activate calcineurin as a signal to promote gut mucosal immunity. BMC Immunol. 8:19.

[133] Galdeano CM, LeBlanc AM, Vinderola G, Bibas Bonet ME, Perdigón G (2007). Proposed model: mechanisms of immunomodulation induced by probiotic bacteria. Clin Vaccine Immunol. 14:485–492.

[134] Perdigón G, Locascio M, Medici M, Pesce de Ruiz Holgado A, Oliver G (2003). Interaction of bifidobacteria with the gut and their influence in the immune function. Biocell, 27:1–9.

[135] Augenlicht LH, Mariadason JM, Wilson A, Arango D, Yang W, Heerdt BG (2002). Short chain fatty acids and colon cancer. J. Nutr. 132:3804S–3808S.

[136] Macgarr SE, Ridlon JM, Hylemon PB (2005). Diet, anaerobic bacterial metabolism, and colon cancer: a review of the literature. J. Clin. Gastroenterol. 39:98–109.

Chapter 10

# Highlights in Probiotic Research

Gülden Başyiğit Kılıç

Additional information is available at the end of the chapter

http://dx.doi.org/10.5772/50004

## 1. Introduction

For centuries, lactic acid bacteria (LAB) have been used for the preservation of food for human consumption. LAB are a large group of fermentative, anaerobe facultative, aerotolerant microorganisms which are usually present in the gut of humans and other animals, raw vegetables, meat and meat products, and cereals (Carr et al., 2002). In animals, their numbers may vary with the species, the age of the host, or the location within the gut (De Vries et al., 2006). In the food industry, lactic acid bacterial strains are widely employed either as starter cultures or as non-starter lactic acid bacteria. Furthermore, owing to their probiotic properties, several LAB strains are used as adjunctive cultures in foods and feed (Sanders, 2000; Leroy & de Vuyst, 2004).

The term "probiotic" originated from the Greek word "probios" meaning "for life" (as opposed to "antibiotic," which means "against life") (Longdet et al., 2011). Probiotics are microbial food supplements which, when administered in adequate amounts, confer health benefits to consumers by maintaining or improving their intestinal microbial flora (Salminen et al., 1998; Reid et al., 2003). The US Food and Drug Administration uses other terms for live microbes for regulatory purposes (Sanders, 2008); live microbes used in animal feeds are called "direct-fed microbials" (FDA, 1995), and, when intended for use as human drugs, they are classified as "live biotherapeutics" (Vaillancourt, 2006). Probiotics are mainly members of the genera *Lactobacillus* and *Bifidobacterium* and are normal residents of the complex ecosystem of the gastrointestinal tract (GIT) of humans.

The research of novel probiotic strains is important in order to satisfy the increasing request of the market and to obtain functional products in which the probiotic cultures are more active and with better probiotic characteristics than those already present on the market (Verdenelli, et al., 2009). According to a recent market research report 'Probiotics Market (2009-2014)', the global probiotics market generated US $15.9 billion in 2008 and is expected to be worth US $ 32.6 billion by 2014 with a compound annual growth rate of 12.6 percent from 2009 to 2014 (FB 1046, 2009).

Several aspects, including general, functional and technological characteristics, have to be taken into consideration while selecting probiotic strains (Sanders & Huis in't Veld 1999; Šušković et al., 2001). This chapter includes selection criteria of bacteria as probiotics, technological usage of probiotics, new approaches for enhancing the performance of probiotics, and health effects of probiotic bacteria.

## 2. Selection of probiotic bacteria

Probiotics are living, health-promoting microorganisms that are incorporated into various kinds of foods. Although there has been a growing interest in using LAB isolated both from naturally fermented products and humans for health benefits (Lim & Im, 2009), the strains should preferably be of human origin and possess a Generally-Recognized-As-Safe status (Rönkä et al., 2003).

In order to exhibit their beneficial effects, probiotic bacteria need to survive during the food-manufacturing process and in human ecosystem conditions; therefore it is important to investigate bacterial behavior under conditions which mimic the GIT (Zago et al., 2011; Lo Curto et al., 2011). Stresses to microorganisms begin in the mouth, with the lysozyme-containing saliva; continue in the stomach, which has a pH between 1.5 and 3.0; and go on to the upper intestine, which contains bile (Corzo & Gilliland, 1999). Acid and bile tolerances are two fundamental properties that indicate the ability of a probiotic microorganism to survive the passage through the GIT, resisting the acidic conditions in the stomach and the bile acids at the beginning of the small intestine (Prasad et al., 1998; Park et al., 2002). To evaluate the probiotic survival in the GIT, several *in vitro* static models of digestion have been developed (Kitazawa et al., 1991; Charteris et al., 1998). One of them is the gastric–small intestinal system TIM-1 (Minekus et al., 1995), which consists of four serial compartments simulating the stomach and the three segments of the small intestine: the duodenum, jejunum, and ileum. Another one, the TIM-2 model, is a more sophisticated *in vitro* model of fermentation in the proximal large intestine. It consists of a series of linked glass vessels containing flexible walls which allow simulation of peristalsis (De Preter et al., 2011). The simulator of the human intestinal microbial ecosystem (SHIME) was developed to simulate the entire human gastrointestinal system (Molly et al., 1993). SHIME consists of a series of five temperature- and pH-controlled vessels that simulate the stomach; small intestine; and ascending, transverse and descending colon, respectively. The SHIME harbors a microbial community resembling that from the human colon both in fermentation activity and in composition (De Preter et al., 2011). Yet another model of the digestive system has been developed by such as TNO to mimic human physiological conditions in the stomach and small intestine (Blanquet et al., 2001). The major limitations of those systems is that digestion products are not removed during the incubation, and they may have a potential inhibitory effect on enzyme activities and on probiotic survival (Pitino et al., 2010). Furthermore, such systems ignore key GIT physical processes, including the temporal nature of gastric and duodenal processing, structure of food, pattern of mixing, particle size reduction and shear, which all affect the digestion rate (Shah 2000; Sumeri et al., 2008).

Effects of probiotics are strain specific. Strain identity is important in order to link a strain with a specific health effect, as well as to enable accurate surveillance and epidemiological studies (Ganguly et al., 2011). It is very important to be able to identify specifically and unambiguously the particular probiotic LAB strains from clinical fecal and intestinal biopsy specimens and from food samples (Tilsala-Timisjärvi & Tapanialtossava, 1998). Identification of bacterial species and strains from commercialized probiotics has been conducted mostly using molecular methods (Holzapfel et al., 2001; Schillinger et al., 2003; Huys et al., 2006; Sheu et al., 2009).

Verdenelli et al. (2009) investigated the probiotic potential of 11 *Lactobacillus* strains isolated from the faeces of elderly Italians. For this purpose, the researchers identified the *Lactobacillus* strains and examined them for resistance to gastric acidity and bile toxicity, adhesion to HT-29 cells, antimicrobial activities, antibiotic susceptibility and plasmid profile. They also examined the survival of the strains as they moved through the human intestine in a 3-month human feeding trial. According to the results, *L. rhamnosus* IMC 501 and *L. paracasei* IMC 502 present favourable strain-specific properties for their utilisation as probiotics in functional foods. Both *in vitro* and *in vivo* studies confirm the high adhesion ability of *L. rhamnosus* IMC 501 and *L. paracasei* IMC 502, used in combination, indicating that the two bacterial strains could be used as health-promoting bacteria.

Başyiğit Kılıç & Karahan (2010) isolated one hundred seven strains of human originated LAB identified by 16S rRNA analysis and examined them for resistance to acidic pH, bile salts and antibiotic susceptibility. They found that *L. plantarum* (AA1–2, AA17–73, AC18–88, AK4–11, and AK7–28), *L. fermentum* (AB5–18, BB16–75, and AK4–180), *Enterococcus faecium* (AB20–98 and BK11–50) and *E. durans* (AK4–14 and BK9–40) are potentially good probiotic candidates for use as health-promoting bacteria. In another study, the *L. plantarum* strains were examined for resistance to gastric acidity in simulated gastric juice at pH 2.0, 2.5, 3.0 and 3.5; 0.4% phenol; production of $H_2O_2$; adhesion to Caco-2 cell line; and antimicrobial activities. The researchers determined that the artificial gastric juice, even at pH 2.0, did not significantly change the viability of the cultures, and all *L. plantarum* strains showed good resistance to 0.4% phenol. They also reported antimicrobial activity and good adhesion of *L. plantarum* strains to Caco-2 cells. The researchers concluded that all of the strains showed probiotic properties, but *L. plantarum* AB6-25, AB7-35, AA13-59, AB16-65, BC18-81 and AK4-11 were the best potential probiotic strains for human use, given their ability to survive in gastric conditions, strong resistance to phenol, and the ability to adhere to the Caco-2 cell line (Başyiğit Kılıç et al., 2011a).

Lo Curto et al. (2011) investigated the survival of three commercial probiotic strains (*L. casei* subsp. *shirota, L. casei* subsp. *immunitas, L. acidophilus* subsp. *johnsonii*) in the human upper GIT. They used a dynamic gastric model (DGM) of digestion followed by incubation under duodenal conditions. The DGM is a computer-controlled gastric model which incorporates the chemical, biochemical, physical environment and processes of the human stomach; the model is based on kinetic data derived from the Echo planar-MRI and data on the rates of GI digestion obtained from human studies (Marciani et al., 2001; 2003; 2005; 2006). The researchers used water and milk as food matrices, and survival was evaluated in both logarithmic and stationary phases. The researchers found that the % of recovery in the logarithmic phase ranged from 1.0% to 43.8% in water for all tested strains, and from 80.5% to 197% in milk. They observed higher survival rates in the stationary phase for all strains. *L. acidophilus* subsp. *johnsonii* showed the highest survival rate in both water (93.9%) and milk (202.4%).

The safety of probiotic bacteria must be carefully assessed, with particular attention to transferable antibiotic resistance (Mathur & Singh, 2005). In the last decade, increasing concern has arisen about the safe use of LAB cultures for food and feed applications, in light of the latest knowledge about their possible role as an antibiotic-resistant gene reservoir. Particular concern is due to evidence of widespread occurrence in this bacterial group of conjugative plasmids and transposons (Clementi & Aquilanti, 2011). It is known that lactobacilli have a high natural resistance to bacitracin, cefoxitin, ciprofloxacin, fusidic acid, kanamycin, gentamicin, metronidazole, nitrofurantoin, norfloxacin, streptomycin, sulphadiazine, teicoplanin, trimethoprim/sulphamethoxazole, and vancomycin (Danielsen & Wind, 2003).

One of the primary benefits associated with probiotic bacterial cultures is that they can exclude pathogenic bacteria from the small and large intestine (Kos et al., 2008). Another benefit is that in food products, antimicrobial activity of probiotic bacteria may contribute to an improvement in the quality of fermented foods. This may result from control of spoilage and pathogenic bacteria, extension of shelf life, and improvement of sensory quality (Wei et al., 2006; Siripatrawan & Harte, 2007). Kos et al. (2008) used overnight cultures and cell-free supernatants of the three probiotic strains *L. acidophilus* M92, *L. plantarum* L4, and *E. faecium* L3 for determining the antagonistic effect against *Listeria monocytogenes, Salmonella typhimurium, Yersinia enterocolitica,* and *Acinetobacter calcoaceticus*. The researchers determined that probiotic strains *L. acidophilus* M92, *L. plantarum* L4, and *E. faecium* L3 demonstrated anti-*Salmonella* activity. *L. acidophilus* M92 was also shown to have antilisterial activity, as demonstrated by *in vitro* competition test.

Production of antimicrobial compounds, which may take part in the inhibition of intestinal pathogens, is another criterion for classifying a potentially probiotic bacteria (Hutt et al., 2006). The inhibition of pathogenic microorganisms by selected probiotic strains may occur via a) production of antibiotic-like substances, b) bacteriocins and bacteriocin-like inhibitory substances such as acidophilin and reuterin, c) lowering of pH by producing organic acids such as acetic, lactic and phenyllactic acid, d) production of hydrogen peroxide and short chain fatty acids, e) decreasing the redox potential, and f) consumption of available nutrients (Holzapfel et al., 1995; Ouwehand, 1998; Tharmaraj & Shah, 2009).

The ability of LAB to adhere to epithelial cells and mucosal surfaces is thought to be an important property of many bacterial strains used as probiotics (FAO/WHO, 2001). Cell adhesion is a complex process involving contact between the bacterial cell membrane and interacting surfaces. Difficulties experienced in studying bacterial adhesion *in vivo*, especially in humans, have stimulated interest in the development of *in vitro* models for preliminary screening of potentially adherent strains (Duary et al., 2011). Attachment and colonization of the gut epithelium prolongs the time for microorganisms to influence the immune system and microbiota of the host (Forestier et al., 2001). HT-29 and Caco 2 cells, the two colonic adenocarcinomas, are derived from human intestinal epithelium. Because they have structural and functional features of normal human enterocytes, they have been extensively used as *in vitro* models in the study of human enterocytic function (Moussavi & Adams, 2009).

The ability of probiotic bacteria to adhere to Caco-2 cells can be determined by plate counting or real time PCR (Matijasic et al., 2003; Candela et al., 2005). Nawaz et al. (2011) used both of these methods and did not find a statistically significant difference. Gaudana et al. (2010) investigated the ability of four different isolates (*L. plantarum* CS23, *L. rhamnosus* CS25, *L. delbrueckii* M and *L. fermentum* ASt1) and two standard strains (*L. plantarum* ATCC 8014 and *L. rhamnosus* GG) to stimulate three types of cells (Caco-2 cells, human peripheral blood mononuclear cells [PBMC] and THP-1 cells). The researchers reported that child faecal isolate CS23 showed high binding ability, high tolerance to acidic pH and bile salts, and significant immunomodulation; therefore they concluded that CS23 can be a good potential probiotic candidate. Duary et al. (2011) determined the colonization potentials of five human faecal *L. plantarum* isolates to the Caco-2 cells. Based on direct adhesion to epithelial cells, *L. plantarum* Lp91 was the most adhesive strain to the Caco-2 cell lines, with adhesion values of approximately 10.2%. They also mentioned that the percentage of adhesion to Caco-2 and HT-29 cell lines was higher among the strains isolated from the human faecal samples and buffalo milk than that which had been isolated from cheese.

## 3. Technological usage of probiotics

The use of starter cultures in the production of fermented food is necessary for guaranteeing safety and standardizing properties. LAB functions primarily to drop the pH of the batter; lower pH a) promotes product safety by inactivating pathogens, b) creates the biochemical conditions to attain the final sensory properties through modification of the raw materials, and c) improves the product stability and shelf life by inhibiting undesirable changes brought about by spoilage microorganisms or abiotic reactions (Ammor & Mayo, 2007).

Functional starter cultures are defined as microbes that possess at least one inherently functional property aimed at improving the quality of the end product (De Vuyst, 2000). The use of probiotics in food has reinforced the acclaimed healthy properties and given rise to an increased consumption of these products in Europe and the USA (Kristo et al., 2003). Probiotics have been evaluated as functional starter cultures in various types of fermented food products such as yoghurt, cheese, dry sausage, salami, and sourdough. They have also been studied in therapeutic preparations to assess their positive effects on physico-chemical properties of foods and their impact on the nutritional quality and functional performance of the raw material (Knorr, 1998; Rodgers, 2008).

Fermented dairy products are widely-accepted, healthy food products and valued components of diets. The incorporation of probiotic bacteria as adjuncts in various fermented milk products is currently an important topic with industrial and commercial consequences. A number of dairy products containing probiotic bacteria are currently on the market. Fermented milk and cheeses have been described as the most suitable carriers, because they enhance the transit tolerance of bacteria (Saarela et al., 2000; Lourens-Hattingh & Viljoen, 2001). Some strains of *Lactobacillus* and *Bifidobacterium* have been shown to tolerate acidic stress when ingested with milk products (Mater et al., 2005). Lactobacilli (e.g. *L. acidophilus, L. casei* subsp. *casei, L. gasseri, L. paracasei, L. reuteri* and *L. rhamnosus*) and bifidobacteria (e.g. *Bifidobacterium adolescentis, B. bifidum, B. breve, B. infantis* and *B. longum*) constitute a significant proportion of probiotic lactic acid bacterium cultures used in the dairy industry (Wood & Holzapfel, 1995; Klein et al., 1998). It is also important to determine the technological features of the strains because they could greatly affect food quality. Further, probiotic starter cultures need to be tested for large-scale production feasibility in regard to acidification, proteolysis, and aroma formation. They must accomplish this without losing viability and functionality or creating unpleasant flavor or texture (De Vuyst, 2000; Lacroix & Yildirim, 2007).

Although the number of cells required to produce therapeutic benefits is not known and might vary as a function of the strain and the health effect desired, in general a minimum level of more than $10^6$ viable probiotic bacteria per mililitre or gram of food product is accepted (Ouwehand & Salminen, 1998). The study of new probiotic strains for their technological relevance and use in food products is important for trade and industry. The search for strains which show resistance to biological barriers of the human GIT, and which possess physiological characteristics compatible with probiotic properties among LAB isolated from food, may eventually lead to the discovery of new probiotic strains for functional food products (Bude-Ugarte et al., 2006).

Studies of fermented food products as a source of new isolates are rapidly accumulating. For example, a mixture of human-derived probiotic strains was tested in the manufacture of ice cream; some of the ice cream was sweetened with sucrose and some was sweetened with aspartame (Başyiğit et al. 2006). The results showed that neither frozen conditions during the storage period nor the type of sweeteners used had any undesired effect on the survival of the probiotic cultures. Georgieva et al. (2009) studied technologically relevant properties of eight candidate probiotic *L. plantarum* strains isolated from cheeses. Researchers tested their capacity to survive over extended shelf-times at refrigerated temperatures and their growth viability in the presence of preservatives widely used in food processing. The researchers determined that the cultures' acidifying and coagulating abilities and enzyme activity make them appropriate for diverse food applications, but especially for dairy products. In another study, the survival of the probiotic strains *L. fermentum* (AB5-18 and AK4-120) and *L. plantarum* (AB16-65 and AC18-82), all derived from human faeces, was investigated in Turkish Beyaz cheese production (Başyiğit Kılıç et al., 2009). The researchers determined the viability of probiotic bacteria in Turkish Beyaz cheese during 4 months of ripening and the bacteria's effect on chemical properties of the cheese. The results of the study revealed that the test probiotic culture mix was successful for cheese production and did not adversely affect cheese quality during ripening.

Essid et al. (2009) characterized 17 strains of *L. plantarum* isolated from traditional Tunisian salted meat products to select the most suitable for use as starters for fermenting meat.

Critical characteristics included acidification and enzymatic activities responsible for final sensory properties; also important were safety characteristics, including antagonistic activity against spoilage strains and antibiotic resistance. The researchers determined that all strains of *L. plantarum* had good acidifying activity; however they showed some differences in antimicrobial, proteolytic and enzymatic activities. Başyiğit Kılıç et al. (2011b) investigated the technological properties of twenty *L. plantarum* strains to evaluate their potential usage as starter cultures in the dairy industry. During two months in cold storage, there were no significant changes in the number of bacteria or the pH of the skim milk inoculated with *L. plantarum* strains. The authors suggested that *L. plantarum* AC3-10 and AB6-25 can be used in industrial yogurt manufacture, based on their technological properties such as proteolitic activity, acidifying ability, and production of flavour compounds.

Floros et al. (2012) tested 19 facultatively heterofermentative lactobacilli from Feta, Kasseri, and Graviera cheeses for potential probiotic strains. Data from this study revealed that isolates B1, G16, G22, E22, E35, and H30 from Feta; PB2.2 from Kasseri; and 631 from Graviera have promising probiotic properties *in vitro*. β-galactosidase, low proteolytic and coagulation activities, and antibacterial activities make them promising candidates as adjunct cultures for the food industry. In another study, yoghurt was produced using a mixture of potential probiotic *L. plantarum* AB6-25, AC18-82, AK4-11 and a commercial starter culture. The yoghurt was divided into four experimental batches to which were added 0.25%, 0.5%, 1%, and 1.5% β-glucan. The survivability of these potential probiotic strains and the physico-chemical properties of the yoghurts were analyzed during a 21-day storage period. The highest *L. plantarum* count was found in the yoghurt containing 0.25% β-glucan. The study found the best physico-chemical properties to be in the 0.25% and 0.5% β-glucan containing yoghurts. Therefore, the researchers suggested using 0.25% and 0.5% β-glucan in yoghurts produced using these potential probiotic bacteria and commercial starter culture (Başyiğit Kılıç, 2012).

Wang et al. (2010) identified and established the functional and technological characteristics of potential probiotic *Lactobacillus* strains isolated from two sources: the faeces of breast-fed infants and traditional Taiwanese pickled cabbages. The authors selected the strains *L. reuteri* F03, *L. paracasei* F08 and *L. plantarum* C06 for producing probiotic fermented milk, due to their acid and bile tolerance and ability to adhere to Caco-2 cells. The milks were fermented with these 3 strains separately, and rats were fed a daily dose of $10^8$ CFU/day for 14 days. After the consumption of the *Lactobacillus*-fermented milk, the rats showed increased faecal lactobacilli counts, while the counts of coliform and *C. perfringens* were significantly decreased. On the other hand, Başyiğit Kılıç et al. (2010) investigated the effects of a probiotic culture mix (*L. fermentum, L. plantarum* and *E. faecium*) and alfa-tocopherol administration on the microbial flora in rat GIT and faeces during a 14-day feeding period. The results indicated that the probiotic culture and alfa-tocopherol administration had no significant effects on the microbial flora of the rat intestinal tract during the 14 days of intake. Minelli et al. (2004) reported that in rats administered milk fermented with L. casei, the faecal *E. coli* counts remained stable, but *Clostridia* counts decreased significantly. Yang et al. (2005) also reported decreased faecal coliform counts as one of advantages of *Lactobacillus* and *Bifidobacterium* proliferation in the rat gut. Such potentially probiotic bacteria colonizing the intestinal mucosa provide a barrier effect against pathogens by using a variety of mechanisms, such as occupation of niches, competition for nutrients, and production of antimicrobials (Ouwehand et al., 2001).

## 3.1. Methods to increase survival and viability of probiotics

Researchers have long been encouraged to find new, efficient methods of improving the viability of probiotics in food products (especially fermented types), since viability can be affected by the acidic-bile conditions of the gastrointestinal tract (Mortazavian et al., 2007). The latest developments focus on fermentation technologies for producing probiotic bacteria; new approaches for enhancing the performance of these fastidious organisms during fermentation, downstream processing, and utilization in commercial products; and improving functionality in the gut. Processes to optimize survival and functionality in the gut include sublethal stress applications during cell production and new fermentation technologies, such as immobilized cell biofilm-type fermentations, are promising in this respect (Lacroix & Yildirim, 2007).

### *3.1.1. Immobilized cell biofilm*

Cell immobilization in fermentations is an attractive and rapidly expanding research area because of its technical and economic advantages, compared to a free cell system (Stewart & Russell, 1986). The immobilization method is cheap, simple and easy (Kourkoutas et al., 2006). The technology of cell immobilization allows an increase in cell stability and a decrease of the lethal effect on the microbial cells, providing protection from the conditions of the environment (Champagne et al., 1994; Grosso & Fávaro-Trindade, 2004). Thus immobilization techniques could provide protection to acid-sensitive LAB and increase their survival rate during the shelf life of the yoghurt and during their passage through the gastrointestinal tract (Cui et al., 2000; Fávaro-Trindade & Grosso, 2002). Kushal et al. (2006) determined that the process of co-immobilization of probiotic strains of *L. acidophilus* NCDC 13 and *B. bifidum* NCDC 255 resulted in better protection of the viability of the cultures during transit through the gastrointestinal tract. In another study conducted by Kourkoutas et al. (2006), *L. casei* cells were immobilized on apple pieces and the immobilized biocatalysts were used separately as adjuncts in producing probiotic fermented milk. The results showed that the immobilized biocatalyst was able to ferment after storage for 15, 98 and 129 days at 4 °C, while no infection was reported during storage periods. Denkova et al. (2007) determined that the immobilization of the cells of *L. acidophilus* A., *L. helveticus* H., *L. casei* subsp. *casei* C. and *L. plantarum* 226-15 in chitosan resulted in preparations with high concentration of viable cells. The immobilized LAB in the chitosan gel beads was resistant to the model conditions of digestion: low and neutral values of pH, enzyme presence, and high concentrations of bile salts.

### *3.1.2. Encapsulation*

Encapsulation is the process of forming a continuous coating around an inner matrix that is wholly contained within the capsule wall as a core of encapsulated material (Kailasapathy,

2002). Encapsulation occurs naturally when bacterial cells grow and produce exopolysaccharides. The microbial cells are entrapped within their own secretions that act as a protective structure or a capsule, reducing the permeability of material through the capsule, and making it less exposed to adverse environmental factors. Many LAB synthesise exopolysaccharides, but they produce insufficient amounts to encapsulate themselves fully (Shah, 2002). Encapsulating probiotics in hydrocolloid beads has been investigated as a means of improving their viability and survival in food products and in the intestinal tract (Picot & Lacroix, 2004). Other benefits of encapsulation include reduction of cell injury, protection of probiotics from bacteriophages (Steenson et al., 1987), increased survival during freeze-drying and freezing (Kim & Yoon, 1995), and greater stability during storage (Kebary et al., 1998). Several methods of encapsulation have been used on probiotics in fermented milk products and biomass production: emulsion or two phase systems, the extrusion or droplet method, and spray drying and spray coating (Mortazavian et al., 2007). The common materials used for microencapsulation of probiotics are alginate and its derivatives, starch, mixtures of xanthan-gelan, carrageenan and its mixtures, gelatin, cellulose acetate phethalate, chitosan, and miscellaneous compounds such as whey proteins, soybean oil, gums, wax, and calcium chloride (Rao et al., 1989, Picot & Lacroix, 2004, Chandramouli et al., 2004).

Hou et al. (2003) demonstrated that encapsulation of *L. delbrueckii* spp. *bulgaricus* increased their bile tolerance, and viability was elevated by approximately four log units after encapsulation within artificial sesame oil emulsions. Encapsulation in spray dried whey protein microcapsules improved survival of *B. breve* R070 but not that of *B. longum* R023 during refrigerated storage in yoghurt (Picot & Lacroix, 2004). Ding & Shah (2007) stated that encapsulation improved the survival of probiotic bacteria including *L. rhamnosus, B. longum, L. salivarius, L. plantarum, L. acidophilus, L. paracasei, B. lactis* type Bl-O4, and *B. lactis* type Bi-07 when exposed to acidic conditions, bile salts, and mild heat treatment. Capela et al. (2006) found improved viability of probiotic organisms encapsulated in 3% v/w sodium alginate in freeze-dried yogurt after 6 months of storage at 4 and 21°C. Ozer et al. (2009) studied the viability of encapsulated bacteria in white-brined cheese; the researchers used *B. bifidum* BB-12 and *L. acidophilus* LA-5 that had been encapsulated in Na-alginate by either an extrusion or an emulsion technique. Both encapsulation techniques were found to be effective in keeping the numbers of probiotic bacteria higher than the level of the therapeutic minimum. While the counts of non-encapsulated probiotic bacteria decreased approximately by 3 logs, the decrease was more limited in the cheeses containing microencapsulated cells (approximately 1 log). Khater et al. (2010) tested the ability of twelve non-encapsulated and encapsulated lactic acid and bifidobacteria strains to assimilate cholesterol and to survive at a low pH and fairly high bile concentrations. The results obtained declared that encapsulation effectively protected the microorganisms from the hostile environment in the GIT, thus potentially preventing cell loss. The assimilative reductions of cholesterol by non-encapsulated and encapsulated strains were clearly different, varying from 32.6% to 89.3% and 27.9% to 85.1% respectively. Kim et al. (2008) stated that encapsulation reduces the ability of LAB to assimilate cholesterol.

## 4. Effects of probiotics on human health

Probiotics have the potential for contributing greatly to human and animal health via a wide range of applications. Historically, probiotics have been used in food for humans and animals without any side effects, while providing for the balance of intestinal flora (Holzapfel & Wood, 1998). The health-promoting effects of probiotics have been widely explored and include stabilization of the indigenous microbial population, boosting of the immune system, inhibition of the growth of pathogenic organisms, prevention of diarrhea from various causes, alleviation of lactose intolerance, increased nutritional value of foods, reduction of serum cholesterol levels, antimutagenicity and anticarcinogenicity, reduction of the risk of inflammatory bowel conditions, improvement of digestion of proteins and fats, synthesis of vitamins, and detoxification and protection from toxins (Klaenhammer, 1998; Perdigon et al., 2002; Gaudana et al., 2010).

Anderson & Gilliland (1999) conducted two controlled clinical studies to test effects of yoghurt on heart-related health. They reported an average reduction of serum cholesterol by 2.9% with regular consumption of yoghurt containing *L. acidophilus* and a 6-10% decrease in cardiac complications due to hypercholesterolemia. A study by Ouwehand et al. (2002) found that a multi-strain probiotic mixture composed of *L. reuteri, L. rhamnosus* and *Propionibacterium freudenreichii* proved effective in both increasing the number of bowel movements and decreasing mucin secretion in elderly subjects. The probiotic mixture was more effective than *L. reuteri* alone, although unfortunately it is difficult to draw conclusions about mixtures versus individual probiotics, since only one component of the mixture was tested and its dose was over 10 times lower than the total bacterial dose in the mixture. Agarwal & Bhasin (2002) have reported that the strain *L. casei* DN-114001 reduced diarrhoeal morbidity by 40% in children.

Isolauri *et al.* (1999) found significant improvement when a supplement of either *L. rhamnosus* or *B. lactis was given* to children from 4 to 6 years of age who had atopic eczema. Another study involving pregnant women and newborns suggested that consumption of probiotic *L. rhamnosus* GG reduced the rate of newborns having atopic dermatitis (Kalliomaki et al., 2001). In an Australian study, 178 newborns of women with allergies who received either *L. acidophilus* LAVRI-A1 or placebo daily for the first 6 months of life showed no difference in atopic dermatitiS. However, at 12 months, the rate of sensitization was significantly higher in the probiotic group. These results suggested that the probiotic treatment had increased the risk of subsequent cow's milk sensitization (Taylor et al., 2007).

Can (2003) used an experimental animal model to study the effects of a probiotic mixture and *L.* GG on immune responses in allergy. The OVA specific IgE levels of the study groups which were administred probiotics and reference strain were found lower than the skim milk fed groups. A double-blind, randomized, placebo controlled trial study was conducted by Abrahamsson et al. (2007) on 188 subjects with allergic disease, in which the mothers received *L. reuteri* ATCC 55730 daily from gestational week 36 until delivery, and their babies continued with the probiotic until 12 months. Probiotic supplemented babies showed less IgE-associated eczema during the second year. Several probiotic effects are mediated through immuneregulation, particularly through establishing and maintaining a balance between pro-and anti-inflammatory cytokines (Isolauri et al., 2001). TNF-a and IL-6 are pro-inflammatory cytokines, which are produced by the host in response to bacterial colonisation or invasion and hence are central to the host defense mechanism against pathogens (Solis-Pereyra et al., 1997). Though lipopolysaccharide of Gram-negative bacteria is known to stimulate their production, Miettinen et al. (1996) have reported an increase in IL-6 and TNF-a production in human PBMC exposed to lactobacilli and thereby suggested the use of probiotics as vaccine vectors and for the purpose of stimulating non-specific immunity. Kailasapathy & Chin (2000) proved that the synthesis of cytokines is increased as the probiotics adhere to the intestinal epithelium.

Ziarno et al. (2007) studied cholesterol assimilation by commercial starter cultures, reporting *L. acidophilus* monocultures to assimilate cholesterol by 49-55%. In another study involving hypercholesterolemic mice, the probiotic potential of *L. plantarum* PHO4 was established by Nguygen et al. (2007). The mice were fed with $10^7$ CFU per day over two weeks. These mice had 7 to 10% lesser serum cholesterol and triglycerides than the control mice deprived of the probiotic feed.

Many probiotic species have been identified to be effective in children suffering from rotaviral diarrhea (Saavedra, 2000). Longdet et al. (2011) investigated the probiotic efficacy of *L. casei* isolated from human breast milk in the prevention of shigellosis in albino rats infected with clinical strains of *Shigella dysenteriae*. The results showed that the experimental rats infected with *S. dysenteriae* but not treated suffered from shigellosis, while the test groups infected and treated with the *L. casei* showed no sign of the disease as well as no clinical effect on the liver.

Senol et al. (2011a) investigated the protective effect of a probiotic mixture of 13 different bacteria and a-tocopherol on 98% ethanol-induced gastric mucosal injury. Levels of gastric mucosal pro-and anti-inflammatory cytokines, malondialdehyde, and secretory immunglobulin A were measured. Results showed that probiotic pretreatment significantly suppressed the ethanol-induced increase of gastric mucosal interleukin-4 levels. Pretreatment with either probiotic or a–tocopherol inhibited the ethanol-induced increase of mucosal malondialdehyde concentration. Probiotic pretreatment enhanced the gastric mucosal secretory immunoglobulin A concentration. The researchers indicated that the probitic mixture and a-tocopherol reduced ethanol-induced gastric mucosal lipid peroxidation, suggesting that these probiotics may be beneficial for helping heal gastric lesions induced by lower ethanol concentration. In another study, the role of a probiotic mixture, including 13 different bacteria, in the prevention of aspirin-induced gastric mucosal injury was investigated. Pretreatment with the probiotic mixture reduced aspirin-induced gastric damage and exerted a tendency toward downregulation of proinflammatory cytokines elicited by aspirin. Researchers also found that the probiotic mixture increased sIgA production approximately 7.5-fold in the stomach, and significantly reduced the malondialdehyde increase in the gastric mucosa elicited by aspirin. Additionally, pretreatment with the probiotic mixture alleviated aspirin-induced reduction of mast cell count in the gastric mucosa. Probiotic mixture pretreatment attenuates the aspirin-induced gastric lesions by reducing the lipid peroxidation, enhancing mucosal sIgA production, and stabilizing mucosal mast cell degranulation into the gastric mucosa (Senol et al., 2011b).

## 5. Final remarks

Significant data have been accumulated on probiotics and their beneficial health effects. Furthermore, more insights and key findings on the impact of processing and storage on probiotic viability and stability have been gained. A variety of microorganisms, typically food grade LAB, have been evaluated for their probiotic potential and are applied as adjunct cultures in various types of food products or in therapeutic preparations. In addition, further studies are needed to determine if preventive probiotic strategies are safe with regard to development of probiotic infections. Cooperation amongst food technologists, medical and nutrition scientists, and anticipation of future consumer demands are crucial for future success in probiotics.

## Author details

Gülden Başyiğit Kiliç
*Mehmet Akif Ersoy University, Department of Food Engineering,*
*Faculty of Engineering-Architecture, Burdur, Turkey*

## 6. References

Abrahamsson, T.R., Jakobsson, T., Bottcher, M.F., Fredrikson, M., Jenmalm, M.C. & Bjorksten, B. (2007). Probiotics in prevention of IgE-associated eczema: a double-blind, randomized, placebo-controlled trial, *Journal of Allergy and Clinical Immunology* 119: 1174 1180.

Agarwal, K.N. & Bhasin, S.K. (2002). Feasibility studies to control acute diarrhoea in children by feeding fermented milk preparations, Actimel and Indian Dahi, *European Journal of Clinical Nutrition* 56: 56-59.

Ammor, M.S. & Mayo, B. (2007). Selection criteria for lactic acid bacteria to be used as functional starter cultures in dry sausage production: An update, *Meat Science* 76: 138–146.

Anderson, J.W. & Gilliland, S.E. (1999). Effect of fermented milk (yogurt) containing *Lactobacillus acidophilus* L1 on serum cholesterol in hypercholesterolemic humans, *Journal of the American College of Nutrition* 18(1): 43-50.

Başyiğit, G., Kuleaşan, H. & Karahan, A.G. (2006). Viability of human derived probiotic lactobacilli in ice-cream produced with sucrose and aspartame, *Journal of Industrial Microbiology and Biotechnology* 33: 796-800.

Başyiğit Kılıç, G., Kuleaşan, H., Eralp, I. & Karahan, A.G. (2009). Manufacture of Turkish Beyaz cheese added with probiotic strains, *LWT - Food Science and Technology* 42: 1003-1008.

Başyiğit Kılıç, G., Kılıç, B., Kuleaşan, H. & Karahan, A.G. (2010). Effect of Probiotics and $\alpha$-tocopherol applications on microbial flora of rat gastrointestinal tract, *Journal of Animal and Veterinary Advances* 9(14): 1972-1977.

Başyiğit Kılıç, G. & Karahan, A.G. (2010). Identification of lactic acid bacteria isolated from the fecal samples of healthy humans and patients with dyspepsia and determination of their pH, bile and antibiotic tolerance properties, *Journal of Molecular Microbiology and Biotechnology* 18: 220-229.

Başyiğit Kılıç, G., Kuleaşan, H. & Çakmak, V.F. (2011a). Determination of probiotic properties of *L. plantarum* strains isolated from the human fecal samples, *Proceedings of Novel Approches in Food Industry, International Food Congress*, 26-29 May 2011, Çeşme-İzmir, Turkey, pp. 56.

Başyiğit Kılıç, G., Kuleaşan, H., Akpınar, D., Çakmak, V.F. (2011b). Characterization of technological properties of human originated probiotic *L. plantarum* strains, *Proceedings of International Scientific Conference on Probiotics and Prebiotics-IPC*, Slovakia, pp. 14.

Başyiğit Kılıç, G. (2012) (unpublished data). Determination of probiotic and technological properties of some *Lactobacillus plantarum* strains, TUBITAK 109 O 623, Ongoing Project.

Blanquet, S., Marol-Bonnin, S., Beyssac, E., Pompon, D., Renaud, M. & Alric, M. (2001). The "biodrug" concept: an innovative approach to therapy, *Trends in Biotechnology* 19(10): 393–400.

Bude-Ugarte, M., Guglielmotti, D., Giraffa, G., Reinheimer, J.A. & Hynes, E. (2006). Non starter lactobacilli isolated from soft and semi hard Argentinean cheeses: genetic characterization and resistance to biological barriers, *Journal of Food Protection* 69: 2983-2991.

Can, R. (2003). The effects of probiotics on allergy, *PhD Thesis*, Süleyman Demirel University, Medicinal Faculty, Departmant of Microbiology, Isparta, Turkey, p. 75 (unpublished).

Candela, M., Seibold, G., Vitali, B., Lachenmaier, S., Eikmanns, B.J. & Brigidi, P. (2005) Real-time PCR quantification of bacterial adhesion to Caco-2 cells: competition between bifidobacteria and enteropathogens, *Research in Microbiology* 156(8): 887-895.

Capela, P., Hay, T.K.C. & Shah, N.P. (2006). Effect of cryoproyectants, prebiotics and microencapsulation on survival of probiotic organisms in yoghurt and freeze-dried yoghurt, *Food Reseach International* 39: 203-211.

Carr, F.J., Chill, D. & Maida, N. (2002). The lactic acid bacteria: a literature survey, *Critical Reviews on Microbiology* 28: 281-370.

Champagne, C.P., Lacroix C. & Sodini-Gallot, I. (1994). Immobilized cell technology for the dairy industry, *Critical Reviews in Biotechnology* 14: 109-134.

Chandramouli, V., Kalasapathy, K., Peiris, P. & Jones, M. (2004). An improved method of microencapsulation and its evaluation to protect *Lactobacillus* spp. in simulated gastric conditions, *Journal of Microbiological Methods* 56: 27-35.

Charteris, W.P., Kelly, P.M., Morelli, L. & Collins, J.K. (1998). Development and application of an *in vitro* methodology to determinate the transit tolerance of potentially probiotic *Lactobacillus* and *Bifidobacterium* species in the upper human gastrointestinal tract, *Journal of Applied Microbiology* 84: 759-768.

Clementi, F. & Aquilanti, L. (2011). Recent investigations and updated criteria for the assessment of antibiotic resistance in food lactic acid bacteria, *Anaerobe* 17: 394-398.

Corzo, G. & Gilliland, S.E. (1999). Bile salt hydrolase activity of three strains of *Lactobacillus acidophilus, Journal of Dairy Science* 82: 472-480.

Cui, J.H., Goh, J.S., Kim, P.H., Choi, S.H. & Lee B.J. (2000). Survival and stability of bifidobacteria loaded in alginate poly-l-lysine microparticles, *International Journal of Pharmacy Research* 210: 51-59

Danielsen, M. & Wind, A. (2003). Susceptibility of *Lactobacillus* spp. to antimicrobial agents, *International Journal of Food Microbiology* 82: 1–11.

Denkova, Z., Slavchev, A., Blazheva, D. & Krastanov, A. (2007). The effect of the immobilization of probiotic lactobacilli in chitosan on their tolerance to a laboratory model of human gut, *Biotechnology & Biotechnology Equipment* 21 (4),442-450.

De Preter, V., Hamer, H.M., Windey, K. & Verbeke, C. (2011). The impact of pre-and/or probiotics on human colonic metabolism: Does it affect human health? *Molecular Nutrition & Food Research* 55: 46–57.

De Vries, M.C., Vaughan, E.E., Kleerebezem, M. & De Vos, W.M. (2006). *Lactobacillus plantarum* – survival, functional and potential probiotic properties in the human intestinal tract, *International Dairy Journal* 16: 1018–1028.

De Vuyst, L. (2000). Technology aspects related to the application of functional starter cultures, *Food Technology and Biotechnology* 38(2): 105–112.

Duary, R.K., Rajput, Y.S., Batish, V.K. & Grover, S. (2011). Assessing the adhesion of putative indigenous probiotic lactobacilli to human colonic epithelial cells, *The Indian Journal of Medical Research* 134: 664-671.

Essid, I., Medini, M. & Hassouna, M. (2009). Technological and safety properties of *Lactobacillus plantarum* strains isolated from a Tunisian traditional salted meat, *Meat Science* 81: 203–208.

Fávaro-Trindade, C.S. & Grosso, C.R.F. (2002). Microencapsulation of *L. acidophilus* (La-05) and *B. lactis* (Bb-12) and evaluation of their survival at the pH values of the stomach and in bile, *Journal of Microencapsulation* 19: 485-494.

FAO/WHO (2001). Evaluation of health and nutritional properties of powder milk and live lactic acid bacteria. Rome, Italy: Food and Agriculture Organization of the United Nations and World Health Organization Expert Consultation Report.

FAO/WHO (2002). Guidelines for the evaluation of probiotics in food, *Report of a Joint FAO/WHO Working Group on Drafting Guidelines for the Evaluation of Probiotics in Food*, London Ontario, Canada.

Fávaro-Trindade, C.S. & Grosso, C.R.F. (2002). Microencapsulation of *L. acidophilus* (La-05) and *B. lactis* (Bb-12) and evaluation of their survival at the pH values of the stomach and in bile, *Journal of Microencapsulation* 19: 485-494.

FB 1046, 2009. Probiotic Market- Advanced Technologies and Global Market (2009 - 2014), By: markets and markets.com. Publishing Date: September 2009. Report Code: FB 1046. [accessed on March 26, 2012]. URL: www.marketsandmarkets.com/Market-Reports/probiotic-market-advanced-technologies-and-global-market-69.html.

Floros, G., Hatzikamari, M., Litopoulou-Tzanetaki, E. & Tzanetakis, N. (2012). Probiotic and technological properties of facultatively heterofermentative lactobacilli from Greek traditional cheeses, *Food Biotechnology* 26: 85–105.

Forestier, C., De Champs, C., Vatoux, C. & Joly, B. (2001). Probiotic activities of *Lactobacillus casei rhamnosus*: in vitro adherence to intestinal cells and antimicrobial properties, *Research in Microbiology* 152(2): 167-173.

Ganguly, N.K., Bhattacharya, S.K., Sesikeran, B., Nair, G.B., Ramakrishna, B.S., Sachdev, H.P.S., Batish, V.K., Kanagasabapathy, A.S., Muthuswamy, V., Kathuria, S.C., Katoch, V.M. Satyanarayana, K., Toteja, G.S., Rahi, M., Rao, S., Bhan, M.K., Kapur, R. & Hemalatha, R. (2011). CMR-DBT guidelines for evaluation of probiotics in food, *The Indian Journal of Medical Research* 134(1): 22–25.

Gaudana, S.B., Dhanani, A.S., Bargchi, T. (2010). Probiotic attributes of *Lactobacillus* strains isolated from food and of human origin, *British Journal of Nutrition* 103: 1620–1628.

Georgieva, R., Iliev, I., Haertlé T., Chobert, J.M., Ivanova, I. & Danova, S. (2009). Technological properties of candidate probiotic *Lactobacillus plantarum* strains, *International Dairy Journal* 19: 696–702.

Grosso, C.R.F. & Fávaro-Trindade, C.S. (2004). Stability of free and immobilized *Lactobacillus acidophilus* and *Bifidobacterium Lactis* in acidified milk and of immobilized *B. lactis* in yoghurt, *Brazilian Journal of Microbiology* 35: 151-156.

Hobbs, C. (2000). Pro-life therapy with probiotics. Health World Online 2000. URL: healthy.net/asp/templates/article.asp?id=953.

Holzapfel, W. H., Geisen, R. & Schillinger, G. (1995). Biological-preservation of foods with reference to protective cultures, bacteriocins, and food grade enzymes, *International Journal of Food Microbiology* 24: 343-362.

Holzapfel, W.H. & Wood, B.J.B. (1998). *The genera of lactic acid bacteria*, Blackie Academic and Professional, London.

Holzapfel, W.H., Haberer, P., Geisen, R., Bjorkroth, J. & Schillinger, U. (2001). Taxonomy and important features of probiotic microorganisms in food and nutrition, *American Journal of Clinical Nutrition* 73: 365-373.

Hou, R.C., Lin M.Y., Wang, M.M. & Tzen, J.T. (2003). Increase of viability of entrapped cells of *Lactobacillus delbrueckii* spp. *bulgaricus* in artificial sesame oil emulsions, *Journal of Dairy Science* 86: 424-428.

Hutt, P., Shchepetova, J., loivukene, K., Kullisaar, T. & Mikelsaar, M. (2006). Antagonistic activity of probiotic lactobacilli and bifidobacteria against entero- and uropathogens, *Journal of Applied Microbiology* 100:1324-1332.

Huys, G., Vancanneyt, M., D'Haene, K., Vankerckhoven, V., Goossens, H. & Swings, J. (2006). Accuracy of species identity of commercial bacterial cultures intended for probiotic or nutritional use, *Research Microbiology* 157: 803–810.

Isolauri, E., Arvola, T., Sutas, Y. & Salminen S. (2001). Probiotics: effects on immunity, *American Journal of Clinical Nutrition* 73(2): 444-450.

Kailasapathy, K. & Chin, J. (2000). Survival and therapeutic potential of probiotic organisms with reference to *Lactobacillus acidophilus* and *Bifidobacterium* spp., *Immunology and Cell Biology* 78: 80–88.

Kailasapathy, K. (2002). Microencapsulation of probiotic bacteria: technology and potential applications, *Current Issues on Intestinal Microbiology* 3: 39-48.

Kalliomaki, M., Salminen, S., Arvilommi, H., Kero, P., Koskinen, P. & Isolauri, E. (2001). Probiotics in primary prevention of atopic disease: a randomised placebo-controlled trial, *Lancet* 357: 1076-1079.

Kebary, K.M.K., Hussein, S.A. & Badawi, R.M. (1998). Improving viability of *Bifidobacteria* and their effect on frozen ice milk, *Egyption Journal of Dairy Science* 26: 319-337.

Khater, K.A.A., Ali, M.A. & Ahmed, E.A.M. (2010). Effect of encapsulation on some probiotic criteria, *Journal of American Science* 6(10): 836-845.

Kim, K.I. & Yoon, Y.H. (1995). A study on the preparation of direct vat lactic acid bacterial starter, *Korean Journal of Dairy Science* 17: 129-134.

Kim, S.J., Cho, S.Y., Kim, S.H., Song, O.J., Shin, S., Chu, D.S. & Park, H.J. (2008). Effect of microencapsulation on viability and other characteristics in *Lactobacillus acidophilus* ATCC 43121, *LWT Food Science and Technology* 41: 493-500.

Kitazawa, H., Toba, T., Itoh, T., Kumano, N., Adachi, S. & Yamaguchi, T. (1991). Antitumoral activity of slime-forming encapsulated *Lactococcus lactis* subsp. *cremoris* isolated from Scandinavian ropy sour milk, "viili", *Animal Feed Science andTechnology* 62: 277-283.

Klaenhammer, T.R. (1998). Functional activities of *Lactobacillus* probiotics: genetic mandate, *International Dairy Journal* 8: 497–505.

Klein, G., Pack, A., Bonaparte, C. & Reuter, G. (1998). Taxonomy and physiology of probiotic lactic acid bacteria*International Journal of Food Microbiology*41: 103–125.

Kos, B., Šušković, J., Beganović, J., Gjuračić, K., Frece, J., Iannaccone, C. & Canganella, F. (2008). Characterization of the three selected probiotic strains for the application in food industry, *World Journal of Microbiology and Biotechnology* 24(5): 699–707.

Kourkoutas, Y., Kanellaki, M. & Koutinas, A.A. (2006). Apple pieces as immobilization support of various microorganisms, *LWT Food Science and Technology* 39: 980–986.

Kristo, E., Biliaderis, C.G. & Tzanetakis, N. (2003). Modelling of rheological, microbiological and acidification properties of a fermented milk product containing a probiotic strain of *Lactobacillus paracasei, International Dairy Journal*, 13: 517–528.

Kushal, R., Anand S.K. & Chander H. (2006). *In vivo* demonstration of enhanced probiotic effect of co-immobilized *Lactobacillus acidophilus* and *Bifidobacterium bifidum, International Journal of Dairy Technology* 59: 265-271.

Lacroix, C. & Yildirim, S. (2007). Fermentation technologies for the production of probiotics with high viability and functionality, *Current Opinion in Biotechnology* 18: 176–183.

Leroy, F. & deVuyst, L. (2004). Lactic acid bacteria as functional starter cultures for the food fermentation industry, *Trends in Food Science and Technology* 15: 67-78.

Lim, S.M. & Im, D.S. (2009). Screening and characterization of probiotic lactic acid bacteria isolated from Korean fermented foods, *Journal of Microbiology and Biotechnology* 19(2): 178-186.

Lo Curto, A., Pitino, I., Mandalari G., Dainty J.R., Faulks, R.M. & Wickham John, M.S. (2011). Survival of probiotic lactobacilli in the upper gastrointestinal tract using an *in vitro* gastric model of digestion, *Food Microbiology* 7: 1359-1366.

Longdet I.Y., Kutdhik, R.J. & Nwoyeocha, I.G. (2011). The probiotic efficacy of *Lactobacillus casei* from human breast milk against Shigellosis in Albino rats, *Advances in Biotechnology & Chemical Processes* 1: 12-16.

Lourens-Hattingh, A. & Viljoen, B.C. (2001). Review: yoghurt as probiotic carrier in food, *International Dairy Journal* 11: 1–17.

Marciani, L., Ramanathan, C., Tyler, D.J., Young, P., Manoj, P., Wickham, M.S.J., Fillery Travis, A., Spiller, R.C. & Gowland, P.A. (2001). Fat emulsification measured using NMR transverse relaxation, *Journal of Magnetic Resonance Imaging* 153: 1-6.

Marciani,L., Wickham, M.S.J., Wright, J., Bush, D., Faulks, R.M., Fillery Travis, A., Gowland, P.A. & Spiller, R.C. (2003). Magnetic resonance imaging (MRI) insights into how fat emulsion stability alters gastric emptying, *Gastroenterology* 124: A581-1.

Marciani, L., Bush, D., Wright, P., Wickham, M.S.J., Pick, B., Wright, J., Faulks, R.M., Fillery Travis, A.J., Spiller, R.C. & Gowland, P.A. (2005). Monitoring of gallbladder and gastric coordination by EPI. *Journal of Magnetic Resonance Imaging* 21: 82-85.

Marciani, L., Wickham, M.S.J., Bush, D., Faulks, R.M., Wright,J., Fillery-Travis, A., Spiller, R. & Gowland, P. (2006). Magnetic resonance imaging of the behaviour of oil-in water emulsions in the gastric lumen of man, *British Journal of Nutrition* 95: 331-339.

Mater, D.D.G., Bretigny, L., Firmesse, O., Flores, M.J., Mogenet, A., Bresson, J.L. & Corthier, G. (2005). *Streptococcus thermophilus* and *Lactobacillus delbrueckii* subsp. *bulgaricus* survive gastrointestinal transit of healthy volunteers consuming yogurt, *FEMS Microbiology Letters* 250: 185-187.

Mathur, S. & Singh, R. (2005). Antibiotic resistance in food lactic acid bacteria--a review, *International Journal of Food Microbiology* 105(3): 281-295.

Matijasic, B.B., Narat, M. & Zoric, M. (2003) Adhesion of two *Lactobacillus gasseri* probiotic strains on Caco-2 cells, *Food Technology and Biotechnology* 41(1): 83-88. Miettinen, M., Vuopio-Varkila, J. & Varkila, K. (1996). Production of human tumor necrosis factor a, interleukin-6, and interleukin-10 is induced by lactic acid bacteria, Infection and Immunity 64: 5403–5405.

Minekus, M., Marteau, P., Havenaar, R. & Huisin't Veld, J.H.J. (1995). A multi compartmental dynamic computer-controlled model simulating the stomach and smal lintestine, *Atla* 23: 197–209.

Minelli, E.B., Benini, A., Marzotto, M., Sbarbati, A., Ruzzenente, O. & Ferrario, R., (2004). Assessment of novel probiotic Lactobacillus casei strains for the production of functional dairy foods, *International Dairy Journal* 14: 723-736.

Molly, K.,Vande Woestyne, M. & Verstraete, W. (1993). Development of a5-step multi-chamber reactor as a simulation of the human intestinal microbial ecosystem, *Applied Microbiology and Biotechnology* 39: 254–258.

Mortazavian, A., Razavi, S.H., Ehsani, m.R. & Sohrabvandi, S. (2007). Principles and methods of microencapsulation of probiotic microorganisms, *Iranian Journal of Biotechnology* 5:1.

Moussavi, M. & Adams, M.C. (2009). An *in vitro* study on bacterial growth interactions and intestinal epithelial cell adhesion characteristics of probiotic combinations, *Current Microbiology* 60: 327-35.

Nawaz, M., Wang, J., Zhou, A., Ma, C., Wu, X. & Xu, J. (2011) Screening and characterization of new potentially probiotic lactobacilli from breast-fed healthy babies in Pakistan, *African Journal of Microbiological Researches* 5(12): 1428-1436.

Nguygen, T.D.T., Kang, J.H. & Lee, M.S. (2007). Characterization of *Lactobacillus plantarum* PH04, a potential probiotic bacterium with cholesterol lowering effects, *International Journal of Food Microbiology* 113(3): 358-361.

Ouwehand, A. C. (1998). Anti-microbial components from lactic acid bacteria, *in* Salminen, S. & Von wright, A. (ed). *Lactic Acid Bacteria: Microbiology and Functional Aspects*, 2nd Edition, Marcel Dekker Inc., NewYork, pp. 139-160.

Ouwehand, A.C., Salminen, S.J. (1998). The health effects of cultured milk products with viable and non-viable bacteria, *International Dairy Journal* 8: 749-758.

Ouwehand, A.C., Tuomola, E.M., Tölkkö, S. & Salminen, S. (2001) Assessment of adhesion properties of novel probiotics strains to human intestinal mucus, *International Journal of Food Microbiology* 64: 119-126.

Ouwehand, A.C., Lagstrom, H., Suomalainen, T. & Salminen, S. (2002). Effect of probiotics on constipation, fecal azoreductase activity and fecal mucin content in the elderly, *Annals of Nutrition and Metabolism* 46(3–4): 159–162.

Ozer, B., Kirmaci, H. A., Senel, E., Atamer, M. & Hayaloglu. A. (2009). Improving the viability of Bifidobacterium bifidum BB-12 and Lactobacillus acidophilus LA-5 in white-brined cheese by microencapsulation, *International Dairy Journal* 19: 22-29.

Park, Y.S., Lee, J.Y., Kim, Y.S. & Shin, D.H. (2002). Isolation and characterization of lactic acid bacteria from feces of newborn baby and from dongchimi, *Journal of Agricultural and Food Chemistry* 50: 2531–2536.

Perdigon, G., Maldonado Galdeano, C. & Valdez, J.C. & Medici, M. (2002). Interaction of lactic acid bacteria with the gut immune system, *European Journal of Clinical Nutrition* 56(4): 21–26.

Picot, A., & Lacroix, C. (2004). Encapsulation of bifidobacteria in whey protein-based microcapsules and survival in simulated gastrointestinal conditions and in yoghurt, *International Dairy Journal* 14 (6): 505-515.

Pitino, I., Randazzo, C.L., Mandalari, G., Lo Curto, A., Faulks, R.M., Le Marc, Y., Bisignano, C., Caggia, C. & Wickham, M.S.J. (2010). Survival of *Lactobacillus rhamnosus* strains in the upper gastrointestinal tract, *Food Microbiology* 27: 1121-1127.

Prasad, J., Gill, H., Smart, J. & Gopal, P.K. (1998). Selection and characterization of *Lactobacillus* and *Bifidobacterium* strains for use as probiotic, *International Dairy Journal* 8: 993–1002.

Rao, A.V., Shiwnarin, N. & Maharij, I. (1989). Survival of microencapsulated *Bifidobacterium pseudolongum* in simulated gastric and intestinal juices, *Canadian Institution of Food Science and Technology Journal* 22: 345-349.

Reid, G., Jass, J., Sebulsky, M.T. & McCormick, J.K. (2003). Potential uses of probiotics in clinical practice, *Clinical Microbiological Reviews* 16(4): 658–72.

Rodgers, S. (2008). Novel applications of live bacteria in food services: probiotics and protective cultures, *Trends in Food Science and Technology* 19: 188-197.

Rönkä E., Malinen E., Saarela M., Rinta-Koski M., Aarnikunnas J., Palva A. (2003). Probitic and milk technological properties of *Lactobacillus brevis, International Journal Food Microbiology* 83: 63–74.

Saarela, M., Mogensen, G., Fonden, R., Matto, J. & Mattila-Sandholm, T. (2000). Probiotic bacteria: safety, functional and technological properties, *Journal of Biotechnology* 84: 197–215.

Saavedra, J. (2000). Probiotics and infectious diarrhea, *American Journal of Gastroenterology* 95: 16-18.

Salminen, S., von Wright, A., Morelli L., Marteau P., Brassart D., de Vos, W.M., Fondén, R., Saxelin, M., Collins, K., Mogensen, G., Birkeland, S.E. & Mattila-Sandholm T. (1998). Demonstration of safety of probiotics – a review, *International Journal Food Microbiology* 44: 93–106.

Sandeep, B., Gaudana, Akhilesh, S., Dhanani & Tamishraha, B. (2010). Probiotic attributes of *Lactobacillus* strains isolated from food and of human origin, *British Journal of Nutrition* 103: 1620–1628.

Sanders, M.E. & Huis in't Veld, J.H.J. (1999). Bringing a probiotic containing functional food to the market: microbiological, product regulatory and labeling issues, *in* Konings, W.N., Kuipers, O.P. & Huis in't Veld, J.H.J. (ed.), *Proceedings of the 6$^{th}$ symposium on lactic acid bacteria: genetics, metabolism and applications*, Kluwer Academic Publishers, Antonie van Leeuwenhoek, Veldhoven, The Netherlands, pp. 293–316.

Sanders, ME. (2000). Considerations for use of probiotic bacteria to modulate human health, *Journal of Nutrition* 130: 384-390.

Sanders, M.E. (2008). Probiotics: Definition, sources, selection, and uses, *Clinical Infectious Diseases* 46: 58–61.

Schillinger, U., Yousif, N.M., Sesar, L. & Franz, C.M. (2003). Use of group-specific and RAPD-PCR analyses for rapid differentiation of Lactobacillus strains from probiotic yogurts, *Current Microbiology* 47: 453–456.

Shah, N.P. (2000) Probiotic bacteria: selective enumeration and survival in dairy foods, *Journal of Dairy Science* 83: 894-907.

Shah, N. (2002). The exopolysaccharides production by starter cultures and their influence on textural characteristics of fermented milks, *Proceeding of Symposium on New Developments in Technology of Fermented Milks*, International Dairy Federation, Comwell Scanticon, Kolding, Denmark, p. 5.

Sheu, S.J., Hwang, W.Z., Chen, H.C., Chiang, Y.C. & Tsen, H.Y. (2009.) Development and use of tuf gene-based primers for the multiplex PCR detection of *Lactobacillus acidophilus, Lactobacillus casei* group, *Lactobacillus delbrueckii,* and *Bifidobacterium longum* in commercial dairy products, *Journal of Food Protection* 72: 93–100.

Senol, A., Isler, M., Karahan, A.G., Kilic, G.B., Kuleasan, H., Kaya, S., Keskin, M., Goren, İ., Saritas, U., Aridogan, C.B. & Delibas, N. (2011a). Preventive effect of probiotics and alpha-tocopherol on the ethanol-induced gastric mucosal injury in rats, *Journal of Medicinal Food* 14(1-2): 173-179.

Senol, A., Isler, M., Karahan, A.G., Basyigit Kilic, G., Kuleasan, H., Goren, İ., Saritas, U., Kaya, S., Ciris, M., Akturk, O., Aridogan, C.B., Demirin, H. & Cakmakci, M.L. (2011b). Effect of the probiotics on aspirin-induced gastric mucosal lesions, *Journal of Turkish Gastroenterology* 22 (1):18-26.

Stewart, G.G. & Russell, I. (1986). One hundred years of yeast research and development in the brewing industry, *Journal of the Institute of Brewing* 92: 537–558.

Siripatrawan, U. & Harte, B.R. (2007). Solid phase microextraction/gas chromatography/mass spectrometry integrated with chemometrics for detection of *Salmonella typhimurium* contamination in a packaged fresh vegetable. *Analytica Chimica Acta* 581: 63–70.

Solis-Pereyra, B., Aattouri, N. & Lemonnier, D. (1997). Role of food in the stimulation of cytokine production, *American Journal of Clinical Nutrition* 66: 521–525.

Steenson, L.R., Klaenhammer, T.R. & Swaisgood, H.E. (1987). Calcium alginate-immobilized cultures of lactic streptococci are protected from attack by lytic bacteriophage, *Journal of Dairy Science* 70: 1121-1127.

Sumeri, I., Arike, L., Adamberg, K. & Paalme, T. (2008). Single bioreactor gastrointestinal tract simulator for study of survival of probiotic bacteria, *Applied Microbiology and Biotechnology* 80: 317-324.

Šušković, J., Kos, B., Goreta, J. & Matošić, S. (2001). Role of lactic acid bacteria and bifidobacteria in symbiotic effect, *Food Technology and Biotechnology* (39): 227–235.

Taylor, A.L., Dunstan, J.A. & Prescott, S.L. (2007). Probiotic supplementation for the first 6 months of life fails to reduce the risk of atopic dermatitis and increases the risk of allergen sensitization in high-risk children: a randomized controlled trial, *Journal of Allergy and Clinical Immunology* 119: 184-191.

Tharmaraj, N. & Shah, N.P. (2009). Antimicrobial effects of probiotics against selected pathogenic and spoilage bacteria in cheese-based dips, *International Food Research Journal* 16: 261-276.

Tilsala-Timisjärvi, A. & Tapanialtossava, T. (1998). Strain-specific identification of probiotic *Lactobacillus rhamnosus* with Randomly Amplified Polymorphic DNA-Derived PCR Primers, *Applied and Environmental Microbiolog* 64(12): 4816–4819.

Vaillancourt, J. (2006). Regulating pre- and probiotics: a U.S. FDA perspective, *Proceedings of ending the war metaphor: the changing agenda for unraveling the host-microbe relationship. in* Institute of Medicine of the National Academies. Washington, DC: National Academies Press, pp. 229–237.

Verdenelli, M.C., Ghelfi, F., Silvi, S., Orpianesi, C., Cecchini, C. & Cresci, A. (2009). Probiotic properties of *Lactobacillus rhamnosus* and *Lactobacillus paracasei* isolated from human faeces, *European Journal of Nutrition* 48: 355–363.

Wang, C., Lin, P., Ng, C. & Shyu, Y. (2010). Probiotic properties of *Lactobacillus* strains isolated from the feces of breast-fed infants and Taiwanese pickled cabbage, *Anaerobe* 16: 578-585.

Wei, H., Wolf, G. & Hammes, W.P. (2006). Indigenous microorganisms from iceberg lettuce with adherence and antagonistic potential for use as protective culture, *Innovative Food Science Emerging Technologies* 7: 294–301.

Wood, B.J.B. & Holzapfel W.H. (1995). *The Lactic Acid Bacteria, Vol. 2, The Genera of Lactic Acid Bacteria*, Blackie Academic & Professional, London, pp 55-124.

Yang, S.C., Chen, J.Y., Shang, H.F., Cheng, T.Y., Tsou, S.C. & Chen, J.R. (2005). Effects of symbiotics on intestinal microflora and digestive enzyme activies in rats, *World Journal of Gastroenterology* 11:7413-7417.

Zago, M., Fornasari, E., Carminati, D., Burns, P., Suàrez, V., Vinderola, G., Reinheimer, J. & Giraffa, G. (2011). Characterization and probiotic potential of *Lactobacillus plantarum* strains isolated from cheeses, *Food Microbiology* 28: 1033-1040.

Ziarno, M., Sekul, E. & LafrayaAguado, A. (2007). Cholesterol assimilation by commercial yoghurt starter cultures, *ACTA Scientiarum Polonorum-Food Science and Human Nutrition* 6(1): 83-94.

Chapter 11

# Dose Effects of LAB on Modulation of Rotavirus Vaccine Induced Immune Responses

Lijuan Yuan, Ke Wen, Fangning Liu and Guohua Li

Additional information is available at the end of the chapter

http://dx.doi.org/10.5772/50379

## 1. Introduction

Probiotics can influence both mucosal and systemic immune responses and function as adjuvants by promoting proinflammatory cytokine production, enhancing both humoral and cellular immune responses. Adjuvant effects of several probiotic lactic acid bacteria (LAB), mostly *Lactobacillus* strains, including *L. rhamnosus* GG, *L. acidophilus* NCFM, *L. acidophilus* CRL431, *L. acidophilus* La-14, *L. fermentum* CECT5716, *L. casei* DN-114 001, and *Bifidobacterium lactis* Bl-04 have been reported in studies of influenza, polio, rotavirus and cholera vaccines and rotavirus and Salmonella typhi Ty21a infections (Boge et al., 2009; Davidson et al., 2011; Isolauri et al., 1995; Kaila et al., 1992; Mohamadzadeh et al., 2008; Olivares et al., 2007; Paineau et al., 2008; Winkler et al., 2005; Zhang et al., 2008b). The word adjuvant in the phrase "probiotic adjuvant" is not used in its traditional definition in which adjuvant implies a substance included in the vaccine formulation to aid the immune response to the vaccine antigen. Instead, probiotic adjuvants enhance the immunogenicity of vaccines when orally administered repeatedly around the time of vaccination and separately from the vaccine. By skewing the balance of pro- and anti-inflammatory innate immune responses and T helper (Th) 1 and regulatory T (Treg) cell adaptive immune responses in the context of vaccination, probiotic adjuvants act as "signal zero" to reduce Treg cell suppression and unleash effector T cell activation (Rowe et al., 2012).

Although the strain-specific effects of LAB in up- or down-regulating inflammatory immune responses have been well recognized, dose effects of probiotics on innate and adaptive immune responses are not clearly understood. The same *Lactobacillus* strain is oftentimes reported by different research groups to have opposite immune modulating effects. We hypothesized that the dose effect is at least one of the reasons for the conflicting reports. Understanding dose effects of probiotics has significant implications in their use as immunostimulatory (adjuvants) versus immunoregulatory agents.

In this chapter, we discuss findings from our serial studies of gnotobiotic pigs on the dose effects of the *L. acidophilus* NCFM strain (LA) on innate and adaptive immune responses induced by an oral attenuated human rotavirus (HRV) vaccine (AttHRV). We studied the effects of low dose (total 2.11 x $10^6$ CFU) and high dose (total 2.22 x $10^9$ CFU) LA on the intestinal and systemic (1) rotavirus-specific IFN-γ producing CD4+ and CD8+ T cell responses; (2) CD4+CD25+FoxP3+ and CD4+CD25-FoxP3+ Treg cell responses and the regulatory cytokine TGF-β and IL-10 production; (3) rotavirus-specific antibody-secreting cell (ASC) and serum antibody responses; and (4) plasmacytoid dendritic cell (pDC) and conventional DC (cDC) frequencies, activation status, TLR expression and cytokine production profile. The protective effect of rotavirus vaccine against virus shedding and diarrhea was assessed in AttHRV-vaccinated gnotobiotic pigs fed with high, low, or no LA and challenged with the virulent HRV.

These studies clearly demonstrated differential immune modulating effects of high dose versus low dose LA on DC and T cell responses, and consequently different effects on the protection conferred by the AttHRV vaccine in gnotobiotic pigs challenged with virulent HRV. Low dose LA enhanced the protection against rotavirus diarrhea in AttHRV-vaccinated pigs whereas high dose LA had negative effects on the effectiveness of the vaccine. Thus, the same probiotic strains at different doses can exert qualitatively different modulating effects on immune responses induced by rotavirus vaccines and possibly other vaccines as well.

## 2. Dose effects of LA on T cell responses

Probiotics have been reported to exert adjuvant properties by inducing pro-Th1 cytokines and promote Th1 type immune responses. For example, *L. lactis* and *L. plantarum* induced production of IL-12 and IFN-γ by splenocytes when the LAB and an allergen were co-administered intranasally to mice (Repa et al., 2003). *L. fermentum* strain CECT5716 enhanced the Th1 responses induced by an influenza vaccine in addition to enhancing virus neutralizing antibody responses (Olivares et al., 2007). Eleven different probiotic strains were tested for cytokine production in human peripheral blood mononuclear cells (MNC) and each tested bacterium was shown to induce production of TNF-α and some strains also induced production of IL-12 and IFN-γ (Kekkonen et al., 2008). Previous studies of gnotobiotic pigs showed that a mixture of LA NCFM strain and *L. reuteri* strain enhanced both Th1 (IL-12, IFN-γ) and Th2 (IL-4 and IL-10) cytokine responses to virulent HRV infection (Azevedo et al., 2012). LA NCFM strain enhanced the HRV-specific IFN-γ producing CD8+ T cell response to a rotavirus vaccine in gnotobiotic pigs, indicating adjuvanticity of the LA strain (Zhang et al., 2008b).

Dose effects of probiotics on modulating T cell immune responses have not been well studied. To address this question, we examined the dose effects of LA NCFM (NCK56) strain on IFN-γ producing CD4+ and CD8+ T cell immune responses induced by an oral rotavirus vaccine in gnotobiotic pigs (Wen et al., 2012). The animal treatment groups included (1) high dose LA plus AttHRV vaccine (HiLA+AttHRV), (2) low dose LA plus

AttHRV (LoLA+AttHRV), (3) AttHRV only, (4) high dose LA only (HiLA), (5) low dose LA only (LoLA), and (6) mock-inoculated control (Mock). Gnotobiotic pigs were orally inoculated at 5 days of age with the AttHRV vaccine at 5 x$10^7$ fluorescent focus-forming units (FFU) per dose. A booster dose was given 10 days later at the same dose and route. Subsets of the pigs were euthanized at post-inoculation day (PID) 28 to assess immune responses and the rest were challenged with the homotypic virulent Wa (G1,P1A[8]) strain HRV at a dose of 1 x $10^5$ FFU to assess protection from post-challenge day (PCD) 1 to 7. The 50% infectious dose and 50% diarrhea dose of the virulent HRV in gnotobiotic pigs are approximately 1 FFU (Ward et al., 1996). The AttHRV inoculation causes virus shedding in about 6% pigs, but it does not cause any illness (Ward et al., 1996). Pigs in the high dose LA groups were fed daily with $10^3$ to $10^9$ CFU/dose of LA for 14 days with 10-fold incremental dose increases every other day from 3-16 days of age. The accumulative total LA dose was 2.22 x $10^9$ CFU. Pigs in the low dose LA groups were fed with $10^3$, $10^4$, $10^5$, $10^6$, and $10^6$ CFU/dose of LA every other day from 3-11 days of age. The accumulative total LA dose was 2.11 x $10^6$ CFU.

## 2.1. Low dose LA, but not high dose LA, enhanced HRV-specific IFN-γ producing T cell responses

The magnitude of HRV-specific IFN-γ producing T cell responses in pigs was differentially modulated by low versus high dose LA at both prechallenge and postchallenge (PID 28 and PCD 7). AttHRV-vaccinated and low dose LA fed pigs (LoLA+AttHRV) had significantly higher frequencies of HRV-specific IFN-γ+CD8+ T cells in ileum (11- and 5-fold higher pre- and postchallenge, respectively), spleen (3.8- and 2.1-fold higher pre- and postchallenge, respectively) and blood (3- and 20-fold higher pre- and postchallenge, respectively) compared to the AttHRV only pigs (Table 1). The LoLA+AttHRV pigs also had significantly higher frequencies of HRV-specific IFN-γ+CD4+ T cells in blood (3-fold higher for both pre- and postchallenge) compared to the AttHRV only pigs. In contrast, high dose LA did not significantly alter the HRV-specific IFN-γ producing CD4+ and CD8+ T cell responses in the HiLA+HRV pigs compared to AttHRV only pigs.

| | Frequencies of IFN-γ+CD8+ T cells among CD3+ cells | | | | | |
|---|---|---|---|---|---|---|
| | PID 28 | | | PCD 7 | | |
| | Ileum | Spleen | Blood | Ileum | Spleen | Blood |
| HiLA+AttHRV | 0.05 | 0.34 | 0.05 | 0.11 | 0.16 | 0.06 |
| LoLA+AttHRV | 1.21 | 0.46 | 0.24 | 0.56 | 0.46 | 0.98 |
| AttHRV only | 0.11 | 0.12 | 0.08 | 0.11 | 0.22 | 0.05 |

(Summarized from Wen et al., 2012)

**Table 1.** Effect of low dose vs. high dose LA on IFN-γ producing CD8+ T cell responses

## 2.2. High dose LA significantly increased frequencies of intestinal and systemic CD4+CD25-FoxP3+ Treg cells whereas low dose LA decreased TGF-β and IL-10 producing Treg cell responses

Frequencies and cytokine production of Treg cells in pigs were differentially modulated by low versus high dose LA at both prechallenge and postchallenge. HiLA+AttHRV pigs had significantly higher frequencies of CD4+CD25-FoxP3+ Treg cells (ranging from 6- to 86-fold higher) in all the tissues compared to LoLA+AttHRV and AttHRV only pigs pre- and postchallenge (Table 2).

Because Treg cells exert regulatory functions through mechanisms involving TGF-β and IL-10, we also compared the effects of high and low dose LA on frequencies of the Treg cell subsets that produced TGF-β or IL-10 among the AttHRV-vaccinated pigs.

| | Frequencies of CD4+CD25-FoxP3+Treg cells among total MNC | | | | | |
|---|---|---|---|---|---|---|
| | PID 28 | | | PCD 7 | | |
| | Ileum | Spleen | Blood | Ileum | Spleen | Blood |
| HiLA+AttHRV | 2.96 | 10.34 | 9.00 | 1.56 | 3.78 | 3.65 |
| LoLA+AttHRV | 0.24 | 0.54 | 0.10 | 0.08 | 0.22 | 0.05 |
| AttHRV only | 0.23 | 0.59 | 0.51 | 0.27 | 0.26 | 0.17 |

(Summarized from Wen et al., 2012)

**Table 2.** Effect of low dose vs. high dose LA on frequencies of Treg cells

Low dose LA reduced frequencies of TGF-β producing CD4+CD25+FoxP3+ and CD4+CD25-FoxP3+ Treg cells compared to high dose LA and AttHRV only pigs in all tissues pre- and postchallenge (Table 3; data for CD25+ Treg cells are not shown). Low dose LA also reduced pre- and postchallenge frequencies of IL-10 producing CD4+CD25+FoxP3+ and CD4+CD25-FoxP3+Treg cells compared to high dose LA and AttHRV- only (except for CD4+CD25-FoxP3+ Treg cells in ileum and spleen postchallenge) (Table 3). High dose LA induced 2.6-fold and 20-fold, respectively higher frequencies of IL-10 producing CD4+CD25-FoxP3+ Treg cells in ileum and spleen postchallenge compared to AttHRV only.

| | PID 28 | | | PCD 7 | | |
|---|---|---|---|---|---|---|
| | Ileum | Spleen | Blood | Ileum | Spleen | Blood |
| | Frequencies of TGF-β+ cells among CD4+CD25-FoxP3+ Treg cells | | | | | |
| HiLA+AttHRV | 4.51 | 1.13 | 0.92 | 4.87 | 10.05 | 1.39 |
| LoLA+AttHRV | 0.32 | 0.15 | 0.31 | 0.00 | 0.17 | 0.38 |
| AttHRV only | 4.19 | 1.19 | 1.55 | 4.34 | 0.79 | 1.53 |
| | Frequencies of IL-10+ cells among CD4+CD25-FoxP3+ Treg cells | | | | | |
| HiLA+AttHRV | 4.21 | 2.97 | 0.88 | 10.68 | 17.70 | 2.51 |
| LoLA+AttHRV | 0.92 | 0.15 | 0.00 | 5.22 | 6.79 | 0.00 |
| AttHRV only | 4.47 | 3.01 | 1.77 | 4.08 | 0.90 | 1.44 |

(Summarized from Wen et al., 2012)

**Table 3.** Effect of low dose vs. high dose LA on CD25-FoxP3+ Treg cell cytokine production

These data clearly demonstrated that low dose LA promoted IFN-γ producing T cell and down-regulated Treg cell responses, whereas high dose LA induced a strong Treg cell response and promoted the regulatory cytokine production by tissue-residing Treg cells postchallenge in gnotobiotic pigs. Studies of other lactobacilli strains have reported similar findings. A mixture of *L. plantarum* CEC 7315 and CEC 7316 at high dose ($5\times10^9$ CFU/day) resulted in significant increases in the percentages of activated potential T-suppressor and NK cells, while at low dose ($5\times10^8$ CFU/day) increased activated T-helper cells, B cells and antigen presenting cells (APCs) in institutionalized seniors (Mane et al., 2011). High concentration ($\geq 1\times10^6$ colony forming unit [CFU]/ml) of a combination containing LA and *Bifidobacterium* or *B. infantis* attenuated mitogen-induced immune responses by inhibiting cell proliferation and arresting the cell cycle at the G0/G1 stage in both mitogen-stimulated spleen cells and peripheral blood MNC. However, low concentration ($\leq 1\times10^6$ CFU/ml) promoted a shift in the Th1/Th2 balance toward Th1-skewed immunity by enhancing IFN-γ and inhibiting IL-4 response (Li et al., 2011). The differences between the "low dose" and "high dose" LAB in these studies are small, yet the immunomodulatory effects are qualitatively different.

Dose effects may explain some of the controversies that result from the same probiotic strain used by different research groups in animal studies showing opposite immunomodulatory functions. For example, administration of *L. casei* suppressed pro-inflammatory cytokine expression by CD4+ T cells and up-regulated IL-10 and TGF-β levels in rats (So et al., 2008a; So et al., 2008b). On the contrary, another study found that *L. casei* was a pure Th1 inducer in mice. In addition to the difference in animal species, the *L. casei* doses used by the different studies differed significantly, with much higher doses used in the studies of rats (So et al., 2008a; So et al., 2008b). In the studies of rats, the amount of *L. casei* was $5 \times 10^9$ or $2 \text{ x}10^{10}$ CFU/dose per rat, three times per week for 11-12 weeks. In the study of mice, the amount of *L. casei* was $2 \text{ x } 10^8$ CFU/dose per mouse, twice per week for 8 weeks (Van Overtvelt et al., 2010). Thus, different dose and frequency of administration of the same LAB strains may result in totally different *in vivo* effects.

The dose effects of LA on immune responses to the AttHRV vaccine in pigs may also partly explain why the efficacies of oral rotavirus vaccines are significantly reduced in low-income countries compared to developed countries. The two licensed rotavirus vaccines, RotaTeq and Rotarix have a protective efficacy of >85% against moderate to severe rotavirus gastroenteritis in middle and high-income countries (O'Ryan et al., 2009). However, the protective efficacy of RotaTeq vaccine is only 39.3% against severe rotavirus gastroenteritis in sub-Saharan Africa (Armah et al., 2010) and 48.3% in developing countries in Asia (Zaman et al., 2010). Rotarix vaccine showed a similar disparity in efficacy in low-income countries in Africa (O'Ryan & Linhares, 2009). In addition to other factors that contribute to the reduction in rotavirus vaccine efficacy (e.g., higher titers of maternal antibodies, malnutrition), during the initial colonization of human infants, exposure to high doses of commensal bacteria (common in countries with lower hygiene standards) would have a suppressive effect on IFN-γ producing T cell responses and promote Treg cell responses, thus leading to the lowered protective immunity after rotavirus vaccination.

## 3. Dose effects of LA on antibody and B cell responses

Probiotics are known to modulate both humoral and cellular immune responses. Probiotics can induce antigen-specific and non-specific IgA antibody responses at mucosal surfaces (Perdigon et al., 2001; Wells & Mercenier, 2008) to prevent invasion by pathogenic microorganisms. Oral administration of *L. acidophilus* L-92 strain led to a significant increase of IgA production in Peyer's patches in mice (Torii et al., 2007). *L. casei* CRL 431 strain increased induction of intestinal IgA secreting cells in mice (Galdeano & Perdigon, 2006). *L. acidophilus* La1 strain and bifidobacteria enhanced specific serum IgA titers to *S. typhi* strain Ty21a and also total serum IgA in humans (Link-Amster et al., 1994). *L. rhamnosus* GG enhanced rotavirus-specific IgA ASC responses in humans and promoted recovery from rotavirus diarrhea (Kaila et al., 1992). In our earlier study of gnotobiotic pigs, a mixture of LA strain and *L. reuteri* strain did not alter virus-specific intestinal and systemic antibody and ASC responses, but they significantly enhanced total intestinal IgA secreting cell responses and total serum IgM and intestinal IgM and IgG titers in rotavirus infected pigs (Zhang et al., 2008a).

The first reported adjuvant effect of probiotic LAB in vaccination was from a human clinical trial in which *L. rhamnosus* GG was shown to enhance rotavirus-specific IgM secreting cells and rotavirus IgA seroconversion in infants receiving a live oral rhesus-human rotavirus reassortant vaccine (Isolauri et al., 1995). In recent years, an increasing number of human clinical trials have demonstrated adjuvant effects of probiotics in enhancing vaccine-induced antibody responses. In a double-blind randomized controlled trial, *L. rhamnosus* GG or *L. acidophilus* CRL 431 increased serum poliovirus neutralizing antibody titers and poliovirus-specific IgA and IgG titers 2- to 4-fold in adult human volunteers vaccinated with the live oral polio vaccine (de Vrese et al., 2005). In another human clinical trial, six out of the seven probiotic strains tested enhanced cholera-specific IgG antibody concentration in serum; for the *B. lactis* Bl-04 and *L. acidophilus* La-14 strains the increase was more significant (Paineau et al., 2008). Daily consumption of a fermented dairy drink (*L. casei* DN-114 001 and yoghurt ferments, Actimel) was shown to increase virus specific antibody responses to the intramuscular inactivated influenza vaccine in individuals of over 70 years of age (Boge et al., 2009). In a randomized, double-blind placebo-controlled pilot study, *L. rhamnosus* GG significantly improved the development of serum antibody responses to the H3N2 strain influenza virus (84% receiving *L. rhamnosus* GG versus 55% receiving placebo had a protective titer 28 days after vaccination) in healthy adults receiving the live attenuated influenza vaccine (FluMist, Medimmune Vaccines, Gaithersburg, MD, USA) (Davidson et al., 2011). Thus, specific strains of probiotics can act as adjuvants to enhance humoral immune responses following not only mucosal (oral or intranasal) but also parenteral vaccination. Yet, dose effects of probiotics on antibodies responses have not been well studied.

In our studies, we demonstrated that high dose LA did not significantly alter the HRV-specific antibody responses whereas low dose LA had negative effects on the antibody responses. The effect of high and low dose LA NCFM strain on HRV-specific serum IgA and IgG antibody levels and HRV-specific ASC and memory B cell responses in the intestinal and systemic lymphoid tissues of gnotobiotic pigs induced by rotavirus vaccination were examined. The animal treatment groups were the same as listed above in the studies of T cell responses. High dose LA did not significantly alter the HRV-specific antibody responses in serum and ASC responses in any tissue at PID 28 and PCD 7 in the AttHRV-vaccinated pigs (Figs 1 and 2), except to reduce the IgG ASC response in ileum of the mock-vaccinated pigs postchallenge (Fig. 2c). In contrast, low dose LA significantly reduced the HRV-specific IgA antibody titers at PID 7 and 14 (Fig. 1a) and reduced or significantly reduced IgG ASC responses in blood pre- and postchallenge as well as IgA ASC responses in spleen and blood postchallenge (Fig. 2a and 2b). Low dose LA also significantly reduced the IgA and IgG ASC responses in ileum of the mock-vaccinated pigs postchallenge (Fig. 2c).

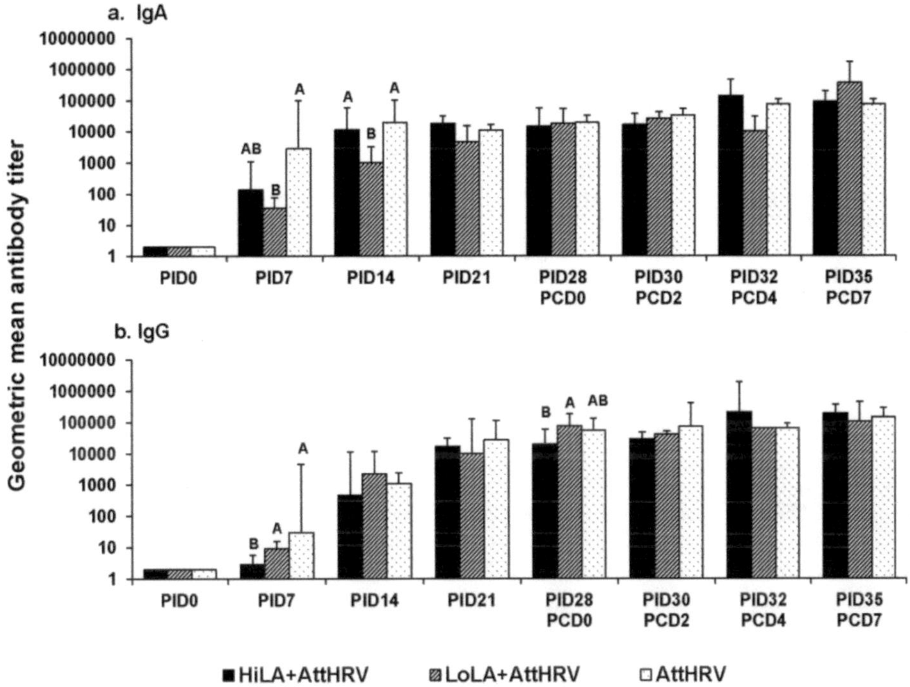


**Figure 1.** Rotavirus-specific serum IgA and IgG antibody responses in Gn pigs vaccinated with AttHRV with or without high or low dose LA feeding. Rotavirus-specific serum IgA (a) and IgG (b) antibody were measured by an indirect isotype-specific antibody ELISA. Error bars indicate the standard error of the mean. Different capital letters (A, B) indicate significant difference among different pig groups at the same time point (Kruskal Wallis Test, $p<0.05$, n=3-27), whereas shared letters or no letters on top indicate no significant difference.

The negative effects of low dose LA on the HRV-specific serum antibody responses and ASC responses induced by the AttHRV vaccine were undesirable for the vaccine's immunogenicity; however it is consistent with the strong pro-Th1 effect of the low dose LA. The skewed balance toward a Th1 type immune response in the low dose LA group may have resulted in the weakened antibody responses. In the subsequent studies, we evaluated the effects of a low dose and an intermediate dose of *L. rhamnosus* GG on the effector T cell, antibody and ASC responses induced by the AttHRV vaccine and we found that *L. rhamnosus* GG enhanced the production of a balanced Th1 and Th2 immune responses to the AttHRV vaccine and significantly increased the virus-specific IFN-γ producing T cell responses, the antibody responses and the protection rate of the AttHRV vaccine (manuscripts under preparation).

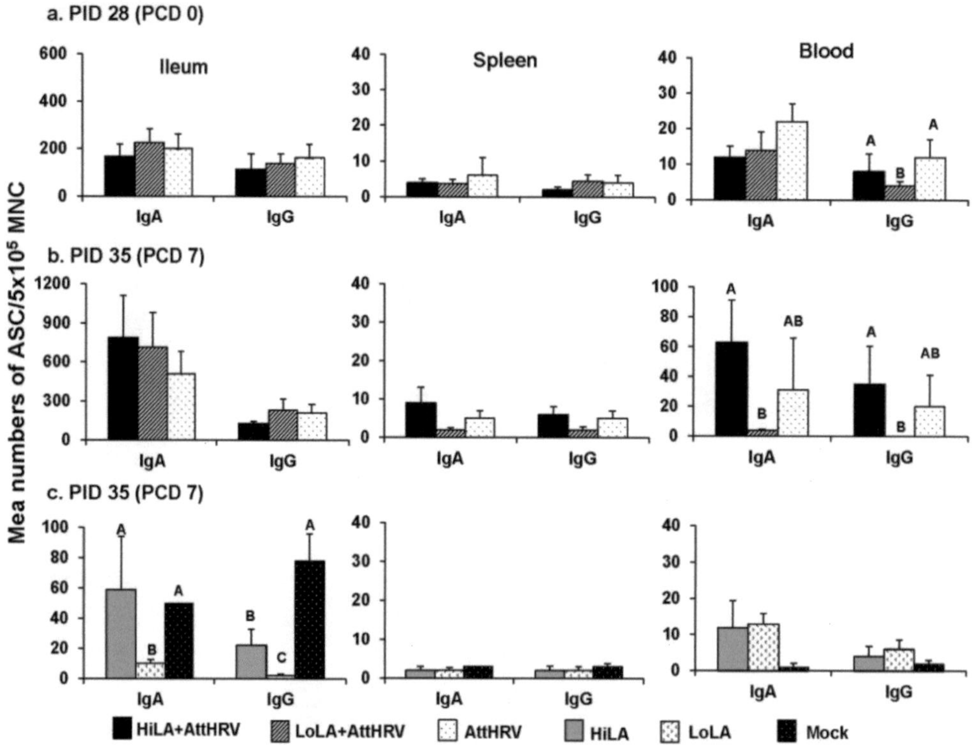


**Figure 2.** Rotavirus-specific IgA and IgG ASC responses in Gn pigs. Rotavirus-specific IgA and IgG ASC in the MNC isolated from ileum, spleen and blood of AttHRV-vaccinated pigs on PID 28 (PCD 0) (a) and PID 35 (PCD 7) (b) and of mock-vaccinated pigs on PID 35 (PCD 7) (c) were enumerated by using an ELISPOT assay and are presented as the mean numbers of virus-specific IgA and IgG ASC per $5 \times 10^5$ MNC. Error bars indicate the standard error of the mean. Different capital letters (A, B, C) on top of the bars indicate significant difference among the treatment groups for the same isotype in the same tissue (Kruskal Wallis Test, $p<0.05$, n=3-14), whereas shared letters or no letters on top indicate no significant difference. Note the y axis scale difference (HiLA, high dose LA; LoLA, low dose LA).

## 4. Dose effects of LA on DC responses

The nature and consequences of a CD4+ T cell response (Th1, Th2, Th17, or Treg type) largely depend on the immune functions of DCs, which are the professional antigen presenting cells that can prime and differentiate naive T cells. Both cDC and pDC are responsible for presenting microbial and dietary antigens to the adaptive immune systems, thereby influencing polarization of the adaptive immune response (Konieczna et al., 2012).

The pDC most effectively sense virus infections and are characterized by their capacity to produce large quantities of IFN-$\alpha$ and the pro-inflammatory cytokines IL-6 and TNF-$\alpha$. These cytokines promote cDC maturation (Summerfield & McCullough, 2009). MHC II expression in professional APCs is tightly regulated. The MHC II of immature DCs are expressed at low levels at the plasma membrane, but abundantly in endocytic compartments. In the presence of inflammatory cytokines such as IFN-$\gamma$, DCs are activated; they stop capturing antigens and markedly increase MHC II expression on their plasma membrane. These MHC II are loaded with peptides derived from antigens captured at the site of inflammation. The mature DCs migrate to lymphoid tissues and up-regulate the co-stimulatory molecules (CD80/86) necessary to activate naïve T cells (Villadangos et al., 2001).

It is known that probiotics can modify the distribution, the phenotype and the function of DC subsets (Grangette, 2012). Both species-specific and strain-specific immunomodulatory effects of different LAB on DCs have been described in a large number of studies and was reviewed previously (Meijerink & Wells, 2010). Among the differential effects, several lactobacilli strains, including *L. acidophilus, L. gasseri, L. fermentum, L. casei, L. plantarum, L. johnsonii,* and *L. rhamnosus* have been reported to stimulate human or murine DCs to produce increased levels of proinflammatory cytokines (IL-2, IL-12, TNF-$\alpha$) that favored Th1 and cytotoxic T cell polarization, and decreased levels of the regulatory cytokine TGF-$\beta$ (Chiba et al., 2010; Christensen et al., 2002; Mohamadzadeh et al., 2005; Van Overtvelt et al., 2010; Vitini et al., 2000; Weiss et al., 2010; Yazdi et al., 2010). Such immune stimulating effects are characteristics of adjuvants. However, studies of dose effects of lactobacilli on DC responses are scarce, with most consisting of *in vitro* experiments, and there is a dearth of comparative studies linking *in vitro* and *in vivo* results. *L. rhamnosus* Lcr35 was shown to induce a dose-dependent immunomodulation of human DCs. Lcr35 at $10^7$ CFU/ml (10 multiplicity of infection), but not $10^4$ CFU/ml induced the semi-maturation of the DCs and a strong pro-inflammatory response (Evrard et al., 2011). LA NCFM induced a concentration dependent production of IL-10, and low IL-12p70 in monocyte derived DCs (Konstantinov et al., 2008). Immature DCs incubated with the LA NCFM at a bacterium to cell ratio of 1000:1 ("high dose") produced significantly higher IL-10 compared with the ratio of 10:1. In contrast, IL-12p70 was up-regulated at a lower concentration of the bacterium (10:1).

In our studies, dose effects of LA on pDC and cDC responses after rotavirus vaccination were examined in gnotobiotic pigs. The animal treatment groups were the same as listed earlier in the studies of T cell and B cell immune responses. Porcine pDC (CD172a+CD4+) and cDC (CD172a+CD11R1+) were defined as previously described (Jamin et al., 2006). The frequencies and tissue distribution, MHC II and costimulatory (CD80/86) molecular, TLR (2, 3, 9) and cytokine (IL-6, IL-10, IFN-$\alpha$, TNF-$\alpha$) expression by pDC and cDC in ileum, spleen and blood of gnotobiotic pigs vaccinated with the AttHRV and fed with high dose, low dose or no LA were determined using multi-color flow cytometry.

The low dose LA group had significantly higher frequencies of pDC in ileum and spleen and cDC in spleen and blood compared to the high dose LA and AttHRV only groups (Fig. 3a). The low dose LA group had overall lower MHC II expression on pDC and cDC in all tissues and lower CD80/86 expression in blood, but significantly higher CD80/86 expression on cDC in ileum, compared to the high dose LA and AttHRV only groups (Fig. 3b). High dose LA did not have a significant effect on DC frequencies or activation marker MHC II and CD80/86 expression, except for the significantly increased CD80/86 expression on pDC in ileum compared to the AttHRV only group (Fig. 3b).

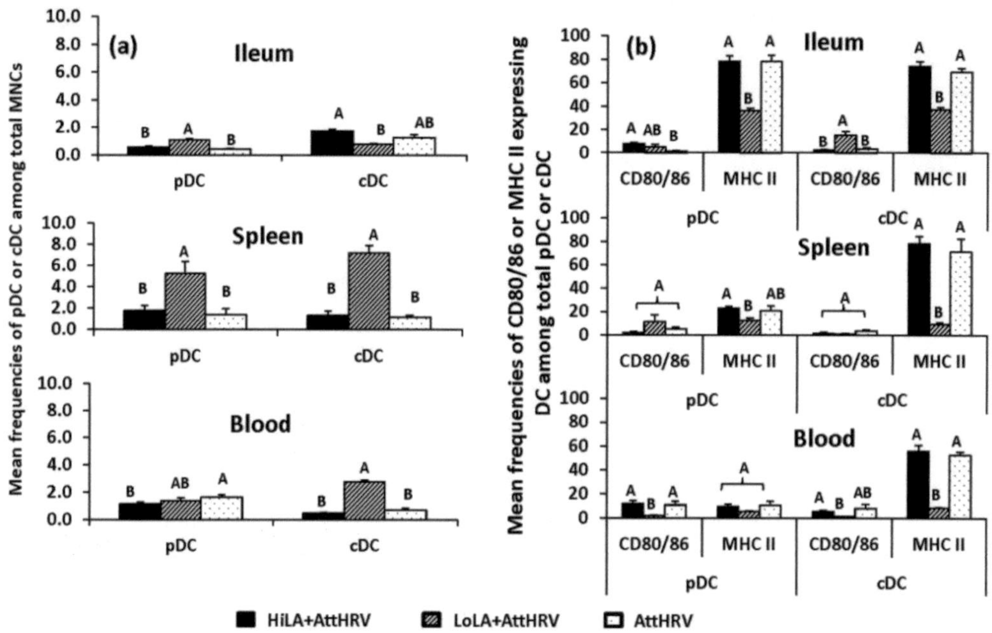


**Figure 3.** Frequencies of pDC and cDC (a) and the activation marker (CD80/86 and MHC II) expression (b) in intestinal and systemic lymphoid tissues of Gn pigs vaccinated with AttHRV vaccine with high dose, low dose or no LA at PID 28. MNC were stained freshly without *in vitro* stimulation before being subjected to flow cytometry analyses. Data are presented as mean frequency ± standard error of the mean (n = 3-13). Different letters on top of bars indicate significant differences in frequencies among groups for the same cell type and tissue (Kruskal–Wallis test, $p < 0.05$), while shared letters indicate no significant difference.

The low dose LA group had lower or significantly lower frequencies of TLR3 expression in both pDC and cDC in all tissues and significantly lower TLR2 expression on cDC in spleen compared to the high dose LA and AttHRV only groups (Fig. 4). High dose LA did not have a significant effect on TLR expression in ileum and spleen. In blood, high dose LA group had significantly lower TLR3 expression in cDC (and lower in pDC) compared to the AttHRV only group.

The most striking dose effect of LA on the cytokine production profile in DCs is the significantly increased IL-6 in the low dose LA group (Fig. 5). The low dose LA group had significantly higher frequencies of IL-6 producing pDC and cDC in all tissues compared to the high dose LA and AttHRV only groups. Interestingly, the low dose LA reduced or significantly reduced the other cytokine TNF-$\alpha$, IL-10 and IFN-$\alpha$ production in pDC in ileum and spleen. In contrast to ileum and spleen, the low dose LA increased or significantly increased TNF-$\alpha$, IL-10 and IFN-$\alpha$ production in blood compared to the high dose LA or the AttHRV only group. High dose LA did not have a significant effect on IL-6, TNF-$\alpha$, and IL-10 but lowered or significantly lowered IFN-$\alpha$ production in both pDC and cDC in all tissues compared to the AttHRV only group.

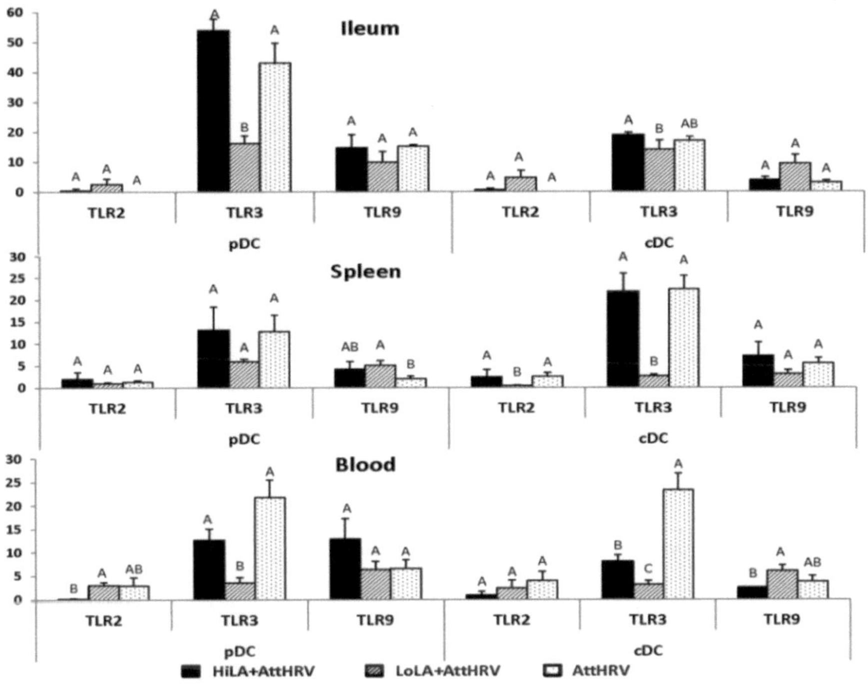


**Figure 4.** TLR expression patterns of pDC and cDC in intestinal and systemic lymphoid tissues of Gn pigs vaccinated with AttHRV vaccine with high dose, low dose or no LA at PID 28. MNC were stained freshly without *in vitro* stimulation before flow cytometry analyses. Data are presented as mean frequency ± standard error of the mean (n = 3-5). Different letters on top of bars indicate significant differences in frequencies among groups for the same TLR in the same tissue (Kruskal–Wallis test, $p < 0.05$), while shared letters indicate no significant difference.

Therefore, the effects of high versus low dose LA on the frequencies, maturation status and functions of DCs were strikingly different. Low dose LA significantly increased frequencies of both DC subsets, but these DCs were immature because they expressed lower frequencies of activation makers CD80/86 and MHC II and had reduced TNF-$\alpha$, IL-10 and IFN-$\alpha$ production compared to the high dose LA and AttHRV only groups. Low dose LA promoted a strong IL-6 response in all tissues and increased all the other cytokine TNF-$\alpha$, IL-10 and IFN-$\alpha$ production in blood for both pDC and cDC. High dose LA did not have such a significant modulating effect on the DC responses compared to the low dose (with a few exceptions). These findings are consistent with the differential effects of low dose versus high dose LA on the adaptive immune responses. The differential modulating effects of high versus low dose LA are intriguing. The biological and immunological implications of these effects and the underlying mechanisms require further investigation. From these data, it is clear that the same probiotic strain at different doses can exert qualitatively different modulating effects on DCs and consequently on adaptive immune responses induced by rotavirus vaccines. It has been reported that the effect of low dose microbe-associated pattern molecular (MAPM), such as lipopolysaccharide, was strikingly different as compared to that of high dose on macrophage cell functions: low dose lipopolysaccharide induced a strong inflammatory response in macrophages (Maitra et al., 2011). It is plausible that a similar interaction occurs between the MAPM from LA and DCs in the gut. Future studies are needed to identify the molecular mechanisms of the dose responses of different MAPM.

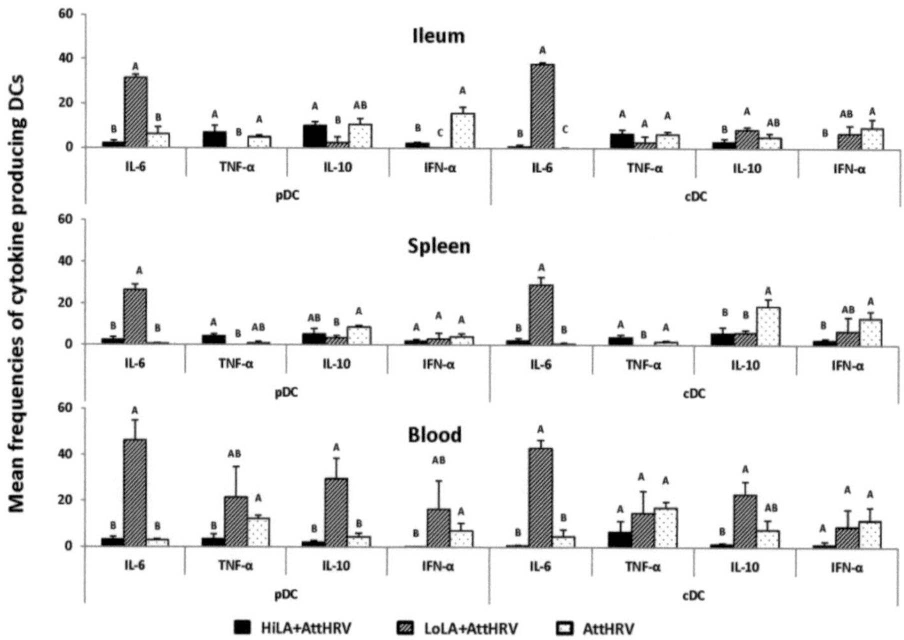


**Figure 5.** Cytokine production profiles of pDC and cDC in intestinal and systemic lymphoid tissues of Gn pigs vaccinated with AttHRV vaccine with high dose, low dose or no LA at PID 28. MNC were stained freshly without *in vitro* stimulation before flow cytometry analyses. Data are presented as mean frequency ± standard error of the mean (n = 3-8). Different letters on top of bars indicate significant differences in frequencies among groups for the same cytokine in the same tissue (Kruskal–Wallis test, $p < 0.05$), while shared letters indicate no significant difference.

## 5. Dose effects of LA on protection conferred by the oral AttHRV vaccine against virulent HRV challenge

To examine the effects of low and high dose LA on improving the protection conferred by the AttHRV vaccine, subsets of gnotobiotic pigs from each treatment group were challenged with the virulent HRV Wa strain at PID 28. Clinical signs and virus shedding were monitored for 7 days postchallenge (Table 4).

After challenge, although the proportion of pigs that developed virus shedding and diarrhea did not differ significantly among the three AttHRV vaccinated pig groups, the LoLA+AttHRV group had the shortest mean durations of fecal virus shedding and diarrhea and the lowest mean cumulative fecal consistency score among all the treatment groups. The durations of diarrhea in the LoLA+AttHRV pigs were significantly shorter compared to the AttHRV only and the mock-vaccinated control pigs. The durations of virus shedding in the LoLA+AttHRV pigs were significantly shorter compared to the HiLA+AttHRV and the mock control pigs. The mean cumulative fecal consistency scores in all the pigs in the LoLA+AttHRV and AttHRV only groups (8.4 and 9.0, respectively) were significantly lower than the control group, indicating significant protection against the severity of diarrhea. Thus, low dose LA slightly, but clearly improved the protection conferred by the AttHRV vaccine against rotavirus diarrhea. In contrast, high dose LA reduced the protection conferred by the AttHRV vaccine as indicated by the significantly longer mean duration of virus shedding (3.8 versus 1.3 days) and higher mean cumulative fecal scores compared to the AttHRV only pigs.

| Treatments | n | Clinical signs | | | Fecal virus shedding (by CCIF and/or ELISA) | | |
|---|---|---|---|---|---|---|---|
| | | % with diarrhea [*, a] | Mean duration days [**] | Mean cumulative score [**, c] | % shedding virus [*] | Mean duration days [**] | Mean peak titer (FFU/ml) [**, d] |
| HiLA+AttHRV | 13 | 92 $^{A}$ | 4.3 (0.7[b]) $^{AB}$ | 12.5 (1.4) $^{AB}$ | 31 $^{B}$ | 3.8 (0.3) $^{A}$ | 2.0 $^{B}$ |
| LoLA+AttHRV | 8 | 88 $^{A}$ | 2.4 (0.7) $^{B}$ | 8.4 (1.3) $^{B}$ | 36 $^{B}$ | 1.0 (0.0) $^{B}$ | 5.6 $^{B}$ |
| AttHRV only | 12 | 67 $^{A}$ | 4.6 (0.5) $^{A}$ | 9.8 (1.4) $^{B}$ | 50 $^{B}$ | 1.3 (0.2) $^{B}$ | 4.9 $^{B}$ |
| Mock control | 9 | 100 $^{A}$ | 5.6 (0.3) $^{A}$ | 14.4 (1.0) $^{A}$ | 100 $^{A}$ | 4.7 (0.7) $^{A}$ | 4558 $^{A}$ |

[Δ] The data was partially presented previously (Wen et al., 2012).

[a] Pigs with daily fecal scores of ≥2 were considered diarrheic. Fecal consistency was scored as follows: 0, normal; 1, pasty; 2, semiliquid; and 3, liquid.

[b] Standard error of the mean.

[c] Mean cumulative score calculation included all the pigs in each group.

[d] FFU, fluorescent focus forming units. Geometric mean peak titers were calculated among pigs that shed virus.

[*] Proportions in the same column with different superscript letters (A, B) differ significantly (Fisher's exact test, $p \leq 0.05$).

[**] Means in the same column with different superscript letters (A, B, C) differ significantly (Kruskal Wallis Test, $p \leq 0.05$).

**Table 4.** Clinical signs and rotavirus fecal shedding in Gn pigs after virulent HRV challenge[Δ]

We reported previously that protection rates against rotavirus diarrhea are correlated with virus-specific intestinal IgA ASC and IFN-γ producing T cell responses at PID 28 in Gn pigs (Yuan et al., 1996; Yuan et al., 2008). A balanced Th1 and Th2 type response is needed for the optimal protective immunity against rotavirus. Although low dose LA further reduced the duration of diarrhea in the AttHRV-vaccinated pigs postchallenge, neither low nor high dose LA significantly altered protection rate against rotavirus challenge (proportions of pigs that were infected and developed diarrhea after challenge). Because virus-specific intestinal

IgA ASC responses probably play a more important role in rotavirus protective immunity than the IFN-γ producing CD8+ T cell responses (Yuan et al., 1996; Yuan et al., 2008), the effect of LA on virus-specific ASC responses also need to be taken into consideration regarding the differences in the protection conferred by the AttHRV vaccine with high or low dose LA. Although the low dose LA enhanced IFN-γ producing CD8+ T cell responses, it had negative effects on the serum antibody and ASC responses induced by the AttHRV vaccine. To improve the AttHRV vaccine efficacy, a different dose of LA (possible an intermediate dose) or a different probiotic strain (i.e. LGG) may be optimal to promote a balanced Th1 and Th2 response without increasing Treg cell responses.

## 6. Conclusion

Differential modulating effects on innate and adaptive immune responses by low dose versus high dose of the same LA NCFM strain were clearly demonstrated in gnotobiotic pigs. Low dose LA significantly enhanced the Th1 type effector T cell responses and decreased Treg cell functions in AttHRV-vaccinated pigs. Meanwhile, low dose LA resulted in a suppressed Th2 response, as evidenced by significantly reduced virus-specific ASC responses and serum antibody titers compared to the AttHRV only group. The dose effects of LA on IFN-γ producing T cell and CD4+CD25-FoxP3+ Treg cell immune responses were similar between the intestinal and systemic lymphoid tissues. Thus the same probiotic strain used in different doses can either increase or reduce mucosal and systemic immune responses induced by vaccines. These findings have significant implications in the use of probiotic lactobacilli as immunostimulatory versus immunoregulatory agents. Probiotic products are increasingly used to improve health, alleviate disease symptoms, and enhance vaccine efficacy. Our findings suggest that probiotics can be ineffective or even detrimental if not used at the optimal dosage for the appropriate purposes, highlighting the importance of not only strain but also dose selection in probiotic studies.

The gnotobiotic pig model is a valuable animal model for study of probiotic-virus-host interaction because of the many similarities between human and porcine intestinal physiology and mucosal immune system (Meurens et al., 2012). The gnotobiotic status prevents confounding factors from commensal microflora that are present in conventionally reared animals or in humans. Unlike gnotobiotic mice, gnotobiotic pigs are devoid of maternal antibodies, thus providing an immunologically naïve background that allows clear identification of the immune responses to a single vaccine in hosts colonized with a qualitatively and quantitatively defined probiotic bacterial strain (Butler, 2009; Yuan & Saif, 2002). Although data from studies of gnotobiotic animal models may not be generalized directly to normal animals or humans, gnotobiotic animals provide a medium in which investigating the complex interrelationships of the host and its associated microbes become possible (Coates, 1975). Our findings provide a good starting point for identification of the optimal dosage of a probiotic strain. But nonetheless, the optimal dosage needs to be confirmed in conventionalized gnotobiotic pigs and in human clinical trials in order to achieve the appropriate adjuvant effect for rotavirus and other vaccines.

## Author details

Lijuan Yuan, Ke Wen, Fangning Liu and Guohua Li
*Department of Biomedical Sciences and Pathobiology,*
*Virginia Polytechnic Institute and State University, USA*

## Acknowledgement

Studies of probiotics' immune modulating effects on the rotavirus vaccine in gnotobiotic pigs were supported by a grant (R01AT004789) from the National Center of Complementary and Alternative Medicine (NCCAM), National Institutes of Health, Bethesda, MD.

## 7. References

Armah, G.E., Sow, S.O., Breiman, R.F., Dallas, M.J., Tapia, M.D., Feikin, D.R., Binka, F.N., Steele, A.D., Laserson, K.F., Ansah, N.A., Levine, M.M., Lewis, K., Coia, M.L., Attah-Poku, M., Ojwando, J., Rivers, S.B., Victor, J.C., Nyambane, G., Hodgson, A., Schodel, F., Ciarlet, M. & Neuzil, K.M. (2010). Efficacy of pentavalent rotavirus vaccine against severe rotavirus gastroenteritis in infants in developing countries in sub-Saharan Africa: a randomised, double-blind, placebo-controlled trial. *Lancet*. 376, 9741, (Aug 21, 2010), 606-614.

Azevedo, M.S., Zhang, W., Wen, K., Gonzalez, A.M., Saif, L.J., Yousef, A.E. & Yuan, L. (2012). Lactobacillus acidophilus and Lactobacillus reuteri modulate cytokine responses in gnotobiotic pigs infected with human rotavirus. *Benef Microbes*. 3, 1, (Mar 1, 2012), 33-42.

Boge, T., Remigy, M., Vaudaine, S., Tanguy, J., Bourdet-Sicard, R. & van der Werf, S. (2009). A probiotic fermented dairy drink improves antibody response to influenza vaccination in the elderly in two randomised controlled trials. *Vaccine*. 27, 41, (Sep 18, 2009), 5677-5684.

Butler, J.E. (2009). Isolator and other neonatal piglet models in developmental immunology and identification of virulence factors. *Anim Health Res Rev*. 10, 1, (Jun, 2009), 35-52.

Chiba, Y., Shida, K., Nagata, S., Wada, M., Bian, L., Wang, C., Shimizu, T., Yamashiro, Y., Kiyoshima-Shibata, J., Nanno, M. & Nomoto, K. (2010). Well-controlled proinflammatory cytokine responses of Peyer's patch cells to probiotic Lactobacillus casei. *Immunology*. 130, 3, (Jul, 2010), 352 362.

Christensen, H.R., Frokiaer, H. & Pestka, J.J. (2002). Lactobacilli differentially modulate expression of cytokines and maturation surface markers in murine dendritic cells. *J Immunol*. 168, 1, (Jan 1, 2002), 171-178.

Coates, M.E. (1975). Gnotobiotic animals in research: their uses and limitations. *Lab Anim*. 9, 4, (Oct, 1975), 275-282.

Davidson, L.E., Fiorino, A.M., Snydman, D.R. & Hibberd, P.L. (2011). Lactobacillus GG as an immune adjuvant for live-attenuated influenza vaccine in healthy adults: a randomized double-blind placebo-controlled trial. *Eur J Clin Nutr*. 65, 4, (Apr, 2011), 501-507.

de Vrese, M., Rautenberg, P., Laue, C., Koopmans, M., Herremans, T. & Schrezenmeir, J. (2005). Probiotic bacteria stimulate virus-specific neutralizing antibodies following a booster polio vaccination. *Eur J Nutr*. 44, 7, (Oct, 2005), 406-413.

Evrard, B., Coudeyras, S., Dosgilbert, A., Charbonnel, N., Alame, J., Tridon, A. & Forestier, C. (2011). Dose-dependent immunomodulation of human dendritic cells by the probiotic Lactobacillus rhamnosus Lcr35. *PLoS One*. 6, 42011), e18735.

Galdeano, C.M. & Perdigon, G. (2006). The probiotic bacterium Lactobacillus casei induces activation of the gut mucosal immune system through innate immunity. *Clin Vaccine Immunol*. 13, 2, (Feb, 2006), 219-226.

Grangette, C. (2012). Bifidobacteria and subsets of dendritic cells: friendly players in immune regulation! *Gut*. 61, 3, (Mar, 2012), 331-332.

Isolauri, E., Joensuu, J., Suomalainen, H., Luomala, M. & Vesikari, T. (1995). Improved immunogenicity of oral D x RRV reassortant rotavirus vaccine by Lactobacillus casei GG. *Vaccine*. 13, 3, (Feb, 1995), 310-312.

Jamin, A., Gorin, S., Le Potier, M.F. & Kuntz-Simon, G. (2006). Characterization of conventional and plasmacytoid dendritic cells in swine secondary lymphoid organs and blood. *Vet Immunol Immunopathol*. 114, 3-4, (Dec 15, 2006), 224-237.

Kaila, M., Isolauri, E., Soppi, E., Virtanen, E., Laine, S. & Arvilommi, H. (1992). Enhancement of the circulating antibody secreting cell response in human diarrhea by a human Lactobacillus strain. *Pediatr Res*. 32, 2, (Aug, 1992), 141-144.

Kekkonen, R.A., Kajasto, E., Miettinen, M., Veckman, V., Korpela, R. & Julkunen, I. (2008). Probiotic Leuconostoc mesenteroides ssp. cremoris and Streptococcus thermophilus induce IL-12 and IFN-gamma production. *World J Gastroenterol*. 14, 8, (Feb 28, 2008), 1192-1203.

Konieczna, P., Groeger, D., Ziegler, M., Frei, R., Ferstl, R., Shanahan, F., Quigley, E.M., Kiely, B., Akdis, C.A. & O'Mahony, L. (2012). Bifidobacterium infantis 35624 administration induces Foxp3 T regulatory cells in human peripheral blood: potential role for myeloid and plasmacytoid dendritic cells. *Gut*. 61, 3, (Mar, 2012), 354-366.

Konstantinov, S.R., Smidt, H., de Vos, W.M., Bruijns, S.C., Singh, S.K., Valence, F., Molle, D., Lortal, S., Altermann, E., Klaenhammer, T.R. & van Kooyk, Y. (2008). S layer protein A of Lactobacillus acidophilus NCFM regulates immature dendritic cell and T cell functions. *Proc Natl Acad Sci U S A*. 105, 49, (Dec 9, 2008), 19474-19479.

Li, C.Y., Lin, H.C., Lai, C.H., Lu, J.J., Wu, S.F. & Fang, S.H. (2011). Immunomodulatory Effects of Lactobacillus and Bifidobacterium on Both Murine and Human Mitogen-Activated T Cells. *Int Arch Allergy Immunol*. 156, 2, (May 16, 2011), 128-136.

Link-Amster, H., Rochat, F., Saudan, K.Y., Mignot, O. & Aeschlimann, J.M. (1994). Modulation of a specific humoral immune response and changes in intestinal flora mediated through fermented milk intake. *FEMS Immunol Med Microbiol*. 10, 1, (Nov, 1994), 55-63.

Maitra, U., Gan, L., Chang, S. & Li, L. (2011). Low-dose endotoxin induces inflammation by selectively removing nuclear receptors and activating CCAAT/enhancer-binding protein delta. *J Immunol*. 186, 7, (Apr 1, 2011), 4467-4473.

Mane, J., Pedrosa, E., Loren, V., Gassull, M.A., Espadaler, J., Cune, J., Audivert, S., Bonachera, M.A. & Cabre, E. (2011). A mixture of Lactobacillus plantarum CECT 7315 and CECT 7316 enhances systemic immunity in elderly subjects. A dose-response, double-blind, placebo-controlled, randomized pilot trial. *Nutr Hosp*. 26, 1, (Jan-Feb, 2011), 228-235.

Meijerink, M. & Wells, J.M. (2010). Probiotic modulation of dendritic cells and T cell responses in the intestine. *Benef Microbes*. 1, 4, (Nov, 2010), 317-326.

Meurens, F., Summerfield, A., Nauwynck, H., Saif, L. & Gerdts, V. (2012). The pig: a model for human infectious diseases. *Trends Microbiol*. 20, 1, (Jan, 2012), 50-57.

Mohamadzadeh, M., Duong, T., Hoover, T. & Klaenhammer, T.R. (2008). Targeting mucosal dendritic cells with microbial antigens from probiotic lactic acid bacteria. *Expert Rev Vaccines*. 7, 2, (Mar, 2008), 163-174.

Mohamadzadeh, M., Olson, S., Kalina, W.V., Ruthel, G., Demmin, G.L., Warfield, K.L., Bavari, S. & Klaenhammer, T.R. (2005). Lactobacilli activate human dendritic cells that skew T cells toward T helper 1 polarization. *Proc Natl Acad Sci U S A*. 102, 8, (Feb 22, 2005), 2880-2885.

O'Ryan, M. & Linhares, A.C. (2009). Update on Rotarix: an oral human rotavirus vaccine. *Expert Rev Vaccines*. 8, 12, (Dec, 2009), 1627-1641.

O'Ryan, M.L., Hermosilla, G. & Osorio, G. (2009). Rotavirus vaccines for the developing world. *Curr Opin Infect Dis*. 22, 5, (Oct, 2009), 483-489.

Olivares, M., Diaz-Ropero, M.P., Sierra, S., Lara-Villoslada, F., Fonolla, J., Navas, M., Rodriguez, J.M. & Xaus, J. (2007). Oral intake of Lactobacillus fermentum CECT5716 enhances the effects of influenza vaccination. *Nutrition*. 23, 3, (Mar, 2007), 254-260.

Paineau, D., Carcano, D., Leyer, G., Darquy, S., Alyanakian, M.A., Simoneau, G., Bergmann, J.F., Brassart, D., Bornet, F. & Ouwehand, A.C. (2008). Effects of seven potential probiotic strains on specific immune responses in healthy adults: a double-blind, randomized, controlled trial. *FEMS Immunol Med Microbiol*. 53, 1, (Jun, 2008), 107-113.

Perdigon, G., Fuller, R. & Raya, R. (2001). Lactic acid bacteria and their effect on the immune system. *Curr Issues Intest Microbiol*. 2, 1, (Mar, 2001), 27-42.

Repa, A., Grangette, C., Daniel, C., Hochreiter, R., Hoffmann-Sommergruber, K., Thalhamer, J., Kraft, D., Breiteneder, H., Mercenier, A. & Wiedermann, U. (2003). Mucosal co-application of lactic acid bacteria and allergen induces counter-regulatory immune responses in a murine model of birch pollen allergy. *Vaccine*. 22, 1, (Dec 8, 2003), 87-95.

Rowe, J.H., Ertelt, J.M. & Way, S.S. (2012). Foxp3(+) regulatory T cells, immune stimulation and host defence against infection. *Immunology*. 136, 1, (May, 2012), 1-10.

So, J.S., Kwon, H.K., Lee, C.G., Yi, H.J., Park, J.A., Lim, S.Y., Hwang, K.C., Jeon, Y.H. & Im, S.H. (2008a). Lactobacillus casei suppresses experimental arthritis by down-regulating T helper 1 effector functions. *Mol Immunol*. 45, 9, (May, 2008a), 2690-2699.

So, J.S., Lee, C.G., Kwon, H.K., Yi, H.J., Chae, C.S., Park, J.A., Hwang, K.C. & Im, S.H. (2008b). Lactobacillus casei potentiates induction of oral tolerance in experimental arthritis. *Mol Immunol*. 46, 1, (Nov, 2008b), 172-180.

Summerfield, A. & McCullough, K.C. (2009). The porcine dendritic cell family. *Dev Comp Immunol*. 33, 3, (Mar, 2009), 299-309.

Torii, A., Torii, S., Fujiwara, S., Tanaka, H., Inagaki, N. & Nagai, H. (2007). Lactobacillus Acidophilus strain L-92 regulates the production of Th1 cytokine as well as Th2 cytokines. *Allergol Int*. 56, 3, (Sep, 2007), 293-301.

Van Overtvelt, L., Moussu, H., Horiot, S., Samson, S., Lombardi, V., Mascarell, L., van de Moer, A., Bourdet-Sicard, R. & Moingeon, P. (2010). Lactic acid bacteria as adjuvants for sublingual allergy vaccines. *Vaccine*. 28, 17, (Apr 9, 2010), 2986-2992.

Villadangos, J.A., Cardoso, M., Steptoe, R.J., van Berkel, D., Pooley, J., Carbone, F.R. & Shortman, K. (2001). MHC class II expression is regulated in dendritic cells independently of invariant chain degradation. *Immunity*. 14, 6, (Jun, 2001), 739-749.

Vitini, E., Alvarez, S., Medina, M., Medici, M., de Budeguer, M.V. & Perdigon, G. (2000). Gut mucosal immunostimulation by lactic acid bacteria. *Biocell*. 24, 3, (Dec, 2000), 223-232.

Ward, L.A., Rosen, B.I., Yuan, L. & Saif, L.J. (1996). Pathogenesis of an attenuated and a virulent strain of group A human rotavirus in neonatal gnotobiotic pigs. *J Gen Virol*. 77 (Pt 7), (Jul, 1996), 1431-1441.

Weiss, G., Rasmussen, S., Zeuthen, L.H., Nielsen, B.N., Jarmer, H., Jespersen, L. & Frokiaer, H. (2010). Lactobacillus acidophilus induces virus immune defence genes in murine dendritic cells by a Toll-like receptor-2-dependent mechanism. *Immunology*. 131, 2, (Oct, 2010), 268-281.

Wells, J.M. & Mercenier, A. (2008). Mucosal delivery of therapeutic and prophylactic molecules using lactic acid bacteria. *Nat Rev Microbiol*. 6, 5, (May, 2008), 349-362.

Wen, K., Li, G., Bui, T., Liu, F., Li, Y., Kocher, J., Lin, L., Yang, X. & Yuan, L. (2012). High dose and low dose Lactobacillus acidophilus exerted differential immune modulating effects on T cell immune responses induced by an oral human rotavirus vaccine in gnotobiotic pigs. *Vaccine*. 30, 6, (Feb 1, 2012), 1198-1207.

Winkler, P., de Vrese, M., Laue, C. & Schrezenmeir, J. (2005). Effect of a dietary supplement containing probiotic bacteria plus vitamins and minerals on common cold infections and cellular immune parameters. *International journal of clinical pharmacology and therapeutics*. 43, 7, (Jul, 2005), 318-326.

Yazdi, M.H., Soltan Dallal, M.M., Hassan, Z.M., Holakuyee, M., Agha Amiri, S., Abolhassani, M. & Mahdavi, M. (2010). Oral administration of Lactobacillus acidophilus induces IL-12 production in spleen cell culture of BALB/c mice bearing transplanted breast tumour. *Br J Nutr*. 104, 2, (Jul, 2010), 227-232.

Yuan, L. & Saif, L.J. (2002). Induction of mucosal immune responses and protection against enteric viruses: rotavirus infection of gnotobiotic pigs as a model. *Vet Immunol Immunopathol*. 87, 3-4, (Sep 10, 2002), 147-160.

Yuan, L., Ward, L.A., Rosen, B.I., To, T.L. & Saif, L.J. (1996). Systematic and intestinal antibody-secreting cell responses and correlates of protective immunity to human rotavirus in a gnotobiotic pig model of disease. *J Virol*. 70, 5, (May, 1996), 3075-3083.

Yuan, L., Wen, K., Azevedo, M.S., Gonzalez, A.M., Zhang, W. & Saif, L.J. (2008). Virus-specific intestinal IFN-gamma producing T cell responses induced by human rotavirus infection and vaccines are correlated with protection against rotavirus diarrhea in gnotobiotic pigs. *Vaccine*. 26, 26, (Jun 19, 2008), 3322-3331.

Zaman, K., Dang, D.A., Victor, J.C., Shin, S., Yunus, M., Dallas, M.J., Podder, G., Vu, D.T., Le, T.P., Luby, S.P., Le, H.T., Coia, M.L., Lewis, K., Rivers, S.B., Sack, D.A., Schodel, F., Steele, A.D., Neuzil, K.M. & Ciarlet, M. (2010). Efficacy of pentavalent rotavirus vaccine against severe rotavirus gastroenteritis in infants in developing countries in Asia: a randomised, double-blind, placebo-controlled trial. *Lancet*. 376, 9741, (Aug 21, 2010), 615-623.

Zhang, W., Azevedo, M.S., Gonzalez, A.M., Saif, L.J., Van Nguyen, T., Wen, K., Yousef, A.E. & Yuan, L. (2008a). Influence of probiotic Lactobacilli colonization on neonatal B cell responses in a gnotobiotic pig model of human rotavirus infection and disease. *Vet Immunol Immunopathol*. 122, 1-2, (Mar 15, 2008a), 175-181.

Zhang, W., Azevedo, M.S., Wen, K., Gonzalez, A., Saif, L.J., Li, G., Yousef, A.E. & Yuan, L. (2008b). Probiotic Lactobacillus acidophilus enhances the immunogenicity of an oral rotavirus vaccine in gnotobiotic pigs. *Vaccine*. 26, 29-30, (Jul 4, 2008b), 3655-3661.

Chapter 12

# *Bifidobacterium* in Human GI Tract: Screening, Isolation, Survival and Growth Kinetics in Simulated Gastrointestinal Conditions

Nditange Shigwedha and Li Jia

Additional information is available at the end of the chapter

http://dx.doi.org/10.5772/50457

## 1. Introduction

Many species of lactic acid bacteria (LAB), *Bacillus*, and fungi such as *Saccharomyces* and *Aspergillus* have been used over the years in the food industry. A few have gained the probiotic status – defined as live microorganisms, which when administered in adequate amounts confer a health benefit on the host (Joint FAO/WHO, 2002) – and most of this belong to *Lactobacillus* (e.g., *L. bulgaricus, L. acidophilus, L. rhamnosus, L. casei, L. johnsonii, L. reuteri,* etc.), *Streptococcus* (e.g., *S. thermophilus,* etc.), and *Bifidobacterium* (e.g., *B. bifidum, B. longum, B. breve, B. infantis*) genera. Bifidobacteria is the predominant species of bacteria in the normal intestinal flora of healthy breast-fed newborns where they constitute more than 95% of the total population (Yildirim & Johnson, 1998). Numerous *Bifidobacterium* strains have gained recognition as probiotics because of their various therapeutic health benefits, including resistance to enteric pathogens (*Clostridium spp., Salmonella spp., Candida spp., Escherichia coli spp. and Listeria monocytogenes*), aid in lactose digestion and/or help to regulate digestion, anti–colon cancer effect, the immune system modulation, anti-allergy, and hepatic encephalopathy (Jia *et al.*, 2010), and also for having a protective effect against acute diarrhoea (Liepke *et al.*, 2002). The food industry recognized the market potential of the numerous strain-specific positive health benefits of the bifidobacteria cultures, namely in beverages. Bifidobacteria can also be administered as capsules or tablets or incorporated into food as dietary adjuncts and into baby foods (Lourens-Hattingh & Viljoen, 2001; Patrignani *et al.*, 2006). In addition, bifidobacteria lower inositol phosphate content during bread making (Palacios *et al.*, 2008).

Several investigators have speculated that the survival of most bifidobacteria is not exceptionally high in most dairy products due to low pH and/or exposure to oxygen.

Nevertheless, problems may arise as a consequence of the difficulties of isolation and cultivation of bifidobacteria. Only a few studies have been published concerning the isolation and characterization of plasmids from bifidobacteria. The human gastrointestinal (GI) tract is the largest tube, running through the body and which include mouth and/or oral cavity, oesophagus, stomach, small intestine and large intestine. (Figure 1).

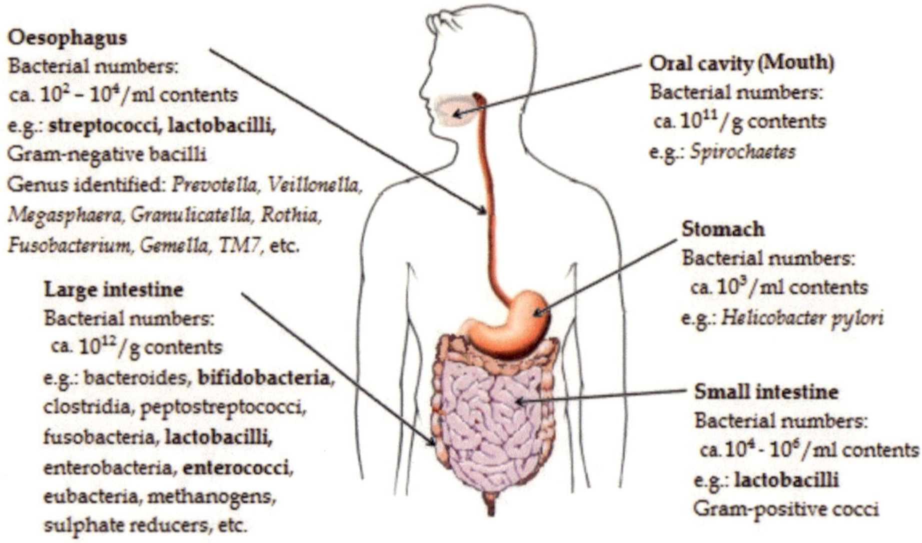


**Figure 1.** The human gastrointestinal tract and its microbiota.

### 1.1. The oral cavity

Ingested foodstuff first comes into contact with the oral cavity, which is composed of different niches of microbial population. In the oral cavity, bacteria are the main group of microorganisms, although viruses and yeasts can also be found. The main ecological habitants of the mouth are the mucosa of lips, cheeks and palate, the tongue, the tooth surface, the saliva, and the tonsillar area. The population of microorganisms in each section is mainly dependent on the presence of oxygen and nutrients as well as the flow rate of the saliva (see Figure 2). The major species in the oral cavity are lactic acid bacteria of the genera *Streptococcus*, *Lactobacillus* and *Bifidobacterium*. In dental plaque and oral infections, many anaerobic species have been isolated, mainly *Prevotella* and *Porphyromonas* species, as well as *Eubacterium*, *Actinomyces* and *Veillonella* (Hartemink, 1999).

The main source of nutrients and energy for oral bacteria is the ingested food, especially carbohydrates, which are rapidly metabolized to lactic and acetic acids by the predominant LAB, leading to a rapid drop in the pH of the saliva after ingestion of carbohydrates. The surplus carbohydrates can be incorporated into exopolysaccharides by a large number of bacteria and be used as energy storage compounds, or as attachment factors (Hartemink, 1999).

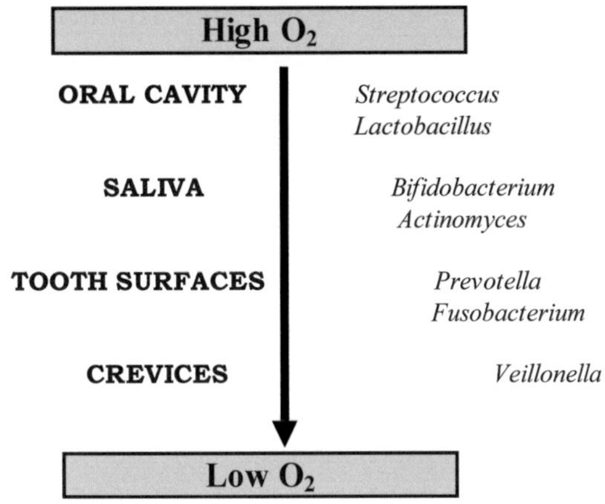


**Figure 2.** Relationship between bacterial species, oxygen tension and habitat in the oral cavity.

## 1.2. The oesophagus

In quantitative terms, the oesophagus and stomach carry the lightest microbial loads in the human GI tract. The predominant culturable bacteria are facultative anaerobes, originating in the oral cavity, such as streptococci and lactobacilli, which occur in relatively small numbers (*ca.* $10^2 - 10^3$ $cm^{-2}$ or $ml^{-1}$ of the mucosal surface or lumenal aspirate, respectively) (Macfarlane & Dillon, 2007). The majority of oesophageal bacteria (including the largely $\alpha$-haemolytic *Streptococcus* species) are cultivable and are almost $10^4$ bacteria per $mm^2$ mucosal surface of the distal oesophagus (Pei *et al.*, 2004). While the bacterial biota in the distal oesophagus is likely to be similar to that of the oropharynx (Kazor *et al.*, 2003), many other species of *Pseudomonas tolaasii, Pseudomonas influorensces, Pseudomonas syringae, Pseudomonas putida,* uncultured *Duganella, Stenotrophomonas maltophilia, Janthinobacterium lividum, Lactobacillus paracasei, Propionibacterium acnes, Pseudomonas Antarctica / meridiana,* and *Brevundimonas bulata* exist in the oesophagus (Pei *et al.*, 2004). Other selected members of the bacterial genera found in human distal oesophagus are given in Figure 1.

## 1.3. The stomach

In general, the human stomach has a remarkably low pH. The normal resting gastric juice's pH is below 3.0, which prevents virtually all bacterial growth, and which is bactericidal for most transient species, especially the LABs. During and shortly after a meal, the pH may increase to values around 6.0. This will allow passing bifidobacteria to survive the gastric juice prior to proceeding onto the small intestine (to battle the bile salts). The resident flora of the gut lumen is highly acidic tolerant and consists mainly of lactobacilli and streptococci.

In the stomach mucosa, the pH is much higher, and bacterial populations may be higher, as well. In addition to lactobacilli and streptococci, some other bacterial species and yeasts may be present (Hartemink, 1999). The gastric juice plays a significant role in digestion of proteins, by activating digestive enzymes, making ingested proteins unravel so that digestive enzymes can alter protein down to individual amino acids. Fermentation of ingested carbohydrates in the stomach hardly occurs.

### 1.4. The small intestine

When the partially digested food enters the small intestine, it is mixed with intestinal secretions, such as bile, pancreatic enzymes and bicarbonates. The bile in particular has a strong bactericidal effect. Together with a strong-fluid secretion by the intestinal mucosa, this also prevents extensive colonization of the small intestine. Colonization usually takes place in crypts and blind loops. In this lower part of the small intestine, the movement is slightly reduced, the bile is diluted, the pH becomes more neutral, and the oxygen tension drops rapidly. This favours the growth and/or transit of different bacteria, initially mainly aerotolerant species, and in the ileum also strict anaerobes as revealed in Figure 3 (Hartemink, 1999). There is not much carbohydrate fermentation in the small intestine in healthy humans, due to the flow rate and the little bacterial mass.

In studies undertaken in pigs, it has been reported that the conditions in the small intestine differed widely. The pH is much higher, and the bile secretion is less abundant, which results in an extensive bacterial growth in the small intestine. This also results in substantial fermentation of ingested carbohydrates. The human body is projected to produce between 20 to 30 g of bile salts per day to replace the loss occurring in the excreta (250 to 500 mg), and these are typically stored in the gall bladder (Glickman, 1980).

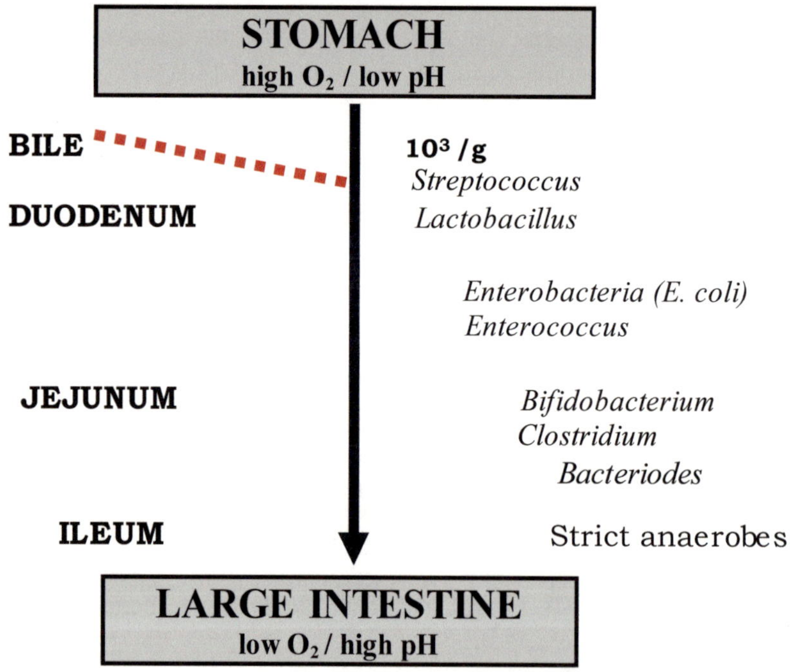


**Figure 3.** Appearance of bacterial species, oxygen tension and habitat in the small intestine.

## 1.5. The large intestine

In the large intestine, the flow rate of the digesta decreases considerably. In addition, the bile is even more diluted, and the pH is close to neutral. Total logarithmic counts may reach up to $10^{11}$ bacteria/gram contents. Higher numbers have been reported, but it is physically impossible to achieve a number over $10^{12}$ bacteria/gram faecal dry weight, taking into account the average balance of faeces and the dimensions of an average bacterium. It is estimated that over 400 different bacterial species reside in the human large intestine. Of these, about 200 have been validly described, but often non-identifiable strains are reported.

In addition to the resident bacteria, transient bacteria are often isolated. The dominant floras in the large intestine are relatively stable, and they include *Bifidobacterium, Bacteriodes* and anaerobic cocci. Large variations also exist in the less dominant species, especially among the facultative or aerotolerant species like *E. coli* and lactobacilli. The numbers of the dominant species are also comparable in different population. Differences in counts are more often due to the methodology used, rather than actual differences. As in individuals, the counts of less dominant species differ widely between different populations. Among the dominant bacterial groups are members of the genera *Bacteriodes, Bifidobacterium, Coprococcus, Peptostreptococcus, Eubacterium* and *Ruminococcus*. Members of the following genera are often isolated and are available in lower numbers: *Fusobacterium, Streptococcus, Lactobacillus, Enterococcus, Veillonella, Megasphaera, Propionibacterium* and *Enterobacteriaceae*.

It is indispensable to emphasize here that the principal function of the GI tract includes breakdown and absorption of food components and water. In general, degradation takes place in the upper part of the GI tract, whereas the major sites of absorption are in the lower part of small intestines and the large intestines. Degradation and absorption are enhanced by the excretion of the large number of digestive enzymes, such as glycosidases, lipases, peptidases and proteinases. The colon receives digesta from the intestinal ileum approximately 5 h after food ingestion. Thereafter, rate of motion slows progressively from the caecum towards the distal colon. Concurrent with this is an increase in water absorption; thus gut contents in the proximal colon are more or less liquid in nature but have a faecal like appearance distally (Hartemink, 1999).

For most of the world's population, the standard gut transit time is 60 h, with a variety of 23 – 168 h. The colon itself has a capacity of approximately 500 ml with about 220 g contents. In general, stools weight correlates inversely with transit time. Studies with healthy volunteers have indicated that speeding up colonic transit times from 67 to 25 h resulted in an increase in stool weight from 148 to 285 g/day. Conversely, when transit time increased, stool weight decreases from 182 to 119 g/day (Hartemink, 1999). The differences are mainly due to changes in the water content of the faecal mass.

The structure of faeces is highly variable. Bacteria may constitute up to 55% of the total solids, whilst fibre and other non-digested, non-fermented compounds represent less than 17% of the weight of which about 24% is soluble material. Faecal water content may be as high as 70% of the total weight. Stool size is influenced by both dietary and endogenous factors. Endogenous factors mainly operate through hormones on the intestinal motility. The well-known endogenous factors include: decreased peristaltic movements during exercises and menstrual cycle. Dietary factors, like non-digestible fibres and polyalcohols (sorbitol), may retain water and thus increases stool bulk. High amounts of these factors may cause diarrhoea, due to the increased osmotic pressure.

## 2. Screening and isolation procedures of bifidobacteria strains

Molecular methods have shown that the average percentage of bifidobacteria in the GI tract of humans is approximately 3% of total microbiota, or they occur at a concentration of $10^9$ – $10^{10}$ CFU/g of faeces (Jia *et al.*, 2010). As to achieve intestinal colonization in humans or animals, bifidobacteria have to endure inhibitory substances secreted by the host, such as gastric acid in the stomach and bile salts (in the small intestine). Although, both the gastric pH (pH < 3) and bile salts are strongly bacteriocidal, some resistant bifidobacteria can handle the low pH's ranges of the stomach and also survive the effects of bile salts in the small intestine of humans. These can be isolated and screened for their leading roles as probiotics.

### 2.1. Isolation and cultivation of bifidobacteria resistant to acidic pH and bile salts

#### *2.1.1. Isolation via stress-shock procedure*

Selection of acid and bile resistant bifidobacteria has been based on the stressing isolation method developed by Chung *et al.*, (1999). Faecal samples are collected from infants and/or adults. The tube containing the faecal sample is promptly screened for the isolation of resistant strains, as follows: Faecal samples (0.8 g each) are inoculated into 8 ml of Transga-lactooligosaccharide-propionate (TP) medium as an enrichment medium for the bifidobacteria. After an anaerobic incubation for 12 h at 37 °C, 0.8 ml of the incubated cultures is transferred into fresh TP medium with pH adjusted to 2.0 and incubated anaerobically for another 12 h at 37 °C. After the acid exposure, an aliquot (0.8 ml) of the incubation medium is transferred into fresh TP medium supplemented with 1.5% ox-gall, and the incubation continued for another 2 h at 37 °C. The resulting incubation medium is serially diluted and plated on TP agar, to select colonies of the resistant bifidobacteria strains. To isolate reference strains, serially diluted *Bifidobacterium* cells grown in the regular TP medium are plated on TP agar medium. In most of the isolation studies, *B. adolescentis, B. longum, B. infantis, B. bifidum* and owner identified *Bifidobacterium* strains (commonly called "own isolates" in microbiology) are used as reference strains. The reference strains are utilized for the convenience of comparison to the resistant strains. Microscopic analysis (1000 × with immersion oil) is routinely performed to confirm *Bifidobacterium* morphology.

In addition, *Bifidobacterium* cells are examined for their biochemical and morphological characteristics according to the Bergey's Manual of Determinative Bacteriology. The cultures are grown in Man, Rogosa, and Sharpe (MRS) medium under anaerobic conditions, in a microprocessor-controlled anaerobic chamber. Cultures are incubated for 18 h at 37 °C and stored at 3 – 5 °C between transfers. For the fermentation test, 0.5 ml of 10% substrate solutions (which were membrane filtered through 0.45 μM filter), are added to 9.5 ml of Peptone Yeast-extract Fildes (PYF) basal medium (Mitsuoka, 1990). After 2.5 days of strictly anaerobic incubation, the pH of the growth medium is measured. Tubes showing pH values below 5.5 are considered to be positive for fermentation. The presence of acetate and lactate in the fermented PYF containing glucose medium is assayed by using gas chromatography (GC) or high performance liquid chromatography (HPLC).

### *2.1.2. Isolation and screening via stress-shock*

Briefly, faecal samples of 3 to 5 days old new-born babies are collected and taken to the laboratory for immediate analysis and isolation of bifidobacteria. About 2 g of each faeces sample is placed in a sterile test-tube (30 ml) and closed tightly with a rubber-stopper. For optimal survival of these highly sensitive anaerobic bacteria, the samples are treated within 15 min after faeces emission, or else the samples are kept in an anaerobic environment until analysis (maximum of 10 h). Screening for the isolation of resistant strains is as follows: faecal samples (2 g each) are inoculated into 10 ml test-tubes of Raffinose-Bifidobacterium (RB) broth (pH 6.8). After an anaerobic incubation for 12 h at 38.5 ºC, 1 ml of the incubated culture is transferred into 10 ml of fresh RB medium with pH adjusted to 3.0 and incubated anaerobically for 2 h at 38.5 ºC. After the acid exposure, an aliquot (1 ml) of the incubation medium is transferred into 10 ml of fresh RB medium supplemented with 1% ox-gall, and the incubation continues for another 2 h at 38.5 ºC. The resulting incubation medium is serially diluted (10-folds) in a pre-reduced Ringer solution with 5 – 10% glycerol for the inhibition of the cellulolytic activity of the fungus. An aliquot of 100 μl from each dilution is plated directly on RB and MRS agars using the surface streak method and incubated anaerobically at 38.5 °C for 3 – 4 days to determine colonies of the resistant *Bifidobacterium* strains.

Likewise, the isolate designated *B. longum* GB-03 was isolated from a pharmaceutical product called Golden Bifid (containing a combination of unspecified *Bifidobacterium* spp., *Streptococcus thermophilus* and *Lactobacillus bulgaricus*) using a similar approach. The first step is crucial to reveal that a single piece (0.5 g) has to be dissolved in 0.2 ml test-tube of sterilized distilled water before being inoculated into 10 ml test tube of fresh RB-medium.

## 3. Morphological identification of bifidobacteria by phase contrast microscopy (PCM)

In the morphological analysis of bifidobacteria population, *in situ,* in human faeces and/or other foods products, microscopes have been used to determine the degree of heterogeneity of these probiotic's populations. The morphology of bifidobacteria determined microscopically has been used as an aid to phenotypic differentiation within the group, while the effect of medium type, low pH and high bile salt concentrations on the bifidobacterial cell morphology has also been studied by this method. Individual

*Bifidobacterium* strains are characterized phenotypically, including morphology identification by phase contrast microscopy (PCM).

Bifidobacteria are gram-positive, anaerobic, rods of various shapes (short, regular, thin cells with pointed ends, coccoidal regular cells, long cells with slight bends or protuberances) or a variety of branching (pointed, slightly bifurcated, club-shaped or spatulated extremities), single or chains of various arrangements (in star-like aggregates or disposed in "V" or "Y" or else "palisade" arrangements) (Scardovi, 1986).

As a pattern to characterize the heterogeneous population of bifidobacteria associated with human origin and other sources, the PCM examinations and two different media (RB & modified MRS) were used to demonstrate a better phenotypical correlation of the natural isolates to the reference strains on RB, MRS and modified MRS media as shown in Figures 4.1 – 4.12). These media are unique and appear to be still the most predominant in culturing the bifidobacteria strains.

Isolates of bifidobacteria are normally cultured anaerobically on appropriate agars at 38 ºC for 3 – 4 days. For gram-staining, a loopful of the culture is streaked on microscope slides (46 × 25 mm) and the staining technique followed thoroughly. Subsequently, the slide is observed under phase contrast microscopy, preferably at 1000 × magnification by oil immersion and can be photographed as well, using the images advanced software package if available.

## 3.1. Morphological characterization of *Bifidobacterium* reference strains

The basic morphologies, namely short, regular, thin cells with pointed ends, coccoidal regular cells, and long cells with slight bends or protuberances are discernible among the 2 *Bifidobacterium* reference strains (*B. adolescentis* and *B. infantis)* shown in Figures 4.1, 4.2, 4.3, 4.4 and 4.11) on modified MRS and RB media. From these micrographs alone, it is obvious to validate that individual variations of the average phenotypic morphologies of bifidobacteria are present as described earlier. The PCM also provided a rapid and clear visualization of the basic bifidobacteria cell morphology, while at the same time, allowed only broad comparisons amongst the bifid structures within a mixture of 2 other LABs (*Streptococcus thermophilus* & *Lactobacillus bulgaricus*) (see Figure 4.12).

The typical colonies of bifidobacteria are altogether round and white on RB and modified MRS media. Colonies are usually picked off of a suitable plate and may be kept sub-cultured 2 – 3 times on a freshly prepared agar as to obtain pure culture without contamination. The morphologies of the 2 reference strains and their relationship to each will now be discussed separately. When the strain of *B. adolescentis* is resuscitated and cultured on modified MRS medium (Figure 4.1) or RB medium (Figure 4.2), it may be differentiated clearly from the *B. infantis* (Figures 4.3 & 4.4) on the basis of morphology. The *B. infantis* was also resuscitated and cultured under the similar conditions. As it can be observed from Figure 4.2, *B. adolescentis* on RB displayed long and thick rod-shaped and regular coccoidal cells. The cells of *B. adolescentis* strain on RB were almost paired and assembled, a feature which was highlighted by PCM. The existence of distinct "V"- and/or "Y"-shapes and some long cells with protuberances or slight curvature of this isolates when grown on RB agar, is a powerful diagnostic feature, particularly when distinguishing this specie from closely related *B. minimum* when grown on Trypticase-Phytone-Yeast extract (TPY) agar stabs (Biavati *et al.,* 1982). In addition, curved cells with smooth and rounded ends are the most one dominating in the micrograph. These features were not compatible with descriptions of this particular species' morphology as described by Reuter (1963), but were common to other species of the genus.

The *B. infantis* strain displayed slender, often short rod-shaped and of the typical club-shaped extremities, which cells of these species are reported to exhibit (see Figures 4.3 & 4.4). The morphology of this strain is almost the same when grown on both the MRS and RB solid growth media. Furthermore, *B. infantis* showed a distinct tendency for chain formation on RB medium. These cells often occurred in "V" and "Y"-shapes and were similar to that of many other species of the genus. Nevertheless, it was also possible to differentiate between this strain and the closely related *B. longum* GB-03 (own isolate, Fig. 4.6) on the basis of small variety of club-shaped extreme morphology.

## 3.2. Morphological differentiation of isolates of bifidobacteria

Morphological consistency is greater among the *Bifidobacterium* isolate (*B. longum* GB-03 and *B. bifidum* WN-04) as shown in Figures 4.5 to 4.11) than the *Bifidobacterium* reference strains. Cell shapes ranged from long and thick–rods with protuberances to long and thin–rods with blunted ends and slightly bifurcated club-shaped extremities, with a number of variations on these basic shapes. Two morphological groups and their potential significance are discussed separately below.

### *3.2.1. Long and thick–rods with protuberances cell morphology*

Figures 4.6 and 4.8 display both isolates of *B. longum* GB-03 and *B. bifidum* WN-04 on RB medium, which consisted of long and thick cells with slight bends. The regular morphology of these cells and the star-like aggregates arrangement (Figure 4.6) was evident under the PCM when grown on RB agar. Also, the presences of sparsely distributed single cells were also evident under the PCM (Figure 4.8). The morphology of these cells was consistent with any of the *Bifidobacterium* reference strains discussed previously. The isolates' morphologies resembled the reference strain of *B. infantis* which are never elongated but have a penchant for group formation (Figure 4.4).

Although no conclusions could be drawn on the basis of morphology alone, the presence of "V"-shaped rods, protuberances with a large variety of bending in *B. bifidum* WN-04 isolate appeared to resemble the reference strains of *B. bifidum,* especially the "amphora-like" cells that are characteristic (Sundman & Bjorksten, 1959). On the RB media, PCM analysis allowed a better correlation of the natural isolates to the reference strains. Speciation of *B. longum* GB-03 (in Figure 4.6) conversely appeared to favour the reference strain of *B. longum,* especially the ultra-elongated and relatively thin cellular elements with slightly irregular contours (Reuter, 1963).

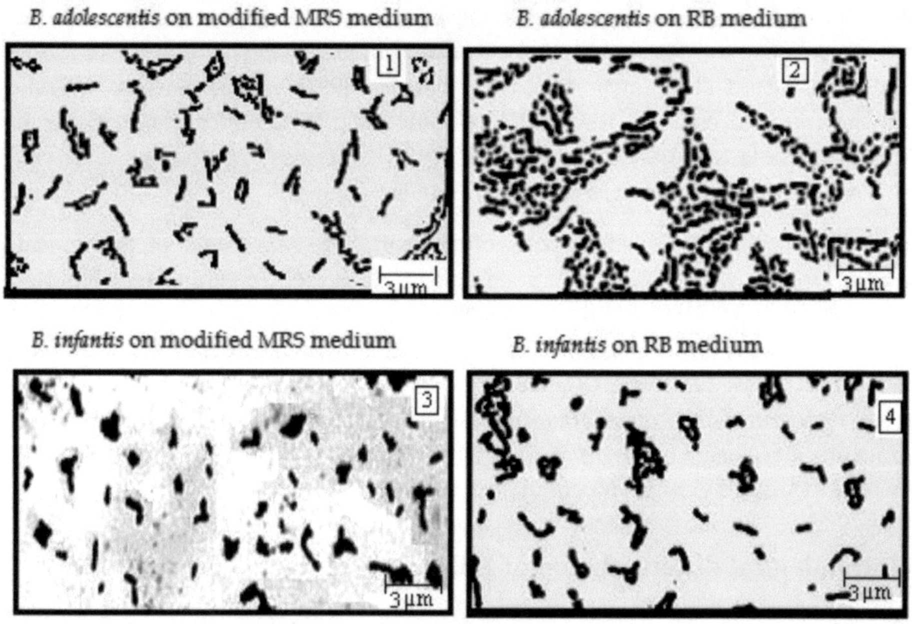


Phase Contrast Micrographs of *Bifidobacterium* reference strains: **Fig. 4.1**, *B. adolescentis* on modified MRS; **Fig. 4.2**, *B. adolescentis* on RB; **Fig. 4.3**, *B. infantis* on modified MRS and **Fig. 4.4**, *B. infantis* on RB, taken at 1000 × magnifications.

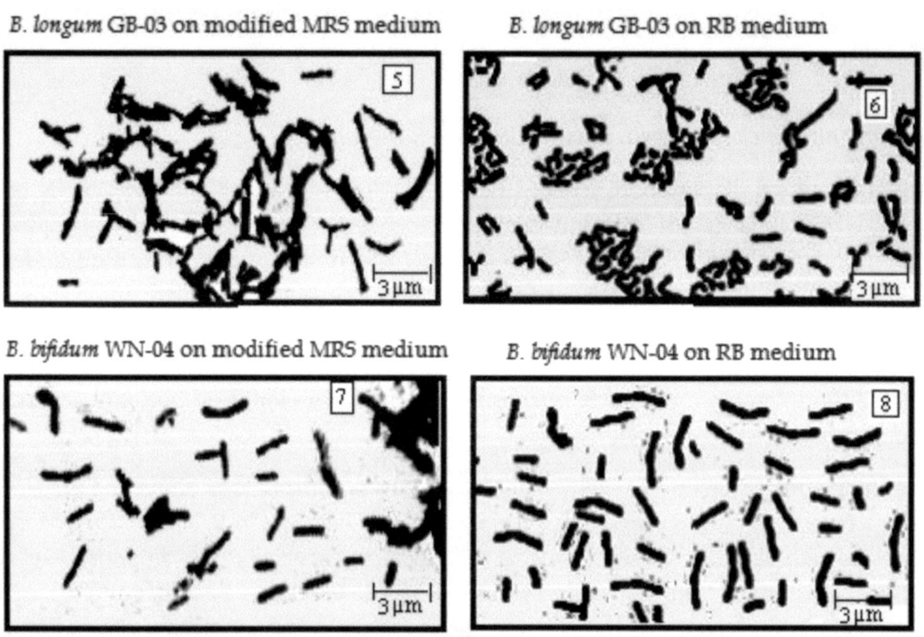


Phase Contrast Micrographs of the isolate strains: **Fig. 4.5**, *B. longum* GB-03 on modified MRS; **Fig. 4.6**, *B. longum* GB-03 on RB; **Fig. 4.7**, *B. bifidum* WN-04 on modified MRS and **Fig. 4.8**, *B. bifidum* WN-04 on RB, taken at 1000 × magnifications.

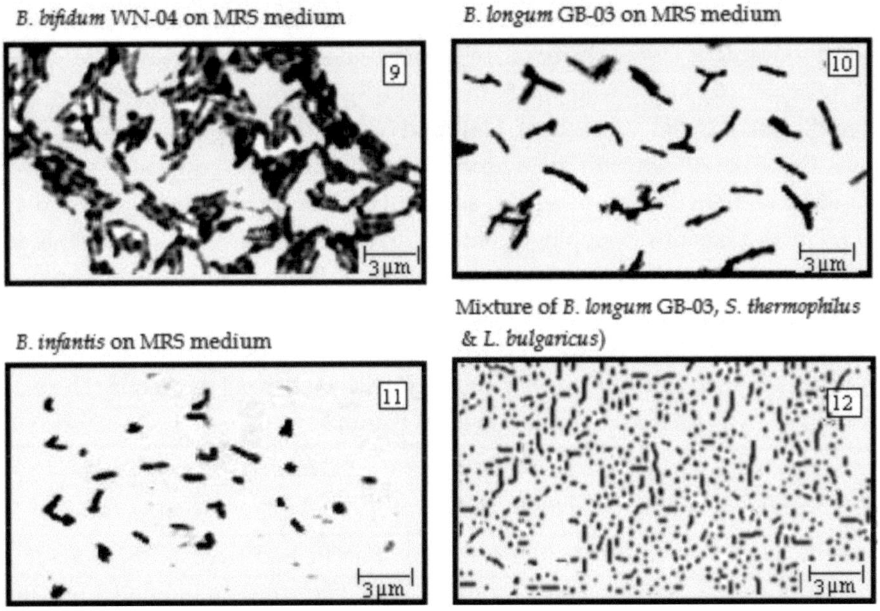


**Figure 4.** Phase Contrast Micrographs of *Bifidobacterium* strains: **Fig. 4.9**, *B. bifidum* WN-04 on unmodified MRS; **Fig. 4.10**, *B. longum* GB-03 on unmodified MRS; **Fig. 4.11**, *B. infantis* on unmodified MRS and **Fig. 4.12**, *B. longum* GB-03 and an assortment of other 2 Lactic Acid Bacteria (*Streptococcus thermophilus* & *Lactobacillus bulgaricus*), taken at 1000 × magnification.

### *3.2.2. Long and thin–rods with blunted ends cell morphology*

This was the most common type of morphology encountered among the *Bifidobacterium* isolates of *B. longum* GB-03 and *B. bifidum* WN-04 on the unmodified MRS agar (Figures 4.5, 4.7, 4.9 & 4.10). Since only the general cell structure was used to differentiate this species from the other bacteria, PCM proved sufficient for this purpose. Variations of morphology within these small groups were visible under PCM as indicated by the following examples. The isolate of *B. longum* GB-03 in Figure 4.12 exemplified the diversity of rods and coccus cells morphology including bifid structures also; with the absence of any coccus build cells when grown on RB agar in Figure 4.6. By comparison with the reference strains, the cells morphology of *B. longum* GB-03 isolate is more peculiar to that displayed by *B. infantis* (Figure 4.4) and the isolate of *B. bifidum* WN-04 (Figures 4.7 & 4.8). All the *Bifidobacterium* isolates displayed long and short club-shaped rods, most of which were long and thin with blunted ends and of conventional "V" and/or "Y"-shaped cells.

## 3.3. Confirmation of identity of *Bifidobacterium* strains

### *3.3.1. Fructose-6-Phosphate Phosphoketolase (F6PPK) verification test*

F6PPK is certainly a key enzyme in the "bifidus pathway" and it allows the discrimination of the specific feature on expression of fructose-6 phosphate in cellular extracts that assigned the bifidobacteria to the genus level (Sgorbati, 1979).

The procedure to test for the F6PPK activity in the *Bifidobacterium* strains is still practised as described by Scardovi (1986). In brief, cells harvested from 10 ml RB or MRS broth are washed twice with 50 mM phosphate buffer (pH 6.5). The cells are disrupted by sonication in the cold, and 0.25 ml of each of NaF and Na iodoacetate solution and fructose-6-phosphate (Na Salt: 70% purity) are added to the sonicate. The reaction is stopped by the addition of 1.5 ml of hydroxylamine HCl, and 1 ml each of trichloroacetic acid and 4 M HCl. Finally, 1.0 ml of a colour-developing agent ($FeCl_3.6H_2O$ 5% (w/v) in 0.1 M HCl) is added. A tube without fructose-6-phosphate serves as a blank, to facilitate the visual comparison. The formation of acetyl phosphate from fructose-6-phosphate, shown by the reddish violet colour formed by the ferric chelate of its hydroxamate is an indicator for F6PPK. This is the distinctive and key enzyme of the "bifid shunt" that characterizes the genus. There are three subtypes of F6PPK in bifidobacteria as shown in Figure 5.

### *3.3.2. Determination of acetic and lactic acids*

One possible method of validating the presence of acetic and lactic acids in the fermented milk by bifidobacteria can be assayed by using High Performance Liquid Chromatography (HPLC). Samples for this analysis are prepared by using a modified method described by Dubey & Mistry, (1996).

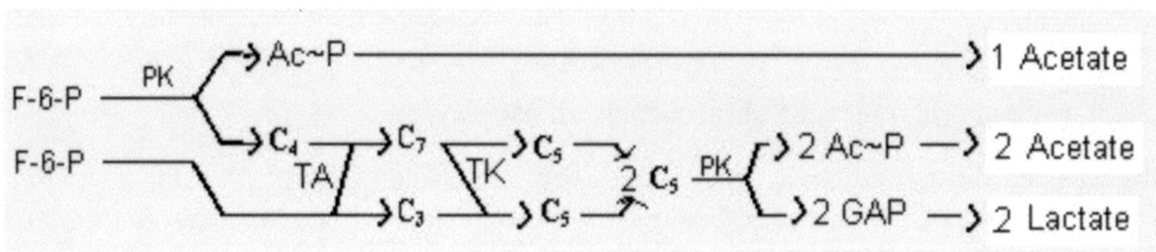


**Figure 5.** Fermentation of hexose for carbohydrate metabolism (the "bifid shunt"), based on Schlegel (1993), where PK, phosphoketolase; TA, transaldolase; TK, transketolase, Ac~P, acetyl phosphate; GAP, glyceraldehydes-3-phosphate.

The strains were maintained anaerobically by propagation in MRS broth (peptone: 10 g/l; meat extract: 8 g/l; yeast extract 5 g/l; D(+)glucose: 20 g/l; di-potassium hydrogen phosphate: 2 g/l; di-ammonium hydrogen citrate: 2 g/l; Tween-80: 1 ml/l; sodium acetate: 5 g/l; magnesium sulfate: 0.2 g/l; manganese sulfate: 0.04 g/l, supplemented with 0.05% (w/v) cysteine-hydrochloride).

The production of acetic and lactic acids, spore formation, aerobic and anaerobic growth, gram reactions, motility, gas production from lactose and carbohydrates fermentation tests are some of the confirmation tests that proves highly diagnostic personality characteristics of different *Bifidobacterium* spp as summarized in Table 1. Furthermore, the taxonomy of bifidobacteria has changed ever since they were first isolated. They had been assigned to the genera *Bacillus, Bacteroides, Nocardia, Lactobacillus* and *Corynebacterium* among others, before being recognized as a separate genus in 1974.

Many of the *Bifidobacterium* species groupings are heterogeneous and the entire genera have been re-examined using DNA-DNA hybridization. A point is made here that, instant phenotypic characterization of most bacteria within their respective genera relies on biochemical tests such as the proportion of acetic and lactic acid relative to the end product of metabolism; the ratio of acetic and lactic acid produced; some key carbohydrate fermentations; colonies and phenotypic morphologies; and the presence of fructose-6-phosphate phosphoketolase (F6PPK), a key enzyme in the bifidus pathway.

| **Characteristics** | ***Bifidobacterium* Strains** | | | |
|---|---|---|---|---|
| | ***B. bifidum*[(1)]** | ***B. longum*[(1)]** | ***B. infantis*[(1)]** | ***B. adolescentis*[(2)]** |
| Spore forming | – | – | – | – |
| Motility | – | – | – | – |
| Gram reaction | + | + | + | + |
| Morphology: rods, pleiomorphic | + | + | + | + |
| Anaerobic growth | + | + | + | + |
| Aerobic growth | – | – | – | – |
| Gas from lactose | – | – | – | – |
| Catalase | – | – | – | – |
| F6PPK | + | + | + | + |
| Acetic and lactic production (ratio 3:2) | + | + | + | + |
| Carbohydrates Fermentation Test | | | | |
| Cellobiose | – | + | – | + |
| Fructose | + | + | + | + |
| Fructooligosacharides | – | + | + | + |
| Galactose | + | + | + | + |
| Glucose | + | + | + | + |
| Isomaltose | – | + | + | + |
| Lactose | + | + | + | + |
| Maltitol[(3)] | – | – | – | – |
| Mannose | – | – | – | – |
| Melezitose | – | – | – | – |
| Raffinose | + | + | + | + |
| Stachyose | + | + | + | + |
| Trehalose | – | – | – | – |
| Xylose | – | + | – | + |

Legends on Table 1: [(1)] Obtained from American Type Culture Collection, Rockville, USA. [(2)] Obtained from China General Microorganisms Culture Collection Center, Beijing, China. [(3)] Maltitol is still widely used as a non-cariogenic sweetener and sugar substitute but is as yet not used as a possible prebiotic. + positive results or fermentation; – negative results or no fermentation observed. F6PPK (fructose-6-phosphate phosphoketolase).

**Table 1.** Phenotypic characteristics of some of the pH- and bile salts-resistant bifidobacteria tested.

## 4. Common media used in isolation and detection of bifidobacteria

Many different media for bifidobacteria are outlined in Table 2.

| Medium | Selectivity based on* | Used for |
|---|---|---|
| Acetylglucosamine-Lactose (AL) agar | lactose, acetylglucosamine | faeces |
| AMC-agar | nal, polymyxin B, kan, iac, TTC, LiCl, prop | *B. longum* |
| Bifidobacterium selective (BS) agar | LiCl, neo, paro, prop | faeces |
| Bifidobacterium selective medium (BBM-agar) | nal, rifampicine, raffinose | faeces |
| Bifidus Blood Agar | aniline blue, blood | faeces |
| Bif-medium | human whey, nal, paro, aztreonam, netilmycin | dairy products |
| Bifidobacterium Iodoacetate Medium (BIM-25 agar) | kan, nal, iac, neo, polymyxin B | sewage |
| BS-agar | LiCl, neo, paro, prop | faeces |
| China Blue (CB) agar | specific impact of china blue | faeces |
| GL-agar | galactose, LiCl | dairy products |
| Liver Cysteine Lactose (LCL) agar | lactose, liver infusion | faeces |
| LP agar | lactose, LiCl, prop | dairy products |
| Modified Rogosa agar | neo, paro, prop, LiCl | dairy products |
| MPN-agar | lactose, nal | faeces |
| MRS-LP-agar | prop, LiCl | dairy products |
| Neomycin Paromomycin Lithium Nalidixic acid (NPLN) agar | LiCl, nal, neo, paro, prop | faeces, dairy products |
| Propionate or Beerens agar | propionic acid, pH 5.0 | faeces |
| Raffinose-Bifidobacterium (RB) Agar | raffinose, LiCl, propionate | faeces, dairy products |
| RCM (modified) | low pH | dairy products |
| RCM + stain | Loeffler's methylene blue stain | dairy products |
| Rogosa agar | low pH | faeces, dairy products |
| Rogosa (modified) | neo, paro, prop, LiCl | dairy products |
| Rogosa-N | low pH, nal | faeces |
| Tomato Casein Peptone Yeast Agar (TCPY) | tomato juice | faeces |
| Transgalactosyloligosaccharide (TOS-Agar) | TOS | faeces, dairy products |
| TOS-Agar (modified) | TOS, nal, neo, paro | dairy products |
| TPYd-agar | dicloxacillin | dairy products |
| TTC-agar | TTC | faecal contamination |
| VF-agar (modified) | LiCl, prop, neo, sodium lauryl sulfate | dairy products |
| YN-6 agar | lactose, nal, neo, bromocresol green | faeces, sewage |

Legends on Table 2: *iac = iodoacetic acid, kan = kanamycin, LiCl = lithiumchloride, nal = nalidixic acid, neo = neomycin, paro = paromomycin, prop = propionate, TOS = transgalactosyl oligosaccharides, TTC = 2,3,5-triphenyl-tetrazoliumchloride

**Table 2.** Popular media used for the enumeration of bifidobacteria from faeces, dairy- and pharmaceutical products, (Adapted from prebiotic effect on non-digestible oligo- and polysaccharides by Hartemink, 1999).

Media used for the detection of bifidobacteria can be classified in 5 different groups. These are non-selective medium (such as MRS and Rogosa), medium without antibiotics but with elective carbohydrate, medium with antibiotics, medium with propionate, and medium with elective substance and/or low pH (Table 3).

| Medium | Group* |
|---|---|
| Acetylglucosamine - Lactose (AL) agar | 2 |
| Bifidobacterium selective (BS) agar | 3, 5 |
| Bifidobacterium selective medium (BBM) agar | 2, 3 |
| Bifidus Blood agar | 5 |
| Bifidobacterium Iodoacetate Medium (BIM-25) agar | 3 |
| China Blue agar | 5 |
| Liver Cysteine Lactose (LCL) agar | 2 |
| Rogosa agar | 1 |
| Modified Rogosa agar | 3, 5 |
| MPN-agar | 2, 3 |
| MRS | 1 |
| MRS agar with LiCl and antibiotics (MRS-NN) | 3, 5 |
| Neomycin Paromomycin Lithium Nalidixic acid (NPLN) agar | 3, 4 |
| Propionate agar or Beerens agar | 4 |
| Raffinose-Bifidobacterium (RB) agar | 2, 4 |
| Reinforced Clostridial agar with Cephalothin and blood (RCB) | 3 |
| Tomato Casein Peptone Yeast agar (TCPY) | 5 |
| Tomato Casein Peptone Yeast agar (TPCY) with azide | 5 |
| Tomato Casein Peptone Yeast agar (TPCY with sorbic acid | 5 |
| Tomato Casein Peptone Yeast agar (TPCY with antibiotics | 3 |
| Transgalactosyloligosaccharide (TOS - agar) | 2 |
| TTC-agar | 5 |
| x-Gal medium | 5 |
| YN-6 agar | 2, 3, 5 |

Legends on Table 3: * group: 1 = non selective medium, 2 = medium without antibiotics but with elective carbohydrate, 3 = medium with antibiotics, 4 = medium with propionate, 5 = medium with elective substance and/or low pH

**Table 3.** Media used for the detection of bifidobacteria from faeces (Source: Hartemink, 1999).

Combinations and media belonging to more than one group are also used. From the large number of media used, it can be concluded that there is no standard medium for the detection of bifidobacteria. *Bifidobacterium* spp. in the GI tract of humans are normally present in an adequate amounts and estimated to be between $10^9$ and $10^{10}$ colony forming units (CFU) per gram wet weight or around 3% of total microbiota (Jia *et al.*, 2010). However, the selectivity of independent media for the quantification of bifidobacteria is thoroughly examined and tested with different baby faeces.

The experimental results of 3 media (PROP, RB and NPLN) tested on bifidobacteria show a wide variation in counts for the different samples (see Figure 6). Absolute counts are highest for the faecal samples on NPLN, followed by RB in 8 of 9 samples. PROP showed the lowest counts. However, as it can be observed from the same Figure 6, the principal difference between these 3 media is exceedingly little, actually less than one log unit.

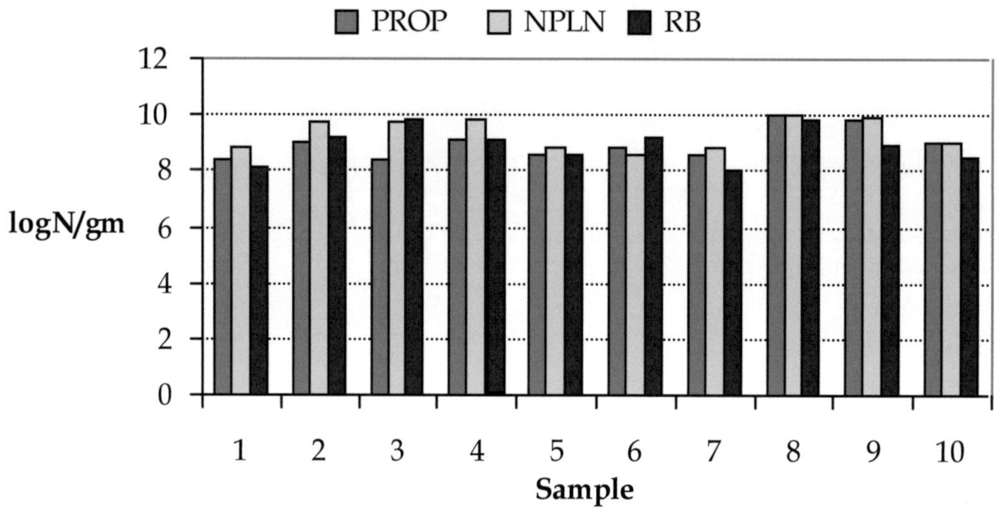


**Figure 6.** Counts (log CFU/gm wet weight) on PROP, NPLN and RB media in babies' faeces.

Selectivity is also determined by microscopic observations of all different colony morphologies on all countable (between 10 and 150 colonies/plate) plates (see Table 4). Based on morphologies, selectivity is highest for babies' faeces with NPLN with 29% false positive colonies (growth, but no bifidobacterial morphology). PROP showed 39% false positive and RB with 50% false positives. False negatives (non-typical colonies, but bifid morphology) can be determined on RB, as this is the only medium for which typical colonies are described. However, no false negatives were observed in this work.

| **Medium** | **Babies' faeces** | | |
|---|---|---|---|
| | **morphology** | | |
| | **n** | **typical** | **non-typ.** [b] |
| RB pos [c] | 24 | 12 | 12 (50) |
| RB neg | 4 | 0 | 4 (0) |
| PROP | 18 | 11 | 7 (39) |
| NPLN | 28 | 20 | 8 (29) |

Legends on Table 4: [b] number in brackets is the percentage of false positive (typical colony, non-typical morphology) or false negatives (non-typical colony, typical morphology) of the colonies tested. [c] pos = colonies showing characteristics for bifidobacteria, neg = colonies not showing characteristics for bifidobacteria. Bifidobacteria characteristics were defined as yellow-green colonies with a yellow halo. This attribute could only be determined on RB, as no characteristics were defined for other media.

**Table 4.** Selectivity of media for bifidobacteria.

Most false positive colonies are reported to be different cocci (mono-, diplo- or streptococci), spore-forming rods and short rods. No yeast is observed on any of the media tested. Based on the actual counts, selectivity can only be determined for RB, as the colonies of bifidobacteria and non-bifidobacteria cannot be determined for the other media and not all colonies are tested for their morphology. Selectivity as percentages of non-typical colonies ranges from around 5 – 7%.

Colonies of different shapes can be tested microscopically. Bacterial morphology is determined, and typical and non-typical morphology is also determined. Typical morphology of bifidobacteria is branched or bifid-shaped rods. For the determination of bifidobacteria, none of the 3 media tested was decidedly selective. In this study, the occurrence of false positive or false negative colonies was determined. The lowest incidence of potential false positive colonies was observed on NPLN, but in all 3 media, the number of non-bifidobacteria capable of growing on the selective media was remarkably high. When many different species are capable of growing on the medium, an increase of one of these species may result in serious mistakes in calculating bifidobacteria. NPLN and RB gave slight higher counts than PROP. The incidence of false positive, based on morphologies on RB was comparable with that on the PROP and slightly higher than that on NPLN. The incidence of competitive flora was relatively low (less than 10% of the total colonies on the plates), as bifidobacteria are one of the main groups of intestinal bacteria in humans.

PROP medium has been described as the best medium for the determination of bifidobacteria by Silvi *et al.*, (1996), but they also concluded that the total bifidobacterial counts were significantly lower on PROP than on the other media tested. Similarly, Favier *et al.*, (1997) concluded that PROP underestimated bifidobacteria in some of their samples. Both studies used human faeces as the test substrate. Several other studies, in which PROP agar is used, also show significantly lower bifidobacterial counts than most other studies (Favier *et al.*, 1997).

NPLN, which has been described as the medium of choice to choose bifidobacteria in dairy products, showed many cocci. This was in accordance with results observed by Silvi *et al.*, (1996). In the same study, BIM-25 was tested, and this medium was found to be non-specific. All these 3 media performed reasonably well for human faeces and bifidobacteria can reliably be counted. The typical colonies morphological trait and the basic cellular-morphology of bifidobacteria were demonstrated well by RB media, with reference to NPNL and PROP medium. The RB medium presented strains with double thickness diameter and more bifurcated cellular morphology under phase contrast microscopy.

## 5. Experimental procedures for the enumeration of bifidobacteria and determining microbial inactivation by low acidic pH or bile salts

LAB or bifidobacteria strains can be selected or isolated from commercial or alleged "own isolates" strains, from freeze-dried cultures which are resuscitated to stationary phase in MRS broth at a ratio of 2% of the volume of the fresh broth. Decimal dilutions are put onto Raffinose–Bifidobacterium (RB) agar plates whose pH had to be adjusted to 6.8 – 7.0 with 2

N NaOH. The agar plates are then incubated anaerobically at 38.5 ºC for 3 – 4 d and number of colony forming units (CFU)/mL are determined. Two hundred microliters of each strain containing about $10^8$ CFU/mL is aseptically transferred into test tubes containing 9 mL of diluted MRS medium with pH adjustments of 3.0, 3.5, 4.0, or 4.5, using 2 N HCl. These suspensions are incubated anaerobically at 38.5 ºC and numbers of survivors are determined after various times as shown in Figure 7 (A). Cells were harvested by centrifugation at 5 ºC, were washed with phosphate-buffered saline (PBS) and were re-suspended in diluted MRS medium without pH adjustment. After thorough mixing on a vortex mixer, the concentration of surviving cells is determined by anaerobic pour plate counts, using 2 plates of RB agar per dilution, and incubated at 38.5 ºC for 3 – 4 days.

Similarly, treatments for the bile salts are carried out at the final concentrations of 0.15%, 0.30%, 0.45%, and 0.60% ox-gall in diluted MRS medium (pH 6.8), exposed to appropriate times as to low pH and incubated anaerobically at 38.5 ºC (see Figure 7 (B). The cells are harvested by centrifugation, washed with PBS, re-suspended in diluted MRS medium without pH adjustment, and mixed using a vortex mixer as described for acidic pH conditions before. Numbers of CFU of bifidobacteria surviving the lytic effect of bile salts are also determined by anaerobic pour plate counts on RB agar after anaerobic incubation for 3 – 4 d at 38.5 ºC.

### 5.1. Characterizations for $D_{(acid)}$-, $D_{(bile)}$-, $z_{(acid)}$, and $z_{(bile)}$-values

$D_{(acid)}$-value is defined as the time (in min) required at a specified acidic pH to reduce the number of cells by 90%, while $D_{(bile)}$-value is defined as the time (in min) required at a specified concentration of bile salts to reduce the number of cells by 90%. In fact, the $D_{(bile)}$-value of any LAB or bifidobacterial strain is directly proportional to the bile salt concentrations, while the $D_{(acid)}$-value is inversely proportional to the acidic pH.

The $z_{(acid)}$ and/or $z_{(bile)}$-values, on the other hand, is defined as a decrease in pH value (pH < 4.5) or an increase in bile salt concentration (% ox-bile) required to reduce the *D*-values by 1 log cycle, however, respectively. The $D_{(acid)}$- and $D_{(bile)}$-values can be directly calculated from the absolute values of the reciprocal of the slopes of the linear-regression equations, using a Microsoft Office–Excel software. It is essential to emphasize that, the regression lines must be applied to all the treatments, for which restriction of the *R*-squared ($R^2$) value is pragmatic above 0.8920. Moreover, the $D_{(acid)}$- and/or $D_{(bile)}$-values can also be calculated algebraically from the regression equation derived using the method of least-squares to be able to produce the $z_{(acid)}$- and/or $z_{(bile)}$-values for the probiotic strains.

In order to determine the $z_{(acid)}$- and/or $z_{(bile)}$-values, the formula is exactly the same as that for heat resistance, replacing T (temperature) with pH values or bile salts (BS) concentrations as described by Equations (1) and (2), respectively. In both of these cases, the effect of acidic conditions and bile salts is determined from the reduction in concentration of colony-forming units. One has to pay attention that the dynamic $z_{(acid)}$- and/or $z_{(bile)}$-values are calculated for a period of exponential destruction of microbial cells (following the logarithmic order of death), using both Equations (1) and (2).

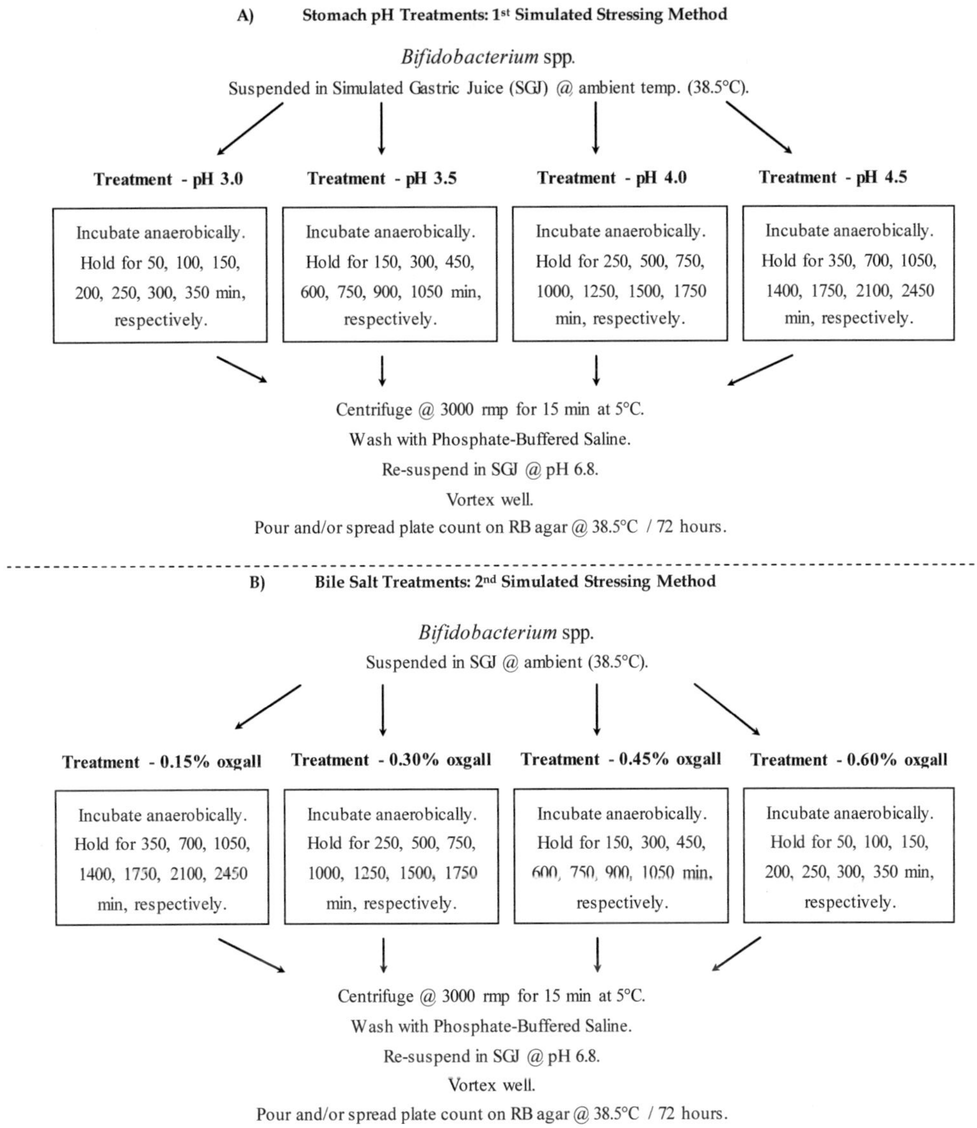


**Figure 7.** Schematic diagram of treatment groups for the selected bifidobacteria. A): Influence of gastric acidity and its residence time. B): Influence of bile salt(s) and its residence time.

$$z_{(\text{acid})} = \frac{-(\text{acid pH}_2 - \text{acid pH}_1)}{\text{Log}_{10}\left(\dfrac{D_{(\text{acid 1})}}{D_{(\text{acid 2})}}\right)} \quad (1)$$

$$z_{(\text{bile})} = \frac{\text{BS}_2 - \text{BS}_1}{\text{Log}_{10}\left(\dfrac{D_{(\text{bile 1})}}{D_{(\text{bile 2})}}\right)} \quad (2)$$

where *z* is the acid pH value (in Equation 1) or bile salts value (in Equation 2) required for a ten-fold reduction in *D*-values; $pH_1$ is the acidic value of pH 1; $pH_2$ is the acidic value of pH 2; $BS_1$ is the concentration of bile salts 1 (%); $BS_2$ is the concentration of bile salts 2 (%); $D_{(acid1)}$ or $D_{(bile1)}$ is a *D*-value obtained at either $pH_1$ or $BS_1$; and $D_{(acid2)}$ or $D_{(bile2)}$ is a *D*-value obtained at either $pH_2$ or $BS_2$.

### 5.2. Survival of bifidobacteria in simulated acidic pH of human stomach

Figure 8 shows the survival of selected *Bifidobacterium* strains exposed to various acidic pH levels. The bifidobacterial counts (range: 5.5 – 6.7 log CFU/ml) of all four strains at pH 3.5 after 5 h are an indication of resistance that may perhaps simulate gastric conditions. In fact, with *B. bifidum*, *B. infantis*, and *B. longum* the counts were > 2.0 log CFU/ml after exposure for 5 h, which indicates that these strains are relatively resistant at pH 3.0. However, as observed from all the experimental results in Figure 8, the *B. adolescentis* strain is more sensitive than the other three strains to all the acid treatments. For instance, numbers of *B. adolescentis* are 3.4 log CFU/ml after 10 h at pH 3.5, but below the level of exposure thereafter, while a similar count (~3.4 log CFU/ml) is observed with the other three strains after 12.5 h. This 2.5 h difference in survival at pH 3.5 is approximately the time chosen by Olejnik *et al.*, (2005) to control acid resistance, as these times simulate residence time in the stomach. On this basis, this specific strain of *B. adolescentis* is considered a less-resistant strain with respect to gastric acidity.

Many other researchers have found pH 2.0 and pH 3.0 to be lethal and sublethal pH values respectively for lactic acid bacteria (LAB), including bifidobacteria (Khalil *et al.*, 2007). It is vital to stress that probiotics are able to confer health benefits despite the brief exposure to exact acidic conditions following ingestion. Although a log-scale reduction of viability may occur, it may still mean that a sufficient number of bifidobacteria survive the gut, depending on the dose. Moreover, the exposure to acid does not mean that the potential health benefits are lost. Some cells may die, and some may be injured. However, these cells may recover later, and they may also have beneficial effects on health. The mechanism may be mediated, for example, through the components of the cell wall of the probiotics which will then be available in both dead and living cells. It should be also borne in mind that, the food matrix in which the probiotics are consumed is likely to have a strong effect on the survival of the bacteria in the gut.

In Figure 8, for example, it is possible to say that the high survival counts of *B. bifidum*, *B. infantis*, and *B. longum* exposed to pH 4.5 after 41 h is a representation of culture stability curves of the acid resistant strains (log CFU/ml) in a food matrix. While the pH of 4.5 does not represent gastric acid conditions, it is a typical representation of fermented products, and in such products, the survival counts are expected to last for much longer periods during shelf life, especially at the refrigeration temperature of 4 °C.

### 5.3. Survival of bifidobacteria in simulated bile salts nature

Resistance to bile salts is considered an intrinsic property for probiotic strains to survive the conditions in the small intestine. The physiological bile salt concentration in the GI tract of humans is estimated to be 0.3 – 0.4% w/v (Jia *et al.*, 2010). As shown in Figure 9, the linear regressions of the loss of CFU did express satisfactorily that *B. bifidum, B. infantis,* and *B. longum* are more-resistant strains to the bile salts. These 3 strains survived well in 0.45% (w/v) bile salts, with more than 4.5 log CFU/ml present after 10 h. Their capacity to survive high bile salt concentrations suggests the existence of defence mechanisms and confirms that strains showing antagonistic effect against enteric pathogens should be able to compete successfully with the pathogens in the GI tract. It is clear that *B. adolescentis* is again the less-resistant strain encountered with only 2.8 log CFU/ml surviving after 10 h in 0.45% bile salts (see Figure 9). Therefore, *B. adolescentis* is considered the less-resistant strain, while *B. bifidum, B. infantis,* and *B. longum,* in that order, are considered the more-resistant. It is well known that, the bile salt hydrolytic (BSH) activity may be the contributing factor towards the resistance of the LABs and to the toxicity of conjugated bile salts in the duodenum, and therefore, is an essential colonization factor.

### 5.4. The feasibility of $D_{(acid)}$-, $D_{(bile)}$-, $z_{(acid)}$- and $z_{(bile)}$-values for selection of probiotic strains and for determining the mechanisms of resistance to acid and bile salts stress

Table 5, shows that accurate tabulation of the $D_{(acid)}$- or $D_{(bile)}$-values and their respective $z_{(acid)}$- or $z_{(bile)}$-values is tremendously helpful in evaluating the resistance and susceptibility of probiotics to acidic pH and high bile salt concentrations, respectively. Both the estimated $D_{(acid)}$- and $D_{(bile)}$-values validated that the most acid- and bile-resistant strain is *B. bifidum* followed by *B. infantis, B. longum,* and final *B. adolescentis.* It is also possible to observe in Table 5, that, increasing the bile salt concentration from 0.15 to 0.60% had a greater impact on survival than decreasing the pH values from 4.5 to 3.0, with the $D_{(bile)}$-values of *B. bifidum* decreasing from 17.40 to 1.40 min and the $D_{(acid)}$-values decreasing from 23.80 to 1.10 min. Similar trends are observed with all other *Bifidobacterium* strains. However, decreases of depicted $z_{(acid)}$-value in the pH value (pH<4.5) or increases of depicted $z_{(bile)}$-value in the bile salt concentrations (% ox-bile) are expected to cause a 1-log reduction in their respective *D*-values. In practice, $z_{(acid)}$- or $z_{(bile)}$-value measures how the sensitivity of probiotic strains is to small changes in [$H^+$] and/or [$OH^-$] or bile salts. As for probiotics to gain intestinal colonization in humans or animals for their proclaimed therapeutic health benefits, obviously, they have to tolerate inhibitory substances secreted by the host, such as gastric acids (in the stomach) and bile salts (in the small intestine).

Of all ions, $H^+$ and $OH^-$ are the most mobile, and minor changes in their concentrations show significant effects on microorganisms. Most organisms survive better when these ions are present in approximately equal concentrations, that is, pH 7.0. Although many bacteria tolerate higher pH values, only a few are acid tolerant or acidophilic. In addition, many other bacteria are tolerant of small pH variations, especially in the pH range of 6.0 to 9.0. For instance, if the pH of the medium changes rapidly, there may be a transient change in the intracellular pH, and this is usually readjusted to the original pH within 30 min. Consequently, any damage produced by adverse pH is not actually due to the $H^+$ and/or

$OH^-$, but to the effect of these ions on the proportion of undissociated weak acids or bases, which penetrate more readily into the bacterial cell than the ionized forms. In contrast, bile salts are biological detergents synthesized in the liver from cholesterol, conjugated to either glycine or taurine, and are then secreted into the intestine where they facilitate fat absorption. Bile salts are well known to be toxic for many cells as they disrupt the lipid bilayer structure of the cellular membranes. Many earlier studies revealed that the autochthonous gastrointestinal microbiota must develop strategies to protect themselves against bile salts.

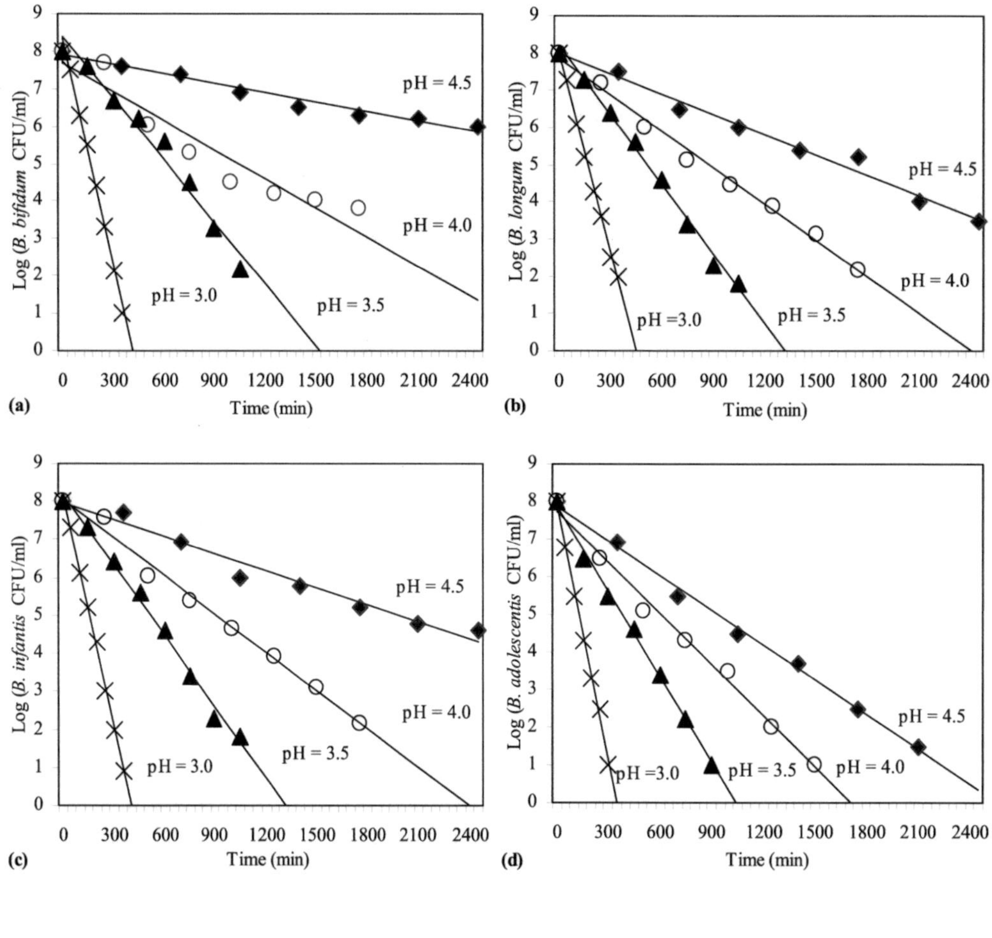


**Figure 8.** Linear regressions of the loss of CFU for the selected bifidobacteria strains when exposed to simulated gastric acidity of pH 3.0, pH 3.5, pH 4.0 and pH 4.5, respectively: (a) *B. bifidum*, (b) *B. longum*, (c) *B. infantis*, (d) *B. adolescentis*.

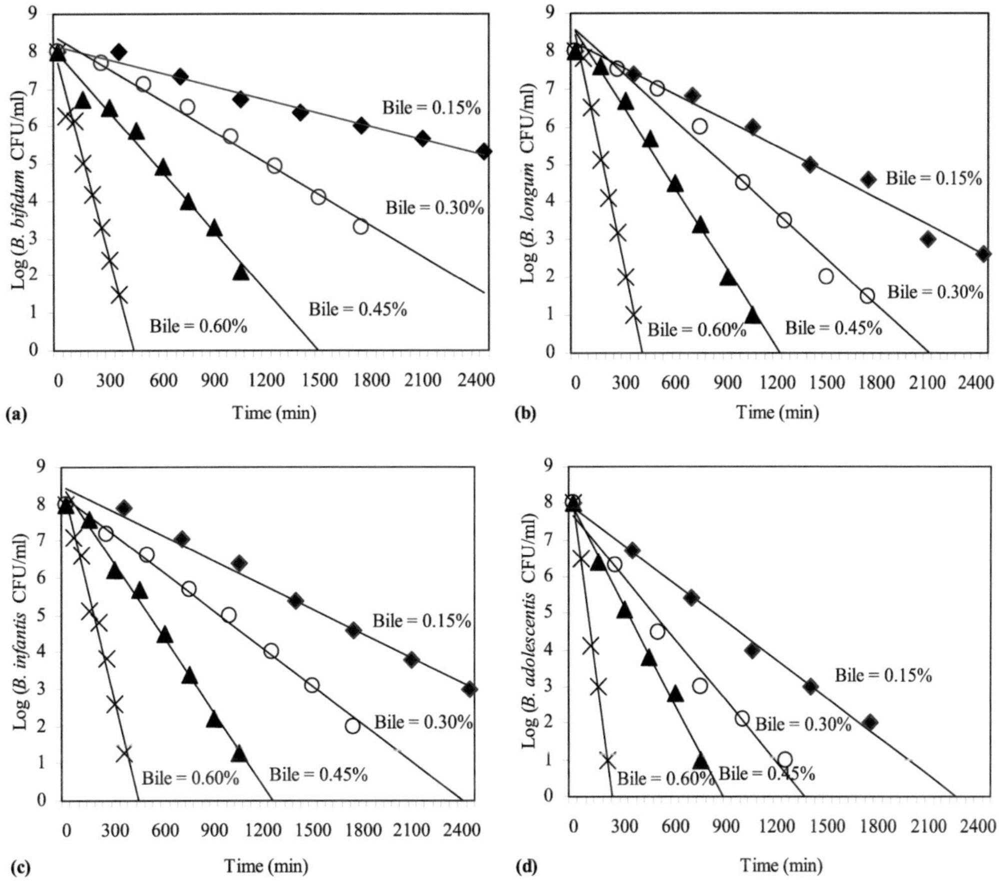


**Figure 9.** Linear regressions of the loss of CFU for the selected bifidobacteria strains when exposed to high bile salt (oxgall) concentrations of 0.60%, 0.45%, 0.30% and 0.15%, respectively: (a) *B. bifidum*, (b) *B. longum*, (c) *B. infantis*, (d) *B. adolescentis*.

The individual $z_{(acid)}$- or $z_{(bile)}$-values calculated from their $D_{(acid)}$- and $D_{(bile)}$-values ranged from 1.11 – 1.55 pH units and 0.40 – 0.49%, respectively (Table 5). Although the combination of both the low acidic pH and bile salts is not assessed, it is assumed that at pH < 3.0, and 0.60% of ox-bile, the combined effects could be more synergistic and even greater in magnitude for probiotic bacteria to survive. Additionally, the $D_{(acid)}$- and $D_{(bile)}$-values reveal a modern and efficient sorting order of the more-resistant probiotic strains to these two distinct hostile GI tract conditions in humans. Many authors have investigated the effect of bile on survival of LAB. For example, Kim *et al.*, (1999) examined the effect of bile concentration in the range of 0 – 0.4% on survival of *Lb. lactis* and found bile to be toxic at concentrations over 0.04%. Shimakawa *et al.*, (2003) reported that 0.2% oxgall in the growth medium inhibited growth of *B. breve* strain Yakult. Others detected that all bacterial cells were killed by 0.2% bile and higher (Olejnik *et al.*, 2005). However, Khalil *et al.*, (2007) reported higher resistance to bile salts, with viability of strains apparently increasing when exposed to high levels of oxgall (0.4%).

| pH | $D_{(acid)}$-value (min) | | | |
|---|---|---|---|---|
| | *B. bifidum* | *B. infantis* | *B. longum* | *B. adolescentis* |
| 4.5 | 23.80 | 14.10 | 12.00 | 7.60 |
| 4.0 | 8.40 | 6.00 | 5.70 | 3.98 |
| 3.5 | 3.00 | 2.70 | 2.60 | 2.05 |
| 3.0 | 1.10 | 1.10 | 1.20 | 1.10 |
| | $z_{(acid)}$-value (in pH units) | | | |
| | 1.11 | 1.55 | 1.35 | 1.55 |
| **Bile Salts** | $D_{(bile)}$-value (min) | | | |
| 0.15% | 17.40 | 10.50 | 9.60 | 6.80 |
| 0.30% | 7.40 | 5.20 | 4.70 | 3.20 |
| 0.45% | 3.20 | 2.55 | 2.30 | 1.58 |
| 0.60% | 1.40 | 1.30 | 1.10 | 0.75 |
| | $z_{(bile)}$-value (% ox-bile concentration) | | | |
| | 0.40 | 0.48 | 0.49 | 0.46 |

**Table 5.** Selected *Bifidobacterium* strains and their calculated $D_{(acid)}$-, $D_{(bile)}$-, $z_{(acid)}$- and $z_{(bile)}$-values.

As compared to previous studies, the practicality of $D_{(acid)}$-, $D_{(bile)}$-, $z_{(acid)}$- and $z_{(bile)}$-values as new kinetic-measurements applied in this study, are indeed, quick to identify comparably higher survival of bifidobacteria cells (> 4.1 log CFU/ml after 2.5 h) at elevated bile salt concentrations of 0.6% (w/v), thereby confirm also that the individual *Bifidobacterium* strains are resistant to harsh intestinal conditions in the following order: *B. bifidum* > *B. infantis* > *B. longum* > *B. adolescentis*. A number of researchers reported that *B. infantis* had the highest survival rates followed by *B. bifidum, B. breve and B. longum,* when exposed to bile salt at concentrations ranging from 0 to 3 g/l. In contrast, the literature contains also one preliminary report that *B. longum* exhibited the highest tolerance to bile salts followed by *B. bifidum* and *B. infantis,* which was almost the exact opposite in order of their tolerance to acidic pH. These contrasting observations may reflect the strain-specific resistance to acid or bile salts stress. It also indicates that tolerance is strain- rather than species-specific. Likewise, the source of isolation of the probiotic strains is particularly influential too.

## 6. Conclusion

Apart from the isolation, enumeration, unequivocal taxonomical characterization, screening and selection of tolerant strains of bifidobacteria to gastric acid and bile salts studies, the assessment of the tolerant bifidobacteria to bile salts and low pH has been made possible by use of *D*- and *z*-value concept. After log-conversion, inactivation followed first-order kinetic law whereby validating the kinetic assumptions of the latter concept. The projected $z_{(acid)}$- and $z_{(bile)}$-values were all fairly similar for the bifidobacteria strains and suggested the effect of increasing the bile salt concentration or decreasing the pH on the $D_{(acid)}$- and $D_{(bile)}$-values. This approach is useful for measuring the resistance and sensitivity of lactic acid bacteria or bifidobacteria to these two hostile gastrointestinal conditions. The approach pursued in this chapter would be extremely useful for predicting the suitability of bifidobacteria and/or other LAB as probiotics for use in real life situations. While the mechanisms of probiotic survival in the GI tract could be more complex, the practical utility of the $D_{(acid)}$- and/or $D_{(bile)}$- and their $z_{(acid)}$- and $z_{(bile)}$-values is significant.

## Author details

Nditange Shigwedha* and Li Jia
*University of Namibia & Meat Corporation of Namibia, Namibia*

## Acknowledgement

We are grateful to Ms. Masa Vidovic and the entire staff of InTech for helpful advices and the opportunity given to us. We would also like to thank Mr. Richard Shigwedha for the good cooperation.

* Corresponding Author

## 7. References

Biavati, B.; Scardovi, V. & Moore, W.E.C. (1982). Electrophoresis Patterns of Protein in the Genus *Bifidobacterium* and Proposal of Four New Species. *International Journal of Systematic Bacteriology*, Vol.32, No.3, (July 1982), pp. 358-373

Chung, H.S.; Kim, Y.B.; Chun, S.L. & Ji, G.E. (1999). Screening and Selection of Acid and Bile Resistant Bifidobacteria. *International Journal of Food Microbiology*, Vol.49, No.1-2, (March 1999), pp. 25-32

Dubey, U.K. & Mistry, V.V. (1996). Effect of Bifidogenic Factors on Growth Characteristics of Bifidobacteria in Infant Formulas. *Journal of Dairy Science*. Vol.79, No.7, (July 1996), pp. 1156-1163

Favier, C.; Neut, C.; Mizon, C.; Cortot, A.; Colombel, J.F. & Mizon, J. (1997). Fecal $\beta$-D-Galactosidase Production and Bifidobacteria are Decreased in Crohn's Disease. *Digestive Diseases and Sciences*, Vol.42, No.4, (April 1997), pp. 817-822

Glickman, R.M. (1980). Intestinal Fat Absorption, In: *Nutrition and Gastroenterology*, M. Winick, (Ed.), 29-41, Wiley, New York, USA

Hartemink, R. (1999). *Prebiotic Effects of Non-Digestible Oligo- and Polysaccharides*, Ponsen & Looijen, ISBN 90-5808-051-X, Wageningen, The Netherlands

Jia, L.; Shigwedha, N. & Mwandemele, O.D. (2010). Use of $D_{acid}$-, $D_{bile}$-, $z_{acid}$- and $z_{bile}$-Values in Evaluating Bifidobacteria with Regard to Stomach pH and Bile Salt Sensitivity. *Journal of Food Science*, Vol.75, No.1, (January/February 2010), pp. M14-M18

Joint FAO/WHO. (2002). Guidelines for the Evaluation of Probiotics in Food. Ontario, Canada: 30 April - 1 May 2002, Available from http://www.who.int/foodsafety/publications

Kazor, C.E.; Mitchell, P.M.; Lee, A.M.; Stokes, L.N.; Loesche, W.J.; Dewhirst, F.E. & Paster, B.J. (2003). Diversity of Bacterial Populations on the Tongue Dorsa of Patients with Halitosis and Healthy Patients. *Journal of Clinical Microbiology*, Vol.41, No.2, (February 2003), pp. 558-563

Khalil, R.; Mahrous, H.; El-Halafawy, K.; Kamaly, K.; Frank, J. & El Soda, M. (2007). Evaluation of the Probiotic Potential of Lactic Acid Bacteria Isolated from Faeces of Breast-Fed Infants in Egypt. *African Journal of Biotechnology*, Vol.6, No.7, (April 2007), pp. 939-949

Kim, W.S.; Ren, J. & Dunn, N.W. (1999). Differentiation of *Lactococcus lactis* Subspecies *lactis* and Subspecies *cremoris* Strains by their Adaptive Response to Stresses. *FEMS Microbiology Letters*, Vol.171, No.1, (February 1999), pp. 57-65

Liepke, C.; Adermann, K.; Raida, M.; Magert, H-J.; Forssmann, W-G. & Zucht, H-D. (2002). Human Milk Provides Peptides Highly Stimulating the Growth of Bifidobacteria. *European Journal of Biochemistry*, Vol.269, No.2, (January 2002), pp. 712-718

Lourens-Hattingh, A. & Viljoen, B.C. (2001). Growth and Survival of a Probiotic Yeast in Dairy Products. *Food Research International,* Vol.34, No.9, (November 2001), pp. 791-796

Macfarlane, S. & Dillon, J.F. (2007). Microbial Biofilms in the Human Gastrointestinal Tract. *Journal of Applied Microbiology,* Vol.102, No.5, (May 2007), pp. 1187-1196

Mitsuoka, T. (1990). Bifidobacteria and their Role in Human Health. *Journal of Industrial Microbiology & Biotechnology,* Vol.6, No.4, (April 1990), pp. 263-267

Olejnik, A.; Lewandowska, M.; Obarska, M. & Grajek, W. (2005). Tolerance of *Lactobacillus* and *Bifidobacterium* Strains to Low pH, Bile Salts and Digestive Enzymes. *Electronic Journal of Polish Agricultural Universities,* Vol.8, No.1, Available from http://www.ejpau.media.pl/volume8/issue1/art-05.html

Palacios, M.C.; Haros, M.; Rosell, C.M. & Sanz, Y. (2008). Selection of Phytate-Degrading Human Bifidobacteria and Application in Whole Wheat Dough Fermentation. *Food Microbiology,* Vol.25, No.1, (February 2008), pp. 169-176

Patrignani, F.; Lanciotti, R.; Mathara, J.M.; Guerzoni, M.E. & Holzapfel, W.H. (2006). Potential of Functional Strains, Isolated from Traditional Maasai Milk, as Starters of the Production of Fermented Milks. *International Journal of Food Microbiology,* Vol.107, No.1, (March 2006), pp. 1-11

Pei, Z.; Bini, E.J.; Yang, L.; Zhou, M.; Francois, F. & Blaser, M.J. (2004). Bacterial Biota in the Human Distal Esophagus. *Proceedings of the National Academy of Sciences (PNAS),* Vol.101, No.12, (March 2004), pp. 4250-4255.

Reuter G. 1963. Vergleichede Untersuchungen über die Bifidus-Flora im Säuglings- und Erwachsenenstuhl. *Zentralbl Bakteriol Parasitenkd Infektionskr Hyg Abt I Orig,* Vol.191, (1963), pp. 486-507

Scardovi, V. (1986). Genus *Bifidobacterium,* In: *Bergey's Manual of Systematic Bacteriology,* P.H.A. Sneath, N.S. Mair, M.E. Sharpe, & J.G. Holt (Eds.), 1418-1434, Williams & Wilkins, Baltimore, USA

Schlegel, H.G. (1993). *General Microbiology,* Cambridge University Press, ISBN 0-521-43372-X, Melbourne, Australia

Sgorbati, B. (1979). Preliminary Quantification of Immunological Relationships among the Transaldolases of the Genus *Bifidobacterium. Antonie van Leeuwenhoek,* Vol.45, No.4, (April 1979), pp. 557-564

Shimakawa, Y.; Matsubara, S.; Yuki, N.; Ikeda, M. & Ishikawa, F. (2003). Evaluation of *Bifidobacterium breve* Strain Yakult-Fermented Soymilk as a Probiotic Food. *International Journal of Food Microbiology,* Vol.81, No.2, (March 2003), pp. 131-136

Silvi, S.; Rumney, C.J. & Rowland, I.R. (1996). An Assessment of the Three Selective Media for Bifidobacteria in Faeces. *Journal of Applied Microbiology,* Vol.81, No.5, (November 1996), pp. 561-564

Sundman, V.; Björksten, K.A.F. & Gyllenberg, H.G. (1959). Morphology of the Bifid Bacteria (Organisms Previously Incorrectly Designated *Lactobacillus bifidus*) and Some Related Genera. *Journal of General Microbiology,* Vol.21, No.2, (October 1959), pp. 371-384

Yildirim, Z. & Johnson, M.G. (1998). Characterization and Antimicrobial Spectrum of Bifidocin B, a Bacteriocin Produced by *Bifidobacterium bifidum* NCFB 1454. *Journal of Food Protection*, Vol.61, No.1, (January 1998), pp. 47-51

# Chapter 13

# Lactic Fermentation and Bioactive Peptides

Anne Pihlanto

Additional information is available at the end of the chapter

http://dx.doi.org/10.5772/51692

## 1. Introduction

Fermented milk products have naturally high nutritional value, and as an extra benefit many health-promoting effects, such as improvement of lactose metabolism, reduction of serum cholesterol and reduction of cancer risk [1]. The beneficial health effects associated with some fermented dairy products may, in part, be attributed to the release of bioactive peptide sequences during the fermentation process. Numerous peptides and peptide fractions, having bioactive properties have been isolated from fermented dairy products. These activities include immunomodulatory, cytomodulatory, hypocholesterolemic, antioxidative, antimicrobial, mineral binding, opioid and bone formation activities. Many recent articles and book chapters have reviewed the release of various bioactive peptides from milk proteins through microbial proteolysis [2-5].

Many industrially utilized dairy starter cultures are highly proteolytic. The use of bioactive peptides producers microbial cultures (starter and non-starter) may allow the development new fermented dairy products. The proteolytic system of lactic acid bacteria e.g. *Lactococcus (L.) lactis, Lactobacillus (Lb.) helveticus* and *Lb. delbrueckii* ssp. *bulgaricus,* is already well characterized. This system consists of a cell wall-bound proteinase and a number of distinct intracellular peptidases, including endopeptidases, aminopeptidases, tripeptidases and dipeptidases [6]. *Lb. helveticus* are known to have high proteolytic activities [7], causing the release of oligopeptides from digestion of milk proteins [8]. These oligopeptides can be a direct source of bioactive peptides following hydrolysis by gastrointestinal enzymes. Rapid progress has been made in recent years to elucidate the biochemical and genetic characterization of these enzymes. The fact that the activities of peptidases are affected by growth conditions makes it possible to manipulate the formation of peptides to a certain extent [9].

Cardiovascular disease (CVD) is the single leading cause of death for both males and females in technologically advanced countries in the world. In lesser-developed countries it generally ranks among the top five causes of death. The World Health Organization estimates that by 2020, heart disease and stroke will have surpassed infectious diseases to become the leading cause of death and disability worldwide [10]. Consequently, there has been an increased focus on improving diet and lifestyle as a strategy for CVD risk reduction.

Elevated blood pressure is one of the major independent risk factors for CVD [11]. Angiotensin I-converting enzyme (ACE) plays a crucial role in the regulation of blood pressure as it promotes the conversion of angiotensin I to the potent vasoconstrictor angiotensin II as well as inactivates the vasodilator bradykinin. By inhibiting these processes, synthetic ACE inhibitors (ACEI) have long been used as antihypertensive agents. In recent years, some food proteins have been identified as sources of ACEI peptides and are currently the best-known class of bioactive peptides [12, 13]. These nutritional peptides have received considerable attention for their effectiveness in both the prevention and the treatment of hypertension.

Oxidant stress, the increased production of reactive oxygen species (ROS) in combination with outstripping endogenous antioxidant defense mechanisms, is another significant causative factor for the initiation or progression of several vascular diseases. ROS can cause extensive damage to biological macromolecules like DNA, proteins and lipids. Specifically, the oxidative modification of LDL results in the increased atherogenicity of oxidized LDL. Therefore, prolonged production of ROS is thought to contribute to the development of severe tissue injury [14]. Some peptides derived from hydrolyzed food proteins exert antioxidant activities against enzymatic (lipoxygenase-mediated) and nonenzymatic peroxidation of lipids and essential fatty acids [15]. The antioxidant properties of these peptides have been suggested to be due to metal ion chelation, free radical scavenging and singlet oxygen quenching.

This review centers on liberation during fermentation, of bioactive peptides with properties relevant to cardiovascular health including the effects on blood pressure and oxidative stress. The focus is mainly to those peptides with in vivo blood pressure lowering effects. Moreover, bioavailability of peptides and aspects of necessary further information is given.

## 2. Release and identification of peptides

### 2.1. Peptides in cheese

Proteolysis in cheese has been linked to its importance for texture, taste and flavour development during ripening. Changes of the cheese texture occur due to breakdown of the protein network. It contributes directly to taste and flavour by the formation of peptides and free amino acids as well as by liberation of substrates for further catabolic changes and thereby formation of volatile flavour compounds. Besides sensory quality aspects of proteolysis, formation of bioactive peptides as a result of proteolysis during cheese ripening has been reported. Cheese contains phosphopeptides as natural constituents [16, 17], and secondary proteolysis during cheese ripening leads to the formation of other bioactive peptides, such as those with ACEI activity. The findings by Meisel et al. [18] showed that inhibitory activity increased as proteolysis developed, however, the bioactivity decreased when proteolysis during ripening exceeded a certain level. Another link between potential antihypertensive peptides and proteolysis was found in Parmesan cheese [19]. A bioactive peptide derived from $\alpha_{s1}$-casein was isolated from 6-month old cheese, but it was degraded further during maturation and was not detectable after 15 month of ripening. ACEI peptide fractions having different potencies have been isolated from various Italian cheeses, e.g. Crescenza (37% inhibition), mozzarella (59% inhibition), Gorgonzola (80% inhibition) and Italico (82% inhibition) [20]. ACEI peptides have also been found in enzyme-modified cheeses [21], in a low-fat cheese made in Finland [22] and Manchego cheeses manufactured with different starter cultures [23]. Mexican Fresco cheese manufactured with *Enterococcus faecium* or a *L. lactis* ssp. *lactis-Enterococcus faecium* mixture showed the largest number of fractions with ACEI activity among tested lactic acid strains [24]. Pripp et al. [25] investigated the relationship between proteolysis and ACE inhibition in Gamalost, Castello, Brie, Pultost, Norvegia, Port Salut and Kesam. The traditional Norwegian cheese Gamalost had per unit cheese weight higher ACE inhibition potential than Brie, Roquefort and Gouda-type cheese. However, ACE inhibition expressed as $IC_{50}$ per unit peptide concentration from ethanol soluble fraction assessed by the OPA-assay was highest for Kesam, a Quark-type cheese with a low degree of proteolysis.

When β-casomorphins were looked from commercial cheese products, no peptides were found or their concentration in the cheese extract was below 2 μg/ml [26]. They further noted that the enzymatic degradation of β-casomorphins was influenced by a combination of pH and salt concentration at the cheese ripening temperature. Therefore, if formed in cheese, β-casomorphins may be degraded under conditions similar to Cheddar cheese ripening. Precursors of β-casomorphins, on the other hand, have been identified in Parmesan cheese [19]. β-Casomorphins were found at a higher level in the mould cheeses (166–648 mg/100 g), whereas the opioid peptides with antagonistic activity (casoxin-6) were identified at a higher level in the semi-hard cheeses (136–276 mg/100 g) and a low quantity of casomorphins (4–100 mg/100 g) [27]. Immunomodulating properties in water-soluble extracts from traditional French Alps cheeses, Abondance and Tomme de Savioe have been observed [28]. However, no correlation between peptide composition and *in vitro* immunomodulation of T-lymphocyte cells could be established.

A limited number of bioactive peptides have been isolated and identified in Gouda, Manchego, Festivo and Crescenza cheeses (Table 1). Several ACEI peptides have been identified from N-terminal of $\alpha_{s1}$-casein of Gouda, Festivo, Cheddar and Fresco cheeses [22, 24, 29, 30]. In addition, peptides from β–casein, Tyr-Pro-Phe-Pro-Gly-Pro-Ile-Pro-Asn (β-cn, f(60–68)); and Met-Pro-Phe-Pro-Lys-Tyr-Pro-Val-Gln-Pro-Phe (β-cn, f(109–119)) from Gouda [29] and Tyr-Gln-Glu-Val-Leu-Gly-Pro-Val-Arg-Gly-Pro-Phe-Pro-Ile-Ile-Val (β-cn, f(193-209)) from Cheddar [30] have been identified. Antihypertensive peptides Val-Pro-Pro (VPP) (β-cn, f(84–86)) and Ile-Pro-Pro (IPP) (β-cn, f(74–76) and κ-cn, f(108–110)), have also been identified and quantified in different cheese varieties [31-33]. In some varieties physiologically relevant amounts was observed, however, a large variation exists between samples of the same cheese variety, as well as between different varieties. The concentrations of VPP and IPP were in the range of 0-224 mg/kg and 0-95 mg/kg, respectively, indicating that some cheese varieties contain similar concentrations of VPP and IPP to fermented milk products. Milk pretreatment, cultures, scalding conditions, and ripening time were identified as the key factors influencing the concentration of these two naturally occurring bioactive peptides in cheese. Thus, it is necessary to develop a reproducible cheese-making process with selected cultures to produce higher concentrations of these peptides that could be used for clinical trials.

| Cheese variety | Milk protein fragment | Peptide sequence | ACE-inhibition $IC_{50}$ μM | Ref |
|---|---|---|---|---|
| Gouda | $\alpha_{s1}$-cn f( 1-9) | RPKHPIKHQ | 13.4 | 29 |
| | $\alpha_{s1}$-cn f(1-13) | RPKHPIKHQGLPQ | ND | |
| | β-cn f(68-66) | YPFPGPIPN | 14.8 | |
| | β-cn f(109–119) | MPFPKYPVQPF | ND | |
| Manchego | ovine $\alpha_{s1}$-cn f(102-109) | KKYNVPQL | 77.2 | 23 |
| | ovine $\alpha_{s1}$-cn f(205-208) | VRYL | 24.1 | |
| Cheddar (with probiotics) | $\alpha_{s1}$-cn f(1-9) | RPKHPIKHQ | ND | 30 |
| | $\alpha_{s1}$-cn f(1-7) | RPKHPIK | | |
| | $\alpha_{s1}$-cn f(1-6) | RPKHPI | | |
| | $\alpha_{s1}$-cn f(24-32) | FVAPFPEVFGK | | |
| | β-cn f(193-209) | YQEPVLGPVRGPFPIIV | | |
| Swiss cheese varieties | β-cn, f(84–86) | VPP | 9 | 31-34 |
| | β-cn, f(74–76) and κ-cn, f(108–110) | IPP | 5 | |
| Fresco cheese | $\alpha_{s1}$-cn f(1-15) | RPKHPIKHQGLPQEV | ND | 24 |
| | $\alpha_{s1}$-cn f(1-22) | RPKHPIKHQGLPQEVLNEN | | |
| | $\alpha_{s1}$-cn f(14-23) | LLR | | |
| | $\alpha_{s1}$-cn f(24-34) | EVLNENLLRF | | |
| | β-cn f(193-205) | FVAPFPEVFGK | | |
| | β-cn f(193-207) | YQEPVLGPVRGPF | | |
| | β-cn f(193-209) | YQEPVLGPVRGPFPI | | |
| | | YQEPVLGPVRGPFPIIV | | |

ND: Not described
IC50: Peptide concentration that shows 50% inhibition of ACE activity
One letter amino acid codes used

**Table 1.** Examples of identified bioactive peptides in different cheese varieties

## 2.2. Fermented milk

During fermentation process, lactic acid bacteria hydrolyze milk proteins, mainly caseins, into peptides and amino acids which are used as nitrogen sources necessary for their growth. Hence, bioactive peptides can be generated by starter and non-starter bacteria used in the manufacture of fermented dairy products (Table 2). Proteolytic system of *Lb. helveticus, Lb. delbrueckii* ssp *bulgaricus, L. lactis* ssp. *diacetylactis, L. lactis* ssp. *cremoris,* and *Streptococcus (Str.) salivarius* ssp. *thermophilus* strains have demonstrated to hydrolyze milk proteins and release ACEI peptides. Among lactic acid bacteria, *Lb. helveticus* has high extracellular proteinase activity and the ability to release large amount of peptides in fermented milk. As a result, among various kinds of fermented milk, antihypertensive effect related to ACEI peptides were found in milk produced by *Lb. helveticus*. Two ACEI peptides have been purified from sour milk and identified as VPP and IPP [34].

| Organisms | ACE-inhibition | Identified peptides | | Dose | Response (Δ SBP mmHg) | Ref. |
|---|---|---|---|---|---|---|
| | $IC_{50}$ mg/ml | Sequence | $IC_{50}$ μM | | | |
| *Lb. helveticus and Str. thermophilus* | ND | VPP<br>IPP | 9<br>5 | 5 ml/kg | -21.8 ±4.2 after 6 h | 34 |
| *Lb. helveticus* | | VPP<br>IPP | 9<br>5 | 27 ml/day | -21 after 4 weeks | 67 |
| *Lb.helveticus* CPN4 | ND | YP | 720 | 10 ml/kg | 32.1 ±7.4 after 6 h | 42 |
| *Lb. helveticus* CHCC637<br>*Lb. helveticus* CHCC641 | 0.16<br>0.26 | | | 10ml/kg | -12 after 4-8 h<br>-11 after 4-8 h | 37 |
| *Lact. delbrueckii* ssp. *bulgaricus*<br>*Str. salivarius* ssp *thermophilus* and *L.lactis* biovar *diacetylactis* | | SKVYPFPGPI<br>SKVYP | 1.7 mg/ml<br>1.5 mg/ml | | ND | 43 |
| *Lb. jensenii* | 0.52 | LVYPFPGPIHNSLPQN<br>LVYPFPGPIH | 71<br>89 | 0.2 kg/kg | approx -12 after 2 h | 38 |
| *Enterococcus faecalis* CECT 5727 | 0.053 | LHLPLP<br>LVYPFPGPIPNSLPQNIPP | 5.5<br>5.2 | 2 mg/kg<br>6 mg/kg | -21.87 ±4.51 after 4h[1)]<br>approx -15 after 4 h | 44 |
| *Lb. delbrueckii* subsp. *bulgaricus* SS1<br>*L. lactis* subsp. *cremoris* FT4 | ND | NIPPLTQTPV<br>LNVPGEIVE<br>DKIHPF | 173.3<br>300.1<br>256.8 | | ND | 36 |
| Mixed lactic acid bacteria (*Lb. casei, acidophilus, bulcaricus, Str. themophilus, Bifidobacterium*) and protease | 0.24 | GTW<br>GVW | 464.4<br>240.0 | 5 mg/ml | SBP -22 after 8 weeks | 76 |

One letter amino acid codes used
ND Not described
1) Pure synthetic peptides were used in the study

**Table 2.** ACE-inhibitory and antihypertensive activity in spontaneously hypertensive rats of peptides produced by fermentation of milk

Pihlanto-Leppälä et al. [35] studied the potential formation of ACEI peptides from cheese whey and caseins during fermentation with various commercial dairy starters used in the manufacture of yogurt, ropy milk and sour milk. No ACEI activity was observed in these hydrolysates. Further digestion of the above samples with pepsin and trypsin resulted in the release of several strong ACEI peptides derived primarily from $\alpha_{s1}$-casein and β-casein. The formation of ACEI peptides was demonstrated in two dairy strains, *Lb. delbrueckii* ssp. *bulgaricus* and *L. lactis* ssp. *cremoris,* after fermentation of milk separately with each strain for 72 hours [36]. The most inhibitory fractions of the fermented milk mainly contained β-casein-derived peptides with inhibitory concentration ($IC_{50}$) values ranging from 8.0 to 11.2 µg/ml. Fuglsang et al. [37] tested a total of 26 strains of wild-type lactic acid bacteria, mainly belonging to *L. lactis* and *Lb. helveticus,* for their ability to produce a milk fermentate with ACEI activity. All tested strains produced ACEI substances in varying amounts, and two of the strains exhibited high ACE inhibition and a high OPA index, which correlates well with peptide formation. In another study 25 lactic acid strains of *Lactobacillus, Lactococccus* and *Leuconsotoc* were used [38]. The strains were tested alone or in combination and the highest activities were observed in *Lb. jensenii, Lb. acidophilus* and *Leuc. mesenteroides* strains and all strains showed correlation between ACE inhibition and degree of proteolysis. In a recent study, milk was fermented to defined pH values with 13 strains of lactic acid bacteria. The highest ACEI activity was obtained with two highly proteolytic strains of *Lb. helveticus* and with the *Lactococcus* strains. Fermentation from pH 4.6 to 4.3 with these strains slightly increased the ACEI activity, whilst fermentation to pH 3.5 with *Lb. helveticus* reduced the ACEI activity [39]. Moreover, four different *Enterococcus faecalis* strains, isolated from raw milk, produced fermented milk with potent ACEI activity [40]. In a recent research it was found that *L. lactis* strains isolated from artisanal dairy starters or commercial starter cultures are potential for the production of fermented dairy products with ACEI properties. Especially, a strain isolated from artisanal cheese presented the lowest $IC_{50}$ (13µg/ml) [41].

Bioactive peptides isolated from skim milk and whey fermented using a range of organisms are summarized in Table 2. The majority of identified peptides are casein-derived ACEI peptides having $IC_{50}$ values ranging from 5 to 500 µM. The best characterized ACEI and antihypertensive peptides liberated with *Lb. helveticus* alone or in combination with *Saccharomyces cerevisiae* are the tripeptides IPP, and VPP. Yamamoto et al. [42] identified an ACEI dipeptide (Tyr-Pro) from a yogurt-like product fermented with *Lb. helveticus* CPN4 strain. This peptide sequence is present in all major casein fractions, and its concentration was found to increase during fermentation, reaching a maximum concentration of 8.1 µg/ml in the product. Ashar and Chand [43] identified an ACEI peptide from milk fermented with *Lb. delbrueckii* ssp. *bulgaricus.* The peptide showed the sequence Ser-Lys-Val-Tyr-Pro-Phe-Pro-Gly-Pro-Ile from β–casein with an $IC_{50}$ value of 1.7 mg/ml. In combination with *Str. salivarius* ssp. *thermophilus* and *L. lactis* biovar. *diacetylactis,* a peptide structure with a sequence of Ser-Lys-Val-Tyr-Pro was obtained from β-casein with an $IC_{50}$ value of 1.4 mg/ml. Both peptides were markedly stable to digestive enzymes, acidic and alkaline pH, as well as during storage at 5 and 10 ºC for four days. Two β-casein-derived peptides were identified from water soluble fraction of milk fermented with *Lb. jensenii.* The identified peptides were Leu-Val-Try-Pro-Phe-Pro-Gly-Pro-Ile-His-Asn-Ser-Leu-Pro-Gln-Asn, and

Leu-Val-Try-Pro-Phe-Pro-Gly-Pro-Ile-His [38]. Quirós et al. [44] identified two peptides in fermented milk with *Enterococcus faecalis* that corresponded to β -casein fragments Lys-His-Leu-Pro-Leu-Pro and Lys-Val-Tyr-Pro-Phe-Pro-Gly-Pro-Ile-Pro-ASn-Ser-Leu-Pro-Gln-Asn-Ile-Pro-Pro, with potent ACEI activity.

Many kinds of proteolytic enzymes have been reported from lactic acid bacteria, and have been reviewed extensively [6, 45]. The components of the proteolytic systems of lactic acid bacteria are divided into three groups, including the extracellular proteinase that catalyzes casein breakdown to peptides, peptidases that hydrolyze peptides to amino acids and a peptide transport system. The extracellular proteinase activity was almost correlated with ACEI activity in the fermented milk, suggesting that the proteolysis of casein by the extracellular proteinase is the most important parameter in the processing of active components [46]. The importance of the proteinase was also supported by the fact that a proteinase negative mutant was not able to generate antihypertensive peptides in the fermented milk, whereas the wild-type strain had the ability to release strong antihypertensive peptides in the fermented milk [47]. The enzymatic process generating the antihypertensive peptides VPP and IPP in *Lb. helveticus* has been elucidated. By the proteolytic action of the extracellular proteinase long peptide with amino acid residue including VPP and IPP sequences were generated. Next the long peptide would be hydrolyzed to shorter peptides by intracellular peptidases. A key enzyme that can catalyze C-terminal processing of Val-Pro-Pro-Phe-Leu and Ile-Pro-Pro-Leu-Thr to VPP and IPP has been purified from *Lb. helveticus* CM4. The endopeptidase has sequence homology in amino terminal sequence to a previously reported pepO-gene product [48]. Kilpi et al. [49] found out higher ACE inhibition in milk fermentation using peptidase-deletion mutants compared to the wild-type of *Lb. helveticus* strain. Unlike with the wild type strain, ACEI remained constant during the course of fermentation with the proline-specific peptidase mutant. The mutant strains had also different peptide profiles than the wild-type strain.

## 2.3. Other

Various types of fermented soybean foods are consumed in Asian countries such as Korea, China, Japan, Indonesia and Vietnam. Soybeans are traditionally fermented primarily by *Bacilli* species during the early stage of fermentation followed by *Aspergillus* species, which predominate during the remaining fermentation period [50]. ACEI peptides have been found in many traditional Asian fermented soy foods, such as soybean paste, soy sauce, natto and tempeh. ACEI peptide His-His-Leu was isolated from Korean fermented soybean paste [51]. Rye gluten sourdoughs fermented with *Lb. reuteri* and added protease were found to contain the lactoripeptides VPP, IPP [52]. Moreover, our recent studies showed that fermentation of rapeseed or flaxseed meals with *Bacillus subtils* or *Lb. helveticus* strains produced ACEI activity [53].

## 2.4. Other activities

It is reasonable to expect that lactic acid bacteria produce scavengers for hydroxyl radical, which can be metabolic compounds produced by bacteria or degradation products of milk proteins. The results have demonstrated that the antioxidant production is commonly higher within the group of obligately homofermentative lactobacilli, than within the facultatively or obligately heterofermentative strain groups. Also heterofermentative *Lactobacillus* sp. have been reported to exhibit antioxidative activity. *Lb. acidophilus, Lb. bulgaricus, Str. thermophilus* and *Bifidobacterium longum* exhibited antioxidative activity by various mechanisms, like metal ion chelating capacity, scavenging of reactive oxygen species (ROS), reducing activity and superoxide dismutase activity [54, 55]. Peptides liberated during fermentation can be partially responsible for the reported antioxidative properties. An antioxidative peptide derived from κ-casein was detected in milk after fermentation with *Lb. delbrueckii* subs. *bulgaricus* [56]. Moreover, Hernández-Ledesma et al. [57] found a moderate ABTS radical scavenging capacity in commercial fermented milk from Europe. Further studies of this radical scavenging activity in different HPLC fractions showed low TEAC values. Virtanen et al. [58] found that fermentation with *Leuc. mesenteroides* ssp. *cremoris, Lb. jensenii* and *Lb. acidophilus* strains produced compounds that showed both radical scavenging activity and inhibition of lipid peroxidation.

Inflammation plays a key role in the development of cardiovascular disease. It often begins with inflammatory changes in the endothelium, which begins to express the adhesion molecule VCAM-1. VCAM-1 attracts monocytes, which then migrate through the endothelial layer under the influence of various proinflammatory chemoattractants [59]. Accordingly, fermentation by lactic acid may be able to release components that possess immunomodulatory properties. Most of the studies have been done with synthetic peptides derived from enzymatic treatment of milk proteins using different *in vitro* models. Leblanc et al. [60] investigated the effect of peptides released during the fermentation of milk by *Lb. helveticus* on the humoral immune system and on the growth of fibrosacromas. The study showed that bioactive components were released during fermentation that contributed to the immunoenhancing and antitumor properties. Antimutagenic compounds were produced during fermentation by *Lb. helveticus,* and release of peptides is one possible explanation [61]. The permeate fraction obtained from milk fermented by *Lb. helveticus* was able to modulate the *in vitro* proliferation of lymphocytes by acting on the production of cytokines [62]. Tompa et al. [63] found that peptide fractions form *Lb. helveticus* BGRA43 fermented milk have anti-inflammatory potential. Matar et al. [64] fed milk fermented with a *Lb. helveticus* strain to mice for three days and detected significantly higher numbers of IgA secreting cells in their intestinal mucosa, compared with control mice fed with similar milk incubated with a non-proteolytic variant of the same strain. The immunostimulatory effect of fermented milk was attributed to peptides released from the casein fraction.

## 3. Antihypertensive effects in vivo

The search for *in vitro* ACEI is the most common strategy followed in the selection of potential antihypertensive peptides derived from food proteins. *In vitro* ACEI activity is generally measured by monitoring the conversion of an appropriate substrate by ACE in the presence and absence of inhibitors. The antihypertensive effects have been assessed by *in vivo* experiments using spontaneously hypertensive rats (SHR) as an animal model to study human essential hypertension [7]. Following a positive response in animal studies human studies may be carried out to ascertain the ACEI potential

### 3.1. Animal studies

A great number of studies have addressed the effects of both short-term and long-term administration of potential antihypertensive peptides using this animal model. Fermented milks with different $IC_{50}$-values ranging from from 0.08 to 1.88 mg/ml have been shown to decrease blood pressure in SHR from 10 to 32 mmHg (Table 2).

The first antihypertensive effect of milk casein-derived peptides was first demonstrated by casein hydrolysate formed by purified proteinase from *Lb. helveticus* CP790 and milk fermented with the same bacteria [65]. The authors concluded that peptides deliberated from casein by extracellular proteinases were responsible for the antihypertensive effect. The active substances were liberated during fermentation of milk with *Lb. helveticus* and *Saccharomyces cerevisiae* and were identified to be IPP and VPP. Oral administration of fermented milk or pure tripeptides were shown to produce strong antihypertensive effect in SHR after single-dose [34, 66]. Thereafter, several animal studies have been conducted to characterize the long-term effects of lactotripeptides or fermented milk containing them. These studies were mainly conducted with SHR but also Goto-Kakizaki (GK) rats and double transgenic rats (dTGR) with malignant hypertension have been used. The development of hypertension was attenuated significantly in rats receiving fermented milk product containing lactotripeptides, attenuation in systolic blood pressure was 12-21 mmHg in SHR, 10 mmHg in high salt-fed GK rats and 19 mmHg in dTGR in comparison to control group [67-69]. Pure tripeptides did not produce as strong antihypertensive effect as the milk products containing them. In addition, minerals alone did not attenuate the development of blood pressure as much as the fermented milk products [68]. These studies indicate that the bioavailability of peptides may be better from milk in comparison of water or is improved by other milk components.

After the blood pressure monitoring has been completed the effect of long-term intake of lactotripeptides on vascular function has been assessed [68,70,71]. Jauhiainen et al. [70], showed improved endothelium-dependent relaxation in mesenteric arteries and aortas of rats that had received minerals and lactotripeptide. Endothelial function of mesenteric arteries was strongly impaired in all groups of salt-loaded GK rats, and significantly improved endothelium-dependent relaxations were observed after treatment with different fermented milk products [68]. Protection of endothelial function after incubation with tripeptides IPP and VPP for 24 h was found in a study with isolated SHR mesenteric arteries [71].

Evidence from ACE inhibition was gained by Masuda et al. [72], who found that after receiving a single-dose of Calpis™ sour milk, ACE activity was decreased in SHR aorta. The lactotripeptides were detected in solubilized fraction from the abdominal aorta of SHR but not from WKY given the sour milk. Moreover, in SHR, plasma rennin activity increased after long-term treatment of fermented milk product containing the lactotripeptides [67]. In addition, treatment with fermented milk containing lactotripeptides and plant sterols decreased serum ACE activity [73]. In salt-loaded GK rats, fermented milk with lactotripeptides decreased serum ACE and aldosterone levels [68].

Besides the most extensively studied lactotripeptides, also other fermented milk products and peptides have been found. Different strains of lactic acid bacteria, such as *Lb. helveticus* CPN4, *Lb. bulgaricus, Lb. jensenii* and *Str. thermophilus,* have been also shown to provoke liberation of peptides with antihypertensive activity in SHR [36, 37, 41]. Two peptides, corresponding to β -casein fragments Leu-Val-Tyr-Pro-Phe-Pro-Gly-Pro-Ile-Pro-Asn-Ser-Leu-Pro-Gln-Asn-Ile-Pro-Pro and Leu-His-Leu-Pro-Leu-Pro, have been isolated in fermented milk with *Enterococcus faecalis* and their antihypertensive effect in SHR, after acute and long-term administration has been proved. The administration of 2 mg/kg of peptide Leu-His-Leu-Pro-Leu-Pro resulted in a significant decrease of the SBP in SHR 4 h post-administration [74,75]. Fermentation of milk with one or more lactic acid bacteria strains followed by hydrolysis using food-grade enzymes liberated tripeptides (Gly-Thr-Trp and Gly-Val-Trp). Oral administration of this fermented whey lowered significantly SBP in SHR from 9 to 15 weeks of age. Bioactive substances, tripeptides and γ-aminobutyric acid (GABA), contributed to lowering blood pressure of SHR [76].

Some of ACE-inhibitory peptide fractions from cheese have shown *in vivo* activities. A water-soluble peptide preparation isolated from Gouda ripened for 8 months was found to have the most potent antihypertensive activity (maximum decrease in SBP = 24.7 (± 0.3) mmHg ($P \leq 0.01$) after 6 h) when administered to SHR by gastric intubation at doses between 6.1 and 7.5 mg/kg body weight. Three peptide fractions were isolated from water-soluble extract by hydrophobic chromatography using different concentrations of acetonitrile. The fractions eluting between 15% and 30%, 30–45% and 60–75% acetonitrile decreased SBP in SHR by 15.0, 29.3 and 18.8 mmHg ($P \leq 0.01$), respectively, 6 h after gastric intubation. The peptide fraction eluting between 30% and 45% acetonitrile was shown to contain the sequences (αs1-cn f(1–9)) Arg-Pro-Lys-His-Pro-Ile-Lys-His-Gln and (β-cn f(60–68)) Tyr-Pro-Phe-Gly-Pro-Ile-Pro-Asn (Table 1), which, respectively, decreased SBP in SHR by 9.3 (± 4.8) and 7.0 (± 3.8) mmHg 6 h after gastric intubation [29].

Several sequences have been proposed as responsible for the antihypertensive activity of soy protein hydrolysates and fermented products, but only the peptide His-His-Leu derived from fermented soy paste was assayed in pure form in SHR, where a decrease of 32 mm Hg of SBP was reached at a dose of 100 mg/kg. Moreover, the synthetic tripeptide His-His-Leu resulted in a significant decrease of ACE activity in the aorta [77]. Soybean-derived products contain isoflavones, which are thought to possess a favourable effect in reducing cardiovascular risk factors as well as vascular function [78]. However, on the basis of *in vitro* results and literature review, Wu and Muir [79] have indicated that the contribution of isoflavones to a blood-pressure-lowering effect in soybean ACEI peptides may be negligible. Similarly, it has been reported that the reduction of hypertension of a fermented product from soy milk was contributed mainly by peptides of 800–900 Da but it could be also attributable to GABA [80]. Moreover, fermented soy product, miso, with added tripeptides

(VPP and IPP) from casein was reported to act as antihypertensive agents in SHR [81]. Recently, Nakahara et al. [82] used the Dahl salt-sensitive rats as a model of salt-sensitive hypertension to evaluate the antihypertensive effect of a peptide-enriched soy sauce-like seasoning. The results of these tests have highlighted an important lack of correlation between the *in vitro* ACEI activity and the *in vivo* action. This fact has provided doubts on the use of the *in vitro* ACEI activity as the exclusive criteria for potential antihypertensive substances, since physiological transformations may occur *in vivo*, and because other mechanisms of action than ACE inhibition might be responsible for the antihypertensive effect.

### 3.2. Effects in clinical studies

Evidence of the beneficial effects of bioactive peptides has to be based on clinical data. Most research has been focused in lactotripeptides, VPP and IPP, and their antihypertensive properties. About twenty human studies have been published linking the consumption of products containing lactotripeptides with significant reductions in both SBP and DBP. Oral administration of these tri-peptides included in different formulas, fermented milk, dried product, fruit juice, etc., products. However, recent studies have provided some conflicting results. Most clinical trials have assessed BP-lowering effects at multiple points over time. Most of the BP studies with lactotripeptides have been done in Japanese subjects, and several studies have been done in Finnish subjects [83-88]. Generally, maximum duration of treatment was 8 weeks at doses between 3 and 52 mg/day (Table 3). From these data, it becomes apparent that the largest part of the total BP reduction takes place in the first 1–2 weeks of treatment. Thereafter, a further gradual lowering is seen, but to a lesser extent than in the first period [84-86]. The first significant effects of lactotripeptides on BP in hypertensive subjects were observed after 1–2 weeks of treatment with dosages as low as 3.8 mg/d. Maximum BP-lowering effects of lactotripeptides approximate 13 mmHg SBP and 8 mmHg DBP active treatment v. placebo, and are likely reached after 8–12 weeks of treatment. Lactotripeptides exert a gradual effect on BP lowering after start of intake and return of BP after end of treatment as well [85, 86, 89]. The highest effective dosage of lactotripeptides was evaluated in a safety study, and consisted of 52.5 mg/d [88]. After 10 weeks of active treatment, mean SBP in subjects with hypertension decreased by 4.1 mmHg and DBP by 1.8 mmHg. The next highest dose of lactotripeptides that was tested amounted to 13.0 mg/d [89]. After 4 weeks of active treatment, SBP in subjects with mild hypertension decreased by 11.2 mmHg compared to placebo, and DBP tended to decrease by 6.5 mmHg. In none of the trials with normotensives were statistically significant BP changes found [90-92]. Even at the highest dosage of lactotripeptides used in normotensives, which included a total of 29.2 mg/d during a period of 7 d, no BP lowering effects by lactotripeptides were observed [93]. Thus lactotripeptides only seem to be active at elevated BP values. Evidence indicates that effectiveness is positively associated with BP level, which is in line with existing data for BP-lowering medication [94].

| Design | Duration (weeks) | Study population | Treatment IPP mg/d | VPP mg/d | Source of peptides | Formula | BP changes mmHg SBP | DBP | Ref. |
|---|---|---|---|---|---|---|---|---|---|
| R, p-c, s-bld, parallel | 8 | 30 eldery hypertensive patients | 1.1 | 1.5 | *Lb. helv + Str. cer* | 1 x 95 ml milk drink | -14.1 | -6.9 | 83 |
| R, p-c, d-bld, parallel | 8 | 64 subjects with SBP 140-159 and DBP 90-99 mmHg | 1.58 | 2.24 | *Lb. helv + Str. cer.* | 2 x 150 g milk drink | -13 | -8.4 | 84 |
| R, p-c, d-bld, parallel | 8 | 32 subjects with SBP 140 - 180 and DBP 90-105 mmHg | 1·60 | 2·66 | *Lb. helv + Str. cer.* | 1 x 120 g milk drink | -12.1 | -5.8 | 85 |
| R, p-c, d-bld, parallel | 8 | 18 hypertensive and 26 normotensive subjects | 1.1 | 1.5 | *Lb. helv + Str. cer.* | 2 x 100 g milk drink | -7.6 | -2 | 91 |
| R, p-c, d-bld, parallel | 8 | 30 subjects with SBP 140-180 and DBP 90-105 mmHg | 1.52 | 2.53 | *Lb. helv + Str. cer.* | 2 x 160 g milk drink | -13.2 | -7.8 | 92 |
| R, p-c, d-bld, parallel 1) | 21 | 39 subjects with SBP 133-176 and DBP 86-108 mmHg | 2.25 | 3.0 | *Lb. helv* LBK-16H | 2 x 150 ml milk drink | -6.7 | -3.6 | 86 |
| R, p-c, d-bld, parallel<br>Cross-over 2) | 10<br>7 | 60 Finnish subjects with SBP 140-180 and DBP 90-110 mmHg | 2.4-2.7 | 2.4-2.7 | *Lb. helv* LBK-16H | 1 x 150 ml milk drink | -2.3<br>-12.3 | -0.5<br>-3.7 | 87 |
| R, p-c, d-bld, parallel | 10 | 94 hypertensive patients | 30 | 22.5 | *Lb. helv* LBK-16H | 2 x 150 ml milk drink | 4.1 | 1.8 | 88 |
| R, p-c, d-bld, parallel | 1 | 20 healthy volunteers normal blood pressure (<130 mmHg SBP and <85 mmHg DPB). | 11.5 | 17.7 | *Lb. helv* CM4 | 1 x 14 tablets | 2.6 | 2 | 93 |
| R, p-c, d-bld, parallel | 8 | 135 Dutch subjects with untreated high-normal BP or mild hypertension | 4.2 | 5.8 | Fermentation | 1 x 200 ml yoghurt drink | -0.5 | -1.2 | 97 |
| R, p-c, d-bld, crossover | 4 | 70 Caucasian subjects with prehypertension or stage 1 hypertension | 15 | - | Hydrolysis by endopeptidase | 2 x 7.5 mg capsules | -3.8 | -2.3 | 102 |

1) Results reported as changes in SBP and DBP after each month of treatment for all subjects (intention-to-treat analysis), and as mean changes over the total intervention period among subjects who had BP measurements for each month (per protocol analysis); 2) First part of the study was carried out in parallel design and second part of the study was carried out in crossover design.

**Table 3.** Hypotensive effects of fermented milks with bioactive peptides in humans

The results have been included in two meta-analysis [95, 96], which described decreases around 5 mmHg for SBP and 2.3 mmHg for DBP. In general, the effects described in Japanese studies on lactotripeptides are larger than those reported in Finnish studies. However, it is unlikely that genetic differences can account for these differential effects. Moreover, clinical trials in Dutch and Danish subjects have described controversial results since no effect on blood pressure was found [97, 98]. In a recent meta-analysis with a total of 18 trials, it was found a reduction of 3.73 mm Hg for SBP and 1.97 mm Hg for DBP but it was highlighted that the effect was more evident in Asian subjects that in Caucasian ones [99]. The relevance of these findings in genetics or dietary patterns should be further investigated. Comparative studies on antihypertensive medication in different races/ethnic groups have demonstrated that pharmacokinetic parameters and haemodynamic effects are essentially the same in Chinese and Japanese subjects compared with Caucasian subjects [100].

Hypertension is a complex multifactor disorder that is thought to result from an interaction between environmental factors and genetic background. Subject characteristics such as age and race/ethnicity can affect BP, including the BP response to specific antihypertensive medication. For certain antihypertensive drugs, it has been reported that a polymorphism found in humans can affect the clinical effectiveness, and similarly, these differences could be also affecting clinical trials of functional ingredients [101]. Although ACE inhibition has been postulated as the underlying mechanism of these lactotripeptides, results about the inhibition of this enzyme are not conclusive in humans. Several studies have shown that rennin or ACE activity was not affected by the oral administration of the tripeptides [95, 102]. Therefore, other mechanisms could be implicated in the observed blood pressure reduction. It has been found that the intake of fermented milk containing these peptides may decrease sympathetic activity, leading to a diminished heart rate variability, heart rate and total peripheral resistance, although differences did not reach statistical significance [98].

## 4. Bioavailability

Bioavilability of bioactive peptides is an important target to establish the relationship between *in vitro* and *in vivo* activities. The likelihood of any bioactive peptide released during fermentation mediating a physiological response is dependent on the ability of that peptide to reach an appropriate target site. Therefore, peptides may need to be resistant to further degradation by proteolytic and peptidolytic enzymes in digestive tract. Thereafter peptides should be absorbed and enter systemic circulation. Resistance to hydrolysis is one of the main factors influencing the bioavailability of bioactive peptides. The effects of digestive enzymes on bioactive peptides, in particular ACEI peptides derived from different food matrices, have been evaluated *in vitro* gastrointestinal simulated systems. The common purpose of these experiments was to assess the effects of the peptidases of the stomach and the pancreas on the preservation of the ACEI activity of different hydrolysates. Studies have shown that the ACEI is low after fermentation but increases during hydrolysis that simulates gastrointestinal digestion [35,103]. The ACEI peptides in rapeseed hydrolysate exhibited good stability in an *in vitro* digestion model using human gastric and duodenal fluids [104]. The digestion of some peptides have been reported. For example, Ile-Val-Tyr was hydrolysed by pepsin, trypsin and chymotrypsin alone or in combination and $IC_{50}$-value did not change significantly during digestion [105]. Proline- and hydroxyproline-containing peptides are usually resistant to degradation by digestive enzymes. Tripeptides containing C-terminal proline-proline are generally resistant to proline-specific peptidases [106]. In some cases, pancreatic digestion is needed to produce active peptide. For instance, the active form of peptide Lys-Val-Leu-Pro-Val-Pro-Glu is generated by hydrolysis of the glutamine residue at the C-terminal during pancreatic digestion [107]. The results are not completely predictive of the resistance of the bioactive peptides because they do not mimic all the physiological factors affecting food digestion, as pH variations, the relative amounts of the enzymes, the interactions with other molecules, and the ratio peptidase/tested compound. These variations may affect the rate of enzymatic degradation of the bioactive peptides under study, therefore affecting the estimated bioavailability of these bioactive peptides. Moreover, commercial enzymes appear to digest whey proteins more efficiently compared with human digestive juices when used at similar enzyme activities [108]. This could lead to conflicting results when comparing human *in vivo* protein digestion with digestion using purified enzymes of non-human species.

Peptides have been reported to have poor permeation across biological barriers (e.g. intestinal mucosa) [109]. Peptides can be transported by active transcellular transport or by passive processes. Although substantial amino acid absorption occurs in the form of di- and tripeptides at the apical side of enterocytes, efflux of intact peptides via the basolateral membrane into the general circulation seems to be negligible [110]. The intestinal absorption of peptides have been performed using *in vitro* tests with monolayer of intestinal cell lines, simulating intestinal epithelium, as well as analysis of peptides and derivatives in blood samples after animal and clinical studies. Foltz et al. [111] investigated the transport of IPP and VPP by using three different absorption models and demonstrated that these tripeptides are transported in small amounts intact across the barrier of the intestinal epithelium. The major transport mechanisms of IPP and VPP were demonstrated to be paracellular transport and passive diffusion [112]. Another ACEI peptide, Leu-His-Leu-Pro-Leu-Pro resisted gastrointestinal simulation but was degraded to His-Leu-Pro-Leu-Pro by cellular peptidases before crossing Caco-2 cell monolayer. The pentapeptide was rapidly transported through Caco-2 cell monolayers through paracellular route [113].

Vascular endothelial tissue peptidases and soluble plasma peptidases further contribute to peptide hydrolysis. As a consequence, for most peptides, the plasma half-life is limited to minutes as shown for endogenous peptides such as angiotensin II and glucagon-like peptide 1 [114]. In order to exert antihypertensive effect ACEI peptides need to resist different peptidases such as ACE. In this regard ACEI peptides can be classified into three groups: the inhibitor type, of which the $IC_{50}$-value is not affected by preincubation with ACE; the substrate type, peptides that are hydrolysed by ACE to give peptides with a weaker activity; the pro-drug type inhibitor, peptides that are converted to true inhibitors by ACE or other proteases/peptidases. Only peptides belonging to pro-drug or inhibitor type exert antihypertensive properties after oral administration. There are some examples showing that peptides are absorbed and can exert *in vivo* activities. As regard to casein-derived IPP,

Jauhiainen et al. [115] used radiolabelled tripeptide and showed that it absorbed partly intact from the gastrointestinal tract after a single oral dose to rats. Considerable amounts of radioactivity were found from several tissues, e.g., liver, kidney and aorta. The excretion of IPP was slow; even after 48 hours the radiolabelled peptide had not been completely excreted. IPP did not bind to albumin or other plasma proteins *in vitro*. Considering this and the long-lasting retention of the radioactivity in the tissues, accumulation of IPP may occur in sufficient concentrations to cause blood pressure lowering effects e.g., by ACE-inhibition in the vascular wall. In another study the absolute bioavailability of the tripeptides in pigs was below 0.1%, with an extremely short elimination half-life ranging from 5 to 20 min [116]. In humans, maximal plasma concentration did not exceed picomolar concentration [117].

The improvement of limited absorption and stability of peptides has been a goal when evaluating their effectiveness. For example, some carriers interact with the peptide molecule to create an insoluble entity at low pH which later dissolves and facilitates intestinal uptake, by enhancing peptide transport over the non-polar biological membrane [118]. Bioavailability of bioactive tripeptides (VPP, IPP, LPP) was improved by administering them with a meal containing fiber, as compared to a meal containing no fiber. High methylated citrus pectin was used as a fiber [119]. Ko et al. [120] applied emulsification, microencapsulation and lipophilization to enhance the antihypertensive activity of a hydrolysate of tuna cooking juice. Among these treatments, lipophilization was the most effective, followed by microencapsulation and lecithin emulsification, getting for each of them a stronger effect than the obtained with the double untreated dosage. Antihypertensive effect of ovokinin (Phe-Arg-Ala-Asp-Pro-Phe-Leu) increased four-times compared to the untreated dosage after administration with egg yolk [121]. In this case, phospholipids were identified as responsible for enhancing the antihypertensive effect, particularly phosphatidylcholine, that could improve intestinal absorption or by protecting ovokinin of peptidases. Among drug delivery systems, emulsions have been used to enhance oral bioavailability or promoting absorption through mucosal surfaces of peptides and proteins [118]. Individually, various components of emulsions have been considered as candidates for improving bioavailability of peptides.

## 5. General conclusions

The interest on foods possessing health-promoting or disease-preventing properties has been increasing. An increasing number of foods sold in developed countries bears nutrition and health claims. Fermented milk with putative antihypertensive effect in humans could be an easy applicable lifestyle intervention against hypertension. In fact, much work has been done with dietary antihypertensive peptides and evidence of their effect in animal and clinical studies. Moreover, there are numerous available patents of products containing antihypertensive bioactive peptides. However, certain aspects, such as identification of the active form in the organism and the different mechanisms of action that contribute in the antihypertensive effect still need to be further investigated. Recent advances on specific analytical techniques able to follow small amounts of the peptides or derivatives from them in complex matrices and biological fluids will allow performing these kinetic studies in model animals and humans. Similarly, advances in new disciplines such as nutrigenomic and nutrigenetic will open new ways to follow bioactivity in the organism by identifying novel and more complex biomarkers of exposure and/or of activity. There is still poor knowledge on the resistance of peptides to gastric degradation, and low bioavailability of peptides has been observed. This reinforces the need of various strategies to improve the oral bioavailability of peptides.

More emphasis has been put on the legal regulation of the health claims attached to the products. Authorities around the world have developed systematic approaches for review and assessment of scientific data. Evidence on the beneficial effects of a functional food product should be enough detailed, extensive and conclusive for the use of a health claim in the product labeling and marketing. Besides being based on generally accepted scientific evidence, the claims should be well understood by the average consumer. First, it is necessary to identify and quantify the active sequences. Antihypertensive peptides are only minor constituents in highly complex food matrices and, therefore, a monitoring of the large-scale production by hydrolytic or fermentative industrial process is mandatory. Second, extensive investigations to prove the antihypertensive effect in humans as well as the minimal dose to show this effect are necessary to fulfill the requirements of the legislation concerning functional foods. Japan was the pioneer with the Foods for Special Health Use (FOSHU) legislation in 1991. Europe adopted a joint Regulation on Nutrition and Health Claims made on Foods in 2006 being the European Food Safety Authority (EFSA). At present, EFSA have concludes that the evidence is insufficient to establish a cause and effect relationship between the consumption of the tripeptides VPP and IPP and the maintenance of normal blood pressure. Bearing in mind that 'essential hypertension' consists of disparate mechanisms that ultimately lead to elevations in systemic BP, it is most probably that that products containing lactotripeptides offer a valuable option as a non-pharmacological, nutritional treatment of elevated blood pressure for some groups of people.

## Author details

Anne Pihlanto
*MTT Agrifood Finland, Biotechnology and Food Research, Jokioinen, Finland*

## 6. References

[1] Shah N (2007) Functional cultures and health benefits. Int. dairy j. 17:1262-1277
[2] Takano T (2002) Anti-hypertensive activity of fermented dairy products containing biogenic peptides. Anton. leeuw. 82: 333-340.
[3] Korhonen HJT, Pihlanto-Leppälä A (2004) Milk-derived bioactive peptides : formation and prospects for health promotion. In: Edited by Colette Shortt and John O'Brien.

Handbook of functional dairy products. Functional foods and nutraceuticals series 6.0: p. 109-124.

[4] FitzGerald RJ, Murray BA (2006) Bioactive peptides and lactic fermentations. Int. j. dairy technology 59: 118-125.

[5] Jäkälä P, Vapaatalo H (2010) Antihypertensive peptides from milk proteins. Pharmaceuticals 3: 251-272.

[6] Christensen JE, Dudley EG, Pederson JA, Steele JL (1999) Peptidases and amino acid catabolism in lactic acid bacteria. Anton. leeuw. 76: 217-246.

[7] Luoma S, Peltoniemi K, Joutsjoki V, Rantanen T, Tamminen M, Heikkinen I, Palva A (2001) Expression of six peptidases from Lactobacillus helveticus in Lactococcus lactis. Appl. environ. microb. 67: 1232–1238.

[8] Foucaud C, Juillard V (2000) Accumulation of casein-derived peptides during growth of proteinase-positive strains of Lactococcus lactis in milk: their contribution to subsequent bacterial growth is impaired by their internal transport. J dairy res. 67: 233-240.

[9] Williams AG, Noble J, Tammam J, Lloyd D, Banks JM (2002) Factors affecting the activity of enzymes involved in peptide and amino acid catabolism in non starter lactic acid bacteria isolated from Cheddar cheese. Int. dairy j. 12: 841–852.

[10] Lopez AD, Murray CC (1998) The global burden of disease, 1990–2020. Nat. med. 4:1241–1243.

[11] Harris T, Cook EF, Kannel W, Schatzkin A, Goldman L (1985) Blood pressure experience and risk of cardiovascular disease in the elderly. Hypertension 7:118–24.

[12] Pihlanto A, Korhonen H (2003) Bioactive peptides and proteins. Adv. food res. 47: 175–276.

[13] Hernández-Ledesma B, del Mar Contreras M, Recio I (2011) Antihypertensive peptides: production, bioavailability and incorporation into foods. Adv. colloid interface sci. 165:23-35

[14] Van Gaal LF, Mertens IL, De Block CE (2006) Mechanisms linking obesity with cardiovascular disease. Nature 444: 876-880.

[15] Pihlanto A (2006) Antioxidative peptides derived from milk proteins. Int. dairy j. 16: 1306–1314.

[16] Roudot-Algaron F, Bars DL, Kerhoas L, Einhorn J, Gripon JC (1994) Phosptiopeptides from Comté Cheese: Nature and origin. J. food sci. 59: 544–547.

[17] Singh TK, Fox PF, Healy A (1997) Isolation and identification of further peptides from diafiltration retentate of the water-soluble fraction of Cheddar cheese. J. dairy res. 64:433-443.

[18] Meisel H, Goepfert A, Günter S (1997) ACE-inhibitory activities in milk products. Milchwissenschaft 52: 307–311.

[19] Addeo F, Chianes L, Salzano A, Sacchi R, Cappuccio U, Ferranti P, Malorni A (1992) Characterization of the 12% tricholoroacetic acid-insoluble oligopeptides of Parmigiano Reggiano cheese. J. dairy res. 59: 401 411.

[20] Smacchi E, Gobbetti M (1998) Peptides from several Italian cheeses inhibitory to proteolytic enzymes of lactic acid bacteria, Pseudomonas fluorescens ATCC 948 and to the angiotensin I-converting enzyme. Enzyme microb.tech. 22: 687–694.

[21] Haileselassie SS, Lee B H, Bibbs BF (1999) Purification and identification of potentially bioactive peptides from enzyme modified cheese. J. dairy sci. 82: 1612–1617.

[22] Ryhänen E-L, Pihlanto-Leppälä A, Pahkala E (2001) A new type of ripened low-fat cheese with bioactive properties. Int. dairy j. 11: 441–447.

[23] Gomez JA, Ramos M, Recio I (2002) Angiotensin-converting enzyme-inhibitory peptides in Manchego cheeses manufactured with different starter cultures. Int. dairy j. 12: 697–706.

[24] Torres-Llanez MJ, González-Córdova AF, Hernandez-Mendoza A, Garcia HS, Vallejo-Cordoba B (2011) Angiotensin-converting enzyme inhibitory activity in Mexican Fresco cheese. J. dairy sci. 94:3794–3800.

[25] Pripp AH, Sørensen R, Stepaniak L and Sørhaug T (2006) Relationship between proteolysis and angiotensin I-converting enzyme inhibition in different cheeses LWT 39: 677–683.

[26] Muehlenkamp MR, Warthesen JJ (1996) β-casomorphins: Analysis in cheese and susceptibility to proteolytic enzymes from Lactococcus lactis ssp. cremoris. J. dairy sci. 79: 20-26.

[27] Sienkiewicz- Szlapka E, Jarmolowska B, Krawczuk S, Kostyra E, Kostyra H, Iwana M (2009) Contents of agonistic and antagonistic opioid peptides in different cheese varieties. Int. dairy j. 19: 258-263.

[28] Durrieu C, Degraeve P, Chappaz S, Martial-Gros A (2006) Immunomodulating effects of water-soluble extracts of traditional French Alps cheeses on a human T-lymphocyte cell line. Int. dairy j. 16: 1505-1514.

[29] Saito T, Nakamura T, Kitazawa H, Kawai Y, Itoh T (2000) Isolation and structural analysis of antihypertensive peptides that exist naturally in Gouda cheese J. dairy sci. 83: 1434–1440

[30] Ong L, Shah NP (2008) Release and identification of angiotensin-converting enzyme-inhibitory peptides as influenced by ripening temperatures and probiotic adjuncts in Cheddar cheeses. LWT - Food sci. technol. 41: 1555-1566.

[31] Bütikofer U, Meyer J, Sieber, R, Wechsler D (2007) Quantification of the angiotensin-converting enzyme-inhibiting tripeptides Val-Pro-Pro and Ile-Pro-Pro in hard, semi-hard and soft cheeses. Int. dairy j. 17: 968-975.

[32] Bütikofer U, Meyer J, Sieber R, Walther B, Wechsler D (2008) Occurrence of the angiotensin-converting enzyme–inhibiting tripeptides Val-Pro-Pro and Ile-Pro-Pro in different cheese varieties of Swiss origin. J. dairy sci. 91:29–38.

[33] Meyer J, Bütikofer U, Walther B, Wechsler D, Sieber R (2009) Hot topic: Changes in angiotensin-converting enzyme inhibition and concentrations of the tripeptides Val-Pro-Pro and Ile-Pro-Pro during ripening of different Swiss cheese varieties. J. dairy sci. 92: 826-836.

[34] Nakamura Y, Yamamoto N, Sakai K, Okubo A, Yamazaki S, Takano T (1995) Purification and characterization of angiotensin I-converting enzyme inhibitors from sour milk. J. dairy sci. 78: 777-783.

[35] Pihlanto-Leppälä A, Rokka T, Korhonen H (1998) Angiotensin I converting enzyme inhibitory peptides derived from bovine milk proteins. Int. dairy j. 8:325-331.

[36] Gobbetti M, Ferranti P, Smacchi E, Goffredi F, Addeo F (2000) Production of angiotensin-I-converting-enzyme-inhibitory peptides in fermented milks started by *Lactobacillus delbrueckii* subsp. *bulgaricus* SS1 and *Lactococcus lactis* subsp. *cremoris* FT4. Appl. environ. microb. 66: 3898–3904.

[37] Fuglsang A, Rattray FP, Nilsson D, Nyborg CB (2003) Lactic acid bacteria: inhibition of angiotensin-converting enzyme in vitro and in vivo. Anton. leeuw. 83: 27–34.

[38] Pihlanto A, Virtanen T, Korhonen H (2010) Angiotensin I converting enzyme (ACE) inhibitory activity and antihypertensive effect of fermented milk. Int. dairy j. 20: 3-10.

[39] Nielsen MS, Martinussen T, Flambard B, Sørensen KI, Otte J (2009) Peptide profiles and angiotensin-I-converting enzyme inhibitory activity of fermented milk products: Effect of bacterial strain, fermentation pH, and storage time Int. dairy j. 19: 155-165.

[40] Muguerza B, Ramos M, Sánchez E, Manso MA, Miguel M, Aleixander, A, Lopez-Fandino R (2006) Antihypertensive activity of milk fermented by *Enterococcus faecalis* strains isolated from raw milk. Int. dairy j. 16: 61–69.

[41] Rodríguez-Figueroa JC, Reyes-Díaz R, González-Córdova AF, Troncoso-Rojas R, Vargas-Arispuro I, Vallejo-Cordoba B (2010) Angiotensin-converting enzyme inhibitory activity of milk fermented by wild and industrial *Lactococcus lactis* strains. J. dairy sci. 93:5032–5038.

[42] Yamamoto N, Maeno M, Takano T (1999) Purification and characterization of an antihypertensive peptide from a yogurt-like product fermented by *Lactobacillus helveticus* CPN4. J. dairy sci. 82: 1388-1393.

[43] Ashar MN, Chand R (2004) Antihypertensive peptides purified form milks fermented with *Lactobacillus belbrueckii* ssp. *bulgaricus*. Milchwissenschaft 59:14-17.

[44] Quiros A, Ramos M, Muguerza B, Delgado MA, Miguel M, Aleixandre A, Recio I (2007) Identification of novel antihypertensive peptides in milk fermented with *Enterococcus faecalis.* Int. dairy j. 17: 33–41.

[45] Kunji ER, Mierau I, Hagting A, Poolman B, Koning, WN (1996) The proteolytic systems of lactic acid bacteria. Anton. leeuw. 70: 187-221.

[46] Yamamoto N, Ishida Y, Kawakami N, Yada H (1991) *Lactobacillus helveticus* bacterium having high capability of producing tripeptide, fermented milk product, and process for preparing the same. EU Patent, 1016709A1.

[47] Yamamoto N, Akino A, Takano T (1993) Purification and specificity of a cell-wall associated proteinase from *Lactobacillus helveticus* CP790. J. biochem, 114: 740-745.

[48] Ueno K, Mizuno S, Yamamoto N (2004) Purification and characterization of an endopeptidase has an important role in the carboxyl terminal processing of antihypertensive peptides in *Lactobacillus helveticus* CM4. Lett. appl. microbiol. 39: 313-318

[49] Kilpi E, Kahala M, Steele JM, Pihlanto A, Joutsjoki V (2007) Angiotensin I-converting enzyme inhibitory activity in milk fermented by wild-type and peptidase-deletion derivatives of *Lactobacillus helveticus* CNRZ32. Int. dairy j. 17: 976-984.

[50] Kwon DY, Daily JW, Kim HJ, Park S (2010) Antidiabetic effects of fermented soybean products on type 2 diabetes. Nutr. res. 30:1–13.

[51] Shin ZI, Yu R, Park SA, Chung DK, Ahn CW, Nam HS, Kim KS, Lee HJ (2001) His-His-Leu, an angiotensin I converting enzyme inhibitory peptide derived from Korean soybean paste, exerts antihypertensive activity in vivo. J. agric. food chem. 49: 3004-3009.

[52] Hu Y, Stromeck A, Loponen J, Lopes-Lutz D, Schieber A, Gänzle MG (2011) LC-MS/MS quantification of bioactive angiotensin I-converting enzyme inhibitory peptides in rye malt sourdoughs. J. agric. food chem. 59: 11983-11989.

[53] Pihlanto A, Johansson T, Mäkinen S (2012) Inhibition of angiotensin I-converting enzyme and lipid peroxidation by fermented rapeseed and flaxseed meal. Eng. life sci. 12 DOI: 10.1002/elsc.201100137

[54] Lin MY, Yen CL (1999) Antioxidative ability of lactic acid bacteria J. agric. food chem. 47: 1460-1466.

[55] Ou CC, Lu TM, Tsai JJ, Yen, JH, Chen, HW, Lin MY (2009) Antioxidative effect of lactic acid bacteria: Intact cells vs. intracellular extracts. J. food drug anal. 17: 209-216.

[56] Kudoh Y, Matsuda S, Igoshi K, Oki T (2001) Antioxidative peptide from milk fermented with *Lactobacillus delbrueckii* subsp. *bulgaricus* IFO13953. Nippon Shokuhin Kagaku Kaishi 48: 44-55.

[57] Hernández-Ledesma B, Miralles B, Amigo L, Ramos M, Recio, I (2005) Identification of antioxidant and ACE-inhibitory peptides in fermented milk. J. sci. food agric. 85: 1041-1048.

[58] Virtanen T, Pihlanto A, Akkanen S, Korhonen H (2007) Development of antioxidant activity in milk whey during fermentation with lactic acid bacteria. J. appl. microbial. 102: 106–115.

[59] Libby P (2006) Inflammation and cardiovascular disease mechanisms. Am. j. clin. nutr. 83: 456S– 460S.

[60] LeBlanc AM, Matar C, Valdéz JC, LeBlanc N, Perdigón G (2002) Immunomodulatory effects of peptidic fractions issued from milk fermented with *Lactobacillus helveticus*. J. dairy sci. 85: 2733–2742.

[61] Matar C, Nadathur SS, Bakalinsky AT, Goulet J (1997) Antimutagenic effects of milk fermented by Lactobacillus helveticus L89 and a protease-deficient derivative. J. dairy sci. 80: 1965-1970.

[62] Laffineur E, Genetet N, Leonil J (1996) Immunomodulatory activity of β-casein permeate medium fermented by lactic acid bacteria. J. dairy sci. 79: 2112–2120.

[63] Tompa G, Laine A, Pihlanto A, Korhonen H, Rogel, I, Marnila P (2011) Chemiluminescence of non-differentiated THP-1 promonocytes: developing an assay for screening anti-inflammatory milk proteins and peptides. Luminescence 26: 251-258,

[64] Matar C, Valdez JC, Medina M, Rachid M, Perdigon G (2001) Immunomodulating effects of milks fermented by *Lactobacillus helveticus* and its non-proteolytic variant. J. dairy res. 68: 601-609.

[65] Yamamoto N, Akino A, Takano T (1994) Antihypertensive effect of the peptides derived from casein by an extracellular proteinase from *Lactobacillus helveticus* CP790. J. dairy sci. 77: 917-922.

[66] Nakamura Y, Yamamoto N, Sakai K, Takano T (1995) Antihypertensive effect of sour milk and peptides isolated from it that are inhibitors to angiotensin I-converting enzyme. J. dairy sci. 78: 1253-1257.

[67] Sipola M, Finckenberg P, Korpela R, Vapaatalo H, Nurminen ML (2002) Effect of long-term intake of milk products on blood pressure in hypertensive rats. J. dairy res. 69: 103–111.

[68] Jäkälä P, Hakala A, Turpeinen A, Korpela R, Vapaatalo H (2009) Casein-derived bioactive tripeptides Ile-Pro-Pro and Val-Pro-Pro attenuate the development of hypertension and improve endothelial function in salt-loaded Goto-Kakizaki rats. J. funct. foods 1: 366–374.

[69] Jauhiainen T, Pilvi T, Cheng ZJ, Kautiainen H, Müller DN, Vapaatalo H, Korpela R, Mervaala E (2010) Milk products containing bioactive tripeptides have an antihypertensive effect in double transgenic rats (dTGR) harbouring human renin and human angiotensinogen genes. J. nutr. metab. doi:10.1155/2010/287030.

[70] Jauhiainen T, Collin M, Narva M, Cheng ZJ, Poussa T, Vapaatalo H, Korpela R (2005) Effect of long-term intake of milk peptides and minerals on blood pressure and arterial function in spontaneously hypertensive rats. Milchwissenschaft 60: 358–362.

[71] Jäkälä P, Jauhiainen T, Korpela R, Vapaatalo H (2009) Milk protein-derived bioactive tripeptides Ile-Pro-Pro and Val-Pro-Pro protect endothelial function in vitro in hypertensive rats. J. funct. foods 1: 266–273.

[72] Masuda O, Nakamura Y, Takano T (1996) Antihypertensive peptides are present in aorta after oral administration of sour milk containing these peptides to spontaneously hypertensive rats. J. nutr. 126: 3063–3068.

[73] Jäkälä P, Pere E, Lehtinen R, Turpeinen A, Korpela R, Vapaatalo H (2009) Cardiovascular activity of milk casein-derived tripeptides and plant sterols in spontaneously hypertensive rats. J. physiol. pharmacol. 60: 11–20.

[74] Miguel M, Recio I, Ramos M, Delgado MA, Aleixandre A (2006) Antihypertensive effect of peptides obtained from *Enterococcus faecalis*-fermented milk in rats J. dairy sci. 89: 3352–3359.

[75] Quiros A, Ramos M, Muguerza B, Delgado MA, Miguc M, Alcixandre A, Recio I (2007) Identification of novel antihypertensive peptides in milk fermented with Enterococcus faecalis Int. dairy j. 17: 33–41.

[76] Chen GW, Tsai JS, Pan BS (2007) Purification of angiotensin I-converting enzyme inhibitory peptides and antihypertensive effect of milk produced by protease-facilitated lactic fermentation. Int. dairy j. 17: 641–647.

[77] Shin ZI, Yu R, Park SA, Chung DK, Ahn CW, Nam HS, Kim KS, Lee HJ (2001) His-His-Leu, an angiotensin I converting enzyme inhibitory peptide derived from Korean soybean paste, exerts antihypertensive activity in vivo. J. agric. food chem. 49: 3004-3009.

[78] Sacks FS, Lichtenstein A, Van Horn L, Harris W, Kris-Etherton P, Winston M (2006) Soy protein, isoflavones, and cardiovascular health. Circulation 113: 1034-1044.

[79] Wu J, Muir AD (2008) Hypotensive and physiological effect of angiotensin converting enzyme inhibitory peptides derived from soy protein on spontaneously hypertensive rats J. agric. food chem. 56: 9899–9904.

[80] Tsai JS, Lin YS, Pan BS, Chen TJ (2006) Antihypertensive peptides and γ-aminobutyric acid from prozyme 6 facilitated lactic acid bacteria fermentation of soymilk. Process biochem. 41: 1282–1288.

[81] Inoue K, Gotou T, Kitajima H, Mizuno S, Nakazawa T, Yamamoto N (2009) Release of antihypertensive peptides in miso paste during its fermentation, by the addition of casein. J. biosci. bioeng. 108: 111–115.

[82] Nakahara T, Sano A, Yamaguchi H, Sugimoto K, Chikata H, Kinoshita E, Uchida R (2010) Antihypertensive effect of peptide-enriched soy sauce-like seasoning and identification of its angiotensin I-converting enzyme inhibitory substances, J. agric. food chem.58: 821-827.

[83] Hata Y, Yamamoto M, Ohni M, Nakajima K, Nakamura Y, Takano T (1996) A placebo-controlled study of the effect of sour milk on blood pressure in hypertensive subjects. Am. j. clin. nutr. 64: 767-771.

[84] Kajimoto O, Kurosaki T, Mizutani J, Ikeda, N, Kaneko K, Yabune M, Nakamura Y (2002) Antihypertensive effects of liquid yogurts containing 'lactotripeptides (VPP, IPP)' in mild hypertensive subjects. J. nutr. food 5: 55–66.

[85] Hirata H, Nakamura Y, Yada H, Moriguchi S, Kajimoto O, Takahashi T (2002) Clinical effects of new sour milk drink on mild or moderate hypertensive subjects. J. new. rem. clin 51: 61–69.

[86] Seppo L, Jauhiainen T, Poussa T, Korpela R. (2003) A fermented milk high in bioactive peptides has a blood pressure-lowering effect in hypertensive subjects. Am. j. clin. nutr. 77: 326-330.

[87] Tuomilehto J, Lindstrom J, Hyyrynen J, Korpela R, Karhunen ML, Mikkola L, Jauhiainen T, Seppo L, Nissinen A (2004) Effect of ingesting sour milk fermented using Lactobacillus helveticus bacteria producing tripeptides on blood pressure in subjects with mild hypertension. J. hum. hypertens. 18: 795–802.

[88] Jauhiainen T, Vapaatalo H, Poussa T, Kyrönpalo S, Rasmussen M, Korpela R (2005) Lactobacillus helveticus fermented milk lowers blood pressure in hypertensive subjects in 24-h ambulatory blood pressure measurement. Am. j. hypertens. 18: 1600–1605.

[89] Arihara K, Kajimoto O, Hirata H, Takahashi R, Nakamura Y (2005) Effect of powdered fermented milk with Lactobacillus helveticus on subjects with high-normal blood pressure or mild hypertension. J. am. coll. nutr. 24: 257–265.

[90] Kajimoto O, Aihara K, Hirata H, Takahashi R, Nakamura Y (2001) Hypotensive effects of the tablets containing lactotripeptides (VPP, IPP). J. nutr. food 4: 51–61.

[91] Itakura H, Ikemoto S, Terada S, Kondo K (2001) The effect of sour milk on blood pressure in untreated hypertensive and normotensive subjects. J. jap. soc. clin. nutr. 23: 26–31.

[92] Kajimoto O, Nakamura Y, Yada H, Moriguchi S, Hirata H, Takahashi T (2001) Hypotensive effects of sour milk in subjects with mild or moderate hypertension. J. jpn. soc. nutr. food sci. 54: 347–354.

[93] Yasuda K, Aihara K, Komazaki K, Mochii M, Nakamura Y (2001) Effect of large high intake of tablets containing 'lactotripeptides (VPP, IPP)' on blood pressure, pulse rate and clinical parameters in healthy volunteers. J. nutr. food 4: 63–72.

[94] Law MR, Wald NJ, Morris JK, Jordan JE (2003) Value of low dose combination treatment with blood pressure lowering drugs: analysis of 354 randomised trials. Br. med. j. 326: 1427–1431.

[95] Pripp AH (2008) Effect of peptides derived from food proteins on blood pressure: a meta-analysis of randomized controlled trials. Food nutr. res. 5: 1–9.

[96] Xu JY, Qin LQ, Wang PY, Li W, Chang C (2008) Effect of milk tripeptides on blood pressure: a meta-analysis of randomized controlled trials. Nutrition 24: 933–940.

[97] Engberink MF, Schouten EG, Kok FJ, van Mierlo LA, Brouwer IA, Geleijnse JM (2009) Lactotripeptides show no effect on human blood pressure: results from a double-blind randomized controlled trial. Hypertension 51:399-405.

[98] Usinger L, Jensen LT, Flambard B, Linneberg A, Ibsen H (2010) The antihypertensive effect of fermented milk in individuals with prehypertension or borderline hypertension. J.hum. hypertens. 24: 678–683

[99] Cicero AFG, Gerocarni B, Laghi L, Borghi C (2011) Blood pressure lowering effect of lactotripeptides assumed as functional foods: a meta-analysis of current available clinical trials. J. hum. hypertens. 25: 425–436.

[100] Vaidyanathan S, Jermany J, Yeh C, Bizot MN, Camisasca R (2006) Aliskiren, a novel orally effective renin inhibitor, exhibits similarpharmacokinetics and pharmacodynamics in Japanese and Caucasian subjects. Br. j. clin. pharmacol. 62: 690–698.

[101] Arsenault J, Lehoux J, , Lanthier L Cabana J, Guillemette G, Lavigne P, Leduc R, Escher E (2010) A single-nucleotide polymorphism of alanine to threonine at position 163 of the human angiotensin II type 1 receptor impairs losartan affinity. Pharmacogenet Genomics 20: 377–388.

[102] Boelsma E, Kloek J (2010) IPP -rich milk protein hydrolysate lowers blood pressure in subjects with stage 1 hypertension, a randomized controlled trial. J. nutr. 9:52 doi:10.1186/1475-2891-9-52.

[103] Hernández-Ledesma B, Amigo L, Ramos M, Recio I (2004) Application of high-performance liquid chromatography–tandem mass spectrometry to the identification of biologically active peptides produced by milk fermentation and simulated gastrointestinal digestion. J. chromatogr. A 1049: 107–114.

[104] Mäkinen S, Johansson T, Vegarud G, Pihlava JM, Pihlanto A (2012) Angiotensin I-converting enzyme inhibitory and antioxidant properties of rapeseed hydrolysates. J. funct. foods (in press)

[105] Matsui T, Li CH, Osajima Y (1999) Preparation and characterization of novel bioactive peptides responsible for angiotensin I-converting enzyme inhibition from wheat germ. J. pept. sci. 5: 289–297.

[106] Vermeirssen V, Van Camp J, Verstraete W (2004) Bioavailability of angiotensin I converting enzyme inhibitory peptides. Br. j. nutr. 92: 357–366.

[107] Maeno M, Yamamoto N, Takano T (1996) Identification of an antihypertensive peptide from casein hydrolysate produced by a proteinase from Lactobacillus helveticus CP790. J. dairy sci. 79: 1316–1321.

[108] Eriksen EK, Holm H, Jensen E, Aaboe R, Devold TG, Jacobsen M, Vegarud GE (2010) Different digestion of caprine whey proteins by human and porcine gastrointestinal enzymes. Br. j. nutr. 104:374-381.

[109] Pauletti GM, Gangwar S, Knipp GT, Nerurkar MM, Okumu FW, Tamura T, Siahaan TJ (1996) Structural requirements for intestinal absorption of peptide drugs. J. control release. 41:3–17.

[110] Daniel H (2004) Molecular and integrative physiology of intestinal peptide transport. Annu. rev. physiol. 66:361–384.

[111] Foltz M, Cerstiaens A, van Meensel A, Mols R, van der Pijl PC, Duchateau GSMJE, Augustijns P (2008) The angiotensin converting enzyme inhibitory tripeptides Ile-Pro-Pro and Val-Pro-Pro show increasing permeabilities with increasing physiological relevance of absorption models. Peptides 29: 1312–1320.

[112] Satake M, Enjoh M, Nakamura Y, Takano T, Kawamura Y, Arai S, Shimizu M (2002) Transepithelial transport of the bioactive tripeptide Val-Pro-Pro, in human intestinal Caco-2 cell monolayers. Biosci. biotech. bioch. 66: 378–384.

[113] Quiros A, Davalos A, Lasuncion MA, Ramos M, Recio I (2008) Bioavailability of the antihypertensive peptide LHLPLP: Transepithelial flux of HLPLP Int. dairy j. 18: 279-286.

[114] Deacon CF, Nauck MA, Toft-Nielsen M, Pridal L, Willms B, Holst JJ (1995) Both subcutaneously and intravenously administered glucagon-like peptide I are rapidly degraded from the NH2-terminus in type II diabetic patients and in healthy subjects. Diabetes. 44:1126–1131.

[115] Jauhiainen T, Wuolle K, Vapaatalo H, Kerojoki O, Nurmela K, Lowrie C, Korpela R (2007) Oral absorption, tissue distribution and excretion of a radiolabelled analog of a milk-derived antihypertensive peptide, Ile-Pro-Pro, in rats. Int. dairy j. 17:1216–1223.

[116] van der Pijl PC, Kies AK, Ten Have GA, Duchateau GS, Deutz NE (2008) Pharmacokinetics of proline-rich tripeptides in the pig. Peptides. 29:2196–2202.

[117] Foltz M, Meynen EE, Bianco V, van Platerink C, Koning TMMG, Kloek J (2007) Angiotensin converting enzyme inhibitory peptides from a lactotripeptide-enriched milk beverage are absorbed intact into the circulation. J. nutr. 137:953–958.

[118] Shaji J, Patole V (2008) Protein and peptide drug delivery: Oral approaches. J. pharm. sci. 70:269-277.

[119] Kies AK, Van Der Pijl P (2012) Peptide availability USA Patent Application 20120040895.

[120] Ko WC, Cheng ML, Hsu KC, Hwang YS (2006) Absorption-enhancing treatments for antihypertensive activity of oligopeptides from tuna cooking juice: In vivo evaluation in spontaneously hypertensive rats. J. food sci. 71:13-17.

[121] Fujita H, Sasaki R, Kurahashi K, Yoshikawa M (1995) Potentiation of the antihypertensive activity of orally administered ovokinin, a vasorelaxing peptide derived from ovalbumin, by emulsification in egg phosphatidyl-choline. Biosci. biotech. bioch.59:2344–2345.

Section 5

# Livestock Feed

Chapter 14

# Lactic Acid Bacteria in Tropical Grass Silages

Edson Mauro Santos, Thiago Carvalho da Silva,
Carlos Henrique Oliveira Macedo and Fleming Sena Campos

Additional information is available at the end of the chapter

http://dx.doi.org/10.5772/50703

## 1. Introduction

Lactic Acid Bacteria (LAB) have applications in many industrial areas and play an important role in the preservation process of moist forages for animal feeding (silage).

The basic principle silage is to store the surplus forage keeping its stability and nutritional value until it is required to feed the animals. This process takes place in anaerobic conditions, where the lactic acid produced by the LAB inhibits the proliferation of spoilage microorganisms, which are less tolerant to acidic conditions. Thus, as the pH values decline, the silage losses decline as well due to the greater conversion of plant soluble carbohydrates (the main substrate for LAB) in lactic acid, with 96.9% rate of energy recovery (Mc Donald et al., 1991). The major soluble carbohydrates present in forage crops are fructose, glucose, sucrose and frutosanas, according to Woolford, (1984), sucrose and frutosanas are rapidly hydrolyzed in their monomers at forage harvest.

Lactic fermentation produces lactic acid as the main product. Therefore, homofermentative bacteria such as *Lactobacillus plantarum* are desirable in the silage fermentation process, once 87% of their metabolites become lactic acid. On the other hand, in the heterofermentative process, additional substances like ethanol, acetate and $CO_2$, are formed. Microbial inoculants used as additives include homofermentative LAB, heterofermentative LAB, or both combined. The specificity between the forage specie and its epiphytic micro flora implicates the need for studies related with isolation and identification of the main microorganism groups present in the forage used for silage.

In this chapter we will discuss the characteristics of tropical grasses, the main LAB species found in these grasses and how the LAB's are used to improve the quality of tropical grass' silages.

## 2. Tropical grass characteristics

The forage characteristics that contribute to a good fermentation are: dry matter content, autochthonous plant microbiota and, most importantly, the quantity of soluble carbohydrates. Corn and sorghum are the most appropriate grasses to make silages due to their high soluble carbohydrate contents and dry matter production. However, some studies have shown that different grasses can be utilized if they are ensilage at the right developmental stage or if appropriate additives are used (Zanine et al., 2010).

The decline in pH values inhibit the spoilage microorganism proliferation, which allows the silage nutritive values to be preserved. Thus, the best silage forages are the ones with high soluble carbohydrates contents, which should be sufficient to promote the fermentation and produce enough acid to preserve the silage. According to Ferreira (2002), the minimum soluble carbohydrates contents recommended to ensure adequate fermentation of good silage, varies between 6% and 12% of the dry mass. McDonald et al. (1991) found that, since the soluble sugar level is adequate, dry mass contents higher than 25% are sufficient to ensure a good silage production. The buffering capacity is another factor affecting the silage final product. It reflects the capacity to resist change in the pH values, determined by buffering substances, represented in plants by inorganic bases such as potassium (K) and calcium (Ca), protein, ammonia ($N\text{-}NH_3$), organic salts (malate, citrate).

Several factors affect the fermentation pattern and consequently the silage quality, including dry matter content, amount of soluble carbohydrates readily available and initial LAB population (Pereira et al., 2006). These inherent plant characteristics may vary according to species and maturity stage. Corn (*Zea mays* L.) and sorghum (*Sorghum bicolor* L. Moench), followed by millet (*Pennisetum glaucum*) and sunflower (*Helianthus annuus*) seems to be the most adapted species for silage due to the high soluble carbohydrates content, low buffering capacity, satisfactory dry matter productivity and quality of the silage produced. Although, sorghum silage nutritional value is considered lower than that of corn, it has shown an important role in forage production in Brazil and in the world as well, standing out as a resistant species to adverse environmental factors, such as drought stress (Miranda et al., 2010). This grass provides silage at low costs and the plant regrowth can be used (Rezende et al., 2011), because they keep the root system active.

As corn and sorghum have ideal characteristics for silage, a factor that drew the researcher's attention was the ideal harvest moment, considering the maturity stage and silage quality. Faria Júnior et al. (2011), working with the effect of seven grain maturity stages on the quality of sorghum BRS 610 silage, observed that the most appropriate stage for ensiling is the milk and soft dough stages, due to its higher silage fermentation quality and nutritional value.

Pearl millet silage presents high crude protein content as an intrinsic characteristic, when compared with corn and sorghum silage. Crude protein values varying from 8.51% to 10.68% were observed by Amaral et al. (2008). The storage system efficiency must not be defined only by the silage nutritional value, but also include the losses that occur from the plant harvest to the animal feeding (Neumann et al., 2007).

Sugarcane (*Saccharum officinarum* L.) is an important grass due to its tolerance to drought periods and high production potential of dry matter and soluble carbohydrates per hectare. The sugarcane silage confection has been unusual, being used more for animal feeding in its natural form, after cutting and chopping, but it can be recommended when desires to store the sugarcane in its higher nutritional value stage (the dry season) for use throughout the year (Molina et al., 2002). However, according to Santos et al. (2006), sugar cane silage becomes justifiable only when there is a surplus or when accidental burning of sugar cane fields happen, always taking into account the difficulty of achieving a good fermentation pattern due to intense alcoholic fermentation (8% to 17% of dry matter of ethanol) caused by yeast (Kung Jr. & Stanley, 1982), leading to losses of up to 30% of dry matter (Ferreira et al., 2007), accumulation of cell wall components and reduction in the *in vitro* dry matter digestibility. Furthermore, sugar cane silage has low aerobic stability, as result of high residual carbohydrate and lactic acid contents (McDonald et al., 1991). On the other hand, the adoption of the silage method represents a chance to keep the sugarcane nutritional value and allows better logistics for their manufacture and use, what implies the hand labor rationalization, concentrating the sugar cane harvest process in a particular time of year or time period, resulting in easier daily farm handling and maximizing the machinery use.

Thus, there has been a growing number of research projects, especially in Brazil, seeking additives that inhibit yeast growth in sugar cane silages (Valeriano et al., 2009). Nevertheless, some studies have shown that grasses can also be stored if they are ensiled at the ideal stage of development, or if the suitable additives are applied (Zanine et al., 2010).

Tropical weather grasses have high production in favorable seasons and a sharp decline in the less favorable ones. In this context, the surplus silage can be an option to increase the dry matter supply to the animals in unfavorable times. Such examples of tropical forages with a potential for silage are: *Brachiaria brizantha* (cv. Marandu), *Brachiaria decumbens* (cv. Basilisk), *Brachiaria humidicula*, *Panicum maximum* Jacq. (Cv. Colonião, Tobiatã, Tanzânia, Mombaça, Vencedor, Centauro, Massai), *Pennisetum purpureum* Schum. (Cv. Napier, Taiwan, Merker, Porto Rico, Cameroon, Mott), *Cynodon dactylon* (Tifton) and the hybrid of *Cynodon dactylon* x *C. nlemfuensis* (Coastcross). (Patrizi et al., 2004; Santos et al., 2006; Ribeiro et al., 2008; Oliveira et al., 2007; Zopollatto et al., 2009; Lopes & Evangelista, 2010). When compared to the others, elephant grass stands out in silages researches because of present high productivity and higher soluble carbohydrates concentration.

According to Evangelista et al. (2004), the tropical grasses present low dry matter contents, high buffering capacity and low soluble carbohydrates in growth stages in which they present good nutritive values, endangering the conservation through ensilage, once secondary fermentations are possible to occur. Bacteria from the *Clostridium* genus are favored by humid environments with high pH values and temperature. These bacteria are responsible for large losses because they produce $CO_2$ and butyric acid instead of lactic acid.

The grasses are colonized by a large number of LAB. In the most of the cases different species occur simultaneously in the same culture (Daeschel et al., 1987). According to Pahlow et al. (2003), in literature review studies, the species more commonly found in plants are *Lactobacillus plantarum, Lactobacillus casei, Pediococcus acidilactici, Enterococcus faecium.* Some heterofermentative lactic bacteria species can also be found in plants.

The lactic acid bacteria from the autochthonous microbiota are essential for the silage fermentation. However, no bacteria group varies as much as this one regarding number, with a detection limit of $10^1$ to $10^5$ CFU $g^{-1}$ in alfalfa forage, $10^6$ in perennial grasses and $10^7$ in corn and sorghum (Pahlow et al., 2003).

The Table 1 shows contents of dry matter, crude protein, soluble carbohydrates and LAB number of mombaça grass (*Panicum maximum*) and *Brachiaria decumbens* with different regrowth ages. It is observed that in none of regrowth ages, neither grass showed dry matter content exceeding 30% and only the grasses cuted over 50 days after regrowth presented LAB population greater than 5 log CFU/g. On the other hand, there is a sharp drop in crude protein content with increasing regrowth age.

| Signal grass (Brachiaria decumbens.) | | | | |
|---|---|---|---|---|
| AGE (days) | DM (%) | CP (%) | SC (%) | LAB (log CFU/g) |
| 30 | 20.99 | 9.65 | 2.62 | 3.93 |
| 40 | 21.23 | 6.97 | 2.92 | 4.81 |
| 50 | 21.94 | 5.86 | 3.13 | 5.37 |
| 60 | 22.35 | 5.30 | 2.73 | 5.32 |
| 70 | 23.67 | 4.37 | 2.53 | 5.51 |
| Mombaça grass(Panicum maximum Jacq. cv. Mombaça) | | | | |
| AGE (days) | DM (%) | CP (%) | SC (%) | LAB (log CFU/g) |
| 30 | 17.75 | 7.43 | 3.34 | 4.35 |
| 40 | 19.63 | 7.30 | 4.12 | 4.56 |
| 50 | 21.50 | 6.47 | 4.18 | 5.16 |
| 60 | 23.38 | 4.94 | 5.43 | 5.55 |

**Table 1.** Dry matter (DM), crude protein (CP) and soluble carbohydrates (SC) and number of lactic acid bacteria (LAB) in signal grass and mombaça grass silage with different regrowth ages (Sousa et al., 2006).

Santos et al. (2011) studying the regrowth age influence in the LAB population observed that silages made with older plants presented LAB populations higher than the silages made with younger plants. According to Knicky (2005), it can be attributed to the increase in soluble carbohydrates and dry matter content, as well as to the decrease of anionic substances such as salts of organic acids, nitrate, sulfates, and so on. Pereira et al. (2005) found an increase in LAB population in elephant grass with the increase in regrowth age.

Meeske et al. (1999) found population of approximately 1 log CFU/g of fresh forage in *Digitaria eriantha*. Cai et al. (1998), analyzing Guinea grass (*Panicum maximum*) indigenous microbiota, found values lower than 3 log CFU/g of fresh forage. Pereira et al. (2007) reported initial LAB population of 4.92 log CFU/g in elephant grass plants.

Table 2 presents a data compilation of chemical composition and other parameters considered determinants of tropical grass silages quality, such as buffering capacity, soluble carbohydrates and pH values.

| | Corn | Sorghum | Pearl millet | Sugar Cane | Elephant grass | Buffel grass | *Brachiaria brizantha* | *Brachiaria decumbens* |
|---|---|---|---|---|---|---|---|---|
| n* | 6 | 6 | 6 | 7 | 5 | 4 | 6 | 6 |
| DM | 30.68 | 30.20 | 31.21 | 25.25 | 20.75 | 37.15 | 38.36 | 30.9 |
| OM | 96.91 | 92.79 | 90.9 | 97.45 | 90.91 | 90.60 | 92.89 | 92.25 |
| CP | 7.22 | 8.04 | 11.09 | 2.80 | 7.81 | 5.03 | 9.67 | 7.01 |
| MM | 5.81 | 4.45 | 9.1 | 2.68 | 9.53 | 9.92 | 5.29 | 7.53 |
| EE | 2.16 | - | - | 0.82 | 3.33 | 1.8 | 1.16 | 2.51 |
| NDF | 50.32 | 61.36 | 60.64 | 46.88 | 72.44 | 73.94 | 70.05 | 75.47 |
| ADF | 26.57 | 37.27 | 35.68 | 28.24 | 44.11 | 50.60 | 38.64 | 38.26 |
| NFC | 32.49 | - | - | 44.21 | 9.99 | 14.05 | 8.74 | 14.12 |
| LIGNIN | 4.72 | 6.2 | 4.24 | 4.72 | 6.24 | 8.4 | 4.67 | 5.9 |
| IVDMD | 59.19 | 52.87 | - | 53.87 | 60.90 | 37.4 | 58.77 | 51.61 |
| pH | 5.60 | 5.93 | 3.62 | 4.76 | 5.6 | - | - | - |
| N-NH3 | 0.785 | - | 1.28 | 1.20 | - | - | - | - |
| ETHANOL | - | - | - | 2.12 | - | - | - | - |
| YEASTS | 5.30 | - | - | 2.71 | - | - | - | - |
| BUFFERING CAPACITY | - | 19.98 | - | 10.80 | - | - | - | - |
| STARCH | 21.31 | - | - | 5.50 | - | - | - | - |

**Table 2.** Chemical characterization of tropical grass used for silage. *Number of researches; DM = dry matter (%); OM = organic matter (%); CP = crude protein (%); EE = ether extract (%); NDF = neutral detergent fiber (%); NFC = non-fibrous carbohydrates (%); IVDMD = *in vitro* dry matter digestibility (%); $N-NH_3$= ammonia nitrogen (% TN); ADF = acid detergent fiber (%); MM = mineral matter (%).

(Pariz, C.M. et al., 2011; Silva, T.C. et al., 2011; Viana, M.C.M. et al., 2011; Hu, W. et al., 2009; Martinez , J.C. et al., 2009; Valeriano, A.R, 2009; Benett, C.G.S. 2008; Reis, J.A.G. et al., 2008; Ribeiro, J.L. et al., 2008; Moreira, J.N. et al., 2007; Pedroso, A.F. et al., 2007; Velho, J.P. et al., 2007; Valadares Filho, S.C. et al., 2006; Velho, J.P. et al., 2006; Kollet, J.L. et al., 2006; Aroeira, L.J.M. et al., 2005; Bernardino, F.S. 2005; Moraes, E.H.B.K. et al., 2005; Santos, G.R.A. et al., 2005; Silva, A.V. et al., 2005; Patrizi, W.L. et al., 2004; Dairy, J. et al., 2003; Santos, M.V.F. et al., 2003; Landell, M.G.A. et al., 2002; Neumann, M. et al., 2002; Rodrigues, P.H.M. et al., 2002).

It is observed that tropical grasses have characteristics influenced by several factors, ranging from species choice to maturity stage at harvest. These factors are primordial in silage confection, because if handled properly, they will favor the LAB development, resulting in higher quality silage.

To understand how the factors related to the grass management will influence the LAB population dynamics consequently the fermentation, it is necessary to know the characteristics related to metabolism and the main tropical grass species.

## 3. Characteristics of lactic acid bacteria (LAB) present in tropical grasses

Lactic acid bacteria are gram-positive. They are negative catalase, do not present motility and do not produce spores. The final fermentation product is lactic acid, however, some groups produce considerable amount of $CO_2$, ethanol and other metabolites, these being called heterofermentative. Particularly, *Lactobacillus plantarum* are the larger silage fermentative bacteria (Ohmomo et al., 2002). *Lactococcus, Streptococcus* and *Enterococcus* are very important in the fermentation initial stage, because they keep an acidic environment, which then becomes, predominantly colonized by Lactobacillus.

Fermentation can be considered the anaerobic decomposition of organic compounds to organic products, which may be metabolized by the cells without the oxygen intervention. Under anaerobiosis conditions, phosphorylation occurs at the substrate level in which an organic acid donates electrons to a $NAD^+$, so that in microorganisms the $NAD^+$ needs to be regenerated and it occurs through various oxidation-reduction pathways, involving pyruvate or its derivatives, like acetyl-CoA. Pyruvate is a key molecule of fermenting microorganisms, from that, it can be formed by several compounds such as: acetaldehyde (ethanol), acetyl-CoA, lactate, acetoacetate (butyrate, isopropanol), acetoin (2, 3-butanediol, diacetyl), acetate, oxaloacetate, succinate, and propionate.

The homofermentative LAB are characterized by a faster fermentation rate, reduced proteolysis, higher lactic acid concentrations, lower acetic and butyric acids contents, lower ethanol content, and higher energy and dry matter recovery. Heterofermentative bacteria utilize pentoses as substrate for acetic and propionic acids production, which are effective at controlling fungi, at low pH values. The facultative heterofermentative use the same hexoses pathway of homofermentative, but they are able to ferment pentoses, as they have aldolase and fosfocetolase enzymes. The facultative heterofermentative may produce lactic and acetic acids when the substrate is a pentose, or lactic acid, ethanol and $CO_2$ when hexose is the substrate, due to the need of oxidation of two NAD molecules produced in the glycolytic pathway (White, 2000).

Table 3 summarizes the main lactic acid bacteria found in silages including some *Lactobacillus* with heterofermentative metabolism and some *Leuconostoc* species which have heterofermentative metabolism also.

For species of *Lactobacillus* genus were defined three groups based on the presence or absence of aldolase and fosfocetolase enzymes (Kandler and Weiss, 1986). These groups are as follows:

| *Lactobacillus* | | *Enterococcus* | *Leuconostoc* | *Pediococcus* |
|---|---|---|---|---|
| *L. plantarum* | *L. brevis* | *E. faecalis* | *L. dextranicum* | *P. acidilactici* |
| *L. casei* | *L. buchneri* | *E. faecium* | *L. citrovorum* | *P. pentosaceus* |
| *L. curvatus* | *L. fermentum* | *E. lactis* | *L. mesenteroides* | *P. cerevisae* |
| *L. acidophilus* | *L. viridescens* | | | |

**Table 3.** Main lactic acid bacteria found in silages. (Woolford, 1984)

**Group 1:** Homofermentative, which ferment hexoses homolacticly almost exclusively to lactic acid (>85%), however, they are unable to ferment pentoses, due to the fosfocetolase enzyme lack;

| Homofermentative *Lactobacillus* | |
|---|---|
| 1A. *Lactobacillus delbrueckii* subsp. *Delbrueckii* | 9. *L. helveticus* |
| 1B. *Lactobacillus delbrueckii* subsp. *lactis* | 10. *L. jensenii* |
| 1C. *Lactobacillus delbrueckii* subsp. *bulgaricus* | 11. *L. ruminis* |
| 2. *L. acidophilus* | 12. *L. salivarius* |
| 3. *L. amylophilus* | 13. *L. sharpeae* |
| 4. *L. amylovorus* | 14. *L. vitulinus* |
| 5. *L. animalis* | 15. *L. yamanashiensis* |
| 6. *L. crispatus* | |
| 7. *L. farciminis* | |
| 8. *L. gasseri* | |

**Group 2**: Facultative heterofermentative that use the same hexoses pathway as the one of group 1, but are able to ferment pentoses, since they have aldolase and fosfocetolase enzymes;

| Facultative heterofermentative *Lactobacillus* | |
|---|---|
| 16. *L. agilis* | 20b. *L. coryniformis* subsp. *Torquens* |
| 17. *L. alimentarius* | 21. *L. curvatus* |
| 18. *L. bavaricus* | 22. *L. homohiochii* |
| 19a. *L. casei* subsp. *Casei* | 23. *L. maltaromicus* |
| 19b. *L. casei* subsp. *pseudo-plantarum* | 24. *L. murinus* |
| 19c. *L. casei* subsp. *rhamnosus* | 25. *L. plantarum* |
| 19d. *L. casei* subsp. *tolerans* | 26. *L. sake* |
| 20a. *L. coryniformis* subsp. *coryniforms* | |

**Group 3**: Obligately heterofermentative, which ferment hexoses, forming lactic acid, ethanol (or acetic acid) and $CO_2$, being able to still ferment pentose to form lactic and acetic acids.

| Mandatory heterofermentative *Lactobacillus* | |
|---|---|
| 27. *L. bifermentans* | 36. *L. halotolerans* |
| 28. *L. brevis* | 37. *L. hilgardii* |
| 29. *L. buchneri* | 38. *L. kandleri* |
| 30. *L. collinoides* | 39. *L. kefir* |
| 31. *L. confusus* | 40. *L. minor* |
| 32. *L. divergens* | 41. *L. reuteri* |
| 33. *L. fermentum* | 42. *L. sanfrancisco* |
| 34. *L. fructivorans* | 43. *L. vaccinostercus* |
| 35. *L. fructosus* | 44. *L viridescens* |

The homofermentative LAB presence in silage is extremely necessary. $CO_2$ generation results in carbon loss, ie, nutrient losses in plant materials. Therefore, homofermentative bacteria such as *Lactobacillus plantarum*, are desirable in the fermentation of silage.

Several lactic acid bacteria have antimicrobial peptides known as bacteriocins which are responsible for inhibiting the growth of or related species which have similar nutritional requirements. The bacteriocins action mechanism involves interaction with specific receptors on the cell membrane to its insertion resulting in proton-motive force dissipation and pores formation, which may cause cell viability loss (Montville and Chen, 1998; Ennahar et al., 2000).

According Lücke (2000), gram-negative bacteria are less susceptible to the action of bacteriocins from lactic acid bacteria due to the presence of outer membrane, which limits the access of peptides to the target site. In addition, the gram-negative bacteria are more sensitive to organic acid produced by LAB compared with the gram-positive bacteria (Ennahar et al., 2000).

Table 4 presents the lactic acid bacteria percentages isolated from sorghum plant in a study conducted by Tjandraatmadja et al. (1991). Likewise, *Lactobacillus plantarum* was the predominant specie and it kept 100 days after ensiling. It was observed the presence of *Lactobacillus fermentum* and *Lactobacillus brevis* heterofermentative bacteria in large quantities at the end of the ensiling process. It demonstrates that these bacteria are active during the fermentation process.

Evaluating the microbiological composition of silages obtained from three different grass species, Tjandraatmadja et al. (1994) found that *Lactobacillus plantarum* and *Pediococcus spp.* are the predominant species, observing one more time the presence of significant amounts of *Lactobacillus brevis* and *Lactobacillus fermentum* (Table 5). Santos et al. (2006) observed that

*Lactobacillus plantrum* was the predominant species in mombaça grass (*Panicum maximum*) and signal grass (*Brachiaria decumbens*).

| | Days after ensiling | | | |
|---|---|---|---|---|
| Species | 0 | 4 | 8 | 100 |
| *Lactobacillus plantarum* | 35 | 84 | 87 | 44 |
| *Leuconostoc spp.* | 59 | 0 | 0 | 0 |
| *Lactobacillus fermentum* | 6 | 6 | 4 | 7 |
| *Lactobacillus brevis* | 0 | 10 | 9 | 49 |

**Table 4.** Percentage of lactic acid bacteria species isolated from sorghum silage (Tjandraatmadja et al., 1991).

| Species | Days after ensiling | | |
|---|---|---|---|
| | *P. maximum* | *D. decumbens* | *S. sphacelata* |
| *Lactobacillus plantarum* | 21 | 39 | 47 |
| *Lactobacillus coryneformis* | 6 | 21 | 0 |
| *Leuconostoc spp.* | 27 | 12 | 0 |
| *Enterococcus faeceium* | 0 | 10 | 4 |
| *Enterococcus faecalis* | 3 | 0 | 3 |
| *Pediococcus spp.* | 30 | 12 | 31 |
| *Lactobacillus brevis* | 7 | 6 | 11 |
| *Lactobacillus fermentum* | 6 | 0 | 4 |

**Table 5.** Main lactic acid bacteria (%) isolated from grasses (*Panicum maximum* cv Hami; *Digitaria decumbens; Setaria sphacelata* cv Kazungula) (Tjandraatmadja et al., 1994).

It is evident that *Lactobacillus plantarum* and the species from the *Pediococcus* genus are prevalent in forage plants. The species from *Leuconostoc* genus are present in plants. However, according to Chunjian et al. (1992) and Tjandraatmadja et al. (1991) they disappear early in the ensiling process.

According Lücke (2000), gram-negative bacteria are less susceptible to the action of bacteriocins from lactic acid bacteria due to the presence of outer membrane, which limits the access of peptides to the target site. In addition, the gram-negative bacteria are more sensitive to organic acid produced by LAB compared with the gram-positive bacteria (Ennahar et al., 2000).

Santos et al. (2011) conducted a study aiming to characterize and quantify microbial populations in signal grass harvested at different regrowth ages. The six lactic acid bacteria strains isolated from signal grass were characterized according Gram staining, catalase enzyme reaction, and bacilli form, submitted to growth and identification tests. The microbial isolates identification was performed by carbohydrates fermentation in API 50 CH kit (BioMéurix - France).

Regarding the predominant bacteria identification in signal grass plants, it is observed in Table 6 that all isolates had the form of short bacilli with rounded ends, arranged in pairs or in short chains (3-4 cells). All of them showed negative reaction to the catalase enzyme test and were gram-positive. None of the strains grew at pH 9.6 and 6.5% NaCl, but all grew at pH 7.2 and 4% NaCl at 45°C.

| | Isolated strain | | | | | | |
|---|---|---|---|---|---|---|---|
| | EB1 | EB2 | EB3 | EB4 | EB5 | EB6 | *Lactobacillus plantarum* |
| Test | | | | | | | |
| form | bacillus | bacillus | bacillus | bacillus | bacillus | bacillus | bacillus |
| Arranjement | DB* | DB | DB | DB | DB | DB | DB |
| Gram | + | + | + | + | + | + | + |
| Catalasis | - | - | - | - | - | - | - |
| | Growth at different pH | | | | | | |
| 7,2 | + | + | + | + | + | + | + |
| 9,6 | - | - | - | - | - | - | - |
| | Growth at different salt concentartion (NaCl) | | | | | | |
| NaCl 4% | + | + | + | + | + | + | + |
| NaCl 6,5% | - | - | - | - | - | - | - |
| | Growth at different temperatures (T °C) | | | | | | |
| 15 °C | + | + | + | + | + | + | + |
| 45 °C | + | + | + | + | + | + | + |

**Table 6.** Morphology and biochemical characteristics of the isolates EB1, EB2, EB3, EB4, EB5, EB6, signal grass plant (*Brachiaria decumbens* cv. Basiliski). *DB: diplobacillus. (Santos et al., 2011).

According with the carbohydrate fermentation pattern (Table 7), the isolates EB1, EB2, EB5 e EB6 were identified as *Lactobacillus plantarum* with 99.9% of similarity.

The *Lactobacillus plantarum* specie, identified as dominant in signal grass plants (*Brachiaria decumbens* cv. Basiliski) (Santos et al., 2011) has been isolated and characterized as major species in several cultures. Lin et al. (1992) evaluated the corn and alfalfa autochthonous microbiota and found that from the total lactic acid bacteria isolated, over 90% were homofermentative lactic bacteria, being *Lactobacillus plantarum* the predominant specie. Tjandraatmadja et al. (1994), in studies on tropical grasses silage, found *Lactobacillus plantarum* and *Pediococcus spp.* as the predominant species.

| | Isolated strain | | | | Lactobacillus plantarum |
|---|---|---|---|---|---|
| | EB1 | EB2 | EB5 | EB6 | |
| Glycerol | - | - | - | - | - |
| Erythritol | (+) | (+) | (+) | (+) | - |
| D-arabinose | - | - | - | - | - |
| L-arabinose | + | + | + | + | + |
| Ribose | + | + | + | + | + |
| D-xylose | - | - | - | - | - |
| L-xylose | - | - | - | - | - |
| Adonitol | - | - | - | - | - |
| β-methyl D-xyloside | - | - | - | - | - |
| Galactose | + | + | + | + | + |
| D-glucose | + | + | + | + | + |
| D-frutose | + | + | + | + | + |
| D-mannose | + | + | + | + | + |
| L-sorbose | - | - | - | + | - |
| Rhamnose | (+) | (+) | (+) | (+) | - |
| Dulcitol | - | - | - | - | - |
| Inositol | - | - | - | - | - |
| Mannitol | + | + | + | + | + |
| Sorbitol | + | + | + | + | + |
| α-methyl D-mannose | - | - | - | - | + |
| α-methyl D-glycoside | - | - | - | - | - |
| N-acetyl-glucosamine | + | + | + | + | + |
| Amygdaline | + | + | + | + | + |
| Arbulin | + | + | + | + | + |
| Esculin | + | + | + | + | + |
| Salicin | + | + | + | + | + |
| Cellobiose | + | + | + | + | + |
| Maltose | + | + | + | + | + |
| Lactose | + | + | + | + | + |
| Melibiose | + | + | + | + | + |
| Saccharose | + | + | + | + | + |
| Trehalose | + | + | + | + | + |
| Inulin | - | - | - | - | - |
| Melezitose | + | + | + | + | + |
| D-raffinose | + | + | + | + | + |
| Amidon | - | - | - | - | - |
| Glycogene | - | - | - | - | - |
| Xylitol | - | - | | - | - |
| β-gentibiose | + | + | + | + | + |
| D-turanose | + | + | + | + | + |

| | Isolated strain | | | | Lactobacillus |
|---|---|---|---|---|---|
| | EB1 | EB2 | EB5 | EB6 | plantarum |
| L-lyxose | - | - | - | - | - |
| D-tagatose | - | - | - | - | - |
| D-fucose | - | - | - | - | - |
| L-fucose | - | - | - | - | - |
| D-arabitol | (+) | (+) | (+) | (+) | - |
| L-arabitol | - | - | - | - | - |
| Gluconate | + | + | + | + | + |
| 2 Cetogluconate | - | - | - | - | - |
| 5 Cetogluconate | - | - | - | - | - |

**Table 7.** Carbohydrate fermentation pattern of the isolates EB1, EB2, EB5, and EB6, signal grass plants (*Brachiaria decumbens* cv. Basiliski). + Intense fermentation, - no fermentation; (+) less intense fermentation (Santos et al., 2011).

In another study, Rocha (2003), evaluating the lactic acid bacteria populations in elephant grass plants cv. Cameroon (*Pennisetum purpureum* Schum) identified the isolates as *Lactobacillus casei ssp. Pseudoplantarum,* using the carbohydrate fermentation profile as an identification criterion. Santos et al. (2011) observed the *Lactobacillus plantarum* as LAB predominant specie in signal grass (*Brachiaria decumbens* Stapf). Based on the reported above, it is observed that there were differences between the LAB dominant species among the cultures evaluated, however *Lactobacillus plantarum* has been identified as the predominant specie for most plants.

## 4. Lactic acid bacteria and their effects on silage fermentation

A suitable acidification is essential for the silage successful preservation, especially when the crop moisture is relatively high, condition which favors the proliferation of spoilage microorganisms. The acidity prevents the development of spoilage microorganisms because they are less tolerant to the acidic conditions than lactic acid bacteria (Woolford, 1984; McDonald et al., 1991).

Among the fermentation stages, aerobic remains during the filling and some hours after the silage closing. The growth of aerobic microorganisms such as yeasts, fungi and bacteria, favored by high concentrations of oxygen ($O_2$) with the plant respiration process, promotes the $O_2$ reduction, initiating the active fermentation process. Thus, occurs a sharp drop in silage pH due to the formation of organic acids from sugars, in which initially actuate the heterofermentative bacteria and enterobacteriaceae, that becomes, then, dominated by homofermentative until the pH falls to below 5.0.

In the stability phase, when only the lactic acid bacteria are active, the anaerobic and acidic pH conditions preserve the silage until the opening time. When the silo is opened, it typically happen the molds and yeasts growth. The inhibition of the fungi multiplication through the contact with $O_2$ is called aerobic stability (Santos et al., 2006).

According to Ohmomo et al. (2002) in the early fermentation stage, Lactococus species, such as Lactococcus lactis, Enterococcus faecalis, Pediococcus acidilactici, Leuconostoc mesenteroides, and Lactobacillus species such as Lactobacillus plantarum, Lactobacillus cellobioses grow together with aerobic microorganisms like yeasts, molds and aerobic bacteria, due to the presence of air between the plant particles. At the same time, it is the plant respiration process. To promote the fermentation, an anaerobic environment is formed making the population to become predominantly composed by LAB, basically Lactococcus and Lactobacillus.

At the final fermentation stage, *Lactobacillus* becomes prevalent, due to their tolerance to the acidity. However, the silage LAB is pretty well diversified, depending on plant material properties, silage technology and silo type. The LAB predominance change from *Lactococcus* to *Lactobacillus* usually occurs in the final fermentation stage. According to Langston et.al (1960), these chemical changes is resulted from bacterial or plant enzymes action making the conversion of carbohydrates into other components such as gas and organic acids, as well as the partial protein breakage resulting in formation of non-protein structures.

The LAB use as microbial inoculants have been widely documented in research (Penteado et al., 2007; Ávila et al., 2009a; Ávila et al., 2009b; Jalč et al., 2009; Reich & Kung Jr., 2010).

Zopollatto et al. (2009) in a meta-analysis study (1999-2009) found a data limitation on the effect of microbial additives in silage quality. They observed that the number of conduced studies is not enough to provide conclusive positions regarding the effects of additives, emphasizing also the data scarcity in certain areas, such as dairy cattle performance. The results documented by these authors show that the magnitude of the response, especially on animal performance, is low. Thus, the justification for the use of additives should be evaluated considering the losses reduction in silage and the higher plant nutritional value preservation. Furthermore, they found that the response intensity varies with plant species and microorganism studied, suggesting a specificity between these components.

However, studies conducted in the 1980s and 1990s had already shown that the fermentation responses differ between strains of the same species (Wooflford & Sawczyc 1984, Hill, 1989; Fitzsimons et al., 1992). Hill (1989) found that inoculating corn silage with two *Lactobacillus plantarum* strains isolated from corn and grass, the dominant strain after ensiling was the isolated from corn. The same was observed for the grass silage, where the dominant lactic bacteria strain of were the one isolated from grass.

Many inconclusive results observed in silage fermentation studies may be related to this principle, which must have been overlooked. The specificity between the forage specie and its epiphytic microflora implicates in the need for studies related with isolation and identification of the main microorganism groups present in the forage used for silage. Ávila et al. (2009b) isolated *Lactobacillus buchneri* strains from sugar cane (*Saccharum officinarum* L.) and found that *L. buchneri* UFLA SIL 72 addition reduced the fungi population and the ethanol concentration in silages. Santos et al. (2007) observed reduction in ammonia concentration and enterobacteria population in mombaça grass silage (*Panicum maximum*) inoculated with *Lactobacillus plantarum*, which were isolated from the epiphytic microflora.

Thereby the silage inoculants can facilitate or accelerate the ensiling process, but they do not replace the fundamental factors (plant maturity, dry matter content, oxygen exclusion), which are essential for producing good quality silage. Among these factors the regrowth age is the one that influences all the silage characteristics, from fermentation to the nutritional value, considering the losses.

Meeske & Basson (1998) evaluated the effect of inoculant containing *Lactobacillus acidophilus, Lactobacillus delbruekii ssp. bulgaricus* and *Lactobacillus plantarum* on corn silage and found no inoculants effect on pH values and the lactic acid production. According to the authors, the high LAB concentrations present in the plant before ensiling led to such results. Furthermore, the amount of bacteria from *Clostridium* genus present in greater numbers in the treatment without inoculants had no effect on the protein content decrease of the untreated silage. It was not detected the butyric acid formation.

The high residual soluble carbohydrates content in silage, mainly the ones made of corn, sorghum and sugarcane, favors the aerobic deterioration process by fungi and yeasts, causing losses after the silo opening. However, the organic acids produced by fermentation, mainly acetic acid, have fungicidal effect and can mitigate the deterioration, increasing silage aerobic stability (Ranjit & Kung Jr. 2000; Kung Jr. & Ranjit, 2001). Therefore, inoculants containing heterofermentative LAB (e.g. *Lactobacillus buchneri*) have been used to increase the silage aerobic stability.

Ávila et al. (2009a) evaluated the aerobic stability of mombaça grass silage (*Panicum maximum* Jacq. cv. Mombaça) inoculated with two *Lactobacillus buchneri* strains, one provinient from a commercial inoculant and another isolated from sugarcane (*Saccharum officinarum* L.) silage. It was observed an increase in dry matter content after silo opening, while the carbohydrate ratio did not change due to the low residual concentration, characteristic of grass silage. The ammonia ($NH_3$) concentrations were above the 12% of the total-N recommended by Molina et al. (2002) for good quality silage, indicating high proteolysis during fermentation, due to low soluble carbohydrates supply, what makes possible a rapid decline of pH values.

Table 8 present few studies evaluating the effect of LAB on the silage fermentation. It is observed that there is a pattern of responses, as discussed previously, and its effect depends of the crop used, the microorganism strain and its concentration at the inoculation time. Although significant, the effects are of low magnitude, which leads to reflect about the use of inoculants without the microbiological principles and characteristics of forage plants knowledge.

Kleinschimit and Kung Jr. (2006), in a meta-analysis study (43 experiments), evaluated the *Lactobacillus buchneri* effect on fermentation and aerobic stability of corn, grasses and small grains silages. In general, the inoculation reduced pH, lactic acid concentration and mold counts. At the same time increases in acetic acid concentrations and aerobic stability were detected in all silage types. The increase in aerobic stability was more pronounced in corn silage. Furthermore, it was observed an increase in the propionic acid and ethanol concentrations, on the other hand decreases in soluble carbohydrates concentrations were found in grass and small grains silages. It was observed correlation between acetic acid concentration and fungi population reduction.

| Crop | Microrganism | pH[1] | $NH_3$[2] | LA[3] | AA[4] | PA[5] | BA[6] | ET[7] | AE[8] | DML[9] | DMR[10] |
|---|---|---|---|---|---|---|---|---|---|---|---|
| | | | % total N | % DM | | | | | hours | % | |
| Grass | LP | -- | -- | ++ | -- | | -- | ns | ns | -- | ++ |
| Corn | LB | ++ | ns | -- | ++ | ns | | ns | + | | ns |
| Grass | | -- | ns | -- | ++ | ++ | | ++ | + | | -- |
| Corn | PA/ LP | -- | | -- | -- | -- | -- | | | | ++ |
| Wheat | LB | ++ | ns | -- | ++ | | | | ++ | ++ | |
| | LP | ns | -- | ++ | ns | | | | ns | -- | |
| | LP/ LB | ns | -- | ns | ++ | | | | ++ | -- | |
| Sorghum | LB | ns | ns | -- | ++ | | | | ++ | ++ | |
| | LP | ns | -- | ns | ns | | | | -- | -- | |
| | LP/ LB | ns | -- | -- | ++ | | | | ++ | -- | |
| Wheat | LP/EF | -- | | ++ | ns | | ns | | ns | | |
| | L.Pe | -- | | ++ | ns | | ns | | ns | | |
| Wheat | LB | ns | | ns | ++ | | | | + | ++ | |
| | LP | ns | | -- | ++ | | | | + | -- | |
| Sugar cane | LB | ns | | ns | ++ | ++ | ns | -- | + | | |
| Sunflowerl | SF/ PA/ LP | ns | ns | ns | ns | ns | ns | | ns | ns | |
| | LP/ L. | ns | ns | ns | ns | ns | ns | | ns | ns | |
| | SF/ LP | -- | -- | ns | -- | ++ | ns | | ns | ns | |
| Potato + WB* | LB | -- | -- | ++ | ++ | ns | -- | | ++ | | |
| | LPa/ LL/ PA | -- | -- | ++ | -- | -- | -- | | -- | | |

**Table 8.** Effect of inoculants with lactic acid bacteria on the fermentation of the silage. *Potato by-product + 30% of wheat bran; [1]lactic acid, [2]acetic acid, [3]propionic acid, [4]butyric acid, [5]ethanol, [6]aerobic stability, [7]dry matter losses, [8]dry matter recovery. ns = not significant, + = numerical increase, - = decreasing numbers; + + = significant increase ($P < 0.05$) / - = significant decrease ($P < 0.05$). (Filya et al., 2000; Rodrigues et al., 2001; Weinberg et al., 2002; Filya, 2003; Kleinschimit & Kung Jr., 2006; Rowghani & Zamiri, 2009; Ávila et al., 2009b; Nkosi et al., 2010; Santos et al., 2011). LP = *Lactobacillus plantarum,* EF = *Enterococcus faecium,* LPe = *Lactobacillus pentosus,* SF = *Streptococcus faecium,* PA = *Pediococcus acidilacti, L = Lactobacillus sp.,* LB = *Lactobacillus buchneri;* Pac = *Propionibacterium acidipropionici;* LPar = *Lactobacillus paracasei paracasei* LL = *Lactococcus lactis.*

In concluded studies, the inoculation with *Lactobacillus buchneri* changed silages fermentation pattern, decreasing the lactate/acetate ratio, without compromising the processes efficiency, because the dry matter values recovery remained above 90%, as the minimum value recommended for this variable in these plants. The authors also suggest the existence of culture-specific effect.

Evaluating barley silage inoculated with *Lactobacillus buchneri*, Taylor et al. (2002) observed a decrease in yeasts and molds number, contrasting with an increase in aerobic stability. Changes in dry matter consumption and milk production were not affected.

The homofermentative LAB are used in order to improve the fermentation of the silage by increasing the concentration of lactic acid, which reduces the ammonia and the loss of dry matter. The heterofermentative LAB, for its turn, promote improvements, especially after the opening of the silo, increasing the aerobic stability of silage by inhibiting the growth of molds and yeasts. Thus, many research papers have recommended the use of inoculant combining the above two groups of LAB, due to its greater efficiency compared to the isolated use.

## 5. Use of additives and management practices aimed at the development of lactic bacteria in tropical grass silages

For an appropriate fermentation process with lactic acid predominance, it is necessary to provide ideal conditions for the LAB to develop and predominate in the silage environment. In order to attend these conditions it is used some additives, which can absorb moisture or provide soluble carbohydrates, making this way a more propitious environment to the LAB growth. Some management practices may also be employed with the same purpose.

The key point in the management of grass for silage is undoubtedly the harvest time. Grass harvested in advanced maturity stage present high LAB population, however high tissues lignification is an intrinsic characteristic also, what reduces its nutritional value. In contrast, young grasses have good nutritional value, however it also have unfavorable characteristics to the fermentation process, such as high humidity, low LAB population and high buffering capacity. In case of young grasses it can be used various additives. In case of mature grasses it can be settled a point in which the dry matter content and the LAB populations are suitable and the nutritive value is not compromised.

Research conducted with tropical grasses, evaluating the addition of a wide variety of additives, show that the increase in forage dry matter content or soluble carbohydrates supply favors lactic fermentation and, in most cases, reduces the silage losses. Among many, it has been used wheat bran, corn, fruit pulp and biodiesel industry by-products, sugar cane molasses and even tropical fruits such as jackfruit (Zanine et al., 2006; Pardo et al., 2008; Santos et al., 2008; Rêgo et al., 2010; Andrade & Melotti, 2004; Zanine et al., 2010; Silva et al., 2011). It is important to remind that these additives should be used respecting the level recommended by the authors, otherwise the effects can endanger the fermentative process.

Andrade & Melotti (2004) evaluated the effect of 20 additives on the silage quality made of elephant grass with 80 days (Tables 9 and 10).

In this study, it is observed that cotton fiber, sweeping residue, corn meal, elephant grass hay and guandu hay were used as additives, absorbing moisture (90.91% of dry matter) .The sweeping residue and molasses were used to supply carbohydrates (97.65%).

Looking at $N\text{-}NH_3$ results, it seems that the use of urea, cotton fiber, elephant grass hay, guandu hay, corn meal and molasses with urea, resulted in increased protein degradation during fermentation process. However, no changes were observed in the lactic acid concentration.

| Treatment | DM | pH | N-NH3 | Lactic acid | Acetic acid | Butyric acid |
|---|---|---|---|---|---|---|
| | % | | % total N | %DM | | |
| Control (without aditive) | 15.58f | 4.15b | 12.39d | 2.40a | 0.30b | 0.00b |
| Urea 0.5 % | 15.49f | 5.36a | 35.76abc | 1.05a | 1.81a | 0.57a |
| Cotton fiber (10%) | 23.25b | 5.33a | 36.07ab | 1.8a | 0.66b | 1.73a |
| Elephant grass hay (10%) | 25.88a | 4.26b | 25.63bcd | 2.48a | 0.46b | 0.12b |
| Guandu hay(10%) | 25.78a | 4.21b | 8.33d | 1.38a | 0.58b | 0.14b |
| Drying for 6 hours | 19.84cd | 4.08b | 15.17d | 1.81a | 0.30b | 0.02b |
| Sugar waste (2%) | 16.50de | 4.09b | 13.68d | 4.69a | 0.66b | 0.00b |
| Corn Meal (2%) | 16.90de | 4.00b | 13.68d | 2.47a | 0.28b | 0.00b |
| Corn Meal (4%) | 20.39c | 4.00b | 12.94d | 4.96a | 1.15a | 0.08b |
| Corn Meal (6%) | 21.60c | 4.04b | 12.01d | 4.41a | 0.33b | 0.00b |
| Corn Meal (2%) / Urea (0.5%) | 17.96de | 4.19b | 36.67ab | 5.31a | 0.53b | 0.04b |
| Corn Meal (4%)/ Urea (0.5%) | 20.26c | 4.29b | 49.36a | 1.96a | 0.85b | 0.05b |
| Corn Meal (6%) / Urea (0.5%) | 20.43c | 4.20b | 46.86a | 2.25a | 0.38b | 0.01b |
| Dried Molasses (1%) | 16.95de | 4.04b | 10.52d | 3.60a | 0.22b | 0.00b |
| Dried Molasses (2%) | 17.58de | 3.92b | 10.27d | 3.29a | 0.23b | 0.00b |
| Dried Molasses (3%) | 16.67de | 3.89b | 9.43d | 3.98a | 0.35b | 0.00b |
| Dried Molasses (1%) Urea (0.5%) | 17.20de | 4.18b | 34.93abc | 1.25a | 0.46b | 0.04b |
| Dried Molasses (2%) Urea (0.5%) | 18.20de | 4.09b | 32.43abc | 5.24a | 0.44b | 0.04b |
| Dried Molasses (3%) Urea (0.5%) | 17.55ed | 3.97b | 11.50d | 4.84a | 0.36b | 0.00b |
| Biosilo inoculant | 15.88f | 4.06b | 15.24d | 2.61a | 0.50b | 0.03b |
| CV (%) | 7.04 | 5.55 | 34.87 | 50.62 | 62.54 | 137.65 |

**Table 9.** Dry matter (DM) content and fermentation pattern of elephant grass, Napier, ensiled with different additives (Andrade & Melotti, 2004). DM = dry matter (%), CP = crude protein (% DM), N-NH3 = ammonia nitrogen/total nitrogen (%), lactic acids, acetic and butyric acids: values in % of the silage DM. Equal means in column do not differ (P>0.05): CV = coefficient of variation.

The lowest in vitro dry matter digestibility was obtained with the use of guandu hay. On the other hand the highest one was obtained using corn meal and urea (Table 10). Compared to the control treatment, only the urea and cotton fiber had higher dry matter loss (11.0 and 10.5%, respectively).

According to the authors, it is not recommended the inclusion of urea, hay and cotton fiber in elephant grass silage. Additives rich in nonstructural carbohydrates, such as corn meal and molasses can be used, however, further studies are required to establish suitable levels for better fermentation. The microbial inoculant 'Biosilo' does not benefit the elephant grass silage.

| Treatment | IVDMD (%DM) | DML (%) |
|---|---|---|
| Control (without aditive) | 41.62abcde | 6.80b |
| Urea 0.5 % | 34.47abcde | 11.00a |
| Cotton fiber (10%) | 27.62de | 10.50a |
| Elephant grass hay (10%) | 34.12abcde | 9.80b |
| Guandu hay(10%) | 26.36e | 7.00b |
| Drying for 6 hours | 41.71abcde | 6.70b |
| Sugar waste (2%) | 42.89abcd | 6.85b |
| Corn Meal (2%) | 41.36abcde | 6.70b |
| Corn Meal (4%) | 45.68abc | 7.20b |
| Corn Meal (6%) | 41.81abcde | 5.70b |
| Corn Meal (2%) /Urea (0.5%) | 50.30ab | 6.60b |
| Corn Meal (4%)/ Urea (0.5%) | 51.31a | 7.10b |
| Corn Meal (6%) /Urea (0.5%) | 41.82abcde | 7.10b |
| Dried Molasses (1%) | 40.03abcde | 6.80b |
| Dried Molasses (2%) | 46.84abc | 6.65b |
| Dried Molasses (3%) | 45.25abc | 6.80b |
| Dried Molasses (1%) Urea (0.5%) | 43.73abc | 6.90b |
| Dried Molasses (2%) Urea (0.5%) | 47.15bc | 7.10b |
| Dried Molasses (3%) Urea (0.5%) | 49.65ab | 6.85b |
| Biosilo inoculant | 32.52de | 7.00b |
| CV (%) | 13.70 | 18.5 |

**Table 10.** *In vitro* dry matter digestibility (IVDMD) and dry matter losses (DML) of elephant grass, Napier, ensiled with different additives (Andrade and Melotti, 2004). Equal means in column do not differ (P>0.05), CV = coefficient of variation.

In more recent studies, evaluating the effect of four additives in sugar cane silage (sugarcane with 1.5% of urea; 0.5% of urea + 4% of corn; 0.5% of urea + 4% of dried cassava, 1.5% of starea and sugar cane control), Lopes & Evangelista (2010) concluded that the additive 0.5% urea + 4% corn, provides better results to the sugar cane silage.

Ávila et al. (2006), using combinations of different additives types (citrus pulp, wheat bran, and corn meal) with various doses (3, 6, 9 and 12%), found that Tanzania grass has low soluble carbohydrates contents and citrus pulp was the additive which contributed to increase the forage carbohydrate concentration and to reduce the buffering capacity. It provides an increase in the relation soluble carbohydrate x buffering capacity and better conditions for the fermentation process, resulting in better quality silages.

Besides the additives, some management practices from the harvest time to the silo sealing can influence the LAB development. When the grass is chopped at harvest time, the LAB population tends to increase due to reactivation of dormant and non-culturable cells. Thus, as faster the time between cutting the grass and sealing the silo, better will be the fermentation conditions.

The well done compaction and sealing is one of the secrets for good silage. It serves to expel the air from inside the forage mass, considering that air presence affects the fermentation process, implicating in losses caused by undesirable microorganisms. According to Senger et al. (2005) the original material must present compression level exceeding 650 kg/m$^3$ of green matter, reducing the quality losses of the ensiled material.

Furthermore, the particle size influences the compression and consequently the silo density. Igarasi (2002) observed an inverse relationship between particle size and silage density, suggesting that as smaller the particle size greater the density, and thus there will be more oxygen remaining among the plant particles.

Neumann et al. (2007) evaluating the effect of particle size (small: 0.2 to 0.6 cm or large: 1.0 to 2.0 cm) and cutting height of corn plants (low: 15 cm or higher: 39 cm) on silage fermentation dynamics and opening period, found that small sized particles provide greater compression efficiency and consequently reduces temperature and pH gradients in the silo opening time. The temperature differential between silage and environment is greater on the top, what is related with the time that the silo remain opened and exposed to the external environment and also the lower compression efficiency. It causes an increase in ammoniac nitrogen content and elevation of silage pH values, indicating changes in silage nutritional value.

The plant moisture content and the particle size after chopping are directly related to the compression. Excessively wet forage provides favorable conditions for butyric fermentation and, favors nutrients losses through leaching, and proteins degradation. On the other hand, forage with high dry matter content hinders compaction and air expulsion in the ensiling process. Amaral et al. (2007) found that increase in compression of 100 to 160 kg MS/m$^3$ increased effluent production from 2.2 to 9.8 kg/t of green matter.

Summarizing, as faster and more efficient the process of harvest, chopping, compaction and sealing, greater is the amount of LAB present in silage, and thus lower the losses.

## 6. Conclusions

The increase in lactic acid fermentation is a big challenge for tropical grass silages confection, determining the success of this technology. It is really important to know the species of lactic acid bacteria prevalent in tropical grasses as well as their metabolism in order to obtain maximum use with its utilization.

The use of lactic acid bacteria as microbial inoculants in tropical grasses silage still shows some inconsistency in the results obtained in research works. More research that evaluates their effects on the fermentation parameters, dry matter losses and mainly on the quality, regarding nutrient intake and animal performance is required.

However, tropical grass silages represent a promising technology for livestock in areas threatened by periodic droughts. Furthermore, in tropical countries like Brazil, this practice has been quite taken by the producers.

## Author details

Edson Mauro Santos and Carlos Henrique Oliveira Macedo
*Department of Animal Science, Federal University of Paraiba, Areia, PB, Brazil*

Thiago Carvalho da Silva
*Department of Animal Science, Federal University of Viçosa, Viçosa, MG, Brazil*

Fleming Sena Campos
*Department of Animal Science, Federal University of Bahia, Salvador, BA, Brazil*

## 7. References

Amaral, R.C., Bernardes, T.F., Siqueira, G.R. & Reis, R.A. (2007). Características fermentativas e químicas de silagens de capim-marandu produzidas com quatro pressões de compactação. *Revista Brasileira Zootecnia*, Vol.36, No.3, (May 2007), pp. 532-539, ISSN 1806-9290

Amaral, P.N.C., Evangelista, A.R., Salvador, F.M. & Pinto, J.C. (2008). Qualidade e valor nutritivo da silagem de três cultivares de milheto. *Ciência agrotécnica*, Vol.32, No.2, (March 2007), pp. 611-617, ISSN 1981-1829

Andrade, S.J.T. & Melotti, L. (2004). Efeito de alguns tratamentos sobre a qualidade da silagem de capim-elefante cultivar Napier (*Pennisetum purpureum*, Schum). *Brazilian Journ al of Veterinary Research and Animal Science*, Vol.41, No.6, (February 2004), pp. 409-415, ISSN 1678-4456

Ávila, C.L.S., Pinto, J.C., Figueiredo, H.C.P., Morais, A.R., Pereira, O.G. & Schwan, R.F. (2009a) Estabilidade aeróbia de silagens de capim-mombaça tratadas com *Lactobacillus buchneri*. *Revista Brasileira de Zootecnia*, Vol. 38, No. 5 (May 2009), pp.779-787, ISSN 1806-9290

Ávila, C.L.S., Pinto, J.C., Tavares, V.B. & Santos, I.P.A. (2006). Avaliação dos conteúdos de carboidratos solúveis do capim-tanzânia ensilado com aditivos. *Revista Brasileira de Zootecnia*, Vol.35, No.3, (Abril 2006), pp. 648-654, ISSN 1806-9290

Aroeira, L.J.M.; Paciullo, D.S.C.; Lopes, F.C.F.; Morenz, M.J.F.; Saliba, E.S.; Silva, J.J. & Ducatti , C. (2005). Disponibilidade, composição bromatológica e consumo de matéria seca em pastagem consorciada de *Brachiaria decumbens* com *Stylosanthes guianensis*. *Pesquisa agropecuária brasileira*, Vol.40, No. 4, (April 2005), pp.413-418, ISSN 0100-204X

Benett, C.G.S.; Buzetti, S.; Silva, K.S.; Bergamaschine, A.F. & Fabricio, J.A. (2008). Produtividade e composição bromatológica do capim-Marandu fontes e doses de nitrogênio. *Ciência e agrotecnologia.*, Vol. 32, No. 5, (September/October 2008), pp. 1629-1636, ISSN 1413-7054

Bernardino, F.S.; Garcia, R.; Rocha, F.C.; Souza, A.L. & Pereira, O.G. (2005). Produção e características do efluente e composição bromatológica da silagem de capim-elefante contendo diferentes níveis de casca de café. *Revista Brasileira de Zootecnia,* Vol.34, No.6, (November/December.2005), pp.2185-2191, ISSN 1806-9290

Cai, Y., Benno Y. Ogawa, M., Ohmomo, S., Kumai, S. & Nakase, T. (1998). Influence of Lactobacillus spp. from an inoculant and of weissella and *Leuconostoc* spp from forage crops on silage fermentation. *Applied and Environmental Microbiology,* Vol. 64, No. 8 (August 1998), pp. 2982-2987, ISSN 1098-5336

Chunjian, L., Bolsen, K.K., Brent, B. E. & Fung, D. Y. C. (1992). Epiphytic lactic acid bacteria succession during the pre-ensiling periods of alfafa and maize. *Journal of Applied Bacteriology,* Vol. 73, No. 5 (November 1992), pp. 375-387, ISSN 1364-5072

Daeschel, M.A., Anderson, R.E. & Fleming, H.P. (1987). Microbial ecology of fermenting plant materials. *FEMS Microbiology Reviews,* Vol. 46, No. 3 (September 1987), pp. 357-367, ISSN 0168-6445

Ennahar, S., Sashihara, T., Sonomoto, K., Ishizaki, A. (2000). Class IIa bacteriocins: biosynthesis. Structure and activity. *FEMS Microbiology Reviews,* Vol. 24, No.1 (January 2000), pp.85-106, 2000 ISSN 0168-6445

Evangelista, A.R., Abreu, J.G., Amaral, P.N.C., Salvador, F.M. & Santana, R.A.V. (2004). Produção de silagem de capim-marandu (*Brachiaria brizantha* Stapf cv. Marandu). *Ciência agrotécnica,* Vol.28, No.2, (July 2003), pp. 443-449, ISSN 1413-7054

Faria Júnior, W.G., Gonçalves, L.C., Ribeiro Júnior, G.O., Carvalho W.T.V., Maurício, R.M., Rodrigues, J.A.S., Faria, W.G., Saliba, E.O.S., Rodriguez, N.M. & Borges, A.L.C.C. (2011). Effect of grain maturity stage on the quality of sorghum BRS-610 silages. *Arquivo Brasileiro de Medicina Veterinária e Zootecnia,* Vol.63, No.5, (June 2011), pp. 1215-1223, ISSN 1678-4162

Ferreira, D.A., Gonçalves, L.C., Molina, L.R., Castro Neto, A.G. & Tomich, T.R. (2007). Características de fermentação da silagem de cana-de-açúcar tratada com uréia, zeólita, inoculante bacteriano e inoculante bacteriano/enzimático. *Arquivo Brasileiro de Medicina Veterinária e Zootecnia,* Vol.59, No.2, (January 2007), pp. 423-433, ISSN 0102-0935

Filya, I. (2003). The effect of *lactobacillus buchneri* and *Lactobacillus plantarum* on the fermentation, aerobic stability and ruminal degradability of low dry matter corn and sorgum silage. *Journal of Dairy Science,* Vol. 86, No.11 (November 2003), pp. 3575-3581, ISSN 1525-3198

Filya, I, Ashbell, G., Hen, Y. & Weinberg, Z.G. (2000). The effect of bacterial inoculants on the fermentation and aerobic stability of whole crop wheat silage. *Animal Feed Science and Technology,* Vol. 88, No. 1-2 (November 2000), pp.39–46, 0377-8401

Fitzsimons, A., Duffner, F., Curtin, D., Brophy, G., O'Kiely, P. & O'Connel, M. (1992). Assessment of *Pediococcus acidilactici* as a potential silage inoculant. *Applied and Environmental Microbiology,* Vol. 58, No. 9 (September 1992), pp. 3047-3052, ISSN 1098-5336

Hu, W.; Schmidt, R.J.; Mcdonell, E.E.; Klingerman, C.M. & Kung Jr., L. (2009). The effect of *Lactobacillus buchneri* 40788 or *Lactobacillus plantarum* MTD-1 on the fermentation and aerobic stability of corn silages ensiled at two dry matter contents. *Journal of Dairy Science,* Vol. 92, No. 8, (August 2009), pp.526-535, ISSN 0022-0302

Hill, H. A. (1989). Microbial ecology of lactobacilli in silage. *Proceedings of the 2nd Forage Symposium,* Pioneer Hi-Bred International, Johnston, IA, pp. 47-64

Igarasi, M.S. (2002). Controle de perdas na ensilagem de capim Tanzânia (*Panicum maximum* Jacq. Cv. Tanzânia) sob os efeitos do teor de matéria seca, do tamanho de partícula, da estação do ano e da presença de inoculante microbiano. *Dissertação* (Mestrado em Ciência Animal e Pastagens) - Escola Superior de Agricultura Luiz de Queiroz, Piracicaba, pp. 152

Jalčl, D., Laukova, A., Simonova, M., Váradyová, Z. & Homolka, P. (2009). The use of bacterial inoculants for grass silage: their effects on nutrient composition and fermentation parameters in grass silages. *Czech Journal of Animal Science,* Vol. 54, No. 2 (February), pp.84-91

Kandler, O. & Weiss, N. (1986). Lactobacillus. In: *Bergey's manual of systematic bacteriology.* Sneath, P.H.A., Mair, N.S., Sharpe, M. E. &Holt, J. G. Baltimore: Williams and Wilkins.

Knicky, M. (2005). Possibilities to improve silage conservation. In: http://pub.epsilon.slu.se/834/1/Thesis_for_epsilon2.pdf. (Consultado em 21/11/2005).

Kleinschimit, D.H. & Kung Jr., L. (2006). A meta-analysis of the effects of Lactobacillus buchneri on the fermentation and aerovic stability of corn and grass and small-grains silages. *Journal of Dairy Science,* Vol. 89, No. 10 (October 2006), pp. 4005-4013, ISSN 1525-3198

Kollet, J.L.; Diogo, J. M.S. & Leite, G.G. (2006). Rendimento forrageiro e composição bromatológica de variedades de milheto (*Pennisetum glaucum* (L.) R. BR.). *Revista Brasileira de Zootecnia,* Vol.35, No.4, (July/August 2006), pp.1308-1315, ISSN 1806-9290.

Kung Jr., L. & Ranjit N.K. (2001).The effect of *Lactobacillus buchneri*and other additives on the fermentation and aerobic stability of barley silage. *Journal of Dairy Science,* Vol. 84, No. 5 (May 2001), pp.1149-1155, ISSN 1525-3198

Kung Jr., L., Stanley, R.W. (1982). Effect of stage of maturity on the nutritive value of whole-plant sugarcane preserved as silage. *Journal of Animal Science,* Vol. 54, No. 4 (April 1982), pp.689-696

Landell, M.G.A.; Campana, M.P.; Rodrigues, A.A. et al. (2002). A variedade IAC86-2480 como nova opção de cana-de-açúcar para fins forrageiros: manejo de produção de uso na alimentação animal. Campinas: Instituto Agronômico, 2002. 39p. (Série Tecnologia APTA, *boletim técnico* IAC; 193).

Langston, C.W. & BOUMA, C. (1960). A study of the microorganisms from grass silage. II. The lactobacilli. Applied Microbiology., Vol. 8, No. 4 (July 1960), 223-234, ISSN 1098-5336

Lin, C., Bolsen, K.K., Brent, B.E., Hart, R.A., Dickerson, J.T., Feyerherm, A.M., Aimutis, W.R. (1992). Epiphytic microflora on alfafa and whole-plant corn. *Journal of Dairy Science,* Vol. 75, No. 9 (September 1992), pp. 2484-2493, ISSN 1525-3198

Lopes, J. & Evangelista, A.R. (2010). Características bromatológicas, fermentativas e população de leveduras de silagens de cana-de-açúcar acrescidas de ureia e aditivos absorventes de umidade. *Revista Brasileira de Zootecnia,* Vol.39, No.5, (May 2009), pp. 984-991, ISSN 1806-9290

Lücke, F. K. (2000). Utilization of microbes to process and preserve meat. *Meat Science,* Vol. 56, No. 2 (October 2000), pp. 105-115, ISSN 0309-1740

Martinez , J.C. (2009). Efeito de *Lactobacillus* na Fermentação e Estabilidade Aeróbica de Silagem de Milho. *Revista inter rural,* Vol. 3, No. 27, (November 2009), pp. 21-22, ISSN 0103-9458

McDonald, P.J., Henderson, A.R. & Heron, S.J.E. (1991). *The biochemistry of silage* (2ª Ed.) Mallow Chalcombe Publications, ISBN 0948617225

Meeske, R., Basson, H.M. & Cruywagen, C. W. (1999). The effect of a lactic acid bacterial inoculant with enzymes on the fermentation dynamics, intake and digestibility of *Digitaria eriantha* silage. *Animal Feed Science Technolog,* Vol. 81, No. 3 (October 1999), pp.237-248, ISSN 0377-8401

Meeske, R., Basson, H. M. (1998). The effect of a lactic acid bacterial inoculant on maize silage. *Animal Feed Science Technology,* Vol. 70, No. 3 (February 1998), pp.239-274, ISSN 0377-8401

Miranda, N.O., Góes, G.B., Andrade Neto, R.A. & Lima, A.S. (2010). Sorgo forrageiro em sucessão a adubos verdes na região de Mossoró, RN. *Revista Brasileira de Ciências Agrárias,* Vol.5, No.2, (June 2010), pp. 202-206, ISSN 1981-0997

Molina, L. R., Ferreira, D. A., Gonçalves, L. C., Castro Neto, A. G. & Rodrigues, N. M. (2002). Padrão de fermentação da silagem de cana-de-açúcar (*Saccharum officinarum* L.) submetida a diferentes tratamentos. *Anais da Reunião da sociedade brasileira de zootecnia,* Recife- PE, July, 2002

Montville, T.J. & Chen, Y. (1998). Mechanistic action of pediocin and nisin: recent progress and unresolved questions. *Applied Microbiology Biotechnology,* Vol. 50, No. 5 (November 1998), pp. 511-519, ISSN 1432-0614

Moreira, J.N.; Lira, M.A.; Santos, M.V.F. & Araújo, G.G.L. (2007). Potencial de produção de capim buffel na época seca no semiárido pernambucano. *Revista Caatinga,* Vol.20, No.3, (July/September 2007), pp.22-29, ISSN 0100-316X

Moraes, E.H.B.K.; Paulino, M.F.; Zervoudakis, J.T.; Valadares Filho, S.C. & Moraes, K.A.K. (2005). Avaliação qualitativa da pastagem diferida de *Brachiaria decumbens* Stapf., sob pastejo, no período da seca, por intermédio de três métodos de amostragem. *Revista Brasileira de Zootecnia,* Vol.34, No.1, (January/February 2005), pp. 30-35, ISSN: 1806-9290

Neumann, M., Muhlbach, P.R.F., Nornberg, J.L., Ost, P.R. & Lustosa, S.B.C. (2007). Efeito do tamanho de partícula e da altura de corte de plantas de milho na dinâmica do processo fermentativo da silagem e no período de desensilagem. *Revista Brasileira de Zootecnia,* Vol.36, No.5, (March 2007), pp.1603-1613, ISSN 1806-9290

Neumann, M.; Restle, J.; Alves Filho, D.C.; Brondani, I.L.; Pellegrini, L.G. & Freitas, A.K. (2002). Avaliação do valor nutritivo da planta e da silagem de diferentes híbridos de sorgo (*Sorghum bicolor,* L. Moench). *Revista Brasileira de Zootecnia,* Vol.31, No.1, (January/February 2002), pp.293-301, ISSN 1806-9290

Nkosi, B.D., Meeske, R., van der Merwe, H.J., Groenewald, I.B. (2010). Effects of homofermentative and heterofermentative bacterial silage inoculants on potato hash silage fermentation and digestibility in rams. *Animal Feed Science and Technology,* Vol. 157, No. 3 (May 2010), p.195–200, ISSN 0377-8401

Ohmomo, S., Tanaka, O., KItamoto, H. K. & Cai, Y. (2002). Silage and microbial performance, old history but new problem. *JARQ,* Vol. 40, No. 2 (April 2002), pp. 59-71, ISSN

Oliveira, J.S., Santos, E.M., Zanine, A.M., Mantovani, H.C., Pereira, O.G. & Rosa, L.O. (2007). Populações microbianas e composição química de silagem de capim-mombaça (Panicum maximum) inoculado com *Streptococcus bovis* isolado de rúmen. *Archives of Veterinary Science,* Vol. 12, No. 2, pp.35-40, ISSN 1517-784X

Oliveira, J.S., Ferreira, R.P., Cruz, C.D., Pereira, A.V., Botrel, M.A., Von Pinho, R.G., Rodrigues, J.A.S., Lopes, F.C.F. & Miranda, J.E.C. (2002). Adaptabilidade e Estabilidade em Cultivares de Sorgo. *Revista Brasileira Zootecnia,* Vol.31, No.2, (August 2002), pp. 883-889, ISSN 1806-9290

Pariz, C.M.; Azenha, M.V.; Andreotti, M.; Araújo, F.C.M.; Ulian, N.A. & Bergamaschine, A.F. (2011). Produção e composição bromatológica de forrageiras em sistema de integração lavoura-pecuária em diferentes épocas de semeadura. *Pesquisa agropecuária brasileira,* Vol.46, No.10, (October 2011), pp. 1392-1400, ISSN 0102-0935

Pardo R.P., Castello Branco van Cleef, E H, da Silva Filho J C, Castro Neto, P, & Neiva Júnior, A.P. (2008). Diferentes níveis de torta de nabo forrageiro (*Raphanus sativus*) como aditivo na silagem de capim elefante. *Livestock Research for Rural Development,* Vol. 20, No. 10 (October 2008), ISSN 0121-3784

Patrizi, W.L.; Madruga Jr.; C.R.F.; Minetto, T.P.; Nogueira, E. & Morais, M.G. (2004). Efeito de aditivos biológicos comerciais na silagem de capim-elefante (*Pennisetum purpureum* Schum). *Arquivo Brasileiro de Medicina Veterinária e Zootecnia,* Vol.56, No. 3, (June. 2004), pp. 392-397, ISSN 0102-0935

Pahlow, G., Muck, R.E. & Driehuis, F. (2003). Microbiology of ensiling. In: Silage Science and Technology. Madison. *Proceedings...* Madison: ASCSSA-SSSA, Agronomy 42, pp.31-93

Pedroso, A.F.; Nussio, L.G.; Loures, D.R.S.; Paziani, S.F.; Igarasi, M.S.; Coelho, R.M.; Horii, J. & Rodrigues, A.A. (2007). Efeito do tratamento com aditivos químicos e inoculantes bacterianos nas perdas e na qualidade de silagens de cana-de-açúcar. *Revista Brasileira de Zootecnia,* Vol.36, No.3 (May/June 2007), pp.558-564, ISSN 1806-9290

Penteado, D. C. S., Santos, E. M., Carvalho, G. G. P., Oliveira, J. S., Zanine, A. M., Pereira, O. G. & Ferreira, C. L. L. F. (2007). Inoculação com Lactobacillus plantarum da microbiota em silagem de capimmombaça. *Archivos de Zootecnia,* Vol.56, No.214, (July 2007), pp. 191-202, ISSN 0004-0592

Pereira, O.G., Rocha, K.D. & Ferreira, C.L.LF. (2007). Composição química, caracterização e quantificação da população de microrganismos em capim-elefante cv. Cameroon (*Pennisetum purpureum,* Schum.) e suas silagens. *Revista Brasileira de Zootecnia,* Vol.36, No.6, (agosto 2007), pp. 1742-1750, ISSN 1806-9290

Pereira, O. G., Santos, E. M., Ferreira, C. L. L. F., Mantovani, H. C. & Penteado, D. C. S. (2006). Populações microbianas em silagem de capim-mombaça de diferentes idades de rebrotação. *Anais da XLIII Reunião anual da sociedade brasileira de zootecnia*. João Pessoa-PB, julho 2006

Pereira, O.G., Sousa, L.O. & Penteado, C.S. (2005). Populações microbianas, pH e relação nitrogênio amoniacal/N total em silagens de capim-elefante com diferentes idades de rebrotação. *Anais da XLII Reunião anual da sociedade brasileira de zootecnia*. UFG-Goiânia, July 2005

Ranjit, N. K. & Kung Jr, L. (2000). The effect of *Lactobacillus buchneri*, *Lactobacillus plantarum*, or a chemical preservative on the fermentation and aerobic stability of corn silage. *Journal of Dairy Science*. Vol.83, No.5, (agosto 2000), pp.526-535, ISSN 2131-2144

Reich, L.J. & Kung Jr, L. (2010). Effects of combining *Lactobacillus buchneri* 40788 with various lactic acid bacteria on the fermentation and aerobic stability of corn silage. *Animal Feed Science and Technology*. Vol.159, No.34, (october 2010), pp.105-109, ISSN 0377-8401

Reis, J.A.G.; Reis, W.; Macedo, V.P. & Sousa, M.M. (2008). Diferentes níveis de uréia adicionados à cana-de-açúcar (*Saccharum officinarum* L.) no momento de sua hidrólise alcalina. *PUBVET*, Vol.2, No.4, (January. 2008), pp. 1- 12, ISSN 1982-1263

Rezende, P. M., Alcantara, H.P., Passos, A.M.A., Carvalho, E.R., Baliza, D.P. & Oliveira, G.T.M. (2011). Rendimento forrageiro da rebrota do sorgo em sistema de produção consorciado com soja. *Revista Brasileira de Ciências Agrárias*, Vol.6, No.2, (April 2011), pp. 362-368, ISSN 1981-0997

Ribeiro, J.L.; Nussio, L.G.; Mourão, G.B.; Mari, L.J.; Zopollatto, M. & Paziani, S.F. (2008). Valor nutritivo de silagens de capim-Marandu submetidas ao efeito de umidade, inoculação bacteriana e estação do ano. *Revista Brasileira de Zootecnia*, Vol. 37, No.7, (July 2008), pp. 1176-1184, ISSN 1806-9290

Rocha, K. D. (2009). Silagens de capim-elefante cv. Cameroon, de milho e de sorgo produzidas com inoculantes ênzimo-bacterianos: populações microbianas, consumo e digestibilidade. *Dissertação* (Mestrado em zootecnia). Universidade Federal de Viçosa-MG, pp. 93

Rowghani, E. & Zamiri, M.J. (2009). The effects of a microbial inoculant and formic acid as silage additives on chemical composition, ruminal degradability and nutrient digestibility of corn silage in sheep. *Iranian Journal of Veterinary Research*. Vol.10, No.2, (September 2009), pp.110-118, ISSN 1728-1997

Rodrigues, P.H.M.; Senatore, A.L.; Andrade, S.J.T.; Ruzante, J.M.; Lucci, C.S. & Lima, F.R. (2002). Efeito da adição de inoculantes microbianos sobre a composição bromatológica e perfil fermentativo da silagem de sorgo produzida em silos experimentais. *Revista Brasileira de Zootecnia*, Vol.31, No.6, (November/Decemper 2002), pp. 2373-2379, ISSN 1806-9290

Santos, E. M., Pereira, O. G., Rasmo, G., Ferreira, C. L. L. F. ; Oliveira, J.S., Silva, T.C. & Rosa, L.O. (2011). Microbial populations, fermentation profile and chemical composition of signalgrass harvsted of different rgrowth ages. *Revista Brasileira de Zootecnia*. Vol.40, No.4, (October 2010), pp.747-755, ISSN 1806-9290

Santos, E.M., Zanine, A.M., Dantas, P.A.S., Dórea, J.R.R., Silva, T.C., Pereira, O.G., Lana, R.P. & Costa, R.G. (2008). Composição bromatológica, perdas e perfil fermentativo de silagens de capim-elefante com níveis de jaca. *Revista Brasileira de Saúde e Produção Animal*, Vol.9, No.1, (March 2008), pp. 71-80, ISSN 1519-9940

Santos, E.M. (2007). Populações microbianas e perfil fermentativo em silagens de capins tropicais e desempenho de bovinos de corte alimentados com dietas contendo silagens de capim Mombaça. *Tese* (Doutorado) Universidade Federal de Viçosa. pp. 126

Santos, E.M., Zanine, A.M. & Oliveira, J.S. (2006). Produção de silagem de gramíneas tropicais. *Revista Electrónica de Veterinaria*, Vol.7, No.7, (July 2006), pp., ISSN 1695-7504

Santos, E. M., Pereira, O. G., Ferreira, C. L. L. F., Mantovani, H. C., Penteado, D. C. S., Oliveira, J.S. & Sousa, L.O. (2006). Isolamento, identificação e caracterização de *Lactobacillus* predominantes em gramíneas tropicais. *Anais da XLIII Reunião anual da sociedade brasileira de zootecnia*. João Pessoa-PB, July 2006

Santos, G.R.A.; Guim, A.; Santos, M.V.F.; Ferreira, M.A.; Lira, M.A.; Dubeux Jr, J.C.B. & Silva, M.J. (2005). Caracterização do pasto de capim-buffel diferido e da dieta de bovinos, durante o período seco no sertão de Pernambuco. *Revista Brasileira de Zootecnia*, Vol.34, No.2, (March/April 2005), pp.454-463, ISSN 1806-9290

Santos, M.V.F.; Dubeux Jr, J.C.B.; Silva, M.C.; Santos, S.F.; Ferreira, R.L.C.; Mello, A.C.L.; Farias, I. & Freitas, E.V. (2003). Produtividade e composição química de gramíneas tropicais na Zona da Mata de Pernambuco. *Revista Brasileira de Zootecnia*, Vol.32, No.4, (July/August 2003), pp.821-827, ISSN 1806-9290

Senger, C.C.D., Muhlbach, P.R.F., Sánchez, L.M.B., Peres Netto, D. & Lima, L.D (2005). Composição química e digestibilidade 'in vitro' de silagens de milho com distintos teores de umidade e níveis de compactação. *Ciência Rural*, Vol.35, Nº.6, (December 2005), pp. 1393-1399, ISSN 0103-8478

Silva, T.C.; Edvan, R.L.; Macedo, C.H.O.; Santos, E.M.; Silva, D.S. & Andrade, A.P. (2011). Características morfológicas e composição bromatológica do capim buffel sob diferentes alturas de corte e resíduo. *Revista Trópica*. Vol. 5, No. 2, (July 2011), pp. 30-39, ISSN 1982-4831

Silva, T.C., Dorea, J. R.R., Dantas, P.A.S., Santos, E.M., Zanine, A.M. & Pereira, O.G. (2011). Populações microbianas, perfil fermentativo e composição bromatológica de silagens de capim-elefante com níveis de jaca. *Archivos de Zootecnia*. Vol.60, No.4, (Setember 2009), pp.247-255, ISSN 0004-0592

Silva, A.V.; Pereira, O.G.; Garcia, R.; Valadares Filho, S.C.; Cecon, P.R. & Ferreira, C.L.L.F. (2005). Composição bromatológica e digestibilidade *in Vitro* da matéria seca de silagens de milho e sorgo tratadas com inoculantes microbianos. *Revista Brasileira de Zootecnia*, Vol.34, No.6, (November/December 2005), pp.1881-1890, ISSN 1806-9290

Sousa, L.O., Santos, E.M., Penteado, D.C.S., Pereira, O.G., Carvalho, G.G.P. & Oliveira, J.S. (2006). Composição bromatológica de silagem de capim-mombaça inoculada com *lactobacilus plantarum* da microbiota epifítica. *Anais do VI Congresso nacional de zootecnia – Zootec*. Recife-PE, November, 2006

Taylor, C.C., Ranjit, N.J. & Mills, J.A. (2002). The effect of treating whole-plant barley with *Lactobacillus buchneri* 40788 on silage fermentation, aerobic stability, and nutritive value for dairy cows. *Journal of Dairy Science,* Vol.85, No.18, (November 2002), pp. 1793-1800, ISSN 1839–1854

Tjandraatmadja, M., Norton, B.W. & Macrae, I.C. (1994). Ensilage characteristics of three tropical grasses as influenced by stage of growth and addition of molasses. *World Journal of Microbiology and Biotechnology*. Vol.10, No.9, (August 1994), pp.74-81, ISSN 0959-3993

Tjandraatmadja, M., Norton, B. W. & Macrae, I. C. (1991). Fermentation patterns of forage sorghum ensiled under different environmental conditions. *World Journal of Microbiology and Biotechnology*. Vol.7, No.4, (July 1991), pp.206-218, ISSN 0959-3993

Valadares Filho, S.C.; Magalhães, K.A. & Rocha Jr, V.R. (2006). *Tabelas brasileiras de composição de alimentos para bovinos.* 2.ed. UFV, ISBN 859060413-6,Viçosa, Minas Gerais

Valeriano, A.R.; Pinto, J.C.; Ávila, C.L.S.; Evangelista, A.R.; Tavares, V.B. & Schwan, R.F. (2009). Efeito da adição de *Lactobacillus* sp. na ensilagem da cana-de-açúcar. *Revista Brasileira de Zootecnia,* Vol.38, No.6, (June 2009), pp. 1009-1017, ISSN 1806-9290

Velho, J.P.; Mühlbach, P.R.F.; Genro, T.C.M.; Velho, I.M.P.H.; Nörnberg, J.L.; Orqis, M.G. & Kessler, J.D. (2006). Alterações bromatológicas nas frações dos carboidratos de silagens de milho "safrinha" sob diferentes tempos de exposição ao ar antes da ensilagem. *Revista Brasileira de Zootecnia,* Vol.35, No.4, (July/August 2006), pp.1621-1628, ISSN 1806-9290

Velho, J.P.; Mühlbach, P.R.F.; Nörnberg, J.L.; Velho, I.M.P.H.; Genro, T.C.M. & Kessler, J.D. (2007). Composição bromatológica de silagens de milho produzidas com diferentes densidades de compactação. *Revista Brasileira de Zootecnia,* Vol.36 No.5, (September/October 2007), pp. 1532-1538, ISSN 1806-9290

Viana, M.C.M.; Freire, F.M.; Ferreira, J.J.; Macêdo, G.A.R.; Cantarutti, R.B. & Mascarenhas, M.H.T. (2011). Adubação nitrogenada na produção e composição química do capim Braquiária sob pastejo rotacionado. *Revista Brasileira de Zootecnia,* Vol.40, No.7, (July 2011), pp.1497-1503, ISSN 1806-9290

Zanine, A.M., Santos, E.M., Dorea, J.R.R., Dantas, P.A.S., Silva, T.C. & Pereira, O. G. (2010). Evaluation of elephant grass with adition of cassava scrapings. *Revista Brasileira de Zootecnia.* Vol.39, No.12, (April 2010), pp.2611-2616, ISSN 1806-9290

Zanine, A.M.; Santos, E.D.; Ferreira, D.J., Pereira, O.G., & Almeida, J.C.C. (2006) Efeito do farelo de trigo sobre as perdas, recuperação da matéria seca e composição bromatológica da silagem de capim mombaça. *Brazilian Journal of Veterinary Research and Animal Science.* Vol.53, No.6, (February 2006), pp.803-809, ISSN 1413-9596

Zhang, T., LI, L., Wang, X., Zeng, Z., Hu, Y. & Cui, Z. (2009). Effects of Lactobacillus buchneri and Lactobacillus plantarum on fermentation, aerobic stability, bacteria diversity and ruminal degradability of alfalfa silage. *World Journal of Microbiology and Biotechnology.* Vol.25, (January 2009), pp.965-971, ISSN 0959-3993

Zopollatto, M., Daniel, J.L.P. & Nussio, L.G. (2009). Aditivos microbiológicos em silagens no Brasil: revisão dos aspectos da ensilagem e do desempenho de animais. *Revista Brasileira de Zootecnia*. Vol.38, No.spe, (February 2006), pp.170-189, ISSN 1806-9290

Weinberg, z.g., Ashbell, G., Hen, Y., Azrieli, A., Szakacs, G. & Filya, I. (2002). Ensiling whole-crop wheat and corn in large containers with *Lactobacillus plantarum* and *Lactobacillus buchneri*. *Journal of Industrial Microbiology & Biotechnology*. Vol.28, No.19, (February 2002), pp.7-11, ISSN 1476-5535

White, D. (2000). *The physiology and biochemistry of prokaryotes* (2). Oxford University Press, ISBN 0195125797, USA

Woolford, M. K. (1984). *The silage fermentation*. Marcel Dekker, ISBN 0824770390, New York

Chapter 15

# Ruminal Digestibility and Quality of Silage Conserved via Fermentation by Lactobacilli

Yang Cao, Yimin Cai and Toshiyoshi Takahashi

Additional information is available at the end of the chapter

http://dx.doi.org/10.5772/50816

## 1. Introduction

The utilization of whole crop rice (WCR) as an animal feed has proven economically viable, not only as a way of disposing of rice straw residues but also as a real alternative for feeding livestock in regions where rice is the main crop (Han et al., 1974). As a result, in Japan and other rice-producing countries, rice is no longer grown exclusively for human consumption but increasingly as a valuable forage crop. Forage rice is in fact believed to be an ideal alternative crop, not only in helping farmers adjust grain rice production but also in preserving the soil, leading to long-term utilization of the paddy field. Yet a major drawback of forage rice is that it yields low-quality silage, due to poor digestibility of nutrients, mostly crude proteins (Cai et al., 2003). Several processes have been developed to improve the fermentation and nutritional value of whole-crop silage from forage paddy rice. Breeding programs are carried out, and newly developed rice varieties with increased yield and amount of digestible nutrients are being grown and tested. Also, harvesting, preparation, and storage techniques are constantly being improved. However, WCR is usually insufficient in sugars and lactic acid bacteria (LAB), and may produce silages rich in ethanol rather than lactic acid and volatile fatty acid (VFA) (Cai et al., 2003). This could be attributed to the structure of the rice plant; the hollow stem may increase the air in a silo, facilitating yeast WCR is usually insufficient in sugars and lactic acid bacteria (LAB) (Cai et al., 2003), and may produce silages rich in ethanol rather than lactic acid and VFA (Yamamoto et al., 2004). This could be attributed to the structure of the rice plant; the hollow stem may increase the air in a silo, facilitating yeast growth especially in the early ensiling period. Furthermore, most of the processes used to date still rely on heavy chemical treatments with ammonia and sodium hydroxide and were reported to reduce the palatability of silage to ruminants (Cai et al., 2003; Enishi et al., 1998). Of the many factors that can affect silage fermentation, the type of microorganisms that dominate the process often dictates the final quality of the silage. For instance, homolactic fermentation by LAB is more desirable than other types of fermentation because the theoretical recoveries of dry matter and energy are greatest. During this type of fermentation, LAB utilizes water-soluble carbohydrates to produce lactic acid, the primary acid responsible for decreasing the pH in silage. In contrast, other fermentations are less efficient. Natural populations of LAB on plant material are often low in number and heterofermentative. Thus, the concept of using a microbial inoculant to silage involves adding fast-growing homofermentative LAB in order to dominate the fermentation, thereby producing higher-quality silage. Some of the commonly used homofermentative LAB in silage inoculants include *Lactobacillus plantarum, Lactobacillus acidophilus, Pediococcus acidilactici,* and *Enterococcus faecium.* Commercially available microbial inoculants contain one or more of these bacteria that have been selected for their ability to dominate the fermentation.

Food by-product such as tofu cake is high in crude protein and fatty acids (Xu et al. 2001). Not only could the by-products be utilized as a source of nutrients for ruminants, but using them to replace imported commercial feedstuffs could save energy in transportation, and possibly reduce the environmental impact of burning them as waste of burying them landfills. Preparing total mixed ration (TMR) silage is one practice whereby food by-products are stored and utilized as animal feeds in Japan (Imai, 2001). Our previous study (Cao et al., 2009b) showed that TMR silage with 30% dried tofu cake had the higher lactic acid content than that with rice bran or green tea waste. Alli et al. (1984) reported that as molasses can provide fermentable sugars for the production of organic acids, it has been used extensively as a fermentation aid, and that silage prepared with molasses may show a lower pH, higher residual water-soluble carbohydrates levels, greater quantities of lactic acid, lower levels of volatile basic nitrogen, and decreased dry matter (DM) loss compared to silage without molasses. Weinberg et al. (2003) also reported a high lactic acid content in silage ensiled with straw and molasses.

However, even if there is plenty of glucose as a substrate, if insufficient in lactic acid bacteria, to preparing good quality silage is difficult. Cai et al. (1999) reported that the factors involved in fermentation quality include chemical composition, particularly the water-soluble carbohydrates content of the silage material and the physiological properties of epiphytic bacteria. Conservation of forage crops by ensiling is based on natural fermentation in which epiphytic LAB convert sugars into lactic acid under anaerobic conditions. As a result, the pH decreases and the forage is preserved. The fermentation quality of silage is influenced by the size, diversity and activity of epiphytic LAB. The population density of LAB has been reported to range from $10–10^3$ CFU/g fresh matter (FM) on standing forage crops to $10^3–10^7$ CFU/g FM on chopped ones entering the silo. Generally, When LAB reaches at least $10^5$ (CFU/g of FM), silage can be well preserved; if LAB values below $10^5$ high-quality silage fermentation may need to be controlled using certain inoculants. *Lactobacillus plantarum* Chikuso-1 is shown to have great potential as an inoculant for WCR silage (Cai et al., 1999, 2003; Cai, 2001; Cao et al., 2011).

Furthermore, Commercial processing of vegetables results in many residues, but most are burned and dumped into landfills or used as compost, which is a waste of resources and leads to possible environmental problems due to unsuitable disposal. Demand for efficient use of food by-products is increasing due to economic and environmental concerns. Residues from vegetables such as white cabbage, Chinese cabbage, red cabbage, and lettuce are high in nutrients such as vitamins, minerals, and vegetable fiber, and large quantities of these vegetables are produced annually in many countries, including Japan. However, these vegetable residues perish easily because of their high moisture content. Technologies to create high-quality animal feed from vegetable residues and to provide long-term storage of the resulting silage need to be developed. In regions where vegetable residues are the main food by-product, the use of vegetable-residue silage as animal feed has proven economically viable, not only as a way of disposing of vegetable residues but also as an alternative livestock feed. To prepare fermented by-product mix as ruminant feed, it is important to investigate the digestive characteristics of these vegetable residues.

The purpose of this experiment was to examine the effects of lactic acid bacteria inoculant on the quality of fermentation and ruminal digestibility and fermentation characteristics of ruminant fed silage, and to determine the chemical and microorganism composition of vegetable residues, and the influence of lactic acid bacteria addition and moisture adjustment on fermentation quality and the in vitro ruminal fermentability of silages.

## 2. Materials and methods

Experiments were conducted with permission from the Committee of Animal Experimentation, and according to the animal care and use institutional guidelines for animal experiments at the Faculty of Agriculture, Yamagata University.

### 2.1. Silage preparation

Whole crop rice (Haenuki) was cultivated using transplant cultivating methods in a paddy field on an experimental farm at Yamagata University, Japan, and harvested at the full-ripe stage with a length of 2 cm. As shown in Table 1, TMR was prepared using compound feed (Kitanihon-Kumiai Feed, Yamagata, Japan), WCR, dried beet pulp, a vitamin-mineral supplement (Snow Brand Seed, Iwate, Japan), dried tofu cake (Zenno, Tsuruoka, Japan), molasses (sugarcane; Dai-Nippon Meiji Sugar Co., Tokyo, Japan) and LAB (*Lactobacillus plantarum* Chikuso-1, Snow Brand Seed, Sapporo, Japan). Experimental treatments included control silage without additive, or with molasses, (4% FM basis), LAB (5 mg $kg^{-1}$ FM basis), and molasses+LAB. Moisture was adjusted with water to approximately 65%. Silages were prepared using a small-scale system of silage fermentation. Approximately 1 kg TMR was packed into plastic film bags (Hiryu BN-12 type, 270 mm × 400 mm, Asahikasei, Tokyo, Japan), and the bags were sealed with a vacuum sealer (SQ303, Sharp, Osaka, Japan). Three silos per treatment were prepared and stored in a room at 20–25°C for 60 days.

Residues of white cabbage (*Brassica campestris* L. *var. capitata*), Chinese cabbage (*Brassica rapa* L. *var. glabra Regel*), red cabbage (*Brassica oleracea* var. *capitata F. rubra*), and lettuce (*Lactuca sativa* L.) (three samples of each) were collected from a local commercial vegetable factory (Fujiyama factory, Matsuya Foods Company, Limited, Fujinomiya,

Shizuoka, Japan). The vegetable residue comprised the outside leaf part of white cabbage, Chinese cabbage, red cabbage, and lettuce with no added bacteria. Experimental treatments included control silage without additive or with LAB, beet pulp (DM, 90.7%; organic matter, 94.9%; crude protein, 8.4%; ether extract, 0.7%; acid detergent fiber, 25.6%; neutral detergent fiber, 52.1%; WSC, 2.1% of DM), and beet pulp+LAB. The strain FG1 (Lactobacillus plantarum Chikuso-1; Snow Brand Seed, Sapporo, Japan) isolated from a commercial inoculant was used. The de Man Rogosa Sharpe agar broth was inoculated with strain FG1 and incubated overnight. After incubation, the optical density of the suspension at 700 nm was adjusted to 0.42 using sterile 0.85% NaCl solution. The inoculum size of LAB was 1 mL of suspension per kilogram of FM. The inoculated LAB number was $1.0 \times 10^5$ CFU/g of FM. The addition ratio of beet pulp was 300 g per kg of FM. Silages were prepared using a small-scale system of silage fermentation (Cai et al., 1999). Approximately 100-g portions of forage material, chopped into about 20-mm length, were mixed well and packed into plastic bags (Hiryu KN type, 180 × 260 cm; Asahikasei, Tokyo, Japan). The bags were sealed with a vacuum sealer (BH 950; Matsushita, Tokyo, Japan). The plastic-bag silos were stored at a room at 25°C. There were ten bag silos per treatment. One bag of silo per treatment was opened on d 60. Samples were dried in a forced-air oven at 60°C for 48 h, ground to pass through a 1-mm screen with a Wiley mill (ZM200, Retch GmbH & Co. KG, Haan, Germany), and used for chemical analysis and in vitro digestibility measurements.

## 2.2. Chemical analyses

The TMR silages, vegetable and its silages were dried in a forced draft oven at 60°C for 48 h and ground into a 2-mm powder with a sample mill (Foss Tecator; Akutalstuku, Tokyo, Japan). Moisture, ash, crude protein, ether extract, and crude fiber contents were determined by general methods. Analyses of neutral detergent fiber and acid detergent fiber contents were made following Van Soest et al. (1991). Heat-stable amylase and sodium sulfite were used in the neutral detergent fiber procedure, and the results were expressed without residual ash. Nonfibous carbohydrate was calculated by the formula as: Nonfibous carbohydrate = organic matter – crude protein – nonfibous carbohydrate – ether extract. The fermentation products of silages were determined using cold-water extracts. Wet silage (50 g) was homogenized with 200 ml sterilized distilled water and stored at 4°C overnight (Cai et al. 1999). The pH of the silages was determined using a glass electrode pH meter (D-21, Horiba, Kyoto, Japan). Lactic acid was analyzed using the methods of Cai et al. (1999). Ammonia- N was determined as described by (Cai et al. 2003). To measure total VFA, silage and ruminal fluid were steam-distilled and titrated using sodium hydroxide. Dried VFA salt was separated and quantified using gas chromatography (G-5000A, Hitachi, Tokyo, Japan) equipped with a thermal conductivity detector and a stainless column (Unisole F-200, 3.2 mm × 2.1 m). The analytical conditions were as follows: column oven temperature, 140°C; injector temperature, 210°C; detector temperature, 250°C. V-score, which was used to assess silage quality, was determined from the proportion ammonia-N in the total nitrogen and VFA contents in the silage.

## 2.3. Cultures and incubations

Two adult wethers (average initial body weight, 78.5 kg) fitted with rumen cannulae were used as donors of ruminal fluid. The wethers were fed a basal diet of 50% reed canary grass (Phalaris arundinacea L.) hay and 50% commercial feed concentrate (Koushi-Ikusei-Special, Kitanihon-Kumiai-Feed, Miyagi, Japan) at maintenance energy level (2.0% DM of their body weight) and had free access to clean drinking water. They were fed once daily at 09:00 h. Wethers were cared for in accordance with the animal care and use institutional guidelines for animal experiments at the Faculty of Agriculture, Yamagata University (Tsuruoka, Japan).

Rumen fluid was collected through the rumen cannulae 2 h after feeding and diverted to plastic bottles. The fluid was filtered through four layers of cheesecloth and combines on an equal volume basis. The combined filtrate was mixed with $CO_2$-bubbled McDougall's artificial saliva (pH 6.8) at a ratio of 1:4 (vol/vol). Then 50 mL buffered rumen fluid was transferred to 128-mL serum bottles containing 0.5 g sample, and flushed with $O_2$-free $CO_2$. Tubes were capped with a butyl rubber stopper and sealed with an aluminum cap. Incubations were performed in triplicate at 39°C for 6 h (Mohammed et al., 2004) in a water bath with a reciprocal shaker (100 strokes/min).

## 2.4. Analysis of fermentation products

To terminate fermentation at the end of incubation, 25 μL of formaldehyde solution (35%) were injected into serum bottles, which were immediately sealed and cooled at room temperature. Gas samples were collected by air syringe from the serum bottles and injected into a gas chromatograph (GC323, GL Sciences, Tokyo, Japan) equipped with a thermal conductivity detector and a stainless steel column (WG-100 SUS, 1.8 m × 6.35 mm OD), and the methane production in each serum bottle was measured. The analytical conditions were as follows: column oven temperature at 50°C, injector temperature at 50°C, and detector temperature at 50°C.

## 2.5. *In vitro* DM digestibility, and methane and VFA production

Separate sub-samples of the supernatant were taken to determine the pH and VFA concentration. The bottles were rinsed with warm water to remove all solid residues, which were then oven-dried at 60°C and stored for further analyses. In total, 2 g of dried residue were oven-dried at 135°C and stored to determine DM digestibility.

## 2.6. Statistical analyses

Analyses were performed using the general linear model procedure (SAS institute, Cary, NC, USA). Data on fermentative characteristics, in vitro DM digestibility, and ruminal methane and VFA production of TMR silages were subjected to one-way analysis of variance Tukey's test was used to identify differences ($P < 0.05$) between means. Data on the fermentative characteristics of each vegetable silage opened on d 30 were analyzed by one-way analysis of variance. Data on chemical compositions, fermentative characteristics, in vitro ruminal DM digestibility, and fermentation products after 6-h incubation of silages opened on d 60 were analyzed using a completely randomized design with a 4 × 4 (vegetable residues × additive treatment) factorial treatment structure. The general linear model procedure of SAS version 9.0 (SAS Institute, Inc., Cary, NC) was used for the analysis, and the model included the main effects of vegetable residues and additive treatment, and their interactions. Sealing time and ensiling duration were excluded from the model because the 60-d silages, wastes processed, and silos were made only one time. The Tukey test was used to identify differences ($P < 0.05$) between means.

## 3. Results and discussion

### 3.1. Chemical composition of materials and silage

The contents of DM, crude protein, ether extract, nonfibous carbohydrate, ash, and neutral detergent fiber in molasses were 72.7, 4.3, 0.7, 83.6, 11.4, and 0%, respectively (Table 1). The contents of organic matter and crude protein in WCR were 86.5 and 5.3%, respectively. The contents of crude protein, nonfibous carbohydrate, and neutral detergent in the tofu cake were 30.1, 15.8, and 37.7%, respectively. And the chemical composition of vegetable residues is shown in Table 2. The DM of the four types of vegetable residues was less than 6%. Their OM contents were more than 70% of DM, and the crude protein and neutral detergent fiber contents were approximately 20% and 30% of DM, respectively. The water-soluble carbohydrates (including sucrose, glucose, and fructose) contents ranged from 8.4 to 21.7% of DM; the highest and lowest values were observed in white cabbage and lettuce residues, respectively.

Preparing TMR silage is one practice whereby food by-products are stored and utilized as ruminant feeds in Japan. It can avoid energy costs associated with drying, and may improve odors and flavors of unpalatable feed resources through fermentation in a silo (Cao et al., 2009a; Imai 2001; Wang & Nishino 2008). We have performed a number of experiments to investigate the effects of food by-products including tofu cake, rice bran, and

| | WCR[1] | Concentrate[2] | Beet pulp | TC[3] | Molasses[4] |
|---|---|---|---|---|---|
| DM[5] (%) | 36.0 | 88.2 | 90.7 | 91.3 | 72.7 |
| CP[6] (% DM) | 5.3 | 16.7 | 8.4 | 30.1 | 4.3 |
| EE[7] (% DM) | 2.2 | 3.8 | 0.7 | 12.2 | 0.7 |
| NFC[8] (% DM) | 32.1 | 60.1 | 33.7 | 15.8 | 83.6 |
| Ash (% DM) | 13.5 | 5.1 | 5.1 | 4.3 | 11.4 |
| ADF[9] (% DM) | 30.2 | 8.7 | 25.6 | 22.2 | – |
| NDF[10] (% DM) | 48.0 | 14.4 | 52.1 | 37.7 | 0 |

[1]Whole crop rice; [2]Formula feed ("Koushi Ikusei Special Mash" made by Zenno with 120g $kg^{-1}$ crude protein in fresh matter); [3]Tofu cake; [4]Wang & Goetsch (1998); [5]Dry matter; [6]Crude protein; [7]Ether extract; [8]Non-fibrous carbohydrate (100 – crude protein – ether extract – neutral detergent fiber – ash); [9]Acid detergent fiber; [10]Neutral detergent fiber.

**Table 1.** Chemical composition of WCR, concentrate, beet pulp and tofu cake used in total mixed ration silages (Cao et al., 2010b)

| Item | White cabbage | Chinese cabbage | Red cabbage | Lettuce |
|---|---|---|---|---|
| DM, % | 3.6 ± 0.02 | 2.2 ± 0.01 | 5.2 ± 0.02 | 2.0 ± 0.01 |
| OM, % of DM | 81.6 ± 0.30 | 70.8 ± 0.08 | 87.3 ± 0.24 | 75.4 ± 0.09 |
| CP, % of DM | 21.9 ± 0.01 | 20.6 ± 0.21 | 22.8 ± 0.32 | 21.1 ± 0.41 |
| Ether extract, % of DM | 2.8 ± 0.07 | 1.5 ± 0.02 | 1.7 ± 0.14 | 4.9 ± 0.07 |
| ADF, % of DM | 31.0 ± 0.20 | 26.5 ± 1.49 | 31.6 ± 0.42 | 29.4 ± 0.14 |
| ADL, % of DM | 2.3 ± 0.19 | 5.8 ± 0.02 | 3.2 ± 0.03 | 5.9 ± 0.30 |
| NDF, % of DM | 32.8 ± 0.56 | 27.7 ± 0.24 | 33.6 ± 0.40 | 31.4 ± 0.01 |
| Sucrose, % of DM | 12.1 ± 0.15 | 1.1 ± 0.02 | 6.3 ± 0.10 | 3.1 ± 0.13 |
| Glucose, % of DM | 4.3 ± 0.04 | 3.6 ± 0.07 | 2.5 ± 0.01 | 1.3 ± 0.04 |
| Fructose, % of DM | 5.3 ± 0.12 | 3.8 ± 0.05 | 3.7 ± 0.10 | 4.0 ± 0.18 |

[1]Values are means ± SD.

**Table 2.** Chemical composition1 of vegetables residues (Cao et al., 2011)

| | Treatment | | | | SEM[3] |
|---|---|---|---|---|---|
| | Control | M[1] | LAB[2] | M+LAB | |
| Ingredient | | | | | |
| LAB (mg kg$^{-1}$ FM[4]) | - | - | 5 | 5 | |
| M (% FM) | - | 4 | - | 4 | |
| Whole crop rice (% DM[5]) | 30 | 30 | 30 | 30 | |
| Concentrate[6] (% DM) | 25 | 25 | 25 | 25 | |
| Vitamin-mineral supplement[7] (% DM) | 1.5 | 1.5 | 1.5 | 1.5 | |
| Dried beet pulp (% DM) | 13.5 | 13.5 | 13.5 | 13.5 | |
| Tofu cake (% DM) | 30 | 30 | 30 | 30 | |
| Chemical composition | | | | | |
| DM (%) | 35.9 | 36.2 | 36.4 | 37.0 | 1.98 |
| Organic matter (% DM) | 92.6 | 92.5 | 92.6 | 92.3 | 0.25 |
| Crude protein (% DM) | 15.3 | 14.6 | 15.4 | 15.1 | 0.38 |
| Ether extract (% DM) | 5.1 | 5.3 | 5.5 | 5.6 | 0.24 |
| Nitrogen free extract (% DM) | 57.1 | 58.4 | 58.4 | 57.5 | 0.40 |
| Crude fiber (% DM) | 15.7 | 14.9 | 14.2 | 14.5 | 0.88 |
| NFC[8] (% DM) | 30.9 | 32.0 | 29.5 | 32.0 | 1.54 |
| Crude ash (% DM) | 7.4 | 7.5 | 7.4 | 7.7 | 0.25 |
| Acid detergent fiber (% DM) | 19.0 | 20.0 | 18.9 | 20.2 | 1.21 |
| Neutral detergent fiber (% DM) | 41.4 | 40.6 | 42.2 | 39.6 | 1.12 |

[1]Molassess; [2]Lactic acid bacteria (Lactobacillus plantarum); [3]Standard error of means; [4]Fresh matter; [5]Dry matter; [6]Formula feed ("Koushi Ikusei Special Mash" made by Zenno; total digestible nutrients, 70.0%; crude protein, 12.0% in fresh matter); [7]Commercial vitamin-mineral supplement product (Snow brand seed, Iwate, Japan); [8]Non-fibrous carbohydrate (100 – crude protein – ether extract – neutral detergent fiber – ash).

**Table 3.** Ingredient and chemical composition of total mixed ration silages (Cao et al., 2010b) wet green tea waste on fermentation quality of WCR-containing TMR silage. Our previous study (Cao et al., 2009a, b) showed that silages with 30% tofu cake had higher lactic acid content, compared to those with rice bran and green tea waste. Therefore, we prepared TMR silage using tofu cake, and in order to investigate if adding LAB or molasses can further increase lactic acid content of the silages with tofu cake, LAB and molasses were added into these silages in this study. LAB can increase the lactic acid content of a silage (Cai, 2001; Cai et al., 2003), and was well used to prepare silage. Molasses is a fermentable carbohydrate (Maiga & Schingoethe, 1997) and many researchers (Alli et al., 1984) have reported its successful use with grass silage. In addition, molasses is a food by-product of sugar beet and sugarcane production. Molasses with high water-soluble carbohydrates is used as a major energy source for meat or milk production (Araba et al., 2002; Granzin & Dryden, 2005; Sahoo & Walli, 2008; Wang & gotsch, 1998).

## 3.2. Fermentation quality

As indicated by the low pH value (around 4.0) and ammonia-N/total N content (2.83–2.97%), high lactic acid content (2.49–2.87%), and V-score (99.8) for the silages, the four TMR silages were well preserved (Table 4). Although the levels of moisture, pH, acetic acid, propionic acid, butyric acid, and ammonia-N/total N and V-score did not differ significantly, lactic acid contents for the silages with LAB and molasses+LAB were higher ($P = 0.005$) than the control and molasses silages.

It is well established that LAB play an important role in silage fermentation, and LAB values have become a significant factor in predicting the adequacy of silage fermentation and in determining whether to apply bacterial inoculants to silage materials. Generally, when LAB reaches at least $10^5$ (CFU/g of FM), silage can be well preserved (Cai 2001; Cai et al., 1999; Cai et al., 2003). However, the LAB values below $10^5$ and aerobic bacteria values above $10^6$ present in most WCR suggest that high-quality silage fermentation may need to be controlled using certain inoculants. The inoculant strain used in this study was Lactobacillus plantarum Chikuso-1; this strain promotes lactic acid fermentation and can grow well in low-pH environments. Therefore, silage prepared using this strain can promote the propagation of LAB, decrease pH, inhibit the growth of clostridia and aerobic bacteria, and improve silage quality (Cai et al., 1999).

| | Treatment | | | | SEM[3] | *P*-value |
|---|---|---|---|---|---|---|
| | Control | M[1] | LAB[2] | M+LAB | | |
| pH | 3.99 | 3.92 | 4.01 | 4.03 | 0.0391 | 0.585 |
| Lactic acid (% FM[4]) | 2.49[a] | 2.52[a] | 2.84[b] | 2.87[b] | 0.0880 | 0.005 |
| Acetic acid (% FM) | 0.09 | 0.09 | 0.10 | 0.09 | 0.0032 | 0.934 |
| Propionic acid (% FM) | 0.003 | 0.003 | 0.001 | 0.001 | 0.0009 | 0.574 |
| Butyric acid (% FM) | 0.002 | 0.003 | 0.003 | 0.003 | 0.0001 | 0.776 |
| ammonia-N/total-N (%) | 2.97 | 2.91 | 2.92 | 2.83 | 0.1775 | 0.956 |
| V-score | 99.8 | 99.8 | 99.8 | 99.8 | 0.0090 | 0.970 |

[1]Molassess; [2]Lactic acid bacteria (Lactobacillus plantarum); [3]Standard error of means; [4]Fresh matter; Means within a row with different letters (a, b) differ ($P < 0.05$).

**Table 4.** Fermentative characteristics of total mixed ration silages (Cao et al., 2010b)

The pH, lactic acid, acetic acid, and ammonia-N were affected not only by vegetable, but also by addition and vegetable × addition. Comparison among the four types of vegetable silages revealed that the pH was the lowest ($P < 0.001$) in silage with white cabbage residue, followed by red cabbage, Chinese cabbage, and lettuce silages. The lactic acid content was highest in white cabbage silage ($P < 0.001$), whereas the acetic acid content was highest in lettuce silage ($P < 0.001$), followed by red cabbage, white cabbage, and Chinese cabbage silages. Propionic and butyric acids were not detected among the four types of vegetable silages; the ammonia-N concentration of white cabbage, red cabbage, and lettuce silages was lower ($P < 0.001$) than that of Chinese cabbage silage. Comparison of the treated silages showed that all silages treated with LAB or BP had lower ($P < 0.001$) pH values and ammonia-N concentrations, but higher ($P < 0.001$) lactic acid contents compared with the control silage.

Alli et al. (1984) assessed the effects of molasses on the fermentation of chopped whole-plant Leucaena. Silages were treated with molasses at a rate of 2.25% or 4.5% fresh weight at the time of ensiling, which led to increased rates of lactic acid production, lower pH, decreased DM loss, and reduced levels of ammonia-N compared to Leucaena to which no molasses was added. In the present experiment, WCR-containing TMR silages were treated with or without molasses at the rate of 4% fresh weight. Although the addition of molasses did not significantly influence the chemical composition, it increased DM and Non-fibrous carbohydrate contents by 3.06 and 3.56%, respectively. Adding molasses did not increase lactic acid content significantly, Adding LAB and molasses+LAB, however, increased lactic acid content significantly. This is probably because that, even if no molasses, there was enough fermentable sugars in TMR silage, and the LAB may have converted more fermentable sugars to lactic acid (Cai 2001; Cai et al., 2003). This study showed the advantage of LAB over molasses. Alli et al. (1984) reported that adding molassesf reduced the ammonia-N of silage, although in present study, ammonia-N did not differ among the four silages.

Furthermore, our previous study (Cao et al., 2011) has determined the effect of LAB inoculant and beet pulp addition on silage fermentation quality and *in vitro* ruminal DM digestion of vegetable residues. The silage treated with LAB or beet pulp had a lower pH and a higher lactic acid content than the control silage (Table 4). After 6 h of incubation, the LAB-inoculated silage had the highest DM digestibility and the lowest methane production. Weinberg et al. (2003) and Filya et al. (2007) reported that LAB inoculants affected the in vitro digestibility of alfalfa hay and corn silage, respectively, after 48 h incubation. In the present study, LAB inoculants not only increased ($P < 0.01$) the silage DM digestibility after 6 h in vitro incubation but also decreased ruminal methane production, which decreases the energy loss of feed (Cao et al., 2010a). Furthermore, LAB inoculants improve the fermentation quality of vegetable silage, which might decrease the degradation of crude protein in the silage; therefore, LAB-treated silage had a high concentration of ruminal ammonia-N. Adding beet pulp to vegetable silage did not affect the DM digestibility after 6 h in vitro incubation, but did increase the production of acetic acid, propionic acid, and even methane, while decreasing the production of butyric acid and ammonia-N. We cannot explain the mechanism of these effects. More research is needed to elucidate the probiotic effect of adding LAB or beet pulp to vegetable silage in ruminants.

| Item | Moisture | pH | Lactic acid | Acetic acid | Ammonia-N |
|---|---|---|---|---|---|
| | % | | g/kg of FM | | |
| Vegetable residue means | | | | | |
| White cabbage | 82.4 | 3.59[c] | 20.9[a] | 1.5[c] | 0.26[b] |
| Chinses cabbage | 82.9 | 4.26[a] | 10.8[b] | 1.3[c] | 0.41[a] |
| Red cabbage | 80.8 | 3.74[b] | 12.7[b] | 2.0[b] | 0.26[b] |
| Lettuce | 83.9 | 4.27[a] | 10.4[b] | 3.3[a] | 0.28[b] |
| Additive treatment means | | | | | |
| Control | 95.4[a] | 4.46[a] | 8.2[c] | 2.4[a] | 0.52[a] |
| LAB | 95.1[a] | 3.95[b] | 10.8[b] | 1.9[b] | 0.26[b] |
| BP | 69.1[b] | 3.77[c] | 17.6[a] | 2.5[a] | 0.23[bc] |
| LAB+BP | 70.3[b] | 3.68[c] | 18.3[a] | 1.4[b] | 0.20[c] |
| Significance of main effects and interactions | | | | | |
| Vegetable residues (V) | 0.249 | <0.001 | <0.001 | <0.001 | <0.001 |
| Additive treatment (A) | <0.001 | <0.001 | <0.001 | <0.001 | <0.001 |
| V × A | 0.841 | <0.001 | <0.001 | <0.001 | <0.001 |

[1]Means within columns with different letters (a-c) differ ($P < 0.05$); [2]Propionic and Butyric acids were not detected; [3]Lactobacillus plantarum (Chikuso-1, Snow Brand Seed, Sapporo, Japan); [3]Beet pulp.

**Table 5.** Fermentation profile of vegetable residue silages prepared with LAB and BP after 60 days of storage (Cao et al., 2011)

The pH, lactic acid, acetic acid, and ammonia-N were affected not only by vegetable, but also by addition and vegetable × addition （Table 5）. Comparison among the four types of vegetable silages revealed that the pH was the lowest ($P < 0.001$) in silage with white cabbage residue, followed by red cabbage, Chinese cabbage, and lettuce silages. The lactic acid content was highest in white cabbage silage ($P < 0.001$), whereas the acetic acid content was highest in lettuce silage ($P < 0.001$), followed by red cabbage, white cabbage, and Chinese cabbage silages. Propionic and butyric acids were not detected among the four types of vegetable silages; the ammonia-N concentration of white cabbage, red cabbage, and lettuce silages was lower ($P < 0.001$) than that of Chinese cabbage silage. Comparison of the treated silages showed that all silages treated with LAB or BP had lower ($P < 0.001$) pH values and ammonia-N concentrations, but higher ($P < 0.001$) lactic acid contents compared with the control silage.

The factors involved in fermentation quality include not only the physiological properties of epiphytic bacteria but also the chemical composition of the silage material (Cai et al., 1999). In this study, the four types of vegetable residues had relatively high water-soluble carbohydrates contents; the epiphytic LAB transformed water-soluble carbohydrates into organic acids during the ensiling process, and as a result, the pH was reduced, which inhibited the growth of some microorganisms, such as bacilli, coliform bacteria, aerobic bacteria, yeasts, and molds. When silage was treated with LAB or BP, the fermentation tended to ensure rapid and vigorous results with the faster accumulation of lactic acid (Table 3) and lower pH values at earlier stages of ensiling, and it also inhibited the production of acetic acid and ammonia-N during silage fermentation and thus improved vegetable-residue conservation. The transitional behavior of the VFA in the silage during fermentation indicated sharp decreases in pH, and corresponding increases in lactic acid contents at earlier stages (7 d of ensiling) were typical of a good fermentation process and were in agreement with previous studies. Subsequently, a steady reduction in pH depicted stability, while lactic acid contents gradually stabilized after a decrease during storage. Some studies (Cai, 2001; Cai et al., 1999; Cai et al., 2003) have shown that the development of LAB peaks in the first 7 d in parallel with the rise in lactic acid concentration of silage, and this is followed by decreases in LAB numbers; however, no apparent decrease in LAB numbers was observed in this study. The d 60 LAB-treated silage had higher organic matter and crude protein, but lower water-soluble carbohydrates than did control silages. During silage fermentation, LAB could effectively utilize water-soluble carbohydrates to produce sufficient lactic acid to reduce pH and inhibit the growth of harmful bacteria; therefore, the resulting silage was of good quality. Furthermore, the moisture content of silage material is also a major factor influencing silage fermentation (Cai et al., 2003). An intrinsic characteristic of vegetable residues is their very high moisture content (95 to 98% of FM), and this is a major limitation to its use as livestock feed. Although dried vegetable residues can easily be incorporated into rations, the energy cost associated with drying wet vegetable residues has been increasing. Moreover, the risk of effluent production is high because of the low DM content. Therefore, pressed vegetable residues have been preferred for adjusting moisture with other feed to ensile. Cai et al. (1999) reported that high-moisture silage is more beneficial to lactic acid fermentation and has less risk of heat damage than low-moisture silage. In this study, according to our preliminary experiment and taking into consideration the cost of feed, the moisture of BP-treated silage was adjusted to 70%, and most beet pulp-treated silages had lower pH and ammonia-N and higher lactic acid content compared with control silage. It is possible that this is because the addition of BP not only adjusted the moisture content of the vegetable residues but also increased the water soluble carbohydrates content; therefore, silages with added BP could greatly contribute to better lactic acid fermentation. Furthermore, we used a small-scale system of silage fermentation; all silages stored well and maintained high quality without aerobic deterioration in this study.

## 3.3. *In vitro* DM digestibility, and methane and VFA production

After in vitro 6 h incubation, DM digestibility, total VFA, acetic acid, isovaleric acid, valeric acid, and the acetic to propionic acid ratio did not differ significantly among the treatments (Table 6). However, methane production for the LAB silage and the molasses silage tended ($P = 0.065$) to decrease and increase, respectively, propionic acid for the LAB silage tended ($P = 0.061$) to increase, and butyric acid for the control silage was higher ($P = 0.008$) than the other silages.

| | Treatment | | | | SEM[3] | *P*-value |
|---|---|---|---|---|---|---|
| | Control | M[1] | LAB[2] | M+LAB | | |
| DM[4] digestibility (%) | 42.2 | 44.5 | 44.8 | 44.5 | 1.03 | 0.313 |
| Methane production (L $kg^{-1}$ DDM[5]) | 10.5 | 11.2 | 9.6 | 10.2 | 0.30 | 0.065 |
| Total VFA (mmol 100 $ml^{-1}$) | 5.3 | 5.8 | 5.9 | 5.7 | 0.16 | 0.340 |
| Acetic acid (A) (mol %) | 37.0 | 38.9 | 38.0 | 38.8 | 0.55 | 0.142 |
| Propionic acid (P) (mol %) | 39.9 | 40.4 | 41.8 | 40.5 | 0.35 | 0.061 |
| Butyric acid (mol %) | 19.9[a] | 17.8[b] | 17.0[b] | 17.8[b] | 0.34 | 0.008 |
| Isovaleric acid (mol %) | 0.4 | 0.3 | 0.3 | 0.3 | 0.06 | 0.844 |
| Valeric acid (mol %) | 2.8 | 2.5 | 2.5 | 2.6 | 0.10 | 0.416 |
| A/P | 0.9 | 1.0 | 0.9 | 1.0 | 0.02 | 0.271 |

[1]Molassess; [2]Lactic acid bacteria (Lactobacillus plantarum); [3]Standard error of means; [4]Dry matter; [5]Digestible dry matter; Means within a row with different letters (a, b) differ ($P < 0.05$).

**Table 6.** *In vitro* dry matter digestibility, methane production and volatile fatty acid concentration after 6 hours incubation of total mixed ration silages (Cao et al., 2010b)

DM digestibility, VFA, and ammonia-N concentrations of vegetable-residue silage after 6 h incubation in vitro are shown in Table 7. Although DM digestibility was not influenced by vegetable, it was influenced by addition and by vegetable × addition; VFA, and ammonia-N were influenced by vegetable, addition, and vegetable × addition. DM digestibility did not differ among silages. However, ruminal $CH_4$ production of white and Chinese cabbage silages was lower ($P < 0.001$) than that of red cabbage and lettuce silages, and the total VFA production of red cabbage and lettuce residue silages was higher ($P = 0.014$) than that of Chinese cabbage silage. The acetic acid production of the lettuce silage was higher ($P < 0.001$) than that of the white cabbage silage. The propionic acid production of white cabbage was the highest ($P < 0.001$) among the four vegetable residues, followed by lettuce, which showed higher propionic acid production ($P < 0.001$) than did red or Chinese cabbage; the last two silages did not differ in this regard. Red cabbage had higher ($P < 0.001$) butyric acid production than Chinese cabbage and lettuce, and butyric acid production was higher in white cabbage ($P < 0.001$) than in Chinese cabbage. The A:P ratio of white cabbage silages was the lowest ($P < 0.001$) among the four types of vegetable silages. The highest and lowest ($P < 0.001$) ammonia-N production was found in white cabbage and lettuce silages, respectively. The LAB-treated silage had a higher ($P < 0.001$) DM digestibility than BP- and beet pulp+LAB-treated silages; it also had the highest ammonia-N production. Together with the control silage, LAB treated silage had lower ($P < 0.001$) total VFA, acetic acid, and propionic acid production, but higher ($P < 0.001$) butyric acid production and acetic acid:propionic acid ratio ratio compared with beet pulp- and LAB+beet pulp-treated silages.

In vitro DM digestibility was higher in silage with LAB than without LAB because LAB reduces DM loss in silage fermentation (Cai 2001; Cai et al., 2003). Furthermore, although there are some reports that adding molasses has no effect on DM digestibility (Granzin & Dryden 2005; Wang & Goetsch 1998), many more studies (Shellito et al., 2006; Sahoo & Walli 2008) have reported that diets with molasses have higher ruminal DM digestibility.

In the present experiment, there was a non-significant increasing trend in DM digestibility with molasses, LAB and molasses+LAB. Ruminal methane production and the molar proportion of propionic acid for silage with LAB decreased by 8.6% and increased by 4.8%, respectively. These might be because that adding LAB increased lactic acid content in the silage, when the silage containing high lactic acid content was incubated in vitro, there are two known mechanisms for the conversion of lactic acid or pyruvic acid to propionic acid, and when lactate acid is secondarily fermented by lactate-utilizing bacteria such as Megasphaera elsdenii, Selenomonas ruminantium, and Veillonella parvula, propionate is generally produced as a major product (Dawson et al., 1997) and this can reduce methanogenesis because electrons are used during propionate formation. But adding molasses, which has a high sugar content, may augment methane production in the rumen (Hindrichsen et al., 2005), perhaps because of which, molasses per se canceled (compensated) the effect of lactic acid content on methane production. A further research is necessary about the effect of molasses and the complex effects of molasses and LAB concerning the methane production in TMR silage. Furthermore, it is not yet clear why adding molasses or LAB decreased in vitro ruminal butyric acid in this study.

| Item | DM digestibility | Total VFA | Acetic acid | Propionic acid | A/P[1] | Ammonia-N |
|---|---|---|---|---|---|---|
| | % | mmol/L | | | | mg/L |
| Vegetable residue means | | | | | | |
| White cabbage | 44.9 | 43.0$^{ab}$ | 24.8$^{b}$ | 9.1$^{a}$ | 2.7$^{b}$ | 112.7$^{a}$ |
| Chinese cabbage | 38.6 | 41.9$^{b}$ | 28.0$^{ab}$ | 7.4$^{bc}$ | 3.8$^{a}$ | 88.3$^{b}$ |
| Red cabbage | 44.3 | 44.8$^{a}$ | 27.5$^{ab}$ | 6.7$^{c}$ | 4.3$^{a}$ | 91.8$^{b}$ |
| Lettuce | 41.6 | 44.7$^{a}$ | 29.2$^{a}$ | 7.6$^{b}$ | 4.0$^{a}$ | 75.1$^{c}$ |
| Additive treatment means | | | | | | |
| Control | 41.3$^{ab}$ | 40.1$^{b}$ | 24.1$^{b}$ | 6.0$^{b}$ | 4.2$^{a}$ | 122.4$^{b}$ |
| LAB | 47.5$^{a}$ | 42.2$^{b}$ | 25.7$^{b}$ | 6.5$^{b}$ | 4.1$^{a}$ | 164.1$^{a}$ |
| BP | 39.8$^{b}$ | 45.6$^{a}$ | 29.3$^{a}$ | 9.1$^{a}$ | 3.2$^{b}$ | 40.3$^{c}$ |
| BP+LAB | 40.6$^{b}$ | 46.4$^{a}$ | 30.4$^{a}$ | 9.2$^{a}$ | 3.4$^{b}$ | 41.2$^{c}$ |
| Significance of main effects and interactions | | | | | | |
| Vegetable residues (V) | 0.056 | 0.014 | 0.014 | <0.001 | <0.001 | <0.001 |
| Additive treatment (A) | 0.001 | <0.001 | <0.001 | <0.001 | <0.001 | <0.001 |
| V × A | <0.001 | 0.009 | 0.01 | <0.001 | <0.001 | <0.001 |

Means within columns with different letters (a-c) differ ($P < 0.05$).

[1]Acetic acid/propionic acid ratio.

[2]Digestible dry matter.

[3]Lactobacillus plantarum (Chikuso-1, Snow Brand Seed, Sapporo, Japan).

[4]Beet pulp.

**Table 7.** Measurements of dry matter digestibility, methane production, VFA concentration and ammonia-N after 6-h in vitro incubation with rumen fluid of vegetable residue silage (Cao et al., 2011)

## 4. Conclusions

The results of the present study show that adding LAB increased the lactic acid content of silage, had the potential to increase DM digestibility and to decrease ruminal methane production.

## Author details

Yang Cao
*College of Animal Science & Veterinary Medicine, Heilongjiang Bayi Agricultural University, Daqing, China*

Yimin Cai
*Japan International Research Center for Agricultural Sciences, Tsukuba, Japan*

Toshiyoshi Takahashi
*Faculty of Agriculture, Yamagata University, Tsuruoka, Japan*

## 5. References

Alli, I., Fairbairn, R., Noroozi, E. & Baker, B. E. (1984). The effects of molasses on the fermentation of chopped whole-plant leucaena. *Journal of the Science of Food and Agriculture*, Vol. 35, No. 3, (March 1984), pp. 285–289, ISSN 0022-5142

Araba, A., Byers, F. M. & Guessous, F. (2002). Patterns of rumen fermentation in bulls fed barley/molasses diets. *Animal Feed Science and Technology*, Vol. 97, No. 1, (May 2002), pp. 53–64, ISSN 0377-8401

Cai, Y. (2001). The role of lactic acid bacteria in the preparation of high fermentation quality. *Japanese Journal of Grassland Science*, Vol. 47. No.5, (December 2001), pp. 527–533, ISSN 0447-5933

Cai, Y., Benno, Y., Ogawa, M. & Kumai, S. (1999). Effect of applying lactic acid bacteria isolated from forage crops on fermentation characteristics and aerobic deterioration of silage. *Journal of Dairy Science*, Vol. 82. No.3, (March 1999), pp. 520–526, ISSN 0022-0302

Cai, Y., Fujita, Y., Murai, M., Ogawa, M., Yoshida, N., Kitamura, R. & Miura, T. (2003). Application of lactic acid bacteria (*Lactobacillus plantarum* Chikuso-1) for silage preparation of forage paddy rice. *Japanese Journal of Grassland Science*, Vol. 49. No. 5, (December 2003), pp. 477–485, ISSN 0447-5933

Cao, Y., Cai Y., Takahashi ,T., Yoshida, N., Tohno, M., Uegaki, R., Nonaka, K. & Terada, F. (2011). Effect of lactic acid bacteria inoculant and beet pulp addition on fermentation characteristics and in vitro ruminal digestion of vegetable residue silage. *Journal of Dairy Science*, Vol. 94, No. 8, (March 2011), pp. 3902–3912, ISSN 0022-0302

Cao, Y., Takahashi, T. & Horiguchi, K. (2009a). Effects of addition of food by-procucts on the fermentation quality of a total mixed ration with whole crop rice and its digestibility, preference, and rumen fermentation in sheep. *Animal Feed Science and Technology*, Vol. 151, No. 1-2, (May 2009), pp. 1–11, ISSN 0377-8401

Cao, Y., Takahashi, T. & Horiguchi, K. (2009b). Effect of food by-products and lactic acid bacteria on fermentation quality and in vitro dry matter digestibility, ruminal methane and volatile fatty acid production in total mixed ration silage with whole-crop rice silage. *Japanese Journal of Grassland Science,* Vol. 55, No. 1, (April 2009), pp. 1–8, ISSN 0477-5933

Cao, Y., Takahashi, T., Horiguchi, K., Yoshida, N. & Cai., Y. (2010a). Methane emissions from sheep fed fermented or non-fermented total mixed ration containing whole-crop rice and rice bran. *Animal. Feed Science and Technology,* Vol. 157, No. 1, (April 2010), pp. 72–78, ISSN 0377-8401

Cao, Y., Takahashi, T., Horiguchi, K. & Yoshida, N. (2010b). Effect of adding lactic acid bacteria and molasses on fermentation quality and in vitro ruminal digestion of total mixed ration silage prepared with whole crop rice. *Grassland Science,* Vol. 56, No. 1, (March 2010), pp. 19–25 ISSN 1744-6961

Dawson, K. A., Rasmussen, M. A. & Allison, M. J. (1997). Digestive disorgers and nutritional toxicity. In: *The Rumen Microbial Ecosystem,* P. N. Hobson, C. S. Stewart, (Ed). 633-660, Blackie Academic and Professional, ISBN10 0751403660, London

Enishi, O., & Shijimaya, K. (1998). Changes of chemical composition of plant parts and nutritive value of silage in male sterile rice plant (*Oryza sativa* L.). *Grassland Science,* Vol. 44, No. 3, (October 1998), pp. 260–265, ISSN 1744-6961

Filya, I., Muck, R. E. & Contreras-Govea, F. E. (2007). Inoculant effects on alfalfa silage: Fermentation products and nutritive value. *Journal of Dairy Science,* Vol. 90, No. 11, (November 2007), pp. 5108–5114, ISSN 0022-0302

Granzin, B. C. & Dryden, G. M. (2005). Monensin supplementation of lactating cows fed tropical grasses and cane molasses or grain. *Animal Feed Science and Technology,* Vol. 120, No. 1-2 (May 2005) pp. 1–16, ISSN 0377-8401

Han, Y. W., & Anderson, A. W. (1974). The problem of rice straw waste. A possible feed through fermentation. *Economic Botany,* 28, pp. 338–344, ISSN 0013-0001

Hindrichsen, I. K., Wettstein, H. R., Machmuller, A., Jörg, B. & Kreuzer, M. (2005). Effect of the carbohydrate composition of feed concentratates on methane emission from dairy cows and their slurry. *Environmental Monitoring and Assessment,* Vol. 107, No. 1-3, pp. 329–350, ISSN 0167-6369

Imai, A. (2001). Silage making and utilization of high moisture by-products. *Japanses Journal of Grassland Science,* Vol. 47, No. 3, (August 2001), pp. 307–310, ISSN 0447-5933

Maiga, H. A. & Schingoethe, D. J. (1997). Optimaizing the utilization of animal fat and ruminal bypass proteins in the diets of lactating dairy cows. *Journal of Dairy Science,* Vol. 80, No.2, (February 1997), pp. 343–352, ISSN 0022-0302

Sahoo, B. & Walli, T. K. (2008). Effects of formaldehyde treated mustard cake and molasses supplementation on nutrient utilization, microbial protein supply and feed efficiency in growing kids. *Animal Feed Science and Technology,* Vol. 142, No. 3-4, (May 2008), pp. 220–230, ISSN 0377-8401

Shellito, S. M., Ward, M. A., Lardy, G. P., Bauer ,M. L. & Caton, J. S. (2006). Effects of concentrated separator by-product (desugared molasses) on intake, ruminal fermentation, digestion, and microbial efficiency in beef steers fed grass hay. *Journal of Animal Science*, Vol. 84, No. 6, (Joune 2006), pp. 1535–1543, ISSN 0021-8812

Van Soest, P. J., Robertson, J. B. & Lewis, B. A. (1991). Methods for dietary fiber, neutral detergent fiber, and non-starch polysaccharides in relation to animal nutrition. *Journal of Dairy Science*, Vol. 74, No. 10, (October 1991), pp. 3583–3597, ISSN 0022-0302

Wang, Z. S. & Goetsch, A. L. (1998). Intake and digestion by Holstein steers consuming diets based on litter harvested after different numbers of broiler growing periods or with molasses addition before deep-stacking. *Journal of Animal Science*, Vol. 76, No. 3, (March 1998), pp. 880–887, ISSN 0021-8812

Wang, F. J. & Nishino, N. (2008). Effect of aerobic exposure after silo opening on feed intake and digestibility of total mixed ration silage containing wet brewers grains or soybean curd residue. *Grassland Science*, Vol. 54, No. 3, (September 2008), pp. 164–166, ISSN 1744-6961

Weinberg, Z. G., Ashbell, G. & Chen, Y. (2003). Stabilization of returned dairy products by ensiling with straw and molasses for animal feeding. *Journal of Dairy Science*, Vol. 86, No. 4, (April 2003), pp. 1325–1329, ISSN 0022-0302

Xu, C. C., Suzuki, H. & Toyokawa, K. (2001). Characteristics of ruminal fermentation of sheep fed tofu cake silage with ethanol. *Animal Science Journal*, Vol. 72, No. 4, (July 2001), pp. 299–305, ISSN 1344-3941

Yamamoto, Y., Deguchi, Y., Mizutani, M., Urakawa, S., Yamada, H,. Hiraoka, H., Inui, K., Kouno, S. & Goto, M. (2004). Improvement of fermentation quality and dry matter digestibility of rice whole crop silage treated with fermented juice of epipytic lactic acid bacteria and mechanical processing. *Japanese Journal of Grassland Science*, Vol. 49, No. 6, (February 2004), pp. 665–668, ISSN 0447-5933

Section 6

# Fish & Seafood Products

Chapter 16

# Lactic Acid Bacteria and Their Bacteriocins: A Promising Approach to Seafood Biopreservation

Mahdi Ghanbari and Mansooreh Jami

Additional information is available at the end of the chapter

http://dx.doi.org/10.5772/50705

## 1. Introduction

The growing interest in a correct life style, including alimentation, and the parallel attention on food quality have contributed to orientate consumers towards fishery products which are considered safe, of high nutritional value and capable of influencing human health in a positive way [1]. The diverse nutrient composition of seafood makes it an ideal environment for the growth and propagation of spoilage micro-organisms and common food-borne pathogens [2]. It has been estimated that as much as 25% of all food produced is lost post-harvest owing to microbial activity [1,2]. It has been mentioned that as many as 30% of people in industrialized countries suffer from a food borne disease each year and in 2000 at least two million people died from diarrhoeal disease worldwide. It is clear that indigenous bacteria present in marine environment as well as the result of post contamination during process are responsible for many cases of illnesses [3,4]. In the last years, the traditional processes applied to seafood like salting, smoking and canning have decreased in favor of mild technologies involving lower salt content, lower cooking temperature and vacuum (VP) or modified atmosphere packing (MAP). The treatments are usually not sufficient to destroy microorganisms and in some cases psychrotolerant pathogenic and spoiling bacteria can develop during the extended shelf-life of these products [2,5]. As several of these products are eaten raw, it is therefore essential that adequate preservation technologies are applied to maintain its safety and quality. Among alternative food preservation technologies, particular attention has been paid to biopreservation to extent the shelf-life and to enhance the hygienic quality, minimizing the impact on the nutritional and organoleptic properties of perishable food products such as seafood [1,6]. Biological preservation refers to the use of a natural or controlled microflora and/or its antimicrobial metabolites to extend the shelf life and improve the safety of food. Lactic acid bacteria (LAB) are particularly interesting candidates for this technique [1,2,6,7]. Indeed, they are frequently naturally present in food products and are often strong competitors, by producing a wide range of antimicrobial metabolites such as organic acids, diacetyl, acetoin, hydrogen peroxide, reuterin, reutericyclin, antifungal peptides, and bacteriocins [8-10). Hence, the last two decades have seen intensive investigation on LAB and their metabolites to discover new LAB strains that can be used in food preservation [1,7,11-13].

## 2. Bacterial hazards associated with fish and fish products

From the viewpoint of microbiology, fish and related products are a risky foodstuff group. Pathogenic bacteria associated with seafood can be categorized into three general groups [14]: 1) Bacteria (indigenous bacteria) that belong to the natural microflora of fish (*Clostridium botulinum*, pathogenic *Vibrio* spp., *Aeromonas hydrophila*); 2) Enteric bacteria (non-indigenous bacteria) that are present due to faecal contamination (*Salmonella* spp., *Shigella* spp., pathogenic *Escherichia coli*, *Staphylococcus aureus*); and 3) bacterial contamination during processing, storage, or preparation for consumption (*Bacillus cereus, Listeria monocytogenes, Staphylococcus aureus, Clostridium perfringens, Clostridium botulinum, Salmonella* spp.).

*Vibrio parahaemolyticus* has been isolated from sea and estuary waters on all continents with elevated sea water temperatures. *V. parahaemolyticus* is frequently isolated from fish, molluscs, and crustaceans throughout the year in tropical climates and during the summer months in cold or temperate climates [15]. Fish food associated with illnesses due to consumption of *V. parahaemolyticus* includes fish-balls, fried mackerel (*Scomber scombrus*), tuna (*Thunnus thynnus*), and sardines (*Sardina pilchardus*). These products include both raw and undercooked fish products and cooked products that have been substantially recontaminated [9,15]. The most affected by the pathogens are Japan, Taiwan, and other Asian coastal regions, though cases of disease have been described in many countries and on many continents [9,16]. Cases of diseases caused by *V. parahaemolyticus* are occasional in Europe. During 20 years, only two cases of gastroenteritis were recorded in Denmark. The interest in this organism has been widened by the finding that similar organisms, *V. alginolyticus* and group of F *Vibrio* sp. also cause serious disease in humans [17]. *V. cholerae* is often transmitted by water but fish or fish products that have been in contact with contaminated water or faeces from infected persons also frequently serve as a source of infection [1,9,19]. The organism would be killed by cooking and recent cases of cholera in South America have been associated with the uncooked fish marinade seviche (*Cilus gilberti*) [18].

*E. coli* is a classic example of enteric bacteria causing gastroenteritis. *E. coli* including other coliforms and bacteria as *Staphylococcus* spp. and sometimes enterococci are commonly used as indices of hazardous conditions during processing of fish. Such organisms should not be present on fresh-caught fish [9,20,21]. The contamination fish derived food with pathogenic *E. coli* probably occurs during handling of fish and during the production process [20,22]. An outbreak of diarrhoeal illness caused by ingestion of food contaminated with enterotoxigenic *E. coli* was described in Japan [23]. The illness was strongly associated with eating tuna paste. Brazilian authors [24] isolated 18 enterotoxigenic strains of *E. coli* (ETEC) from 3 of 24 samples of fresh fish originating from Brazilian markets; 13 of them produced a thermolabile enterotoxin. Infection with verocytotoxin _ producing strains of *E. coli* (VTEC) after ingestion of fish was recorded in Belgium [25]. An outbreak caused by salted salmon roe contaminated, probably during the production process, with enterohaemorrhagic *E. coli* (EHEC) O157 occurred in Japan in 1998 [22]. The roe was stored frozen for 9 months but it appears that O157 could survive freezing and a high concentration of NaCl and retained its pathogenicity for humans [26].

*Aeromonas* spp. has been recognized as potential foodborne pathogens for more than 20 years. Aeromonads are ubiquitous in fresh water, fish and shellfish and also in meats and fresh vegetables [27]. The epidemiological results so far are, however, very questionable. The organism is very frequently present in many food products, including raw vegetables, and very rarely has a case been reported. Up to 8.1% of cases of acute enteric diseases in 458 patients in Russia were caused by *Aeromonas* spp. [28]. In this study, *Aeromonas* spp. isolates with the same pathogenicity factors were isolated from river water in the Volga Delta, from fish, raw meat, and from patients with diarrhoea. Most *Aeromonas* spp. isolates are psychrotrophic and can grow at refrigerator temperatures [29]. This could increase the hazard of food contamination, particularly where there is a possibility of cross-contamination with ready-to-eat food products.

*Salmonella* has been isolated from fish and fishery product, though it is not psychrotrophic or indigenous to the aquatic environment [30]. The relationship between fish and *Salmonella* has been described by several scientists; some believe that fish are possible carriers of *Salmonella* which are harbored in their intestines for relatively short periods of time and some believe that fish get actively infected by *Salmonella* [31]. Most outbreaks of food poisoning associated with fish derive from the consumption of raw or insufficiently heat treated fish and cross-contamination during processing and the U.S. Food and Drug Administration's (FDA) data showed that *Salmonella* was the most common contaminant of fish and fishery products [31]. The highest *Salmonella* incidence in fishery products was determined in Central Pacific and African countries while it was lower in Europe and including Russia, and North America [32]. The most common serovar found in the world was *S.* sub Weltvreden [30, 31]. In seafood the commonest serotype encountered was *S.* sub Worthington followed by *S.* sub Weltevreden.

Enterotoxins produced by *Staphylococcus aureus* are another serious cause of gastroenteritis after consumption of fish and related products. In 3 of 10 samples of fresh fish, higher counts of *Staph. aureus* were detected than permitted by Brazilian legislation [20, 33]. In the southern area of Brazil, *Staph. aureus* was isolated from 20% of 175 examined samples of fresh fish and fish fillets (*Cynoscion leiarchus*). *Staph. aureus* has also been detected during the process of drying and subsequent smoking of eels in Alaska in 1993 [34]. During the process, *S. aureus* populations increased to more than $10^5$ CFU $g^{-1}$ of the analyzed sample, after 2 to 3 days of processing. Subsequent laboratory studies showed that a pellicle (a dried skin-like surface) formed rapidly on the strips when there was rapid air circulation in the smokehouse and that bacteria embedded in/under the pellicle were able to grow even when heavy smoke deposition occurred.

In ready-to-eat products, cooking, preservation ingredients, and storage atmosphere inhibit the Gram-negative organisms, resulting in a longer shelf life. Such conditions favor the growth of psychotropic pathogens such as *Listeria monocytogenes*, allowing them to grow to dangerous levels [9,35,36]. *L. monocytogenes* is a serious threat to consumer health and safety and has been implicated in several deadly outbreaks around the world [1,2]. This organism is halotolerant (up to 28% w/v for short periods), resistant to freezing temperatures, can grow and multiply during refrigeration, where other competing organisms cannot, and is able to survive at low water activity (aw) [9,14,37]. *L. monocytogenes* is widely distributed in the general environment including fresh water, coastal water and live fish from these areas. Contamination or recontamination of seafood may also take place during processing [37-39]. Moreover, *L. monocytogenes* is a psychrotrophic pathogen with the ability to grow from under 0 to 45°C [40]. This ability to grow at storage temperatures means that this bacterium is the main hazard in this kind of product. The pathogenic bacteria *L. monocytogenes* may grow on fresh seafood. *Listeria* has been found in farmed rainbow trout [41]. The outbreak of listeriosis related to vacuum packed gravad and cold-smoked fish was described in at least eight human cases for 11 months in Sweden [42]. Cold-smoked and gravad rainbow trout (*Oncorhynchus mykiss*) and salmon (*Salmo salar*) have been focused on during recent years as potential sources of infection with *L. monocytogenes* and there are several report on isolation of this food borne pathogen from fish-processing plants environments [14,37,39,43-45]. Seafood treatment is necessary to prevent food-borne illness. However, the pervasive nature of *L. monocytogenes* makes it difficult for processors to fully eliminate the organism from the environment.

Development of new-generation foods, which are mildly processed, contain few or no preservatives, are packaged in vacuum or modified atmospheres to ensure long shelf life and rely primarily on refrigeration for preservation, has raised concerns of potential increases in botulism risk caused by psychrotrophic nonproteolytic group II *Clostridium botulinum* [46]. An average of 450 outbreaks of foodborne botulism with 930 cases have been reported annually worldwide [47]. The main habitat of clostridia is the soil but they are also found in sewage, rivers, lakes, sea water, fresh meat, and fish [48,49]. Most critical are the hygienic conditions for handling the product after smoking. There is a risk of botulism due to the growth of *C. botulinum* type E in smoked fish. The bacterium becomes a hazard when processing practices are insufficient to eliminate botulinal spores from raw fish, particularly improper thermal processing [21]. The growth of *C. botulinum* and toxin production then depends on appropriate conditions in food before eating: the temperature, oxygen level, water activity, pH, the presence of preservatives, and competing microflora [21]. A problem with *C. botulinum* has been encountered with some traditional fermented fish products. These rely on a combination of salt and reduced pH for their safety. If the product has insufficient salt, or fails to achieve a rapid pH drop to below 4.5, *C. botulinum* can grow. There was no evidence that the fish had been mishandled, but a low salt environment in the viscera allowed the bacterium to multiply and to produce toxin. *C. perfringens*, an important cause of both food poisoning and non-food-borne diarrhoeas in humans, was found in a number of fish owing to contamination with sewage, which is the main source of this organism [21].

## 3. Biopreservation

Seafood products are known to be especially susceptible to both microbiological and biochemical spoilage pathways. The development of effective processing treatments to extend the shelf life of fresh fish products is a must [2]. Additionally, the consumers' demand for high-quality and minimally processed seafood has recently captivated great attention [5, 9]. However, an increase in foodborne illness outbreaks is concomitant with the increase in consumer demand for less processed foods [1]. These trends highlight the importance of studying new microbial stress factors to extend the shelf-life of foods. Until now, approaches to reduce the risk of outbreaks of food poisoning have relied on the search for addition of more efficient chemical preservatives or on the application of more drastic physical treatments such as heating, refrigeration, high hydrostatic pressure (HHP), ionising radiation, pulsed-light, ozone, ultrasound, etc [1,5,50]. In spite of some possible advantage, these types of treatments have many drawbacks and limitation in seafood products: the proven toxicity of many of the commonest chemical preservatives (e.g. nitrites) (3), the alteration of the organoleptic and nutritional properties of seafood by physical treatments due to their delicate nature (e.g. freezing damage, discolouration in case of HHP and ionising radiation) [50,51] and especially recent consumer trends in purchasing and consumption, with demands for healthy seafood products that have been subjected to less extreme treatments (less heat and chill damage), with lower levels of salts, fats, acids, and sugars and/or the complete or the partial removal of chemically synthesized additives [1,2,7]. To harmonize consumer demands with the necessary safety standards, traditional means of controlling microbial spoilage and safety hazards in seafood are being replaced by an alternative solution that is gaining more and more attention: "biopreservation technology" [2,9,13,52,53]. It consists in inoculating food with microorganisms, or their metabolites, selected for their antibacterial properties and may be an efficient way of extending shelf life and food safety through the inhibition of spoilage and pathogenic bacteria without altering the nutritional quality of raw materials and food products [54, 55].

Lactic acid bacteria (LAB) possess a major potential for use in biopreservation because they are safe to consume, and during storage they naturally dominate the microbiota of many foods. Certain LAB species and strains isolated from seafood have been shown to exert strong antagonistic activity against spoilage and pathogenic microorganisms such as *Listeria, Clostridium, Staphylococcus,* and *Bacillus* spp [56-58]. The antagonistic and inhibitory properties of LAB are due to the competition for nutrients and the production of one or more antimicrobially active metabolites such as organic acids (lactic and acetic acid), hydrogen peroxide, and antimicrobial peptides (bacteriocins) [10]. Certain LAB are able to grow at refrigeration temperatures and are tolerant to modified-atmosphere packaging, low pH, high salt concentrations, and the presence of certain additives such as lactic acid, acetic acid, and ethanol. Because of these benefits, LAB can be used as protective cultures to restrict the growth of undesired organisms such as certain spoilage and pathogenic bacteria, with the subsequent benefits in terms of food safety [9,10,58]. Moreover, these microorganisms may have additional functional properties and, in some circumstances, they can be beneficial for the consumers [6]. LAB represent the microbial group most commonly used as protective cultures, as they are present in all fermented foods and have a long history of safe use [8]. Safety for the consumers is an aspect of great importance, in particular for some seafood products which are not cooked before consumptions, but also for other types of foods.

# 4. The role of lactic acid bacteria in biopreservation technology

## 4.1. Characterization and classification

Lactic acid bacteria (LAB) encompass a heterogeneous group of microorganisms having as a common metabolic property the production of lactic acid as the majority end - product from the fermentation of carbohydrates [59]. LAB are Gram (+), usually nonmotile, non - sporulating, catalase - negative, acid - tolerant, facultative anaerobic organisms and have less than 55 mol% G+C content in their DNA [60-62]. Except for a few species, LAB members are nonpathogenic organisms with a reputed generally recognized as safe status (GRAS). Taxonomic revisions of these genera and the description of new genera mean that LAB could, in their broad physiological definition, comprise around 20 genera [10]. However, from a practical, food-technology point of view, the following genera are considered the principal LAB: *Aerococcus, Carnobacterium, Enterococcus, Lactobacillus, Lactococcus, Leuconostoc, Oenococcus, Pediococcus, Streptococcus, Tetragenococcus, Vagococcus,* and *Weissella* [61]. The classification of lactic acid bacteria into different genera is largely based on morphology, mode of glucose fermentation, growth at different temperatures, configuration of the lactic acid produced, ability to grow at high salt concentrations, and acid or alkaline tolerance [62, 63]. An important characteristic used in the differentiation of the LAB genera is the mode of glucose fermentation under standard conditions. In this regard, the accepted definition is that given by Hommes and Vogel [64]: obligately homofermentative LAB are able to ferment hexoses almost exclusively to lactic acid by the Embden–Meyerhof–Parnas (EMP) pathway while pentoses and gluconate are not fermented as they lack phosphoketolase; facultatively heterofermentative LAB degrade hexoses to lactic acid by the EMP pathway and are also able to degrade pentoses and often gluconate as they possess both aldolase and phosphoketolase; finally, obligately heterofermentative degrade hexoses by the phosphogluconate pathway producing lactate, ethanol or acetic acid and carbon dioxide; moreover, pentoses are fermented by this pathway [62]. Several strains of groups 1 and 2 and some of the hetero fermentative group 3 are either used in fermented foods, but group 3 are also commonly associated with food spoilage. (For a more detailed discussion concerning the metabolic pathways, see [59].

## 4.2. Antimicrobial components from LAB

### *4.2.1. Bacteriocins*

Bacteriocins are ribosomally synthesized peptides, that exert their antimicrobial activity against either strains of the same species as the bacteriocin producer (narrow range), or to more distantly related species (broad range) [1,2,7]. It has been estimated that between 30% and 99% of all bacteria and archaea produce bacteriocins; their production by LAB is very significant from the point of view of their potential applications in food systems and thus, unsurprisingly, these have been most extensively investigated [6,10,12,60,65,66]. It has been noted that the activity of bacteriocins is frequently directed against bacteria that are related to the bacteriocin - producing strain or against bacteria found in similar environments [67]. It has also been noted that some bacteriocins can also play a role in cell signaling. Microorganisms that produce bacteriocins also possess immunity mechanisms to confer self - protection, that is, to protect bacteriocin producers from committing "suicide" [10,68,69]. Besides concern about antibiotic resistance, increasing consumer awareness of potential health risks associated with chemical preservatives has increased interest in bacteriocins. Bacteriocins are naturally produced so they are more easily accepted by consumers [54]. Bacteriocins are usually classified combining various criteria. The main ones being the producer bacterial family, their molecular weight and finally their amino acid sequence homologies and/or gene cluster organization [59,70]. Based on a relatively recent approach [69,71,72] bacteriocins produced by LAB have been categorized into two major classes: the lanthionine - containing bacteriocins or lantibiotics (class I) and the largely unmodified linear peptide antimicrobials (class II).

### *4.2.2. Organic acid production*

An important role of meat LAB starter cultures is the rapid production of organic acids; this inhibits the growth of unwanted flora and enhances product safety and shelf life. The types and levels of organic acids produced during the fermentation process depend on the LAB strains present, the culture composition, and the growth conditions [74]. Fermentation of the carbohydrates, glucose, glycogen, glucose-6-phosphate and small amounts of ribose, in meat and meat products, produces organic acids by glycolysis (Embden-Meyerhof Parnas pathway, EMP pathway) or the Hexose Monophosphate, HMP pathway. L (+) lactic acid is more inhibitory than its D (-) counterpart [68]. The antimicrobial effect of organic acids lies in the reduction of pH, and in the action of undissociated acid molecules [75]. It has been proposed that low external pH causes acidification of the cytoplasm. The lipophilic nature of the undissociated acid allows it to diffuse across the cell membrane collapsing the electrochemical proton gradient. Alternatively, cell membrane permeability may be affected, disrupting substrate transport systems [72]. The LAB in particular are able to reduce the pH to levels where putrefactive (e.g. clostridia and pseudomonads), pathogenic (e.g. *Salmonella* s and *Listeria* spp.) and toxinogenic bacteria (*Staphylococcus aureus. Bacillus cereus, Clostridium botulinum*) will be either inhibited or killed [7]. Also, the undissociated acid, on account of its fat solubility, will diffuse into the bacterial cell, thereby reducing the intracellular pH and slowing down metabolic activities, and in the case of Enterobacteriaceae such as *E. coli* inhibiting growth at around pH 5.1.

### 4.2.3. *Other antimicrobials of LAB*

Hydrogen peroxide is produced from lactate by LAB in the presence of oxygen as a result of the action of flavoprotein oxidases or nicotinamide adenine dinucleotide (NADH) peroxidise [76]. The antimicrobial effect of $H_2O_2$ may result from the oxidation of sulfhydryl groups causing denaturing of a number of enzymes, and from the peroxidation of membrane lipids thus increasing membrane permeability [8]. Most undesirable bacteria such as *Pseudomonas* spp. and *S. aureus* are many times sensitive to $H_2O_2$. Carbon dioxide ($CO_2$) is mainly produced by heterofermentative LAB. $CO_2$ plays a role in creating an anaerobic environment which inhibits enzymatic decarboxylations, and the accumulation of $CO_2$ in the membrane lipid bilayer may cause a dysfunction in permeability [8]. $CO_2$ can effectively inhibit the growth of many food spoilage microorganisms, especially Gram-negative psychrotrophic bacteria [77]. Diacetyl, an aroma component, is produced by strains within all genera of LAB by citrate fermentation. It is produced by heterofermentative lactic acid bacteria as a by-product along with lactate as the main product [8]. Diacetyl is a high value product and is extensively used in the dairy industry as a preferred flavour compound. Diacetyl also has antimicrobial properties. Diacetyl was found to be more active against gram-negative bacteria, yeasts, and molds than against gram-positive bacteria. Diacetyl is thought to react with the arginine-binding protein of gram-negative bacteria and thereby interfering with the utilization of this amino acid [78].

## 5. LAB in fish and fish products

LAB are not considered as genuine microflora of the aquatic environment, but certain genera, including *Carnobacterium, Lactobacillus, Enterococcus,* and *Lactococcus,* have been found in fresh and sea water fresh fish [61,63,79-83]. The number of lactobacilli in the gastro-intestinal tract of Arctic char was smaller in those reared in sea water than in fresh water, while the number of *Leuconostoc* and enterococci remained the same [84]. It is well documented that lactobacilli are part, not dominant, of the native intestinal microbiota of Arctic charr (*Salvelinus alpinus* L.), Atlantic cod, Atlantic salmon (*Salmo salar* L.), and brown trout (*Salmo trutta*) [82,85]. Several studies have shown the presence of other lactic acid bacteria, specially carnobacteria such as *Carnobacterium maltaromaticum* and *Carnobacterium divergence* within the intestinal content of salmonid species like Arctic charr (*Salvelinus alpinus*), Atlantic salmon (*Salmo salar*), rainbow trout (*Oncorhyncus mykiss*) [63,86-89], Atlantic cod [89], common wolffish (*Anarhichas lupus* L.) [85], brown trout [82] and also wild pike [63,82]. Bacteria of the genus *Enterococcus* have been isolated from the intestine of common carp (*Cyprinus carpio*) and brown trout [80,82].

LAB dominating in spoiled vacuum-packaged cold-smoked fish products include the genera of *Lactobacillus, Leuconostoc, Lactococcus* and *Carnobacterium* [9]. Magnússon & Traustadóttir

[91] reported the complete dominance of homofermentative lactobacilli in vacuum-packaged cold-smoked herring. In vacuum packaged cold-smoked salmon and herring, *Lactobacillus curvatus* has been found in majority together with lower numbers of *Lactobacillus sakei, Lactobacillus plantarum, Lactococcus* spp. and *Leuconostoc mesenteroides* [58,]. Paludan-Müller, Huss, & Gram [92] identified *Carnobacterium piscicola* as the dominant microorganism isolated from spoiled vacuum-packaged cold-smoked salmon. Leroi et al. [93] also isolated carnobacteria during the first stage of storage of vacuum-packaged cold-smoked salmon, whereas *Lactobacillus farciminis, Lactobacillus sakei,* and *Lactobacillus alimentarius* were isolated at advanced storage times. Other studies have also confirmed that most bacteria in vacuum-packaged "gravad" fish products stored at refrigeration temperatures are carnobacteria [94] and *L. sakei,* and to a lesser extent *Leuconostoc* spp., *L. curvatus,* and *Weissella viridescens* [95]. Gancel et al [90] have isolated 78 strains belonging to the genus *Lactobacillus* from fillets of vacuum packed smoked and salted herring *(Clupea harengus).* LAB has been found to occur in marinated herring, herring fillets and cured stockfish [58]. In marinated or dried fish, the lactic acid bacteria flora maybe quite diverse since the presence of *Lactobacilli* and *Pediococci* has been reported [90]. Thai fermented fishery products were screened for the presence of LAB by Ostergaard et al. [96]. LAB was found to occur in the low salted fermented products in the range of $10^7$-$10^9$ cfu/g. The high salt product "hoi dorng" had a lower LAB count of $10^3$-$10^5$ cfu/g. Olympia et al [97] have isolated $10^8$ LAB/g from a Philippine low salt rice-fish product burong bangus. Several studies have been mentioned that some species of *Carnobacteriuim* such as *C. divergens* and *C. maltaromaticum* are present in seafood and are able to grow to high concentrations in different fresh and lightly preserved products such as modified atmosphere-packed (MAP) [98-100], chilled MAP [101,102], high-pressure processing treated seafood products [103] and vacuum-packed cold smoked or sugar-salted ('gravad') seafood [53,93,95]. These studies clearly highlight the ability of LAB fish isolates to grow on different harsh condition rather than other organisms. Obviously many investigation have been shown that carnobacteria are common in chilled fresh and lightly preserved seafood, but at higher storage temperatures (15–25°C) other species could be dominate the spoilage microbial community of seafood.

## 6. Application of LAB in seafood

Treating catfish fillets with of 0.50% sodium acetate, 0.25% potassium sorbate with 2.50% lactic acid culture completely inhibited growth of Gram negative bacteria, improved catfish odor and appearance during 13 days storage [110]. Einarsson & Lauzon [111] treated shrimps with various bacteriocins from lactic acid bacteria and reported shelf life extension except carnocin UI49. Total mesophilic and psychotropic bacteria and MRS counts of the samples treated with carnocin UI49 were not different than those of controls at 4.5°C. In a study with five strains of lactic acid bacteria (four *Lactobacillus* and one *Carnobacterium*) on fermented salmon fillets, *L. sake* LAD and *L. alimentarius* BJ33 was regarded as suitable starters for fermentation of salmon fillets [112] based on starter growth (increase of more than 1log in 3 days) and acidification of muscle (e.g. pH reduction of approximately 0.7

units in 5 days) as well as sensory evaluation. Kisla & Ünlütürk [113] studied the microbial shelf life of rainbow trout treated with nisin-containing aqueous solution of *Lactococcus lactis* subsp. lactis NCFB 497and lactic acid. They reported the dipping of rainbow trout fillets into a lactic culture did not prolonged the shelf life due to the low inoculum level and type of lactic culture used. Elotmani & Assobhei [114] evaluated the inhibition of the microbial flora of sardine by using nisin and a lactoperoxidase system (LP), observing the efficiency of the nisin–LP combination in inhibiting fish spoilage flora. In another study growth of *L. monocytogenes* was significantly inhibited ($P < 0.05$) by *L. sakei* Lb706 in rainbow trout fillets stored under vacuum at 4°C during 10 days of storage while bacteriocin negative Lb706-B did not affect the growth of *L. monocytogenes*. In the presence of the sakacin A-producing strain of *L. sakei* (Lb706), the growth of *L. monocytogenes* was significantly inhibited ($P < 0.05$) in the first 3 days of storage at 10°C, after which its count increased to $10^7$ CFU $g^{-1}$ [115]. Altieri et al. [106] succeeded in inhibiting *Pseudomonas* spp. and *P. phosphoreum* in VP fresh plaice fillets at low temperatures by using a *Bifidobacterium bifidum* starter, and extending the shelf-life, especially under MAP. Bifidobacteria combined with sodium acetate (SA) extended refrigerated shelf-life of catfish fillets at 4°C [116]. The application of two *Lactobacillus sakei* CECT 4808 and *L. curvatus* CECT 904T protective cultures on refrigerated vacuum-packed rainbow trout (*Oncorhynchus mykiss*) fillets resulted in extension of shelf-life by 5 days by significantly improved in the counts of all microbiological spoilage indicator organisms (Enterobacteriaceae, *Pseudomonas* spp., H2S-producing bacteria, yeasts and moulds) and also significantly improved in all examined chemical parameters and off-odour [117].

Under biopreservation, combined coating of *Lactobacillus casei* DSM 120011 and *Lactobacillus acidophilus* 1M in *Streptomces* sp. NIOF metabolites, played effective role in lowering the biochemical and microbiological changes, extended shelf-life and safety of stored fish under low temperature as reported by Daboor & Ibrahim [118]. Tahiri et al. [119] suggest that selection of protective strains to improve the sensory quality of seafood products should focus on specific spoilage microorganism's inhibition. This approach was chosen by Matamoros et al., [120] who have isolated seven strains from various marine products on the basis of their activity against many spoiling and pathogenic, Gram positive and Gram negative marine bacteria. Among strains, two *Le. gelidum,* and two *Lc. piscium* demonstrated promising effect in delaying the spoilage of tropical shrimp and of VP CSS. However, no correlation with the classical quality indices measured was evidenced. A recent study demonstrated that this protective effect could be due to the inhibition of *B. thermosphacta* identified as one of the major spoiler organisms in cooked shrimp stored under MAP [121]. The inoculation of Tilapia (*Oreochromis niloticus*) fillets with *Lactobacillus casei* DSM 120011 and *Lactobacillus acidophilus* 1M at 2% concentration decreased both total volatile basic nitrogen (TVB-N), trimethylamine nitrogen (TMA-N) and thiobarbituric acid (TBA) values and improved the biochemical quality criteria, microbial aspects and safety of frozen fish fillets during 45 and 90 days storage. [122].

For Shirazinejad et al. [123] 2.0% lactic acid combined with nisin indicated the highest reduction in population of *Pseudomonas* spp. and $H_2S$ producing bacteria during storage time of Chilled Shrimp. Fall et al. [121] evidenced the in situ inhibition of *B. thermosphacta*, a major spoiling bacterium, by *L. piscium* that could explain the protective effect observed in shrimp. Additionally, those strains also showed an inhibitory effect on *L. monocytogenes* [124] and *Staph. aureus*. Recently, Sudalayandi & Manja [109] succeeded to preserve fresh fish through controlling spoilage bacteria and amines of Indian mackerel fish chank for two days at 37°C by inoculating them with different strains of LAB such as *Pediococcus acidilactici, Pediococcus pentosaceous, Streptococcus thermophilus, Lactococcus lactis, Lactobacillus plantarum, Lactobacillus acidophilus* and *Lactobacillus helveticus*. Using bacteriocin-like metabolite producer and non-producer strains of *Pediococcus* spp. [125] only slightly improved sensory quality of Horse Mackerel during cold storage. It was concluded that *Pediococcus* strains used in this study were not proper for preserving horse mackerel fillets especially at low storage temperatures. EntP-producing enterococci isolated from farmed turbot, under a spray-dried format exhibited antilisterial, antistaphylococcal, and antibacilli activities in turbot fillets either vacuum-packaged or subjected to modified-atmosphere packaging [2].

LAB Protective cultures have not been applied in many other seafood products except for cold smoked salmon (CSS), as they are normally flora of such products at the end of storage, and *L. monocytogenes* control. The effectiveness of bacteriocins to control growth of *L. monocytogenes* in vacuum packed cold smoked salmon has also been demonstrated by several researchers. Among them, Sakacin P has been found to be very potent against *L. monocytogenes* and is one of the most extensively studied bacteriocins [126-131]. Leroi et al. [132] succeeded in increasing the sensory use-by-date of CSS slices by inoculating them with strains of *Carnobacterium* sp. However the results varied depending on the batch treated. Addition of nisin to $CO_2$ packed cold smoked salmon resulted in a 1 to 2 $log_{10}$ reduction of *L. monocytogenes* [11]. Using a strain of *C. maltaromaticum,* Paludan-Müller et al. [92] only slightly extended the shelf-life of smoked salmon. Budu-Amoako et al. [133] tested nisin combined with heat as anti Listerial treatment in cold- packed lobster meat, finding decimal reductions of inoculated *L. monocytogenes* of 3 to 5 logs, whereas heat or nisin alone resulted in decimal reductions of 1 to 3 logs.

Duffes et al. [65] isolated *C. divergens* and *C. maltaromaticum* strains that exhibited listericidal activity in a model experiment with cold-smoked fish. They found that *C. piscicola* V1 inhibited *L. monocytogenes* by the in situ production of bacteriocins in vacuum-packed cold-smoked salmon stored at 4°C and 8°C. In contrast, another related species, namely, *C. divergens* V41 and its divercin V41, only exhibited a bacteriostatic effect on the target microorganism. Two strains of *C. maltaromaticum* isolated from CSS demonstrated their efficiency to limit the growth of *L. monocytogenes* in VP CSS during 31 days of storage at 5°C [134]. In a study using vacuum-packed cold smoked rainbow trout, the combination of nisin and sodium lactate injected into smoked fish decreased the count of *L. monocytogenes* from 3.3 to 1.8 $log_{10}$ over 16 days of storage at 8°C [135]. Sakacin P was added to vacuum-packed cold smoked salmon, a lightly processed high-fat (15–20%) product, together with a sakacin P-producing *L. sakei* culture in order to study the effect on the growth of *L. monocytogenes*. In this product, the combination of purified sakacin P and a live culture was found to be bactericidal against *L. monocytogenes*. The addition of sakacin P alone inhibited the growth of *L. monocytogenes* on this product for about 1 week [126]. Silva et al. [136] used a bacteriocin-producing *Carnobacterium* strain under a spray-dried format. This strain survived the process and retained antilisterial ability, although it lost activity against other Gram-positive targets such as *Staph. aureus.* Some authors have evaluated the antimicrobial activity of nisin combined with other bacteriocins. Bouttefroy & Milliere [137] tested combinations of nisin and curvaticin 13 produced by *L. curvatus* SB13 for preventing the regrowth of bacteriocin-resistant cells of *L. monocytogenes,* finding that this combination induced a greater inhibitory effect than the use of a single bacteriocin. Aasen et al. [131] studied the interactions of the bacteriocins sakacin P and nisin with food constituents in cold-smoked salmon, chicken cold cuts, and raw chicken. They stated that owing to the amphiphilic nature of these peptides, they can be adsorbed to food macromolecules and undergo proteolytic degradation, which may limit their use as preservation agents. More than 80% of the added sakacin P and nisin were rapidly adsorbed by proteins in the food matrix that had not been heat-treated, less than 1% of the total activity remaining after 1 week in cold- smoked salmon. In heat-treated foods, they found that, bacteriocin activity was stable for more than 4 weeks. No important differences were observed between sakacin P and nisin, but less nisin was adsorbed by muscle proteins at low pH. The growth of *L. monocytogenes* was completely inhibited for at least 3 weeks in both chicken cold cuts and cold-smoked salmon by the addition of sakacin P (3.5 μ/g), despite proteolytic degradation in the salmon.

In the presence of the bacteriocinogenic strain *C. maltaromaticum* CS526 isolated from surimi, the population of *L. monocytogenes* in CSS decreased from $10^3$ to 50 CFU $g^{-1}$ after 7 days at 4°C [138]. This activity could be linked to the production of the bacteriocin piscicocin CS526, since a non-bacteriocin producing strain had a lower effect on the growth of the pathogenic bacteria [138, 139]. The growth of the protective *Carnobacterium* strains did not modify the sensory characteristic of the product. One of these strains showing the strongest inhibition activity produces a bacteriocin, named Carnobacteriocin B2 that was involved in the antilisterial activity [105]. Three strains of bacteriocin producing *Carnobacterium* have been tested with the agar diffusion test method against a wide collection of *L. monocytogenes* (51 strains) isolated from seafood. All of the *Listeria* strains were sensitive. The inhibition was confirmed in co-culture with a mix of *L. monocytogenes* strains in sterile CSS [140]. One of these strains, *C. divergens* V41 showed its ability to maintain *L. monocytogenes* at the initial inoculating level of 20 CFU $g^{-1}$ during 28 days of storage at 4°C and 8°C. The effect of this strain on sensory characteristics and physico-chemical parameters revealed that it did not spoiled the product [56].

A bacteriocinogenic strain of *L. sakei* isolated from CSS allowed a 4 log reduction of *Listeria innocua* after 14 days of storage at 4°C. A reduction of 2 log units after 24 h at 5°C was also demonstrated with that strain in CSS juice towards *L. monocytogenes* [141]. Mix of bacteriocin-producing LAB like *L. casei, L. plantarum* and *C. maltaromaticum* were successfully used to limit the growth of *L. innocua* in CSS [142]. *C. maltaromaticum* had no effect on the inhibition of the Gram positive spoilage bacteria *B. thermosphacta* in cooked shrimps [143]. The anti-listerial activity of 3 LAB strains used individually or as co-cultures was assayed on cold-smoked salmon artificially contaminated with *L. innocua* and stored under vacuum at 4°C [142]. The association of *L. casei* T3 and *L. plantarum* PE2 was the most effective, probably due to a competition mechanism against the pathogen. In their study Tomé et al. [144] have also selected a strain of *Enterococcus faecium* among five bacteriocinogenic LAB strains for its ability to induce a decrease of the population of *L. innocua* inoculated in CSS. However in these studies the inhibition activities were not confirmed on *L. monocytogenes*. For Matamoros et al. [145] two LAB strains, *Lactococcus piscium* EU2241 and *Leuconostoc gelidum* EU2247 were efficient to limit the growth of both pathogenic bacteria *L. monocytogenes* and *S. aureus* in a challenge test in cooked shrimp stored under VP from 2 to 3 log CFU $g^{-1}$ units after 4 weeks at 8°C followed by 1 week at 20°C. The strain of *Leuconostoc* produced a bacteriocin-like compound but its activity was slight lower than the *Lactococcus* strain that was non-bacteriocinogenic. In another study, the application of *C. divergens* M35 towards *L. monocytogenes* in CSS resulted in a maximal decrease of 3.1 log CFU $g^{-1}$ of the pathogenic bacteria after 21 days of storage at 4°C whereas a non bacteriocinogenic strain had no effect [119].

## 7. Conclusion and future prospective

The presence of LAB in many processed seafood product is now well documented and the bio-protective potential of many strains and/or their bacteriocin has been highlighted in the last years. In situ production is readily cost-effective provided that the bacteriocin producers are technologically suitable. To date, only nisin and pediocin PA - 1 have been applied commercially in food applications where they are used to protect against spoilage and pathogenic organisms. However, other bacteriocins could be at least as effective for food processors as it is possible to apply them with hurdle approaches, particularly in light of consumer demands for minimally processed, safe, preservative - free foods. Control of pathogenic bacteria has widely focused on *L. monocytogenes* considered as the main risk in ready-to-eat seafood. However, in these minimally processed products, the new combination of hurdles can give selective advantages to enhance food safety and quality, particularly effective against other pathogenic bacteria like clostridia, vibrio or staphylococci. These goals can be facilitated through the incorporation of live bacteriocin - producing strain(s) or through the use of bacteriocins as concentrated preparations, either through direct addition to the seafood or in an immobilized form on packaging as well as in conjunction with other factors such as high pressure or pulse electric fields, to achieve more effective preservation of foods. The great results obtained with protective culture, bacteriocins for improving safety and quality of seafood products clearly indicate that the application of LAB protective culture and/or their bacteriocins in seafood product can suggest several important benefits; 1) extended shelf life of seafood during storage time, 2) decrease the risk for transmission of foodborne pathogens in lightly preserved seafood products, 3) ameliorate the economic losses due to seafood spoilage, 4) reduce the application of chemical preservatives and drastic physical treatments such as heating, refrigeration, etc. causing better preservation nutritional quality of food, 5) good option for industry due to cost effective way and finally 6) a good response to consumer demands for minimally processed, safe, preservative - free foods. At present the new techniques and disciplines emerging in the post – genomic era, such as genomics, proteomics, metabolomics, and system biology, open new avenues for interpretation of biological data. In combination with classical and molecular techniques, these new methods will be invaluable in the rational optimization of LAB function in order to obtain safer traditional and new seafood products.

## Author details

Mahdi Ghanbari* and Mansooreh Jami
*University of Zabol, Faculty of Natural Resources, Department of Fishery; Zabol, Iran*
*BOKU — University of Natural Resources and Life Sciences, Department of Food Sciences and Technology, Institute of Food Sciences, Vienna, Austria*

## 8. References

[1] Cortesi ML, Panebianco A, Giuffrida A, Anastasio A (2009) Innovations in seafood preservation and storage. Veterinary Research Communications.; Supplement 1: p. S15-S23.
[2] Campos A, Castro P, Aubourg SP, Velázquez JB (2012) Use of Natural Preservatives in Seafood. In McElhatton A, Sobral. Novel Technologies in Food Science, Integrating Food Science and Engineering Knowledge Into the Food Chain.: © Springer Science+Business Media;. p. 325-360.
[3] Feldhusen F (2000) The role of seafood in bacterial foodborne diseases. Method. Microbiol. 2: 1651-1660.
[4] ICMSF (2011) Fish and Seafood Products. In International Commission on Microbiological Specifications for Foods (ICMSF). Microorganisms in Foods.: Springer Science+Business Media, LLC. p. 107-133.
[5] Alzamora S, Welti-Chanes J, Guerrero S (2012) Rational Use of Novel Technologies:A Comparative Analysis of the Performance of Several New Food Preservation Technologies for Microbial Inactivation. In McElhatton A, Sobral PJA(). Novel Technologies in Food Science, Integrating Food Science and Engineering Knowledge Into the Food Chain.: © Springer Science+Business Media, LLC.
[6] Soomro AH, Masud T, Anwaar K (2002) Role of lactic acid bacteria (LAB) in food preservation and human health—A review. Pak. J. Nut. 1: 20-24.
[7] Gálvez A, Abriouel H, López R, Omar N (2007) Bacteriocin-based strategies for food biopreservation. Int. J. Food Microbiol. 120: 51-70.

---

* Corresponding Author

[8] Holzapfel WH, Geisen R, Schillinger U (1995) A review paper: biological preservation of foods with reference to protective cultures, bacteriocins and food-grade enzymes. Int. J. Food Microbiol. 24: p. 343-362.

[9] Calo-Mata P, Arlindo S, Boehme K, Miguel T, Pascola A, Barros-Velazquez J (2008) Current Applications and Future Trends of Lactic Acid Bacteria and their Bacteriocins for the Biopreservation of Aquatic Food Products. Food Biopro. Tech.1: 43–63.

[10] Collins B, Cotter P, Hill , Paul Ross R (2010) Applications of Lactic Acid Bacteria - Produced Bacteriocins. In Mozzi F, Raya R, GM V. Biotechnology of Lactic Acid Bacteria Novel Applications.: Blackwell Publishing. 89-109.

[11] Nilsson L (1997) Control of *Listeria monocytogenes* in cold-smoked salmon by biopreservation: Danish Institute for Fisheries Research and The Royal Veterinary and Agricultural University of Copenhagen, Denmark, Ph. D Dissertation.

[12] Cleveland J, Montville T, Nes I, Chikindas M (2001) Bacteriocins: Safe, natural antimicrobials for food preservation. Int. J. Food Microbiol. 71: p. 1 – 20.

[13] Dortu C, Thonart P (2009) Bacteriocins from lactic acid bacteria: interest for food products biopreservation. Biotech. Agr. Society Environ. 13: p. 143-154.

[14] Beaufort A, Rudelle S, Gnanou-Besse N, Toquin MT, Kerouanton A, Bergis H (2007). Prevalence and growth of *Listeria monocytogenes* in naturally contaminated cold-smoked salmon. Lett. Appl. Microbiol. 44: 406-411.

[15] Baffone W, Pianei A, Bruscolini F, Barbieri E, Cierio B (2000) Occurrence and expression of virulence-related properties of Vibrio species isolated from widely consumed seafood products. Int. J. Food Microbiol. 54: 9-18.

[16] InternationalDiseaseSurveillanceCenter (IDSC). 1999. *Vibrio parahaemolyticus,* Japan 1996-1998, Infectious Agents Surveillance Report (IASR), 20:1-2.

[17] Joseph SW, Colwell RR, Kaper JB (1982) *Vibrio parahaemolyticus* and related halophilic Vibrios. Crit. Rev. Microbiol. 10(1): 77-124.

[18] Kam KM, Leung TH, Ho YY, Ho NK, Saw TA (1995) Outbreak of *Vibrio cholerae* 01 in Hong Kong related to contaminated fish tank water. Public Health. 109(5): p. 389-395.

[19] Colwell RR (1996) Global climate and infectious diseases: the cholera paradigm. Sci. 274: 2025-2031.

[20] Ayulo AM, Machado R, Scussel V (1994) Entero toxigenic *Escherichia coli* and Staphylococcus aureus in fish and seafood from the southern region of Brazil. Int. J. Food Microbiol. 24:171–178.

[21] Chattopadhyay P (2000) Fish – catching and handling. In Robinson RK. Encyclopedia of Food Microbiol. London: Academic Press; 153 p.

[22] Asai Y, Murase T, Osawa R, Okitsu T, Suzuki R, Sata S, Terajima J,Izumiya H, Watanabe H (1999) Isolation of Shiga toxin-producing *Escherichia coli* O157:H7 from processed salmon roe associated with the outbreaks in Japan, 1998, and a molecular typing of the isolates by pulsed-field gel electrophoresis. Kansenshogaku Zasshi. 73: 20-24.

[23] Mitsuda, T; Muto, T; Yamada, M; Kobayashi, N; Toba, M; Aihara, Y; Ito, A; Yokota, S (1998) Epidemiological study of a food-borne outbreak of enterotoxigenic *Escherichia coli* O25:NM by pulsed-field gel electrophoresis and randomly amplified polymorphic DNA analysis. J. Clin. Microbiol. 36: 652-656.

[24] Vieira RHSF, Rodrigues DP, Gocalves FA, Menezes FGR, Aragao JS, Sousa OV (2001) Microbicidal effect of medicinal plant extracts (*Psidium guajava* Linn.and *Carica papaya* Linn.) upon bacteria isolated from fish muscle and known to induce diarrhea in children. Rev. Inst. Med. trop. S. Paulo 43 (3): 145-148.

[25] Pierard D, Crowcroft N, de Bock S, Potters D, Crabbe G,VLF, Lauwers S (1999) A case-control study of sporadic infection with O157 and non-O157 verocytotoxin-producing *Escherichia coli*. Epid. Infec. 122: 359-365.

[26] Semanchek JJ, Golden DA (1998) Influence of growth temperature on inactivation and injury of *Escherichia coli* O157:H7 by heat, acid, and freezing. J. Food Prot. 61: 395-401

[27] Isonhood JH, Drake M (2002) *Aeromonas* species in foods. J. Food Prot. 65: 575–582.

[28] Pogorelova NP, Zhuravleva LA, Ibragimov FKH, Iushchenko GV (1995) Bacteria of the genus *Aeromonas* as the causative agents of saprophytic infection. Zh Mikrobiol Epidemiol Immunobiol. 4: 9-12.

[29] Fernandes CF, Flick GJ, Thomas TB (1998) Growth of inoculated psychrotrophic pathogens on refrigerated fillets of aquacultured rainbow trout and channel catfish. J. Food Prot. 61(3): 313-317.

[30] Novotny L, Halouzka R, Matlova L, Vavra O, Dvorska L, Bartos M (2010) Morphology and distribution of granulomatous inflammation in freshwater ornamental fish infected with mycobacteria. J. Fish Dis. 33: 947-955.

[31] Olgunoğlu IA (2012) *Salmonella* in Fish and Fishery Products. In Mahmoud BSM. *Salmonella* - A Dangerous Foodborne Pathogen.: InTech; 2012.

[32] Heinitz ML, Ruble RD, Wagner DE, Tatini SR (2000) Incidence of *Salmonella* in fish and seafood. J. Food Prot. 63(5): 579-592.

[33] Vieira RHSF, Rodrigues DP, Gocalves FA, Menezes FGR, Aragao JS, Sousa OV (2001) Microbicidal effect of medicinal plant extracts (*Psidium guajava* Linn. and *Carica papaya* Linn.) upon bacteria isolated from fish muscle and known to induce diarrhea in children. Revista do Instituto de Medicina Tropical de São Paulo. 43:145-148.

[34] Eklund MW, Peterson ME, Poysky FT, Paranjpye RN, Pelroy GA (2004) Control of bacterial pathogens during processing of cold-smoked and dried salmon strips. J. Food Prot. 67: 347-351.

[35] Francis GA, O'Beirne D (1998) Effects of the indigenous microflora of minimally processed lettuce on the survival and growth of L. monocytogenes. Int. J. Food Sci. Tech. 33: 477-488.

[36] Alves VF, De Martinis ECP, Destro MT, Vogel BF, Gram L (2005) Antilisteral activity of a *Carnobacterium piscicola* isolated from brazilian smoked fish (Surubim (*Pseudoplatystoma* sp.)) and its activity against a persistent strain of *Listeria monocytogenes* isolated from surubim. J. Food Prot.11: 2068-2077.

[37] Zunabovic M, Domig K, Kneifel W (2011)Practical relevance of methodologies for detecting and tracing of *Listeria monocytogenes* in ready-to-eat foods and manufacture environments - A review. LWT- Food Sci. Tech. 44(2): 351-362.

[38] Huss HH, Jørgensen LV, Vogel BF (2000) Control options for *Listeria monocytogenes* in seafoods. Int. J. Food Microbiol. 62(3): 267-74.

[39] Gudmundsdöttir S, Gudbjörnsdottir B, Lauzon H, Einarsson H, Kristinsson KG, Kristjansson M (2005) Tracing *Listeria monocytogenes* isolates from cold smoked salmon and its processing environment in Iceland using pulsed-field gel electrophoresis. Int. J. Food Microbiol. 101: 41-51.

[40] Bayles DO, Annous BA, Wilkinson BJ (1996) Cold stress proteins induced in *Listeria monocytogenes* in response to temperature downshock and growth at low temperatures. Appl. Environ. Microbiol. 62: 1116-1119.

[41] Miettinen H, Wirtanen G. Prevalence and location of *Listeria monocytogenes* in farmed rainbow trout (2005) Int. J. Food Microbiol. 104: 135-143.

[42] Tham W, Ericsson H, Loncarevic S, Unnerstad H, Danielsson-Tham ML (2000) Lessons from an outbreak of listeriosis related to vacuum-packed gravad and cold-smoked fish. Int. J. Food Microbiol. 62(3): 173-175.

[43] Fonnesbech Vogel B, Huss HH, Ojeniyi B, Ahrens P, Gram L (2001) Elucidation of *Listeria monocytogenes* contamination routes in cold-smoked salmon processing plants detected by DNA-based typing methods. Appl. Environ. Microbiol. 67(6): p. 2586-2595.

[44] Hoffman AD, Gall KL, Norton DM, Wiedmann M (2003) *Listeria monocytogenes* contamination patterns for the smoked fish processing environment and for raw fish. J. Food Prot. 66: p. 652-670.

[45] Thimothe J, Kerr Nightingale K, Gall K, Scott VN, Wiedmann M (2004) Tracking of *Listeria monocytogenes* in smoked fish processing plants. J. Food Prot. 67: 328–341.

[46] Peck MW(1997) Clostridium botulinum and the safety of refrigerated processed foods of extended durability. Trend.Food Sci. Tech. 8: 186-192.

[47] Hatheway CL (1995) Hath Botulism: the present status of the disease. Curr. Top. Microbiol. Imm. 195.

[48] Haagsma J. (1991) The distribution of Pathogenic anaerobic bacteria and the environment. Sci. Technic. Rev. Office Int. des. 10: 49-764.

[49] Sramova H, Benes C (1998) Occurrence of botulism in the Czech Republic (in Czech). Zpravy CEM (SZU Praha). 7: 395-397.

[50] Zhou GH, Xu XL, Liu Y (2010) Preservation technologies for fresh meat. Meat Sci. 86: 119-128.

[51] Devlieghere F, Vermeiren L, Debevere J (2004) New preservation technologies: Possibilities and limitations. Int. Dairy J.14: 273-285.

[52] Rodgers S (2001) Preserving non-fermented refrigerated foods with microbial cultures - a review. Trends. Food Sci. Tech. 12: 276-284.

[53] Pilet MF, Leroi F (2011)Applications of protective cultures , bacteriocins and bacteriophages in fresh seafood and seafood product. In Lacroix C. Protective cultures,

antimicrobial metabolites and bacteriophages for food and beverage biopreservation.: © 2011 Woodhead Publishing Limited.

[54] Galvez A, Abriouel H, Benomar N, Lucas R (2010) Microbial antagonists to food-borne pathogens and biocontrol. Cur. Opin. Biotech. 21: 142-148.

[55] Garcia P, Rodriguez L, Rodriguez A, Martinez B (2010) Food biopreservation:Promising strategies using bacteriocins, bacteriophage and endolysins. Trends. Food Sci. Tech. 373-382.

[56] Brillet A, Pilet MF, Prévost H, Cardinal M, Leroi F (2005) Effect of inoculation of inoculation of *Carnobacterium divergens* V41, a biopreservative strain against *Listeria monocytogenes* risk, on the microbiological, and sensory quality of cold-smoked salmon. Int. J. Food Microbiol. 104: 309-324.

[57] Pinto AL, Fernandes M, Pinto C, Albano H, Castilho F, Teixeira P, Gibbs PA (2009) Characterization of anti- *Listeria* bacteriocins isolated from shellfish. Int. J.Microbiol. 129: 50-58.

[58] Leroi F (2010) Occurrence and role of lactic acid bacteria in seafood products. Food Microbiol. 27: 698-709.

[59] Mozzi F, Raya RR, Vignolo GM, editors (2010) Biotechnology of Lactic Acid Bacteria: Novel Applications: Blackwell Publishing.

[60] Stiles E (1996) Biopreservation by lactic acid bacteria. Antonie van Leeuwenhoek. 70: 331-345.

[61] Ghanbari M, Rezaei M, Jami M, Nazari M (2010) Isolation and characterization of *Lactobacillus* species from intestinal content of Beluga(*Huso huso*) and persian sturgeon (*Acipenser persicus*). Iran. J. Vet. Res. 10(2): 152-157.

[62] Mayo B, Aleksandrzak - Piekarczyk T, Fernández M, Kowalczyķ M, Álvarez - Martín P, Bardowski J (2010) Updates in the Metabolism of Lactic Acid Bacteria. In Mozzi F, Raya RR, Vignolo GM, editors. Biotechnology of Lactic Acid Bacteria: Novel Applications.: Blackwell Publishing.

[63] Ringo E, Gatesoupe F (1998) Lactic acid bacteria in fish: a review. Aquacult. 160: 177-203.

[64] Hammes WP, Vogel RF (1995) The genus *Lactobacillus* Glasgow: Blackie Academic & Professional.

[65] Duffes F, Corre C, Leroi F, Dousset X, Boyaval P (1999) Inhibition of *Listeria monocytogenes* by in situ produced and semipurifi ed bacteriocins of Carnobacterium spp. on vacuum-packed, refrigerated. J. Food Prot. 62: 394–1403.

[66] Campos C, Rodríguez O, Calo-Mata P, Prado M, Barros-Velazquez J (2006) Preliminary characterizationof bacteriocins from *Lactococcus lactis* , *Enterococcus faecium* and *Enterococcus mundtii* strains isolated from turbot (*Psetta maxima* ). Food Res. Int. 39: 356-64.

[67] Drider D, Fimland G, Hechard Y, McMullen L, Prevost H (2006) The continuing story of class IIa bacteriocins. Microbiology and Molcular Biology Reviews. 70: 564 – 582.

[68] Ouwehand A, Vesterlund S (2004) Antimicrobial Components from Lactic Acid Bacteria. In Salminen S, Wright v, Ouwehand A. Lactic Acid Bacteria Microbiological and Functional Aspects.: Marcel Dekker, Inc.

[69] Cotter PD, Hill C, Ross RP (2005) Bacteriocins: Developing innate immunity for food . Nat. Rev. Microbiol 3, 777 – 788 .

[70] Nes I, Yoon S, Diep D (2007) Ribosomally Synthesiszed Antimicrobial Peptides (Bacteriocins) in Lactic Acid Bacteria: A Review. Food Sci. Biotech.. 16(5): 675-690.

[71] Cotter PD, Draper LA, Lawton EM, McAuliffe O, Hill C, Ross RP (2006) Overproduction of wild-type and bioengineered derivatives of the lantibiotic lacticin 3147 . Appl. Environ. Microbiol. 72, 4492 – 4496

[72] Gillor O, Etzion A, Riley M (2008) The dual role of bacteriocins as anti- and probiotics. Appl. Microbiol. Biotech.. 81: 591-606.

[73] Nes I (2011) History, Current Knowledge, and Future Directions on Bacteriocin Research in Lactic Acid Bacteria. In Drider D, RS, (eds.). Prokaryotic Antimicrobial Peptides: From Genes to Applications.: Springer Science+Business Media, LLC p. 3-12.

[74] Lindgren SE, Dobrogosz WJ (1990) Antagonistic activities of lactic acid bacteria in food and feed fermentations. FEMS Microbiol. Lett. 87(1-2): 149-164.

[75] Podolak PK, Zayas JF, Kastner CL, Fung DYC (1996) Inhibition of *Listeria monocytogenes* and *Escherichia coli* O157:H7 on beef by application of organic acids. J. Food Prot. 59: 370-373.

[76] Ammor MS, Mayo B (2007) Selection criteria for lactic acid bacteria to be used as functional cultures in dry sausage production: An update. Meat Sci. 76: 138–146.

[77] Devlieghere F, Debevre J (2000) Influence of dissolved carbon dioxide on the growth of spoilage bacteria. Lebensmittel- und Wissenschaft-Technologie. 33: 531-537.

[78] Lanciotti E, Santini C, Lupi E, Burrini D (2003) Actinomycetes, cyanobacteria and algae causing tastes and odours in water of the River Arno used for the water supply of Florence. J. Water Sup. Res. Tech. 52(7): 489–500.

[79] Kvasnikov EI, Kovalenko NK, Materinskaya LG (1997) Lactic acid bacteria of freshwater fish. Microbiol. 46: 619-624.

[80] Cai Y, Suyanandana P, Saman P (1999) Classification and characterization of lactic acid bacteria isolated from the intestines of common carp and freshwater prawns. The J. Gen. Appl. Microbiol. 45: 177-184.

[81] Huss HH, Jeppesen VF, Johansen C, Gram L (1995) Biopreservation of fish products a reviewof recent approaches and results. J. Aquat. Food Prod. Tech. 4: 5-26.

[82] González CJ, Encinas JP, García-López ML, Otero A (2000) Characterization and identification of lactic acid bacteria from freshwater fishes. Food Microbiol. 17: 383-391.

[83] Bucio A, Hartemink R, Schrama JW, Verreth J, Rombouts FM (2006) Presence of lactobacilli in the intestinal content of freshwater fish from a river and from a farm with a recirculation system. Food Microbiol. 23(5): 476-482.

[84] Ringo E, Strom E (1994) Microflora of Arctic char, *Salvelinus alpinus* (L.); gastrointestinal microflora of free-living fish, and effect of diet and salinity on the intestinal microflora. Aquacult. Fish. Manag. 25: 623-629.

[85] Ringo E (2004) Lactic acid bacteria in fish and fish farming. In Salminen S, Wright A, Ouwehand A, editors. Lactic acid bacteria : Microbiological and Functional Aspects. 3rd ed. New-York: CRC Press; 581-610.

[86] Ringø E, Strøm E, Tabachek JA (1995) Intestinal microflora of salmonids: a review. Aqua. Res. 26: 773–789.

[87] Ringø E, Olsen RE (1999) The effect of diet on aerobic bacterial flora associated with intestine of Arctic charr (*Salvelinus alpinus* L.). J. Appl. Microbiol. 86: 12–28.

[88] Spanggaard B, Huber I, Nielsen J, Nielsen T, Appel KF, Gram L (2000) The microflora of rainbow trout intestine. A comparison of traditional and molecular identification. Aquacult. 182: 1–15.

[89] Seppola M, Olsen RE, Sandaker E, Kanapathippillai P, Holzapfel W, Ringø E (2006) Random amplification of polymorphic DNA (RAPD) typing of carnobacteria isolated from hindgut chamber and large intestine of Atlantic cod (*Gadus morhua* L.). Sys. Appl. Microbiol. 29: p. 131–137.

[90] Gancel F, Dzierszinski F, Tailliez R (1997) Identification and characterization of *Lactobacillus* species isolated from fillets of vacuum-packed smocked and salted herring (*Clupea harengus*). J. appl. Microbiol. 82: 722-728.

[91] Magnússon H, Traustadóttir K (1982) The microbial flora of vacuum-packed smoked herring fillets. J. Food Tech. 17: 695–702.

[92] Paludan-Müller C, Dalgaard P, Huss H, Gram L (1998) Evaluation of the role of *Carnobacterium piscicola* in spoilage of vacuum and modified atmosphere-packed-smoked salmon stored at 5°C. Int. J. Food Microbiol. 39: 155-166.

[93] Leroi F, Joffraud JJ, Chevalier F, Cardinal M (1998) Study of the microbial ecology of cold smoked salmon during storage at 8°C. Int. J. Food Microbiol. 39: 111-121.

[94] Leisner J, Laursen B, Prevost H, Drider D, Dalgaard P (2007) Carnobacterium:positive and negative effects in the environment and in foods. FEMS Microbiol. Rev. 13: 592-613.

[95] Leisner JJ, Millan JC, Huss HH, Larsen LM (1994) Production of histamine and tyramine by lactic acid bacteria isolated from vacuum-packed sugar-salted fish. J. Appl. Bacter. 76: 417–423.

[96] Østergaard A, Ben Embarek PK, Wedel-Neergaard C, Huss HH, Gram L (1998) Characterization of anti-listerial lactic acid bacteria isolated from Thai fermented fish products. Food Microbiol. 15: 223–233.

[97] Olympia M, Ono H, Shinmyo A, Takano M (1992) Lactic acid bacteria in fermented fishery, burong bangus. J. Fer. Bioeng. 73(3): 193-197.

[98] Mauguin S, Novel G (1994) Characterization of lactic acid bacteria isolated from seafood. J. Appl. Bacter. 76 : 616–625.

[99] Emborg J, Laursen BG, Rathjen T, Dalgaard P (2002) Microbial spoilage and formation of biogenic amines in fresh and thawed modified atmosphere-packed salmon (Salmo salar) at 2 degrees C. J. Appl. Microbiol. 92(4). 790-799

[100] Franzetti L, Scarpellini M, Mora D, Galli A (2003) *Carnobacterium* spp. in seafood packaged in modified atmosphere. Annal. Microbiol. 53: 189-193.

[101] Emborg J, Laursen BG, Rathjen T, Dalgaard P (2002) Microbial spoilage and formation of biogenic amines in fresh and thawed modified atmosphere-packed salmon (*Salmo salar*) at 2°C. J. Appl. Microbiol. 92: 790-799.

[102] Dalgaard P, Madsen HL, Samieian N, Emborg J (2006) Biogenic amine formation and microbial spoilage in chilled garfish (*Belone belone*) effect of modified atmosphere packaging and previous frozen storage. J. Appl. Microbiol. 101: 80 - 95.

[103] Lakshmanan R, Dalgaard P (2004) Effect of high-pressure processing on *Listeria monocytogenes*, spoilage microflora and multiple compound quality indices in chilled cold-smoked salmon. J. Appl. Microbiol. 96: 398-408.

[104] Wessels S, Huss HH (1996) Suitability of *Lactococcus lactis* ATCC 11454 as a protective culture for lightly preserved fish products. Food Microbiol. 13: 323-332.

[105] Nilsson L, Ng YY, Christiansen JN, Jorgensen BL, Grotinum D, Gram L (2004) The contribution of bacteriocin to inhibition of *Listeria monocytogenes* by *Carnobacterium piscicola* strains in cold-smoked salmon systems. J. Appl. Microbiol. 96: 133-143.

[106] Altieri C, Speranza B, Del Nobile MA, Sinigaglia M (2005) Suitability of bifidobacteria bacteria and thymol as biopreservatives in extending the shelf life of fresh packed plaice fillets.. J. Appl. Microbiol. 99: 1294-1302.

[107] Yin LJ, Wu CW, Jiang ST (2007) Biopreservative effect of pediocin ACCEL on refrigerated seafood. Fish. Sci. 73: 907-912.

[108] Ringo E (2008) The ability of carnobacteria isolated from fish intestine to inhibit growth of fish pathogenic bacteria. Aqua. Res. 39: 171-180.

[109] Sudalayandi, K.M (2011) Efficacy of lactic acid bacteria in the reduction of trimethylamine-nitrogen and related spoilage derivatives of fresh Indian mackerel fish chunks. Afr. J. Biotech. 10: 42-47.

[110] Kim CR, Hearnsberger JO (1994) Gram negative bacteria inhibition by lactic acid culture and food preservatives on catfish fillets during refrigerated storage. J. Food Sci. 59: 513-516.

[111] Einarsson H, Lauzon HL (1995) Biopreservtaion of brined shrimp (*Pandalus borealis*) by bacteriocins from lactic acid bacteria. Appl. Environ. Microbiol. 61: 669-675.

[112] Morzel M, Fransen NG, Arendt EK (1997) Defined starter cultures for fermentation of salmon fillets. J. Food Sci. 62(6): 1214-1217.

[113] Kışla D, Ünlütürk A (2004) Microbial shelf life of rainbow trout fillets treated with lactic culture and lactic acid. Adv. Food Sci. 26: 17-20.

[114] Elotmani F, Assobhei O (2004) In vitro inhibition of microbial flora of fish by nisin and lactoperoxidase system. Lett. Appl. Microbiol. 38: 60–65.

[115] Aras Husar S, Kaban G, Husar O, Yanik T, Kaya M (2005) Effect of *Lactobacillus* sakei Lb706 on Behavior of *Listeria monocytogenes* in Vacuum-Packed Rainbow Trout Fillets. Tur. J. Vet. Anim. Sci. 29: 1039-1044.

[116] Kim Y, Ohta T, Takahashi T, Kushiro A, Nomoto K, Yokokura T, Okada N, Danbara H (2006). Probiotic *Lactobacillus* casei activates innate immunity via NF-κB and p38 MAP kinase signaling pathways. Microb. Infec.. 8: 994–1005.

[117] Katikou P, Ambrosiadis IGD, Koidis P, Georgakis SA 2(2007) Effect of *Lactobacillus* cultures on microbiological, chemical and odour changes during storage of rainbow trout fillets. J. Sci. Food Agri. 87: 477–484.

[118] Daboor SM, Ibrahim SM (2008) Biochemical and microbial aspects of tilapia (*Oreochromis niloticus* L.) biopreserved by Streptomces sp. metabolites. In 4th International Conference of Veterinary Research Division, National Research Center (NRC); Cairo, Egypt. p. 39-49.

[119] Tahiri M, Desbiens E, Kheadr C, Lacroix IF (2009) Comparison of different application strategies of divergicin M35 for inactivation of *Listeria monocytogenes* in cold-smoked wild salmon. Food Microbiol. 26: p. 783-793.

[120] Matamoros S, Pilet MF, Gigout F, Prévost H, Leroi F (2009) evaluation of seafood-borne psychrotrophic lactic acid bacteria as inhibitors of pathogenic and spoilage bacteria. Food Microbiol. 26: 638-644.

[121] Fall PA, Leroi F, Cardinal M, Chevalier F, Pilet MF (2010) Inhibition of *Brochothrix thermosphacta* and sensory improvement of tropical peeled cooked shrimp by *Lactococcus piscium* CNCM I-4031. Lett. Appl. Microbiol. 50: 357-361.

[122] Ibrahim SM, Salha GD (2009) Effect of antimicrobial metabolites produced by lactic acid bacteria on quality aspects of frozen Tilapia (*Oreochromis niloticus*) fillets. World Journal of Fish and Marine Sciences. 1: 40-45.

[123] Shirazinejad AR, Noryati I, Rosma A, Darah I (2010) Inhibitory Effect of Lactic Acid and Nisin on Bacterial Spoilage of Chilled Shrimp. World Acad. Sci. Eng. Tech.

[124] Fall PA, Leroi F, Chevalier F, C G, Pilet MF (2010) Protective effect of a non-bacteriocinogenic Lactococcus piscium CNCM I-4031 strain against *Listeria monocytogenes* in sterilised tropical cooked peeled shrimp. J. Aquat. Food Prod. Tech. 19: 84-92.

[125] Cosansu S, Mol S, Ucok Alakavuk D, Tosun ŞY (2011) Effects of *Pediococcus* spp. on the quality of vacuum-packed Horse Mackerel during Cold Storage. J. Agri. Sci. 17: 59-66.

[126] Katla T, Moretro T, Aasen IM, Holck A, Axelsson L, Naterstad K (2001) Inhibition of *Listeria monocytogenes* in cold smoked salmon by addition of sakacin P and/or live *Lactobacillus sakei* cultures. Food Microbiol. 18: 431-439.

[127] Blom H, Katla T, Hagen BF, Axelsson L (1997) A model assay to demonstrate how intrinsic factors affect diffusion of bacteriocins. Int. J. Food Microbiol. 38: 103-109.

[128] Brurberg MB, Nes IF, Eijsink VGH (1997) Pheromone-induced production of antimicrobial peptides in *Lactobacillus*. Mol. Microbiol. 26: 347-360..

[129] Eijsink VGH, Skeie M, Middelhoven H, Brurberg MB, Nes IF (1998) Comparative studies of pediocin-like bacteriocins.. Appl. Environ. Microbiol. 64: 3275-3281.

[130] Ganzle MG, Weber S, Hammes WP (1999) Effect of ecological factors on the inhibitory spectrum and activity of bacteriocins. Int. J. Food Microbiol. 46: 207 – 217.

[131] Aasen IM, Moretro T, Katla T, Axelsson L, Storro I (2000) Infuence of complex nutrients, temperature and pH on bacteriocin production by *Lactobacillus* sakei CCUG 42687. Appl. Microbiol. Biotech.. 53: 159-166.

[132] Leroi F, Arbey N, Joffraud J, Chevalier F (1996) 'Effect of inoculation with lactic acid bacteria on extending the shelf-life of vacuum-packed cold-smoked salmon. Int. J. Food Sci. Tech. 1996; 31: p. 497-504.

[133] Budu-Amoako B, Albert RF, Harris J, Delves-Broughton J (1999) Combined effect of nisin and moderate heat on destruction of *Listeria monocytogenes* in cold-pack lobster meat. J. Food Prot. 62: 46–50.

[134] Nilsson L, Gram L, Huss H (1999) Growth control of *Listeria monocytogenes* on cold smoked salmon using a competitive lactic acid bacteria flora. J. Food Prot. 62: 336-342.

[135] Nykanen A, Weckman K, Lapvetelainen A (2000) Synergistic inhibition of *Listeria monocytogenes* on cold-smoked rainbow trout by nisin and sodium lactate. Int. J. Food Microbiol. 61: 63-72.

[136] Silva J, Carvalho AS, Teixeira P, Gibbs PA (2002) Bacteriocin production by spray-dried lactic acid bacteria. Lett. Appl. Microbiol. 34(2): 77-81.

[137] Bouttefroy A, Millière JB (2000) Nisin-curvaticin 13 combinations for avoiding the regrowth of bacteriocin resistant cells of *Listeria monocytogenes* ATCC 15313. Int. J. Food Microbiol. 62: 65-75.

[138] Yamazaki K, Suzuky M, Kawai Y, Inoue N, Montville TJ (2003) Inhibition of *Listeria monocytogenes* in cold-smoked salmon by *Carnobacterium piscicola* CS526 isolated from frozen surimi. J. Food Prot. 66: 1420-1425.

[139] Yamazaki K, Suzuki M, Kawai Y,IN, Montville TJ (2005) Purification and characterization of a novel class IIa bacteriocin, piscicocin CS526, from surimi associated *Carnobacterium piscicola* CS526. Appl. Environ. Microbiol. 71: p. 554-557.

[140] Brillet A, Pilet MF, Prevost H, Bouttefroy A, Leroi F (2004). Biodiversity of *Listeria monocytogenes* sensitivity to bacteriocin-producing Carnobacterium strains and application in sterile cold-smoked salmon. J. Appl. Microbiol. 97: 1029-1037.

[141] Weiss A, Hammes WP (2006) Lactic acid bacteria as protective cultures against *Listeria* spp. on cold-smoked salmon. Eur. Food Res. Tech. 222: 343-346.

[142] Vescovo M, Scolari G, Zacconi C (2006) Inhibition of *Listeria innocua* growth by antimirobial-producing lactic acid cultures in vacuum-packed cold-smoked salmon. Food Microbiol. 23: 689-693.

[143] Laursen BG, Bay L, Cleenwerck I, Vancanneyt M, Swings J, Dalgaard P (2005). *Carnobacterium divergens* and *Carnobacterium maltaromicum* as spoilers or protective cultures in meat and seafood: phenotypic and genotypic characterisation. Sys. Appl. Microbiol. 28: 151-164.

[144] Tomé E, Pereira VL, Lopes CI, Gibbs PA, Teixeira PC (2008) In vitro tests of suitability of bacteriocin-producing lactic acid bacteria, as potential biopreservation cultures in vacuum-packaged cold-smoked salmon. Food Control. 19: 535-543.

[145] Matamoros S, Leroi F, Cardinal M, Gigout F, Kasbi Chadli F, Cornet J, Prevost F, Pilet M.F (2009). Psychrotrophic lactic acid bacteria used to improve the safety and quality of vacuum-packaged cooked and peeled tropical shrimp and cold smoked salmon. J. Food Prot. 72: 365-374.

# Chapter 17

# Selection of *Lactobacillus* Species from Intestinal Microbiota of Fish for Their Potential Use as Biopreservatives

Mahdi Ghanbari, Masoud Rezaei and Mansoureh Jami

Additional information is available at the end of the chapter

http://dx.doi.org/10.5772/50166

## 1. Introduction

Despite recent advances in seafood production, seafood safety is still an important public health issue. It is clear that indigenous bacteria present in marine environment as well as resulting from post contamination during processing are responsible for many cases of illnesses [1-3]. In the last years, traditional processes applied to seafood like salting, smoking and canning have decreased in favor of mild technologies involving lower salt content, lower heating temperature and vacuum (VP) or modified atmosphere packing (MAP, 3-5]. Most of these treatments are usually not sufficient to destroy microorganisms and in some cases psychrotolerant pathogenic such as *Listeria monocytogenes* or spoilage causing bacteria can develop during prolonged shelf-life of these products [2,5,6]. As several of these products are eaten raw, it is therefore essential that adequate precautious and preservation technologies are applied to maintain their safety and quality. Among alternative preservation technologies, particular attention has been paid to biopreservation to extend the shelf-life and to enhance the hygienic quality of perishable food products such as seafood, thereby minimizing the impact on nutritional and organoleptic properties [1,7,8]. In this context, lactic acid bacteria (LAB) possess a major potential in biopreservation strategies, since they are safe to consume, and during storage they naturally dominate the microbiota of many foods [7-11]. Lactic acid bacteria are gram-positive, non-sporulating and catalase negative rods or cocci that ferment various carbohydrates mainly to lactate and acetate [12]. Accordingly, they are commonly associated with nutritious environments like foods, decaying material and the mucosal surfaces of the gastrointestinal and urogenital tract [12- 14], where they enhance the host protection against pathogens [13]. Their antagonistic and inhibitory properties are due to the competition for nutrients and the production of one or more antimicrobially active metabolites such as organic acids (lactic and acetic acid), hydrogen peroxide, and antimicrobial peptides like bacteriocins [8-11,15-17]. Bacteriocins are ribosomally synthesized peptides that exert their antimicrobial activity against either strains of the same species as the bacteriocin producer (narrow range), or to more distantly related species (broad range) [7,15,18]. An important reason for research on LAB based bacteriocins is due to their activity at nanomolar concentrations against number of bacterial pathogens [1,3,5,6,19,20]. Some bacteriocins even exhibit their activities against multidrug-resistant nosocomial pathogens such as methicillin-resistant *Staphylococcus aureus* (MRSA) and vancomycin-resistant enterococci [VRE, 17, 21]. Thus they also may have some big potential in medical and veterinary applications. Fermented food and plant material have been a well-known source for bacteriocin-producing LAB, but isolates from the intestinal of animals and humans has become an increasingly important source for such strains due to an increased awareness of their importance as probiotics. In fish the presence of LAB is meanwhile well documented and the bio-protective potential of some strains and/or their bacteriocin has been highlighted in the last years [4-6,16,18,22-26]. Kvasnikov et al. [12] described the presence of lactic acid bacteria, including *Lactobacillus* in the intestines of various fish species at larval, fry and fingerling stages inhabiting ponds in Ukraine. They give information on the changes in their composition as a function of the season of the year and life-stage of the fish. However, it was discussed that some human activities like artificial feeding in ponds would have had an effect on the bacterial composition and load in some fish, like carp (*Cyprinus carpio*) which showed the highest content of lactic acid bacteria in the intestines. Cai et al. [27] described the lactic acid bacteria in *Cyprinus carpio* collected from the Thajin river in Thailand. They reported the presence of *Enterococcus* spp. and the dominance of *Lactococcus garviae,* an emerging zoonotic pathogen, in *Cyprinus carpio.* Bucio Galindo et al. [23] studied the distribution of lactobacilli in the intestinal content of river fish and reported that various species of lactobacilli were present in relatively high numbers in the intestines of edible freshwater fish from the river, especially in warm season but in low numbers in cold season. There are no reports on the presence of *Lactobacillus* in the intestines of sturgeon fish inhabiting Caspian Sea, whereas other groups of bacteria have been studied in more details. In comparison with other food products of dairy or meat origin, only few bacteriocinogenic LAB strains have been recovered from seafood. The present study focuses on the characterization of antimicrobial compounds produced by the lactobacilli isolates, in addition, their ability to inhibit the growth of relevant food borne pathogens as well as of spoilage bacteria and last but not least, of contaminants in aquaculture.

## 2. Materials and methods

### 2.1. Fish intestine samples

Two species of Persian sturgeon (*Acipenser persicus*) and Beluga (*Huso huso*) were collected from the south coast of Caspian Sea in Iran. Twenty two individuals of these fish in adult stage were selected. The weight and length of the fish were measured before dissection. The fish were sacrificed by physical destruction of the brain, and the number of incidental organisms was reduced by washing the fish skin with 70% ethanol. Then, the ventral surface was opened with sterile scissors. After dissecting the fish, 1 g of the intestinal tract content of each fish was removed under aseptic condition and placed into previously weighed flasks containing storage medium.

## 2.2. Media and culture condition

Intestinal content was homogenized in a storage medium using a vortex mixer. One milliliter was transferred to reduced neutralized bacterial peptone (NBP, Oxoid L34, Hampshire, England) 0.5 g/L, NaCl 8 g/L, cysteine.HCl 0.5 g/L, pH adjusted to 6.7 [29]. Afterwards serial dilutions were spread on plates of selective media and incubated at the following conditions. Columbia blood agar (CAB, Oxoid CM 331) was used as a selective medium to make an estimation of the cultivable total anaerobic counts [29]. All the inoculated plates were incubated anaerobically at 30°C for 48 h. The following two media were used to isolate lactic acid bacteria (LAB). MRS (MRS, Merck, Darmstadt, Germany) with 1.5% agar (M641, HiMedia, Mumbai, India) and pH adjusted to 4.2 (MRS 4.2) and incubated anaerobically at 30°C for 96 h was used as a selective medium for lactic acid bacteria. MRS is an inhibitory medium for *Carnobacterium*. Anaerobic MRS with Vancomycin and Bromocresol green (LAMVAB), incubated at 30°C for 96 h was used as an elective and selective medium for *Lactobacillus* spp. [30]. Anaerobic incubation of the three media was made in an anaerobic Gas-Pack system (LE002, HiMedia, Mumbai, India) with a mixture of 80% $N_2$, 10% $H_2$ and 10% $CO_2$. Colonies were selected either randomly, or in case of less than 10 colonies per each plate, all the samples were counted according to the method described by Thapa et al. [31]. Purity of the isolates was checked again by streaking them onto fresh agar plates of the isolation media, followed by microscopic examinations. Identified strains of lactobacilli were kept in MRS broth with 15% (v/v) glycerol at -20°C.

## 2.3. Characterization procedures for lactic acid bacteria

Eighty four strains were randomly selected for identification procedures based on the phenotypical characteristics. Cell morphology and motility of all isolates were observed using a phase contrast microscope (CH3-BH-PC, Olympus, Japan). Isolates were gram-stained and tested for catalase production test. Preliminary identification and grouping was based on the cell morphology and phenotypic properties such as $CO_2$ production from glucose, hydrolysis of arginine, growth at different temperatures (10, 15 and 45°C), and at different pH (3.9 and 9.6). As well as the ability to grow in different concentrations of NaCl (6.5% (w/v), 10% (w/v) and 18% (w/v)) in MRS broth was checked as well. The configuration of lactic acid produced from glucose was determined enzymatically using d-lactate and l-lactate dehydrogenase test kits (Roche Diagnostic, France). The presence of diaminopimelic acid (DAP) in the cell walls of LAB was determined using thin-chromatography on cellulose plates. Fermentation of carbohydrates was determined using API 50 CHL (API 50 CH is a standardized system, associating 50 biochemical tests for the study of carbohydrate metabolism in microorganisms. API 50 CH is used in conjunction with API 50 CHL Medium for the identification of *Lactobacillus* and related genera) strips according to the manufacturer's instructions (Biomerieux, Marcy l' Etoile, France). The APILAB PLUS database identification software (bioMe´rieux, France) was used to interpret the results. Identification was undertaken according to the method described by Kandler and Weiss [12] and Hammes and Vogel [32].

## 2.4. Statistical analysis

Statistical analysis using Student's t-test (SPSS, Version 11.0) was performed to find significant difference on lactobacilli count between LAMVAB and MRS 4.2. Pearson's correlation coefficient was used to investigate the correlation of lactobacilli count between LAMVAB and MRS 4.2 (SPSS Inc., Version 11.0, Chicago, USA). A significance level of $p<0.05$ was used.

## 2.5. Screening of *Lactobacillus* strains for their inhibitory potential

In a first test series, the ability of each of the *Lactobacillus* isolates to exert an antibacterial effect against *Listeria monocytogenes* ATCC 19115 and *Salmonella* Typhimurium PTCC 1186 were examined by using three methods: the spot-on-lawn method, standardized agar disk diffusion method and the well diffusion method as described by Schillinger and Lucke [33], Benkerroum et al. [34] and Tagg & Mc Given [35]. Throughout, cell-free supernatants (CFS) of strains were obtained by centrifugation at 10,000 ×g for 20 min and then adjusted to pH 6.5 by applying NaOH (to exclude the effect of organic acid) before sterilization by filter (0.2 μm, Sigma, UK). Based on the screening tests, the inhibitory spectrum of potential bacteriocin-producing isolates was assessed against 42 indicator strains using a standardized agar disk diffusion test. The strains were kept frozen in 20% (v/v) glycerol at -20°C. For this purpose, an aliquot of 20 ml CFS was applied on disks (6 mm) and set on agar plates previously inoculated with each individual indicator strain suspension, which corresponded to a $10^5$ CFU/ml. Plates were incubated 24 h at optimum temperatures of the test organism. Antimicrobial activity was detected as a translucent halo in the bacterial lawn surrounding the disks.

## 2.6. Characterization of the inhibitory effect

In order to determine the biological nature of the antimicrobial activity of bacteria, CFS (pH 6.0) of 24-h lactobacilli cultures of two selected isolates (*Lactobacillus casei* AP8 & *Lactobacillus plantarum* H5) incubated at 30°C, were tested for their sensitivity to the proteolytic enzymes. One ml of CFS was treated for 2 h with 1 mg $ml^{-1}$ final concentration of the following enzymes: papain, trypsin, proteinase K, pronase E and $\alpha$-amylase (Sigma, London). To clarify whether the antimicrobial activity detected derives from the production of hydrogen peroxide, 2600 IU/ml of catalase (Sigma, London) were added to 1 ml portions of extracellular extracts of LAB exhibiting antimicrobial activity and incubated for 24 h at ambient temperature. Chemicals were added to the CFS and the samples incubated for 5 h before being tested for antimicrobial activity. To determine the sensitivity of potential bacteriocin activities to the temperatures, samples of CFS were incubated under defined conditions. The effect of pH on bacteriocin activity was determined by adjusting the pH of the CFS (cell free supernatant (pH 6.5) of 24-h lactobacilli cultures incubated at 30°C) with diluted appropriate volumes of HCl and NaOH (Table 3). After incubating for 2 h, the pH of the samples was readjusted to 6.5 followed by sterilization (0.2 µm, Sigma, UK). In all cases, the remaining bacteriocin activity was assessed exemplarily by using strain *L. monocytogenes* ATCC 19115 as the indicator bacterium and by applying the agar disk diffusion plate bioassay. Untreated cell-free supernatants were used as controls and experiments were performed in duplicate.

## 2.7. Growth dynamics and antimicrobial compounds production

The time course of inhibitory substance production was performed by inoculating 10 mL of an overnight culture of selected *Lactobacillus* isolates into 100 mL of MRS broth followed by incubation at 30°C. Cells were subsequently removed by centrifugation at 10,000 ×g for 20 min. At appropriate intervals, changes in pH and optical density (600 nm) of the cultures were measured to monitor bacterial growth using a spectrophotometer (Hitachi U 1100, Tokyo, Japan). Antibacterial activity was evaluated every hour by using serial twofold dilutions of each culture used as a neutralized cell-free supernatant (CFS) tested against *L. monocytogenes* ATCC 19115 based on the agar disk diffusion plate bioassay. In a separate experiment, the inhibitory effect of CFSs of lactobacilli strains on target cells in liquid medium was also examined against *L. monocytogenes* ATCC 19115 as indicator strain. For this purpose, 20 mL of each filter-sterilized bacteriocin-containing cell-free supernatant were added to a 100 mL culture of the indicator organism at early exponential phase (4 h old). These experiments were also repeated with stationary-phase cells. The optical density at 600 nm and viable cell count were determined every hour during an observation period of 20 h. Indicator cells without CFSs were used as control.

## 2.8. Adsorption of bacteriocin to producer cells

Bacteriocin-producing cells were cultured for 18 h at 30 °C. The pH of the cultures was adjusted to 6.0 with 1 M NaOH to allow maximal adsorption of the bacteriocin to the producer cells, according to the method described by Yang et al. [36]. The cells were then harvested (10,000 ×g 20 min, 4 °C) and washed with sterile 0.1 M phosphate buffer (pH 6.5). The pellet was re-suspended in 10 ml of 100 mM NaCl (pH 2.0) and stirred slowly for 1 h at 4 °C. The suspension was then centrifuged (10,000 ×g 20 min, 4 °C), the CFS was neutralized to pH 7.0 with sterile 1 M NaOH followed by testing the bacteriocin activity as described above.

## 2.9. Partial purification and characterization of the bacteriocin

Bacteriocin producer strains were grown in MRS broth, and incubated without agitation for 18 h at 30°C. The cells were harvested (10,000 ×g, 20 min, 4 °C) and the bacteriocin precipitated from the CFS with 60% saturated ammonium sulphate [45]. The precipitate in the pellet and floating on the surface were collected and re-suspended in one-tenth volume 25 mM ammonium acetate buffer (pH 6.5). The sample was stored at -20 °C for one week and activity tests were performed as described above. For the determination of the molecular size of the bacteriocins, precipitated peptides re-suspended in 25 mM ammonium acetate buffer (pH 6.5) were separated by Tricine-SDS-PAGE, according to Schägger and Von Jagow [38]. Low molecular weight markers, ranging from 2.5 to 45 kDa (Pharmacia, Sweden) were used. One half of the gel containing the molecular marker was fixed for 20 min in 5% (v/v) formaldehyde, then rinsed with water and stained with Coomassie Brilliant Blue R250 (Bio-Rad) overnight. The other half of the gel (not stained and extensively pre-washed with sterile distilled water) was overlaid with a culture of $10^6$ cfu/ml *L. monocytogenes* ATCC 19115 embedded in BHI agar. The position of the active bacteriocin was visualized by an inhibition zone around the active protein band [39].

# 3. Results

## 3.1. Isolation of lactobacilli

Intestinal content of 22 fish were analysed for the presence of lactobacilli. To determine the most appropriate medium for isolating lactobacilli from fish intestines, two media (MRS agar, LAMVAB) were used. LAMVAB was highly selective to quantify lactobacilli, as 99% of 143 randomly picked colonies and purified isolates were identified as *Lactobacillus* spp. and confirmed according to [12] (Table 1). Counts of intestinal lactobacilli for Persian sturgeon and beluga were detected at the range of approximately $10^{5.3}$ to $10^{6.4}$ cfu/g, respectively. The physiological and biochemical characterization of *Lactobacillus* isolates and the presumptive *Lactobacillus* species found in two fish species are shown in Table 2. From 84 isolates, 2 metabolic groups of *Lactobacillus* were recovered: facultative and obligate heterofermentatives. *L. sakei* and *L. plantarum* were the most often found isolates (Table 2). MRS 4.2 was suitable to quantify lactobacilli. As 30 randomly picked colonies on the highest dilution were identified as lactobacilli and coccoid forms were not found. Means of counts of 90 samples were not statistically different to LAMVAB counts in the Student's t-test (P=0.29) and were correlated with LAMVAB counts (r = 0.85; P<0.001). The correlation of counts on MRS 4.2 with those on LAMVAB and the absence of coccoids suggests that lactobacilli were the most important acidophilic lactic acid bacteria in the samples analysed. Facultative anaerobic flora recovered in CAB medium provided the highest counts in the samples analysed (Table 1).

| Fish species | No. | CAB (cfu/g) | LAMVAB (cfu/g) | MRS 4.2 (cfu/g) |
|---|---|---|---|---|
| *Acipenser persicus* | 12 | 7.84 | 5.32 | 4.85 |
| *Huso huso* | 10 | 8.21 | 6.45 | 5.64 |

CAB: Columbia blood agar; LAMVAB: *Lactobacillus* spp. Anaerobic MRS with Vancomycin and Bromocresol green; MRS 4.2: deMan, Rogosa and Sharp

**Table 1.** Average bacterial counts of intestinal bacteria (Log cfu/g of intestinal content) for Persian sturgeon and beluga in different media

## 3.2. Screening of *Lactobacilli* strains for antimicrobial activity and bacteriocin production

Eighty four lactobacilli strains previously isolated from two species of Sturgeon fish identified and their cell free supernatant extracts were assayed for antimicrobial activity and

| Presumptive *Lactobacillus* species | *L. sakei* | *L. plantarum* | *L. coryneformis* | *L. alimentarius* | *L. brevis* | *L. casei* | *L. oris* |
|---|---|---|---|---|---|---|---|
| No. of isolates | 30 | 18 | 12 | 10 | 7 | 5 | 2 |
| Diaminopimelic acid | ND | + | ND | ND | ND | ND | *ND* |
| $CO_2$ from glucose | - | - | - | - | + | - | + |
| $NH_3$ from arginine | - | - | - | - | + | - | + |
| 10°C | + | + | + | + | + | + | + |
| 15°C | + | + | + | + | + | + | + |
| 45°C | - | - | - | 2 | - | - | - |
| Glycerol | - | + | - | 1 | - | + | - |
| L-Arabinose | + | + | - | 2 | 2 | - | + |
| Ribose | + | - | - | + | + | + | + |
| D-Xylose | 26 | - | - | - | - | - | + |
| Galactose | 29 | - | - | - | - | + | - |
| Rhamnose | - | - | + | - | 2 | + | - |
| Inositol | - | + | - | - | - | + | + |
| Mannitol | - | + | 5 | - | + | + | - |
| Sorbitol | - | + | - | - | - | + | - |
| 1-Methyl-D-mannoside | - | + | - | - | - | - | + |
| 1-Methyl-D-glucoside | - | + | - | 7 | + | - | + |
| N-Acetyl glucosamine | 28 | + | + | + | + | + | + |
| Amygdaline | 10 | + | - | + | - | + | + |
| Arbutine | 1 | + | - | + | - | + | + |
| Esculine | + | + | + | + | 1 | + | + |
| Salicin | + | + | - | + | - | + | + |
| Cellobiose | 27 | + | - | + | - | + | + |
| Maltose | 19 | + | - | + | + | + | + |
| Lactose | 26 | + | - | + | - | + | + |
| Melibiose | + | + | + | 2 | + | - | + |
| Sucrose | + | + | + | 8 | + | + | + |
| Trehalose | + | + | - | + | - | + | - |
| Melezitose | - | + | - | - | + | + | + |
| D-Raffinose | 29 | - | - | 2 | + | - | - |
| Starch | - | - | - | - | - | + | - |
| Xylitol | - | + | 3 | - | - | - | + |
| 2-Gentiobiose | + | + | - | + | - | + | + |
| D-Turanose | - | - | - | - | + | - | - |
| D-Tagatose | 1 | - | - | + | - | + | - |
| D-Arabitol | - | + | 5 | - | - | - | + |
| Gluconate | + | - | - | + | + | + | + |
| 2-keto-gluconate | - | - | 1 | 2 | - | - | + |
| 5-keto-gluconate | - | - | - | 1 | - | + | + |
| *Lactic acid configuration* | *DL* | *DL* | *DL* | *DL* | *DL* | *DL* | D |

† +: Positive reaction of all the isolates. Numbers are the positive isolates. All isolates fermented D-Glucose, D-Fructose, D-Mannose, however they didn't ferment erythrol, D-Arabinose, L-Xylose, Adonitol, 2-Methyl-xyloside, L-Sorbose, Dulcitol, Inulin, Glycogen, D-Fucose, L-Fucose, L-Arabitol. ND: Not data

**Table 2.** Biochemical characteristics of *Lactobacillus* species isolated from the intestines of Persian sturgeon and beluga

bacteriocin production against *Listeria monocytogenes* ATCC 19115 and *Salmonella* Typhimurium PTCC 1186 by using spot-on-lawn method, standardized agar disk diffusion method and well diffusion method. In each instance, diameters of inhibition were quantified. Fifteen strains (18%) exhibited inhibitory activity against both indicator organisms. Consequently, all candidate isolates (Inhibition zone> 8mm) subjected to different tests such as growth at different temperatures, pH, salt content, antibiotic resistance, etc. Based on the result of aforementioned tests, two strains *Lactobacillus* casei AP8 and *Lactobacillus plantarum* H5, isolated from Persian sturgeon and beluga respectively, were chosen as active strains and were subjected to further examinations.

| Presumptive *Lactobacillus* species | *L. sakei* | *L. plantarum* | *L. coryneformis* | *L. alimentarius* | *L. brevis* | *L. casei* | *L. oris* |
|---|---|---|---|---|---|---|---|
| *Acipenser persicus* | ** | ** | * | ** | - | ** | * |
| *Huso huso* | ** | * | - | * | ** | * | * |

* = Presence of lactobacilli. ** = High number of lactobacilli presence

**Table 3.** *Lactobacillus* species isolated from the intestines of sturgeon fish

### 3.3. Inhibitory spectrum of bacteriocin

As the results in screening test showed that greater inhibition was observed by agar disk diffusion tests of cell-free supernatant extracts, so this method was selected as the best technique for examining the antibacterial activity of *L. casei* AP8 and *L. plantarum* H5 CFSs against forty two Gram-positive and Gram-negative bacteria. The CFS preparations from both strains showed a broad inhibitory spectrum against a wide range of LAB of different species and some food-borne pathogens and spoilage bacteria including *Listeria innocua, L. monocytogenes, Staphylococcus aureus, Aeromonas hydrophila, Aeromonas salmonicida, Bacillus cereus, Bacillus pumilus, Bacillus subtilis, Brochotrix thermosphacta,* Gram-negative *E. coli, Salmonella* and *Pseudomonas, Clostridium perfringens* and *Vibrio parahaemolyticus* (Table 4). Result showed that the Gram-positive bacteria tested were more sensitive to the bacteriocin produced by the isolates than Gram-negative bacteria. The largest spectrum of inhibition was showed by *L. casei* AP8 bacteriocin, which inhibited 33 out of 42 indicator strains.

### 3.4. Characterization of inhibitory effect

Table 5 and table 6 depict the stability of inhibitory substances at different physic-chemical conditions. To determine the biological nature of the antimicrobial activity of bacteria, CFSs were tested for their sensitivity to the proteolytic enzymes. Antimicrobial activities exhibited by *L. casei* AP8 and *L. plantarum* H5 were sensitive to proteolytic enzymes since proteolytic, but not lipolytic or glycolytic enzymes, completely inactivated the antimicrobial effect of both cell-free supernatants, confirming the proteinaceous nature of the inhibitors (Table 3). The effect of several chemicals on the antimicrobial activity was also evaluated. Interestingly, the cell-free extracts remained active after treatment with chemicals such as catalase, SDS, Triton X-100, Tween 20, Tween 80 and EDTA after 5 h of exposure (Table 2). Enhancing the antimicrobial activity in case of *L. casei* AP8 bacteriocin was observed after treating by EDTA and SDS against *L. monocytogenes* ATCC 19115. The stability study of

| Indicator organism | Medium* | Temp. [°C] | Bac AP8 | Bac H5 |
|---|---|---|---|---|
| **Gram Negative Group** | | | | |
| *Aeromonas hydrophilus* MJ 1120 | BHI | 37 | ++ | 0 |
| *Aeromonas hydrophilus* MJ 1240 | BHI | 37 | +++ | + |
| *Aeromonas salmonicida* CC 1546 | BHI | 37 | + | + |
| *Aeromonas salmonicida* RT 7895 | BHI | 37 | ++ | + |
| *Brochothrix thermosphacta* RF 35 | BHI | 37 | ++ | + |
| *Escherichia coli* ATCC 25922 | BHI | 37 | ++ | 0 |
| *Escherichia coli* PTCC 1325 | BHI | 37 | ++ | ++ |
| *Photobacterium damselae ssp. Piscida* | BHI | 37 | 0 | 0 |
| *Pseudomonas aeruginosa* PTCC 1310 | BHI | 37 | ++ | + |
| *Pseudomonas fluorescens* HFC 1236 | BHI | 37 | ++ | 0 |
| *Salmonella enteritidis* ATCC 13076 | BHI | 37 | ++ | ++ |
| *Salmonella spp* SM 162 | BHI | 37 | +++ | ++ |
| *Vibrio anguillarum* MI12 | BHI | 37 | ++ | + |
| *Vibrio parahaemolyticus* MI 23 | BHI | 37 | +++ | 0 |
| *Vibrio parahaemolyticus MI 56* | BHI | 37 | +++ | + |
| **Gram Positive Group** | | | | |
| *Bacillus cereus* ATCC 9634 | BHI | 37 | +++ | +++ |
| *Bacillus coagulans* | BHI | 37 | +++ | ++ |
| *Bacillus licheniformis* PTCC 1331 | BHI | 37 | ++ | 0 |
| *Bacillus subtilis* ATCC 9372 | BHI | 37 | +++ | + |
| *Clostridium perfringens* ATCC 3624 | RCM | 37 | ++ | + |
| *Clostridium sporogenes* PTCC 1265 | RCM | 37 | ++ | + |
| *Lactobacillus acidophilus* ATCC 4356 | MRS | 30 | ++ | + |
| *Lactobacillus alimentarius* AP 10 | MRS | 30 | + | ++ |
| *Lactobacillus brevis* H56 | MRS | 30 | ++ | ++ |
| *Lactobacillus brevis* AP 83 | MRS | 30 | ++ | ++ |
| *Lactobacillus casei* PTCC 1608 | MRS | 30 | 0 | ++ |
| *Lactobacillus casei* RN 78 | MRS | 30 | 0 | 0 |
| *Lactobacillus casei* LB 10 | MRS | 30 | 0 | + |
| *Lactobacillus casei* LB 46 | MRS | 30 | 0 | + |
| *Lactobacillus plantarum* PTCC 1050 | MRS | 30 | 0 | 0 |
| *Lactobacillus plantarum* AP 76 | MRS | 30 | + | 0 |
| *Lactobacillus plantarum* H12 | MRS | 30 | + | 0 |
| *Lactobacillus sakei* AP 43 | MRS | 30 | 0 | + |
| *Lactobacillus sakei* | MRS | 30 | 0 | 0 |
| *Lactococcous sp* | MRS | 30 | + | 0 |
| *Lactobacillus curvatus* | MRS | 30 | 0 | + |
| *Listeria innocua AN 15* | BHI | 37 | ++ | ++ |
| *Listeria monocytogenes* ATCC 7644 | BHI | 37 | +++ | +++ |
| *Listeria monocytogenes* PTCC 1163 | BHI | 37 | ++ | ++ |
| *Listeria monocytogenes* PTCC1297 | BHI | 37 | ++ | ++ |
| *Staphylococcus aureus* ATCC 25923 | BHI | 37 | +++ | + |
| *Staphylococcus aureus* PTCC 1112 | BHI | 37 | +++ | + |

* BHI: brain hearth infusion, MRS: de Man-Rogosa-Sharpe agar and RCM: reinforced clostridial medium. 0 no zone of inhibition; +, 1 mm<zone<5 mm; ++, 5 mm<zone<8 mm; +++, zone>8 mm.; PTCC: Persian Type Culture Collection; ATCC: American Type Culture Collection.

**Table 4.** Antimicrobial activity of potential bacteriocin producing strain *L. casei* AP8 *and L. plantarum* H5 as examined with selected bacterial indicator strains.

inhibitory compounds of *L. casei* AP8 and *L. plantarum* H5 in different conditions indicated the high resistance of these agents. The antimicrobial compounds were able to resist most of these factors to which it was exposed even during prolong incubation period (Table 6). Cell free extracts prepared from both the isolates are found to be thermo-stable. When *L. casei* AP8 bacteriocin was heated at 40-100° C for 30 min, it retained inhibitory activity against *L. monocytogenes* ATCC 19115. However, a loss in activity in the ranges of 35% was observed when heated at 120°C for 15 min (Table 6). The Antilisterial activity of *L. plantarum* H5 bacteriocin was resistant to heat treatments of 40-100°C for 30 min and remained constant after heating at 121ºC for 15 min. Both investigated bacteriocins were most stable at 4°C and - 20°C and able to retain their antilisterial activity for 30 days without any decrease. *L. casei* AP8 bacteriocin was active in a wide range of pH, as full activity was retained at pH values between 3 and 10. *L. plantarum* H5 bacteriocin remained stable after incubation for 2 h at pH values between 2.0 - 12.0.

| Treatment | Concentration | %Residual antimicrobial activity | |
|---|---|---|---|
| | | *L. casei* AP8 | *L. plantarum* H5 |
| **Enzymes** | | | |
| Trypsin | 1 mg/ml$^{-1}$ | 0 | 0 |
| Papain | 1 mg/ml$^{-1}$ | 100 | 100 |
| Proteinase K | 1 mg/ml$^{-1}$ | 0 | 0 |
| Pronase E | 1 mg/ml$^{-1}$ | 0 | 0 |
| α- amylase | 1 mg/ml$^{-1}$ | 100 | 100 |
| Catalase | 1mg/ml$^{-1}$ | 100 | 100 |
| **Organic solvents** | | | |
| Butanol | 10% [v/v] | 100 | 100 |
| Ethanol | 10% [v/v] | 100 | 100 |
| Methanol | 10% [v/v] | 92 | 100 |
| Ethyl ether | 10% [v/v] | 100 | 100 |
| EDTA | 5 mmol l$^{-1}$ | 100 | 83 |
| Sodium deoxycholate | 1mg ml$^{-1}$ | 100 | 100 |
| Sulphobetaine 14 | 1mg ml$^{-1}$ | 92 | 100 |
| SDS | 1% [w/v] | 100 | 100 |
| Tween 20 | 1% [v/v] | 100 | 100 |
| Tween 80 | 1% [v/v] | 92 | 100 |

**Table 5.** Effect of enzymes and chemicals on the antimicrobial activity of two selected strains *L. casei* AP8 and *L. plantarum* H5. For details see text

| Treatment (Storage, Temperature and pH stability) | Residual antimicrobial activity | |
|---|---|---|
| | *L. casei* AP8 | *L. plantarum* H5 |
| 4 ºC, -20ºC/ 30 d | + | + |
| 40-100 ºC/30 min | + | + |
| 121 ºC/10 min | + [-35%] | + |
| 121 ºC/15 min | - | + |
| pH= 2 | - | + |
| pH= range 3-10 | + | + |
| pH= 11 | - | + |
| pH= 12 | - | + |

No inhibition= -; inhibition= +

**Table 6.** Effect of cold storage, different temperatures and pH on inhibitory activity against *Listeria monocytogenes* ATCC 19115. For details see text.

## 3.5. Growth and bacteriocin production

Figure 1 shows the growth and bacteriocin production curves of *L. casei* AP8 and *L. plantarum H5* cultured at 30°C. For *Lactobacillus* casei AP 8 cell growth reached the stationary phase at 12 h of cultivation. Kinetics of bacteriocin production showed that its synthesis and/or secretion started at 4 h growth in the exponential phase of growth and maximum activity was observed at the early stationary phase of growth (1800 AU $ml^{-1}$) and had stabled for 6 h before the bacteriocin activity decreased (Figure 1). The pH values decreased from 6.5 to 3.7 at the end of incubation. For *L. plantarum* H5, bacteriocin activity was detectable in the culture supernatant after 5hr when an absorbance of 0.55 at 600 nm of the culture broth. Production of bacteriocin increased throughout logarithmic growth. In the stationary phase, *L. plantarum* H5 showed maximum bacteriocin activity (3400 AU/mL) and stabilized for 2 hr. But since then, bacteriocin activity declined gradually and stabilized at 1600 AU/ml during the following 4 h. In the stationary phase, extracellular pH was maintained, however, bactericidal activity decreased, excluding a possibility of lactic acid as a bactericidal mechanism.

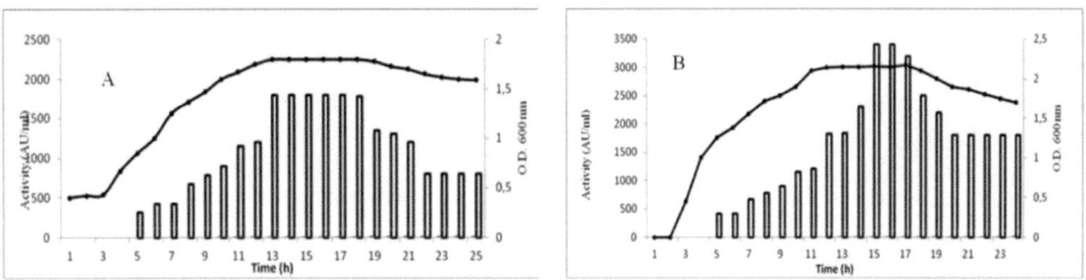


**Figure 1.** Antimicrobial activity [bars] against *L. monocytogenes* ATCC19115 of *L. casei* AP8 [A] and *L. plantarum* H5 [B] observed during growth in MRS medium [•] and expressed in AU/ml. Results are represent the mean of three independent experiments.

To investigate the reduction of viable cells of target organism in presence of inhibitory substances, twenty mL of each filter-sterilized bacteriocin-containing cell-free supernatant were added to 100 mL of *L. monocytogenes* ATCC 19115 (4 h old at 30°C). The optical density at 600 nm and viable cell count were determined every hour during 24 h. In the control samples inoculated with indicator strain the viable cell count reached to $10^{11}$CFU/ml after 24 h incubation at 37°C. The inhibition kinetics using the bacteriocin AP8 (Figure 2) indicated a bactericidal mode of action against *L. monocytogenes*. Addition of the bacteriocin *L. casei* AP8 to early logarithmic-phase cells of indicator strain resulted in grows inhibition after 1h, followed by complete growth inhibition (slow decline) for the remaining time (20 h). In the case of *L. plantarum* H5 bacteriocin the inhibition kinetics showed a bacteriostatic mode of inhibition against indicator strain. Addition of bacteriocin H5 to culture of *L. monocytogenes* showed a growth inhibition after 1 h followed by slow growth. Experiment with stationary-phase cells did not showed any inhibition. No increase in the activity of bacteriocin AP8 and H5 were observed after treatment of the producer cells with 100 mmol/l NaCl at low pH, suggesting that these bacteriocins do not adhere to the surfaces of the producer cells.

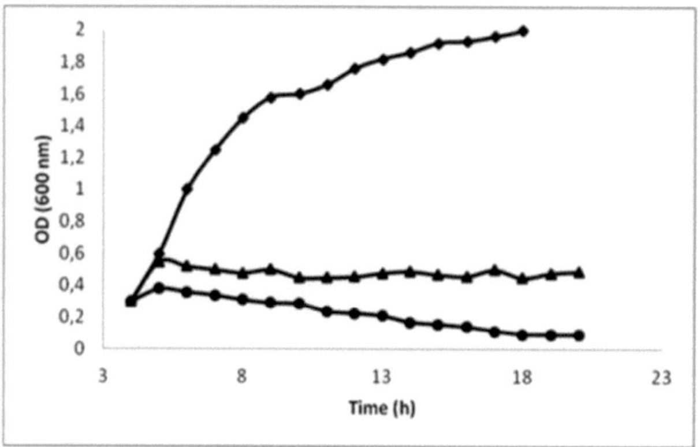


**Figure 2.** Antimicrobial effect of the CFS of *L. casei* AP8 [▲] and *L. plantarum* H5 [•] on the growth of *L. monocytogenes* ATCC 19115 at 30°C. Growth of *L. monocytogenes* ATCC 19115 without added bacteriocins [control, ♦].

### 3.6. Partial purification and molecular size of bacteriocins AP8 and H5

Ammonium sulfate precipitation method with 60% saturated ammonium sulphate is used for partial purification of both bacteriocins. Results showed an increase (10-15%) in the inhibitory activity of both bacteriocins against *L. monocytogenes* ATCC 19115 after precipitation. The SDS-PAGE analysis of the partially purified samples showed peptide bands for bacteriocins AP8 and H5 in size of approximately 5 and 3 kDa respectively (Figure 3).

## 4. Discussion

In this study, we isolated, quantified and characterized *Lactobacillus* species from two species of sturgeon fish inhabiting Caspian Sea to make a bank collection of strain for further research. These fishes are highly valuable species for fisheries and aquaculture in

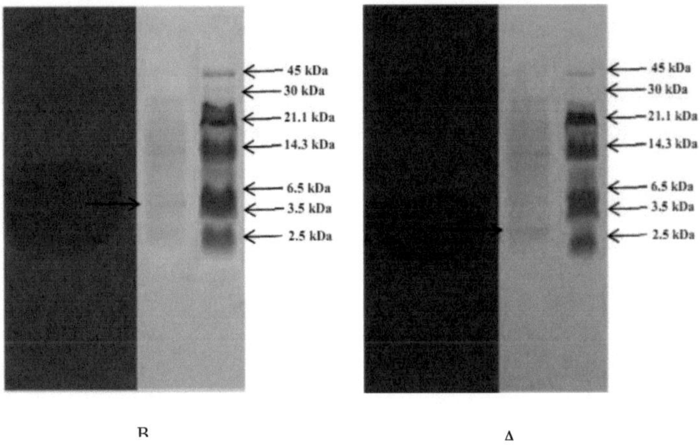


**Figure 3.** Tricine-SDS-PAGE gel of partially purified bacteriocins [precipitated by 60% saturated ammonium sulphate] *L. casei* AP8 [A] and *L. plantarum* H5 [B] along with the standard MW markers. The gel was overlaid with *L. monocytogenes* ATCC 19115 [approx. $10^6$ CFU/ml], embedded in BHI agar, after incubation at 30 °C for 24.

Iran. Presumptive lactobacilli species found in this study were relatively similar to the species described by Bucio Galindo et al. [28]. These authors reported *L. alimentarius, L. coryneformis, L. casei, L. sakei, L. pentosus, L. plantarum, L. brevis and L. oris,* as lactobacilli presented in the intestinal content of studied fish. However, the fish species analysed in that study were different from the two species in this study which were collected from a lake environment. The biochemical characteristics used for identification of *Lactobacillus* may suggest some ideas in relation to the occurrence of the strains in nature. Most of *Lactobacillus* examined in this study (80%) had the capacity to ferment lactose and galactose. Generally, most lactobacilli are able to ferment lactose, by uptake of this disaccharide by a specific permease and splitting it by S-galactosidase for further phosphorylation of galactose and glucose [12]. Because, lactose is only present in milk and milk derivates, it is possible that these strains have evolved from environments related with mammals, as was suggested for other lactose positive *Lactobacillus* [40]. Lactose may be present or was present in the environment as a waste; resulting from livestock production, and disposal effluents from dairy factories. Another component, often fermented by the strains was the amino-sugar N-acetyl-glucosamine, a compound present in peptidoglycans, in blood, chitin and as one of the main constituents of mucus in the gastrointestinal tract [41]. The carbohydrate portion constitutes above 40% of the weight of the mucus [42] or higher values [41].

It could be shown that two strains, *Lactobacillus casei* AP8 and *Lactobacillus plantarum* H5 isolated from intestinal bacterial flora of beluga (*Huso huso*) and Persian sturgeon (*Acipenser persicus*) were able to produce antibacterial substances. According to the findings it was likely that the antibacterial effect was due to the formation of bacteriocin. Results from enzyme inactivation studies demonstrated that antimicrobial activity of isolates AP 8 and H5 was lost or unstable after treatment with all the proteolytic enzymes, confirming the protein status of metabolites and indicating the presence of bacteriocins. Furthermore, treatment with lipolytic or glycolytic enzymes did not affect the activity of antimicrobial compound produced by strain, suggesting that produced bacteriocins do not belong to the controversial group IV of the bacteriocins, which contain carbohydrates or lipids in the active molecule structure [45-47]. It is important to note that, their activities were not due to hydrogen peroxide or acidity, as antimicrobial activity was not lost after treatment with catalase. Both of the presumptive were considered to be heat stable. Although heat stability of antibacterial substances produced by *Lactobacillus* spp. has been well established [39,48,49,50-53] heat stability of *L. casei* AP8 121ºC for 10 min is novel. The result of pH stability were not coherent with previous report that had indicated the tolerance of bacteriocins to acidic pH rather than alkaline [36,54]. The loss of antimicrobial activity of AP8 bacteriocin at pH > 10 might be ascribed to proteolytic degradation, protein aggregation or instability of proteins at this extreme pH [39,48,55]. *L. casei* AP8 bacteriocin showed an increase in the inhibitory activity after treatments with SDS and EDTA, may be due to the ability of these compounds to break down the proteinaceous complex from its large form into smaller more active unite [21]. Similar to Lactocin RN78 and Plantaricin LC74, both bacteriocin *L. casei* AP8 and *L. plantarum* H5 were found to be stable after treatment with organic solvents like butanol, ethanol and methanol confirming their proteinaceous and soluble nature [18,21,56]. Pronounced inhibitory potential against various species of Gram-positive bacteria were shown, including pathogenic and spoilage microorganisms such as *A. hydrophila, A. salmonicida, C. perfringens, B. cereus* and *L. monocytogenes*. Observed effects were consistent with reports about bacteriocins produced by other strains of LAB [1,3,17,19,20,25,41,49,55,57 59,60]. Although bacteriocins from LAB usually are ineffective against Gram-negative bacteria and rather relate to a narrow antimicrobial spectrum [9,51,53], both presumptive bacteriocins AP8 and H5 showed broad antimicrobial activity against several genera of Gram-positive and Gram-negative bacteria. Even representatives of *Pseudomonas, Salmonella, E. coli, A. hydrophila, A. salmonicida and V. anguillarum* could be inhibited. Moreover a high level of inhibitory activity against *Listeria monocytogenes* was observed. Earlier studied have shown that several marine bacteria may produce inhibitory substances against bacterial pathogens in aquaculture systems [1,16,19]. Hence the use of such bacteria releasing antimicrobial substances in now gaining importance in fish farming as a natural alternative to administration of antibiotics [1,61-63]. In kinetic studies, both crude bacteriocins were continuously produced during logarithmic phase followed by optimal production during stationary growth phase, suggesting that these peptides may be secondary metabolites. Similar results were reported for some bacteriocins produced by some LAB isolates [5,64,65] and is contrary for other *Lactobacillus* species bacteriocins [1,16,25,55,60,66]. Bacteriocin H5 showed a decrease in activity towards the end of stationary growth may be due to proteolytic degration, protein aggregation, and feedback regulation as has been observed for Lactacin ST13BR, Lactacin B, Helveticin J and Enterocin1146 [53,55,67]. *L. casei* AP8 crude bacteriocin demonstrated a bactericidal mode of action, as the immediate decrease in the optical density of *L. monocytogenes* was observed in mix culture. In the case of H5 bacteriocin a bacteriostatic mode of action was observed. Crud H5 bacteriocin showed a growth inhibition, followed by decrease activity for remained time, suggesting that indicator organism became resistant to the bacteriocin or bacteriocin was destroyed by proteolytic enzymes [55]. Treating of bacteriocins AP 8 and H5 with NaCl at low pH did not result in increased levels of antilisterial activity, suggesting no adsorption of bacteriocins to their producer cells in agreement with result reported before for *Lactobacillus* strains bacteriocins [55,64,66].

More accurate techniques could be used to determine the molecular mass of molecules, yet the SDS-PAGE technique provides valuable information about the presence of the peptides [3]. In recent years, a large number of new bacteriocins produced by *L. plantarum* have been identified and characterized and the molecular masses of all the bacteriocins produced have been reported in the range of 3-10 kDa [5,39,55]. However, to our knowledge, there is no bacteriocin produced by any *L. casei* strain with a molecular mass of 5 kDa with similar characteristics to strain investigated in this study. Thus, it is possible that this bacteriocin may be a novel bacteriocin produced by *L. casei*. The physiochemical properties of bacteriocins from *L. casei* AP8 and *L. plantarum* H5 were similar to those of other bacteriocins of lactobacilli belonging to the group IIa lactic acid bacteria with respect to molecular weight, heat and pH stability and also sensivity to proteolytic enzymes [9,45,51]. Characteristics unifying all members of class IIa bacteriocins are 1) below 10kDa [1] their potent activity against *Listeria* spp., 2) their resistance to elevated temperatures and extreme pHs, and 3) their cystibiotic feature attributed to the presence of at least one disulfide bridge, which is crucial for antibacterial activity [15,45,51,55]. Class IIa bacteriocins were formerly considered as "narrow"-spectrum antibiotics, with antimicrobial activity directed against related strains. Recently, some class IIa bacteriocins, such as bacteriocin OR-7, enterocin E50-52, and enterocin E760, have been shown to be active against both Gram-negative and Gram-positive bacteria, including *Campylobacter jejuni, Yersinia* spp., *Salmonella* spp., *Escherichia coli* O157:H7, *Shigella dysenteriae, Staphylococcus aureus,* and *Listeria* spp. [15, 54, 45, 51, 3].

## 5. Conclusion

Bacteriocins AP8 and H5 showed a wide spectrum of antibacterial activity against seafood borne pathogens like *Listeria, Clostridium, Bacillus* spp, *S. aureus* and even Gram-negative pathogens like *Pseudomonas, Salmonella* and *E. coli*. Some of these foodborne pathogens can produce toxins resulting in human illness. In addition to the broad inhibition spectrum, theirs technological properties and especially cold, heat and storage stability, indicate that bacteriocins AP8 and H5 have potential for application not only as biopreservative agents to control pathogens in food products that are pasteurized and cook-chilled but also as bioprotect compounds at aquaculture. Accordingly *L. casei* may be of great interest as probiotics strains because of their ability to adhere to intestinal epithelial cells and being of human origin. Several authors have reported the production of bacteriocins by *L. casei* and *L. plantarum* strains from plant, dairy or meat origin. However, very few bacteriocins from *L. plantarum* have been reported to be isolated from fish and also based on our knowledge this is the first report of a *L. casei* bacteriocin isolated from fish.

## Author details

Mahdi Ghanbari and Mansoureh Jami
*University of Zabol, Faculty of Natural Resources, Department of Fishery; Zabol, Iran*
*BOKU -University of Natural Resources and Life Sciences, Department of Food Sciences and Technology, Institute of Food Sciences; Vienna, Austria*

Masoud Rezaei
*Tarbiat Modares University, Faculty of Marine Science, Department of Seafood Science and Technology, Noor, Mazandaran, Iran*

## Acknowledgement

The authors thank the personnel of the Shahid Rajaei Aquaculture Center for their assistance. We are also grateful to Professor Mehdi Soltani from Faculty of Veterinary Medicine, University of Tehran, Dr Ali Shahreki from Medicine Science University of Zahedan, Iran, Professor Masoud Rezaei from Tarbiat Modares University, Iran, Professor Wolfgang Kneifel and Professor Konrad Domig from University of natural Resources and Life Sciences, Vienna, Austria for their helpful opinions. Some part of this research was funded by Tarbiat Modares University, Tehran, Iran.

## 6. References

[1] Indira, K., S. Jayalakshmi, A. Gopalakrishnan, and M. Srinivasan. 2011. Biopreservative potential of marine *Lactobacillus* spp. African Journal of Biotechnology. 5[16]: 2287-2296

[2] Okafor, N. and B. C. Nzeako. 1985. Microbial flora of fresh and smoked fish from Nigerian fresh waters. Food Microbiology. 2: 71-75.

[3] Pinto, A. L., M. Fernandes, C. Pinto, H. Albano, F. Castilho, P. Teixeira, and P. A. Gibbs. 2009. Characterization of anti-Listeria bacteriocins isolated from shellfish: Potential antimicrobials to control non-fermented seafood. International Journal of Food Microbiology. 129: 50–58

[4] Ghanbari, M. 2007. The Use of Gram positive bacilli isolated from intestinal flora of fish to control growth of *Listeria monocytogenes* in cold smoked Roach. M.Sc Thesis. Trabiat Modares University. Iran.

[5] Todorov, S.D., P. Hob, M. Vaz-Velho, and L. M. T. Dicks. 2010. Characterization of bacteriocins produced by two strains of *Lactobacillus plantarum* isolated from Beloura and Chouriço, traditional pork products from Portugal. Meat Science. 84: 334–343

[6] Leroi, F., 2010. Occurrence and role of lactic acid bacteria in seafood products. Food Microbiology. 27, 1-12.

[7] Glvez, A., H. abriouel, R. L. Lopez and N. B. Omar. 2007. Bacteriocin-based strategies for food biopreservation. International Journal of Food Microbiology. 120: 51-70.

[8] Holzapfel, W. H., R. Geisen, and U. Schillinger. 1995. Biological preservation of foods with reference to protective cultures, bacteriocins and food grade enzymes. International Journal of Food Microbiology. 24: 343-362.

[9] Cleveland J., T. J., Montville, I. F. Nes, and M. L. Chikindas. 2001. Bacteriocins: safe natural antimicrobials for food preservation. International Journal of Food Microbiology. 71: 1-20.

[10] Kandler, O. and N. Weiss. 1986. In: Bergey's Manual of Systematic Bacteriology Vol. 2, Baltimore: Williams and Wilkins. 1209–1234.

[11] Maragkoudakis, P.A., G. Zoumpopoulou, C. Miaris, G. Kalantzopoulos, B. Pot and E. Tsakalidou. 2006. Probiotic potential of *Lactobacillus* strains isolated from dairy products. International Dairy Journal. 16: 189-199.

[12] Kandler, O and Weiss, N [1986]. Genus *Lactobacillus* beijerinck 1901, 212$^{AL}$. In: Sneath, PHA; Mair, NS; Sharpe, ME and Holt, JG [Eds.], *Bergey's manual of systematic bacteriology*. [Eds.], Vol. 2, Baltimore: Williams and Wilkins. PP: 1209-1234.

[13] Havenaar, R; Ten Brink, B and Huis in't Veld, JHJ [1992]. Selection of strains for probiotic use. In: Fuller, R [Ed.], *Probiotics: the scientific basis*. [1st Edn.], London, Chapman and Hall. PP: 209-224.

[14]Walstra, P; Geurts, TJ; Noomen, A; Jellema, A and van Boekel, MAJS [1999]. *Dairy technology, principles of milk properties and processes*. Eds., New York, Marcel Dekker, Inc., P: 727.

[15] Belguesmia, Y., K. Naghmouchi, N. E. Chihib, and D. Drider. 2011. Class IIa bacteriocins: current knowledge and perspectives, p 171-195. *In* D. Drider and S. Rebuffat [ed.], Prokaryotic Antimicrobial Peptides: From Genes to Applications, DOI 10.1007/978-1-4419-7692-5_10, © Springer Science+Business Media, LLC 2011.

[16] Campos, C.A., O. Rodríguez, P. Calo-Mata, M. Prado, and J. Barros-Velásquez. 2006. Preliminary characterization of bacteriocins from *Lactococcus lactis*, *Enterococcus faecium* and *Enterococcus mundtii* strains isolated from turbot [*Psetta maxima*]. Food Research International 39: 356–364.

[17] Karska-Wysocki, B., M. Bazo and W. Smoragiewicz. 2010. Antibacterial activity of *Lactobacillus acidophilus* and *Lactobacillus casei* against methicillin-resistant *Staphylococcus aureus* [MRSA]. Microbiology Research 165: 674-686.

[18] Ghanbari, M., M. Rezaei, M. Soltani, G. R. Shahosseini., 2009. Production of bacteriocin by a novel Bacillus sp. strain RF 140, an intestinal bacterium of Caspian Frisian Roach [*Rutilus frisii kutum*]. Iranian Journal of Veterinary Research 10 [3]: pp. 267-272.

[19] Chahad, O. B., M. El Bour, P. Calo-Mata, A. Boudabous, J. Barros-Velazquez. 2011. Discovery of novel biopreservation agents with inhibitory effects on growth of food-borne pathogens and their application to seafood products. Research in Microbiology. doi:10.1016/j.resmic.2011.08.005

[20] Delves-Broughton, J., P. Blackburn, R. J. Evans, and J. Hugenholts. 1996. Application of the bacteriocin, nisin. Antonie van Leeuwenhock. 69: 193-202.

[21] Mojgani, N., and C. Amirinia. 2007. Kineitics of growth and bacteriocin production in *L. casei* RN 78 isolated from a dairy sample in IR Iran. International Journal of Dairy Science. 2[1]: 1-12

[22] Bucio, A., R. Hartemink, J. W. Schrama, J. Verreth, F. M. Rombouts. 2006. Presence of lactobacilli in the intestinal content of freshwater fish from a river and from a farm with a recirculation system. Food Microbiology. 23: 476–482.

[23] Ghanbari, M., M. Rezaei, R. M. Nazari, and M. Jami. 2009. Isolation and characterization of *Lactobacillus* species from intestinal contents of beluga [*Huso huso*] and Persian sturgeon [*Acipenser persicus*]. Iranian Journal of Veterinary Research 10 [2]: 152-157.

[24] Itoi, S., T. Abe, S. Washio, E. Ikuno, Kanomata, and H. Sugita. 2008. Isolation of halotolerant *Lactococcus lactis* subps. lactis from intestinal tract of coastal fish. International Journal of Food Microbiology. 121: 116121.

[25] Noonpakdee, W., P. Jumriangrit, K. Wittayakom, J. Zendo, J. Nakayama, K. sonomoto, and S. panyim. 2009. Two-peptide bacteriocin from *Lactobacillus plantarum* PMU 33 strainisolated from som-fak, a Thai low salt fermented fish product. Asia-Pacific Journal of Molecular Biology and Biotechnology. 17: 19-25

[26] Ringø, E., and F. J. Gatesoup. 1998. Lactic acid bacteria in fish: a review. Aquaculture 160: 177-203.

[27] Cai, YM; Suyanandana, P; Saman, P and Benno, Y [1999]. Classification and characterization of lactic acid bacteria isolated from the intestines of common carp and freshwater prawns. The Journal of General and Applied Microbiology 45: 177-184.

[28] Bucio Galindo, A; Hartemink, R; Schrama, JW; Verreth, JAJ and Rombouts, FM [2006]. Presence of lactobacilli in the intestinal content of freshwater fish from a river and from a farm with a recirculation system. Food Microbiology 23: 476-482.

[29] Hartemink, R and Rombouts, FM [1999]. Comparison of media for the detection of bifidobacteria, lactobacilli and total anaerobes from faecal samples. Journal of Microbiology Methods 36: 181-192.

[30] Hartemink, R; Domenech, VR and Rombouts, FM [1997]. LAMVAB - a new selective medium for the isolation of lactobacilli from faeces. Journal of Microbiology Methods 29: 77-84.

[31] Thapa, N; Pal, J and Tamang, JP [2006]. Phenotypic identification and technological properties of lactic acid bacteria isolated from traditionally processed fish products of the Eastern Himalayas. International Journal of Food Microbiology 107: 33-38.

[32] Hammes, WP and Vogel, RF [1995]. The genus *Lactobacillus*. In: Wood, BJB and Holzapfel, WH [Eds.], *The lactic acid bacteria, the genera of lactic acid bacteria*. [Eds.], Vol. 2, London, Blackie Academic and Professional. PP: 19-54.

[33] Schillinger, U. and F. Lucke, 1989. Antibactrial activity of *Lactobacillus sake* isolated from meat. Applied and Environmental Microbiology 55: 1901-1906.

[34] Benkerroum, N., Y. Ghouati, W. E. Sandine, and Tantaout-Elaraki. 1993. Methods to demonstrate the bactericidal activity of bacteriocins. Letters in Applied Microbiology 17: 78-81.

[35] Tagg J.R., and A.R. Mcgiven. 1971. Assay system for bacteriocins. Applied and Environmental Microbiology 21: 943-943.

[36]Yang, R., M.C. Johnson, and B. Ray, 1992. Novel method to extract large amounts of bacteriocin from lactic acid bacteria. Applied and Environmental Microbiology 58: 3355-3359.

[37] Sambrook, J.E., F. Eritsch, and J. Maniatis.1989. Molecular Cloning: A Laboratory Manual, 2nd ed. Cold Spring harbour Laboratory Press, Cold Spring Harbour, NY.

[38] Schägger, H., and G. Von Jagow. 1987. Tricine-sodium dodecyl sulphate-polyacrylamide gel electrophoresis for the separation of protein in the range from 1 to 100 kDa. Analytical Biochemistry 166: 368–379.

[39] Todorov, S.D., H. Nyati, M. Meincken, and L. M. T. Dicks. 2007. Partial characterization of bacteriocin AMA-K, produced by *Lactobacillus plantarum* AMA-K isolated from naturally fermented milk from Zimbabwe. Food Control 18: 656–664.

[40] 40. Garvie, EI [1984]. Taxonomy and identification of dairy bacteria. In: Davies, FL and Law, BA [Eds.], Advances in the microbiology and biochemistry of cheese and fermented milk, London: Elsevier Applied Science Publishers. PP: 35-65.

[41]Hicks, SJ; Theodoropoulos, G; Carrington, SD and Corfield, AP [2000]. The role of mucins in host-parasite interactions. Part I– Protozoan parasites. Parasitol. Today. 16: 476-481.

[42] Stephen, A.M. [1985]. Effect of food on the intestinal microflora. In: Food and the Gut [Hunter, J.O. and Alun Jones, V., Eds.]. Baillière Tindall. Sussex, England. pp .57-77

[43] Ouwehand, AC; Kirjavainen, PV; Grönlund, MM; Isolauri, E and Salminen, S [1999]. Adhesion of probiotic micro-organisms to intestinal mucus. International Dairy Journal 9: 623-630.

[44]Fuller, R [1989]. Probiotics in man and animals. Journal of Applied Bacteriology 66: 365-378.

[45] Cotter, P.D., C. Hill, and R. P. Ross. R.P. 2005. Bacteriocins: Developing innate immunity for food . Nature Reviews Microbiology 3: 777 – 788 .

[46] Lewus, C. B., S. Sun, and T. J. Montville. 1992. Production of an amylase-sensitive bacteriocin by an atypical *Leuconostoc paramesenteroides* strain. Applied and Environmental Microbiology 58:143-149.

[47] Keppler, K., R. Geisen and H. Holzapfelw. 1994. An amylase sensitive bacteriocin of *Leuconostoc carnosum*. Food Microbiology. 11, 39-45.

[48] Bhattacharya, S., and A. Das. 2010. Study of Physical and Cultural Parameters on the Bacteriocins Produced by Lactic Acid Bacteria Isolated from Traditional Indian Fermented Foods. American Journal of Food Technology *5: 111-120.*

[49] Chung, H. J., and A. E. Yousef. 2010. Synergistic effect of high pressure processing and *Lactobacillus casei* antimicrobial activity against pressure resistant *Listeria monocytogenes*. New Biotechnology 27 [4]: 403-408

[50] Hernandez, D., E. Cardell, and V. Zarate. 2005. Antimicrobial activity of lactic acid bacteria isolated from Tenerife cheese: initial characterization of plantaricin TF711, a bacteriocin-like substance produced by *Lactobacillus plantarum* TF711. Journal of Applied Microbiology 99: 77 84.

[51]Drider, D., G. Fimland, Y. Héchard, L. M. McMullen, and H. Prévost. 2006. The continuing story of Class IIa bacteriocins. Microbiology and Molecular Biology Reviews 70 [2]: 564–582.

[52] Ogunbanwo, S.T., A.I. Sanni, and A.A. Onilude. 2003. Characterization of bacteriocin produced by *Lactobacillus plantarum* F1 and *Lactobacillus brevis* OG1. African Journal of Biotechnology 2[8]: 219-227.

[53] Parente, E. and A. Ricciardi. 1999. Production, recovery and purification of bacteriocins from lactic acid bacteria. Applied Microbiology and Biotechnology 52: 628–638.

[54] Bhunia, A.K., M.C. Johnson, B. Ray, and N. Kalchayanand. 1991. Mode of action of pediocin AcH from *Pediococcus acidilactis* H on sensitive bacterial strains. Journal of Applied Bacteriology 70: 25-33.

[55] Todorov, S. D., and L. M. T. Dicks. 2004. Partial characterization of bacteriocins produced by four lactic acid bacteria isolated from regional South African barely beer. Annals of Microbiology 54[4]: 403-413.

[56] Rekhif, N., A. Atrih, M. Michel and G. Lefebvre. 1995. Activity of plantaricin SA6, a bacteriocin produced by *Lactobacillus plantarum* SA6 isolated from fermented sausages. Journal of Applied Bacteriology 78: 349–58.

[57] Cuozzo, S. A., F. Sesma, J. M. Palacios, A. P. de Ruiz Holgado, and R. R. Raya. 2000. Identification and nucleotide sequence of genes involved in the synthesis of lactocin 705, a two-peptide bacteriocin from *Lactobacillus casei* CRL 705. FEMS Microbiology Letters 185: 157-161

[58] Rammelsberg, M., E. Miiller, and F. Radler. 1990. Caseicin 80: purification and characterization of a new bacteriocin from *Lactobacillus casei.* Archives of Microbiology 154: 249 – 252

[59] Van Reenen, C. A., L. M. T. Dicks and M. L. Chikindas. 1998. Isolation, purification and partial characterization of plantaricin 423, a bacteriocin produced by *Lactobacillus plantarum* . Journal of Applied Microbiology 84: 1131–1137.

[60] Vignolo, G., S. Fadda, M. N. De Kairuz, A. A. P. de Ruiz Holgado, and G. Oliver. 1996. Control of *Listeria monocytogenes* in ground beef by Lactocin 705, a bacteriocin produced by *Lactobacillus casei* CRL 705. International Journal of Food Microbiology. 27: 397402.

[61] Venkat, H. K., N. P. Shau and K. J. Jain. 2004. Effect on feeding *Lactobacillus*-based probiotics on the gut microflora, growth and survival of postlarvae of *Macrobrachium rosenbergii* [de Man]. Aquaculture Research 35: 501-507.

[62] Verschuere, L., G. Rombaut, P. Sorgeloos and W. Verstraete. 2000. Probiotic bacteria as biological control agents in aquaculture. Microbiology and Molecular Biology Reviews. 64: 655-671.

[63] Vijayan, K. K., I. S. B. Singh, N. S. Jayaprakash, S. V. Alavandi, S. S. Pai, R. Preetha, J. J. S. Rajan and T. S. Santiago. 2006. A brackishwater isolate of *Pseudomonas* PS-102, a potential antagonistic bacterium against pathogenic vibrios in penaeid and non-penaeid rearing systems. Aquaculture 251: 192-200.

[64] Todorov, S., B. Onno, O. Sorokine, J. M. Chobert, I. Ivanova, and X. Dousset. 1999. Detection and characterization of a novel antibacterial substance produced by *Lactobacillus plantarum* ST 31 isolated from sourdough. International Journal of Food Microbiology. 48: 167-177

[65] Toméa, E.,V. L. Pereiraa, C. I. Lopesa, P. A. Gibbsc, and P. C. Teixeiraa. 2008. In vitro tests of suitability of bacteriocin-producing lactic acid bacteria, as potential biopreservation cultures in vacuum-packaged cold-smoked salmon. Food Control 19[5]: 535-543

[66] Karthikeyan, V., and S. W. Santosh. 2009. Isolation and partial characterization of bacteriocin produced from *Lactobacillus plantarum* . African Journal of Microbiology Research 3 [5]: 233-239

[67] Daba, H., S. Pandian, J. F. Gosselin, R. E. Simard, J. Huang, C. Lacroix. 1991. Detection and activity of bacteriocin produced by *Leuconostoc mesenteriodes*. Applied and Environmental Microbiology 57: 3450-3455.

[68] Stiles, M. E. and W. H. Holzaphel. 1997. Lactic acid bacteria of foods and their current taxonomy. International Journal of Food Microbiology. 36: 1 –29.

[69] Chung, H. J., and A. E. Yousef. 2009. Screening of Lactobacilli Derived from Fermented Foods and Partial Characterization of *Lactobacillus casei* OSY-LB6A for Its Antibacterial Activity against Foodborne Pathogens. Journal of Food Science and Nutrition *14: 162-167*

Section 7

# New Fields of Application

Chapter 18

# Exploring Surface Display Technology for Enhancement of Delivering Viable Lactic Acid Bacteria to Gastrointestinal Tract

Shirin Tarahomjoo

Additional information is available at the end of the chapter

http://dx.doi.org/10.5772/50783

## 1. Introduction

Anchoring of proteins to the cell surface is a common theme in nature and the processes governed by different surface proteins are bases of many biological phenomena, such as cell-cell recognition, signal transduction, adherence, colonization, and immunoreactions (Westerlund & Korohonen, 1993). The utilization of cellular surface anchoring systems for the display of heterologous proteins on the surface of microbial cells has been developed into an active research area that holds a great promise for a variety of biotechnological applications including the production of whole cell biocatalysts, microbial adsorbents, live vaccines, antibody fragments, and screening of novel proteins (Hansson et al., 2001; Kondo& Ueda, 2004; Lee et al., 2003). Generally construction of these systems is accomplished by the expression of heterologous peptides or proteins as fusions with anchoring domains, which are able to attach to the cell surface (Fig 1.). Anchoring domains are usually cell surface proteins or their fragments. Depending on the characteristics of target and anchor proteins, N-terminal fusion, C-terminal fusion or sandwich fusion strategy can be considered (Lee et al., 2003).

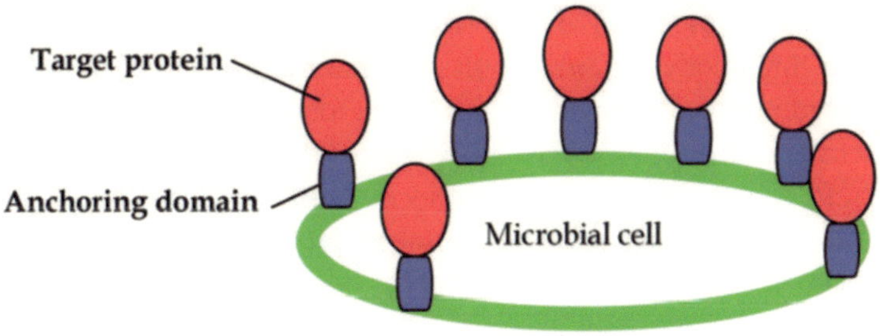


**Figure 1.** Schematic representation of a microbial surface display system

LAB are gram positive, non-spore forming, fastidious, acid tolerant, and strictly fermentative that secret lactic acid as the major end product of sugar fermentations (Axelsson, 1998). LAB are naturally present in media rich in organic nutrients such as food products and digestive tracts. They are a genetically diverse group of bacteria with GC contents varying from 34 to 53%, including rod shaped bacteria such as lactobacilli and also cocci such as lactococci, enterococci, pediococci, and leuconostoc (Stiles & Holzapfel, 1997). Since time immemorial, LAB have been used for the fermentation and preservation of food products, particularly dairy products, fermented meats, and vegetables. Consequently, several strains of LAB have a long record of safe association with humans and human foodstuffs (Mckay & Baldwin, 1990). The display of proteins on the surface of LAB cells can broaden or improve applications of these bacteria. In this chapter, we intend to describe cell surface anchoring domains used in LAB surface display systems. Then applications of surface engineered LAB are depicted and key factors affecting their performances are highlighted. Moreover, we explained comprehensively a novel application of the protein display in LAB, which is potentially useful for enhancement of the delivery of viable LAB to the gastrointestinal tract (GIT).

## 2. Anchoring domains in surface display systems of LAB

The cell surface anchoring domains utilized for the development of LAB surface display systems are associated with the cell surface either covalently or noncovalently. Depending on the applied anchoring domains, two modes of the protein display can be considered, including internal and external mode of the protein display. In the case of the internal mode of protein display, fusions of target proteins to anchoring domains are expressed in LAB cells, and therefore target proteins are displayed on the surface of expression hosts, whereas in the case of the external mode of protein display, expression and display hosts are different from each other. If the association of anchoring domains with the cell surface is of a non covalent nature, they can bind to cells when they are added from the outside. Therefore, fusions of target proteins to anchoring domains are produced in suitable expression hosts capable of their correct folding and modifications. The fusion proteins are then purified and incubated with desired display hosts in order to attach to the cell surface. As a result, the external mode of protein display can retain the nongenetically modified status of cells and is valuable for food and vaccine development.

The surface display systems based on the internal mode of protein display are often associated with the limitations in terms of the translocation of target proteins to the cell surface and the control of surface intensity of target proteins. The mislocalization of target proteins can affect their functions negatively( Dieye et al., 2003; Van Der Vaart et al., 1997; Wan et al., 2002). In contrast, the surface display systems based on the external mode of protein display can ensure the full exposure of target proteins outside of the cell wall and the surface intensity of target proteins can readily be adjusted by selecting appropriate display hosts and suitable concentrations for the fusion proteins in the incubation mixture. However, regarding noncovalent interaction of target proteins with the cell surface, the possibility of dissociation of target proteins from the cell surface should be considered.

## 2.1. Covalent anchors

### *2.1.1. LPXTG motif containing anchors*

The most widely used surface display systems of LAB are based on cell wall anchored proteins that contain an LPXTG motif. These proteins are first synthesized as a preprotein containing an N-terminal signal peptide and a C-terminal cell wall anchor domain. The anchor domain starts at its C-terminus with a short tail of positively charged residues (five to seven amino acids) that remain in the cytoplasm. Upstream of the cytoplasmic domain, a stretch of approximately 30 hydrophobic amino acids is preceded by the highly conserved pentapeptide LPXTG. The charged tail and hydrophobic domain are thought to function as a temporary stop to position the LPXTG motif for proteolytic cleavage. Correct positioning results in cleavage between the threonine and glycine residues followed by amid linkage of the threonine residue to the peptide crossbridge in the peptidoglycan of cell wall, by the action of a sortase (Navarre & Schneewind, 1994). The amino acid composition of the peptide crossbridge varies among the different LAB species and is flexible with respect to the sorting reaction (Strauss et al., 1998; Ton-That et al., 1998). The anchor domain is preceded by a wall associated region of about 50 to 125 residues and is characterized by a high percentage of proline/glycine and/or threonine/serine residues (Fischetti et al., 1990). In surface display genetic constructs, secretion signal peptides are fused to the N-termini of target proteins to transport them to the outside of cell and the LPXTG motif containing anchors are fused to the C-termini of target proteins in order to attach them to the cell surface. The cell surface associated proteinases of *Lactococcus lactis* and *Lactobacilli casei* (PrtPs), M6 protein of *Streptococcus pyogenes*, and Protein A of *Staphylococcus aureus* (SpA) are among LPXTG type anchors, which have mainly been used for the construction of surface display systems in LAB (Maassen et al., 1999; Norton et al., 1996; Piard et al., 1997; Pouwels et al., 1996; Steidler et al., 1998).

Slight deviations from the LPXTG consensus sequence is found in some LAB. Recently Kleerebezem (2003) found that sortase substrates of *Lactobacillus plantarum* contain LPQTXE motifs instead of the LPXTG motifs. In addition, in the carboxy end of cell surface proteinase of *Lactobacillus delbrueckii* (PrtB), a degenerated LPKKT motif is surrounded by two imperfect repeats of 59 residues, which are lysine rich. However, downstream of the LPKKT motif, there is no hydrophobic domain and no charged tail at the extreme C-terminus. It was shown that the C-terminal region of PrtB was able to attach to the cell wall of *L. lactis* and the capacity of attachment was drastically reduced by absence of the duplicated sequences. The high content of total positive charges in the anchoring region of PrtB suggests interactions of the anchor with the negatively charged teichoic acids of the cell wall. The mechanism of PrtB attachment to the cell wall probably implicates electrostatic forces (Germond et al., 2003).

### *2.1.2. Lipoprotein anchors*

Lipoproteins are lipid modified proteins produced as secretory precursors with a signal peptide linked to their amino termini. The C-terminal region of their signal peptide contains a well conserved lipobox motif of four amino acid residues and invariably, the last residue is cysteine. The covalent binding of lipoproteins to the cell membrane is generally achieved via diacylglyceryl modification of the indispensible cysteine residue in the lipobox by a diacylglyceryl transferase. Lipidation of the cysteine residue is a perquisite for cleavage of the signal peptide by a lipoprotein specific signal peptidase (SPase II)(Yamaguchi et al., 1988; Venema et al., 2003). The lipoprotein anchors should be fused at their C-termini to N-termini of target proteins in order to display them on the cell surface. Poquet (1998) identified four lactococcal lipoprotein anchors using nuclease of *S. aureus* as an export specific reporter enzyme. The nuclease activity was shown to require an extracellular location in *L. Lactis* demonstrating its suitability to report the protein export. The enzyme activity was detected in a plate test by the presence of pink halos. Fusions of the lipoprotein anchors to the nuclease expressed in *L. Lactis* were associated with the cell fraction and the recombinant lactococcal cells showed strong nuclease activities indicating the cell surface anchoring function of the lipoproteins. For one of the anchors (NlpI), the surface location of the fusion protein was also confirmed by proteinase K treatment of *L. lactis* cells (Poquet et al., 1998). Basic membrane protein A (BmpA) of *L. Lactis* is a putative lipoprotein that has been used for the protein display on the surface of lactococci (Berlec et al., 2011).

#### *2.1.3. Transmembrane anchors*

The strategy to insert target amino acid sequences in the exterior loop between transmembrane spanning domains (TMSs) can limit the insert size in order not to disturb the membrane protein topology. Therefore, a fusion approach is often preferred in which a target protein is simply linked at its N-terminus to one or more TMSs of a cytoplasmic membrane protein. *L. Lactis* bacteriocin transport accessory protein (LcnD) and *Bacillus subtilis* poly-γ-glutamate synthetase A protein (PgsA) are transmembrane proteins, which were fused to the N-termini of target proteins in order to display them on the cell surface of *L. Lactis* and *L. casei*, respectively (Franke et al., 1996; Narita et al., 2006). In addition, in the same random procedure as described above for the lipoprotein anchors Poquet (1998) identified seven lactococcal gene fragments encoding TMSs that function as membrane anchors in *L. Lactis*.

### 2.2. Noncovalent anchors

#### *2.2.1. S- layer protein anchors*

Some LAB strains possess a surface layer (S-layer) of proteins as the outermost structure of the cell envelope. These S-layers are composed of regularly arranged subunits of a single protein (SLP) and may constitute up to 20% of the total cell protein content. S-layers self assemble in entropy driven process during which multiple noncovalent interactions between individual SLPs and the underlying cell surface take place. These two types of interactions in SLPs can be assigned to two separate domains including the self assembly domain and the cell wall binding domain. These domains have been characterized in SLPs of *Lactobacillus acidophilus* ATCC 4356 (SA), *Lactobacillus crispatus* JCM 5810 (CbsA), and

*Lactobacillus brevis* ATCC 8287 (SlpA). The C-terminal regions of SA and CbsA showed the cell surface anchoring function and the N-terminal regions were involved in the self assembly process. In contrast, the self assembly domain of SlpA was located in the C-terminal region and its cell wall binding domain was found in the N-terminal region. The (lipo)teichoic acids were identified as the cell wall ligands of SA and CbsA. Moreover, the specific cell wall component that interacts with SlpA was shown to be the neutral polysaccharide moiety of the cell wall (Antikainen et al., 2002; Avall-Jaaskelainen et al., 2008; Smit et al., 2002). Avall-Jaaskelainen (2002) decribed the construction of recombinant *L. brevis* strains expressing poliovirus epitope VP1 of 10 amino acid residues inserted in the *slpA* gene. Insertion sites in the *slpA* gene were selected on the basis of the hydrophilicity profile of the SlpA protein. The four most hydrophilic parts of the SlpA protein were selected for testing because it was expected that parts of them were likely to be sites where the epitope would be accessible to the cell surface. One of the insertion sites was at the N-terminus of SlpA and the others were at its C-terminus. Only one site at the C-terminus showed strong colour response in whole cell enzyme linked immunosorbent assay (ELISA) using anti epitope antibody demonstrating that the epitope was accessible on the surface of the recombinant *L. brevis*. In another study, the C-terminal region of SLP of *L. crispatus* K2-4-3 (LcsB) isolated from the chicken intestine was used for the construction of surface display systems. Green fluorescent protein (GFP) was fused to the N-terminus of LcsB. The fusion protein (GFP-LcsB) was expressed in *Escherichia coli*. It was then purified and mixed with various LAB. The binding of the fusion protein to LAB cells was viewed by the fluorescence microscopy. GFP-LcsB was associated with the cell surface of various LAB including *L. delbrueckii, L. brevis, Lactobacillus helveticus, Lactobacillus johnsonii, L. crispatus, Streptococcus thermophilus, L. lactis* and *Lactobacillus salivarius*. GFP alone did not bind to the cells. These results indicated that binding of GFP to the surface of LAB cells is directed by LcsB. However, the fusion protein could not attach to the cell surface of *L. casei*. The reason for this observation requires further studies to elucidate the target ligand of LcsB on the cell surface of LAB (Hu et al., 2011).

### *2.2.2. Lysin motif containing anchors*

The lysine motif (LysM) was first discovered in the lysozyme of *Bacillus* phage ø29 as a C-terminal repeat composed of 44 amino acids separated by 7 amino acids. The cell wall attachment of several bacterial proteins in both gram positive and gram negative organisms occurs through LysMs, often repeated several times in the protein sequence. Many LysM containing proteins are cell wall hydrolases (Buist et al., 2008). The C-terminal region of peptidoglycan hydrolase (AcmA) of *L. lactis* MG 1363 (CpH) contains three 44 amino acid residue lysMs separated by stretches of 21 to 31 amino acids rich in the serine, threonine, and asparagine residues. CpH is able to bind to the cell surface of lactococci and several strains of lactobacilli. Moreover, CpH is able to bind both to the cell surface of LAB treated with sodium dodecyl sulfate (SDS) to remove cell wall associated proteins and LAB treated with trichloroacetic acid (TCA) to remove carbohydrates and (lipo) teichoic acids. These findings suggest that peptidoglycan is the binding ligand of the CpH domain (Buist et al.,

1995). The C-terminus of endolysin Lyb5 of *Lactobacillus fermentum* bacteriophage øPYB5 (Ly5C) contains three LysMs. Each of LysMs is composed of 41 amino acids and they are separated by intervening sequences varying in length and composition. Ly5C fused to GFP was expressed in *E. coli*. After mixing the fusion protein with various cells *in vitro*, GFP was successfully displayed on the surfaces of *L. lactis, L. casei, L. brevis, L. plantarum, L. fermentum, L. delbrueckii, L. helveticus*, and *S. thermophilus* cells. Increases in the fluorescence intensities of TCA treated *L. lactis* and *L. casei* cells compared to those of nontreated cells showed that the cell wall peptidoglycan was the cell surface binding target of Ly5C. Concentration of sodium chloride and pH influenced the binding capacity of the fusion protein, and optimal conditions of these factors were determined empirically in order to obtain high fluorescence intensities of *L. lactis* and *L. casei* cells (Hu et al., 2010). N-terminus of putative muropeptidase (MurO) of *L. plantarum* also contains two LysMs composed of about 43 amino acids separated by 22 amino acid residue sequences. The LysM domain fused to GFP was expressed in *E. coli* and it was able to bind to the cell surface of *L. plantarum* after being mixed with the cells (Xu et al., 2011). Examination of supernatant fractions from broth cultures of *L. fermentum*, revealed the presence of a 27-kDa protein termed Sep. The N-terminus of Sep contains a LysM. Sep fused N-terminally to a six histidine epitope was expressed in *L. fermentum, Lactobacillus rhamnosus*, and *L. lactis*. The protein was found associated with the surface of the expression hosts. However, it was largely present in the supernatant of the cell cultures (Turner et al., 2004).

### *2.2.3. WxL anchors*

The C-terminal cell wall binding domain designated WxL was first identified in proteins of *Lactobacillus* and other LAB based on *in silico* analysis (Kleerebezem et al., 2010). WxL domain contains a WxL motif followed by a proximal well conserved YXXX(L/I/V)TWXLXXXP motif. This domain was found in gene clusters that also encode additional extracellular proteins with C-terminal membrane anchors and LPxTG motif containing anchors, suggesting that they form an extracellular protein complex (Siezen et al., 2006). The C-terminal WxL domains identified in two proteins of *Enterococcus faecalis* were fused at their N-termini to an export reporter enzyme (nuclease of *S. aureus)* and a secretion signal peptide. The fusion proteins expressed in *E. faecalis* were detected in both cell wall and supernatant fractions of the recombinant enterococci. Removal of the WxL domains from the fusion proteins nearly eliminated them in the cell wall. Treatment of the cell wall fractions with SDS disrupted binding of the fusion proteins to these fractions. These results indicated that the fusion proteins had noncovalent interactions with the cell wall of *E. faecalis*. The fusion proteins were able to attach to the cell surface of *E. faecalis* and *L. johnsonii* when they were added exogenously (Brinster et al., 2007).

### *2.2.4. Other anchors*

Basic surface protein A (BspA) is a surface located protein of *L. fermentum* BR11. Sequence comparisons have been shown that BspA is a member of family III of the solute binding proteins. Most solute binding proteins are lipoproteins. However, BspA is not a lipoprotein and is attached to the cell envelope by electrostatic interactions. It has been used as a fusion partner to direct proteins to the cell surface of *L. fermentum* BR11. In these constructs, BspA was fused at its C-terminus to target proteins and the fusion proteins were expressed in *L. fermentum* BR11 (Turner & Giffard, 1999). The C-terminal region of cell associated dextransucrase of *Leuconostoc mesentroides* IBT-PQ (DsrP) contains five repeats of 65 amino acid residues. The domain expressed in *E. coli* was able to bind to the cell surface of *L. mesentroides* IBT-PQ cells after being mixed with the cells (Olvera et al., 2007). The carboxy end of PrtP of *Lactobacillus acidophilus* was used for the protein display on *L. acidophilus* using the internal mode. The association of this domain with the cell surface was mediated by electrostatic interactions (Kajikawa et al., 2011).

## 3. Applications of surface engineering of LAB

Research in the field of surface engineering of LAB has mainly been focused on the construction of vaccine delivery vehicles but other interesting applications have also been reported. In this section, we will describe different areas of biotechnology in which surface display of heterologous proteins on LAB have been investigated.

### 3.1. Development of vaccine delivery vehicles

Vaccination represents one of the most effective public health strategies to combat infectious diseases (Mielcarek et al., 2001). One of the technologies being developed for vaccine production is the use of bacteria as live vectors for the delivery of recombinant vaccine antigens to the immune system. Such vaccines have the potential for the production of protective antigens *in vivo* and are inexpensive to manufacture (Moore et al., 2001).

Most infections affect or initiate infectious processes at mucosal surfaces and mucosal local immune responses can block pathogens at the portal of entry. Live bacterial vaccines can induce mucosal, as well as systemic, immune responses when delivered via mucosal routes, such as oral or intranasal administration (Mielcarek et al., 2001). The mucosal, needle free, administration of vaccines can significantly decrease the need for syringes with their inherent added cost and risk of disease transmission, and it can increase compliance, and consequently the coverage of vaccination programs (Giudice & Campbell, 2006). The first live recombinant bacterial vectors developed were derived from attenuated pathogenic microorganisms. In addition to the difficulties often encountered in the construction of stable attenuated mutants of pathogenic organisms, attenuated pathogens may retain a residual virulence level that renders them unsuitable for the vaccination of partially immunocompetent individuals such as infants, the elderly or immunocompromised patients (Curtiss, 2002). These problems can be addressed by the application of nonpathogenic food grade LAB as antigen delivery vehicles. LAB therefore represent attractive alternatives as antigen carriers and their use has mainly been focused on the construction of mucosal vaccines. The cellular location of antigens can influence the elicited immunological responses. Cell surface anchored antigens are better recognized by the immune system than those produced intracellularly. Furthermore, intracellular production of antigens may limit their *in vivo* release. The vaccines constructed by the cell surface anchoring of antigens are of particulate nature. In contrast to most soluble antigens, which are ignored by the immune systems, particles are recognized as foreign and as danger eliciting effective immune responses (Storni et al., 2005). Tetanus toxin fragment C (TTFC) is an immunogen protective against tetanus. In a pioneer study by Norton (1996) three recombinant strains of *L. lactis* expressing TTFC in three cellular locations, intracellular, secreted or cell surface anchored via lactococcal PrtP were constructed. The recombinant lactococcal cells were used to immunize mice, which were then challenged by the subcutaneous inoculation of tetanus toxin. When compared in terms of the dose of expressed TTFC required to elicit protection against the lethal challenge, the cell surface displayed form of TTFC was significantly (10-20 fold) more immunogenic than the alternative forms of the protein. The result of this study indicated the advantage of antigen display on the cell surface for the construction of the lactococcal vaccines. In addition to TTFC, several other antigens were displayed on the surface of LAB and the protection studies were carried out to evaluate the efficiency of these vaccines (Bermudez-Humaran et al., 2005; Hou et al., 2007; Kajikawa et al., 2007; Lee & Faubert, 2006; Lee et al., 2006; Li etal., 2010; Lindholm et al., 2004; Liu et al., 2009; Medina et al., 2008; Poo et al., 2006; Tang & Li, 2009; Wei et al., 2010; Xin et al., 2003).

F18 fimbrial *E. coli* strains are associated with porcine postweaning diarrhea and pig edema disease. Adherence of F18 fimbrial *E. coli* to porcine intestinal epithelial cells is mediated by the FedF adhesin of F18 fimbriae. For the development of a mucosal vaccine against porcine postweaning diarrhea and edema disease, different expression cassettes for the display of FedF on the cell surface of *L. lactis* were constructed. Preliminary attempts to express the entire FedF protein as a fusion protein in *L. lactis* resulted in inefficient secretion and degradation of the adhesin. Therefore, only those regions of FedF required for binding specificity to porcine intestinal epithelial cells, were used in the construction of cell surface display systems. Initially, recombinant *L. lactis* clones secreting the partially overlapping receptor binding domains of FedF (42 and 62 amino acid residues) were prepared using two different signal peptides. Substantially higher levels of the fusion proteins (four- to six-fold) were secreted by the clones possessing *L. brevis* SlpA signal peptide than by those possessing *L. lactis* Usp45 signal peptide. In order to enhance the secretion of the fusion proteins, a synthetic sequence encoding the propeptide LEISSTCDA was inserted between the signal sequences and the receptor binding domains of FedF. For the construction of surface display systems, the secreted proteins were anchored to the cell wall of *L. lactis* via the CpH protein or the lactococcal PrtP protein. Three groups of expression vectors with *prtP* spacer sequences of 0.6, 0.8 and 1.5 kb were also designed. The spacers inserted between the receptor binding domains and the anchors. Whole cell ELISA for the detection of cell surface exposure of the FedF receptor binding regions showed that the CpH anchor performed significantly better than the PrtP anchor, particularly in a *L. lactis* mutant devoid of the extracellular housekeeping protease, HtrA. Among the cell surface display systems possessing the CpH anchor, only those with the longest PrtP spacer resulted in efficient binding of the recombinant *L. lactis* cells to porcine intestinal epithelial cells (Lindholm et al., 2004).

In another study, pneumococcal surface antigen A (PsaA), a conserved membrane anchored virulence factor, was expressed in different strains of LAB and it was associated with the surface of LAB cells. *L. plantarum* and *L. helveticus* were found to be more effective at inducing mucosal and systemic anti-PsaA immune responses than *L. casei* following intranasal vaccination of mice. Because all three *Lactobacillus* strains expressed almost the same amount of PsaA and were also recovered from mice nasal mucosa in the same period (3 days), the observed differences among their respective antibody responses may reflect their different intrinsic adjuvant properties. PsaA expressed by *L. lactis* at $2\times10^{-8}$ ng/ colony forming unit (CFU), which was about 10% of PsaA amount produced by the *Lactobacillus* strains. The recombinant *L. lactis* remained in the nasal mucosa only 1 day after the inoculation. Therefore, the inability of *L. lactis* expressing PsaA to significantly induce serum IgG or secreted IgA in mice can be explained by the low level of antigen production in this bacterium compared with the *Lactobacillus* strains and also its shorter persistence in the nasal mucosa. Intranasal inoculation of the mice with *L. lactis* expressing PsaA did not exert any effect on *Streptococcus pneumoniae* recovery from the nasal mucosa upon colonization challenge, in comparison with inoculation of saline or the control *L. lactis* carrying the expression vector devoid of the antigen gene. On the other hand, all the recombinant *Lactobacillus* strains showed a significant reduction of *S. pneumoniae* colonization when compared with the saline group ($10^{0.6}$-$10^{1.35}$ CFU). However, only *L. helveticus* expressing PsaA showed a significant reduction of *S. pneumoniae* colonization in relation to control *L. helveticus* (10 CFU) (Oliviera et al., 2006). Among LAB, *L. lactis* and *Lactobacillus* strains have mostly been used for the construction of LAB vaccines. Selection of LAB strains for use as antigen carriers depends on their persistence in the host, capacity to express foreign antigens, and intrinsic adjuvanticity. *L. lactis* does not colonize the internal cavities of man or animals. Therefore, the use of lactobacilli, which are able to colonize the cavities such as the GIT transiently seems more advantages than that of *L. lactis* for developing LAB vaccines because the longer persistence of lactobacilli in the host body may enhance immunological responses. On the other hand, the progress in the genetics of lactobacilli is more recent than that of lactococci. Furthermore, the availability of a commercial powerful gene expression system for *L. lactis*, nisin inducible gene expression system, urged many researchers to establish LAB vaccines based on lactococci. It has been reported that several LAB strains particularly strains from the genera *Lactobacillus* are able to act as adjuvants. This aspect should be considered when selecting a vaccine strain as it is a natural way to potentiate the immune reaction against heterologous antigens produced by recombinant LAB. It might be speculated that a high level of the antigen expression will not be necessary when using immunostimulatory LAB strains. However, studies have not yet been reported for comparison of the adjuvanticity of *L. lactis* and different lactobacilli. *L. casei* and *L. plantarum* are among *Lactobacillus* strains, which can colonize the GIT of human and mice and they show immunostimulatory properties (Pouwels et al., 1998; Wells et al., 1996).

It has been reported that the flagellin of *Salmonella* has significant vaccine potential because it is the only surface antigen of *Salmonella* detected to have a mitogenic stimulatory effect on lymphocytes (Toyota-Hantani et al., 2008). Bacterial flagellins can also induce innate immune responses through their interaction with Toll-like receptor 5 (TLR5) (Ramos et al., 2004). Recombinant *L. casei* cells expressing the flagellin of *Salmonella* Enteritidis (LCF) on their cell surface via *L. casei* PrtP anchor were constructed. Intragastric immunization of mice with the recombinant lactobacilli resulted in a significant level of protective immunity against an oral challenge with *S.* Enteritidis. There was no significant difference in the level of protection after immunization with the recombinant lactobacilli compared with the free flagellin isolated from *S. Enteritidis*, although the amount of flagellin carried by LCF was less than that of the free flagellin. The immunization of mice with the recombinant lactobacilli did not result in antigen-specific antibody responses in either feces or sera but did induce the release of interferon (IFN)-γ on restimulation of primed lymphocytes *ex vivo*. These results suggested that the protective efficacy provided by flagellin expressing *L. casei* was mainly attributable to cell mediated immune responses. When the levels of IFN-γ produced by primed and flagellin restimulated lymphocytes were compared between the recombinant *L. casei* cells expressing flagellin on their cell surface, and a mixture of the purified flagellin and normal *L. casei*, the results indicated that the *Lactobacillus* strain showed adjuvanticity only when the flagellin was expressed on the cell surface (Kajikawa et al., 2007).

Two recombinant *L. acidophilus* strains displaying flagellin of *Salmonella typhimurium* on the cell surface were constructed using different anchor motifs. In one construct, the flagellin gene was fused at its carboxy end to the C-terminal region of PrtP of *L. acidophilus*. In other construct, the flagellin gene was fused in the same way to the anchor region of mucus binding protein (Mub) of *L. acidophilus* containing an LPXTG motif. The density of the flagellin fused protein at the cell surface of *L. acidophilus* displaying the flagellin by the PrtP protein (FliC-PrtP) was higher than that of *L. acidophilus* displaying the antigen by the Mub anchor (FliC-Mub). Both of the recombinant lactobacilli showed TLR5 stimulating activity, which indicated that the surface associated flagellin was recognized by TLR5. The magnitude of the TLR5 stimulating activity of *L. acidophilus* cells expressing FliC-PrtP was higher than that of *L. acidophilus* cells expressing FliC-Mub and this result showed that the magnitude of the TLR5 stimulating activity was dependent on the quantity of surface located flagellin. Moreover, the two recombinant lactobacilli exhibited dissimilar maturation and cytokine production by human myeloid dendritic cells (Kajikawa et al., 2011).

Human papillomavirus type 16 (HPV-16) has been associated with more than 50% of HPV related cervical cancer (CxCa) (Krinbauer et al., 1992). HPV-16 E7 oncoprotein is constitutively expressed in CxCa cells during malignant progression of HPV-16 induced cervical lesions and is therefore considered as an effective target for the cancer immunotherapy. *L. lactis* was engineered to express HPV-16 E7 on the cell surface (LL-E7) using the M6 anchor and its coadministration with another lactococci secreting IL-12 (LL-IL-12) was investigated for the immunization and immunotherapy of HPV-related CxCa. IL-12 is a heterodimeric cytokine that induces Th1 responses, enhances cytotoxic T-lymphocyte (CTL) maturation, promotes natural killer (NK) cell activity and induces IFN-γ production. Mice were vaccinated intranasally with LL-E7 and were then challenged by injection of tissue culture number 1 (TC-1) tumor cells. Thirty five percent of the vaccinated mice remained tumor free. Coadministration of LL-IL-12 with LL-E7 resulted in higher antitumor activities as half of the inoculated mice remained tumor free and the tumor median size in the remaining tumor bearing animals was less than that in LL-E7 immunized mice. Antitumoral activity elicited by covaccination with LL-E7 and LL-IL-12 appeared to be long lasting as when the tumor free animals were rechallenged 3 months later with TC-1 cells, they remained tumor free for up to 6 months. To investigate the therapeutic effects of the coadministration of LL-E7 and LL-IL-12, mice were challenged with the TC-1 tumor cell line prior to the initiation of immunotherapy. Once 100% of the mice had palpable tumor, the immunotherapy was started. Only LL-E7/LL-IL-12 treatment resulted in total tumor regression in 35% of the immunized animals. Moreover, the tumor median size in the remaining tumor bearing mice was lower than that measured in mice treated with LL-E7. In contrast, no tumor regression was observed in mice treated with LL-E7 alone. Mice immunized with LL-E7/LL-IL-12 also exhibited both systemic and mucosal humoral responses, which were induced at higher levels than those in mice vaccinated with LL-E7 (Bermudez-Humaran et al., 2005).

Boiling of *L. lactis* cells in TCA followed by washing and neutralization resulted in nonviable spherical peptidoglycan microparticles , which are deprived of surface proteins and their intracellular content is largely degraded. The proteins IgA1 protease (IgA1p), putative proteinase maturation protein A (PpmA) and streptococcal lipoprotein A (SlrA) were bound to the surface of the lactococcal particles after recombinant production of the antigens as hybrids with the CpH domain. TCA removes (lipo) teichoic acids from the lactococcal cell wall that results in enhancement of the binding capacity for CpH fusions. Mice immunized intranasally with the monovalent lactococcal particle based vaccines were not protected against an intranasal pnemococcal challenge. However, intranasal immunization with a trivalent vaccine containing PpmA, SlrA and IgA1p bound to the lactococcal particles by the CpH domain showed protection against fatal pneumococcal pneumonia in mice (Audouy et al., 2007)

## 3.2. Development of whole cell biocatalysts

Cellulosome is a multi enzyme complex in which various cellulytic enzymes assemble into a macromolecular structure by their attachment to a nonenzymatic central scaffold protein for the efficient degradation of cellulose (Bayer et al., 2004; Demain et al., 2005). Cellulosome of the gram positive thermophile *Clostridium thermocellum* is anchored to the surface of cells, resulting in one of the most efficient bacterial systems for the cellulose hydrolysis. All of the cellulosomal enzymes of *C. thermocellum* contain a twice repeated sequence, usually at their C-termini, called type I dockerin domain. These dockerin domains are considered to bind to the hydrophobic domains termed cohesins, which are repeated nine times in the central scaffold protein (CipA) of the bacterium. CipA also contains a cellulose binding module (CBM3a), allowing the different cellulases to act in synergy on the crystalline substrate, as well as a type 2 dockern domain, which binds the cell surface anchor proteins, ensuring the cellulosome's attachment to the cell surface. Association of the cellulosome with the cell surface yields formation of cellulose-enzyme-microbe ternary complexes and results in enhanced activity and synergy (Begum et al., 1996). The assembly of recombinant cellulosome inspired complexes on the cell surface of surrogate hosts such as LAB is highly desirable. LAB can produce commodity chemicals such as lactic acid and bioactive compounds (De Vuyst & Leory, 2007; Hofvendahl & Han-Hagerald, 2007; Siragusa et al., 2007). The economics of these processes would be greatly improved if LAB could utilize cellulosic substrates, which are cheap and abundant. While most of LAB can not assimilate cellulose, by the display of cellulosome on the surface of these bacteria, the hydrolysis of cellulosic substrates and the fermentation of hydrolysis products to desirable compounds can be carried out in a single step process, which has economical advantages. As a key step in the development of these recombinant LAB, fragments of CipA were functionally expressed on the cell surface of *L. lactis*. The fragments engineered to contain a single cohesin module, two cohesin modules, one cohesin and CBM3a, or only CBM3a. Cell toxicity from over expression of the proteins was circumvented by use of the nisin A (nisA) inducible promoter. Incorporation of the C-terminal anchor motif of the streptococcal M6 protein in the expression cassette resulted in the successful surface display of the fragments. All of the constructs containing cohesin modules were able to bind to an engineered hybrid reporter enzyme, *E. coli* ß-glucuronidase fused to the type I dockerin domain of a cellusomal enzyme. These results demonstrated that the cohesins were displayed on the cell surface of *L. lactis* cells in the functional form. In addition, the cell surface complex formation was dependent on the presence of both cohesin and dockerin modules (Wieczorek & Martin, 2010).

The process for the conversion of starch to lactic acid by LAB includes the enzymatic hydrolysis of starch followed by the fermentation of resultant oligosaccharides to lactic acid by LAB. These steps can be carried out simultaneously by $\alpha$-amylase displaying LAB. Therefore, using these whole cell biocatalysts can result in economical benefits for the conversion of starch to lactic acid. The PgsA anchor protein was fused to the N-terminus of $\alpha$-amylase of *Streptococcus bovis* 148 (AmY). The resulting fusion protein expressed in *L. casei* and it was associated with the membrane and cell wall fractions. However, the status of exposure of AmY outside of the cell wall was not clarified. The constructed whole cell biocatalyst was able to convert starch to lactic acid. Because the lactic acid concentration increased as the total sugar concentration decreased, it was concluded that the lactic acid was produced by simultaneous saccharification and fermentation of starch. The yield of lactic acid was improved by repeated utilizations of the recombinant *L. casei* cells (Narita et al., 2006).

The C-terminal region of peptidoglycan hydrolase (AcmA) of *L. lactis* IL 1403 (CphI) is a homolog of CpH of *L. lactis* MG 1363 and it contains three LysMs. CphI can bind to the surface of various LAB (Tarahomjoo et al., 2008a). We studied the capability of CphI for the development of $\alpha$-amylase displaying LAB using the external mode of protein display. These whole cell biocatalysts are expected to be effective for the direct fermentation of starch to lactic acid. Starch is a large substrate that is not capable of penetrating the cell wall. For this reason, in order to achieve its efficient hydrolysis, the enzyme must be exposed on the outside of the cell wall such that it is accessible to starch. The display systems based on the internal mode of protein anchoring often have limited ability for the translocation of target proteins to the cell surface. In contrast, the display systems based on the external mode of protein anchoring can display the enzyme completely outside of the cell wall. Therefore, a whole cell biocatalyst constructed using the externally added cell surface adhesive $\alpha$-amylase is considered as a suitable selection for our purpose. Moreover, the cell surface adhesive $\alpha$-amylase can readily be recovered together with LAB cells at the end of starch conversion process for reuse. CphI fused to AmY either at its C-terminus (CphI-AmY) or at its N-terminus (AmY-CphI) was expressed in *E. coli*. Both of the fusion proteins were able to bind to the cell surface of *L. lactis* ATCC 19435. Therefore, CphI is considered as a bidirectional anchor protein. However, the number of bound molecules per cell in the case of CphI-AmY was 3 times greater than that in the case of AmY-CphI. The change in the fusion direction may cause conformational alterations in the fusion protein leading to a better accessibility of CphI for the cell surface binding and an increase in the number of bound molecules. Moreover, the specific activity for starch digestion of CphI-AmY was 11 fold higher than that of AmY-CphI. The starch binding domain of AmY is located at its C-terminus. As a result, the fusion of CphI to the N-terminus of AmY may help improve the adsorption of starch onto the enzyme and enhance starch degradation, resulting in a higher specific activity for starch digestion. In addition to *L. lactis* ATCC 19435, *L. plantarum* NRRL B531, *L. lactis* IL1403, *L. casei* NRRL B441, *L. delbrueckii* ATCC 9649 and *L. casei* NRRL B445 were examined in terms of the binding of CphI-AmY. Of the LAB tested, *L. lactis* ATCC 19435 showed the highest binding capability for CphI-AmY, up to $6\times10^4$ molecules per cell. The binding of CphI-AmY to *L. delbrueckii* ATCC 9649 cells was very stable and its dissociation rate constant at 37°C was $7\times10^{-6}$ $s^{-1}$ (the half life of binding ($t_{1/2}$) was 28 h). The binding of this protein to cells of *L. lactis* ATCC 19435 was also stable, with a dissociation rate constant of $5\times10^{-5}$ $s^{-1}$ at 30°C ($t_{1/2}$=4 h). Lactate production by lactic acid bacteria is maximal during the exponential growth phase. Therefore, for a successful application of the constructed whole cell biocatalysts in lactate production, suitable fermentation conditions should be specified to adjust the duration of the exponential growth phase with respect to the dissociation rate of the protein. These half lives are long enough for lactic acid fermentation when the inoculm size is adequet and/or suitable growth conditions with high specific growth rates are used (Tarahomjoo et al., 2008a). A CphI mutant devoid of its N-glycosylation sites was expressed extracellularly in *Pichia pastoris*. This domain was able to bind to the cell surface of *L. casei*. However, its dissociation rate constant from the cell surface was 3.5 fold lower than that of CphI. These results indicated that the protein engineering approaches can be useful for increasing the binding stability of noncovalent anchors (Tarahomjoo et al. 2008b).

### 3.3. Attachment of bacteria to host tissues

Display of adhesins capable of binding to a host tissue on the surface of LAB can provide them with a specific adhesion capability, which can be beneficial when LAB are used as mucosal delivery vehicles of bioactive compounds. Because the displayed adhesins can increase the persistence of bacteria in the host tissue and as a result, the desired effects of the delivered bioactive compounds can be enhanced. The N-terminal region of SlpA was recently shown to mediate adhesion to human intestinal cell lines *in vitro*. The SlpA adhesion mediating domain fused to the N-terminus of lactococcal CpH protein was expressed in *L. lactis*. To increase the surface accessibility of the hybrid protein, a part of PrtP of *L. lactis* subsp. *cremoris* Wg2 was used as a spacer protein, which could extend the SlpA receptor binding region out of the cell surface. The spacer was inserted between the receptor binding domain and the anchor protein. *In vitro* adhesion assay with the human intestinal epithelial cell line Intestine 407 indicated that the recombinant lactococcal cells had gained an ability to adhere to Intestine 407 cells significantly greater than that of parental nonrecombinant *L. lactis* cells (Avall-Jaaskelainen et al., 2003).

It has been reported that *Mycobacterium bovis* bacillus Calmette-Guerin (BCG) exerts antitumor effects against superficial bladder cancer. These antitumor effects are due to the activation of immune responses, which are mediated by the attachment of BCG to the bladder wall through fibronectin. In addition, *L. casei* strain Shirota showed an antitumor activity when it was administered to the mouse model of superficial bladder cancer. However, the *L. casei* cells exhibits no significant binding to fibronectin. Therefore, higher antitumor activity could be expected for the *L. casei* cells genetically engineered to acquire binding capacity to fibronectin. For this purpose, fibronectin binding domain (FbD) of *S. pyogenes* ATCC 21059 fused to the C-terminal region of PrtP 763 of *L. lactis* NCDO 763 was expressed in *L. casei* strain Shirota. The recombinant *L. casei* cells were able to bind to fibronectin. Furthurmore, FbD expressed on the *L. casei* cell surface promoted the adherence to murine fibroblast STO cells, which secret fibronectin in large amounts (Kushiro et al., 2001).

## 3.4. Enhancement of delivering viable bacteria to gastrointestinal tract

Probiotics are live microbial food supplements, which benefit the health of consumers by improving their intestinal microbial balance (Fuller, 1989). Since the viability and activity of a probiotic is essential at the site of action, it must survive the harsh environment of the upper GIT, and it must be able to function in the gut environment (Collins et al., 1998). Most commonly used probiotics are lactobacilli and bifidobacteria (Daly & Davis, 1998). However, several studies indicate that most of these bacteria may not be able to withstand the harsh acidity of the upper GIT (Conway et al., 1987; Lankaputhra & Shah, 1995). In a study by Wang (1999), the enhancement of survival of bifidobacteria grown in the presence of starch granules and mixed with them was reported. However, the exact mechanism underlying the protective effect of starch was not clarified. The adhesion of bacteria to starch was considered as a possible explanation for these observations. The results of their study suggested that starch granules can be used to protect living microbes from environmental stress factors.

Microencapsulation is an approach that has been proposed to protect probiotics from environmental stresses. It segregates cells from adverse environments; thus, it minimizes cell injury (Anal & Singh, 2007). Myllaerinen (1999) has recently developed a

and incubated at 37°C for 2 h with gentle shaking. After centrifugation, the supernatant was examined for the presence of unbound proteins by 12.5% SDS-PAGE and CBB staining.

#### *3.4.1.7. Aggregation of bacteria with starch and microencapsulation*

After performing the cell surface binding assay as described above, the cells were washed and resuspended in PB to a final density of $10^9$ cells/ml. The suspension of the starch granules in PB was mixed with an equal volume of the cell suspension for 30 min. The mixture was then allowed to stand at room temperature for 1 h. The formation of aggregates was studied both visually and with phase contrast microscopy. To determine the percentage of cells adhering to starch, a 0.5 ml sample was taken from below the liquid surface after the sedimentation. Optical density at 540 nm was measured and compared with those of controls including bacteria without starch and starch without bacteria to calculate the starch adhesion percentage as described by Crittenden (2001). For coating of the aggregates with amylose, a 1% solution of amylose in water (amylose from potato, Sigma –Aldrich, Tokyo, Japan) was prepared by heating it to a temperature of 170°C in a pressure heater (Taiatsu Techno, Tokyo, Japan), which was then cooled down to about 37°C. The aggregates were mixed gently with 0.5 ml of the amylose solution and the coating was allowed to form overnight at 4°C.

#### *3.4.1.8. Survival of cells in simulated gastric juice*

Simulated gastric juice (SGJ) was prepared as described by Lian (2003), which was a pepsin solution (3 g/*l*) in saline (0.5% NaCl). The SGJ was prepared freshly and its pH was adjusted to 2.0 or 3.0 with 5 M HCl. The cells ($5 \times 10^7$ cfu) were mixed with 1 ml of the filter sterilized SGJ and incubated at 37°C. At specified time intervals, the gastric juice was removed following centrifugation and the cells were washed once with PB following with two washes with saline. Amylose coated cells were then resuspended in PB containing 30 U/ml $\alpha$-amylase (Megazyme, Bray, Ireland) and incubated at 40°C for 20 min to aid the release of cells from the encapsulating materials. Viable bacteria were enumerated on MRS-agar after incubation for 24 h at 37°C and survival percentage was determined by dividing the final viable population (cfu/ml) with the initial viable population (cfu/ml) of *L. casei* cells exposed to the SGJ.

### *3.4.2. Results*

#### *3.4.2.1. Expression and purification of fusion protein*

To investigate the capability of CphI for the construction of a cell surface adhesive SbD, this domain was fused at its C-terminus to the linker region and SbD of AmY (Fig. 2). The fusion protein (CphI-SbD) was expressed intracellularly in *E. coli* using the T5 promoter at 0.35 g/*l*. The molecular size was 56 kDa as expected and 75% of the protein was present in the soluble form. When the protein was purified under native conditions by the histidine tag affinity chromatography, two additional protein bands corresponding to 73 and 71 kDa were present in the protein preparation (Fig. 3, lane 1). After incubation of the protein preparation with corn starch, no band for CphI-SbD was detected in the supernatant indicating its adsorption to the starch granules (Fig. 3, lane 2). CphI-SbD was further purified using an anion exchange chromatography (Fig. 3, lane 4). The result of starch binding assay showed that the purified fusion protein was in the active form and it was able to adhere to the starch granules (data not shown).

**Figure 2.** Structure of expression cassette

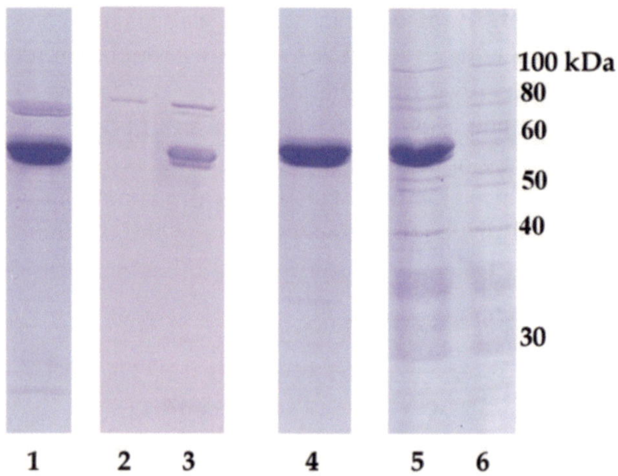


**Figure 3.** Purification of CphI- SbD and its binding to corn starch and *L. casei* cells. Lane 1, nickle chelate column purified protein preparation; lane 2, supernatant after starch binding assay; lane 3, control without starch; lane 4, ion exchange chromatography purified CphI-SbD; lane 5, cells bound to CphI-SbD; lane 6, cells only.

*3.4.2.2. Binding of CphI-SbD to L. casei cells*

*L. casei* cells were incubated with the purified fusion protein and studied in terms of the binding of the fusion protein by SDS-PAGE. As shown in Fig. 3 (lane 5), the fusion protein was associated with the cells. The result of the densitometrical analysis showed that $6{\times}10^4$ molecules of CphI-SbD were bound to each cell of *L. casei.*

*3.4.2.3. Aggregation of bacteria with starch*

For aggregation of bacteria with starch, an optimal ratio between bacteria and starch must be determined. Therefore, we examined the dependence of aggregate formation on the final starch concentration in the mixture at a constant cellular density. Free cells mixed with starch, and starch without cells were used as controls. For each concentration of starch, we compared the volume of sediment formed in the sample containing the mixture of bacteria bound to CphI-SbD and starch with those of the controls visually. The result is shown in Table 1. The volumes of the sediments formed in the controls after 1 h standing at room temperature, were almost the same at all the starch concentrations tested. When the starch concentration was 5 mg/ml, the volume of formed sediment in the sample containing the mixture of bacteria bound to CphI-SbD and starch was markedly larger than those of the controls. The adhesion percentage of *L. casei* cells to the starch granules under these conditions was determined by cosedimentation assay, which measures the reduction in optical density in bacterial suspensions after the addition of starch (Crittenden et al., 2001), and it was 32% for the bacteria bound to CphI-SbD and 4% for the free cells.

| Starch concentration (mg/ml) | BPS[1] | BS[2] | S[3] |
|---|---|---|---|
| 1 | + | + | + |
| 2 | ++ | + | + |
| 5 | +++ | + | + |
| 10 | + | + | + |

[1]: Mixture of *L. casei* cells displaying CphI-SbD and starch
[2]: Mixture of *L. casei* cells and starch
[3]: Starch only

**Table 1.** Comparison of sediment formation at different starch concentrations.

Mixing of the bacteria bound to CphI-SbD with starch at a final concentration of 5 mg/ml resulted in the crosslinking of starch granules and the formed aggregates were observed by phase contrast microscopy (Fig. 4). In contrast, when starch was mixed with free cells or when no bacteria were present in the mixture, the starch granules were found to be separated from each other and no aggregates were observed by phase contrast microscopy (data not shown).

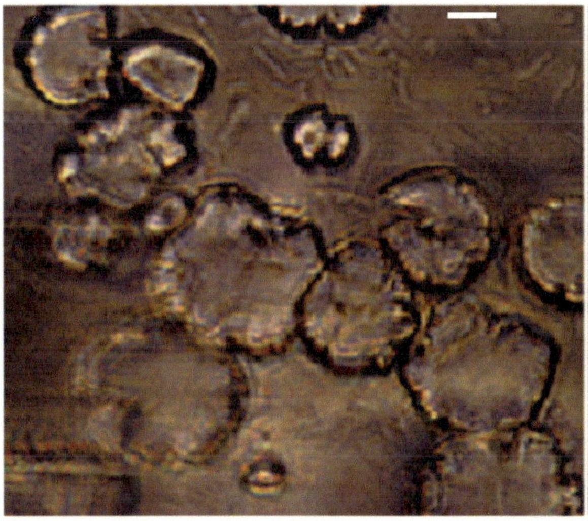

**Figure 4.** Aggregation of *L. casei* cells displaying CphI-SbD with starch granules (scale bar=5 μm)

*3.4.2.4. Survival of encapsulated cells under simulated gastric conditions*

When *L. casei* free cells were exposed to the SGJ at pH 3.0 and 2.0 for 1 h, the survival percentages of the cells were 0.074% and 0.002%, respectively. However, the survival percentages of amylose coated bacterial aggregates after 1 h incubation in the SGJ were 63.9% and 6.03% at pH 3.0 and 2.0, respectively (Fig. 5).

We studied the effects of starch, amylose coating and CphI-SbD on cell survival in the SGJ (pH 3.0, 1 h) (Table 2). It was observed that when free cells were mixed with the starch granules, the survival percentage was 3.1%, and the survival percentage of amylose coated free cells in the SGJ was 7.2%. The survival of the CphI-SbD displaying bacteria (0.093%) was not significantly different from that of free cells (0.074%) and the survival of amylose coated fusion protein displaying bacteria was comparable to that of amylose coated free cells. However, when fusion protein displaying bacteria were aggregated with the starch granules, the survival percentage was 7.7% higher than that of free cells mixed with the starch granules, and when the aggregates were coated with amylose, the survival percentage was 27.6% higher than that of the amylose coated mixture of the starch granules and free cells.

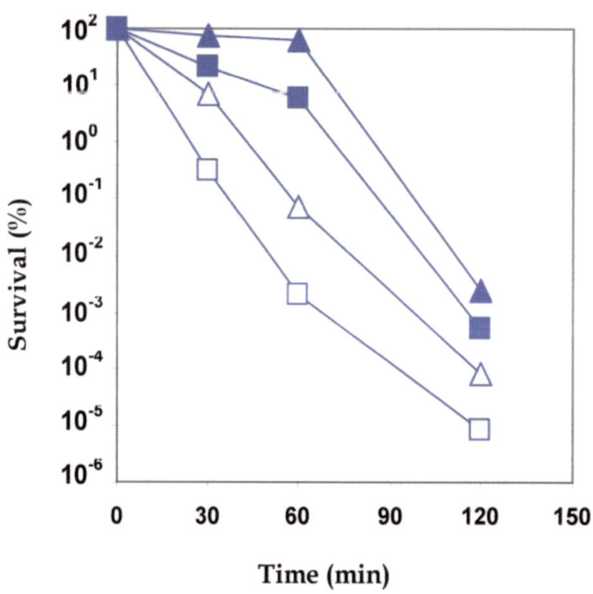


**Figure 5.** Time course of survival of *L. casei* cells under simulated gastric conditions. Open and filled triangles are free cells and amylose coated bacterial aggregates, respectively at pH 3.0; Open and filled rectangles, those at pH 2.0. The Data represent the means of two independent experiments.

| Bacteria | CphI-SbD | Starch | Amylose | Survival (%) |
|---|---|---|---|---|
| + | – | – | – | 0.074 ± 0.012 |
| + | + | – | – | 0.093 ± 0.018 |
| + | – | + | – | 3.1 ± 0.4 |
| + | + | + | – | 10.8 ± 2.0 |
| + | – | – | + | 7.2 ± 1.7 |
| + | + | – | + | 6.6 ± 2.5 |
| + | – | + | + | 36.3 ± 5.4 |
| + | + | + | + | 63.9 ± 2.7 |

**Table 2.** Effects of protective systems components on cell survival. Data represent means± standard deviations of three independent experiments.

### *3.4.3. Discussion*

The objective of this study was to evaluate the possibility for enhancement of the delivery of viable microorganisms to the GIT through the conferment of starch binding ability to them. In this way, the bacteria are entrapped between starch granules to use the protective effect of starch for maintaining the cell viability under adverse conditions. The surface display technology based on the external mode was used to provide starch adhesion ability for the cells while retaining their nongenetically modified status because consumption of genetically modified microorganisms is not favorable for consumers. It was observed that CphI-SbD was able to bind both to the cell surface of *L. casei* and to the starch granules. Therefore, the fusion protein was able to mediate the adhesion of cells to the starch granules. We examined the aggregation of cell's with starch as an alternative protective strategy, which entraps bacteria between starch granules. Compared with the previous method of entrapping bacteria within the porous starch granules prepared by an enzymatic digestion (Myllaerinen et al., 2001), our technique, is much simpler and faster. Moreover, starch granules can be used in their intact forms without any modifications.

When the aggregates of CphI-SbD displaying bacteria with starch were coated with amylose and exposed to the SGJ, there were significant increases in the survival percentage (63.9% at pH 3.0, and 6.03% at pH 2.0) with respect to those of free cells (0.074% at pH 3.0, and 0.002% at pH 2.0). The pHs of the SGJ measured for all the combinations of bacteria, fusion protein, starch and amylose coating (Table 2), in addition to that of the SGJ without any of these components both before and after the incubation at 37°C, were in the range of 3.00- 3.04. Therefore, the observed increases in the survival were not due to the modifications of the pH of SGJ.

Analysis of the effects of fusion protein, starch and amylose coating on the cell survival showed that the binding of fusion protein to *L. casei* cells did not have a significant effect on the cell survival (Table 2). When free cells were mixed with the starch granules, their survival percentage was 3.03% higher than that of free cells, which indicated the protective effect of starch on the cell survival. It was observed that, the entrapment of bacteria between the starch granules with the aid of CphI-SbD (the aggregation of bacteria with the starch granules) enhanced the protective effect of starch, and the survival percentage was increased to 10.8%. The effect of the amylose coating on the survival of CphI-SbD displaying bacteria was comparable to that of the free cells (6.6 and 7.2% respectively), and the observed difference was not statistically significant. Incorporation of the fusion protein in the protective system composed of the starch and amylose resulted in a 27.6% increase in the cell survival percentage, which showed that the simultaneous application of two protective strategies (the aggregation of bacteria with starch and the amylose coating) resulted in the highest cell survival percentage (63.9%). In conclusion, in this study we showed the potential usefulness of the cell surface display technology for protection of cells under adverse gastric conditions.

## 4. Conclusion

Surface display is an attractive technology that can be used to confer new functions to LAB. The effectiveness of these systems depends on the appropriate selection of several factors including the anchoring domains, secretion signals, and host strains. Moreover, a proper strategy for the fusion of anchoring domains to target proteins should be determined to protect the functionality of target proteins. So far, a limited number of surface display systems have been developed. The characterization of anchoring, secretion, and regulatory signals from genome sequences can expand the surface display systems. The low transformation efficiency of LAB is a major obstacle for the construction of surface display systems and the establishment of efficient transformation protocols is therefore necessary.

## Author details

Shirin Tarahomjoo

*Department of Biotechnology, Razi Vaccine and Serum Research Institute, Karaj, Iran*

*Food Industries and Biotechnology Research Center, Amirkabir University of Technology, Tehran, Iran*

## 5. References

Anal, A. K., and Singh, H. (2007). Recent advances in microencapsulation of probiotics for industrial applications and targetd delivery. *Trends Food Sci Technol* 18, 240-251 (2007).

Antikainen, J., Anton, L., Sillanpää, J., Korhonen, T. K. (2002). Domains in the S-layer protein CbsA of *Lactobacillus crispatus* involved in adherence to collagens, laminin and lipoteichoic acids and in self-assembly. *Mol Microbiol* 46, 381-394.

Audouy, S. A. L., van Slem, S., van Roosmalen, M. L., Post, E., Kanninga, R., Neef, J., Estevao, S., Nieuwenhuis,, E. S., Adrian, P. V., Leenhouts, K. And Hermans, P. W. M. (2007). Development of lactococcal GEM based pneumococcal vaccines. *Vaccine* 25, 2497-2506.

Avall-Jääskeläinen, S., Kyla-Nikkila, K., Kahala, M., Miikkulainen, T. and Palav, A. (2002). Surface display of foreign epitopes on the *Lactobacillus brevis* S-layer. *Appl Environ Microbiol* 68, 5943-5951.

Avall-Jääskeläinen, S., Lindholm, A. And Palva, A. (2003). Surface display of the receptor binding region of the Lactobacillus brevis S-layer protein in *Lactococcus lactis* provides nonadhesive lactococci with the ability to adhere to intestinal epithelial cells. *Appl Environ Microbiol* 69, 2230-2236.

Avall-Jääskeläinen, S., Hynönen, U., Ilk, N., Pum, D., Sleytr, U. B. and Palva, A. (2008). Identification and characterization of domains responsible for self-assembly and cell wall binding of the surface layer protein of *Lactobacillus brevis* ATCC 8287. *BMC Microbiol* 8, 165-181.

Axelsson L. (1998). *Lactic acid bacteria: classification and physiology. In Lactic acid bacteria: Microbiology and functional aspects*, Salminen S. and Wright A. W. Marcel Dekker Inc., New York.

Bayer, E. A., Belaich, J. P., Shoham, Y. and Lamed, R. (2004). The cellulosomes: multienzyme machines for degradation of plant cell wall polysaccharides. *Ann Rev Microbiol* 58, 521-554.

Begume, P. and Lemaire, M. (1996). The cellulosome: an extracellular multiprotein complex specialized in cellulose degradation. *Critic Rev Biochem Mol Bio* 31, 201-236.

Berlec, A., Zadravec, P., Jevnikar, Z. and Strukelj, B. (2011). Identification of candidate carrier proteins for surface display on *Lactococcus lactis* by theoretical and experimental analysis of the surface proteom. *Appl Environ Microbiol* 77, 1292-1300.

Bermudez-Humaran, L. G., Cortes-Perez, N. G., Lefever, F., Guimaraes, V., Rabot, S., Alcocer-Gonnzalez, J. M., Gratadoux, J-J., Rodriguez-Padilla, C., Tamez-Guerra, R. S., Corthier, G., Gruss, A. and Langella, P. (2005) A novel mucosal vaccine based on live lactococci expressing E7 antigen and IL-12 induces systemic and mucosal immune responses and protects mice against human papillomavirus type 16-induced tumors. *J Immunol* 175, 7297-7302.

Brinster, S., Furlan, S., Serror, P. (2007). C-terminal WXL domain mediates cell wall binding in *Enterococcus faecalis* and other gram positive bacteria. *J Bacterio* 189, 1244-1253.

Buist, G., Kok, J., Leenhouts, K. J., Babrowska, M., Venema, G., and Haandrikman, A. J. (1995). Molecular cloning and nucleotide sequence of the gene encoding the major peptidoglycan hydrolase of *Lactococcus lactis*, a muramidase needed for cell separation. *J Bacteriol* 177, 1554-1563.

Buist, G., Steen, A., Kok, J. and Kuipers, O. P. (2008). LysM, a widely distributed protein motif for binding to (peptide)glycans. *Mol Microbiol* 68, 838-847.

Collins J.K., Thornton, and G., Sullivan, G. O. (1998). Selection of probiotic strains for human applications. *Int Dairy J* 8, 487-490.

Conway, P. L., Gorbasch, S. L., and Goldin, B. R. (1987). Survival of lactic acid bacteria in the human stomach and adhesion to intestinal cells. *J dairy Sci* 70, 1-12.

Crittenden, R., Laitila, A., Forssell, P., Matto, J., Saarela, M., Mattilla-Sandholm, T., and Myllarinen, P. (2001). Adhesion of bifidobacteria to granular starch and its implications in probiotic technologies. *Appl Environ Microbiol* 67, 3469-3475.

Curtiss, R. (2002) Bacterial infectious disease control by vaccine development. *J Clin Invest* 110, 1061-1066.

Daly, C., and Davis, R. (1998). The biotechnology of lactic acid bacteria with emphasis on applications in food safety and human health. *Agric Food Sci* 7, 219-250 (1998).

Demain, A. L., Newcomb, M. and Wu, J. H. D. (2005). Cellulase, clostridia and ethanol. *Microbiol Mol Biol Rev* 69, 124-154.

De Vuyst, L. and Leory, F. (2007). Bacteriocins from lactic acid bacteria: production, purification, and food applications. *J Mol Microbiol Biotechnol* 13, 194-199.

Dieye, Y., Hoekman, A. J. W., Clier, F., Juillard, V., Boot, H. J. and Piard, J-C. (2003). Ability of *Lactococcus lactis* to export viral capsid antiges: a crucial step for development of live vaccines. *Appl Environ Microbiol* 69, 7281-7288.

Fischetti, V. A., Pancholi, V. and Schneewind, O. (1990). Conservation of a hexapeptide sequence in the anchr region of surface proteins from gram positive cocci. *Mol Microbiol* 4, 1603-1605.

Franke , C. M., Leenhouts, K. J., Haandrikman, A. J., Kok, J., Venema, G. and Venema, K. (1996). Topology of LcnD, a protein implicated in the transport of bacteriocins from *Lactococcus lactis*. *J Bacteriol* 178, 1766-1769.

Fuller, R. (1989). Probiotics in man and animals. *J Appl Bacteriol* 66, 365-378.

Germond, J-E., Delley, M., Gilbert, C. and Atlan, D. (2003). Determination of the domain of the *Lactobacillus delbrueckii* subsp. *bulgaricus* cell surface proteinase PrtB involved in attachment to the cell wall after heterologous expression of the *prtB* gene in *Lactococcus lactis*. *Appl Environ Microbiol* 69, 3377-3384.

Giudice, E. L. and Campbell, J. D. (2006) Needle free vaccine delivery. *Adv Drug Deliv Rev* 58, 68-89.

Hansson, M., Samuelson, P., Gunneriusson, E. and Stahl, S. (2001). Surface display on gram positive bacteria. Comb. *Chem High throuput Screen* 4, 171-184.

Hofvendahl, K. and Han-Hagerdal, B. (2000). Factors affecting the fermentative lactic acid production from renewable resources. *Enz Microbial Technol* 26, 87-107.

Hou, X-L., Yu, L-Y., Liu, J. and Wang, G-H. (2007) Surface-displayed porcine epidemic diarrhea viral (PEDV) antigens on lactic acid bacteria. *Vaccine* 26, 24-31.

Hu, S., Kong, J., Kong, W., Guo, T. and Ji, M. (2010). Characterization of a novel LysM domain from *Lactobacillus fermentum* bacteriophage endolysin and its use as an anchor to display heterologous proteins on the surfaces of Lactic acid bacteria. *Appl Environ Microbiol* 76, 2410-2418.

Hu, S., Kong, J., Sun, Z., Han, L., Kong, W. and Yang, P. (2011). Heterologous protein display on the cell surface of lactic acid bacteria is mediated by the s-layer protein. *Microbial Cell Factories* 10, 86.

Kajikawa, A., Satoh, E., Leer, R. J., Yamamoto, S. and Igimi, S. (2007) Intragastic immunization with recombinant *Lactobacillus casei* expressing flagellar antigen confers antibody-independent protective immunity against *Salmonella enterica* serovar *Enteritidis*. *Vaccine* 25, 3599-3605.

Kajikawa, A., Nordone, S. K., Zhang, L., Stoeker, L. L., LaVoy, A. S., Klaenhammer, T. R. and Dean, G. A. (2011). Dissimilar properties of two recombinant *Lactobacillus acidophilus* strains displaying *Salmonella* FliC with different anchoring motifs. Appl Environ Microbiol 77, 6587-6596.

Kleerebezem, M., Boekhorest, J., van Kranenburg, R., Molenaar, D., Kuipers, O. P., Leer, R., Tarchini, R., Peters, S. A., SAndbrink, H. M., Fiers, M. W., Stiekema, W., Lankhorst, R. M., Bron, P. A., Hoffer, S. M., Groot, M. N., Kerkhoven, R., de Vries, M., Ursing, B., de Vos, W. M. and Siezen, R. J. (2003). Complete genome sequence of *Lactobacillus plantarum* WCFS1. *Proc Natl Acad Sci USA* 100, 1990-1995.

Kleerebezem, M., Hols, P., Bernard, E., Rolain, T., Zhou, M., Siezen, R. J. and Bron, A. (2010). The extracellular biology of the lactobacilli. FEMS Microbiol Rev 34, 199-230.

Kondo, A. and Ueda, M. (2004). Yeast cell-surface display- applications of molecular display. *Appl Microbiol Biotechnol* 64, 28-40.

Krinbauer, R., Booy, F., Cheng, N., Lowy, D. R. and Schiller, J. T. (1992) Papillomavirus L1 major capsid protein self assembles into virus-like particles that are highly immunogenic. *Proc Natl Acad Sci USA* 89, 12180-12184.

Kushiro, A., Takahashi, T., Asahara, T., Tsuji, H., Nomoto, K. and Morotomi, M. (2001). *Lactobacillus casei* acquires the binding activity to fibronectin by the expression of the fibronectin binding domain of *Streptococcus pyogenes* on the cell surface. *J Mol Microbiol Biotechnol* 3, 563-571.

Lankaputhra, W. E. V., and Shah, N. P. (1995). Survival of *Lactobacillus acidophilus* and *Bifidobacterium* spp. in the presence of acid and bile salts. *Cultured Dairy Products J* 30, 2-7.

Lee, J-S., Poo, H., Han, D. P., Hong, S-P., Kim, K., Cho, M. W., Kim, E., Sung, M H. and Kim, C-J. (2006) Mucosal immunization with surface displayed severe acute respiratory syndrome coronovirus spike protein on *Lactobacillus casei* induces neutralization antibodies in mice. *J Virol* 80, 4079-4087.

Lee, P. and Faubert, G. M. (2006) Expression of the *Giardia lamblia* cyst wall protein 2 in *Lactococcus lactis*. *Microbiol* 152, 1981-1990.

Lee, S. Y., Choi, J. H. and Xu, Z. (2003). Microbial cell-surface display. *Trends Biotechnol* 21, 45-52.

Li, Y-J., Ma, G-P., Li, G-W., Qiao, X-Y., Ge, J-W., Tang, L-J., Liu, M. and Liu, L-W. (2010) Oral vaccination with the porcine rotavirus VP4 outer capsid protein expressed by *Lactococcus lactis* induces specific antibody production. *J Biomed Biotech* 2010, 1-9.

Lian W-H., Hsiao H-C., and Chou C-C. (2003). Viability of microencapsulated bifidobacteria in simulated gastric juice and bile solution. *Int J Food Microbiol* 86, 293-301.

Lindholm, A., Smeds, A. and Palva, A. (2004) Receptor binding domain of *Escherichia coli* F18 fimbrial adhesin FedF can be both efficiently secreted and surface displayed in a functional form in *Lactococcus lactis*. *Appl Environ Microbial* 70, 2061-2071.

Liu, J-K., Hou, X-L., Wei, C-H., Yu, L-Y., He, X-J., Wang, G-H., Lee, J-S. and Kim, C-J. (2009) Induction of immune responses in mice after oral immunization with recombinant *Lactobacillus casei* strains expressing enterotoxigenic *Escherichia coli* F41 fimbrial protein. *Appl Environ Microbiol* 75, 4491-4497.

Maassen, C. B. M., Laman, J. D., Heijne den bak-Glashouwer M. J., Tielen, F. J., van Holten-Neelen, J. C. P. A., Hoogteijling, L., Antonissen, C., Leer, R. J., Pouwels, P. H., Boersma,

W. J. A. and Shaw, D. M. (1999). Instruments for oral disease intervention strategies: recombinant *Lactobacillus casei* expressing tetanus toxin fragment C for vaccination or myelin proteins for oral tolerance induction in multiple sclerosis. *Vaccine* 17, 2117-2128.

Mckay, L. L. and Baldwin, K. A. (1990). Applications for biotechnology: present and future improvements in lactic acid bacteria. *FEMS Microbiol Lett* 87, 3-14.

Medina, M., Villena, J., Vintini, E., Hebert, E. M., Raya, R. and Alvarez, S. (2008) Nasal immunization with *Lactococcus lactis* expressing the pnemococcal protective protein A induces protective immunity in mice. *Infect Immun* 76, 2696-2705.

Mielcarek, N., Alonso, S. and Locht, C. (2001) Nasal vaccination using live bacterial vectors. Adv Drug Deliv Rev 51, 55-69.

Moore, R. J., Stewart, D. J., Lund, K. and Hodgson, L. M. (2001) Vaccination against ovine footrot using a live bacterial vector to deliver basic protease antigen. FEMS Microbiol Lett 194, 193-196.

Myllaerinen, P., Forssell, P., Von, W. A., Alander, M., Mattila-sandholm, T. and Poutanen, K. (2001) Starch capsules containing microorganisms and/or polypeptides or proteins and a process for producing them, In: *European patentEP1063976,* Available from: <http://www.freepatentsonline.com/EP1063976A1.html>.

Narita, J., Okano, K., Kitao, T., Ishida, S., Sewaki, T., Sung, M-H., Fukuda, H. and Kondo, A. (2006). Display of $\alpha$-amylase on the surface of *Lactobacillus casei* cells by use of the PgsA anchor protein, and production of lactic acid from starch. 2006. *Appl Environ Microbiol* 72, 269-275.

Navarre, W. W. and Schneewind, O. (1994). Proteolytic cleavage and cell wall anchoring at the LPXTG motif of surface proteins in gram positive bacteria. *Mol Microbiol* 14, 115-121.

Norton, P. M., Brown, H. W. G., Wells, J. M., Macpherson, A. M., Wilson, P. W. and Le Page, R. W. F. (1996). Factors affecting the immunogenicity of tetanus toxin fragment C expressed in *Lactococcus lactis*. *FEMS Immunol Med Microbiol* 14, 167-177.

Oliviera, M. L. S., Areas, A. P. M., Campos, I. B., Monedro, V., Perez-Martinez, G., Miyaji, E. N., Leite, L. C. C., Aires, K. A. and Ho, P. L. (2006) Induction of systemic and mucosal immune response and decrease in *Streptococcus pneumoniae* colonization by nasal inoculation of mice with recombinant lactic acid bacteria expressing pneumococcal surface antigen A. *Microbes Infect* 8, 1016-1024.

Olvera, C., Fernandez-Vazquez, J. L., Ledezma-Candanoza, L. and Lopez-Munguia, A. (2007). Role of the C-terminal region of dextransucrase from *Leuconostoc mesentroides* IBT-PQ in cell anchoring. Microbiol 153, 3994-4002.

Piard, J. C., Hautefort, I., Fischetti, V. A., Ehrlich, S. D., Fons, M. and Gruss A. (1997a). Cell wall anchoring of *Streptococcus pyogenes* M6 protein in various lactic acid bacteria. *J Bacteriol* 179, 3068-3072.

Poo, H., Pyo, H-M., Lee, T-Y., Yoon, S-W., Lee, J-S., Kim, C-J., Sung, M-H. and Lee, S-H. (2006) Oral administration of human papilomavirus type 16 E7 displayed on *Lactobacillus casei* induces E7-specific antitumor effects in C57/BL6 mice. *Int J Cancer* 119, 1702-1709.

Poquet, I., Dusko Ehrlich, S. and Gruss, A. (1998). An export specific reporter designed for gram positive bacteria: application to *Lactococcus lactis*. J Bacteriol 180, 1904-1912.

Pouwels, P. H., Leer, R. J., and Boersma, W. J. A. (1996). The potential of Lactobacillus as a carrier for oral immunization: Development and preliminary characterization of vector systems for targeted delivery of antigens. *J Biotechnol* 44, 183-192.

Pouwels, P. H., Leer, R. J., Shaw, M., den Bak-Glashouwer, M-J. H., Tielen, F. D., Smit, E., Martinez, B., Jore, J. and Conway, P. L. (1998) Lactic acid bacteria as antigen delivery vehicles for oral immunization purposes. *Int J Food Microbiol* 41, 155-167.

Ramos, H. C., Rumbo, M. and Sirad, J-C. (2004) Bacterial flagellins: mediators of pathogenicity and host immune responses in mucosa. *Trends Microbiol* 12, 509-517.

Shigechi, H., Koh, J., Fujita, Y., Matsumoto, T., Bito, Y., Ueda, M., Satoh, E., Fukuda, H., and Kondo, A. (2004). Direct production of ethanol from raw corn starch via fermentation by use of a novel surface-engineered yeast strain codisplaying glucoamylase and $\alpha$-amylase. *Appl Environ Microbiol* 70, 5037-5040.

Siezen, R., Boekhorest, J., Muscariello, L., Molenaar, D., Renckens, B., Kleerebezem, M. (2006). *Lactobacillus plantarum* gene clusters encoding putative cell surface protein complexes for carbohydrate utilization are conserved in specific gram positive bacteria. *BMC Genomics* 7, 126.

Siragusa, S., De Angelis, M., Di Cagno, R., Rizzello, G., Coda, R. and Gobbetti, M. (2007). Synthesis of $\gamma$-aminobutyric acid by lactic acid bacteria isolated from a variety of Italian cheeses. *Appl Environ Microbiol* 73, 7283-7290.

Smit, E., Jager, D., Martinez, B., Tielen. F. J. and Pouwels, P.H. (2002). Structural and functional analysis of the S-layer protein crystallisation domain of *Lactobacillus acidophilus* ATCC 4356: evidence for protein-protein interaction of two subdomains. *J Mol Biol* 324,953-964.

Steidler, L., Viaene, J., Fiers, W. and Remaut, E. (1998). Functional display of a heterologous protein on the surface of *Lactococcus lactis* by mean of cell wall anchor of *Staphylococcus aureus* protein A. *Appl Environ Microbiol* 64, 342-345.

Stiles, M. E., and Holzapfel, W. H. (1997). Lactic acid bacteria of foods and their current taxonomy. *Int J Food Microbiol* 36, 1-29.

Storni, T., Kundig, T. M., Senti, G., and Johansen, P. (2005). Immunity in response to particulate antigen delivery systems. Adv Drug Deliv Rev 57, 333-355.

Tang, L. and Li, Y. (2009) Oral immunization of mice with recombinant *Lactococcus lactis* expressing porcine transmissible gastroenteritis virus spike glycoprotein. *Virus Genes* 39, 238-245.

Tarahomjoo, S., Katakura, Y., Satoh, E. and Shioya, S. (2008a). Bidirectional cell-surface anchoring function of the C-terminal repeat region of peptidoglycan hydrolase of *Lactococcus lactis* IL1403. *J Biosci Bioeng* 105, 116-121.

Tarahomjoo, S., Katakura, Y., and Shioya, S. (2008b). Expression of C-terminal repeat region of peptidoglycan hydrolase of *Lactococcus lactis* IL1403 in methylotrophic yeast *Pichia pastoris*. J Biosci Bioeng 105, 134-139.

Toyota-Hantani, Y., Inoue, M., Ekawa, T., Ohta, H., Igimi, S. and Baba, E. (2008) importance of the major FliC antigenic site of *Salmonella enteritidis* as a subunit vaccine antigen. *Vaccine* 26, 4135-4137.

Turner, M. S. and Giffard, P. M. (1999). Expression of *Chlamydia psittaci* and human immunodeficiency virus derived antigens on the cell surface of *Lactobacillus fermentum* BR11 as fusions to BspA. *Infect Immun* 67, 5486-5489.

Turner, M. S., Hafner, L. M., Walsh, T. and Giffard, P. (2004). Identification and characterization of the novel LysM domain containing surface protein Sep from *Lactobacillus fermentum* BR11 and its use as a peptide fusion partner in *Lactobacillus* and *Lactococcus*. *Appl Environ Microbiol* 70, 3673-3680.

Van Der Vaart, J. M., Biesebeke, R., Chapman, J. W., Toschka, H. Y., Klis, F. M., and Verrips, T. (1997). Comparison of cell wall proteins of *Saccharomyces cervisiae* as anchors for cell surface expression of heterologous proteins. *Appl Environ Microbiol*, 63, 615-620.

Venema, R., Tjalsma, H., van Dijl, J. M., de Jong, A., Leenhouts, K., Buist, G. and Venema, G. (2003). Active lipoprotein precursors in the gram positive eubacterium *Lactococcus lactis*. *J Biol Chem* 278, 14739-14746.

Wan, H-M., Chang, B-Y., and Lin, S-C. (2002). Anchorage of cyclodextrin glucanotransferase on the outer membrane of *Escherichia coli*. *Biotechnol Bioeng* 79, 457-464.

Wang, X., Brown, I. L., Evans, A. J., and Conway, P. L. (1999). The protective effect of high amylose maize (amylomaize) starch granules on the survival of *Bifidobacterium* spp. in the mouse intestinal tract. *J Appl Microbiol* 87, 631-639.

Wei, C-H., Liu, J-K., Hou, X-L., Yu, L-Y., Lee, J-S. and Kim, C-J. (2010) Immunogenicity and protective efficacy of orally or intranasally administered recombinant *Lactobacillus casei* expressing ETEC K99. *Vaccine* 28, 4113-4118.

Wells, J. M., Robinson, K., Chamberlain, L. M., Schofield, K. M. and Le Page, R. W. F. (1996) Lactic acid bacteria as vaccine delivery vehicles. *Antonie van Leeuwen* 70, 317-330.

Westerlund, B., and Korhonen, T.K. (1993). Bacterial proteins binding to the mammalian extracellular matrix. *Mol Microbiol* 9, 687-694.

Wieczorek, A. and Martin, V. J. J. (2010). Engineering the cell surface display of cohesions for assembly of cellulosome inspired enzyme complexes on *Lactococcus lactis*. *Microbial Cell Fact* 9, 69.

Xin, K-Q., Hoshino, Y., Toda, Y., Igimi, S., Kojima, Y., Jounai, N., Ohba, K., Kushiro, A., Kiwaki, M., Hamajima, K., Klinman, D. and Okuda, K. (2003) Immunogenicity and protective efficacy of orally administered recombinant *Lactococcus lactis* expressing surface-bound HIV Env. *Blood* 102, 223-228.

Xu, W., Huang, M., Zhang, Y., Yi, X., Dong, W., Gao, X. and Jia, C. (2011). Novel surface display system for heterogonous proteins on *Lactobacillus plantarum*. *Lett Appl Microbiol* 53, 641-648.

Yamaguchi, K., Fujio, Y. and Inouye, M. (1988). A single amino acid determinant of the membrane localization of lipoproteins in *E. coli*. Cell 53, 423-432.

Chapter 19

# Lactic Acid Bacteria and Mitigation of GHG Emission from Ruminant Livestock

Junichi Takahashi

Additional information is available at the end of the chapter

http://dx.doi.org/10.5772/50358

## 1. Introduction

The gases which bring greenhouse effect are water vapor and trace gases in atmosphere, carbon dioxide ($CO_2$), methane ($CH_4$), nitrous oxide ($N_2$), sulfur hexafluoride ($SF_6$), hydrofluorocarbons (HFCs), and perfluorocarbons (PFCs). Global warming due to increases in the atmospheric concentration of greenhouse gases (GHG) is an important issue. The worldwide trends of carbon dioxide have shown an increase in the greenhouse effect on global warming (Houghton, 1994). However, $CH_4$ is an important greenhouse gas second only to $CO_2$ in its contribution to global warming due to its high absorption ability of infrared in the radiation from sun (IPCC, 1994). The world population of ruminants is important source of methane, contributing approximately 15-18% of the total atmospheric $CH_4$ flux. The control of $CH_4$ emission is a logical option since atmospheric $CH_4$ concentration is increasing at a faster rate than carbon dioxide (Moss, 1993). $CH_4$ emitted from ruminants is mainly generated in the rumen by hydrogenotrophic methanogens that utilize hydrogen to reduce carbon dioxide, and is a significant electron sink in the rumen ecosystem (Klieve and Hegarty, 1999), although acetotrophic methanogens may play a limited role for rumen methanogenesis (McAllister, 1996). Methane contains 892.6 kJ combustible energy per molecule at 25ºC and 1013hPa, while not contributing to the total supply of metabolic energy to ruminants (Takahashi *et al.*, 1997). As reported by Leng (1991), methane production from ruminants in the developing countries may be high since the diets are often deficient in critical nutrients for efficient microbial growth in the rumen. So far, a number of inhibitors of methanogenesis have been developed to improve feed conversion efficiency of ruminant feeds claimed to be effective in suppressing methanogens or overall bacterial activities (Chalupa, 1984). Attempts to reduce methanogenesis by the supplementation of chemicals such as ionophores (monensin and lasalocid), have long been made (Chalupa, 1984; Hopgood and Walker, 1967). However, these ionophores may depress fiber digestion and protozoal growths (Chen and Wolin, 1979). In addition, some resistant bacteria will appear in the rumen from the results of long term use of the ionophores. Therefore, development of manipulators to mitigate rumen methanogenesis must pay attention to secure safety for animals, their products and environment as alternatives of ionophores.

Theoretically, methanogenesis can be reduced by either a decrease in the production of $H_2$, the major substrates for methane formation or an increase in the utilization of $H_2$ and formate by organisms other than methanogens. However, direct inhibition of $H_2$-forming reactions may depress fermentation in microorganisms that produce $H_2$, including main cellulolytic bacteria such as *Ruminococcus albus* and *Ruminococcus flavefaciens* (Belaich *et al.*, 1990; Wolin, 1975). Therefore, a reduction in $H_2$ production by the enhancement of reactions that accept electrons is desirable (Stewart and Bryant, 1988). In the rumen, metabolic $H_2$ is produced during the anaerobic fermentation of glucose. This $H_2$ can be used during the synthesis of volatile fatty acids and microbial organic matter. The excess $H_2$ from NADH is eliminated primarily by the formation of $CH_4$ by methanogens, which are microorganisms from the *Archea* group that are normally found in the rumen ecosystem (Baker, 1999). The stoichiometric balance of VFA, $CO_2$ and $CH_4$ indicates that acetate and butyrate promote $CH_4$ production whereas propionate formation conserves $H_2$, thereby reducing $CH_4$ production (Wolin, and Miller, 1988). By contrast, reductive methanogenesis might contribute to mitigate methane (Immig *et al.*, 1996). Therefore, a strategy to mitigate ruminal $CH_4$ emission is to promote alternative metabolic pathway to dispose the reducing power, competing with methanogenesis for $H_2$ uptake. Oligosaccharides are naturally occurring carbohydrates with a low degree of polymerisation and consequently low molecular weight, being commonly found to perform in the various plant and animal sources. β1-4 Galactooligosaccharides (GOS) are non-digestible carbohydrates, which are resistant to gastrointestinal digestive enzymes, but fermented by specific colonic bacteria. The products of fermentation of GOS in the colon, mainly short chain fatty acids, have a role in the improvement of the colonic environment, energy supply to the colonic epithelium, and calcium and magnesium absorption (Sako, *et al.*, 1999). The indigestibility and stability of GOS to hydrolysis by α-amylase of human saliva, pig pancreas, rat small intestinal contents and human artificial gastric juice has been shown in several *in vitro* experiments (Ohtsuka *et al.*, 1990; Watanuki *et al.*, 1996). This is because GOS have β-configuration, whereas human gastrointestinal digestive enzymes are mostly specific for α-glycosidic bonds. From this point of view, expectedly, GOS will be readily degraded in the rumen as a result of the ruminal enzymes being specific for β-glycosidic bonds. Thus, lactic acid bacteria may consume GOS to promote propionate formation through acrylate pathway, and consequently the competition with methanogens for hydrogen will occur. Thus, the amplifying competition of metabolic $H_2$ with probiotics may be a key factor in the regulation of rumen methanogenesis. However, direct effects of prebiotics and secondary metabolites such as tannin, saponin and natural resin on methanogens and eubacteria in the rumen remain to be elucidated to secure the safety for animals, their products and environment. The mechanism for accreditation of manipulators must be established to mitigate global $CH_4$ emission.

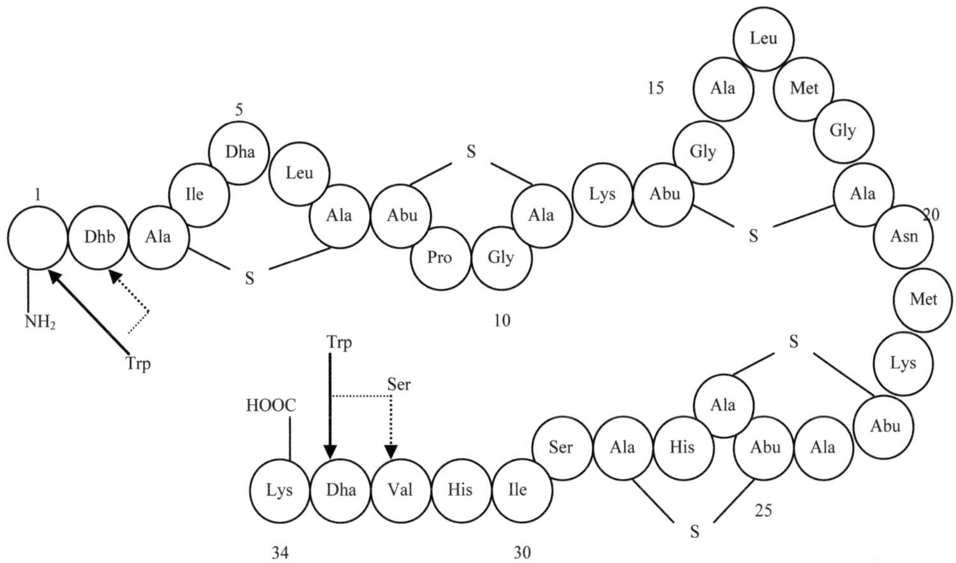


Dha = dehydroalanine, Dhb = dehydrobutyrine, Ala-S-Ala = lanthionine, Abu-S-Ala = β-methyllanthionine. (adapted from Breukink et al., 1998).

**Figure 1.** Primary structure of nisin.

## 2. Possible control of indirect action of lactic acid bacteria as probiotics on rumen methanogenesis

Rumen manipulation with ionophores such as monensin has been reported to abate rumen methanogenesis (Mwenya *et al.*, 2005), However, there is an increasing interest in exploiting prebiotics and probiotics as natural feed additives to solve problems in animal nutrition and livestock production as alternatives of the antibiotics due to concerns about incidences of resistant bacteria and environmental pollution by the excreted active-antibacterial substances (Mwenya *et al.*, 2006). Particular interest concerning bacteriocins which produced by lactic acid bacteria has increased recently.

Bacteriocins, antimicrobial proteinaceous polymeric material substances, are ubiquitous in nature being produced by a variety of Gram-negative and Gram-positive bacteria, and typically narrow spectrum antibacterial substances under the control of plasmid. Nisin is produced by *Lactococcus lactis* ssp. *lactis* which is an amphiphilic peptide composed by 34 amino acids with two structural domains that are connected by a flexible hinge (Breukink *et al.*, 1998; Montville and Chen, 1998), and is classified into the group of lantibiotics. Nisin has a mode of action similar to ionophores, which show antimicrobial activity against a broad spectrum of Gram-positive bacteria and is widely used in the food industry as a safe and natural preservative (Delves-Broughton *et al.*, 1996). It is generally recognized as safe (GRAS) and given international acceptance in 1969 by the joint Food and Agriculture Organization/World Health Organization (FAO/WHO) Expert Committee on Food

Additives. Recent works have indicated that *Lactococcus lactis* subsp. *lactis* produce nisin Z, which has been identified from Korean traditional fermented food "Kimchi" besides nisin A (Park, 2003). They have similar antibacterial ability to mitigate methane emission (Mwenya *et al.*, 2004; Santoso *et al.*, 2004; Sar *et al.*, 2006), to inhibit growth both of *Clostridium amoniphilum*, which is obligate amino-acid fermenting bacteria (Callaway *et al.*, 1997) and lactic acid-producing ruminal Staphylococci and Enterococci (Lauková, 1995). *Leuconostoc mesenteroides* ssp. *mesenteroides, Leuconostoc lactis* and *Lactococcus lactis* ssp. *lactis* were isolated from "Laban" which was a traditional fermented milk product in Yemen and determined the mitigating effect on in vitro rumen methane production. These strains isolated from Laban enhanced propionate production and decreased acetate/propionate ratio. In consequence, they reduced methane production remarkably (Gamo *et al.*, 2002). For *Leuconostoc mesenteroides* ssp. *mesenteroides*, in particular, the mitigating effect was amplified with GOS, which was degradable about 80% within 1 hour incubation in the artificial rumen fluid due to the stimulation of reduction reactions consuming metabolic hydrogen. However, direct involvement of bacteriocin or lower molecular substances produced by the strain on rumen methanogenesis remains to be elucidated.

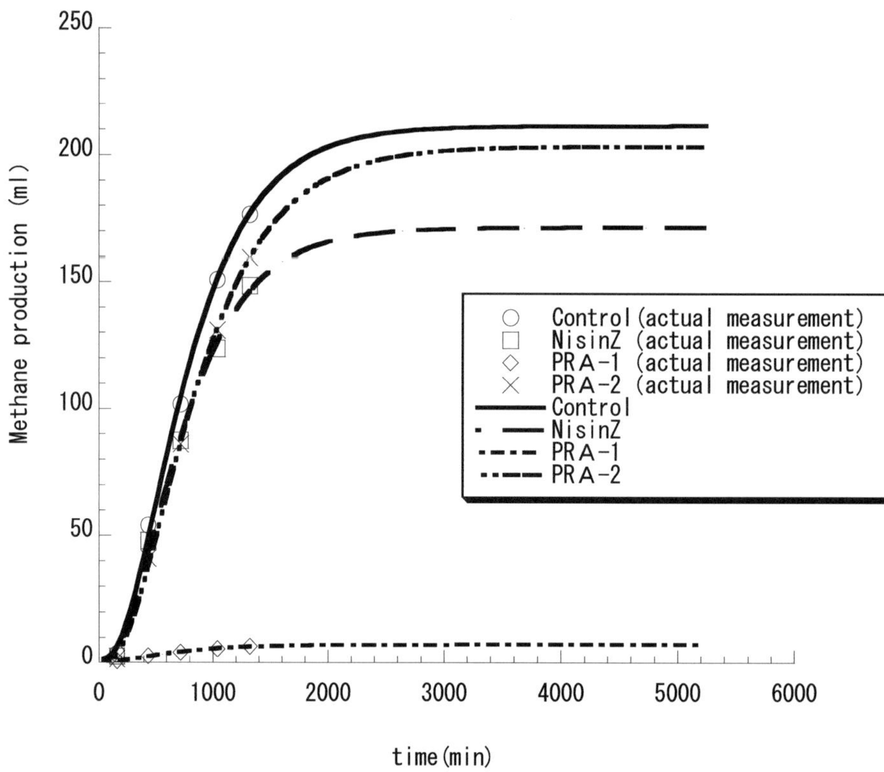


Where, y(ml) =gas produced at time *t* (min), a=first gas production, b=second gas production and c=frctional rate gas production, using Kaleida Graph (Version 3.6, Synergy Software, Reading, PA, USA).

**Figure 2.** Effect of PRA on the cumulative methane production extrapolated by nonlinear regression analysis; $y=a+b\ (1-e^{-ct})^3$.

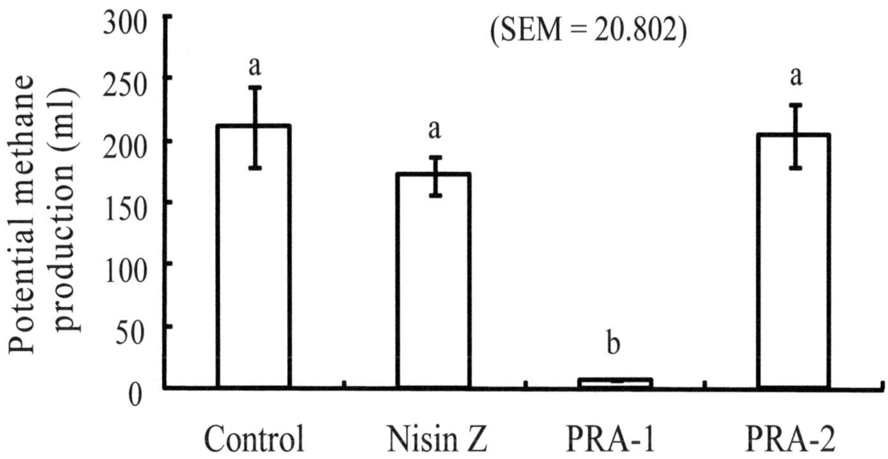


**Figure 3.** Effect of PRA on potential methane production. Control: *Lactococcus lactis* ATCC19435 (non-antibacterial substances), Nisin-A: *Lactococcus lactis* NCIMB702054, PRA-1: *Lactobacillus plantarum* TUA1490L, and PRA-2: *Leuconostoc citreum* JCM9698. Vertical bars represent standard deviation (n = 4). Means with different letters differ significantly (p<0.01).

## 3. Abatement of rumen methanogenesis by direct action of lactic acid bacteria as prebiotics producer

For low molecular compounds, small amounts of volatile fatty acids (acetic acid, formic acid), hydrogen peroxide, β-hydroxy-propionaldehyde (reuterin) are produced by lactic acid bacteria as antibacterial substances in addition to lactic acid. Because lactic acid bacteria themselves don't have a group of catalase, considerable amount of hydrogen peroxide accumulates in the bacterial cells. Many strains of the genus *Lactobacillus* are commonly referred to as having high ability to produce hydrogen peroxide (Jaroni and Brashears, 2000; Aroucheva *et al*, 2001; Gardiner *et al.*, 2002).

Its antimicrobial activity is effective against numerous Gram-positive bacteria. Although it has been reported that nisin suppress rumen methanogenesis, the suppressing efficacy of nisin on rumen methanogenesis may not be sustained, because proteinaceous nisin is degradable in the rumen due to bacterial protease (Sang *et al.*, 2002). Several strains of lactic acid bacteria produce different types of protease resistant antimicrobial substance (PRA). In our research, the strain of lactic acid bacteria that produce PRA were screened on MRS agar plates containing Umamizyme G (protease mixture from *Aspergillus oryzae*, amino Enzyme Inc, Nagoya, Japan) as follows: candidates were inoculated onto MRS agar with or without 1,000 IU $ml^{-1}$ of Umamizyme G and incubated for 24 h at 30 °C. the plates were then overlaid with Bacto Lactobacilli agar AOAC (Becton, Dickinson and Company, NJ. USA) containing an indicator strain, *Lactobacilli sakei* JCM1157T. The agar overlays were incubated for 24 h at 30°C and examined for zones of clearing. Protease degradable anti-microbial substances were decomposed by Umamizyme G, thus a clear zone did not form on the plate with Umamizyme G. Two strains of lactic acid bacteria, *Lactobacillus plantarum* TUA1490L and

*Leuconostoc citreum* JCM9698 that produced almost the same size of clear zone on a Umamizyme G containing plate as that on a plate without Umamizyme G, were selected as PRA producers. *Lactobacillus plantarum* TUA1490L and *Leuconostoc citeum* JCM9698 were selected as PRA-1 and PRA-2 producers. GYEKP medium to prepare inoculants for PRA-1, PRA-2, nisin Z and control were used for the culture of lactic acid bacteria. Each strain of lactic acid bacteria was inoculated into a shaking flask containing GYEKP, and was cultivated for 20 h at 30°C using SILIKOSEN (Shin-Etsu polymer, Tokyo), which was culture plug for aeration cultivation after confirmation of the stationary phase. The cells were removed by centrifugation at 8,000 × g at 4°C and filtration with 0.45 μm membrane filter. The supernatants were used as PRA inoculants in the in vitro gaseous quantification trials. Methane mitigating effects of PRA-1 from *Lactobacillus plantarum* TUA1490L and PRA-2 from *Leuconostoc citeum* JCM9698 isolated from foods were determined in comparison with *Lactococcus lactis* ATCC19435 which did not produce any antibacterial substances as a negative control and *Lactococcus lactis* NCIMB702954 which produced nisin-Z as a positive control using in vitro continuous incubation system equipped with automated infra-red quantification apparatus (Takahashi *et al.*, 2005). Fig.2 shows effects of PRA-1 and PRA-2 produced by *Lactobacillus plantarum* and *Leuconostoc citreum* on cumulative methanogenesis extrapolated by nonlinear regression analyses. PRA-1 remarkably decreased cumulative methane production. For PRA-2, there were no effects on $CH_4$ and $CO_2$ production and fermentation characteristics in mixed rumen cultures. Fig. 3 shows the effect of PRA on potential methane production which estimated from non-linier regression analysis of cumulated methane production. It has been suggested that PRA-1 significantly decreases potential methane production by rumen methanogens (Asa *et al.*, 2010). The PRA maintained their antimicrobial effects after incubation with proteases, while nisin lost its activity. Therefore, the PRA was hypothesized to be a more sustained agent than nisin for the mitigation of rumen methane emission. Fig. 4 shows DGGE band patterns of archaea and eubacteria. All fluorescence brightness of methanogens bands of PRA-1 were remarkably light in color compared with control. Band No. 1 to No.3 in archaea might be *Methanobrevibacter sp.* which is a Gram positive or parasitic methanogens sticking on protozoan surface (Fig.5). PRA-1 increased the fluorescence brightness of the band of the Gram positive bacteria and declined the fluorescence brightness of the band of the Gram negative bacteria. For Gram positive bacteria, *Streptococcus* sp., *Clostridium* sp., *Butyrivibrio* sp. and *Clostridium aminophilum* were increased, whereas *Prevotella* sp., *Prevotella ruminicola, Pseudobutyrivibrio* sp, *Prevotella* sp, *Succinivibrio dextrinosolvens and Schwartzia succinivorans* in Gram negative bacteria were decreased by adding PRA-1.

Natural antimicrobial substances can be used alone or in combination with other novel preservation technologies to facilitate the replacement of traditional approaches (Brijesh, 2009). *Lactobacillus plantarum* produces bacteriocin from many foods including meat and meat products (Garriga *et al.*,1993; Enan *et al.*,1996; Aymerich *et al.*, 2000), milk (Rekhif *et al.*, 1995), cheese (Gonzalez *et al.*,1994), fermented cucumber (Daeschel *et al.*,1994), olives (Jimenez-Diaz *et al.*, 1993; Leal *et al.*, 1998), grapefruit juice (Kelly *et al.*,1996), Turkish fermented dairy products (Aslim *et al.*, 2005), and sourdough (Todorov *et al.*, 1999). PRA-1 was the antibacterial substance produced from a strain of *Lactobacillus plantarum* TUA1490L that was isolated from tomato in Japan. However, methane suppressing activity of PRA-1 was not inactivated by treatments Umamizyme G and protease K. Moreover, aeration cultivation is an essential procedure for activation of PRA-1 to abate methanogenesis. Therefore, possible mechanism of PRA-1 produced by *Lactobacillus plantarum* TUA1490L on rumen methane production might be assumed as resulting from the direct involvement of low molecule substance such as hydrogen peroxide due to the requirement of aeration for the preparation.

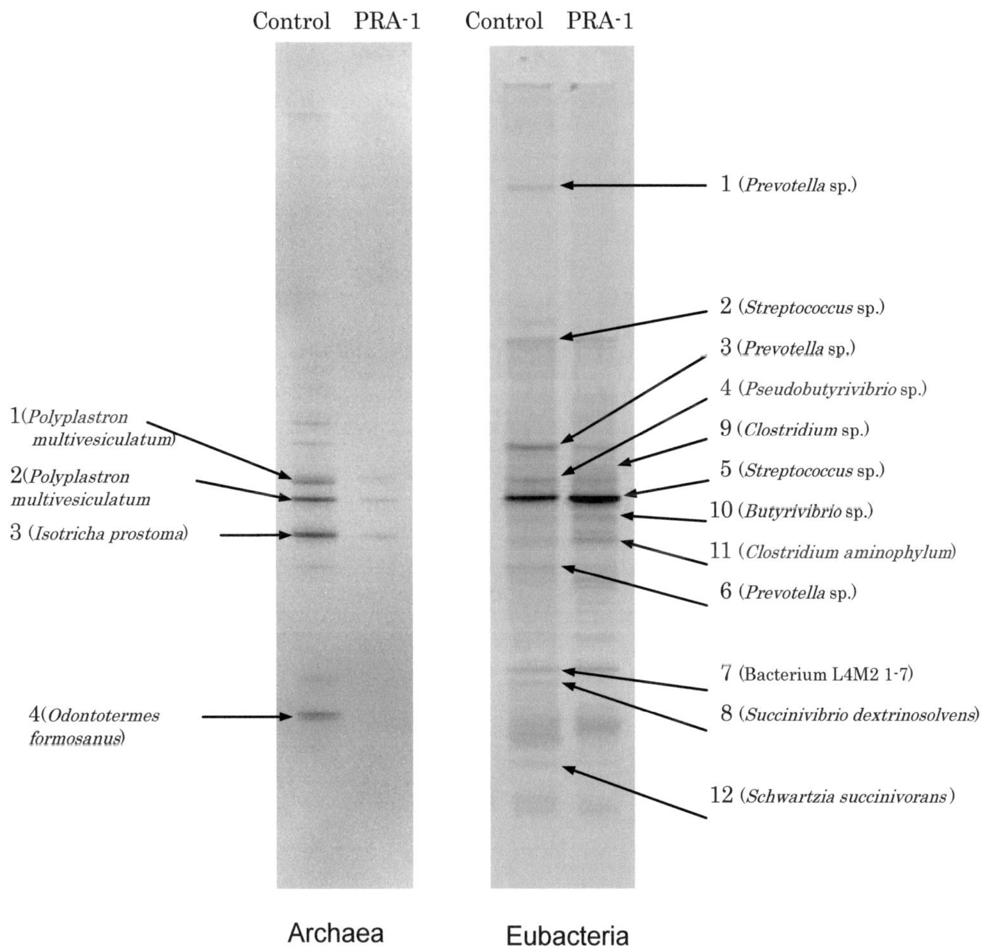


**Figure 4.** DGGE band patterns

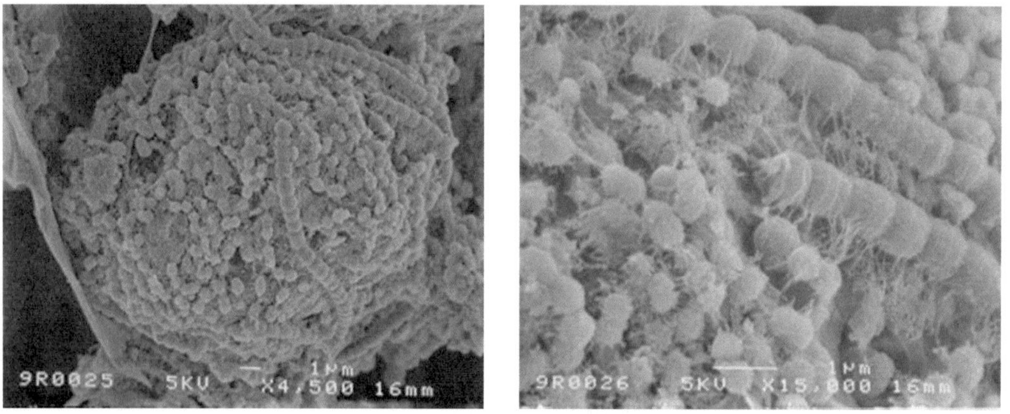


**Figure 5.** Electric scanning microscopy of symbioses of methanogens on the surface of Ciliate Protozoa.

## Author details

Junichi Takahashi
*Graduate School of Animal Science, Obihiro University of Agriculture and Veterinary Medicine, Obihiro, Japan*

## 4. References

Asa, R., A. Tanaka, A. Uehara, I. Shinzato, Y. Toride, N. Usui, K. Hirakawa and J. Takahashi., 2010. Effects of protease-resistant antimicrobial substances produced by lactic acid bacteria on rumen methanogenesis. Asian-Aust. J. Anim. Sci., 23:700-707.

Aslim, B., Z. N. Yuksekdag., E. Sarikaya and Y. Beyatli., 2005. Determination of the bacteriocin-like substances produced by some lactic acid bacteria isolated from Turkish dairy products. LWT, 38: 691-694.

Aroucheva, A., D. Gariti and M. Simons., 2001. Defense factors of vaginal Lactobacilli. Am. J. Obstet. Gynecol., 185: 375-379.

Aymerich. M., T., M. Garriga., J. M. Monfort., I. Nes and M. Hugas., 2000. Bacteriocin-producing lactobacilli in Spanish-style fermented sausages: characterization of bacteriocins. Food Microbiol., 17: 33-45.

Baker, S.K., 1999. Rumen methanogens, and inhibition of methanogenesis. Aust. J. Agric. Res., 50: 1293-1298.

Belaich, J.P., M. Bruschi and J. L. Garcia., 1990. Microbiology and Biochemistry of Strict Anaerobes Involved in Interspecies Hydrogen Transfer. Plenum Press, NY. USA.

Breukink, E., C. Van Kraaij, A. Van Dalen, R.A. Demel, R.J. Siezen, B. De Kruijff, and O.P. Kuipers., 1998. The orientation of nisin in membranes. Biochem., 37: 8153–8162.

Brijesh, K. T., P. V. Vasilis, P. O. D. Colm, M. Kasiviswanathan B. Paula and P. J. Cullen., 2009. Applivation of natural antimicrobials for food preservation. J. Agric. Food Chem., 57: 5987-6000.

Callaway, T. R., Alexandra M. S. Carneiro De Melo and J. B. Russell., 1997. The effect of nisin and monensin on ruminal fermentations in vitro. Curr Microbiol., 35:90-96.

Cadieux, P., J. Burton, G. Gardiner, J. Braunstein, A.W. Bruce, G.Y. Kang and G. Reid., 2002. Lactobacilli strains and vaginal ecology. J. Am. Med. Assoc., 287: 1940-1941.

Chalupa, W., 1984. Manipulation of rumen fermentation. In: Recent Advances in Animal Nutrition. W. Haresign and D. Cole, (eds). Butterworths, London, England, pp.143-160.

Chen, M. and M.J. Wolin, M.J., 1979. Effect of monensin and lasalocid-sodium on the growth of methanogenic and rumen saccharolytic bacteria. Appl. Environ. Microbiol., 38: 72-77.

Daeschel, M. A., M. C. Mckenny and L. C. McDonald., 1990. Bacteriocidal activity of Lactobacillus plantarum C11. Food Microbiol. 7: 91-99.

Delves-Broughton, J., P. Blackburn., R. Evans and J. hugenholtz., 1996. Applications of the bacteriocin, nisin. Antonie van Leeuwenhoek. 69: 193-202.

Enan, G., A. A. El-Essawy., M. Uyttendaele and J. Debevere., 1996. Antibacterialactivity of Lactobacillus plantarum UG1 isolated from dry sausage: characterization, production and bactericidal action of plantarcin UG1. Int. J. Food. Microbiol., 30: 189-215.

Gamo, Y., M. Mii, X.G. Zhou, C. Sar, B. Santoso, I. Arai, K. Kimura and J. Takahashi., 2002. Effects of lactic acid bacteria, yeasts and galactooligosaccharides supplementation on in vitro rumen methane production. In: J. Takahashi and B.A. Young, (eds), Greenhouse Gases and Animal Agriculture. ELSEVIER, Amsterdam, Netherland. pp201-204.

Garriga, M., M. Hugas., T. Aymerich and J. M. monfort., 1993. Bacteriocinogenic activity of lactobacilli from fermented sausages. J. Appl. Bacteriol., 75: 142-148.

González, B., P. Arca., B. Mayo and J. E. Suárez., 1994. Detection, purification and partial characterization of plantaricin C, a bacteriocin produced by a Lactobacillus plantarum strain of dairy origin. Appl. Environ. Microbiol., 6: 2158-2163.

Hopgood, M. F. and D. J. Walker., 1967. Succinic acid production by rumen bacteria. II. Radioisotope studies on succinateproduction by *Ruminococcus flavefaciens.* Aust. J. Biol. Sci., 20: 183-192.

Houghton, J., 1994. Global warming. Lion Publishing plc. Oxford, UK, pp: 29-45.

IPCC (Intergovermental Panel on Climate Change), 1994. Houghton, J.H., Meria filho, J. Bruce, L.Hoesung, B.A.Callander, H. Haites, N.Harris and K.Maskell, (eds), Cambridge University Press. New York. pp25-27.

Immig I., D. Demeyer, D. Fiedler, C. Van Nevel and L. Mbanzamihigo., 1996. Attempts to induce reductive acetogenesis into a sheep rumen. Arch. Tierernahr. 49:363-370.

Jaroni, D. and M. M. Brashears., 2000. Production of Hydrogen Peroxide by *Lactobacillus delbrueckii* subsp. *lactis* as Influenced by Media Used for Propagation of Cells . J. Food Sci., 65: 1033-1036.

Jiménez-Díaz R., R. M. Rios-Sánchez, M. Desmazeaud, J. L. Ruiz-Barba and J. C. Piard., 1993. Plantaricin S and T, two new bacteriocins produced by Lactobacillus plantarum LPCO10 isolated from a green olive fermentation. Appl. Environ. Microbiol. 59: 1416-1424.

Kelly, W. J., R. V. Asmundson and C. M. Huang., 1996. Characterization of plantaricin KW30, a bacteriocin produced by Lactobacillus plantarum. J. Appl. Bacteriol. 81: 657-662.

Klieve, A.V. and R. S. Hegarty., 1999. Opportunities of biological control of ruminant methanogenesis. Aust. J. Agric. Res., 50: 1315-19.

Leal, M. V., M. Baras, J. L. Ruiz-Barba, B. Floriano and R. Jimenez-Diaz., 1998. Bacteriocin production and competitiveness of Lactobacillus plantarum LPCO10 in olive juice broth, a culture medium obtained from olives. Int. J. Food. Microbiol. 43: 129-134.

Leng, R.A., 1991. Improving ruminant production and reducing methane emissions from ruminants by strategic supplementation. EPA/400/1-91/004, US Environmental Protection Agency, Washington, DC, pp6-10.

Lauková, A., 1995. Inhibition of ruminal staphylococci and enterococci by nisin in vitro. Lett. Appl. Microbiol., 20:34-36.

Moss, A.R., 1993. Methane: Global Warming and Production by Animals. Chalcombe, Canterbury, UK, p.105.

McAllister, T. A., E.K. Okine, G.W. Mathison, K.-J. Cheng., 1996. Dietary, environmental and microbiological aspects of methane production in ruminants. Can. J. Anim. Sci. 76: 231-243

Montville, T.J and Y. Chen., 1998. Mechanistic action of pediocin and nisin: recent progress and unresolved questions. Appl. Microbiol. Biotechnol., 50: 511-519.

Mwenya, B., C. Sar, Y. Gamo, T. Kobayashi, R. Morikawa, K. Kimura, H. Mizukoshi, and J. Takahashi., 2004. Effects of *Yucca schidigera* with or without nisin on ruminal fermentation and microbial protein synthesis in sheep fed silage- and hay- based diets. Anim. Sci. J., 75: 525-531.

Mwenya, B. C. Sar, B. Santoso, T. Kobayashi, R. Morikawa, K. Takaura, K. Umetsu, S. Kogawa, K. Kimura, H. Mizukoshi and J. Takahashi., 2005. Comparing the effects of β1-4 galacto-oligosaccharides and L-cysteine to monensin on energy and nitrogen utilization in steers fed a very high concentrate diet. Anim. Feed Sci. Technol. 118: 19-30.

Mwenya, B., C. Sar, B. Pen, R. Morikawa, K. Takaura, S. Kogawa, K. Kimura, K. Umetsu and J. Takahashi., 2006. Effects of feed additives on ruminal methanogenesis and anaerobic fermentation of manure in cows and steers. In: C. Soliva, J. Takahashi and M. Kreutzer (eds), Greenhouse Gases on Animal Agriculture update. International Congress Series. Elsevier, Amsterdam. 1293: 209-212.

Ohtsuka, K., K.Tsuji, Y. Nakagawa, H. Ueda, O. Ozawa, T. Uchida and T. Ichikawa., 1990. Availability of 4`-galactosyllactose (o-beta-D-galactopyranosyl -(1-4)-o- beta-D-galactospyranosyl-(1-4)-D-glucopyranose) in rat. J. Nutr. Sci. Vitaminol., 36: 265-276.

Park, S.E., K. Itoh, E. Kikuchi, H. Niwa and T. Fujisawa., 2003. Identification and characteristics of nisin Z-producing *Lactococcus lactis* subsp. *lactis* isolated from Kimchi. Curr. Microbiol., 46: 385-358.

Rekhif, N, A. Atrih and G Lefebvre., 1995. Activity of plantaricin SA6, a bacteriocin produced by Lactobacillus plantarum SA6 isolated from fermented sausage. J. Appl. Bacteriol., 78: 349-358.

Sako, T., K. Matsumoto and R. Tanaka., 1999. Recent progress on research and applications of non-digestible galacto-oligosaccharides. Inter. Dairy J., 9: 69-80.

Sang, S.L., H. C. Mantovani, and J.B. Russell., 2002. The binding and degradation of nisin by mixed ruminal bacteria. FEMS Microbiol. Ecol., 42:339-345.

Santoso, B., B. Mwenya, C. Sar, Y. Gamo, T. Kobayashi, R. Morikawa and J. Takahashi., 2004. Effect of Yucca schidigera with or without nisis on ruminal fermentation and microbial protein synthesis in sheep fed silage- and hay-based diets. Anim. Sci. J., 75: 525-531.

Sar, C., B. Mwenya, B. Pen, R. Morikawa, K. Takaura, T. Kobayashi and J. Takahashi., 2006. Effect of nisin on ruminal methane production and nitrate/nitrite reduction *in vitro*. Aust. J. Agric. Res., 56: 803-810.

Stewart, C.S., and M.P. Bryant., 1988. The rumen bacteria. In: The Rumen Microbial Ecosystem. P.N. Hobson, (ed). Elsevier Appl. Sci., New York, NY, pp.21-75.

Takahashi, J., A.S. Chaudhry, R.G. Beneke, Suhubdy and B.A. Young., 1997. Modification of methane emission in sheep by cysteine and a microbial preparation. Sci. Total Environ., 204: 117-123

Takahashi, J., B. Mwenya, B. Santoso, C. Sar., K. Umetsu, T. Kishimoto, K. Nishizaki, K. Kimura and O. Hamamoto., 2005. Mitigation of methane emission and energy recycling in animal agricultural systems. Aisan-Aust. J. Anim. Sci., 18:1199-1208.

Todorov, S., B. Onno., O. Sorokine., J. M. Chobert., I. Ivanova and X. Dousset., 1999. Detection and characterization of a novel antibacterial substance produced by *Lactobacillus plantarum* ST 31 isolated from sourdough. Int. J. Food. Microbiol. 48:167-177.

Watanuki, M., Y. Wada and K. Matsumoto., 1996. Digestibility and physiological heat of combustions of β1-4 and β1-6 galacto-oligosaccharides *in vitro*. Annu. Rep. Yakult Cent. Inst. Microbiol. Res., 16: 1-12.

Wolin, M. J., 1975. Interactions between the bacterial species of the rumen. In: Digestion and Metabolism in the Ruminant. I.M. McDonald and A.C.I. Warner (eds). Univ. New England Publ. Unit, Sydney, Australia, pp.135-148.

Wolin, M.J. and T.L. Miller, 1988. Microbe-microbe interactions. In: The rumen microbial ecosystem. Elsevier Applied Science, London, UK, pp. 343-359

Chapter 20

# Lactose and $\beta$-Glucosides Metabolism and Its Regulation in *Lactococcus lactis*: A Review

Tamara Aleksandrzak-Piekarczyk

Additional information is available at the end of the chapter

http://dx.doi.org/10.5772/50889

## 1. Introduction

### 1.1. Lactic acid bacteria

Lactic acid bacteria (LAB) are a group of Gram-positive, non-sporulating, low-GC-content bacteria that comprise 11 bacterial genera, such as *Lactococcus, Lactobacillus, Leuconostoc, Streptococcus* and others (Stiles & Holzapfel, 1997). LAB have a generally regarded as safe (GRAS) Food and Drug Administration (FDA) status, and some strains of different LAB species exhibit also probiotic properties (Gilliland, 1989). They are ubiquitous in many nutrient rich environments, such as milk, meat and plant material, and some of them are permanent residents of mainly mammalian intestinal tracts, while others are able to colonize them temporarily. Due to their ability to produce lactic acid as an end product of sugar fermentation, they are industrially important and are used as starter cultures in various food-fermentation processes. The importance of LAB for humans can be appreciated from the estimated 8.5 billion kg of fermented milk produced annually in Europe, leading to human consumption of $8.5 \times 10^{20}$ LAB (Franz et al., 2010).

Understanding the mechanisms involved in carbohydrate metabolism and its regulation in LAB is essential for improving the industrial properties of these microorganisms. There are several ways to improve the metabolic potential of LAB cells, of which metabolic engineering offers a very efficient and effective tool.

### 1.2. Genus *Lactococcus*

Lactococci are homofermentative, mesophilic LAB that basically inhabit two natural environments, milk and plants, of which plants seem to constitute the primary niche. Occasionally, there have been reports that *L. lactis* was also isolated from soil, effluent water, the skin of cattle (Klijn et al., 1995), insects (leafhoppers, termites) (Bauer et al., 2000;

Latorre-Guzman et al., 1977; Schultz & Breznak, 1978) and fish (Itoi et al., 2008, 2009; Pérez et al., 2011). Adaptation of lactococcal strains from plants to the dairy environment has caused the loss of some functions, resulting in smaller chromosomes and acquisition of genes (often plasmidic) important for growth in milk (Kelly et al., 2010).

Since *Lactococcus lactis* was first described in 1919 (Orla-Jensen, 1919), its taxonomy has changed repeatedly and still is confusing in some aspects. This group of bacteria, previously designated lactic streptococci, was placed in the new *Lactococcus* taxon in 1985 (Schleifer *et al.*, 1985). The current taxonomy of *L. lactis* is based on phenotype and includes four subspecies (*lactis, cremoris, hordniae,* and the newly identified subsp. *tructae*) and one biovar (subsp. *lactis* biovar diacetylactis) (Schleifer et al., 1985; van Hylckama Vlieg et al., 2006; Pérez et al., 2011; Rademaker et al., 2007). Among them, only *L. lactis* subsp. *hordniae* and subsp. *tructae* have never been isolated from dairy products. The *lactis* and *cremoris* phenotypes are distinguished on the basis of several basic criteria, such as: arginine and maltose utilization, decarboxylation of glutamate to $\gamma$-aminobutyric acid (GABA), and 40°C, 4% NaCl and pH 9.2 tolerance. *L. lactis* subsp. *cremoris* strains are reported to be negative for all of these features (Nomura et al., 1999; Schleifer et al., 1985). Moreover, the biovar diacetylactis strains are able to metabolize citrate, which is converted to diacetyl, an important aroma compound. Additionally, numerous genetic studies (DNA–DNA hybridization, 16S rRNA and gene sequence analysis) of *L. lactis* isolates of dairy and plant origin have revealed the existence among them of two main genotypes that have also been called *L. lactis* subsp. *lactis* (*lactis* genotype) and *L. lactis* subsp. *cremoris* (*cremoris* genotype). Furthermore, it has been demonstrated that the genotype and phenotype do not always correspond within one isolate, thus introducing a degree of disorder into the taxonomy of this species (Tailliez et al., 1998). It has been observed that within the group of *cremoris* genotype, strains with both *lactis* (MG1363) and *cremoris* (SK11) phenotypes may occur, and, likewise, within the group of *lactis* genotype there are ones with *lactis* (KF147) as well as biovar diacetylactis (IL594) phenotypes (Bayjanov et al., 2009; Kelly et al., 2010; Nomura et al., 2002; Rademaker et al., 2007; Tanigawa et al., 2010). Hence, the *L. lactis* has an atypical taxonomic structure with two phenotypically distinct groups, such as *L. lactis* subsp. *lactis* and *L. lactis* subsp. *cremoris*, which may belong to two distinct genotype groups. As a result, in order to sufficiently describe the individual strains, it is necessary to specify both the genotype (*cremoris* or *lactis*) and the phenotype (*cremoris, diacetylactis,* or *lactis*).

Strains belonging to *L. lactis* subsp. *lactis* and *L. lactis* subsp. *cremoris* together with a diverse assortment of other LAB are widely used as dairy starters for the production of a vast range of fermented dairy products, including various types of cheeses, sour cream, buttermilk and butter (Daly, 1983; Davidson et al., 1996). In the dairy industry, the *lactis* subspecies are better for making soft cheeses and the *cremoris* subspecies for the hard ones. Overall, it is generally accepted that the *L. lactis* subsp. *cremoris* strains make better quality products than *L. lactis* subsp. *lactis* because of their important contribution to flavour development via their unique metabolic mechanisms (Salama et al., 1991; Sandine, 1988).

During growth in milk, the primary function of *L. lactis* is rapid conversion of lactose to lactic acid, which provides preservation of the fermented product by preventing growth of pathogenic and spoilage bacteria, it supports curd formation, and creates optimal conditions for ripening. Further, due to their proteolytic activity and amino acid conversion, lactococci contribute to the final texture (moisture, softness) and flavour of dairy products (Smit et al., 2005). Many of lactococcal functions vital for successful fermentations are borne on plasmids, which are a common feature in lactococci, even in strains isolated from non-dairy sources (Davidson et al., 1996). For example, specific plasmid-borne genes encode proteins involved in lactose transport and metabolism and in hydrolysis and utilization of casein (Davidson, et al., 1996; McKay, 1983). Hence, there is considerable selective pressure on dairy strains to retain these plasmids, since plasmid-cured derivatives grow poorly in milk. Since plasmids are mobile elements, they can be readily exchanged among different strains (via conjugal transfer) (Gasson, 1990).

Due to its industrial importance *L. lactis* has become the best studied LAB, and although most studies have been performed on a small number of laboratory strains of dairy origin, it is regarded as a model organism for this bacterial group. A number of genome sequences of *L. lactis* strains are available, including strains from *L. lactis* subsp. *lactis*, such as IL1403, KF147 and CV56, as well as strains from *L. lactis* subsp. *cremoris*, such as MG1363, A76, NZ9000 and SK11 (according to http://www.ncbi.nlm.nih.gov/genome/). Among them, *L. lactis* subsp. *lactis* IL1403 (Chopin et al., 1984) and *L. lactis* subsp. *cremoris* MG1363 (Gasson, 1983) are the most important laboratory strains, and they can be distinguished by differences in specific DNA sequences, including those encoding 16S rRNA (Godon et al., 1992), and by their genome organization (Le Bourgeois et al., 1995). These two strains are plasmid-cured derivatives of the dairy starter strains IL594 (IL1403) and NCDO 712 (MG1363) respectively, and due to their industrial importance, their metabolism, physiology and genetics have been extensively studied over the past years. Both belong to *L. lactis* subsp. *lactis* phenotypically, but the parent strain of IL1403 has a citrate permease plasmid (Górecki et al., 2011) and is able to metabolize citrate, placing it with *L. lactis* subsp. *lactis* biovar diacetylactis, whereas MG1363 has a *lactis* phenotype and a *cremoris* genotype (Kelly et al., 2010). Despite their physiological and 16S rRNA gene sequence similarities, they share only about 85% chromosomal sequence identity, which is comparable to the genetic distance between *Escherichia coli* and *Salmonella typhimurium* (McClelland et al., 2001; Salama et al., 1991; Wegmann et al., 2007). A derivative of MG1363 was created by the integration of the *nisRK* genes (involving the "NICE" system for nisin-controlled protein overexpression) into the *pepN* gene, yielding *L. lactis* NZ9000 (Kuipers et al., 1998).

## 2. Lactose metabolism

Most microorganisms have adapted to growth in milk habitat due to acquisition of the ability to the use its most abundant sugar, lactose, as a carbon source. This disaccharide consists of a galactose moiety linked at its $C_1$ via a β-galactosidic bond to the $C_4$ of glucose. Because of the efficiency and economic importance of its fermentation, a large number of studies have focused on the utilization of lactose by LAB.

Uptake of lactose into a bacterial cell can be mediated by several pathways, such as the lactose-specific phosphotransferase system (*lac*-PTS), ABC protein-dependent systems and secondary system transporters like lactose-galactose antiporters and lactose-$H^+$ symport systems (de Vos & Vaughan, 1994). While ABC protein-dependent lactose transport has been demonstrated only in non-LAB, Gram-negative *Agrobacterium radiobacter* (Williams et al., 1992), the *lac*-PTS as well as secondary lactose transport systems have been described for many LAB species.

## 2.1. Lactose-specific phosphotransferase systems (*lac*-PTS)

Although LAB used as starter cultures may also convert pyruvate to a variety of end products, these pathways are not expressed during lactose fermentation, which is homolactic in most strains (Cocaign-Bousquet et al., 2002; Neves et al., 2005). Since the primary function of LAB in dairy fermentations is the conversion of lactose to lactic acid, the industrial strains are primarily selected on the basis of their ability for its rapid, homolactic fermentation (de Vos & Simons, 1988).

Starter lactococcal strains transport lactose exclusively by the most abundant in LAB uptake system for various sugars - the phosphoenolpyruvate-dependent phosphotransferase system (PEP-PTS). The *lac*-PTS has a very high affinity for this sugar and is bioenergetically the most efficient system since one lactose molecule is translocated and phosphorylated in a single step, at the expense of a single ATP equivalent. Concomitantly with transport, PTS catalyzes the phosphorylation of the incoming sugar. Phosphoenolpyruvate is the first phosphoryl donor, which phosphorylates Enzyme I (EI), and then the phosphoryl group is transferred in sequence to HPr, EIIA, EIIB, and finally, via transmembrane porter (EIIC), to the transported sugar (Lorca et al., 2010). After translocation via *lac*-PTS, lactose is hydrolyzed by P-$\beta$-galactosidase to glucose and galactose-6-P. While glucose enters the Embden-Meyerhof-Parnas glycolytic pathway through phosphorylation by glucokinase, galactose-6-P, before it also enters the glycolytic pathway, is further metabolized via the D-tagatose-6-P (Tag-6P) pathway. This involves three enzymes: (i) galactose-6-P isomerase (LacAB); (ii) tagatose-6-P kinase (LacC); and (iii) tagatose-1,6-diphosphate aldolase (LacD). The resulting triosephosphates (glyceraldehydes-3-P and dihydroxyacetone-P) are further metabolized via glycolysis. The operons engaged in this rapid, homolactic lactose fermentation are usually plasmid-located (*lac*-plasmids) and, in addition to the genes for the *lac*-PTS proteins and P-$\beta$-galactosidase, contain genes coding for the enzymes of the Tag-6P pathway. Their transcription is regulated by various repressors, with tagatose-6-P being the molecular inducer in *L. lactis* (van Rooijen et al., 1991).

It is believed that plasmid-encoded ability for rapid lactose fermentation characteristic for dairy strains was recently acquired by wild-type plant strains, as a result of their adaptation to milk-environment (Kelly et al., 2010).

## 2.2. Lactose permease-$\beta$-galactosidase systems

Another strategy developed by LAB for lactose metabolism depends on its uptake via secondary transport systems. These systems transport lactose in an unphosphorylated form via specific permeases belonging to the LacS subfamily (TC No. 2.A.2.2.3) of the 2.A.2 glycoside-pentoside-hexuronide (GPH) family (Saier, 2000). Carriers of the LacS subgroup are chimeric in nature: at their carboxy terminal end they contain an approximately 160 amino acid hydrophilic extension homologous to the EIIA domains of PTS. Thus, lactose transport is controlled by HPr-dependent phosphorylation (Gunnewijk et al., 1999; Gunnewijk & Poolman, 2000a; Gunnewijk & Poolman, 2000b). Due to this additional domain these lactose permeases are larger than the other carriers from the GPH family, which are generally about 500 amino acids in length. Depending on the organism, LacS can mediate lactose transport coupled to proton symport or by antiport with galactose. Following its import, lactose is hydrolyzed by β-galactosidase (David et al., 1992; Vaughan et al., 1996) yielding glucose and galactose. The glucose moiety is further metabolized via glycolysis, whereas the galactose moiety follows different pathways depending on the particular LAB. While some thermophilic strains of LAB (*e.g., Lactobacillus bulgaricus* and *Streptococcus thermophilus*) are known to release the galactose moiety of lactose into the medium, other LAB (*e.g., Lactobacillus helveticus, Leuconostoc lactis* and *Streptococcus salivarius*) metabolize this saccharide via the Leloir pathway (de Vos, 1996; Poolman, 1993; Vaughan et al., 2001). This pathway was one of the first central metabolic pathways to be discovered, by L. F. Leloir and coworkers in the early 1950s. It includes the key enzyme galactokinase (GalK), and hexose-1-P uridylyltransferase (GalT) plus UDP-glucose 4-epimerase (GalE), all of which are involved in the conversion of galactose to glucose-1P. The generated glucose-1P, after conversion to glucose-6P by phosphoglucomutase, enters the glycolytic pathway. Aldose-1-epimerase, a mutarotase (GalM), is an additional, more recently characterized enzyme required for rapid galactose metabolism (Bouffard et al., 1994; Mollet & Pilloud, 1991; Poolman et al., 1990). GalM catalyses the interconversion of the α- and β-anomers of galactose. This enzyme was found to be essential for efficient lactose utilization in *E. coli* since cleavage of this β-galactoside by β-galactosidase yields glucose and β-D-galactose, the latter being the sole substrate for GalK (Bouffard et al., 1994).

The existence of genes encoding components of the lactose permease-β-galactosidase system seems to be limited among the *L. lactis* strains as they have been identified only in the genomes of the dairy-derived strain IL1403 (Bolotin et al., 2001), non-dairy NCDO2054 (Vaughan et al., 1998) and KF147 isolated from mung bean sprouts (Siezen et al., 2010). Remarkably, in addition to galactose genes of the Leloir pathway cluster, these strains contain genes needed for lactose assimilation, such as *lacZ* (β-galactosidase) and *lacA* (thiogalactoside acetyltransferase), arranged in an identical layout. Directly upstream of the aforementioned genes required for lactose hydrolysis and subsequent galactose conversion, there is the gene encoding the LacS permease for sugar uptake.

Some details concerning the role of the lactose permease-β-galactosidase system in lactose utilization have been reported for the slow lactose fermenter - *L. lactis* NCDO2054 (Vaughan et al., 1998), and for the devoid of the *lac*-plasmid, essentially lactose-negative *L. lactis* IL1403 strain (starts to utilize lactose slowly after approximately 40 h of incubation) (Aleksandrzak-Piekarczyk et al., 2005). Since these strains possess the complete lactose permease-β-galactosidase system and an active Leloir pathway, it seems odd that they are barely capable of lactose metabolism. In the case of *L. lactis* NCDO2054, which can accumulate a high intracellular concentration of lactose-6-phosphate by using an efficient *lac*-PTS and possesses low-level P-$\beta$-galactosidase activity, it has been suggested that the slow fermentation of lactose may be due to this rate-limiting P-$\beta$-galactosidase activity and the inhibitory effect of the accumulated lactose-6-phosphate (Bissette & Anderson 1974; Crow & Thomas, 1984). However, other explanations of lactose fermentation problem can be envisaged: (i) lactose transport is inefficient due to low affinity of LacS for lactose or (ii) the strains lack a functional $\beta$-galactosidase. Indeed, the *lacS* gene of *L. lactis* IL1403 is almost identical to that of *L. lactis* NCDO2054, but also to *galP* of the lactose-negative *L. lactis* MG1363 strain (Grossiord et al., 2003). These permeases belong to the same subfamily (TC No. 2.A.2.2.3 according to the Transporter Classification Database: http://www.tcdb.org/; Saier, 2000), which includes transporters specific for galactose uptake, in contrast to LacS permeases of another subfamily (TC No. 2.A.2.2.1) with a proven high lactose-transport rate. The lack of LacS involvement in lactose transport is confirmed by the fact that disruption of *lacS* in *L. lactis* IL1403 had a minor effect on lactose assimilation (Aleksandrzak-Piekarczyk et al., 2005). Another indispensable factor in lactose assimilation, the $\beta$-galactosidase enzyme, is also encoded by the genomes of *L. lactis* IL1403 and NCDO2054 strains. In spite of the high similarity in the protein level of both enzymes, $\beta$-galactosidase of *L. lactis* NCDO2054, in contrast to the one of *L. lactis* IL1403 (Aleksandrzak-Piekarczyk et al., 2005), seems to be highly active and strongly regulated (Griffin et al., 1996). It has been suggested that the *lacZ* gene of *L. lactis* IL1403 may not be expressed or the encoded enzyme may be inactive since this strain does not exhibit $\beta$-galactosidase activity (Aleksandrzak-Piekarczyk et al., 2005). Furthermore, the *in trans* complementation of chromosomal *lacZ* by an active $\beta$-galactosidase in *L. lactis* IL1403 did not improve its ability for lactose assimilation, indicating that the lack of $\beta$-galactosidase activity is not the only obstacle in its ability to efficiently ferment lactose (unpublished personal observations).

Taken together, it seems that in *L. lactis* strains lactose permease-$\beta$-galactosidase systems play a minor role in lactose assimilation or function under certain environmental conditions. It appears that the major obstacle is the galactose-specific LacS permease, which shows only weak affinity for lactose and functions almost only in transport of galactose (Fig. 1). This thesis is confirmed by the study of Solem et al. (2008), in which an efficient lactose transporter (LacS; TC No. 2.A.2.2.1 ) and $\beta$-galactosidase (LacZ), encoded by the *lacSZ* operon, were introduced from lactose-positive *S. thermophilus* into the lactose-negative strain *L. lactis* MG1363, devoid of lactose permease-$\beta$-galactosidase system. As a result, fast-growing lactose-positive mutant strains were obtained. This shows that addition of the LacSZ system containing LacS with a proven high lactose-transport rate can strongly increase the lactose-transport capacity in *L. lactis*.

## 3. Metabolism of $\beta$–glucosides

In addition to dairy environment, plant surfaces and fermenting plant material are also important ecosystems occupied by *L. lactis*. With regard to fermentation, lactococcal strains usually occur there only at the beginning of this process, to be later replaced by microorganisms more resistant to low pH values (Kelly & Ward, 2002; Kelly et al., 1998). The majority of plant-associated strains belong to *L. lactis* subsp. *lactis*, whereas *L. lactis* subsp. *cremoris* is typical for dairy fermentations (Kelly & Ward, 2002; Kelly et al., 1998). In comparison to the dairy environment, fermenting plant material differs highly with respect to chemical composition, exhibiting, for instance, much lower protein concentration and wider availability of carbohydrates other than lactose. The ability of plant-associated *L. lactis* subsp. *lactis* strains to utilize such a large variety of plant carbohydrates is reflected in their genomes and sugar fermentation capabilities. Comparison between milk- and plant-associated lactococcal strains clearly shows that the latter possess a larger number of genes involved in transport and metabolism of carbohydrates, resulting in their increased sugar fermentation capabilities (Siezen et al., 2008).

Besides lactose, the PTS systems can also transport various other carbohydrates, including sugars widely distributed in plants, namely β-glucosides, like e.g. amygdalin, arbutin, cellobiose, esculin, gentobiose and salicin (Tobisch et al., 1997). Except for amygdalin, these sugars are composed of two molecules joined by the β-glucosidic bond, of which at least one is glucose. The best known example of this group is cellobiose, the structural unit of one of the most abundant renewable polymers on earth – cellulose, and also the main product in its enzymatic hydrolysis (Teeri, 1997). Unlike most of other β-glucosides (aryl-β-glucosides e.g., arbutin, amygdalin, esculin, and salicin), which are composed of a single glucose molecule and respective aglycone, cellobiose consists of two glucose molecules linked via a β(1-4) bond.

It is well known from sugar fermentation characteristics that *L. lactis* strains of different origin can utilize a variety of β-glucosides (e.g., Aleksandrzak-Piekarczyk et al., 2011; Bardowski et al., 1995; Fernández et al., 2011; Siezen et al., 2008). The metabolic potential for catabolism of these sugars can be chromosomally encoded by more than one genetic system, as was shown for *L. lactis* IL1403. Eight genes, which encode proteins homologous to EII proteins of β-glucoside-dependent PTS, involved in the uptake and phosphorylation of β-glucosides have been found throughout the *L. lactis* IL1403 chromosome (Bolotin et al., 2001). Three of them encode the three-domain EIIABC PTS components (PtbA, YedF and YleE), another three, EIIC permeases (CelB, PtcC and YidB), one an EIIA component (PtcA) and one an EIIB component (PtcB). CelB, PtcA, PtcB, PtcC and YidB are members of the Lac family (TC No. 4.A.3), which includes several lactose porters of Gram-positive bacteria as well as the *E. coli* and *Borrelia burgdorferi* N,N'-diacetylchitobiose (Chb) porters (according to http://www.tcdb.org/). The involvement of CelB and CelB/PtcC permeases in cellobiose transport has been experimentally confirmed in *L. lactis* IL1403 and MG1363, respectively (Aleksandrzak-Piekarczyk et al., 2011; Campelo et al., 2011). Although *L. lactis* IL1403 has such a large number of β-glucosides-specific PTS systems, CelB is the only permease operative in cellobiose uptake in this strain (Aleksandrzak-Piekarczyk et al., 2011) (Fig. 1), whereas in *L. lactis* MG1363 also another PTS permease, namely PtcC, seems to participate in the transport of this sugar, albeit to a much lesser extent than CelB (Campelo et al., 2011). It has been proposed that the observed low expression of the *ptcC* gene may be the result of repression by carbon catabolite control protein A (CcpA) as mutations in its binding site (catabolite responsive element - *cre*) in the *ptcC* promoter region led to high upregulation of this gene in strain NZ9000 compared to strain MG1363, even under repressive conditions (Linares et al., 2010).

On the other hand, the EIIAB components, namely PtcA and PtcB, seem to be more versatile, being involved in the metabolism of numerous sugars (arbutin, cellobiose, glucose, lactose, salicin) in *L. lactis* (Aleksandrzak-Piekarczyk et al., 2011; Castro et al., 2009; Pool et al., 2006). No other PTS systems dedicated to transport of other *β*-glucosides have yet been described in detail in any *L. lactis* strain. However, according to unpublished preliminary data, the PtbA protein appears to be involved in the transport of arbutin, esculin and salicin, but not cellobiose, in *L. lactis* IL1403 (unpublished personal observation) (Fig. 1). In this strain, inactivation of the *ptbA* gene led to serious defects in growth in medium supplemented with each of these sugars (unpublished).

After translocation by PTS through the bacterial membrane, the P-*β*-glucoside sugar is cleaved by P-*β*-glucosidase into glucose and glucose-6P or the respective aglycon (Tobisch et al., 1997). There are plenty of genes encoding P-*β*-glucosidases present in *L. lactis* chromosomes sequenced so far. Their large number is probably the result of adaptation of these bacteria to life on plants with abundant where *β*-glucosides. However, the data concerning their involvement in *β*-glucosides assimilation are rather scarce in scientific literature. It has only been demonstrated that a P-*β*-glucosidase, BglS, is responsible for hydrolysis of cellobiose, but not of salicin in *L. lactis* IL1403 (Aleksandrzak-Piekarczyk et al., 2005) (Fig. 1). On the other hand, no function has been attributed to another P-*β*-glucosidase encoded by the *bglA* gene, and forming one operon with *ptcC*. According to unpublished results, the disruption of *bglA* did not alter growth of the IL1403 mutant strain in medium supplemented with a wide array of sugars (unpublished personal analysis).

Expression of *β*-glucosides' catabolic genes can be controlled by various regulatory mechanisms. Among them, catabolite repression (Aleksandrzak-Piekarczyk et al., 2005, 2011; Zomer at al., 2007) and transcriptional antitermination through the BglR protein (Bardowski et al., 1994) were shown to be operational in *L. lactis*. The antitermination mechanism allows for expression of *β*-glucoside-specific genes in the absence of a metabolically preferred carbon source, such as glucose (Rutberg, 1997). It is believed that antiterminator proteins act by binding to a ribonucleic antiterminator (RAT) site at a specific mRNA secondary structure to prevent the formation of a hairpin terminator structure that would otherwise terminate transcription (Aymerich & Steinmetz, 1992; Rutberg, 1997). The binding of the antiterminator protein to the mRNA permits transcription through the sequestered terminator sequence into a *β*-glucoside-specific operon that is not normally transcribed. The function of BglR has been studied earlier in *L. lactis* IL1403, and it was shown to be involved in the activation of assimilation of *β*-glucosides such as arbutin, esculin and salicin, except for cellobiose (Bardowski et al., 1994; 1995) (Fig. 1). Inspection of the *L. lactis* IL1403 genome sequence downstream of *bglR* revealed the presence of two genes, *ptbA* and *bglH*, encoding proteins homologous to a putative three-domain EIIABC PTS component specific for the assimilation of *β*-glucosides, and P-*β*-glucosidase, respectively. Upstream of *bglR*, a putative *cre*-box (differing from the *cre* consensus by one nucleotide), a putative promoter sequence and a RAT sequence were identified. This RAT sequence has been reported previously (Bardowski et al., 1994, 1995) to be involved in the autoregulation of BglR. This sequence partially overlapped a putative *rho*-independent terminator, which comprised six nucleotides at the 3' end of the RAT. The *ptbA* gene is located 141 nt downstream of *bglR*. *In silico* sequence analysis revealed that the *ptbA* gene is also preceded by a DNA sequence highly similar to the RAT consensus sequence, suggesting that the regulation of *ptbA* expression may involve the BglR-mediated antitermination mechanism (unpublished personal analysis). Moreover, the short intergenic DNA region (47 nt) between *ptbA* and the next gene (*bglH*), plus the lack of an obvious hairpin structure or a promoter sequence strongly suggest that these two genes might be cotranscribed, and thus undergo common BglR-mediated regulation (unpublished) (Fig. 1).

## 4. Alternative lactose utilization system and its interconnection with cellobiose assimilation

The existence in several lactococcal strains devoid of *lac*-plasmids of cryptic lactose transport and catabolism systems has already been suggested in earlier studies (Anderson & McKay, 1977; Cords & McKay, 1974; de Vos & Simons, 1988; Simons et al., 1993). The presence in *L. lactis* of chromosomally-encoded lactose permease has been proposed since introduction of the *E. coli lacZ* gene into a lactose-deficient *L. lactis* strain restored its ability to utilize lactose (de Vos & Simons, 1988). Moreover, P-$\beta$-galactosidase activities have also been detected in strains cured of their lactose plasmids, suggesting the presence of chromosomally-encoded cryptic *lac*-PTS(s) (Anderson & McKay, 1977; Cords & McKay, 1974). However, it was suggested that these PTSs are not specific for lactose, but rather for the translocation of other sugars (e.g., $\beta$-glucosides), and lactose could be transported alternatively. This hypothesis was supported by observations suggesting that a putative P-$\beta$-glucosidase, involved in cellobiose hydrolysis, is probably also involved in lactose-6-P cleavage in *L. lactis* strain ATCC7962 (Simons et al., 1993). This seems reasonable, as according to http://www.tcdb.org/, PTS lactose transporters belong to the Lac family (TC No. 4.A.3) and porters of this family have broad substrate specificity. Besides lactose, they can also transport aromatic $\beta$-glucosides and cellobiose.

Until recently (Aleksandrzak et al., 2000; Aleksandrzak-Piekarczyk et al., 2005, 2011; Kowalczyk et al., 2008), little information on the organization in *L. lactis* strains of chromosomal alternative lactose utilization genes has been available. It was shown that in *lac*-plasmid-free, and thus lactose-negative *L. lactis* IL1403, the ability to assimilate lactose can be induced in two ways: (i) by the presence of cellobiose or (ii) by inactivation of CcpA (Aleksandrzak et al., 2000; Aleksandrzak-Piekarczyk et al., 2005). The CcpA protein is a member of the LacI-GalR family of bacterial repressors and exists only in Gram-positive bacteria (Weickert & Adhya, 1992). It exerts its regulatory role in carbon catabolite repression (CCR) by binding to DNA sites called *cre*s, which occur in the vicinity of CcpA-regulated genes (Weickert & Chambliss, 1990). In *L. lactis* the known targets of CcpA are the *gal* operon for galactose utilization (Luesink et al., 1998), the *fru* operon for fructose utilization (Barrière et al., 2005), the *ptcABC* operon for cellobiose utilization (Zomer et al., 2007), and *cel-lac* genes for cellobiose and lactose utilization (Aleksandrzak-Piekarczyk et al., 2011). Thus, one could speculate that in *L. lactis* IL1403 cellobiose-inducible chromosomal alternative lactose utilization genes are under the negative control of CcpA, and, therefore, inactivation of the *ccpA* gene could result in their derepression and ability to assimilate lactose by the IL1403 *ccpA* mutant.

Further studies of Aleksandrzak-Piekarczyk et al. (2005, 2011) and Kowalczyk et al. (2008) provided details on interconnected metabolism of $\beta$-glucosides (cellobiose) and $\beta$-galactosides (lactose) and its variable regulation in *L. lactis* IL1403. Several genes have been implicated in coupled cellobiose and lactose assimilation in *L. lactis* IL1403, such as *bglS* and *celB, ptcA* and *ptcB,* encoding proteins homologous to P-$\beta$-glucosidase and EII components of cellobiose-specific PTS, respectively (Fig. 1). It has been shown that in *L. lactis* IL1403 the cellobiose-specific PTS system, comprising of *celB, ptcB* and *ptcA,* is also able to transport lactose because cellobiose-specific permease CelB has also an affinity for lactose, and, moreover, is the only permease involved in lactose uptake (Aleksandrzak-Piekarczyk et al., 2011). Furthermore, internalized lactose-P is hydrolyzed exclusively by BglS – an enzyme with dual P-$\beta$-glucosidase and P-$\beta$-galactosidase activity, and high affinity for cellobiose (Aleksandrzak-Piekarczyk et al., 2005) (Fig. 1). Thus, BglS activity generates glucose and galactose-P molecules. Glucose enters the Embden-Meyerhof-Parnas glycolytic pathway through phosphorylation by glucokinase, whereas galactose-P requires dephosphorylation performed by an unidentified phosphatase or phosphohexomutase, before entering the Leloir pathway (Neves et al., 2010) (Fig. 1). Moreover, this alternative lactose utilization system has been shown to be tightly controlled by CcpA-directed negative regulation (Fig. 1), since inactivation of the *ccpA* gene led to derepression of *bglS, celB, ptcA* and *ptcB* and *L. lactis* IL1403 *ccpA* mutant ability to assimilate lactose (Aleksandrzak-Piekarczyk et al., 2011). In addition to CcpA-mediated repression, the *celB* and *bglS* genes are specifically activated by cellobiose, as its presence leads to an increase in their transcription. This phenomenon has not been observed when other sugars, such as glucose, galactose or salicin, were used as carbon sources (Aleksandrzak-Piekarczyk et al., 2011). Preliminary results suggest that a hypothetical transcriptional regulator, namely YebF, could be engaged in this cellobiose-dependent activation of *celB* and *bglS* (Aleksandrzak-Piekarczyk et al., 2011; unpublished personal analysis) (Fig. 1). The YebF protein belongs to the RpiR family of phosphosugar binding proteins (Sorensen & Hove-Jensen, 1996), and, in addition to its sugar binding domain (SIS), it has a putative helix-turn-helix (HTH) DNA-binding domain. In addition to *yebF* mutant ferment lactose inability (Aleksandrzak-Piekarczyk et al., 2005), inactivation of the *yebF* gene in IL1403 resulted in inability to grow on cellobiose (unpublished personal analysis), suggesting the gene's requirement in both cellobiose and lactose assimilation. Further studies on this phenomenon in *L. lactis* are needed to address it in greater detail.

When cellobiose is available, it activates the cellobiose-specific PTS transport system, comprising CelB, PtcB and PtcA proteins, and *L. lactis* IL1403 is able to grow on cellobiose and lactose. This growth is supported by the activity of cellobiose-inducible BglS protein, which splits lactose-P into galactose-P and glucose. Then, after the dephosphorylation step, galactose is further metabolized through the Leloir pathway, while glucose enters glycolysis. Therefore, inactivation of the *ccpA* gene results in derepression of the cellobiose-specific PTS transport system and also of the *bglS* gene, which in turn enable the IL1403 strain to grow on lactose.

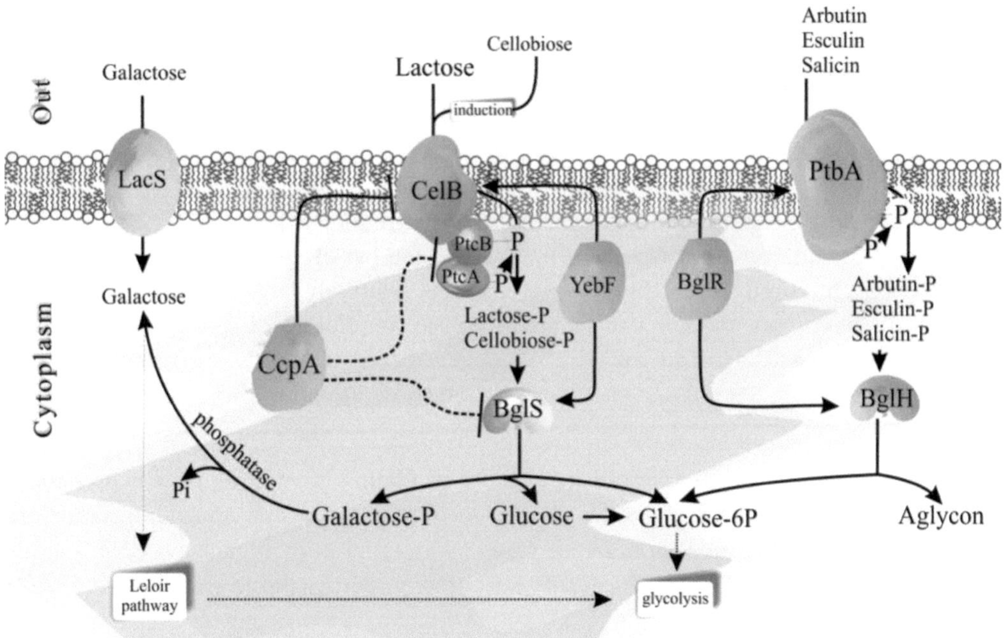


**Figure 1.** Schematic representation of the proposed mechanism of chromosomally-encoded lactose, cellobiose-inducible lactose and β-glucosides metabolism and of its regulation in *L. lactis* IL1403. In this model the key elements are the CelB, PtcB, PtcA, BglS and PtbA proteins. In the presence of glucose, IL1403 is unable to assimilate either lactose or β-glucosides. Under these conditions, these catabolic systems are either repressed by the CcpA protein and/or are not induced by the BglR activator.

Besides cellobiose, other β-glucosides like arbutin, esculin and salicin are transported by the PtbA-mediated PTS system. In the absence of any of these three sugars, *ptbA* expression is not induced by the inactive the phosphorylated BglR antiterminator protein. Once a β-glucoside is available, BglR becomes dephosphorylated and active, inducing the expression of the *ptbA* gene. The PtbA protein transports, with concomitant phosphorylation, arbutin, esculin and salicin, which are then probably hydrolyzed by BglH, a P-β-glucosidase, encoded by a gene located downstream of and in the same operon as the *ptbA* gene.

It is also proposed in this model that LacS is not engaged in lactose internalization and its function is limited to galactose transport.

## 5. Conclusions

Despite the fact that the metabolism of lactose and β-glucosides is very important for the biotechnological processes catalysed by *L. lactis*, thorough studies of the chromosomally encoded features enabling use of these carbon sources were so far rather scarce. The reason for this could be the fact that *L. lactis* demonstrates a very large and complex metabolic capability towards carbohydrates used as carbon and energy sources, and, moreover, that this genetic potential is tightly regulated by various environmental and intracellular factors. It seems that the main obstacle in studies on the complicated mechanisms involved in assimilation of $\beta$–glycoside sugars was the lack of complex data specifying the sequences of genes potentially involved in the metabolism of these sugars and its regulation. Indeed, recent access to the genomic sequences of some these bacteria greatly advanced the research on the metabolism of various $\beta$–glycosides. As expected, the results of sequencing of lactococcal genomes and genes annotations confirmed that there are numerous genes encoding potential $\beta$-glucosides-specific transport systems and $\beta$-glucosidases, sometimes with dual activities. And, to complicate the matter even further, the analysis of the list of genes annotated in *L. lactis* leads to over a hundred transcriptional regulators. A relatively large number of them may be related to carbon metabolism control. These regulators, together with signals modulating their activity, and the controlled genes form a regulatory network that is necessary for sensing the environmental conditions and adjusting the catabolic capacities of the cell.

Detailed knowledge of sugar metabolism and the regulators controlling gene expression in *Lactococcus lactis* may contribute to the improvement of mechanisms controlling significant cellular processes in these bacteria. In the case of industrial microorganisms, acting on the defined regulatory network may drastically affect the properties of the bacteria and have an impact on bioprocesses.

Lastly, is shown as an example that by the use of a simple microbiological screen, it is possible and worthwhile to modify the metabolic potential of lactococcal strains initially unable to assimilate lactose. By inactivation of the *ccpA* gene or induction of particular genes by supplementation of the medium with cellobiose and thus activation of YebF, it is possible to turn on an alternative lactose assimilation pathway in *L. lactis* IL1403. In contrast to plasmid-located *lac*-operons, the *cel-lac* system is within the chromosome, resulting in a stable, highly adapted strain, potentially valuable for the dairy industry.

## Author details

Tamara Aleksandrzak-Piekarczyk
*Institute of Biochemistry and Biochemistry, Polish Academy od Sciences, Warsaw, Poland*

## Acknowledgement

Some of the data presented were funded in part by the NCN grant UMO-2011/01/B/NZ2/05377.

## 6. References

Aleksandrzak, T., Kowalczyk, M., Kok, J. & Bardowski, J. (2000). Regulation of carbon catabolism in *Lactococcus lactis*. In: *Food biotechnology, Proceedings of an International Symposium organized by the Institute of Technical Biochemistry, Technical University of Lodz.*

Bielecki, S., Tramper, J., Polak, J. (eds), Vol. 17, pp. 61–66. Elsevier Science BV. ISBN 978-0-444-50519-4, Zakopane, Poland, May 1999

Aleksandrzak-Piekarczyk, T., Polak, J., Jezierska, B., Renault, P. & Bardowski, J. (2011). Genetic characterization of the CcpA-dependent, cellobiose-specific PTS system comprising CelB, PtcB and PtcA that transports lactose in *Lactococcus lactis* IL1403. *International Journal of Food Microbiology*, Vol. 145, No. 1, pp. 186-94, ISSN 1879-3460

Aleksandrzak-Piekarczyk, T., Kok, J., Renault, P. & Bardowski, J. (2005). Alternative lactose catabolic pathway in *Lactococcus lactis* IL1403. *Applied and Environmental Microbiology*, Vol. 71, No. 10, pp. 6060-9, ISSN 0099-2240

Anderson, DG. & McKay, LL. (1977). Plasmids, loss of lactose metabolism, and appearance of partial and full lactose-fermenting revertants in *Streptococcus cremoris* B1. *Journal of Bacteriology*, Vol. 129, No. 1, pp. 367-77, ISSN 0021-9193

Aymerich, S. & Steinmetz, M. (1992). Specificity determinants and structural features in the RNA target of the bacterial antiterminator proteins of the BglG/SacY family. *Proceedings of the National Academy of Sciences of the United States of America*, Vol. 89, No. 21, pp. 10410-4, ISSN 0027-8424

Bardowski, J., Ehrlich, SD. & Chopin, A. (1994). BglR protein, which belongs to the BglG family of transcriptional antiterminators, is involved in beta-glucoside utilization in *Lactococcus lactis*. *Journal of Bacteriology*, Vol. 176, No. 18, pp. 5681-5, ISSN 0021-9193

Bardowski, J., Ehrlich, SD. & Chopin, A. (1995). A protein, belonging to a family of RNA-binding transcriptional anti-terminators, controls beta-glucoside assimilation in *Lactococcus lactis*. *Developments in Biological Standardization*, Vol. 85, pp. 555-9, ISSN 0301-5149

Barrière, C., Veiga-da-Cunha, M., Pons, N., Guédon, E., van Hijum, SA., Kok, J., Kuipers, OP., Ehrlich, DS. & Renault, P. (2005). Fructose utilization in *Lactococcus lactis* as a model for low-GC gram-positive bacteria: its regulator, signal, and DNA-binding site. *Journal of Bacteriology*, Vol. 187, No. 11, pp. 3752-61, ISSN 0021-9193

Bauer, S., Tholen, A., Overmann, J. & Brune, A. (2000). Characterization of abundance and diversity of lactic acid bacteria in the hindgut of wood- and soil-feeding termites by molecular and culture-dependent techniques. *Archives of Microbiology*, Vol. 173, No. 2, pp. 126-37, ISSN 0302-8933

Bayjanov, JR., Wels, M., Starrenburg, M., van Hylckama Vlieg, JE., Siezen, RJ. & Molenaar, D. (2009). PanCGH: a genotype-calling algorithm for pangenome CGH data. *Bioinformatics (oxford, England)*, Vol. 25, No. 3, pp. 309-14, ISSN 1367-4811

Bissett, DL. & Anderson, RL. (1974). Lactose and D-galactose metabolism in group N streptococci: presence of enzymes for both the D-galactose 1-phosphate and D-tagatose 6-phosphate pathways. *Journal of Bacteriology*, Vol. 117, No. 1, pp. 318-20, ISSN 0021-9193

Bolotin, A., Wincker, P., Mauger, S., Jaillon, O., Malarme, K., Weissenbach, J., Ehrlich, SD. & Sorokin, A. (2001). The complete genome sequence of the lactic acid bacterium

*Lactococcus lactis* ssp. *lactis* IL1403. *Genome Research,* Vol. 11, No. 5, pp. 731-53, ISSN 1088-9051

Bouffard, GG., Rudd, KE. & Adhya, SL. (1994). Dependence of lactose metabolism upon mutarotase encoded in the *gal* operon in *Escherichia coli. Journal of Molecular Biology,* Vol. 244, No. 3, pp. 269-78, ISSN 0022-2836

Campelo, AB., Gaspar, P., Roces, C., Rodríguez, A., Kok, J., Kuipers, OP., Neves, AR. & Martínez, B. (2011). The Lcn972 bacteriocin-encoding plasmid pBL1 impairs cellobiose metabolism in *Lactococcus lactis*. *Applied and Environmental Microbiology,* Vol. 77, No. 21, pp. 7576-85, ISSN 1098-5336

Castro, R., Neves, AR., Fonseca, LL., Pool, WA., Kok, J., Kuipers, OP. & Santos, H. (2009). Characterization of the individual glucose uptake systems of *Lactococcus lactis*: mannose-PTS, cellobiose-PTS and the novel GlcU permease. *Molecular Microbiology,* Vol. 71, No. 3, pp. 795-806, ISSN 1365-2958

Chopin, A., Chopin, MC., Moillo-Batt, A. & Langella, P. (1984). Two plasmid-determined restriction and modification systems in *Streptococcus lactis*. *Plasmid,* Vol. 11, No. 3, pp. 260-3, ISSN 0147-619X

Cocaign-Bousquet, M., Even, S., Lindley, ND. & Loubière, P. (2002). Anaerobic sugar catabolism in *Lactococcus lactis*: genetic regulation and enzyme control over pathway flux. *Applied Microbiology and Biotechnology,* Vol. 60, No. 1-2, pp. 24-32, ISSN 0175-7598

Cords, BR. & McKay, LL. (1974). Characterization of lactose-fermenting revertants from lactose-negative *Streptococcus lactis* C2 mutants. *Journal of Bacteriology,* Vol. 119, No. 3, pp. 830-9, ISSN 0021-9193

Crow, VL. & Thomas, TD. (1984). Properties of a *Streptococcus lactis* strain that ferments lactose slowly. *Journal of Bacteriology,* Vol. 157, No. 1, pp. 28-34, ISSN 0021-9193

Daly, C. (1983). The use of mesophilic cultures in the dairy industry. *Antonie Van Leeuwenhoek,* Vol. 49, No. 3, pp. 297-312, ISSN 0003-6072

David, S., Stevens, H., van Riel, M., Simons, G. & de Vos, WM. (1992). *Leuconostoc lactis* beta-galactosidase is encoded by two overlapping genes. *Journal of Bacteriology,* Vol. 174, No. 13, pp. 4475-81, ISSN 0021-9193

Davidson, BE., Kordias, N., Dobos, M. & Hillier, AJ. (1996). Genomic organization of lactic acid bacteria. *Antonie Van Leeuwenhoek,* Vol. 70, No. 2-4, pp. 161-83, ISSN 0003-6072

De Vos, WM. & Simons, G. (1988). Molecular cloning of lactose genes in dairy lactic streptococci: the phospho-beta-galactosidase and beta-galactosidase genes and their expression products. *Biochimie,* Vol. 70, No. 4, pp. 461-73, ISSN 0300-9084

de Vos, WM. & Vaughan, EE. (1994). Genetics of lactose utilization in lactic acid bacteria. *FEMS Microbiology Reviews,* Vol. 15, No. 2-3, pp. 217-37, ISSN 0168-6445

de Vos, WM. (1996). Metabolic engineering of sugar catabolism in lactic acid bacteria. *Antonie Van Leeuwenhoek,* Vol. 70, No. 2-4, pp. 223-42, ISSN 0003-6072

Fernández, E., Alegría, A., Delgado, S., Martín, MC. & Mayo, B. (2011). Comparative phenotypic and molecular genetic profiling of wild *Lactococcus lactis* subsp. *lactis* strains of the *L. lactis* subsp. *lactis* and *L. lactis* subsp. *cremoris* genotypes, isolated from starter-free cheeses made of raw milk. *Applied and Environmental Microbiology*, Vol. 77, No. 15, pp. 5324-35, ISSN 1098-5336

Franz, CMAP., Cho, G-S., Holzapfel, WH. & Gálvez, A. (2010). Safety of Lactic Acid Bacteria, In: *Biotechnology of Lactic Acid Bacteria: Novel Applications*, Mozzi, F., Raya, RR. & Vignolo, GM., pp. 341-59, Wiley-Blackwell, ISBN 978-0-8138-1583-1, Oxford, UK

Gasson, MJ. (1983). Plasmid complements of *Streptococcus lactis* NCDO 712 and other lactic streptococci after protoplast-induced curing. *Journal of Bacteriology*, Vol. 154, No. 1, pp. 1-9, ISSN 0021-9193

Gasson, MJ. (1990). In vivo genetic systems in lactic acid bacteria. *FEMS Microbiology Reviews*, Vol. 7, No. 1-2, pp. 43-60, ISSN 0168-6445

Gilliland, SE. (1989). Acidophilus milk products: a review of potential benefits to consumers. *Journal of Dairy Science*, Vol. 72, No. 10, pp. 2483-94, ISSN 0022-0302

Godon, JJ., Delorme, C., Ehrlich, SD. & Renault, P. (1992). Divergence of Genomic Sequences between *Lactococcus lactis* subsp. *lactis* and *Lactococcus lactis* subsp. *cremoris*. *Applied and Environmental Microbiology*, Vol. 58, No. 12, pp. 4045-7, ISSN 0099-2240

Górecki, RK., Koryszewska-Bagińska, A., Gołębiewski, M., Żylińska, J., Grynberg, M. & Bardowski, JK. (2011). Adaptative potential of the *Lactococcus lactis* IL594 strain encoded in its 7 plasmids. *Plos One*, Vol. 6, No. 7, pp. e22238, ISSN 1932-6203

Griffin, HG., MacCormick, CA. & Gasson, MJ. (1996). Cloning, DNA sequence, and regulation of expression of a gene encoding beta-galactosidase from *Lactococcus lactis*. *DNA Sequence : the Journal of DNA Sequencing and Mapping*, Vol. 6, No. 6, pp. 337-46, ISSN 1042-5179

Grossiord, BP., Luesink, EJ., Vaughan, EE., Arnaud, A. & de Vos, WM. (2003). Characterization, expression, and mutation of the *Lactococcus lactis galPMKTE* genes, involved in galactose utilization via the Leloir pathway. *Journal of Bacteriology*, Vol. 185, No. 3, pp. 870-8, ISSN 0021-9193

Gunnewijk, MG. & Poolman, B. (2000a). HPr(His approximately P)-mediated phosphorylation differently affects counterflow and proton motive force-driven uptake via the lactose transport protein of *Streptococcus thermophilus*. *The Journal of Biological Chemistry*, Vol. 275, No. 44, pp. 34080-5, ISSN 0021-9258

Gunnewijk, MG. & Poolman, B. (2000b). Phosphorylation state of HPr determines the level of expression and the extent of phosphorylation of the lactose transport protein of *Streptococcus thermophilus*. *The Journal of Biological Chemistry*, Vol. 275, No. 44, pp. 34073-9, ISSN 0021-9258

Gunnewijk, MG., Postma, PW. & Poolman, B. (1999). Phosphorylation and functional properties of the IIA domain of the lactose transport protein of *Streptococcus thermophilus*. *Journal of Bacteriology*, Vol. 181, No. 2, pp. 632-41, ISSN 0021-9193

Itoi, S., Abe, T., Washio, S., Ikuno, E., Kanomata, Y. & Sugita, H. (2008). Isolation of halotolerant *Lactococcus lactis* subsp. *lactis* from intestinal tract of coastal fish. *International Journal of Food Microbiology*, Vol. 121, No. 1, pp. 116-21, ISSN 0168-1605

Itoi, S., Yuasa, K., Washio, S., Abe, T., Ikuno, E. & Sugita, H. (2009). Phenotypic variation in *Lactococcus lactis* subsp. *lactis* isolates derived from intestinal tracts of marine and freshwater fish. *Journal of Applied Microbiology*, Vol. 107, No. 3, pp. 867-74, ISSN 1365-2672

Kelly, W. & Ward, L. (2002). Genotypic vs. phenotypic biodiversity in *Lactococcus lactis*. *Microbiology (reading, England)*, Vol. 148, No. Pt 11, pp. 3332-3, ISSN 1350-0872

Kelly, WJ., Davey GP., & Ward, LJ. (1998). Characterization of lactococci isolated from minimally processed fresh fruit and vegetables. *International Journal of Food Microbiology*, Vol. 45, No. 2, pp. 85-92, ISSN 0168-1605

Kelly, WJ., Ward, LJ. & Leahy, SC. (2010). Chromosomal diversity in *Lactococcus lactis* and the origin of dairy starter cultures. *Genome Biology and Evolution*, Vol. 2, pp. 729-44, ISSN 1759-6653

Klijn, N., Weerkamp, AH. & de Vos, WM. (1995). Detection and characterization of lactose-utilizing Lactococcus spp. in natural ecosystems. *Applied and Environmental Microbiology*, Vol. 61, No. 2, pp. 788-92, ISSN 0099-2240

Kowalczyk, M., Cocaign-Bousquet, M., Loubiere, P. & Bardowski, J. (2008). Identification and functional characterisation of cellobiose and lactose transport systems in *Lactococcus lactis* IL1403. *Archives of Microbiology*, Vol. 189, No. 3, pp. 187-96, ISSN 0302-8933

Kuipers, OP., de Ruyter, PG., Kleerebezem, M. & de Vos WM. (1998). Quorum sensing-controlled gene expression in lactic acid bacteria. *Journal of Biotechnology*, Vol. 64, No. 1, pp. 15-21

Latorre-Guzman, BA., Kado, CI. & Kunkee, R. (1977). *Lactobacillus hordniae*, a new species from the leafhopper (*Hordnia circellata*). *International Journal of Systematic Bacteriology*, Vol. 27, No. 4, pp. 362–70, ISSN 1466-5026

Le Bourgeois, P., Lautier, M., van den Berghe, L., Gasson, MJ. & Ritzenthaler, P. (1995). Physical and genetic map of the *Lactococcus lactis* subsp. *cremoris* MG1363 chromosome: comparison with that of *Lactococcus lactis* subsp. *lactis* IL 1403 reveals a large genome inversion. *Journal of Bacteriology*, Vol. 177, No. 10, pp. 2840-50, ISSN 0021-9193

Linares, DM., Kok, J. & Poolman, B. (2010). Genome sequences of *Lactococcus lactis* MG1363 (revised) and NZ9000 and comparative physiological studies. *Journal of Bacteriology*, Vol. 192, No. 21, pp. 5806-12, ISSN 1098-5530

Lorca, G., Reddy, L., Nguyen, A., Sun, EI., Tseng, J., Yen, M-R. & Saier, MH. (2010). Lactic Acid Bacteria: Comparative Genomic Analyses of Transport Systems, In: *Biotechnology of Lactic Acid Bacteria: Novel Applications*, Mozzi, F., Raya, RR. & Vignolo, GM., pp:73-87. ISBN 978-0-8138-1583-1 Oxford, UK

Luesink, EJ., van Herpen, RE., Grossiord, BP., Kuipers, OP. & de Vos, WM. (1998). Transcriptional activation of the glycolytic las operon and catabolite repression of the *gal* operon in *Lactococcus lactis* are mediated by the catabolite control protein CcpA. *Molecular Microbiology*, Vol. 30, No. 4, pp. 789-98, ISSN 0950-382X

McClelland, M., Sanderson, KE., Spieth, J., Clifton, SW., Latreille, P., Courtney, L., Porwollik, S., Ali, J., Dante, M., Du, F., Hou, S., Layman, D., Leonard, S., Nguyen, C., Scott, K., Holmes, A., Grewal, N., Mulvaney, E., Ryan, E., Sun, H., Florea, L., Miller, W.,

Stoneking, T., Nhan, M., Waterston, R. & Wilson, RK. (2001). Complete genome sequence of *Salmonella enterica* serovar Typhimurium LT2. *Nature*, Vol. 413, No. 6858, pp. 852-6, ISSN 0028-0836

McKay, LL. (1983). Functional properties of plasmids in lactic streptococci. *Antonie Van Leeuwenhoek*, Vol. 49, No. 3, pp. 259-74, ISSN 0003-6072

Mollet, B. & Pilloud, N. (1991). Galactose utilization in *Lactobacillus helveticus*: isolation and characterization of the galactokinase (galK) and galactose-1-phosphate uridyl transferase (galT) genes. *Journal of Bacteriology*, Vol. 173, No. 14, pp. 4464-73, ISSN 0021-9193

Neves, AR., Pool, WA., Kok, J., Kuipers, OP. & Santos, H. (2005). Overview on sugar metabolism and its control in *Lactococcus lactis* - the input from in vivo NMR. *FEMS Microbiology Reviews*, Vol. 29, No. 3, pp. 531-54, ISSN 0168-6445

Neves, AR., Pool, WA., Solopova, A., Kok, J., Santos, H. & Kuipers, OP. (2010). Towards enhanced galactose utilization by *Lactococcus lactis*. *Applied and Environmental Microbiology*, Vol. 76, No. 21, pp. 7048-60, ISSN 1098-5336

Nomura, M., Kimoto, H., Someya, Y. & Suzuki, I. (1999). Novel characteristic for distinguishing *Lactococcus lactis* subsp. *lactis* from subsp. *cremoris*. *International Journal of Systematic Bacteriology*, Vol. 49 Pt 1, pp. 163-6, ISSN 0020-7713

Nomura, M., Kobayashi, M. & Okamoto, T. (2002). Rapid PCR-based method which can determine both phenotype and genotype of *Lactococcus lactis* subspecies. *Applied and Environmental Microbiology*, Vol. 68, No. 5, pp. 2209-13, ISSN 0099-2240

Orla-Jensen, S. (1919). The lactic acid bacteria, pp. 1-196. Host & Son, Copenhagen, Denmark

Pérez, T., Balcázar, JL., Peix, A., Valverde, A., Velázquez, E., de Blas, I. & Ruiz-Zarzuela, I. (2011). *Lactococcus lactis* subsp. *tructae* subsp. nov. isolated from the intestinal mucus of brown trout (*Salmo trutta*) and rainbow trout (*Oncorhynchus mykiss*). *International Journal of Systematic and Evolutionary Microbiology*, Vol. 61, No. Pt 8, pp. 1894-8, ISSN 1466-5034

Pool, WA., Neves, AR., Kok, J., Santos, H. & Kuipers, OP. (2006). Natural sweetening of food products by engineering *Lactococcus lactis* for glucose production. *Metabolic Engineering*, Vol. 8, No. 5, pp. 456-64, ISSN 1096-7176

Poolman, B. (1993). Energy transduction in lactic acid bacteria. *FEMS Microbiology Reviews*, Vol. 12, No. 1-3, pp. 125-47, ISSN 0168-6445

Poolman, B., Royer, TJ., Mainzer, SE. & Schmidt, BF. (1990). Carbohydrate utilization in *Streptococcus thermophilus*: characterization of the genes for aldose 1-epimerase (mutarotase) and UDPglucose 4-epimerase. *Journal of Bacteriology*, Vol. 172, No. 7, pp. 4037-47, ISSN 0021-9193

Rademaker, JL., Herbet, H., Starrenburg, MJ., Naser, SM., Gevers, D., Kelly, WJ., Hugenholtz, J., Swings, J. & van Hylckama Vlieg, JE. (2007). Diversity analysis of dairy and nondairy *Lactococcus lactis* isolates, using a novel multilocus sequence analysis scheme and (GTG)5-PCR fingerprinting. *Applied and Environmental Microbiology*, Vol. 73, No. 22, pp. 7128-37, ISSN 0099-2240

Rutberg, B. (1997). Antitermination of transcription of catabolic operons. *Molecular Microbiology*, Vol. 23, No. 3, pp. 413-21, ISSN 0950-382X

Saier, MH. (2000). Families of transmembrane sugar transport proteins. *Molecular Microbiology*, Vol. 35, No. 4, pp. 699-710, ISSN 0950-382X

Salama, M., Sandine, W. & Giovannoni, S. (1991). Development and application of oligonucleotide probes for identification of *Lactococcus lactis* subsp. *cremoris*. *Applied and Environmental Microbiology*, Vol. 57, No. 5, pp. 1313-8, ISSN 0099-2240

Sandine, WE. (1988). New nomenclature of the non-rod-shaped lactic acid bacteria. *Biochimie*, Vol. 70, No. 4, pp. 519-21, ISSN 0300-9084

Schleifer, KH., Kraus, J., Dvorak, C., Kilpper-Bälz, R., Collins MD. & Fischer, W. (1985). Transfer of *Streptococcus lactis* and related streptococci to the genus *Lactococcus* gen. nov. *Systematic and Applied Microbiology*, Vol. 6, No. 2, pp. 183-95, ISSN 07232020

Schultz, JE. & Breznak, JA. (1978). Heterotrophic bacteria present in hindguts of wood-eating termites [*Reticulitermes flavipes* (Kollar)]. *Applied and Environmental Microbiology*, Vol. 35, No. 5, pp. 930-6, ISSN 0099-2240

Siezen, RJ., Bayjanov, J., Renckens, B., Wels, M., van Hijum, SA., Molenaar, D. & van Hylckama Vlieg, JE. (2010). Complete genome sequence of *Lactococcus lactis* subsp. *lactis* KF147, a plant-associated lactic acid bacterium. *Journal of Bacteriology*, Vol. 192, No. 10, pp. 2649-50, ISSN 1098-5530

Siezen, RJ., Starrenburg, MJ., Boekhorst, J., Renckens, B., Molenaar, D. & van Hylckama Vlieg, JE. (2008). Genome-scale genotype-phenotype matching of two *Lactococcus lactis* isolates from plants identifies mechanisms of adaptation to the plant niche. *Applied and Environmental Microbiology*, Vol. 74, No. 2, pp. 424-36, ISSN 1098-5336

Simons, G., Nijhuis, M. & de Vos, WM. (1993). Integration and gene replacement in the *Lactococcus lactis lac* operon: induction of a cryptic phospho-beta-glucosidase in LacG-deficient strains. *Journal of Bacteriology*, Vol. 175, No. 16, pp. 5168-75, ISSN 0021-9193

Smit, G., Smit, BA. & Engels, WJ. (2005). Flavour formation by lactic acid bacteria and biochemical flavour profiling of cheese products. *FEMS Microbiology Reviews*, Vol. 29, No. 3, pp. 591-610, ISSN 0168-6445

Solem, C., Koebmann, B. & Jensen, PR. (2008). The extent of co-metabolism of glucose and galactose by *Lactococcus lactis* changes with the expression of the *lacSZ* operon from *Streptococcus thermophilus*. *Biotechnology and Applied Biochemistry*, Vol. 50, No. Pt 1, pp. 35-40, ISSN 1470-8744

Sørensen, KI. & Hove-Jensen, B. (1996). Ribose catabolism of *Escherichia coli*: characterization of the *rpiB* gene encoding ribose phosphate isomerase B and of the *rpiR* gene, which is involved in regulation of *rpiB* expression. *Journal of Bacteriology*, Vol. 178, No. 4, pp. 1003-11, ISSN 0021-9193

Stiles, ME. & Holzapfel, WH. (1997). Lactic acid bacteria of foods and their current taxonomy. *International Journal of Food Microbiology*, Vol. 36, No. 1, pp. 1-29, ISSN 0168-1605

Tailliez, P., Tremblay, J., Ehrlich, SD. & Chopin, A. (1998). Molecular diversity and relationship within *Lactococcus lactis*, as revealed by randomly amplified polymorphic

DNA (RAPD). *Systematic and Applied Microbiology*, Vol. 21, No. 4, pp. 530-8, ISSN 0723-2020

Tanigawa, K., Kawabata, H. & Watanabe, K. (2010). Identification and typing of *Lactococcus lactis* by matrix-assisted laser desorption ionization-time of flight mass spectrometry. *Applied and Environmental Microbiology*, Vol. 76, No. 12, pp. 4055-62, ISSN 1098-5336

Teeri, TT. (1997) Crystalline cellulose degradation: new insight into the function of cellobiohydrolases. *Trends in Biotechnology*, Vol. 15, No. 5, pp. 160–67, ISSN 0167-7799

Tobisch, S., Glaser, P., Krüger, S. & Hecker, M. (1997). Identification and characterization of a new beta-glucoside utilization system in *Bacillus subtilis*. *Journal of Bacteriology*, Vol. 179, No. 2, pp. 496-506, ISSN 0021-9193

van Hylckama Vlieg, JE., Rademaker, JL., Bachmann, H., Molenaar, D., Kelly, WJ. & Siezen, RJ. (2006). Natural diversity and adaptive responses of *Lactococcus lactis*. *Current Opinion in Biotechnology*, Vol. 17, No. 2, pp. 183–190, ISSN 0958-1669

van Rooijen, RJ., van Schalkwijk, S. & de Vos, WM. (1991). Molecular cloning, characterization, and nucleotide sequence of the tagatose 6-phosphate pathway gene cluster of the lactose operon of *Lactococcus lactis*. *The Journal of Biological Chemistry*, Vol. 266, No. 11, pp. 7176-81, ISSN 0021-9258

Vaughan, EE., David, S. & de Vos, WM. (1996). The lactose transporter in *Leuconostoc lactis* is a new member of the LacS subfamily of galactoside-pentose-hexuronide translocators. *Applied and Environmental Microbiology*, Vol. 62, No. 5, pp. 1574-82, ISSN 0099-2240

Vaughan, EE., Pridmore, RD. & Mollet, B. (1998). Transcriptional regulation and evolution of lactose genes in the galactose-lactose operon of *Lactococcus lactis* NCDO2054. *Journal of Bacteriology*, Vol. 180, No. 18, pp. 4893-902, ISSN 0021-9193

Vaughan, EE., van den Bogaard, PT., Catzeddu, P., Kuipers, OP. & de Vos, WM. (2001). Activation of silent *gal* genes in the *lac-gal* regulon of *Streptococcus thermophilus*. *Journal of Bacteriology*, Vol. 183, No. 4, pp. 1184-94, ISSN 0021-9193

Wegmann, U., O'Connell-Motherway, M., Zomer, A., Buist, G., Shearman, C., Canchaya, C., Ventura, M., Goesmann, A., Gasson, MJ., Kuipers, OP., van Sinderen, D. & Kok, J. (2007). Complete genome sequence of the prototype lactic acid bacterium *Lactococcus lactis* subsp. *cremoris* MG1363. *Journal of Bacteriology*, Vol. 189, No. 8, pp. 3256-70, ISSN 0021-9193

Weickert, MJ. & Adhya, S. (1992). A family of bacterial regulators homologous to Gal and Lac repressors. *The Journal of Biological Chemistry*, Vol. 267, No. 22, pp. 15869-74, ISSN 0021-9258

Weickert, MJ. & Chambliss, GH. (1990). Site-directed mutagenesis of a catabolite repression operator sequence in *Bacillus subtilis*. *Proceedings of the National Academy of Sciences of the United States of America*, Vol. 87, No. 16, pp. 6238-42, ISSN 0027-8424

Williams, SG., Greenwood, JA. & Jones, CW. (1992). Molecular analysis of the *lac* operon encoding the binding-protein-dependent lactose transport system and beta-galactosidase in *Agrobacterium radiobacter*. *Molecular Microbiology*, Vol. 6, No. 13, pp. 1755-68, ISSN 0950-382X

Zomer, AL., Buist, G., Larsen, R., Kok, J. & Kuipers, OP. (2007). Time-resolved determination of the CcpA regulon of *Lactococcus lactis* subsp. *cremoris* MG1363. *Journal of Bacteriology*, Vol. 189, No. 4, pp. 1366-81, ISSN 0021-9193

Chapter 21

# Lactic Acid Bacteria in Hydrogen-Producing Consortia: On Purpose or by Coincidence?

Anna Sikora, Mieczysław Błaszczyk,
Marcin Jurkowski and Urszula Zielenkiewicz

Additional information is available at the end of the chapter

http://dx.doi.org/10.5772/50364

## 1. Introduction

Hydrogen is both a valuable energy carrier and a feedstock for various branches of the chemical industry. It is thought to be one of the most important energy carriers of the future, an alternative to conventional fossil fuels. Water vapor and heat energy are the sole products of hydrogen burning. Therefore, the use of hydrogen to generate energy does not contribute to ozone depletion, the greenhouse effect, climate changes or acid rains. Hydrogen is a highly efficient energy source; its specific energy equals 33 Wh/g, which is the highest among all fuels. For comparison, the specific energy of methane is 14.2 Wh/g and coal, 9.1 Wh/g. Hydrogen can be used as a fuel in hydrogen fuel cells or burn directly in internal combustion engines. In the chemical industry, hydrogen is used for syntheses of ammonia, alcohols, aldehydes, hydrogen chloride and for the hydrogenation of edible oils, heavy oils or ammonia, for removal of oxygen traces in prevention against metal oxidation and corrosion processes (Nath & Das, 2003; Logan, 2004; Antoni et al., 2007; Piela & Zelenay, 2004).

Conventional methods of hydrogen production, such as gasification of coal, steam reforming of natural gas and petroleum, and electrolysis of water, are based on fossil fuels. Therefore, these methods are regarded as energy expensive and cause environmental pollution (Nath & Das, 2003; Logan, 2004; Nath & Das, 2004).

Considering the limited reserves of fossil fuels, environmental pollution and global warming, there is great interest in biological methods of producing fuels, such as bio-hydrogen, biogas (methane), ethanol or diesel. Among the known biological processes leading to hydrogen production are dark fermentation, photofermentation, direct and indirect biophotolysis, as well as anaerobic respiration of sulphate-reducing bacteria under conditions of sulphate depletion. Taking under account potential applications, microbial hydrogen production has been focused on: (i) photolysis of water using algae and *Cyanobacteria*, (ii) photofermentation of organic compounds by photosynthetic bacteria, and (iii) dark fermentation of organic compounds using anaerobic bacteria.

Members of the *Clostridiales* and *Enterobacteriaceae* are well-recognized hydrogen-producers during the process of dark fermentation. For future applications, dark fermentation seems to be the most promising concept. However, low hydrogen yields and generation of large quantities of non-gaseous organic products remain key problems of dark fermentation. The theoretical maximum hydrogen yield during dark fermentation is 4 moles of $H_2$/mole of glucose (~33% substrate conversion), but the actual yield is only 2 moles of $H_2$/mole of glucose (~17% conversion). Currently, many investigations are focused on improving the hydrogen yield during fermentation as an alternative method of hydrogen production and combining dark fermentation with other processes, like methanogenesis, photofermentation or microbial electrolysis of cells, to achieve more effective substrate utilization (Li & Fang, 2007; Das & Veziroglu, 2008; Hallenbeck & Ghosh, 2009; Lee et al., 2010; Hallenbeck, 2011). Biohydrogen fermentations may be carried out in different batch types, continuous or semi-continuous bioreactors, where mixed microbial consortia develop. In the most effective systems, consortia are selected for growth and dominance under non-sterile conditions and usually show high stability and resistance to transient unfavorable changes in the bioreactor environment. Depending on the bioreactor type and growth conditions, consortia form various structures which ensure retention and accumulation of the active biomass. These include microbial-based biofilms and macroscopic aggregates of microbial cells, such as flocs and granules (Campos et al., 2009; Hallenbeck & Ghosh, 2009). A good understanding of the structure of hydrogen-producing microbial communities, symbiotic relationships within the consortia as well as factors favoring hydrogen production is vital for optimizing the process.

Interestingly, lactic acid bacteria (LAB) are often detected in mesophilic hydrogen-producing consortia as bacteria that accompany hydrogen producers. In this chapter, we discuss the issue of whether LAB are bad or good (positive or negative) components of hydrogen-producing consortia. We present different opinions about the potential significance and the role of LAB in hydrogen-producing communities.

## 2. Hydrogen-producing bacteria

Fermentation is an anaerobic type of metabolic process of low energy gain in which organic compounds are degraded in the absence of external electron acceptors and a mixture of oxidized and reduced products are formed. Products, namely organic compounds and gasses (hydrogen and carbon dioxide), determine the type of fermentation. Main hydrogen yielding fermentations are butyric acid fermentation (saccharolytic clostridial-type fermentation) and mixed-acid fermentation (enterobacterial-type fermentation). The first step of both fermentations is the Embden-Meyerhof pathway or glycolysis in which glucose is converted into pyruvate and NADH is formed.

In the clostridial-type fermentation pyruvate is oxidized to acetyl-CoA by pyruvate:ferredoxin oxidoreductase (PFOR) in the presence of ferredoxin (Fd) (See Equation 1).

$$\text{Pyruvate} + \text{CoA} + \text{Fd} \xrightarrow{(\text{PFOR})} \text{acetyl}-\text{CoA} + \text{FdH} + \text{CO}_2 \tag{1}$$

Reduced ferredoxin is also formed in the reaction with NADH catalyzed by NADH:ferredoxin oxidoreductase (NFOR) (See Equation 2).

$$\text{NADH} + \text{Fd} \xrightarrow{(\text{NFOR})} \text{NAD}^+ + \text{FdH} \tag{2}$$

Hydrogen is released by hydrogenases that catalyze proton reduction using electrons from ferredoxin. The activity of PFOR and NFOR enzymes is thermodynamically regulated by the hydrogen concentration. Partial hydrogen pressure >60 Pa inhibits the NFOR activity and favors formation of non-gaseous end-products from acetyl-CoA including acetate, butyrate, ethanol, butanol and lactate. PFOR is active at hydrogen concentrations up to $3\times10^4$ Pa (Angenent et al., 2004; Girbal et al., 1995; Hallenbeck, 2005; Kraemer & Bagley, 2007; Lee et al., 2011).

The theoretical maximum hydrogen yield during clostridial-type fermentation is 4 moles of hydrogen per mole of glucose, when all of the substrate is converted to acetic acid (See Equation 3).

$$C_6H_{12}O_6 + 2\,H_2O \rightarrow 4\,H_2 + 2\,CO_2 + 2\,CH_3COOH \tag{3}$$

This gives the maximal possible level of hydrogen yield during dark fermentation. When the glucose is converted to butyrate the hydrogen yield drops to 2 moles (See Equation 4).

$$C_6H_{12}O_6 + 2\,H_2O \rightarrow 2\,H_2 + 2\,CO_2 + CH_3CH_2CH_2COOH \tag{4}$$

Formation of other non-gaseous end products of fermentation causes further decrease in hydrogen yields. The scheme of the clostridial-type fermentation is presented in Figure 1 (Papoutsakis, 1984; Saint-Amans et al., 2001).

The described type of fermentation is the most characteristic for spore-forming representatives of the *Clostridium* as well as *Bacillus* genera and others, such as the rumen bacteria e.g. *Ruminococcus albus*. Among the fermentative anaerobes, clostridia have been well known and extensively studied for their capability to produce hydrogen from various carbohydrates (Kalia & Purohit, 2008; Lee et al., 2011). The hydrogen yields of pure *Clostridium* cultures, including *C. acetobutylicum, C. bifermentans, C. butyricum, C. kluyveri, C. lentocellum, C. paraputrificum, C. pasteurianum, C. saccharoperbutylacetonicum, C. thermosuccinogenes,* and *C. thermolacticum* were examined. The optimum hydrogen yields observed for these bacteria varied between 1.1 moles of $H_2$/mole of hexose and 2.6 moles of $H_2$/mole of -hexose, dependent on the organism per se as well as environmental conditions (for review see Lee et al., 2011).

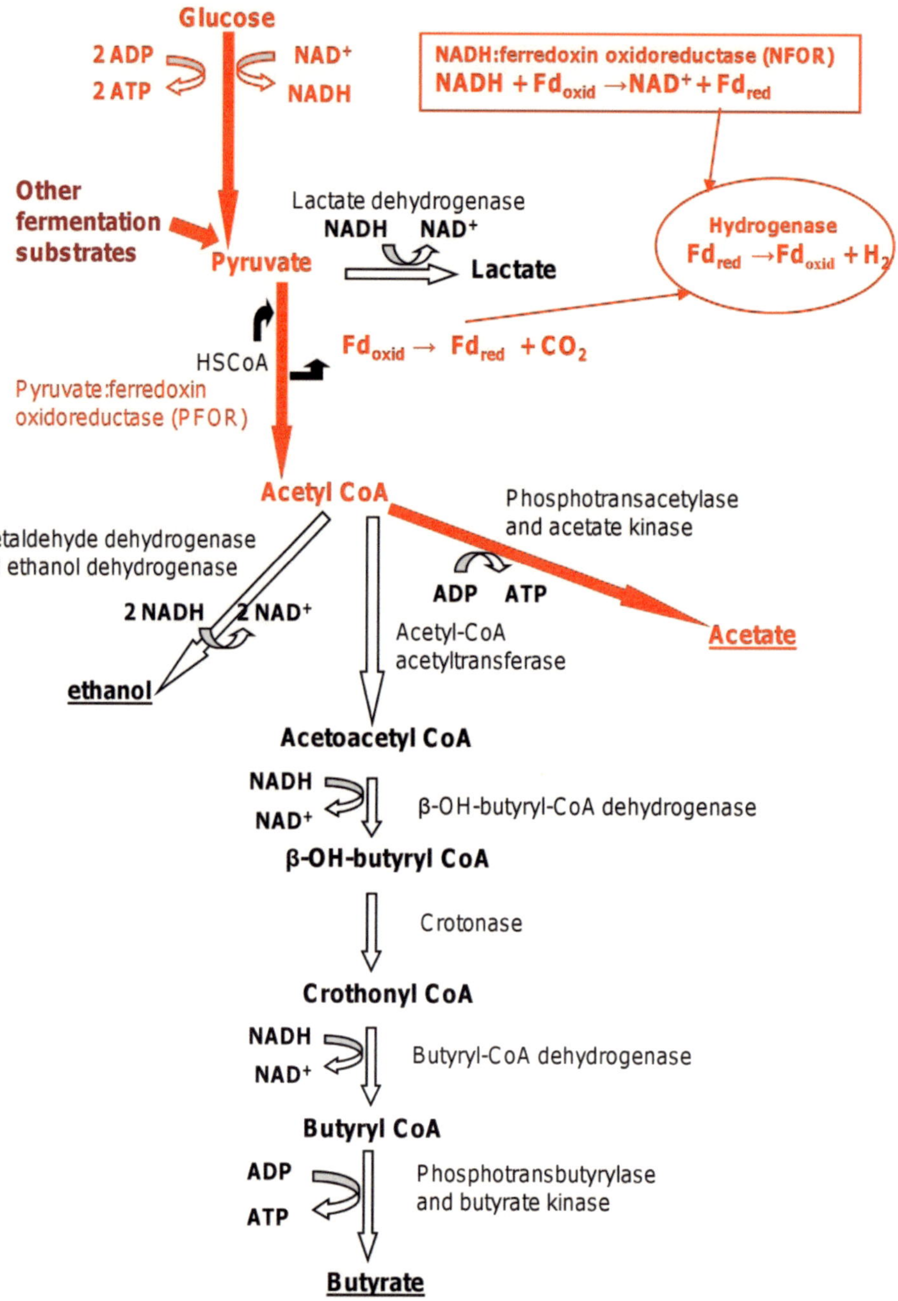


**Figure 1.** The scheme of clostridial-type fermentation. The pathway leading to the theoretical maximum hydrogen yield of 4 moles of hydrogen per 1 mole of glucose, when all of the substrate is converted to acetic acid is labeled in red.

In the mixed acid-fermentation (also known as formic acid fermentation) pyruvate formate-lyase (PFL) converts pyruvate to acetyl-CoA and formic acid (See Equation 5).

$$\text{Pyruvate} + \text{CoA} \xrightarrow{\text{PFL}} \text{acetyl} - \text{CoA} + \text{formic acid} \tag{5}$$

The formic acid can be degraded into hydrogen and carbon dioxide by formate hydrogen-lyase (FHL) (See Equation 6).

$$\text{Formic acid}\left(\text{HCOOH}\right) \xrightarrow{\text{FHL}} H_2 + CO_2 \tag{6}$$

There are two types of mixed-acid fermentations. In the first type ethanol and a complex mixture of acids, particularly acetic, lactic, succinic and formic acids are produced. This pattern is seen in *Escherichia, Salmonella, Proteus* and other genera. The second type is characteristic for *Enterobacter, Serratia, Erwinia* and some species of *Bacillus*. In this type of fermentation, acetoin, 2,3-butanediol, ethanol and lower amount of acids are formed.

The theoretical hydrogen yields during mixed acid fermentation are lower than those described for the clostridial-type fermentation. Hydrogen yields of *Escherichia* spp., as obtained for the pure culture of *E. coli* NCIMB 11943, are in a range of 0.2–1.8 moles of $H_2$/mole of hexose, when glucose or starch hydrolysate are substrates, whereas hydrogen yields determined for pure *Enterobacter* spp. cultures are much higher, ranging from 1.1 moles of $H_2$/mole of hexose to ca. 3.0 moles of $H_2$/mole of hexose (Lee et al., 2011). It is known that in the *Enterobacter*-type fermentation hydrogen is also generated through oxidation of NADH by NFOR in reactions similar to those described for the clostridial-type fermentation (Nakashimada et al., 2002; Sawers, 2005; Maeda et al., 2007).

The pathway of the mixed-acid fermentation is presented in Figure 2.

## 3. Lactic acid bacteria in hydrogen-producing consortia

### 3.1. Lactic acid bacteria – General information

Lactic acid bacteria are Gram-positive bacteria, producing lactic acid as the main product of carbohydrate fermentation. Two types of lactic acid fermentation are distinguished: homolactic and heterolactic fermentation. In homolactic acid fermentation, two molecules of pyruvate that are formed during glycolysis are converted to lactate. In heterolactic acid fermentation, one molecule of pyruvate is converted to lactate; the other is converted to ethanol and carbon dioxide.

At present, nearly 400 LAB species have been recognized. They include bacteria belonging to the order *Lactobacillales* classified into seven families: *Lactobacillaceae* (genera: *Lactobacillus* and *Pediococcus*); *Aerococcaceae* (genus *Aerococcus*); *Carnobacteriaceae* (genera: *Alloiococcus, Carnobacterium, Dolosigranulum, Granulicatella* and *Lactosphaera*); *Enterococcaceae* (genera: *Enterococcus, Tetragenococcus* and *Vagococcus*); *Leuconostocaceae* (genera: *Leuconostoc, Oenococcus* and *Weisella*); *Streptococcaceae* (genera: *Streptococcus, Lactococcus* and *Melissococcus*); *Microbacteriaceae* (genus *Microbacterium*). Extremely varied among lactic acid bacteria is genus *Lactobacillus* which comprises over 145 species. Genera *Bifidobacterium* and *Propionibacterium* (class: *Actinobacteria*) as well as spore forming rods belonging to the order *Bacillales*, family *Sporolactobacillaceae*, genus *Sporolactobacillus* constitute further groups of LAB. With the exception of bacteria belonging to the genera *Lactobacillus*, *Carnabacterium*, *Weissella* and *Sporolactobacillus* which are rods, other species of lactic acid bacteria are cocci (de Vos et al., 2009).

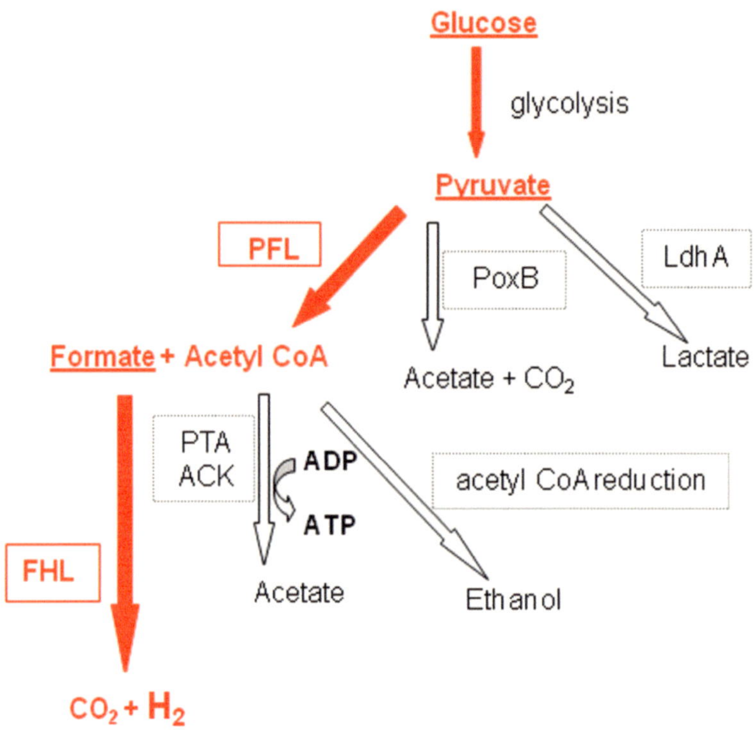


LdhA – lactate dehydrogenase, PoxB – pyruvate oxydase, PTA – phosphotransacetylase, ACK – acetate kinase, PFL – pyruvate formate lyase, FHL – formate hydrogen lyase.

**Figure 2.** The scheme of mixed-acid fermentation (*Escherichia coli*-type). The pathway leading to hydrogen production is shown in red.

LAB are microorganisms ubiquitous in the environment. Due to their high nutritional requirements, they are usually found in environments rich in carbohydrates, amino acids and nucleotides. On the other hand, they show considerable adaptation to the harsh conditions, which allows them to inhabit a range of various niches (Korhonen, 2010).

The digestive tracts of man and animals are among the environments where LAB occur. They have been reported in saliva, the small intestine and colon (Korhonen, 2010). The development of the gastrointestinal microflora in infants is influenced by contact with diverse microflora of the mother and of the closest surrounding. The main species found in both infants and adults are *Lactobacillus ruminis*, *L. salivarius*, *L. gasseri*, *L. reuteri* as well as *Bifidobacterium longum* and *B. breve* (Salminen et al., 2005; Ishibashi et al., 1997). The diversity of lactic acid bacteria colonizing the human digestive system is high; however, the species composition is constantly changing as most of the species colonize the gastrointestinal tract for only a short period (Korhonen, 2010). Microorganisms in the adult intestine outnumber by 10-fold cells constituting the human body. The microbial composition for each individual is unique, depending on age, diet, diseases and environmental factors (Qin J. et al., 2010). LAB have been widely used as probiotic bacteria in the human gastrointestinal tract, contributing to pathogen inhibition and immunomodulation (Zhang et al., 2011).

The natural occurrence of lactic acid bacteria on plants (fruits, vegetables and grains) as well as in milk permitted their use in biotechnology (Makarova et al., 2006). *Lactobacillus, Pediococcus, Leuconostoc* and *Oenococcus* which reside on grapes, enable fruit fermentation and wine production (de Nadra, 2007). Also, LAB can occur naturally or be intentionally added as starter cultures during plant, meat and dairy fermentation (Korhonen, 2010).

In marine environments LAB play a role in the breakdown of organic matter. In the last decade LAB belonging to the following genera: *Amphibacillus, Alkalibacterium, Marinilactibacillus, Paraliobacillus, Halolactibacillus* were isolated from the samples taken from the sea and oceanic as well as from animals that inhabit these ecosystems. These bacteria were named "marine LAB" (Ishikawa et al., 2005).

### 3.2. Lactic acid bacteria – Influence on hydrogen producers

Interestingly, lactic acid bacteria are often detected in mesophilic hydrogen-producing consortia as bacteria that accompany hydrogen producers. The technique most commonly used for analyzing the diversity of hydrogen-producing microbial communities is polymerase chain reaction-denaturing gradient gel electrophoresis (PCR-DGGE), followed by either direct sequencing or cloning and sequencing of DGGE bands. One of the disadvantages of this method is underestimation of the true bacterial diversity due to the fact that only the most prominent DGGE bands are analyzed. Various studies have shown that DGGE bands representing LAB are one of the most dominant bands (Fang et al., 2002; Kim et al., 2006; Li et al., 2006; Wu et al., 2006; Hung et al., 2007; Ren et al., 2007; Jo et al., 2007; Lo et al., 2008; Sreela-or et al., 2011). Another method of analyzing the biodiversity of hydrogen-producing consortia is cloning and sequencing of the 16S rDNA gene amplified on the total DNA isolated from the culture probes. Also with this method, sequences related to lactic acid bacteria have been detected (Yang et al., 2007). An alternative method used by our group for the first time to perform metagenomic analysis of hydrogen-producing microbial communities is 454-pyrosequencing. Our results showed that *Clostridiaceae, Enterobacteriaceae* and heterolactic fermentation bacteria, mainly *Leuconostocaeae,* were the most dominant bacteria in hydrogen-producing consortia under optimal condition for gas production (Chojnacka et al., 2011).

The aim of the chapter is a provocative discussion on the true role of LAB in hydrogen-producing bioreactors and their influence on hydrogen producers. Table 1 presents a set of selected studies which examine the possible influence of lactic acid bacteria on hydrogen production during dark fermentation.

| **Subject of examination** | **Results and suggested influence of LAB on hydrogen producers** | **References** |
|---|---|---|
| **A.** | **Negative role of LAB** | |
| Investigation of the effects of LAB on hydrogen fermentation of bean curd manufacturing waste in a series of co-cultures of *Clostridium butyricum* and two strains of *C. acetobutylicum* with *Lactobacillus paracasei* and *Enterococcus durans*. | Inhibition of hydrogen producers by LAB due to (i) substrate competition (replacement of hydrogen fermentation by lactic acid fermentation); (ii) excretion of bacteriocins. | Noike et al., 2002 |
| Fermentative hydrogen production from molasses in continuous stirred-tank reactors and DG-DGGE (double gradient denaturating gradient gel electrophoresis) analysis of bacterial community structure. | *C. pasteurianum, Lactococcus* sp., *Desulfovibrio ferrireducens, Actinomyces* sp., *Klebsiella oxytoca, Acidovorax* sp., uncultured *Actinobacterium* and *Bacteroidetes* were detected in the bioreactor where the main non-gaseous end products were ethanol, butyric acid and acetic acid. Negative role of *Lactococcus* species: inhibition of hydrogen production by substrate competition (competitive ethanol production). | Ren et al., 2007 |
| DGGE examination of microbial community during unstable hydrogen production from food waste of kimchi in a continuous culture. | Conversion of hydrogen fermentation to lactic acid fermentation due to shifts in the microbial community structure from *Clostridium* spp. to *Lactobacillus* spp. Negative role of LAB: substrate competition. | Jo et al., 2007 |
| Investigation of hydrogen production from food waste in batch fermentation by anaerobic mixed cultures and DGGE analysis of microbial community. | *Clostridium* species (*C. butyricum, C. acetobutylicum, C. beijerinckii, Clostridium* sp.) were the dominant hydrogen producers. Negative role of LAB representatives (*Lactobacillus* sp., *Enterococcus* sp.): inhibition of hydrogen production by substrate competition (competitive ethanol and lactic acid production). | Sreela-or et al., 2011 |

| Subject of examination | Results and suggested influence of LAB on hydrogen producers | References |
|---|---|---|
| **B** | **Role of LAB in hydrogen-producing consortia not discussed** | |
| Fermentative hydrogen production from sucrose-containing wastewater in a well-mixed reactor and DGGE analysis of bacterial community structure of the granular sludge. | *Clostridium* species (*C. pasteurianum, C. tyrobutyricum, C. acidisoli*) and *Sporolactobacillus racemicus* were detected in the bioreactor. A high-rate fermentative hydrogen production was observed. The role of LAB (*Sporolactobacillus racemicus*) in the microbial community is not discussed. | Fang et al., 2002 |
| Fermentative hydrogen production from sucrose in a continuously stirred anaerobic bioreactor seeded with silicone-immobilized sludge and DGGE analysis of bacterial community structure of the granular sludge. | DGGE analysis revealed the presence of representatives of the following genera and species: *Clostridium* (*C. intestinale* and *C. pasteurianum*), *Escherichia coli, Streptococcus* sp., *Klebsiella pneumoniae.* A high-rate fermentative hydrogen production was observed. The role of LAB (*Streptococcus* sp.) in the microbial community is not discussed. | Wu et al., 2006 |
| Fermentative hydrogen production from sucrose in a continuous stirred tank reactor and DGGE analysis of bacterial community structure of the granular sludge. | *Clostridium cellulosi, Clostridium* sp., *Klebsiella ornithinolytica, Prevotella* sp. and *Leuconostoc pseudomesenteroides* were detected in the bioreactor. A high-rate fermentative hydrogen production was observed. The role of LAB (*L. pseudomesenteroides*) in the microbial hydrogen-producing community is not discussed. | Li et al., 2006 |
| Fermentative hydrogen production from sucrose or xylose in a continuous dark fermentation bioreactor and DGGE analysis of the bacterial community structure. | *Clostridium* species (*C. butyricum, C. pasteurianum* on sucrose and *C. celerecrescens* on xylose), *Klebsiella pneumoniae, K. oxytoca, Streptococcus* sp., *Escherichia* sp., *Pseudomonas* sp. *Dialister* sp., *Bacillus* sp., *Bifidobacterium* sp. were detected in the bioreactor. The role of LAB (*Streptococcus* sp. and *Bifidobacterium* sp.) in the microbial community is not discussed. | Lo et al., 2008 |

| **Subject of examination** | **Results and suggested influence of LAB on hydrogen producers** | **References** |
|---|---|---|
| C. | **Positive role of LAB** | |
| Fermentative hydrogen production from glucose in anaerobic agitated granular sludge bed bioreactors and DGGE and FISH analyses of the granular sludge. | The DGGE analysis showed that the bacterial community was mainly composed of *Clostridium* sp., *Klebsiella oxytoca* and *Streptococcus* sp. A high-rate fermentative hydrogen production was observed. The FISH images suggested that *Streptococcus* cells acted as seeds for granule formation. | Hung et al., 2007 |
| Fermentative hydrogen production from cheese whey wastewater by mixed continuous cultures and molecular analysis of the consortium by cloning and sequencing of the 16S rDNA gene amplified on the total DNA isolated from the culture probe. | The most prevalent bacteria, representing approximately 50% of the total sequences analyzed, were representatives of the genus *Lactobacillus*. Remaining sequences belonged to the genera *Olsenella, Clostridium* and *Prevotella*. Decrease in hydrogen production was accompanied by the reductions in the number of detected bacteria from the genus *Lactobacillus*. Authors declare isolation of *Lactobacillus* bacteria capable of hydrogen production in the process of lactose fermentation. | Yang et al., 2007 |
| Fermentative hydrogen production from molasses in packed bed bioreactors and metagenomic analysis of bacterial biofilms and granules by 454-pyrosequencing. | Metagenomic analysis of microbial consortia by 454-pyrosequencing of amplified 16S rDNA fragments revealed that the most dominant bacteria were the representatives of the *Firmicutes* (*Clostridiaceae* and *Leuconostocaeae*) and *Gammaproteobacteria* (*Enterobacteriaceae*). Bacteria of heterolactic fermentation were one of the predominant microbes in hydrogen-producing consortia. The speculation that LAB may favor hydrogen production is discussed. For details see Tables 2-4, Figures 3-5 and description in the text. | Chojnacka et al., 2011 |

**Table 1.** A set of selected studies demonstrating the contribution of LAB in hydrogen-producing cultures and presenting their possible influence on hydrogen production.

Some studies argue that development of LAB in bioreactors may inhibit hydrogen production (Table 1, part A). Cessation of hydrogen generation by LAB was suggested to be due to (i) substrate competition and/or (ii) excretion of bacteriocins inhibiting growth of other bacteria. These observations derive from examinations of both batch (Sreela-or et al., 2011) and continuous (Ren et al., 2007; Jo et al., 2007) mixed cultures as well as co-cultures where one component was a representative of clostridia and the second one of lactic acid bacteria (Noike et al., 2002). Heat treatment was proposed as a method of eliminating lactic acid bacteria (Noike et al., 2002; Baghchehsaraee et al., 2008).

Substrate competition includes changes in the type of fermentation occurring in the bioreactors during long-term continuous processes and replacement of hydrogen fermentation by lactic acid or ethanol fermentation (Noike et al., 2002; Jo et al., 2007; Ren et al., 2007; Sreela-or et al., 2011). In all of the studies decrease in hydrogen production was observed with simultaneous increase of lactic acid and ethanol concentrations in the effluents or fluid phase of the culture.

The hypothesis that bacteriocins may act as inhibitors of hydrogen production was postulated by Noike and co-workers (2002), who showed in a series of co-cultures experiments that cessation of hydrogen production by *C. acetobutylicum* and *C. butyricum* was caused by both the presence of *Enterococcus durans* and *Lactobacillus paracasei* as well as supernatants from their culture media. Moreover, treatment of the supernatants with trypsin recovered normal hydrogen production by selected clostridial strains.

Studies listed in part B of Table 1 determined the presence of lactic acid bacteria in hydrogen-producing consortia; yet, their role in these microbial communities is not discussed. It is noteworthy that (i) those papers discuss efficient systems of biohydrogen production and (ii) studies were performed under optimal conditions for hydrogen production (Fang et al., 2002; Kim et al., 2006; Li et al., 2006; Wu et al., 2006; Hung et al., 2007).

Part C of Table 1 presents the only so far available studies arguing that LAB could play a positive role in hydrogen-producing microbial communities and stimulate hydrogen production.

Hung and colleagues (2007) studied the efficiency of fermentative hydrogen production from glucose in anaerobic agitated granular sludge bed bioreactors under different substrate concentration and hydraulic retention times (HRT). PCR-DGGE and FISH methods were used to analyze the biohydrogen-producing microbial community of the granular sludge. The bacterial community was composed of *Clostridium* sp. (possibly *C. pasteurianum*), *Klebsiella oxytoca* and *Streptococcus* sp. The percentage of *Streptococcus* sp. contributing to the microbial community was dependent on the HRT. The shorter HRT, meaning the faster the flow of the medium and increased dilution rate, the higher the contribution of *Streptococcus* sp. in the bacterial consortium was observed. Formation of granular sludge enables biomass retention. FISH analysis revealed that *Streptococcus* cells are located inside granules surrounded by *Clostridium* cells. Authors postulate that *Streptococcus* cells may act as the seed for sludge granule formation.

According to Yang et al. (2007) some LAB are able to produce hydrogen. They declare isolation of strains from the genus *Lactobacillus* capable of hydrogen production during lactose fermentation.

### 3.3. Fermentative hydrogen production and microbial analysis of bacterial biofilms and granular sludge formed in packed bed bioreactors

We developed an effective system of bacterial hydrogen production based on long-term continuous cultures (from an inoculum of a lake bottom sediment) grown on sugar beet molasses in packed bed reactors filled with granitic stones (Chojnacka et al., 2011). In separate cultures, two consortia of anaerobic fermentative bacteria producing hydrogen-rich gas developed on the stones as biofilms. Furthermore, in one of the cultures a granular sludge was also observed (Figures 3 and 4). Cultures were named, respectively, (i) the culture with stone biofilm only and (ii) granular sludge culture. Both cultures were regularly renewed by removal of an excess of biomass.

Analysis of the surface topography of biofilms from both cultures revealed their porous, irregular structure with many cavities and channels. Bacteria appeared to be suspended in and surrounded by a matrix substance. The granules were white and light cream in color, with a diameter between 0.2 – 2 mm, and of hard structure, resistant to squashing or crumbling. Moreover, the granules were clustered in structures resembling bunches of grapes with a noticeable net of channels. Similar to the bacterial biofilm, the granules consisted of bacterial cells surrounded by a matrix.

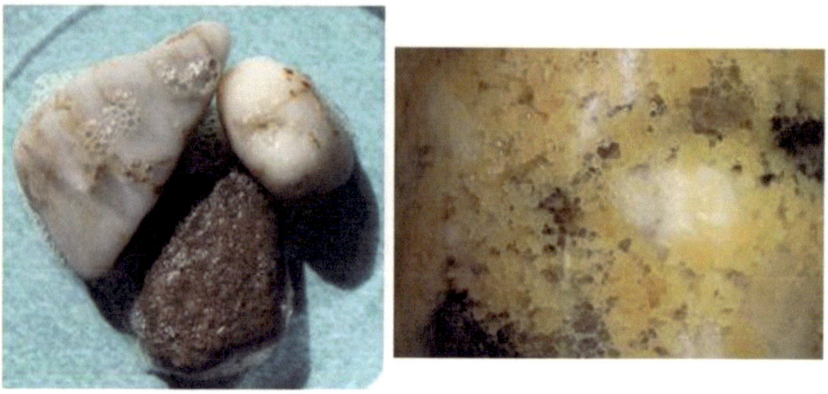

(a) (b)

**Figure 3.** Images of the two structures formed by selected consortia of fermentative bacteria grown in a bioreactor on M9 medium containing molasses: (a) stones covered with bacterial biofilm (b) the granular sludge.

Metagenomic analysis of microbial communities by 454-pyrosequencing of amplified 16S rDNA fragments revealed that the overall biodiversity of hydrogen-producing cultures was quite small. Stone biofilm from the culture without the granular sludge was dominated by *Clostridiaceae* and heterolactic fermentation bacteria, mainly *Leuconostocaeae*. Representatives of *Leuconostocaeae* and *Enterobacteriaceae* were dominant in both the granules and the stone biofilm formed in the granular sludge culture. The granular sludge contained bacteria of heterolactic fermentation, dominated by *Leuconostoc* species as well as unclassified *Streptococcaceae* and unclassified *Enterobacteriaceae*. Surprisingly, sequences representing the *Clostridiaceae* were in a relative minority (Table 2).

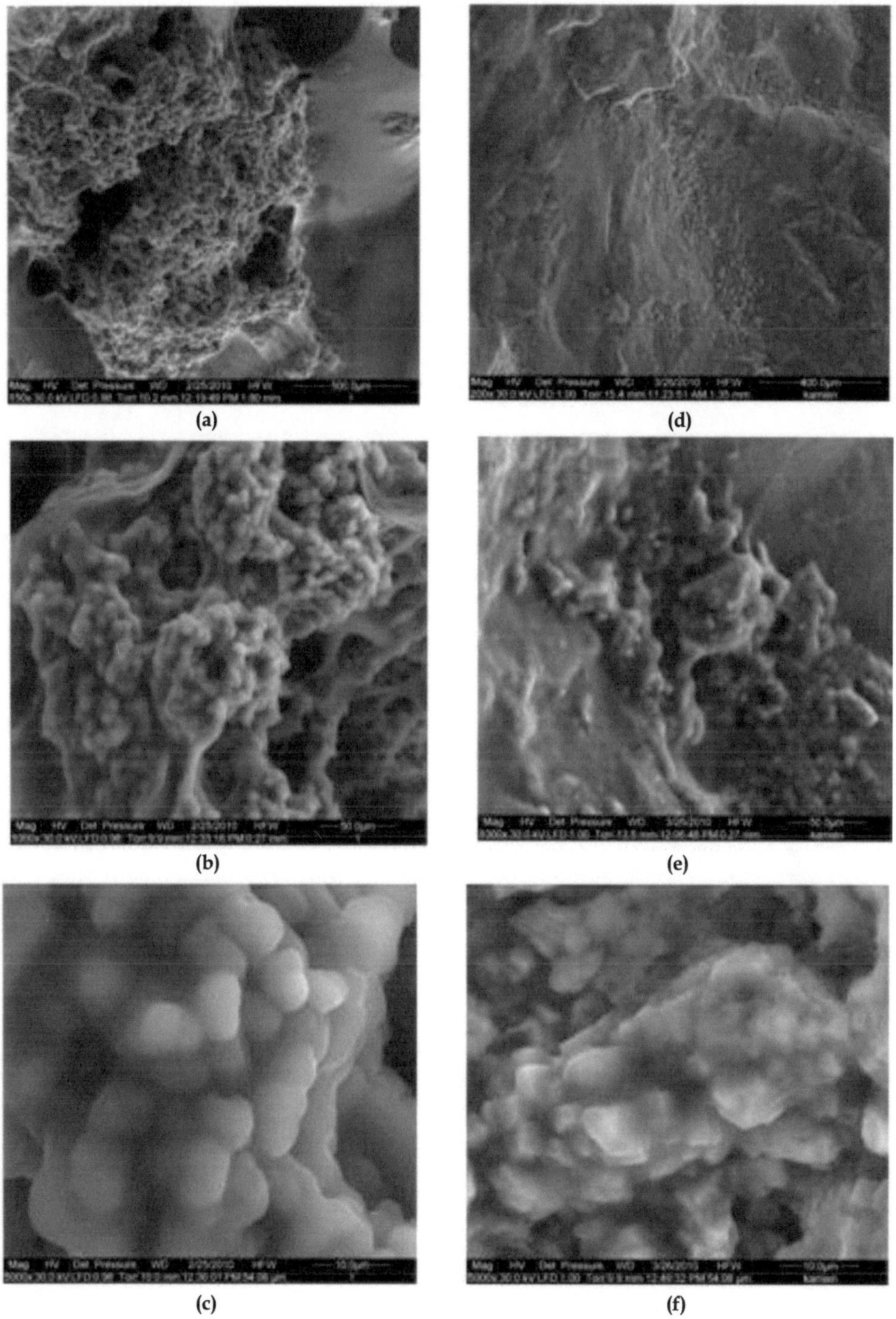


**Figure 4.** Scanning electron micrographs of structures formed by selected consortia of fermentative bacteria grown on M9 medium containing molasses: (a – c) granules; (d – f) bacterial biofilm formed on the granitic stones filling the bioreactor in the granular sludge culture.

| taxon | B | G | Bg |
|---|---|---|---|
| Bacteria | 4596 | 8578 | 26066 |
| Firmicutes | 4410 | 6473 | 10216 |
| Bacilli | 1564 | 5425 | 7586 |
| Bacillales | 119 | 17 | 96 |
| Bacillaceae | 1 | 0 | 11 |
| Sporolactobacillaceae | 94 | 6 | 44 |
| Sporolactobacillus | 94 | 6 | 44 |
| Lactobacillales | 1343 | 5302 | 7272 |
| Enterococcaceae | 5 | 16 | 417 |
| Enterococcus | 2 | 2 | 123 |
| Lactobacillaceae | 96 | 3 | 790 |
| Lactobacillus | 92 | 3 | 725 |
| Leuconostocaceae | 826 | 4661 | 4589 |
| Leuconostoc | 826 | 4634 | 4586 |
| Streptococcaceae | 4 | 37 | 95 |
| Lactococcus | 4 | 17 | 74 |
| Clostridia | 2627 | 579 | 2240 |
| Clostridiales | 2572 | 443 | 2023 |
| Clostridiaceae | 2131 | 134 | 1100 |
| Clostridium | 1182 | 66 | 593 |
| Proteobacteria | 168 | 1974 | 15755 |
| Gammaproteobacteria | 168 | 1970 | 15744 |
| Enterobacteriales | 156 | 1769 | 15528 |
| Enterobacteriaceae | 156 | 1769 | 15528 |
| Enterobacter | 61 | 319 | 2130 |
| Raoultella | 6 | 50 | 23 |
| Pseudomonadales | 6 | 8 | 137 |
| Moraxellaceae | 5 | 0 | 128 |
| Acinetobacter | 2 | 0 | 68 |
| Pseudomonadaceae | 1 | 8 | 9 |
| Pseudomonas | 1 | 7 | 9 |

B – stone biofilm from the culture without granular sludge; Bg – stone biofilm from the granular sludge culture; G – granules from the granular sludge culture.

**Table 2.** Number of reads assigned to respective taxonomic branches of 16S rRNA gene fragments amplified from the total DNA pool from bacterial communities formed in bioreactors.

Results of the metagenomic analysis by 454-pyrosequencing were confirmed by FISH (Fluorescence-In-Situ-Hybridization) analysis (Fig. 5) as well as by isolatating of lactic acid bacteria from the culture (Table 3).

Both, the stone biofilm and granules are composed of bacteria of many different shapes. As judged from fluorescence *in situ* hybridization, the relative abundance of selected bacterial groups varied during the rounds of bioreactor cycles. At the very beginning of biofilm development clostridial and lactobacilli cells were detected only sporadically among gammaproteobacteria (Fig. 5A a-c). In the growing biofilm systematic increase of Firmicutes (especially lactobacilli) cells was observed (Fig. 5Bd-e).

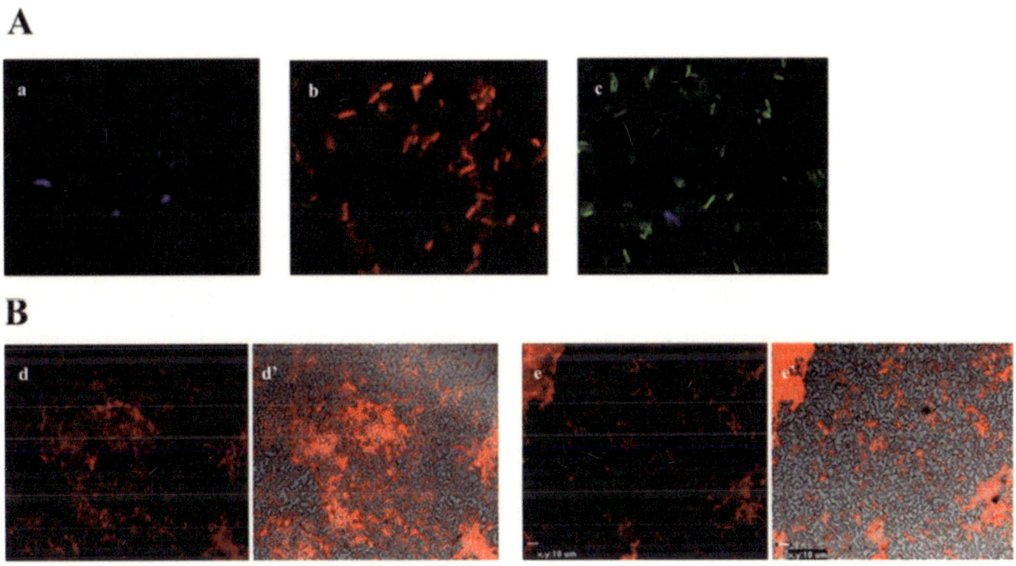


A: young biofilm; a-clostridia/dyLight405, b-lactobacilli/TAMRA; c-gammaproteobacteria/CY3,
B: mature biofilm; d-firmicutes/CY5; e-lactobacilli/TAMRA, d', e'-in combination with phase contrast.

**Figure 5.** FISH image of the hydrogen-producing biofilm from the granular sludge culture described previously (Chojnacka et al., 2011) analyzed by confocal laser fluorescence microscopy. The sample was stained with fluorescently labeled specific probes.

A cultivable approach with the use of media promoting the growth of lactic acid bacteria (MRS, M17) revealed that the bioreactor was inhabited by a vast number of these bacteria. Similarly to the metagenomic data, the majority of growing colonies represented *Leuconostoc* or *Lactobacillus* genera. All in all, six different species listed in Tab. 3 were isolated. It was determined that heterofermentative species (*Leuconostoc, L. brevis, L. rhamnosus*) slightly outnumbered homofermentatives.

| Isolate | Isolation ratio | Homofermenters:heterofermenters ratio |
|---|---|---|
| ***Lactobacillus plantarum*** | 46,7% | 0,89 |
| ***Enterococcus casseliflavus*** | 0,5% | |
| *Leuconostoc mesenteroides* | 46,7% | |
| *Leuconostoc mesenteroides* ssp. *mesenteroides* | 0,5% | |
| *Lactobacillus brevis* | 5,1% | |
| *Lactobacillus rhamnosus* | 0,5% | |

**Table 3.** The species of LAB isolated from the hydrogen-producing bioreactor.

Samples were collected from both, stone (biofilm) and liquid phase of hydrogen-producing culture, and plated on selective media for lactic acid bacteria. Plates were incubated under anaerobic conditions. Obtained colonies were tested for Gram positivity and lack of catalase enzyme. For strains which gave positive results, the V3 fragment of the 16S rRNA gene was amplified. Subsequently, fragments were analyzed using MSSCP technique. Strains with unique or representative gel patterns were chosen for further studies based on amplification and sequencing of 16S rRNA gene. Resulting sequences were identified by comparison to known sequences using the NCBI database. Names of homofermentative species are written in bold.

Formation of granular sludge rich in heterolactic bacteria significantly enhanced hydrogen production. Table 4 presents a list of parameters describing and comparing the two bacterial cultures that were the subject of the study of Chojnacka et al. (2011), under optimal conditions for hydrogen production. Significantly higher total gas production was observed for the culture containing granular sludge than for the biofilm-only culture (9.5 vs. 6.6 $cm^3$/min/working volume of the bioreactor). Furthermore, the percentage contribution of hydrogen was almost 49 and 36 %, whereas of carbon dioxide 47 and 60%, in the former and latter cultures, respectively. The granular sludge culture produced hydrogen at the rate of 6649 $cm^3$/day/working volume of the bioreactor, whereas the biofilm-only culture at the rate of 3393 $cm^3$/day/working volume of the bioreactor. Fermentation gas produced by both cultures contained 0.0004% methane, meaning that it was practically methane-free. Consequently, under optimal conditions, the culture containing granular sludge rich in heterolactic bacteria was two-fold more effective in producing hydrogen than that containing biofilm only: 5.43 moles of $H_2$ vs. 2.8 moles of $H_2$/mole of sucrose from molasses, respectively.

It is known that butyrate is the predominant metabolite during butyric acid fermentation at pH 5.0 – 5.5 (Li and Fang, 2007). The analysis of the non-gaseous fermentation products in both cultures in the study of Chojnacka et al. (2011) revealed that butyric acid was the main metabolite with partial contribution of ethanol. Concentration of butyric acid was almost 1.8-fold higher in the culture containing granular sludge than in the biofilm-only culture. No net production of lactic and propionic acids was observed in the granular sludge culture, whereas these were the second and third most abundant fermentation products in the cultures containing only biofilm. The formic and acetic acids present in the medium were utilized by both cultures. It is noteworthy that in the granular sludge culture rich in heterolactic bacteria showing very good performance in hydrogen production and a high content of butyric acid, the number of *Clostridiales* sequences was significantly lower than in the biofilm-only culture.

Based on our results presented in the study of Chojnacka et al. (2011) we speculate that LAB may possibly play a significant but not fully understood and perhaps underestimated role in the hydrogen producing communities. This hypothesis is based on two observations: (i) the higher the number of LAB in the hydrogen-producing community, the more efficiently hydrogen is produced; (ii) complete consumption of lactic acid, significantly increased concentration of butyric acid as well as larger hydrogen yield in the culture containing granular sludge than in that with just the biofilm.

| Parameter | Culture without granular sludge | Culture containing granular sludge |
|---|---|---|
| **Total gas production (cm³/min/working volume of the bioreactor)** | **6.6** | **9.5** |
| **Composition of fermentation gas (%):** | | |
| Hydrogen | 35.7 % | 48.6 % |
| Carbon dioxide | 60% | 47.1 % |
| Water vapor | ~4.3% | ~4.3 % |
| Methane | 0.0004 % | 0.0004% |
| Others (NH3, H2S, formic, acetic, propionic and butyric acids) | ~1% | ~1% |
| **Hydrogen production (cm³/day/working volume of the bioreactor)** | **3393** | **6649** |
| **Yield of hydrogen (moles $H_2$/mole of sucrose)** | **2.8** | **5.43** |
| **Net production of the non-gaseous end products (mg/L):** | | |
| Lactic acid | 2419 ± 42.6 | 0 |
| Formic acid | 0 | 0 |
| Acetic acid | 0 | 0 |
| Propionic acid | 248 ± 0.7 | 0 |
| Butyric acid | 4331 ± 60.0 | 7641 ± 33.1 |
| Ethanol (%) | 0.06 ± 0.002 | 0.1 ± 0.004 |

**Table 4.** Parameters describing two cultures of hydrogen-producing bacteria under optimal conditions for hydrogen production based on the study of Chojnacka et al. (2011).

### 3.4. Enhancement of hydrogen production by lactic acid

Based on the study of Chojnacka et al. (2011), for the culture containing granular sludge rich in heterolactic bacteria no net production of lactic acid was observed, indicating complete consumption of this metabolite, whereas its concentration in the biofilm-only culture was quite high (Table 4). Noticeable is the fact that molasses - a fermentative substrate in this study, also contains acetic and lactic acids at concentrations of about 800 mg/L each. Furthermore, a significantly higher concentration of butyric acid was detected in the culture containing granular sludge than in biofilm-only culture.

There are studies arguing that lactic acid and acetic acid mixed with the substrate stimulate biohydrogen production. Baghchehsaraee et al. (2009) showed that the addition of lactic acid to a mixed culture grown on starch-containing medium increased both hydrogen production and butyric acid formation. Furthermore, complete consumption of lactic acid produced by the culture was observed. When lactic acid was the only carbon source, the level of hydrogen production was very low (0.5% substrate conversion efficiency). Therefore, authors claimed that the addition of lactic acid to the medium probably alters the metabolic pathways in bacterial cells.

In the study of Kim et al. (2012), the effects of different lactate concentrations on hydrogen production from glucose in batch and continuous cultures were examined. Lactic acid was determined to be a factor increasing the efficiency of hydrogen production in a proper range of concentrations. The key issue was to establish the optimal lactic acid concentration. FISH analyses revealed that *Clostridium* sp. was the dominant hydrogen producer in the examined system.

Matsumoto and Nishimura (2007) examined fermentative hydrogen production from sweet potato sho-chu post-distillation slurry that contained large amounts of organic acids. Hydrogen production was accompanied by a decrease in the concentrations of acetic ad lactic acids and co-production of butyric acids. The authors isolated a clostridial strain, *Clostridium diolis* JPCC H-3, capable of effective hydrogen production from the slurry solution and a mixture of acetic and lactic acids in an artificial medium.

The ability to produce hydrogen from lactic and acetic acids seems to be widely conserved in the genus *Clostridium* and other hydrogen-producing bacteria capable of butyric acid fermentation of carbohydrates. It was shown that the *Clostridium acetobutylicum* strain P262 and *Butyribacterium methylotrophicum* utilized lactate and acetate and converted them to butyrate, carbon dioxide and hydrogen in the absence of carbohydrates in the medium. Cell extracts from bacteria grown on acetate and lactate showed a higher activity of NAD-independent lactate dehydrogenase than these from bacteria grown on carbohydrate-rich medium (Diez-Gonzales et al., 1995; Shen et al., 1996). The authors presented potential biochemical pathways leading to butyrate and hydrogen production from lactate and acetate. Conversion of lactate and acetate to butyrate and symbiotic interactions between LAB and clostridial species in animal intestinal tracts are intensively studied and discussed in section 4. Therefore, also the biochemical routes leading to butyrate and hydrogen production from lactate and acetate are presented in the same section.

In the study of Matsumoto and Nishimura (2007) the process of hydrogen production by *C. diolis* from both the slurry solution and a mixture of acetic and lactic acids in an artificial medium occurred to be pH-dependent and was observed in a range of pH (~5.8 – 7.4). Juang et al. (2011) also observed utilization of lactate and acetate for biohydrogen and butyrate production during their studies on hydrogen and methane production from organic residues of ethanol fermentation from tapioca starch by mixed bacteria culture. Lactate and acetate came from maltose fermentation, the main carbohydrate of ethanol fermentation residues. The optimal hydrogen production was observed at pH 5.5 – 6.0. Jo et al. (2008) showed that conversion of lactate and acetate to butyrate and hydrogen by *Clostridium tyrobutyricum* was inhibited due to pH decrease from 5.5 to 4.6. The pH values were dependent on HRT and organic loading rate. At high organic loading rate accumulation of lactate, pH decrease and a lower efficiency of hydrogen production were observed.

Matsumoto and Nishimura (2007), Jo et al. (2008) and Juang et al. (2011) point to pH values as a critical factor for hydrogen production from lactate and acetate. Various optimal pH for hydrogen production are observed. The differences may depend on the microbial system applied for hydrogen production and the initial substrate. It is speculated that unfavorable changes in pH could be the main reason of inhibiting hydrogen production that could be incorrectly attributed to the presence of lactic acid bacteria in hydrogen-producing consortia. In the study of Chojnacka et al. (2011), the optimal pH was around 5.0. Any change in pH, a decrease below 4.5 or increase above 5.5, caused a significant decline in fermentative gas production. Changes in pH may either be the reason or the results of disturbing the "homeostasis" of hydrogen-producing microbial communities in bioreactors.

## 4. Interactions between LAB and clostridial species in the animal intestinal tract

Microflora of the mammalian intestine is composed of a diverse population of both aerobic and anaerobic bacteria. Symbiotic relationships occur between different intestinal species or groups of species, among which are interactions between LAB and clostridial species. Numerous observations arising from different models describe lactate conversion to butyrate by intestinal bacteria and enhancement of butyrate production by LAB (Hashizume et al., 2003; Duncan et al. 2004; Bourriaud, et al., 2005; Meimandipour et al., 2009; Abbas, 2010; Munoz-Tamayo, et al., 2011).

The microbial community of the human colon contains many bacteria that produce lactic acid including lactobacilli, bifidobacteria, enterococci and streptococci. However, lactate is normally detected only at very low concentration (<5 mM) in feces of healthy individuals due its rapid conversion to short chain fatty acids (SCFAs; acetate, propionate and butyrate) by acid-utilizing bacteria. Therefore, lactate is thought to be a precursor of the formation of various SCFAs (Hashizume et al., 2003; Duncan et al., 2004; Bourriaud et al., 2005; Munoz-Tamayo et al., 2011). Bourriaud and colleagues (2005) performed convincing experiments exploring the lactate metabolism and short fatty acids production. They incubated three human microfloras with media containing $^{13}C$-labelled lactate and detected the labeled products of fermentation by $^{13}C$ NMR spectrometry. Results revealed that butyrate was the major net product of lactate conversion by human fecal microflora. Other SCFAs produced were: propionate, acetate and valerate. Inter-individual differences between the three microfloras were observed. Similar studies performed using $^{2}H$-labelled acetate and $^{13}C$-labelled lactate and gas chromatography-mass spectrometry (GC-MS) analysis showed that acetic and lactic acids are important precursors of butyrate production in human fecal samples (Morrison et al., 2006).

The metabolic pathway of lactate and acetate utilization to produce butyrate proposed for *Eubacterium hallii* and *Anaerostipes caccae* is shown on Figure 6 (Duncan, et al., 2004; Munoz-Tamayo, et al., 2011). The butyrate produced (in moles) is approximately equal to the sum of half of the acetate and lactate coming from the medium. Lactate is converted to pyruvate by lactate dehydrogenase. The next steps are analogous to those ones presented on Figure 1. Pyruvate is oxidized to acetyl coenzyme A (acetyl-CoA), which is further routed to acetate and butyrate. Acetate is produced via acetate kinase, the pathway generating energy in the form of ATP. For butyrate formation, two molecules of acetyl-CoA are condensed to one molecule of acetoacetyl-CoA, and subsequently reduced to butyryl-CoA. Butyrate can be synthesized from two metabolic pathways: phosphotransbutyrylase and butyrate kinase as shown on Figure 1, and butyryl CoA:acetate CoA transferase as shown on Figure 6. The latter mechanism seems to be the dominant in the human colonic ecosystem. Butyryl-CoA:acetate CoA-transferase transports the CoA component to exterior of acetate releasing butyrate and acetyl-CoA (Duncan et al., 2004; Munoz-Tamayo, et al., 2011). Hydrogen can be produced by both PFOR and NFOR complexes and hydrogenases, as described in section 2. The reaction catalyzed by NFOR is assumed to be the main route for $H_2$ production by intestinal microflora (Bourriaud et al., 2005). Similar pathway is proposed for clostridial species (eg. *C. acetobutylicum*; Diez-Gonzales et al., 1995) and other hydrogen and butyrate producing bacteria (eg. *B. methylotrophicum*; Shen et al., 1996), as mentioned in section 3.4. Conversion of lactate and acetate to butyrate and hydrogen is an energetically favorable process (Duncan et al., 2004; Jo et al., 2008).

The known lactate-utilizing butyrate-producing bacteria belong to the *Firmicutes* phylum, which includes the following species: *Megasphaera elsdenii, Anaerostipes caccae, Anaerostipes coli, E. hallii* and species distantly related to *Clostridium indolis*. *A. coli* is a dominant member of the human colonic microbiota recognized for its importance in butyrate production. *M. elsdenii* is one of the main butyrate producers from lactate in ruminants as well as monogastric animals, such as pigs or rodents. *A. caccae, A. coli, E. hallii* and species distantly related to *Clostridium indolis* belong to the clostridial cluster XIVa (*Lachnospiraceae*), known butyrate-producing bacteria of gastrointestinal tracts in mammals. However, only a few butyrate-producing species within the clostridial cluster XIVa are capable of converting lactate to butyrate (Duncan, 2004; Hashizume et al., 2003; Munoz-Tamayo, et al., 2011).

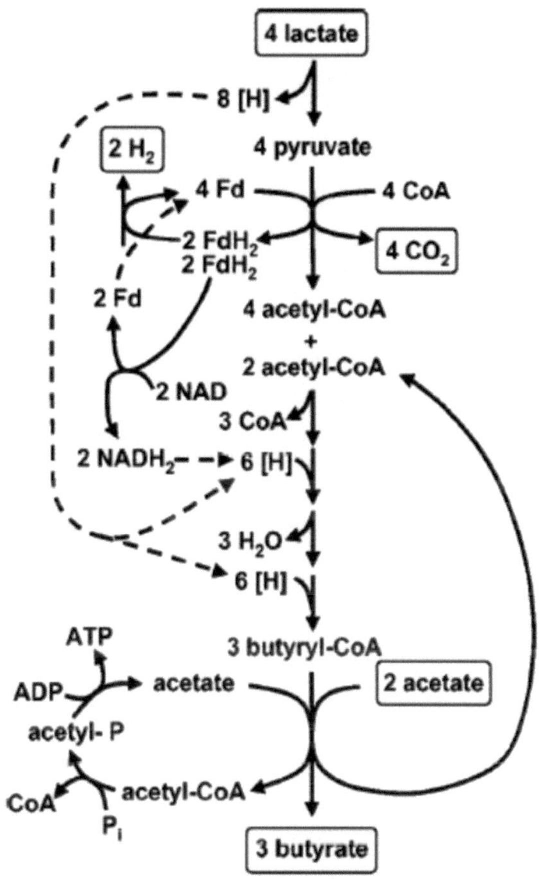


**Figure 6.** Scheme for butyrate production from lactate in *E. hallii* and *A. caccae*, adapted from Duncan et al., 2004.

The issue of stereospecificity of lactate utilization was addressed in the study of Duncan et al. (2004). Three *E. hallii*-related strains (SL6/1/1, SM6/1 and L2-7) and two *A. caccae* strains (L1-92 and P2) were able to use both D and L isomers of lactate during incubation on DL-lactate-containing medium. Interestingly, the addition of glucose to the medium almost completely inhibited lactate utilization by the tested strains. Additional studies showed that *E. hallii* L2-7, when grown with DL-lactate, used all of the supplied lactic acid together with some acetate, producing more than 20 mM of butyrate. Less butyrate, but a noteworthy amount of formate, was produced during growth on glucose or on glucose plus lactate. Interestingly, the highest level of hydrogen production was observed when strains were grown on lactate and the lowest for growth on glucose plus lactate. However, the *Clostridium indolis*-related strain SS2/1 was able to use D-lactate, but not L-lactate, during growth on DL-lactate containing media, which suggests that it lacks both an L-lactate dehydrogenase, capable of producing pyruvate from L-lactate, and a racemase, capable of converting L-lactate into D-lactate. According to Bourriaud and colleagues (Bourriaud et al., 2005), both lactate enantiomers are equally utilized by human intestinal microflora, treated as a whole consortium, not as pure strains.

Conversion of lactate to butyrate is one of the important factors for maintaining homeostasis in gastrointestinal tracts. Accumulation of lactate leads to different intestinal disorders (Hashizume et al., 2003). A number of studies have been performed to confirm the symbiotic interaction between lactic acid bacteria and butyric acid bacteria, mainly the *Clostridiales* representatives isolated from animal gastrointestinal tracts. Co-culture experiments that simulated the relations occurring *in vivo* were carried out. Symbiotic interactions were described to rely on the phenomenon of cross-feeding of lactate and involve conversion of lactate to butyrate by butyrate-producing bacteria stimulated by LAB.

It is noteworthy that results from studies of the gastrointestinal microflora indicate that acidity seems to be a key regulatory factor in lactate metabolism. The pH values may influence both bacterial growth and development of specific groups of bacteria as well as fermentation processes affecting the relative proportions of SCFAs (Belenguer et al., 2006; Meimandipour, et al., 2009; Belenguer et al., 2011). These observations are in agreement with our position concerning the potentially important role of pH in hydrogen-producing consortia discussed in section 3.4.

We postulate that the phenomenon analogous to cross-feeding observed in the gastrointestinal tract might take place in hydrogen-producing bioreactors. Although LAB may seem to be undesirable in such processes as they use some of $H_2$ to produce lactate, their stimulatory effects on hydrogen producers seem to exceed the potentially unbeneficial features. In many studies, it has been explicitly proven that the presence of LAB positively affects the production of butyrate. Most probably, hydrogen producers, mainly species belonging to the *Clostridiales* order, are capable of utilizing lactate as the main precursor of butyrate formation. Further investigations are required.

## 5. Conclusions

Lactic acid bacteria are detected in almost all biohydrogen-producing microbial communities of dark fermentation. Many studies indicate that LAB inhibit hydrogen production due to substrate competition and replacement of hydrogen fermentation by lactic acid and ethanol fermentations, and/or excretion of bacteriocines. On the other hand, some positive interactions between LAB and clostridial species have also been noted. They include hydrogen production from lactate by many clostridial species and symbiotic interactions, called lactate cross-feeding, occurring between LAB and clostridia.

These phenomena rely on the conversion of lactate and acetate to butyrate and hydrogen. Symbiotic interactions between LAB and butyrate-producing bacteria involving clostridia have been described in the gastrointestinal tract. We postulate that similar relations exist in biohydrogen-producing bioreactors. According to our hypothesis, pH may be a critical factor affecting bacterial growth, development of specific groups building hydrogen-producing microbial communities and fermentation processes. Acidity changes in bioreactors might be either the reason or the results of disturbances in the balance between microorganisms constituting hydrogen-producing microbial communities in bioreactors. Still, there are no data on symbiotic interactions between LAB and enterobacteria in hydrogen-producing microbial consortia. All these issues require further investigations.

## Author details

Anna Sikora and Urszula Zielenkiewicz
*Institute of Biochemistry and Biophysics, Polish Academy of Sciences, Poland*

Mieczysław Błaszczyk and Marcin Jurkowski
*Faculty of Agriculture and Biology, Warsaw University of Life Sciences, Poland*

## Acknowledgement

We would like to thank Dominika Brodowska for performance of FISH analyses and Dr. Agnieszka Szczepankowska for editorial assistance.

## 6. References

Abbas, K.A. (2010). The synergistic effects of probiotic microorganisms on the microbial production of butyrate in vitro, *McNair Scholars Research Journal* 2(1): 103-113.

Angenent, L.T., Karim, K., Al-Dahhan, M.H., Wrenn, B.A. & Domiguez-Espinoza, R. (2004). Production of bioenergy and biohemicals from industrial and agricultural wastewater. *TRENDS Biotechnol.* 22 (9): 477-485.

Antoni, D., Zverlov, V.V. & Schwarz, W.H. (2007). Biofuels from microbes. *Appl. Microbiol. Biotechnol.* 77(1): 23–35.

Baghchehsaraee, B., Nakla, G., Karamanev, D., Margaritis, A. & Reid, G. (2008). The effect of heat pretreatment temperature on fermentative hydrogen production using mixed cultures. *Int. J. Hydrogen Energy* 33(15): 4064-4073.

Baghchehsaraee, B., Nakhla, G., Karamanev, D. & Margaritis, A. (2009). Effect of extrinsic lactic acid on fermentative hydrogen production. *Int. J. Hydrogen Energy* 34 (6): 2573–2579.

Belenguer, A., Duncan, S.H., Calder, A.G., Holtrop, G., Louis, P., Lobley, G.E. & Flint H.J. (2006). Two Routes of Metabolic Cross-Feeding between *Bifidobacterium adolescentis* and Butyrate-Producing Anaerobes from the Human Gut. *Appl. Environ. Microbiol* 72(5): 3593–3599.

Belenguer, A., Holtrop, G., Duncan, S.H., Anderson, S.H., Calder, G.A, Harry, J., Flint H.J. & Lobley G.E. (2011). Rates of productionand utilization of lactate by microbial communities fromthe human colon. *FEMS Microbiol Ecol* 77(1): 107–119.

Bourriaud, C., Robins R.J., Martin, L., Kozlowski, F., Tenailleau, E., Cherbut, C. & Michel, C. (2005). Lactate is mainly fermented to butyrate by human intestinal microfloras but inter-individual variation is evident. *J. Appl. Microbiol.* 99(1): 201–212.

Campos, J.L., Arrojo, B., Franco, A., Belmonte, M., Mosquera-Corral, A., Roca, E. & Méndez, R. (2009). Effect of shear stress on wastewater treatment systems performance, *in* F.E. Dumont & J.A. Sacco (eds.), *Biochemical Engineering,* Nova Science Publishers, Inc., New York, pp. 227-244.

Chojnacka, A., Błaszczyk, M.K., Szczęsny, P., Latoszek, K., Sumińska, M., Tomczyk, K., Zielenkiewicz, U. & Sikora, A. (2011). Comparative analysis of hydrogen-producing bacterial biofilms and granular sludge formed in continuous cultures of fermentative bacteria. *Bioresour. Technol.* 102 (21): 10057-10064.

Das, D. & Veziroglu, T.N. (2008). Advances in biological hydrogen production processes. *Int. J. Hydrogen Energy* 33(21): 6046-6057.

de Nadra, M. (2007). Nitrogen metabolism in lactic acid bacteria from fruits: a review, *in* A. Méndez-Vilas (ed.), *Communicating Current Research and Educational Topics and Trends in Applied Microbiology* vol. 1, Microbiology Book Series, Communicating Current Microbiology with an Educational Vocation, pp. 500-510.

de Vos, P., Garrity, G., Jones, D., Krieg, N.R., Ludwig, W., Rainey, F.A., Schleifer, K.H. & Whitman, W.B. (2009). Bergey's Manual of Systematic Bacteriology, second edition, vol. 3 (The *Firmicutes*), Springer, Dordrecht, Heidelberg, London, New York.

Diez-Gonzalez, F., Russell, J.B. & Hunter, J.B. (1995) The role of an NAD-independent lactate dehydrogenase and acetate in the utilizationof lactate by *Clostridium acetobutylicum* P262. *Arch. Microbiol.* 164(1): 36–42.

Duncan, S.H., Petra Louis, P. & Flint, H.J. (2004). Lactate-utilizing bacteria, isolated from human feces, that produce butyrate as a major fermentation product. *Appl. Environ. Microbiol.* 70(10): 5185-5190.

Fang, H.H.P., Liu, H. & Zhang, T. (2002). Characterization of hydrogen-producing granular sludge. *Biotechnol. Bioeng.* 78 (1): 45-52.

Girbal, L., Croux, C., Vasconcelos, I. & Soucaille, P. (1995). Regulation of metabolic shifts in *Clostridium acetobutylicum* ATCC 824. *FEMS Microbial. Rev.* 17 (3): 287-297.

Hallenbeck, P.C. (2005). Fundamentals of the fermentative production of hydrogen. *Water Sci. Technol.* 52(1-2): 21-29.

Hallenbeck, P.C. & Ghosh, D. (2009). Advances in fermentative biohydrogen production: the way forward? *Trends Biotechnol.* 27 (5): 287–297.

Hallenbeck, P.C. (2011). Microbial paths to renewable hydrogen production. *Biofuels* 2(3): 285–302.

Hashizume, K., Tsukahara, T., Yamada, K., Koyama, H. & Ushida K. (2003) *Megasphaera elsdenii* JCM1772 normalizes hyperlactate production in the large intestine of fructooligosaccharide-fed rats by stimulating butyrate production. *J. Nutr.* 133 (10): 3187-3190.

Hung, C.-H., Lee, K.-S., Cheng, L.-H., Huang, Y.-H., Lin, P.-J. & Chang, J.-S. (2007). Quantitative analysis of a high-rate hydrogen-producing microbial community in anaerobic agitated granular sludge bed bioreactors using glucose as substrate. *Appl. Microbiol. Biotechnol.* 75(3): 693–701.

Ishibashi, N., Yaeshima, T. & Hayasawa, H. (1997). Bifidobacteria: their significance in human intestinal health. *Malaysian J. Nutrition* 3 (2): 149-159.

Ishikawa, M., Nakajima, K., Itamiya, Y., Furukawa, S., Yamamoto, Y. & Yamasato, K. (2005). *Halolactibacillus halophilus* gen. nov., sp. nov. and *Halolactibacillus miurensis* sp. nov., halophilic and alkaliphilic marine lactic acid bacteria constituting a phylogenetic lineage in *Bacillus* rRNA group 1. *Intern. J. System. Evolution. Microbiol.* 55 (Pt6): 2427–2439.

Jo, J.H., Jeon, C.O., Lee, D.S., Park & J.M. (2007). Process stability and microbial community structure in anaerobic hydrogen-producing microflora from food waste containing kimchi. *J. Biotechnol.* 131 (3): 300-308.

Jo, J.H., Lee, D.S., Park, D. & Park, J.M. (2008). Biological hydrogen production by immobilized cells of *Clostridium tyrobutyricum* JM1 isolated from a food waste treatment process. *Bioresour. Technol.* 99(14): 6666-6672.

Juang, C.-P., Whang, L.-M. & Cheng, H.-H. (2011). Evaluation of bioenergy recovery processes treating organic residues from ethanol fermentation process. *Bioresour. Technol.* 102 (9): 5394-5399.

Kalia, V.C. & Purohit, H.J. (2008) Microbial diversity and genomics in aid of bioenergy. *J. Ind. Microbiol. Biotechnol.* 35(5): 403-419.

Kim, T.-H., Lee, Y., Chang, K.-H. & Hwang, S.-J. (2012). Effects of initial lactic acid concentration, HRTs, and OLRs on bio-hydrogen production from lactate-type fermentation. *Bioresour. Technol.* 103 (1):, 136-141.

Kim, S.-H., Han, S.-K. & Shin, H.-S. (2006) Effect of substrate concentration on hydrogen production and 16S rDNA-based analysis of the microbial community in a continuous fermenter . *Process Bioch.* 41 (1): 199–207.

Korhonen, J. (2010). Antibiotic resistance of lactic acid bacteria. Dissertations in Forestry and Natural Sciences, Publications of the University of Eastern Finland, Kuopio.

Kraemer, J.T. & Bagley, D.M. (2007) Improving the yield from fermentative hydrogen production. *Biotechnol. Lett.* 29 (5): 685-695.

Lee, H.-S., Vermaas, W.F.J. & Rittmann, B.E. (2010). Biological hydrogen production: prospects and challenges. *Trends Biotechnol.* 28 (5): 262–271.

Lee, D.-J., Show, K.-Y. & Su, A. (2011). Dark fermentation on biohydrogen production: Pure cultures. *Bioresour. Technol.* 102(18): 8393-8402.

Li, C., Zhang, T. & Fang, H.H. (2006). Fermentative hydrogen production in packed-bed and packing-free upflow reactors. *Water Sci. Technol.* 54 (9): 95-103.

Li, C. & Fang, H.H.P. (2007). Fermentative hydrogen production from wastewater and solid wastes by mixed cultures. *Crit. Rev. Env. Sci. Tech.* 37 (1): 1–39.

Lo, Y.-C., Chen, W.-M., Hung, C.-H., Chen, S.-D. & Chang, J.-S. (2008). Dark $H_2$ fermentation from sucrose and xylose using $H_2$-producing indigenous bacteria: Feasibility and kinetic studies. *Water Res.* 42(1-2): 827-842.

Logan, B.E.: Extracting hydrogen and electricity from renewable resources. (2004). *Environ Sci Technol* 38(9): 160A-167A.

Maeda, T., Sanchez-Torres, V. & Wood, T.K. (2007). Enhanced hydrogen production from glucose by metabolically engineered *Escherichia coli. Appl. Microbiol. Biotechnol.* 77(4): 879-90.

Makarova, K., Slesarev, A., Wolf, Y., Sorokin, A., Mirkin, B., Koonin, E., Pavlov, A., Pavlova, N., Karamychev, V., Polouchine, N., Shakhova, V., Grigoriev, I., Lou, Y., Rohksa,r Y., Lucas, S., Huang, K., Goodstein, D. & Hawkins, T. (2006). Comparative genomics of the lactic acid bacteria. *PNAS* 103 (42): 15611–15616.

Matsumoto, M. & Nishimura, Y. (2007). Hydrogen production by fermentation using acetic acid and lactic acid. *J. Biosci. Bioeng.* 103(3): 236-241.

Meimandipour, A., Shuhaimi, M., Hair-Bejo, M., Azhar, K., Kabeir, B.M., Rasti, B. & Yazid, A.M. (2009). *In vitro* fermentation of broiler cecal content: the role of lactobacilli and pH value on the composition of microbiota and end products fermentation. *Lett. Appl. Microbiol.* 49(4): 415-420.

Morrison, D.J., Mackay, W.G., Edwards, C.A., Preston T., Dodson B. & Weaver L.T. (2006). Butyrate production from oligofructose fermentation by the human faecal flora: what is the contribution of extracellular acetate and lactate? *British J. Nutr.* 96 (3): 570–577

Munoz-Tamayo, R., Laroche, B., Walter, E., Dore, J., Duncan, S.H., Flint, H.J. & Leclerc, M. (2011). Kinetic modelling of lactate utilization and butyrate production by key human colonic bacterial species. *FEMS Microbiol. Ecol.* 76 (3): 615–624.

Nakashimada, Y., Rachman, M.A., Kakizono, T. & Nishio N. (2002). Hydrogen production of *Enterobacter aerogenes* altered by extracellular and intracellular redox state. *Int. J. Hydrogen Energy* 27(11-12): 1399-1405.

Nath, K. & Das, D. (2003). Hydrogen from biomass. *Current Science* 85 (3): 265-271.

Nath, K. & Das, D. (2004). Improvement of fermentative hydrogen production: various approaches. *Appl. Microbiol. Biotechnol.* 65(5): 520-529.

Noike, T., Takabatake, H., Mizuno, O. & Ohba, M. (2002). Inhibition of hydrogen fermentation of organic wastes by lactic acid bacteria. *Int. J. Hydrogen Energy* 27(11-12): 1367-1371.

Papoutsakis, E.T. (1984). Equation and calculations for fermentations of butyric acid bacteria. *Biotechnol. Bioeng.* 26(2): 174-187.

Piela P. & Zelenay P. (2004). Researchers redefine the DMFC roadmap. *The Fuel Cell Review.* 1(2): 17-23.

Qin, J., Li, R., Raes, J., Arumugam, M., Burgdorf, K.S., Manichanh, C., Nielsen, T., Pons, N., Levenez, F., Yamada, T., Mende, D.R., Li, J., Xu, J., Li, S., Li, D., Cao, J., Wang, B., Liang, H., Zheng, H., Xie, Y., Tap, J., Lepage, P., Bertalan, M., Batto, J.M., Hansen, T., Le Paslier, D., Linneberg, A., Nielsen, H.B., Pelletier, E., Renault, P., Sicheritz-Ponten, T., Turner, K., Zhu, H., Yu, C., Li, S., Jian, M., Zhou, Y., Li, Y., Zhang, X., Li, S., Qin, N., Yang, H., Wang, J., Brunak, S., Doré, J., Guarner, F., Kristiansen, K., Pedersen, O., Parkhill, J., Weissenbach, J.; MetaHIT Consortium, Bork, P., Ehrlich, S.D. & Wang, J. (2010). A human gut microbial gene catalogue established by metagenomic sequencing. *Nature* 464(4): 59-67

Ren, N., Xing, D., Rittmann, B.E., Zhao, L., Xie, T. & Zhao, X. (2007). Microbial community structure of ethanol type fermentation in bio-hydrogen production. *Environ. Microbiol.* 9 (5): 1112–1125.

Saint-Amans, S., Girbal, L., Andrade, J., Ahrens, K. & Soucaille, P. (2001). Regulation of carbon and electron flow in *Clostridium butyricum* VPI 3266 grown on glucose-glycerol mixtures. *J. Bacteriol.* 183(5): 1748-1754.

Salminen, S., Nurmi, J. & Gueimonde, M. (2005). The genomics of probiotic intestinal microorganisms. *Genome Biology* 6 (7): 225.

Sawers, R.G. (2005). Formate and its role in hydrogen production in *Escherichia coli*. *Biochem. Soc. Trans.* 33(Pt1): 42-46.

Shen, G.-J., Annous, B.A., Lovitt, R.W., Jain, M.K. & Zeikus, J.G. (1996). Biochemical route and control of butyrate synthesis in *Butyribacterium methylotrophicum*. Appl. Microbiol. Biotechnol. 45(3): 355-362.

Sreela-or, C., Imai, T., Plangklang, P. & Reungsang, A. (2011). Optimization of key factors affecting hydrogen production from food waste by anaerobic mixed cultures. *J. Hydrogen Energy*.36(21): 14120–14133).

Wu, S.-Y., Hung, C.-H., Lin, C.-N., Chen, H.-W., Lee, A.-S. & Chang, J.-S. (2006). Fermentative hydrogen production and bacterial community structure in high-rate anaerobic bioreactors containing silicone-immobilized and self-flocculated sludge. *Biotechnol. Bioeng.* 93(5): 934–946.

Yang, P., Zhang, R., McGarvey, J.A. & Benemann, J.R. (2007). Biohydrogen production from cheese processing wastewater by anaerobic fermentation using mixed microbial communities. *In. J. Hydrogen Energy* 32(18): 4761-4771.

Zhang, Z.G., Ye, Z.Q., Yu, L. & Shi, P. (2011). Phylogenomic reconstruction of lactic acid bacteria: an update. *BMC Evolutin. Biol.* 11(1): 1–12.

Chapter 22

# Exopolysaccharides of Lactic Acid Bacteria for Food and Colon Health Applications

Tsuda Harutoshi

Additional information is available at the end of the chapter

http://dx.doi.org/10.5772/50839

## 1. Introduction

Lactic acid bacteria (**LAB**) are used in many fermented foods, particularly fermented dairy products such as cheese, buttermilk, and fermented milks. LAB produce lactic acid, carbon dioxide, and diacetyl/acetoin that contribute to the flavor, texture, and shelf life of fermented foods. Some LAB produce exopolysaccharide (EPS), and generally, EPS play a major role as natural texturizer in the industrial production of yoghurt, cheese, and milk-based desserts. Recently, EPS produced by LAB have received increasing attention, mainly because of their health benefits. In particular, immune stimulation, antimutagenicity, and the antitumor activity of fermented dairy products prepared with EPS-producing LAB or EPS themselves have been investigated [1-4].

EPS are polysaccharides secreted from the cell, or produced on the outer cell by extracellular enzymes. EPS from LAB are divided into two classes, homo- and hetero-EPS. Homo-EPS are composed of one type of monosaccharide, whereas hetero-EPS consist of regular repeating units of 3-8 different carbohydrate moieties synthesized from intracellular sugar nucleotide precursors [5]. The biosynthesis of homo-EPS and hetero-EPS are different. Homo-EPS are made from sucrose using glucansucrase or levansucrase [6-7], and the synthesis of hetero-EPS involves four major steps, sugar transportation, sugar nucleotide synthesis, repeating unit synthesis, and polymerization of the repeating units [8]. The major physiological function of EPS is believed to be biological defenses against various stresses such as phage attack, toxic metal ions, and desiccation [9], and it is very unlikely that bacteria use EPS as an energy source. However, some potentially probiotic LAB strains have been reported to degrade EPS produced by the other LAB strains [10-11].

The term "probiotic" was first proposed by Fuller [12], and its definition was further refined to "Live microorganisms which when consumed in adequate amounts as part of food confer a health benefit on the host" [13]. Probiotic LAB thus represent a class of live food ingredients that exert a beneficial effect on the health of the host. Beneficial microorganisms in the intestine are enhanced by "prebiotics," which are defined as "nondigestible food ingredients that beneficially affect the host by selectively stimulating the growth and activity of one or a limited number of bacterial species already resident in the colon, and thus improving host health" [14].

Most of the current prebiotics are low molecular weight except for inulin. As long carbohydrate chains are metabolized more slowly than the short ones, and polysaccharides thus exert prebiotic effects in more distal colonic regions compared to oligosaccharides, which are more rapidly digested in the proximal colon [15]. Therefore, EPS produced by LAB can be used as prebiotics. This chapter reviews the physicochemical properties, genetics, and bioactivities of the EPS produced by LAB.

## 2. Chemical composition of EPS

### 2.1. Homo-EPS

Some LAB can produce EPS that are either secreted to the environment or attached to the cell surface forming capsules. EPS are classified into two groups: homo-EPS, consisting of a single type of monosaccharide (α-D-glucans, β-D-glucans, fructans, and others represented by polygalactan) and hetero-EPS, composed of different types of monosaccharides, mainly D-glucose, D-galactose, L-rhamnose, and their derivatives [16].

The differences arise between the homopolysaccharides mainly because of the features of their primary structure such as the pattern of main chain bonds, molecular weight, and branch structure. Two important groups of homo-EPS are produced by LAB; (i) α-glucans, mainly composed of α-1,6- and α-1,3-linked glucose residues, namely dextrans, produced by *Leuconostoc mesenteroides* subsp. *mesenteroides* and *Leuconostoc mesenteroides* subsp. *dextranicum* and mutans produced by *Streptococcus mutans* and *Streptococcus sobrinus*; and (ii) fructans, mainly composed of β-2,6-linked fructose molecules, such as levan produced by *Streptococcus salivarius* [17].

The formation of dextran from sucrose has been recorded for *Leuc. mesenteroides* subsp. *mesenteroides*. However, the ability to form dextran is often lost when serial transfers are made in media with increasing salt concentrations. Nevertheless, non-dextran-producing strains of *Leuconostoc* sp. can revert to dextran production when they are inoculated into medium containing tomato or orange juice [18]. In the 1950s, the use of a cell-free enzyme solution permitted dextran synthesis under controlled conditions yielding a polymer of greater purity. A common feature of all dextrans is the preponderance of α-1,6-linkages with branch points at positions 2, 3, or 4 [17]. Some strains of *Leuconostoc amelibiosum* [19] and *Lactobacillus curvatus* [20] are reported to be dextran-producing strains.

Mutan is the glucan synthesized by various serotypes of *Str. mutans*, and differs from dextran in that it contains a high percentage of α-1,3 linkages. Differences in solubility result

| Homo-EPS | Main linkage (branching linkage) | Organism |
|---|---|---|
| Glucans | | |
| Dextran | 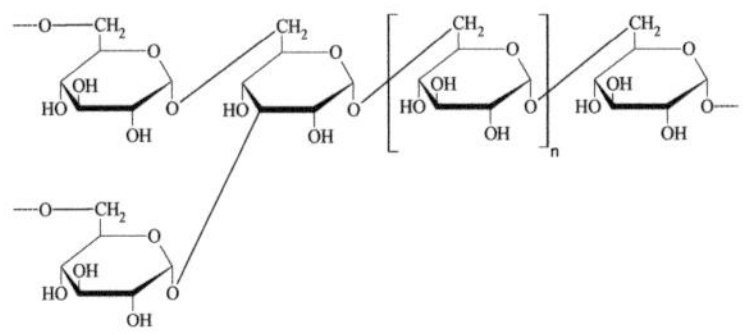<br>α-1,6 (α-1,3) | *Leuc. mesenteroides* subsp. *mesenteroides, Leuc. mesenteroides* subsp. *dextranicum, Leuc. amelibiosum, Lb. curvatus* |
| Mutan | 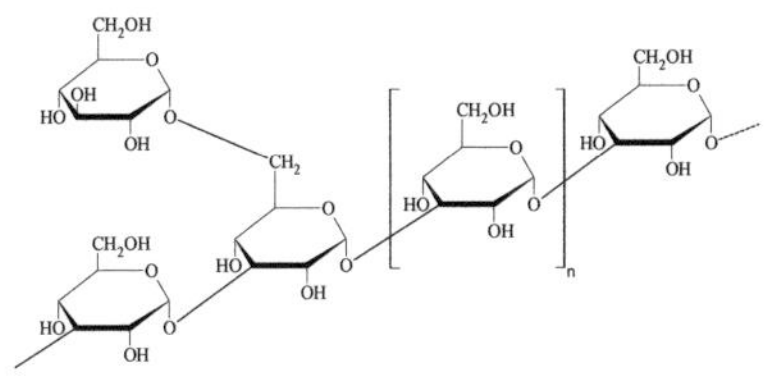<br>α-1,3 (α-1,6) | *Str. mutans* |
| Alternan | 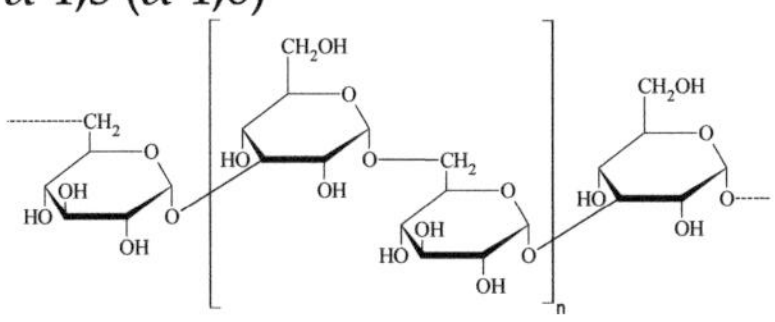<br>α-1,3 and α-1,6 | *Leuc. mesenteroides* |
| Fructans | | |
| Levan | 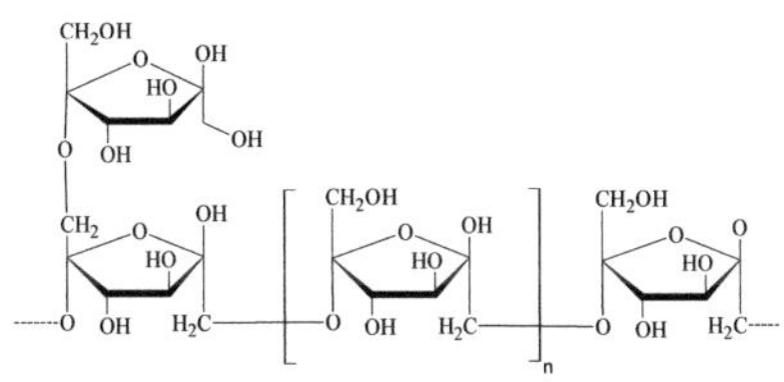<br>β-2,6 (β-2,1) | *Leuc. mesenteroides, Lb. reuteri* |
| Inulin | 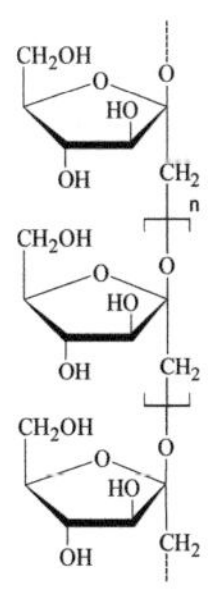<br>β-2,1 (β-2,6) | *Str. mutans* |

**Table 1.** Homo EPS produced by LAB

from the proportions of different types of linkages; water-soluble glucans are rich in $\alpha$-1,6 linkages, while water-insoluble glucans are rich in $\alpha$-1,3 linkages [17]. Ingestion of mutan has been linked with dental caries, as insoluble mutans can adhere to teeth, thus helping microorganisms adhere to the surface of teeth.

Alternan has alternate $\alpha$-1,6 and $\alpha$-1,3 linkages, and this structure is thought to be responsible for its distinctive physical properties including high solubility and low viscosity. These characteristics provide this glucan with a potential commercial application as a low viscosity texturizer in foods. *Leuc. mesenteroides* NRRL B-1355 was first reported to be an alternan-producing strain [21]

Levan is an EPS produced from sucrose. It is fructan composed of β-2,6-linked fructose molecules with some β-2,1-linked branches. Incidentally, inulin is a fructan composed of β-2,1-linked fructose molecules with some β-2,6-linked branches. *Str. salivarius, Leuc. mesenteroides,* and *Lactobacillus reuteri* are known to be levan-producing LAB [22-23]. In addition, the EPS produced by *Lactobacillus sanfranciscensis* TMW 1.392 has been reported to be fructan [11].

## 2.2. Hetero-EPS

The chemical composition of hetero-EPS shows wide variablity. Hetero-EPS are polymerized repeating units mainly composed of D-glucose, D-galactose, and L-rhamnose. The composition of the monosaccharide subunits and the structure of the repeating units are considered not to be species-specific, except in case of *Lactobacillus kefiranofaciens* subsp. *kefiranofaciens*. This species, isolated from kefir grain, a fermented dairy food from the North Caucasus region, produces large amounts of polysaccharides [24]. Hetero-EPS-producing strains of *Streptococcus thermophilus, Lactococcus lactis, Lactobacillus delbrueckii,* and *Lactobacillus helveticus,* among others have been identified (Table 2) [25-49]. Heterofermentative LAB such as *Leuc. dextranicum* are well known homo-EPS producers, while homofermentative LAB are well-studied hetero-EPS producers. Heterofermentative in addition to homofermentative LAB can produce EPS. *Lactobacillus fermentum* is an EPS-producing heterofermentative LAB for which the EPS structure has been determined [50]. Figueroa et al. reported that *Lactobacillus brevis* and *Lactobacillus buchneri* showed ropiness on glucose- or sucrose-containing media, although they did not investigate whether such ropiness derived from hetero-EPS or from other slimy substances [51].

The quantities of hetero-EPS produced by LAB vary greatly. EPS production is 50-350 mg/l for *Str. thermophilus*, 80-600 mg/l for *Lc. lactis subsp. cremoris*, 60-150 mg/l for *Lb. delbrueckii* subsp. *bulgaricus*, 50-60 mg/l for *Lactobacillus casei* [52], and approximately 140 mg/l for *Lactobacillus plantarum* [45, 53]. The highest recorded yields of hetero-EPS are 2775 mg/l for *Lactobacillus rhamnosus* RW-9595M [54] and 2500 mg/l for *Lb. kefiranofaciens* WT-2B [55]. However, the quantities of EPS produced by LAB are much lower than the yields from other industrially important microorganisms such as *Xanthomonas campestris*, which produces 30-50 g/l xanthan gum [56]. Even so, amounts of EPS produced by LAB are sufficient to exploit for in situ applications. LAB are 'generally recognized as safe' (GRAS) microorganisms, and

LAB strain culture would be a useful method to produce EPS for food applications if the LAB could be grown in edible and safe culture media such as whey, and if fermentation conditions were optimized to obtain a high yield.

Fermentation conditions using undefined media have been improved to maximize yields. However, a chemically defined medium containing a carbohydrate source, mineral salts, amino acids, vitamins, and nucleic acid bases is more suitable for investigating the influence of different nutrients on LAB growth and EPS biosynthesis. The total yield of EPS produced by LAB depends on the composition of the medium (carbon and nitrogen sources) and the growth conditions, i.e., temperature, pH, and incubation time.

Under conditions of higher temperatures and slower growth, the production of the polymer per cell in *Lb. delbrueckii* subsp. *bulgaricus* NCFB 2772 was greater in milk [57]. Another study investigated the optimum culture conditions for EPS production by *Lb. delbrueckii* subsp. *bulgaricus* RR in semidefined medium [58], and determined the optimum temperature and pH conditions for EPS production to be 36°C - 39°C and pH 4.5 - 5.5. The optimal temperature for EPS production was approximately 40°C for thermophilic LAB strains, and around 25°C for mesophilic LAB. Gamar et al. [59] reported increased slime production at lower incubation temperatures, and an increase in the final EPS concentration in *Lb. rhamnosus* following incubation at 25°C instead of 30°C. The effects of temperature on EPS production in whey were investigated in *Lb. plantarum* [53], and the yield was found to be higher at 25°C than at either 30°C or 37°C. Moreover, an inverse relationship was observed between EPS production per cell and the growth temperature for *Lactobacillus sake* [49], i.e. the lower the temperature, the higher the EPS production per cell. However, the growth rate in the exponential phase decreased at low temperatures. Therefore, the temperature for the maximal production of EPS is based on a balance of cell density and EPS production per cell. Maximal EPS production by *Lb. sake* was obtained under anaerobic conditions at 20°C, although EPS production per cell was higher at 10°C. Therefore, it is possible that severe environmental conditions trigger EPS production as a protective mechanism.

The effects of alterations to the nitrogen and carbon sources used in EPS production have also been investigated. According to early reports, neither LAB growth nor EPS production was specifically linked to the presence of casein or whey proteins in the growth medium. Garcia et al. [57] reported that EPS production by *Lb. delbrueckii* subsp. *bulgaricus* NCFB 2772 increased during the early growth pase in the presence of hydrolyzed casein in milk, while the addition of hydrolyzed casein to MRS medium did not increase EPS production. This strain produced 25 mg/l EPS when grown on fructose in a defined medium, and 80 mg/l EPS when grown on glucose [60]. The optimum Bacto-casitone concentration for EPS production by *Lb. delbrueckii* subsp. *bulgaricus* RR was investigated in semidefined medium [58]. In this study, there was a significant relationship between the Bacto-casitone concentration and EPS production; the higher the casitone concentration, the higher the EPS yield that was obtained. For *Lb. plantarum* grown in whey, yeast extract was a more effective nitrogen source for EPS production than soy peptide, tryptone, peptone, and Lab-Lemco powder, and glucose was a more effective carbon source than galactose, sucrose, maltose, fructose, and raffinose [53]. EPS production by *Lb. casei* CG11 was investigated in basal minimum medium containing galactose, glucose, lactose, sucrose, maltose, and melibiose; glucose was the most efficient carbon source, and lactose and galactose were the least efficient ones [61]. EPS production by *Lb. rhamnosus* C83 was investigated in a chemically defined medium containing different carbon sources (glucose, fructose, mannose, and maltose) at different concentrations. Mannose at 40 g/l was by far the most efficient carbon source. Furthermore, increased Mg, Mn and Fe concentrations stimulated EPS production in synthetic media [59]. In addition, Macedo et al. [54] reported about the importance of salts in culture media and the strong positive effect of salts and amino acids on *Lb. rhamnosus* RW-9595M growth and EPS production. The addition of salts and amino acids largely increased EPS production (to 2775 mg/l) in whey permeate supplemented with yeast extract, although the addition of amino acids alone had no effect on EPS production.

It has been shown that an optimal ratio between the carbon and nitrogen is absolutely necessary to achieve high EPS yields [62]. The production of EPS by *Str. thermophilus* LY03 is modulated by both the absolute quantities and the ratio of carbon to nitrogen (C/N ratio). The carbon source is converted into lactic acid to produce energy as well as to synthesize the cell wall and EPS, and nitrogen is necessary for the synthesis of essential cell components. Therefore, a higher C/N ratio and sufficient quantities of both carbon and nitrogen increase EPS production.

## 3. EPS biosynthesis by LAB

### 3.1. Homo EPS biosynthesis

Homo EPS are synthesized outside the cell by specific glycosyltransferase (GTF) or fructosyltransferase (FTF) enzymes (commonly named glucansucrases or fructan-sucrases). Homo-EPS producing LAB also use extracellular GTF enzymes to synthesize high-molecular mass $\alpha$-glucans from sucrose. This process uses sucrose as a specific substrate, and the energy required for the process comes from sucrose hydrolysis. There is no energy requirement for EPS-production other than for enzyme biosynthesis because EPS synthesis by GTF or FTF does not involve active transport processes or the use of activated carbohydrate precursors. Therefore, large amounts of sucrose can easily be converted to EPS. *Lb. sanfranciscensis* produces up to 40 g/l levan and 25 g/l 1-kestose during growth in the presence of 160 g/l sucrose [63].

Glucan synthesis reactions catalysed by GTF can be written as follows (Fig. 1):
sucrose + $H_2O$ → glucose + fructose
sucrose + acceptor carbohydrate → oligosaccharide + fructose
sucrose + glucan (n) → glucan (n+1) + fructose

Although GTF enzymes have a high degrees of similarity, lactobacilli produce a broad spectrum of glucans, including polymers with $\alpha$-1,6 linkages (dextran), $\alpha$-1,3 linkages (mutan), and both $\alpha$-1,6 and $\alpha$-1,4 linkages (alternan). The relative molecular weight of glucans from lactobacilli range from $1 \times 10^6$ Da to $5 \times 10^7$ Da [6]. In addition, GTF enzymes are not saturated by their substrate, and transfer reactions exceed the sucrose hydrolysis under sucrose concentrations above 100 mM [64].

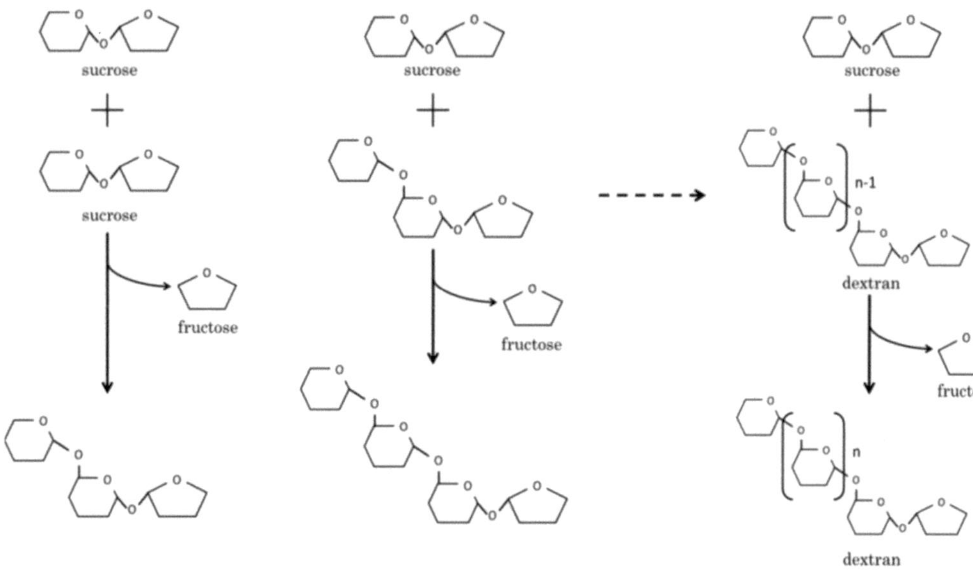


**Figure 1.** The dextran synthesis by GTF (dextran sucrase).

The GTF enzymes of streptococci are generally produced constitutively. In contrast, the GTF enzymes of *Leuconostoc* species are specifically induced by sucrose. For example, GTF expression in *Leuc. mesenteroides* is low in the presence of carbon sources other than sucrose and is increased by the addition of sucrose [5]. GTF expression during sucrose fermentation is 10-15-fold higher than that measured during glucose fermentation in *Leuc. mesenteroides* Lcc4. In fed-batch fermentation with both glucose and sucrose, GTF activity was similar to that obtained with sucrose alone. These results show that GTF expression is low in the presence of glucose alone, and that GTF activity is significantly induced by sucrose. A sucrose concentration of 20 g/l is sufficient to ensure the induction of enzyme synthesis, and higher concentrations (up to 60 g/l) do not lead to a further increase in enzyme synthesis [65].

The fructan synthesis reaction catalyzed by FTF can be written as follows:
sucrose + $H_2O$ → fructose + glucose
sucrose + acceptor carbohydrate → oligosaccharide + glucose
sucrose + fructan (n) → fructan (n+1) + glucose

Fructans generally have a relative molecular weight exceeding $5 \times 10^6$ Da. Similar to GTFs, FTFs are not saturated by their substrate, namely, sucrose, and transfer reactions exceed the rate of sucrose hydrolysis for sucrose concentrations above 200 mM [5]. FTFs such as Lev, Inu, and LevS from lactobacilli exhibit pH optima of between 5.0 and 5.5. The optimum temperature for enzymes from the thermophilic *Lb. reuteri* is higher (50°C) than that of the *Lb. sanfranciscensis* enzyme (35°C – 40°C) [5].

### 3.2. Hetero EPS biosynthesis

Hetero EPS are not synthesized by extracellular enzymes, but are instead synthesized by a complex sequence of interactions involving intracellular enzymes. EPS are made by polymerization of repeating units, and these repeating units are built by a series of addition of sugar nucleotides at the cytoplasmic membrane. Sugars are the starting materials for the synthesize sequence. LAB strains can utilize various monosaccharides and disaccharides as energy sources, via some well-studied sugar uptake systems include primary transport systems, direct coupling of sugar translocation to ATP hydrolysis via a transport-specific ATPase; secondary sugar transport systems, coupling of sugar transport to the transport of ions or other solutes, both as symport and antiport transport systems; and group translocation systems, coupling of sugar transport to phosphorylation via the phosphoenolpyruvate (PEP)-dependent phosphotransferase system (PTS; Fig. 2) [8]. Polysaccharides must be hydrolyzed before uptake. For example, starch is hydrolyzed by $\alpha$-amylase, and the raction products are subsequently hydrolyzed by the enzymes described above.

*Lc. lactis* strains possess a lactose-specific PEP-PTS sugar transport system that imports extracellular lactose, resulting in increased intracellular lactose-6-phosphate. Lactose-6-phosphate is then hydrolyzed, and the galactose-6-phosphate moiety is metabolized by the tagatose-6P pathway (Fig. 2).

*Lb. delbrueckii* subsp. *bulgaricus* and *Str. thermophilus* are generally galactose-negative and take up lactose via a lactose/galactose antiport transport system. The glucose moiety of imported lactose is fermented by these strains, while the galactose moiety is excreted via the lactose/galactose antiport system.

After the addition of a hetero-EPS repeating unit, the unit is exported through the cell membrane and becomes polymerized into the final hetero-EPS. Hence, several enzymes and proteins are involved in the biosynthesis and secretion of heterotype EPS, and the enzymes and proteins involved in these processes may not be unique to hetero-EPS anabolism.

Sugars taken into the cell are converted into sugar nucleotides. Iintracellular monosaccharides are converted to sugar nucleotide substrates for polymerization reactions, including UDP (uridine diphosphate), dNTP (thymidine diphosphate), and GDP (guanosine diphosphate). Such polymerization reactions are catalyzed by glycosyl pyrophosphorylases.

Glu-1P (Gal-1P) + UTP $\rightarrow$ UDP-Glu (UDP-Gal) + pyrophosphate
UDP-glucose is then converted to UDP-galactose by epimerases such as UDP-glucose-4-epimerase. This reaction is reversible.
UDP-glucose $\leftrightarrow$ UDP-galactose

Glycosidic linkages are formed on membranes in the cytoplasm. A sugar moiety is transferred to C55-polyprenyl phosphate, a carrier lipid and component of the membrane, by priming glycosyl transferases. This transfer triggers the addition of a repeating unit to the hetero-EPS molecule. Disruption of the priming glycosyl transferase gene generates non-EPS-producing mutants [66]. Thus, priming glycosyl transferases are thought to be crucial

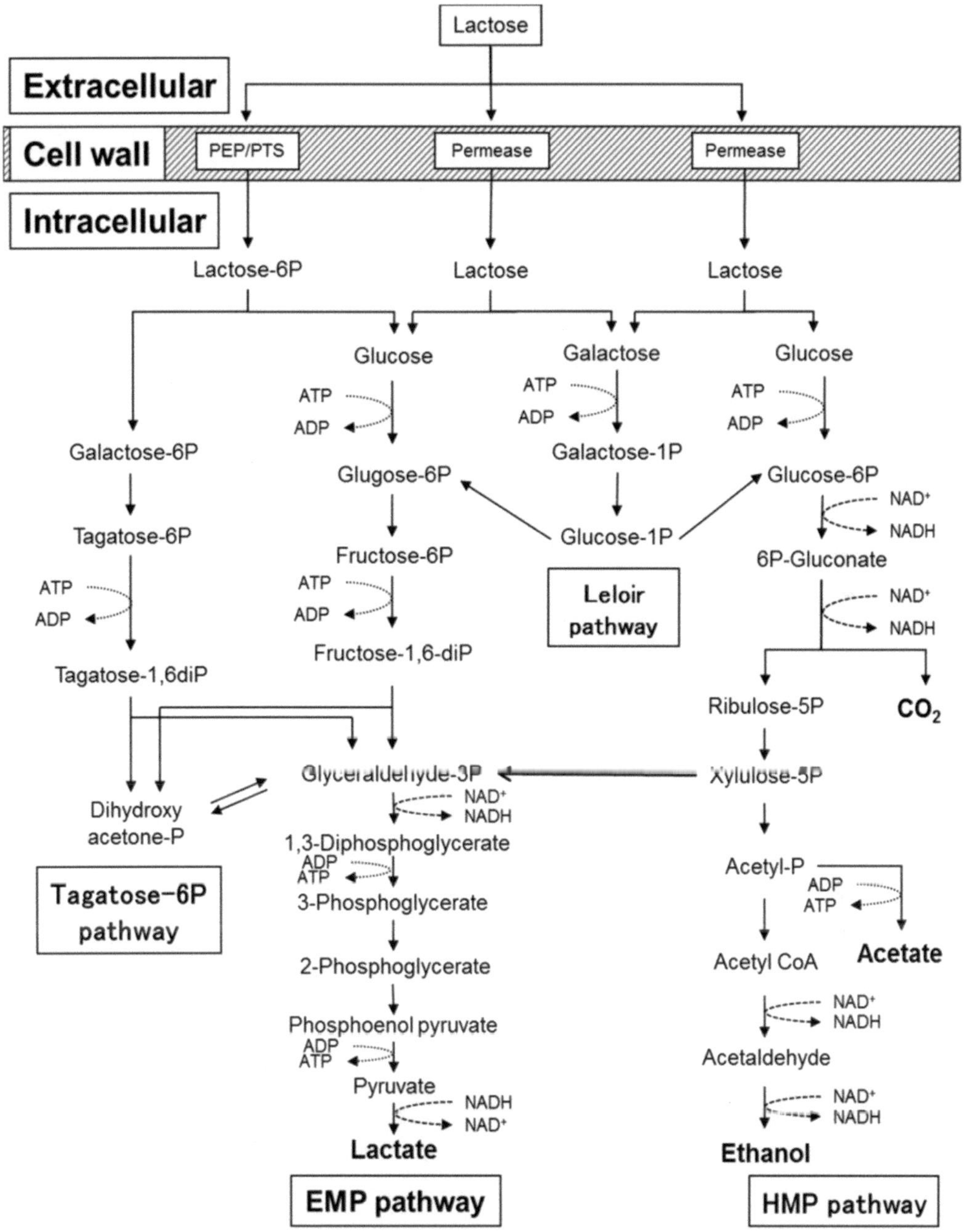


**Figure 2.** Pathway of lactose fermentation in lactic acid bacteria.

for EPS biosynthesis. The addition of the repeating unit is completed by the action of glycosyl transferase on the sugar residue attached to C55-polyprenyl phosphate. Therefore, the type and number of glycosyl transferases available determine the range of repeating units in hetero-EPS. C55-polyprenyl phosphate is also involved in bacterial cell wall biosynthesis, and therefore, cell wall biosynthesis and EPS synthesis compete for this substrate. The repeating unit is exported through the bacterial membrane, and is polymerized to become a hetero-EPS (Fig. 3).

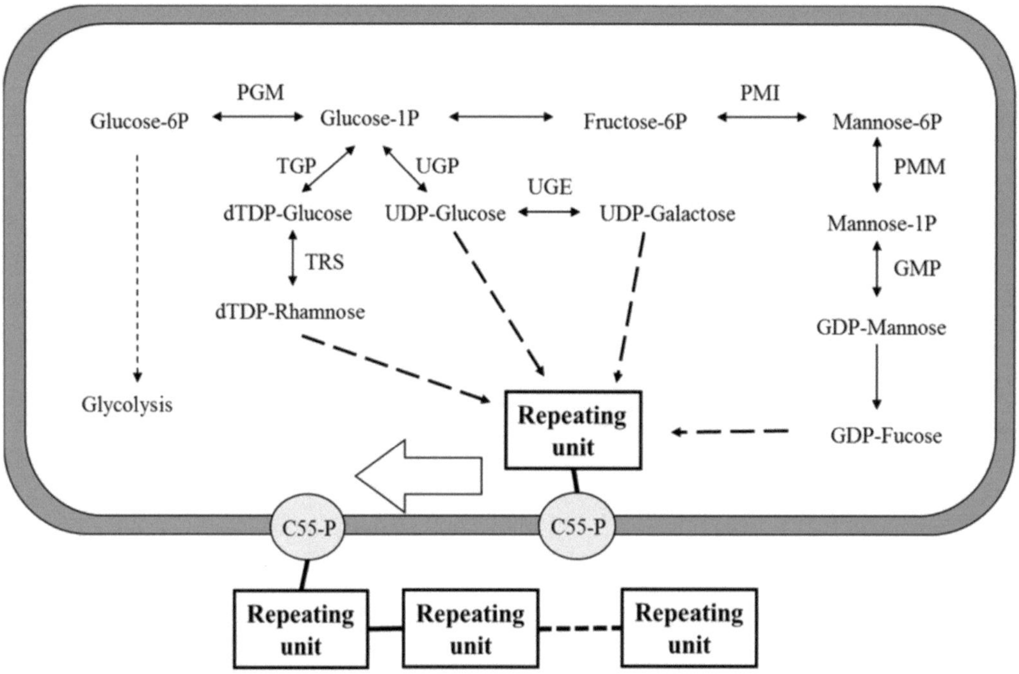


**Figure 3.** Outline of biosynthesis of hetero EPS.
PGM: α-phosphoglucomutase, UGP: UDP-glucose pyrophospholyrase
UGE: UDP-galactose 4-epimerase, TGP: dTDP-glucose pyrophospholyrase
TRS: dTDP-rhamnose synthetic enzyme system, PMI: phosphomannoisomerase
PMM: phosphomannomutase, GMP: GDP-mannose pyrophospholyrase

### 3.3. Instability of EPS production

The instability of hetero-EPS production has been reviewed by de Vuyst et al. [8]. Briefly, a loss in the ability to produce slime may be caused by repeated subculture of bacterial strains or incubation at high temperatures. The loss of plasmids from ropy mesophilic LAB strains is generally the reason for loss of slime production. On the other hand, thermophilic LAB, namely, *Lb. delbrueckii* subsp. *bulgaricus* and *Str. thermophilus*, have been shown to lack plasmids encoding components required for slime production. These species can usually recover the ability to produce slime following loss due to culture conditions. Thus, genetic instability could be a consequence of the actions of mobile genetic elements such as insertion sequences. Recently, the EPS gene cluster in *Lb. fermentum* TDS030603 was reported to be located in chromosomal DNA [67].

Priming glycosyl transferases are thought to be crucial for EPS biosynthesis and disruption of the priming glycosyl transferase gene generates non-EPS-producing mutants. Tsuda et al. generated the EPS-producing mutant strain 301102S from the non-EPS-producing *Lb. plantarum* 301102 following exposure to the mutagens acridine orange and novobiocin [4]. The activities of $\alpha$-phosphoglucomutase (PGM), UDP-glucose pyrophosphorylase (UGP), and UDP-galactose 4-epimerase (UGE) were measured in parental and mutant strains by using the method of Mozzi [68], and were found to be almost the same for both [Tsuda & Miyamoto, unpublished data]. Next, priming glycosyl transferase genes in parental and mutant strains were amplified with the thermal cycler. Primers were designed to amplify a priming glycosyl transferase gene referring to complete *Lb. plantarum* WCFS1 genome sequenced [69]. PCR products were subjected to restriction digestion, which allowed identification of putative priming glycosyl transferase gene. PCR products were also applied to single strand conformation polymorphism (SSCP) analysis for detecting point mutations. However, both parental and mutant strains had the same priming glycosyl transferase gene sequence, and similar levels of activities of the PGM, UGP, and UGE enzymes. Thus, although priming glycosyl transferases are essential, other factors may also be necessary for EPS production, and a mutation affecting EPS production may occure in another gene. Morona et al. reported that an autophosphorylating protein-tyrosine kinase is essential for encapsulation in *Streptococcus pneumoniae* [70]. A point mutation in the gene encoding the autophosphorylating protein-tyrosine kinase affecting the ATP-binding domain resulted in loss of EPS production.

## 4. Polysaccharides and oligosaccharides for colon health

EPS produced by LAB have various functional roles in human or animal health including immunomodulatory properties, antiviral activity, antioxidant activity, and antihypertensive activity [1, 55, 71, 72], and have also been used as food additives for texture improvement. These properties have been extensively reviewed [8, 9, 56, 73, 74]. Besides these properties, prebiotics based on LAB and oligosaccharides have other health benefits. Prebiotics are usually non-digestible oligosaccharides that selectively stimulate the growth and activity of a limited number of bacterial species in the colon, such as bifidobacteria and lactobacilli, and therefore, improve host health. Detrimental bacteria may form substances such as ammonia, hydrogen sulfide, indles, and amines that are noxious to the host. However, beneficial bacteria such as bifidobacteria and lactobacilli inhibit the proliferation of detrimental bacteria, and their cell components stimulate the host immune system [75]. Gastrointestinal microflora consist of approximately $10^{14}$ colony forming units (cfu)/g of various types of both detrimental and beneficial bacteria, and the numbers and composition vary greatly along the gastrointestinal tract. The balance of the gastrointestinal micro flora influences different aspects of host health such as bowel movement, tympanites flatulence, and the absorption of nutrients. Many factors may upset this balance, including stress, consumption of antibiotics, infection, food poisoning, and the natural ageing process. To redress this balance, the growth and activities of beneficial bacteria may be enhanced by specific ingredients in foods.

| Speceis | Strain | Glc | Gal | Rha | Fuc | NAc Gal | GlcA | Gly | Reference |
|---|---|---|---|---|---|---|---|---|---|
| *Streprococcus thermophilus* | CNCMI 733 | 1 | 2 | | | 1 | | | [25] |
| | SFi39 | 1 | 1 | | | | | | [26] |
| | SFi12 | 1 | 3 | 2 | | | | | [26] |
| | LY03 | 1 | 4 | | | | | | [27] |
| | OR901 | | 5 | 2 | | | | | [28] |
| | MR-1C | | 5 | 2 | 1 | | | | [29] |
| *Lactococcus lactis* subsp. *cremoris* | NIZO B891 | 3 | 2 | | | | | | [30] |
| | Ropy352 | 2 | 3 | | | | | | [31] |
| | NIZO B39 | 2 | 3 | 2 | | | | | [32] |
| | SBT 0495 | 2 | 2 | 1 | | | | | [33] |
| *Lactobacillus delbrueckii* subsp. *bulgaricus* | OLL 1073R-1 | 1 | 1.6 | | | | | | [34] |
| | NCFB 2772 | 1 | 2.4 | | | | | | [35] |
| | Lb18 | 1 | 1 | | | | | | [36] |
| | EU23 | 1 | | 1 | | | | | [37] |
| | rr | 1 | 5 | 1 | | | | | [38] |
| | NCFB 2772 | 1 | 7 | 0.8 | | | | | [35] |
| *Lb. helveticus* | TN-4 | 1 | 1 | | | | | | [39] |
| | 766 | 2 | 1 | | | | | | [40] |
| | 2091 | 1 | 2 | | | | | | [41] |
| | Lb161 | 5 | 2 | | | | | | [42] |
| *Lb. rhamnosus* | RW-9595M | 2 | 1 | 4 | | | | | [43] |
| | GG | 1 | 4 | 1 | | | | | [44] |
| *Lb. plantarum* | EP56 | 3 | 1 | | | 1 | | | [45] |
| | EP56 | 3 | 1 | 1 | | | | | [45] |
| *Lb. pentosus* | LPS26 | 1 | | 2 | | | 2 | | [46] |
| *Lb. paracasei* | 34-1 | | 3 | | | 1 | | 1 | [47] |
| *Lb. kefiranofaciens* | $K_1$ | 1 | 1 | | | | | | [48] |
| *Lb. sake* | 0-1 | 3 | | 2 | | | | | [49] |

**Table 2.** Monosaccharide ratio in hetero EPS
Glc: glucose, Gal: galactose, Rha: rhamnose, Fuc: fucose, NAc Glu: N-acetyl glucosamine, NAc Gal: N-acetyl galactosamine, GlcA: glucuronic acid.

Various oligosaccharides have been identified as prebiotics, that can increase the number of *Bifidobacterium* in the host colon. Galacto-oligosaccharides (GOS) and fructo-oligosaccharides (FOS) are considered important prebiotics. Other carbohydrates including gluco-oligosaccharides, isomalto-oligosaccharides, lactulose, mannan-oligosaccharides, and nigero-oligosaccharides are also considered prebiotics. Increased numbers of bifidobacteria and/or lactobacilli in the colon have been shown to have beneficial effects, although the specific mixtures of populations of these genera necessary to provide health-promoting effects has not yet been determined. This is because the beneficial effects are likely to be due to improvement in the balance of coloni micro flora. However, difference do exist in the micro flora among individuals. To function most effectively, prebiotics must be resistant to digestive processes in the stomach and small bowel, so that they can come into contact with the bacteria growing in the large intestine.

The food for specified health use (FOSHU) system was introduced in Japan in 1991. FOSHU refers to foods containing ingredients that provide health benefits and have officially approved physiological effects on the human body. FOSHU is intended to be consumed for the maintenance or promotion of health or for special health uses, for example, to control conditions such as blood pressure or blood cholesterol. To be defined as FOSHU, it is important to assess the safety of the food as well as the effectiveness of health promotion, and this assessment must be approved by the Ministry of Health, Labour and Welfare in Japan. At present (2012), 990 foods are recognized as FOSHU, and of these, 86 provide gastrointestinal health benefit. Foods for balancing gastrointestinal micro flora contain galactosylsucrose, soy oligosaccharides, lactulose, GOS, FOS, isomalto-oligosaccharides, raffinose, xylo-oligosaccharides, mannobiose, and brewer's yeast cell wall as functional ingredients.

## 4.1. GOS

GOS are well-known type of prebiotic oligosaccharides found in human milk. The concentration of oligosaccharides is 100 times higher in human breast milk than in bovine milk [76]. Many studies have shown that breast-fed infants have intestinal microflora dominated by bifidobacteria. The reason for this phenomenon is thought to be that the oligosaccharides in breast milk, including GOS, can reach the upper gut without being digested where the bifidobacteria can utilize them. At present, GOS is produced by the enzymatic treatment of lactose by β-galactosidase. GOS produced in this manner usually have degrees of polymerization (DP) between 2 and 10. Furthermore, the type of glycosidic linkage is determined by the reaction conditions: final products usually possess β-1,2, β-1,3, or β-1,4 linkages. GOS is given a caloric value of 2 kcal/g in Japan and Europe for food-labelling purpose.

The effect of GOS on defecation has been studied in healthy volunteers. Defecation frequency was significantly increased, and faeces became significantly softer after the subjects drank a beverage containing 5.0 g of GOS, on a daily basis. Therefore, consumption of a beverage containing 5.0 g of GOS can improve defecation in individuals with a tendency for constipation [77]. Ishikawa et al. reported that the number of faecal bifidobacteria increased significantly after subjects consumed 2.5 g of GOS/day for 3 weeks [78]. GOS utilization by enterobacteria was further investigated in vitro. The trisaccharide forms of GOS were utilized by *Bifidobacterium, Lactobacillus acidophilus, Lb. reuteri, Bacteroides, Clostridium perfringens, Klebsiella pnumoniae, Enterococcus faecium*, and the tetra-saccharide forms were utilized by *Bifidobacterium adolescentis, Bifidobacterium breve, Bifidobacterium infantis,* and *Ent. faecium*. These results suggest that a higher DP of GOS enhanced selectivity, and that the tetrasaccharide forms of GOS are specifically utilized by bifidobacteria. Similarly, *Bifidobacterium lactis* DR10 utilizes trisaccharide and tetra-saccharide forms of GOS, whereas *Lb. rhamnosus* DR20 prefers disaccharides and monosaccharides [79]. Barboza reported that *Bif. breve* and *Bif. longum* subsp. *infantis* can consume GOS with a DP ranging from 3 to 8 [80]. Furthermore, *Bif. longum* subsp. *infantis* preferentially consume GOS with a DP of 4, and *Bif. adolescentis* utilizes GOS with DP of 3. In addition, the structure of GOS influences its utilization by lactobacilli and bifidobacteria [81]. Trisaccharides of 4'-GOS (β-1,4 linkage) and 6'-GOS (β-1,6 linkage) can be used as the sole carbon source. Almost all lactobacilli and bifidobacteria tested preferred to utilize 4'-GOS, while *Lb. acidophilus, Lb. reuteri,* and *Lb. casei* could utilize both 4'- and 6'-GOS. GOS are used to stimulate beneficial bacteria, but can also be utilized by bacteroides and clostridia [82]. GOS selectivity may be enhanced by altering the structure and increasing the DP.

The use of beneficial bacteria or their enzymes in the synthesis of prebiotics may be a good way to produce prebiotics with high specificity. Rabiu reported that five different GOS were produced using β-galactosidase extracted from five different *Bifidobacterium* species, and that each GOS showed an increased growth rate in producer strains, except for *Bif. adolescentis* [83]. The utilization of these GOS by faecal bacteria was investigated using commercial GOS as control. The number of *Bacteroides* was decreased with GOS from bifidobacteria, whereas both GOS extracts and commercial GOS increased the number of bifidobacteria, lactobacilli, and clostridia.

### 4.2. FOS

FOS is used as a generic term for all β-2,1 linear fructans with a variable DP. Inulin and oligofructose are common forms of FOS that are widely found in nature. Chicory inulin has a DP of 2-60, and the product of its partial enzymatic hydrolysis is oligofructose or FOS with a DP of 2-10.

The effect of FOS intake on intestinal microflora was studied in humans. The number of bifidobacteria in faeces was significantly increased during the FOS intake (1 g/d) period, and a significant increase in stool frequency and a softening effect on stool were observed [84]. FOS increased the level of bifidobacteria in faeces, whereas that of bacteroides, clostridia, and fusobacteria decreased in subjects that were fed FOS (15 g/d) for 15 days [85]. Another study measured the increase in number of *Bifidobacterium* species in faeces by using real-time PCR [86]. The composition of bifidobacteria in the gut microflora was studied by clone library analysis in ten volunteers. All ten volunteers carried *Bif. longum,* and nine of these also carried *Bif. adolescentis*. The consumption of inulin (10 g/d) increased the number of bifidobacteria in faeces with *Bif. adolescentis* showing the highest increase response among *Bifidobacterium* species. Rossi et al. reported that only 8 of 55 *Bifidobacterium* strains fermented inulin in pure cultures, although inulin increased the number of bifidobacteria in faecal culture [87]. They, therefore, suggested that most bifidobacteria were not able to utilize long fructans in the absence of other intestinal bacteria that can hydrolyze fructans, and that fermentation of oligosaccharides in the colon is the result of a complex metabolic sequence carried out by numerous species.

### 4.3. Selection of high-efficiency prebiotics

It is not clear which oligosaccharides are the most suitable substrates for the selective growth of specific beneficial species or strains. Several research group have suggested useful methods to investigate the potential prebiotic activity of oligosaccharides [88-92]. Potential prebiotic activities were determined on the basis of the changes in the growth of beneficial and undesirable bacteria, such as bifidobacteria, lactobacilli, clostridia, and bacteroides. Such methods can evaluate the ability of specific strains to utilize a particular prebiotic, and a comparison of the prebiotic activities of oligosaccharides by using these methods could help in the choice of prebiotics for improving the gastrointestinal microflora on an individual basis. However, it is important to understand that only a limited group of bacteria can be chosen from the gastrointestinal microflora by using these methods, and that polysaccharides and oligo-saccharides are fermented by numerous species in the gastrointestinal tract.

Oligosaccharides produced by beneficial bacteria or their enzymes may enhance the growth of beneficial bacteria. A novel GOS mixture produced using *Bif. longum* NCIBM 41171 galactosidases increased the proportion of bifidobacteria in faeces relative to commercial GOS [93]. In the above-described study, oligosaccharides synthesized by the enzymes from *Bifidobacterium* strains were favored by the producer strains [83]. These studies suggest that the oligosaccharides produced by beneficial bacteria are selectively utilized by the producer strain, because the enzymes required for their degradation are already available. In addition, glycosyltransferases may possess both hydrolytic and transglycosylation activities [94], and glycosidases and glycosyltransferases may coexist in the same strains. Schwab et al. reported the production of novel oligosaccharides [95]. Hetero-oligosaccharides were produced from lactose, mannose, fucose, and N-acetylglucosamine by using crude cell extracts and whole cells of LAB and bifidobacteria. These hetero-oligosaccharides contained mannose, fucose, and N-acetylglucosamine, and could be digested by LAB strains. The prebiotic activities of these oligosaccharides were not investigated; however, a similar approach using probiotic and intestinal beneficial bacteria may lead to the production of highly selective prebiotics.

The dietary fiber, arabinoxylan is the predominant hemicellulose from cereals and exhibits prebiotic activity [96]. The addition of water-unextractable arabinoxylans increased the population of bifidobacteria and bacteroides in a medium inoculated with faecal slurry. Polysaccharides are not usually utilized by microorganisms. Remarkably, however, *Bifidobacterium bifidum* DSM20456 can utilize the EPS produced by *Pediococcus pentosaceus*, *Lb. plantarum*, *Weissella cibaria*, and *Weissella confusa*, and some growth is observed in cas of *Bif. longum*, *Bif. adolescentis*, and *Lb. acidophilus* [97]. For EPS production by LAB, reduced yields were frequently observed after the maximal level had been reached, which might be caused by the enzymes produced by the bacteria [98]. Tsuda and Miyamoto investigated the prebiotic activity of EPS produced by *Lb. plantarum* 301102S [52], a mutant strain derived from *Lb. plantarum* 301102. Oral administration of the parental strain 301102 showed the survivability and proliferation in porcine gastrointestinal tract [99]. The potential prebiotic activities of EPS, GOS, and inulin were measured in 37 LAB strains, and the activity scores of EPS in the strains 301102 and 301102S were highest. This suggests that the EPS produced by the mutant strain is utilized by the same strain 301102S and the parental strain, and that the parental strain has enzymes that can degrade the EPS.

## 5. Conclusion

Poly- and hetero-oligosaccharides produced by LAB may be potential prebiotics. Studies on the production of polysaccharides and oligosaccharides by enzymes in beneficial microorganisms may lead to the production of highly selective prebiotics, although in vitro evaluation may be difficult because of degradation and utilization of polysaccharides by various microorganisms in the gastrointestinal tract. Administration of synbiotic food containing a combination of a probiotic bacterial strain and the prebiotic sugar produced by that strain could be effective in improving human health.

## Author details

Tsuda Harutoshi
*National Institute of Health and Nutrition, Tokyo, Japan*

## 6. References

[1] Chabot S, Yu H.L, de Leseleuc L, Cloutier D, Van Calsteren M.R, Roy D, Lacroix M, Oth D (2001) Exopolysaccharides from *Lactobacillus rhamnosus* RW-9595M stimulate TNF, IL-6 and IL-12 in human and mouse cultured immunocompetent cells, and IFN-γ in mouse splenocytes. Lait 81: 683-698.

[2] Kitazawa H, Harata T, Uemura J, Saito T, Kaneko T, Itoh T (1998) Phosphate group requirement for mitogenic activation of lymphocytes by an extracellular phosphopolysaccharide from *Lactobacillus delbrueckii* ssp. *bulgaricus*. Int. J. Food Microbiol. 40: 169-175.

[3] Sreekumar O, Hosono A (1998) The antimutagenic properties of a polysaccharide produced by *Bifidobacterium longum* and its cultured milk against some heterocyclic amines. Can. J. Microbiol. 44: 1029-1036.

[4] Tsuda H, Hara K, Miyamoto T (2008) Binding of mutagens to exopolysaccharide produced by *Lactobacillus plantarum* mutant strain 301102S. J. Dairy Sci. 91: 2960-2966.

[5] Ganzle M. Michael G., Schwab C (2005) Exopolysaccharide production by intestinal lactobacilli. In: Tannock G. W, editors. Probiotics & Prebiotics: Scientific Aspects. Norfolk: Caister Academic Press. pp. 83-96.

[6] Kralj S., van Geel-Schutten G. H., Dondorff M. M. G., Kirsanovs S, van der Maarel M. J. E. C., Dijkhuizen L (2004) Glucan synthesis in the genus *Lactobacillus*: isolation and characterization of glucansucrase genes, enzymes and glucan products from six different strains. Microbiology 150: 3681-3690.

[7] van Hijum S. A. F. T., Szalowska E., van der Maarel M. J. E. C., Dijkhuizen L (2004) Biochemical and molecular characterization of a levansucrase from *Lactobacillus reuteri*. Microbiology 150: 621-630.

[8] de Vuyst L., de Vin F., Vaningelgem F., Degeest B (2001) Recent developments in the biosynthesis and applications of heteropolysaccharides from lactic acid bacteria. Int. Dairy J. 11: 687-707.

[9] Ruas-Madiedo P., Hugenholtz J., Zoon P (2002) An overview of the functionality of exopolysaccharides produced by lactic acid bacteria. Int. Dairy J. 12:163-171.

[10] Ruijssenaars H, Stingele F, Hartmans S (2000) Biodegradability of food-associated extracellular polysaccharides. Curr. Microbiol. 40: 194-199.

[11] Korakli M, Ganzle M.G, Vogel R.F (2002) Metabolism by bifidobacteria and lactic acid bacteria of polysaccharides from wheat and rye, and exopolysaccharides produced by *Lactobacillus sanfranciscensis*. J. Appl. Microbiol. 92: 958-965.

[12] Fuller R (1989) Probiotics in man and animals. J. Appl. Bacteriol. 66: 365-378.

[13] FAO/WHO (2002) Guidelines for the evaluation of probiotics in food. Report of a joint FAO/WHO working group on drafting guidelines for the evaluation of probiotics in food. London, Ontario, Canada.

[14] Gibson G.R, Roberfroid M.B (1995) Dietary modulation of the human colonic microbiota: Introducing the concept of prebiotics. J. Nutr. 125: 1401-1412.

[15] Rastall R (2003) Enhancing the functionality of prebiotics and probiotics. In: Mattila-Sandholm T, Saarela M, editors. Functional Dairy Products. Florida: CRC Press LLC. pp. 301-315.

[16] Mayo B, Aleksandrzak-Piekarczyk T, Fernandez M, Kowalczyk M, Alvarez-Martin P, Bardowski J (2010) Updates in the metabolism of lactic acid bacteria. In: Mozzi F, Raya R. R, Vignolo G. M, editors. Biotechnology of Lactic Acid Bacteria. Iowa: Blackwell Publishing. pp. 3-33.

[17] Cerning J (1990) Exocellular polysaccharides produced by lactic acid bacteria. FEMS Microbiol. Lett. 87: 113-130.

[18] Pederson C. S, Albury M (1955) Variation among the heterofermentative lactic acid bacteria. J. Bacteriol. 70: 702-708

[19] Dellaglio F., Dicks L. M. T., Torriani S (1995) The genus *Leuconostoc*. In: Wood B. J. B., Holzapfel W. H, editors. The genera of lactic acid bacteria. Glasgow: Blackie Academic and Professional. pp. 235-278.

[20] Minervini F, Angelis M. D, Surico R. F, Cagno R. D, Ganzle M, Gobbetti M (2010) Highly efficient synthesis of exopolysaccharides by *Lactobacillus curvatus* DPPMA10 during growth in hydrolyzed wheat flour agar. Int. J. Food Microbiol. 141: 130-135.

[21] Cote G. L, Robyt F. J (1982) Isolation and partial characterization of an extracellular glucansucrase from *Leuconostoc mesenteroides* NRRl B-1355 that synthesizes an alternating (1→6), (1→3)-$\alpha$-D-glucan. Carbohydr. Res. 101:57-74.

[22] Uchida K (1996) Nyuusannkinn no kouzou to kinntaiseibunn. In: Nyuusannkinn kennkyuu syuudannkai, editors. Nyuusannkinn no kagaku to gijutsu. Tokyo: Gakkai Syuppann Center. pp. 59-88. (In Japanese)

[23] Van Geel-Schutten G. H, Faber E. J, Smit E, Bonting K, Smith M. R, Ten Brink B, Kamerling J. P, Vliegenthart J. F. G, Dijkhuizen L (1999) Biochemical and structural characterization of the glucan and fructan exopolysaccharides synthesized by the *Lactobacillus reuteri* wild-type strain and by mutant strains. Appl. Environ. Microbiol. 65: 3008-3014.

[24] De Vos P, Garrity G. M, Jones D, Krieg N. R, Ludwig W, Rainey F. A, Schleifer K. H, Whitman W. B editors (2009) Bergey's Manual of Systematic Bacteriology Second Edition Volume Three. New York: Springer.

[25] Doco T, Wieruszeski J. M, Fournet B (1990) Structure of an exocellular polysaccharide produced by *Streptococcus thermophilus*. Carbohydr. Res. 198: 313-321.

[26] Lemoine J, Chirat F, Wieruszeski J. M, Strecker G, Favre N, Neeser J. R (1997) Structural characterization of the exocellular polysaccharides produced by *Streptococcus thermophilus* SFi39 and SFi12. Appl. Environ. Microbiol. 63: 3512-

[27] Degeest B, de Vuyst L (2000) Correlation of activities of the enzymes a-phosphoglucomutase, UDP-galactose 4-epimerase, and UDP-glucose pyrophosphorylase with exopolysaccharide biosynthesis by *Streptococcus thermophilus* LY03. Appl. Environ. Microbiol. 66: 3519-3527.

[28] Bubb W. A, Urashima T, Fujiwara R, Shinnai T, Ariga H (1997) Structural characterization of the exocellular polysaccharide produced by *Streptococcus thermophilus* OR901. Carbohydr. Res. 301: 41-50.

[29] Low D, Ahlgren J. A, Horne D, McMahon D. J, Oberg C. J, Broadbent J. R (1998) Role of *Streptococcus thermophilus* MR-1C capsular exopolysaccharide in cheese moisture retention. Appl. Environ. Microbiol. 64: 2147-2151.

[30] Van Casteren W. H. M, de Waaed P, Dijkema C, Schols H. A, Voragen A. G. J (2000) Structural characterization and enzymic modification of the exopolysaccharide produced by *Lactococcus lactis* subsp. *cremoris* B891. Carbohydr. Res. 327: 411-422.

[31] Knoshaug E. P, Ahlgren J. A, Trempy J. E (2007) Exopolysaccharide expression in *Lactococcus lactis* subsp. *cremoris* Ropy352: Evidence for novel gene organization. Appl. Environ. Microbiol. 73: 897-905.

[32] van Casteren W. H. M, Dijkema C, Schols H. A, Beldman G, Voragen A. G. J (2000) Structural characterization and enzymic modification of the exopolysaccharide produced by *Lactococcus lactis* subsp. *cremoris* B39. Carbohydr. Res. 324: 170-181.

[33] Nakajima H, Hirota T, Toba T, Itoh T, Adachi S (1992) Structure of the extracellular polysaccharide from slime-forming *Lactococcus lactis* subsp. *cremoris* SBT 0495. Carbohydr. Res. 224: 245-253.

[34] Kitazawa H, Ishii Y, Uemura J, Kawai Y, Saito T, Kaneko T, Noda K, Itoh T (2000) Augmentation of macrophage functions by an extracellular phosphopolysaccharide from *Lactobacillus delbrueckii* ssp. *bulgaricus.* Food Microbiol. 17:109-118.

[35] Grobben G. J, Smith M. R, Sikkema J, de Bont J. A. M (1996) Influence of fructose and glucose on the production of exopolysaccharides and the activities of enzymes involved in the sugar metabolism and the synthesis of sugar nucleotides in *Lactobacillus delbrueckii* subsp. *bulgaricus* NCFB 2772. Appl. Microbiol. Biotechnol. 46:279-284.

[36] Petry S, Furlan S, Waghorne E, Saulnier L, Cerning J, maguin E (2003) Comparison of the thickening properties of four *Lactobacillus delbrueckii* subsp. *bulgaricus* strains and physicochemical characterization of their exopolysaccharides. FEMS Microbiol. Lett. 221: 285-291.

[37] Harding L. P, Marshall V. M, Elvin M, Gu Y, Laws A. P (2003) Structural characterization of a perdeuteriomethylated exopolysaccharide by NMR spectroscopy: characterization of the novel exopolysaccharide produced by *Lactobacillus delbrueckii* subsp. *bulgaricus* EU23. Carbohydr. Res. 338:61-67.

[38] Gruter M, Leeflang B. R, Kuiper J, Kamerling J. P, Vliegenthart F. G (1993) Structural characterization of the exopolysaccharide produced by *Lactobacillus delbrueckii* subspecies *bulgaricus* rr grown in skimmed milk. Carbohydr. Res. 239:209-226.

[39] Yamamoto T, Nunome T, Yamauchi R, Kato K, Sone Y (1995) Structure of an exocellular polysaccharide of *Lactobacillus helveticus* TN-4, a spontaneous mutant strain of *Lactobacillus helveticus* TY1-2. Carbohydr. Res. 275: 319-332.

[40] Robijn G. W, Thomas J. R, Haas H, van den Berg D. J. C, Kamerling J. P, Vliengenthart J. F. G (1995) The structure of the exopolysaccharide produced by *Lactobacillus helveticus* 766. Carbohydr. Res. 276: 137-154.

[41] Staaf M, Widmalm G, Yang Z, Huttunen E (1996) Structural elucidation of an extracellular polysaccharide produced by *Lactobacillus helveticus*. Carbohydr. Res. 291: 155-164.

[42] Staaf M, Yang Z, Huttunen E, Widmalm G (2000) Structural elucidation of the viscous exopolysaccharide produced by *Lactobacillus helveticus* Lb161. Carbohydr. Res. 326: 113-119.

[43] Van Calsteren M. R, Pau-Roblot C, Begin A, Roy D (2002) Structure determination of the exopolysaccharide produced by *Lactobacillus rhamnosus* strains RW-9595M and R. Biochem. J. 363: 7-17.

[44] Landersjo C, Yang Z, Huttunen E, Widmalm G (2002) Structural studies of the exopolysaccharide produced by *Lactobacillus rhamnosus* strain GG (ATCC 53103). Biomacromolecules 3: 880-884.

[45] Tallon R, Bressollier P, Urdaci M. C (2003) Isolation and characterization of two exopolysaccharides produced by *Lactobaciilus plantarum* EP56. Res. Microbiol. 154: 705-712.

[46] Rodriguez-Carvajal M. A, Sanchez J. I, Campelo A. B, Martinez B, Rodriguez A, Gil-Serrano A. M (2006) Structure of the high-molecular weight exopolysaccharide isolated from *Lactobacillus pentosus* LPS26. Carbohydr. Res. 343: 3066-3070.

[47] Robijn G. W, Wienk H. L. J, van den Berg D. J. C, Haas H, Kamerling J. P, Vliengenthart J. F. G (1996) Structural studies of the exopolysaccharide produced by *Lactobacillus paracasei* 34-1. Carbohydr. Res. 285: 129-139.

[48] Mukai T, Toba T, Itoh T, Adachi S (1990) Structural investigation of the capsular polysaccharide from *Lactobacillus kefiranofaciens* K1. Carbohydr. Res. 204: 227-232.

[49] van den Berg D. J. C, Robijn G. W, Janssen A. C, Giuseppin M. L. F, Vreeker R, Kamerling J. P, Vliegenthart J. F. G, Ledeboer A. M, Verrips C. T (1995) Production of a novel extracellular polysaccharide by *Lactobacillus sake* 0-1 and characterization of the polysaccharide. Appl. Environ. Microbiol. 61: 2840-2844.

[50] Leo F, Hashida S, Kumagai D, Uchida K, Motoshima H, Arai I, Asakuma S, Fukuda K, Urashima T (2007) Studies on a neutral exopolysaccharide of *Lactobacillus fermentum* TDS030603. J. Appl. Glycosci. 54: 223-229.

[51] Figueroa C, Davila A. M, Pourquie J (1995) Lactic acid bacteria of the sour cassava starch fermentation. Lett. Appl. Microbiol. 21: 126-130.

[52] Cerning J (1995) Production of exopolysaccharides by lactic acid bacteria and dairy propionibacteria. Lait, 75: 463-472.

[53] Tsuda H, Miyamoto T (2010) Production of exopolysaccharide by *Lactobacillus plantarum* and the prebiotic activity of the exopolysaccharide. Food Sci. Technol. Res. 16: 87-92.

[54] Macedo M.G, Lacroix C, Gardner N.J, Champagne C.P (2002). Effect of medium supplementation on exopolysaccharide production by *Lacctobacillus rhamnosus* RW-9595M in whey permeate. Int. Dairy J. 12: 419-426.

[55] Maeda H, Zhu X, Suzuki S, Suzuki K, Kitamura S (2004) Structural characterization and biological activities of an exopolysaccharide kefiran produced by *Lactobacillus kefiranofaciens* WT-2B$^T$. J. Agric. Food Chem. 52: 5533-5538.

[56] de Vuyst L, Degeest B (1999) Heteropolysaccharides from lactic acid bacteria. FEMS Microbiol. Rev. 23: 153-177.

[57] Garcia-Garibay M, Marshall V. M. E (1991) Polymer production by *Lactobacillus delbrueckii* ssp. *bulgaricus*. J. Appl. Bacteriol. 70: 325-328.

[58] Kimmel S. A, Roberts R. F, Ziegler G. R (1998) Optimization of exopolysaccharide production by *Lactobacillus delbrueckii* subsp. *bulgaricus* RR grown in a semidefined medium. Appl. Environ. Microbiol. 64: 659-664.

[59] Gamar L, Blondeau K, Simonet J. M (1997) Physiological approach to extracellular polysaccharide production by *Lactobacillus rhamnosus* strain C83. J. Appl. Microbiol. 83: 281-287.

[60] Grobben G. J, Chin-Joe I, Kitzen V. A, Boels I. C, Boer F, Sikkema J, Smith M. R, de Bont J. A. M (1998) Enhancement of exopolysaccharide production by *Lactobacillus delbrueckii* subsp. *bulgaricus* NCFB 2772 with a simplified defined medium. Appl. Environ. Microbiol. 64: 1333-1337.

[61] Cerning J, Renard C. M. G. C, Thibault J. F, Bouillanne C, Landon M, Desmazeaud M, Topisirovic L (1994) Carbon source requirements for exopolysaccharide production by *Lactobacillus casei* CG11 and partial structure analysis of the polymer. Appl. Environ. Microbiol. 60: 3914-3919.

[62] de Vuyst L, Vanderveken F, Van de Ven S, Degeest B (1998) Production by and isolation of exopolysaccharides from *Streptococcus thermophilus* grown in a milk medium and evidence for their growth-associated biosynthesis. J. Appl. Microbiol. 84: 1059-1068.

[63] Korakli M, Pavlovic M, Ganzle M. G, Vogel R. F (2003) Exopolysaccharide and kestose production by *Lactobacillus sanfranciscensis* LTH2590. Appl. Environ. Microbiol. 69: 2073-2079.

[64] Kralj S, van Geel-Schutten G. H, van der Maarel M. H. E. C, Dijkhuizen L (2004) Biochemical and molecular characterization of *Lactobacillus reuteri* 121 reuteransucrase. Microbiology 150: 2099-2112.

[65] Neubauer H, Bauche A, Mollet B (2003) Molecular characterization and expression analysis of the dextransucrase DsrD of *Leuconostoc mesenteroides* Lcc4 in homologous and heterologous *Lactococcus lactis* cultures. Microbiology 149: 973-982.

[66] Dabour N, LaPointe G (2005) Identification and molecular characterization of the chromosomal exopolysaccharide biosynthesis gene cluster from *Lactococcus lactis* subsp. *cremoris* SMQ-461. Appl. Environ. Microbiol. 71: 7414-7425.

[67] Dan T, Fukuda K, Sugai-Bannai M, Takakuwa N, Motoshima H, Urashima T (2009) Characterization and expression analysis of the exopolysaccharide gene cluster in *Lactobacillus fermentum* TDS030603. Biosci. Biotechnol. Biochem. 73: 2656-2664.

[68] Mozzi F, Savoy de Giori G, Font de Valdez G (2003) UDP-galactose 4-epimerase: a key enzyme in exopolysaccharide formation by *Lactobacillus casei* CRL 87 in controlled pH batch cultures. J. Appl. Microbiol. 94: 175-183.

[69] Kleerebezem M. et al. (2003) Complete genome sequence of *Lactobacillus plantarum* WCFS1. Proc. Natl. Acad. Sci. USA. 100: 1990-1995.

[70] Morona J. K, Paton J. C, Miller D. C, Morona R (2000) Tyrosine phosphorylation of CpsD negatively regulates capsular polysaccharide biosynthesis in *Streptococcus pneumonia*. Mol. Microbiol. 35: 1431-1442.

[71] Nagai T, Makino S, Ikegami S, Itoh H, Yamada H (2012) Effects of oral administration of yogurt fermented with *Lactobacillus delbrueckii* ssp. *bulgaricus* OLL1073R-1 and its exopolysaccharides against influenza virus infection in mice. Int. Immunopharmacolgy 11: 2246-2250.

[72] Kodali V. P, Sen R (2008) Antioxidant and free radical scavenging activities of an exopolysaccharide from a probiotic bacterium. Biotechnol. J. 3: 245-251.

[73] Kleerebezem M, van Kranenburg R, Tuinier R, Boels I. C, Zoon P, Looijesteijn E, Hugenholtz J, de Vos W. M (1999) Exopolysaccharides produced by *Lactococcus lactis*: from genetic engineering to improved rheological properties? Ant. van Leeuwenhoek 76: 357-365.

[74] O'Connor E. B, Barrett E, Fitzgerald G, Hill C, Stanton C, Ross R. P (2005) Production of vitamins, exopolysaccharides and bacteriocins by probiotic bacteria. In: Tamime A. Y, editors. Probiotic Dairy Products. Oxford: Blackewell Publishing Ltd. pp. 167-194.

[75] Mitsuoka T (1992) Intestinal flora and aging. Nutr. Rev. 50: 438-446.

[76] Kunz C, Rudloff S, Baier W, Kein N, Strobel S (2000) Oligosaccharides in human milk: structure, functional, and metabolic aspects. Ann. Rev. Nutr. 20: 699-722.

[77] Deguchi Y, Matsumoto K, Ito A, Watanuki M (1997) Effects of β 1-4 galactooligosaccharides administration on defecation of healthy volunteers with constipation tendency. Eiyougakuzasshi 55: 13-22. (in Japanese)

[78] Ishikawa F, Takayama H, Matsumoto K, Ito M, Chonan O, Deguchi Y, Kikuchi-Hayakawa H, Watanuki M (1995) Effects of β 1-4 linked galactooligosaccharides on human fecal microflora. Bifizusu 9: 5-18. (in Japanese)

[79] Gopal P. K, Sullivan P. A, Smart J. B (2001) Utilisation of galacto-oligosaccharides as selective substrates for growth by lactic acid bacteria including *Bifidobacterium lactis* DR10 and *Lactobacillus rhamnosus* DR20. Int. Dairy J. 11: 19-25.

[80] Barboza M, Sela D. A, Pirim C, LoCascio R. G, Freeman S. L, German J. B, Mills D. A, Lebrilla C. B (2009) Glycoprofiling bifidobacterial consumption of galacto-oligosaccharides by mass spectrometry reveals strain-specific, preferential consumption of glycans. Appl. Environ. Microbiol. 75: 7319-7325.

[81] Cardelle-Cobas A, Corzo N, Olano A, Pelaez C, Requena T, Avila M (2011) Galactooligosaccharides derived from lactose and lactulose: Influence of structure on *Lactobacillus*, *Streptococcus* and *Bifidobacterium* growth. Int. J. Food Microbiol. 149: 81-87.

[82] Macfarlane G.T, Steed H, Macfarlane S (2008). Bacterial metabolism and health-related effects of galacto-oligosaccharides and other prebiotics. J. Appl. Microbiol. 104: 305-344.

[83] Rabiu B. A, Jay A. J, Gibson G. R, Rastall R. A (2001) Synthesis and fermentation properties of novel galacto-oligosaccharides by β-galactosidases from *Bifidobacterium* species. Appl. Environ. Microbiol. 67: 2526-2530.

[84] Tokunaga T, Nakada Y, Tashiro Y, Hirayama M, Hidaka H (1993) Effects of fructooligosaccharides intake on the intestinal microflora and defecation in healthy volunteers. Bifizusu 6: 143-150. (in Japanese)

[85] Gibson G. R, Beatty E. R, Wang X, Cummings J. H (1995) Selective stimulation of bifidobacteria in the human colon by oligofructose and inulin. Gastroenterology 108: 975-982.

[86] Ramirez-Farias C, Slezak K, Fuller Z, Duncan A, Holtrop G, Louis P (2009) Effect of inulin on the human gut microbiota: stimulation of *Bifidobacterium adolescentis* and *Faecalibacterium prausnitzii*. Br. J. Nutr. 101: 541-550.

[87] Rossi M, Corradini C, Amaretti A, Nicolini M, Pompei A, Zanoni S, Matteuzzi D (2005) Fermentation of fructooligosaccharides and inulin by bifidobacteria: a comparative study of pure and fecal cultures. Appl. Environ. Microbiol. 71: 6150-6158.

[88] Olano-Martin E, Gibson G. R, Rastall R. A (2002) Comparison of the in vitro bifidogenic properties of pectins and pectic-oligosaccharides. J. Appl. Microbiol. 93: 505-511.

[89] Palframan R, Gibson G. R, Rastall R. A (2003) Development of a quantitative tool for the comparison of the prebiotic effect of dietary oligosaccharides. Lett. Appl. Microbiol. 37: 281-284.

[90] Vulevic J, Rastall R. A, Gibson G. R (2004) Developing a quantitative approach for determining the in vitro prebiotic potential of dietary oligosaccharides. FEMS Microbiol. Lett. 236: 153-159.

[91] Sanz M. L, Gibson G. R, Rastall R. A (2005) Influence of disaccharide structure on prebiotic selectivity in vitro. J. Agric. Food Chem. 53: 5192-5199.

[92] Huebner J, Wehling R. L, Hutkins R. W (2007) Functional activity of commercial prebiotics. Int. Dairy J. 17: 770-775.

[93] Depeint F, Tzortzis G, Vulevic J, I'Anson K, Gibson G. R (2008) Prebiotic evaluation of a novel galactooligosaccharide mixture produced by the enzymatic activity of *Bifidobacterium bifidum* NCIMB 41171, in healthy humans: a randomized, double-blind, crossover, placebo-controlled intervention study. Am. J. Clin. Nutr. 87: 785-791.

[94] van den Broek L. A. M, Hinz S. W. A, Beldman G, Vincken J. P, Voragen A. G. J (2008) *Bifidobacterium* carbohydrases – their role in breakdown and synthesis of (potential) prebiotics. Mol. Nutr. Food Res. 52: 146-163.

[95] Schwab C, Lee V, Sorensen K. I, Ganzle M. G (2011) Production of galactooligosaccharides and heterooligosaccharides with disrupted cell extracts and whole cells of lactic acid bacteria and bifidobacteria. Int. Dairy J. 21: 748-754.

[96] Vardakou M, Palop C. M, Christakopoulos P, Faulds C. B, Gasson M. A, Narbad A (2008) Evaluation of the prebiotic properties of wheat arabinoxylan fractions and induction of hydrolase activity in gut microflora. Int. J. Food Microbiol. 123: 166-170.

[97] Hongpattarakere T, Cherntong N, Wichienchot S, Kolida S, Rastall R. A (2012) In vitro prebiotic evaluation of exopolysaccharides produced by marine isolated lactic acid bacteria. Carbohydr. Polym. 87: 846-852.

[98] Pham P.L, Dupont I, Roy D, Lapointe G, Cerning J (2000) Production of exopolysaccharide by *Lactobacillus rhamnosus* R and analysis of its enzymatic

degradation during prolonged fermentation. Appl. Environ. Microbiol. 66: 2302-2310.

[99] Tsuda H, Hara K, Miyamoto T (2008) Survival and colonization of orally administered *Lactobacillus plantarum* 301102 in porcine gastrointestinal tract. Anim. Sci. J. 7: 274-278.

Chapter 23

# Dynamic Stresses of Lactic Acid Bacteria Associated to Fermentation Processes

Diana I. Serrazanetti, Davide Gottardi, Chiara Montanari and Andrea Gianotti

Additional information is available at the end of the chapter

http://dx.doi.org/10.5772/51049

## 1. Introduction

Despite their negligible mass the microbial agents, starters and non starters, play a profound role in the characterization of the fermented foods in terms of chemical and sensorial properties. In fact, fermented foods may be defined as foods processed through the activity of microorganisms. Fermentation processes take a special place in the evolution of human cuisine, by altering the taste experience of food products, as well as extending the storage period. In particular, foods fermented with lactic acid bacteria (LAB) have constituted an important part of human diet and of fermentation processes (involving various foods, including milk, meat, vegetables and fruits) [1] since ancient times. They have played an essential role in the preservation of agricultural resources and in the improvement of nutritional and organoleptic properties of human foods and animal feed. Moreover, these organisms nowadays are increasingly used as health promoting probiotics, enzyme and metabolite factories and vaccine delivery vehicles [2].

It is interesting to outline how the changes of food characteristics during the fermentation process can be described as dynamic fluctuations of the food environment itself and, at the same time, stress source for the microorganisms involved [3, 4], such as LAB. In fact, whenever autochthonous bacteria are adapted and competitive in their respective environment, the environment can be described as stressful for LAB [5, 4]. The fermentation parameters, including temperature, water activity (Aw), oxygen, pH, as well as the concentration of starter cultures, affect the regulatory mechanism and the response mechanisms of LAB, as well as their effects on the final products properties [4].

When LAB are added to food formulations, several factors that may influence the ability of those microorganisms to survive, growth and become active in the new matrix have to be considered [6]. These factors include: 1) the physiological state of the LAB used as starters (whether the cells are from the logarithmic or the stationary growth phase); 2) the physical conditions of product ripening and storage (eg. temperature); 3) the chemical composition of the matrix (eg. acidity, available carbohydrates content, nitrogen source, mineral content, water activity and oxygen concentration); 4) possible interactions of the starter cultures with probiotics and other microorganisms naturally occurring or added to the system [6].

In figure 1 the main factors affecting the viability and the responses of LAB from production to storage are described [7].

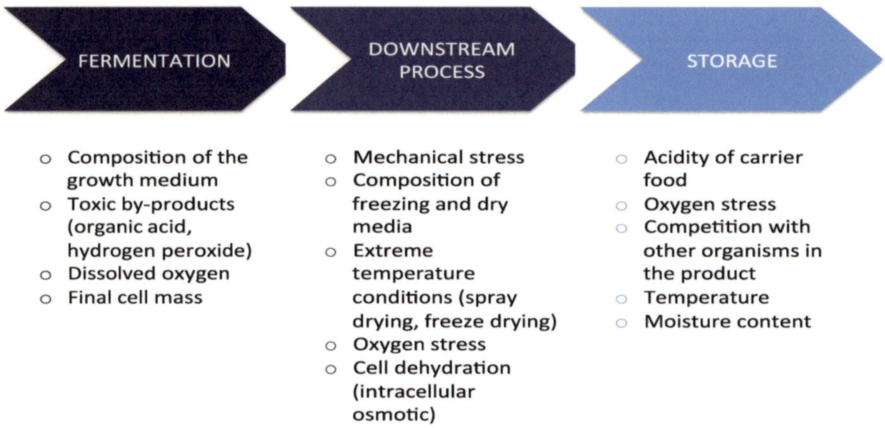


**Figure 1.** Factors affecting the viability and the responses of LAB to the various fermented foods production steps.

To better elucidate what happens to LAB during fermentation processes, we decided to use a model (defined "virtual food") that mimics various steps occurring during processing and that can affect LAB performances or viability.

## 2. Lactic acid bacteria and stress: Basic concepts

"Stress results from interactions between subjects and their environment that are perceived as straining or exceeding their adaptive capacities and threatening their well-being. The element of perception indicates that human stress responses reflect differences in personality, as well as differences in physical strength or general health" [8].

Stress has driven evolutionary changes (the development and natural selection of species over time). Thus, the species that adapted best to the causes of stress (stressors) have survived and evolved into the plant and animal kingdoms we now observe. The same evolutionary process regarded microorganisms. In fact, bacteria, irrespective of natural habitat, are exposed to constant fluctuations in their growth conditions. Consequently they have developed sophisticated responses, modulated by the re-modelling of protein complexes and by phosphorylation dependent signal transduction systems, to adapt and to survive to a variety of insults. To ensure survival to environmental adversities, bacteria may adapt to changes in their immediate vicinity by responding to the imposed stress. These responses are different and vast and depend on the microorganism nature and on the environmental stress and are accomplished by changes in the patterns of gene expression for those genes whose products are required to combat the deleterious [3]. In particular, cellular metabolic pathways are closely related to stress responses and the flux of particular metabolites to understand the hypothetically shifts and implications in the food systems has been studied in LAB [9-13, 4, 14, 15].

LAB are a functionally related group of organisms known primarily for their bioprocessing roles in food and beverages [16]. LAB play a crucial role in the development of the organoleptic and hygienic quality of fermented products. These microorganisms are used as starter cultures in many fermented products (i.e. beer, milk, dough, sausages and wine). Therefore, the reliability of starter cultures in terms of quality and functional properties (important for the development of aroma and texture), but also in terms of growth performance and robustness, has become essential for successful fermentations [17]. There have been some reports describing the physiological stress responses in LAB, particularly *Lactobacillus* species, which have a broad biodiversity [17-21, 13, 22, 4, 14, 15].

LAB evolved specific mechanisms to respond and to survive to environmental stresses and changes (stress-sensing system and defences). In fact, microorganisms could have specific regulators tailored to each of their regulated genes and adapt their expression according to environment. Stress defences are good examples of such integrated regulation systems. Bacterial stress responses rely on the coordinated expression of genes that alter different cellular processes (cell division, DNA metabolism, housekeeping, membrane composition, transport, etc.) and act in concert to improve the bacterial stress tolerance. The integration of these stress responses is accomplished by networks of regulators that allow the cells to react to various and complex environmental shifts. LAB respond to stress in a very specific way dependent on the species, on the strains and on the type of stress. The best-studied stresses are acid, heat, oxidative and cold stresses, although for the latter most of the studies focused on a specific family of proteins instead of analyzing the whole response [4].

Despite the extensive use of LAB, there is a paucity of information concerning the stress-induced mechanisms studied *in vivo* for improving the survival of these organisms during real food processing. A better knowledge of the adaptive responses of LAB is important because the fermentation processes often expose these microorganisms to adverse environmental conditions. LAB should resist to adverse conditions encountered in industrial processes, for example during starter handling and storage (freeze drying, freezing or spray-drying) and during the fermentation environment dynamic changes. These phenomena reinforce the need for robust LAB since they may have to survive and grow in different unfavorable conditions expressing specific functions (for example during stationary phase or storage) [17].

## 3. Principal responses to the most common stresses

**Heat shock response:** The effect of heat shock and the induction of a stress response in *Lactobacillus* spp. have been studied for *Lactobacillus delbrueckii* subsp. *bulgaricus* [23] and *Lactobacillus paracasei* [24, 25], *Lactobacillus acidophilus, Lactobacillus casei* and *Lactobacillus helveticus* [26], *Lactobacillus collinoides* [27] , *Lactobacillus sakei* [28], *Lactobacillus johnsonii* [29], *Lactobacillus rhamnosus* [30], *Lactobacillus plantarum* [31-33] and *Lactobacillus salivarius* [34]. The heat resistance of LAB is a complex process involving proteins with different roles in cell physiology, including chaperone activity, ribosome stability, stringent response mediation, temperature sensing and control of ribosomal functions [31]. The time taken to initiate the stress response is different for different treatments and different strains. The major problem encountered by cells at high temperature is the denaturation of proteins and their subsequent aggregation. In addition Earnshaw et al. [35], , Texeira et al. [36] and Hansen et al. [37] described also as response to heat stress the destabilization of macromolecules as ribosomes and RNA as well as alterations of membrane fluidity.

Heat stress response is characterized by the transient induction of general and specific proteins and by physiological changes. In every strain tested the involvement of Heat Shock Proteins (HSPs such as DnaK, GroEL and GroES during the heat stress was clear) [23-38]. The role of these stress proteins is complex; in fact, the bind substrate proteins in a transient non-covalent manner prevent premature folding and promote the attainment to the correct state *in vivo*. The resistance to heat stress is higher when the cells were previously exposed and adapted to this type of stress in the stationary phase, otherwise, when pre-adapted in exponential phase, the cells are more sensitive. In particular, the storage stability of the culture that was heat shocked after stationary phase was superior to that of culture heat shocked after log phase [34, 23, 30].

**Cold shock response:** It is very important to improve knowledge about LAB behavior in cold environment. In fact, during industrial processes, like in cheese ripening and refrigerated storage of fermented products, these microorganisms are subjected to different temperatures far below the optimal growth temperature. When LAB living cells are exposed to these cold environments, important physiological changes occur, such as decrease in membrane fluidity and stabilization of secondary structures of RNA and DNA, resulting in a reduced efficiency of translation, transcription and DNA replication. The response of microorganisms to these effects is termed cold-shock response during which a number of Cold Induced Proteins (CIPs) are synthesized. The roles of these proteins are at the levels of membrane fluidity, DNA supercoiling and transcription and translation. Few papers have described cold shock proteins and mechanisms in LAB, in particular they have focused on *Lactococcus lactis* and *L. plantarum* [39-42]. Kim et al. [39, 40] tested different LAB to evaluate cold shock effects on cryotolerance. Improved understanding of cold-shock-induced cryotolerance may contribute to the development of environmental conditions that allow improved viability/activity of frozen or freeze-dried commercial LAB starter cultures. The results showed that, as with heat stress, there is also an improvement of the viability of the tested strains as concerning the cryotolerance after a cold shock. The process of freezing appeared to have different effects on different LAB as well as different effects on strains within the same genus. Moreover, the freezing response of the strains depends on the time of the cold shock process and the induction of cryotolerance appears to be dependent on the growth phase in which the cold shock took place [43-47].

Another interesting study regarding LAB response to sub-lethal cold stress was developed by Montanari et al.[14]. These Authors separated and quantified the cell cyclopropane fatty acids lactobacillic (C19cyc11) and dehydrosterculic (C19cyc9) to study the adaptive response to sub-lethal acid and cold stresses in *L. helveticus* and *Lactobacillus sanfranciscensis*. These microorganisms showed different fatty acids composition and environmental adaptation to short term cold and acidic stresses. In *L. helveticus* C19cyc11 dramatically increased after 2 h at 10°C and with the pH decrease, particularly in micro-aerobic conditions, in the presence of tween 80, and in anaerobic conditions. The increase of lactobacillic acid in *L. helveticus* is necessary to maintain the cell membrane in a suitable state of fluidity. Moreover, cyclopropane fatty acids confer resistance to ozonolysis, singlet oxygen and mild oxidative treatments [48, 49], suggesting a cross protection and response of LAB cell membrane to physicochemical stresses. A combined analysis of the genome-wide transcriptome and metabolism was performed with a dairy *Lactococcus lactis* subsp. *lactis* under dynamic conditions similar to the conditions encountered during the cheese-making process. Specific responses to acid and cold stresses were identified, but also the induction of unexpected pathways was determined. In particular, the induction of purine biosynthesis and prophage [50].

**Oxidative stress response:** LAB are facultative anaerobic microorganisms that have in common the reduction of part of pyruvate produced to lactate production in order to regenerate NAD+ from NADH formed during glycolysis. They do not require oxygen for growth and, in fact, a negative effect of oxygen on the development of these bacteria has often been observed. It was generally believed that these bacteria could under no condition use oxygen as the terminal electron acceptor [17]. However, many LAB have NADH oxidase and some can even express a functionally active respiratory chain in the presence of heme [51-57]. Respiration-competent LAB differ from the features of *Escherichia coli* and *Bacillus subtilis*, since they carry limited equipment for respiration. All respiring LAB carry genes encoding electron donor (NADH dehydrogenase) and a single electron acceptor (cytochrome bd oxidase) [58]. Addition of heme to the system activates respiration chain NADH oxidase activity, but none of the tested LAB synthesize heme [01].

When for some reasons the generation of free radicals is higher than the rate of their detoxification the cells are exposed to a constraint called "oxidative stress" [59]. For the food-associated LAB a still fragmented picture of the resistance mechanisms present emerges. Representatives of the different mechanisms have been described in different LAB [60-64]. Apart from the toxic effects of oxygen, aeration can induce important changes in the sugar metabolism of LAB. In fact, the presence of oxygen is a factor that greatly affects the outcome of a fermentation process. In general, LAB tolerate oxygen but grow better under nearly anaerobic conditions. However, in the presence of heme and oxygen LAB start respiration metabolism, by which the cell metabolism is reprogrammed so that pH, oxygen status, growth capacity and survival are markedly altered [56]. In the presence of oxygen and during the fermentation metabolism, $H_2O_2$ is formed. Numerous species of LAB contain peroxidase and/or catalase to prevent and eliminate these deleterious effects [17]. Concerning the prevention of reactive oxygen species (ROS) formation, the scope of the reactions is the eliminations of free oxygen. In a study on *L. helveticus* the fatty acids composition in the cell membrane changed in response to oxidative stress. In fact, the activity of oxygen consuming desaturase system increased to reduce the free radical damage to the cell [19]. Generally, the response to oxidative stress of LAB is similar, but also depends on the species, on the strains and, with regard to catalase action, on the bacterial density [4]. In *L. lactis* several genes have been identified and the respective encoded proteins have been shown to contribute to oxidative stress resistance. Moreover, the induction of these genes is growth phase-dependent (exponential or stationary) and their products confer multi-stress resistance [52]. General stress resistance mechanisms may also confer resistance to oxidative stress. In fact, in a model system several acid resistant mutants of *L. lactis* that appeared also more resistant to oxidative stress were isolated [64].

**Acid stress response:** Understanding the acid resistance mechanism used by LAB to survive to by-products of their own metabolism (i.e. homofermentative *L. lactis* converts 90% of metabolized sugar to lactic acid) and the response available in low-pH foods is of great importance. In LAB one of the most effective mechanisms for resistance in acid stress environment is the glutamate decarboxylase (GAD). In fact, few years ago, it was proposed that amino acid decarboxylase functions to control the pH of the bacterial environment by consuming hydrogen ions as part of carboxylation reaction [65]. LAB are also capable of inducing an Acid Tolerance Response (ATR) in response to mild acid treatments. The system induced includes pH homeostatis, protection and repair mechanisms. Genes and proteins, involved in pH homeostasis and cell protection or repair, play a role in acid adaptation, but this role can also extend to more general acid tolerance mechanisms. A more specific study was developed on the effects of lactic acid stress on *L. plantarum* by transcription profiling [66]. The difference, in terms of stress response, into the dissociated or undissociated forms of lactic acid has been highlighted. The toxicity of organic acids depends on their degree of dissociation and thus on the pH. For LAB end product inhibition by lactic acid could result in a disturbance of the regeneration of cofactor NAD+, especially under anaerobic conditions, in which the cell does not have the possibility of NAD+ regeneration by NADH oxidase. The response at membrane fatty acids level to acid stress was studied in *L. helveticus* and *L. sanfranciscensis* [14]. The relevant proportion of dodecanoic acid in the latter species under acid stress suggests that carbon chain shortening is the principal strategy of *L. sanfranciscensis* to modulate fluidity or chemico-physical properties of the membranes in the presence of acid stress. Moreover, a specific shift in leucine catabolic pathway at pH 3.6 was identified in *L. sanfranciscensis* [15]. In fact, the acid stress induced a metabolic shift toward overproduction of 3-methylbutanoic and 2-methylbutanoic acids, accompanied by sugar reduced consumption and primary carbohydrate metabolite production. The metabolites coming from branched chain amino acids (BCAAs) catabolism increased up to seven times under acid stress. While the overproduction of 3-methylbutanoic acid under acid stress can be attributed to the need to maintain redox balance, the rationale for the production of 2-methylbutanoic acid from leucine can be found in a newly proposed biosynthetic pathway leading to 2-methylbutanoic acid and 3 mol of ATP per mol of leucine. Leucine catabolism to 3-methylbutanoic and 2-methylbutanoic acids suggests that the switch from sugar to amino acid catabolism supports growth of *L. sanfranciscensis* in restricted environments such as sourdough, characterized by acid stress and recurrent carbon starvation.

**Osmotic stress response:** In the various applications in food and feed industry LAB can be exposed to osmotic stress when important amounts of salts or sugars are added to the product [17]. In fact, in most of the food habitats where lactobacilli live, they are confronted with salt [67] and sugar stress [68]. Study on the differences between salt and sugar osmotic stress revealed that the hyperosmotic conditions imposed by sugar stress are much less detrimental and only transient (transient osmotic stress), because the cells are able to balance the extra and the intracellular concentrations of lactose and sucrose [17]. Bacteria need to adapt to this change in their environment in order to survive [69], and they can do it by accumulating (by uptake or synthesis) compatible solutes, generally of organic origin, under hyperosmotic conditions [17]. The compatible solutes are defined as osmoprotectants. The main strategy to adapt to high osmolarity of non-halophilic bacteria is associated with the enhancement of the osmotolerance [68]. Moreover, the osmoprotectants can also stabilize enzymes and provide protection not only against osmotic stress but also against other type of stresses (high temperature, freezing and drying). The intracellular accumulation of compatible solutes prevents the loss of water caused by high external osmolarity and allows the maintenance of turgor [68]. The accumulation of carnitin, betain and proline was determined in LAB grown in MRS and complex diluted MRS medium (DMRS medium) [70]. Moreover, a specific response mechanism to osmotic stress was identified in a sourdough model system [13]. In particular, the growth of *L. sanfranciscensis* under osmotic stress resulted in a relevant accumulation of 3-methylbutanoic acid. Its synthesis is associated with the BCCAs., is NAD+ dependent and produces NADH during the reaction [71]. The accumulation of 3-methylbutanoic acid as predominant metabolite has been also observed in model systems simulating sourdough as a consequence of osmotic, acid or oxidative stress [12, 15].

**High pressure stress response:** High-pressure processing (HPP) or high pressure homogenization (HPH) are non-thermal processes capable of inactivating and eliminating pathogenic and food spoilage microorganisms in specific foods [11, 72], and it represents an exceptional stimulus for most mesophilic bacteria. Several proteins are induced after high pressure treatment and some of these have also been involved in the response to other various stresses [8]. The responses to HHP stress have been studied in particular on *L. sakei* and *L. sanfranciscensis* [73, 18]. These Authors suggested the presence of *de novo* protein synthesis as a consequence of HHP stress [73]. As concerning HPH several interesting studies on the responses on *Lactobacillus* spp., at the level of proteolytic and metabolic activities point of view have been conducted [11, 21, 22, 74]. HPH treatment positively affects the proteolytic activity of some of *Lactobacillus* strains, but the activation and the quantitative and qualitative changes of the metabolic activity appear to be the most promising results. The pre-treatment at different pressure was able to induce relevant changes in term of fermentation dynamics and metabolism with respect to the untreated cells [11]. The same approach was applied on *L. acidophilus* and *L. paracasei* to improve the technological performances of probiotic strains [21, 22, 74]. The sub-lethal treatment with HPH enhanced the capacity of some *in vitro* probiotic features (i.e. hydrophobicity and tolerance to simulated gastric acidity) in a strain dependant way. *L. paracasei* A13 enhanced cellular hydrophobicity and auto-aggregation capacity after HPH treatment at 50 MPa. On the contrary, the HPH treatment decreased these features in the other strains considered. Highest values of hydrophobicity were found for *L. acidophilus* DRU and its bile-resistant derivative *L. acidophilus* DRU+, while lower values were obtained for *L. paracasei* strain [74]. Moreover, the stress responses enable survival under more severe conditions, enhancing resistance to subsequent processing conditions [75]. HPH treatment at 50 MPa can favour the maintenance of cell viability during a refrigerated storage in buttermilk, a suitable medium to maintain the cell viability during refrigeration [76]. The increased viability can be attributed to the increased precocious availability of low molecular weight peptides and free fatty acids such as oleic acid [21, 22].

**Competition and communication:** Food fermentations are typically carried out by mixed cultures consisting of multiple strains or species [77]. Mixed-culture food fermentations are of primary economic importance. The performance of these cultures, consisting of LAB, yeasts, and/or filamentous fungi, is not the simple result of "adding up" the individual single-strain functionalities, but is largely determined by interactions at the level of substrates, exchange of metabolites and growth factors or inhibiting compounds [77].

General microbial interference is an effective non-specific control mechanism common to all populations and environments including foods. It represents the inhibition of the growth of certain microorganisms by other members of the habitat.

The mechanisms involved are common to all genera and include [78]:

1. Nutrient competition,
2. Generation of unfavorable environment,
3. Competition for attachment/adhesion sites.

Most substrates for food fermentations have a highly heterogeneous physicochemical composition, which offers the possibility for the simultaneous occupation of multiple niches by "specialized" strains, for instance, through the utilization of different carbon sources. In these substrates, coexisting strains often interact through trophic or nutritional relations via multiple mechanisms [77].

Carbon sources are often present at high concentrations in food substrates, and therefore competition concerns the rapid uptake of nutrients and conversion into biomass. In dairy fermentations nitrogen is limiting, and initially organisms compete for the free amino acids and small peptides available. While in the later stages of fermentation, they compete for the peptides released by the actions of proteolytic enzymes [77].

In a cell-density-dependent quorum-sensing system, bacteria produce extracellular signaling molecules such as peptides or post-translationally modified peptides that act as inducers for gene expression when concentrations of these molecules exceed a certain threshold value [79]. These changes might eventually lead to competitive advantages for the population, more effective adaptation and responses to changing environmental conditions, or the co-ordination of interactions between bacteria and their abiotic and biotic environments [7]. In fact, microorganisms produce diffusible chemicals for the purpose of communication and it has been reported that the stress caused by the exposure of microbial cells to their own cell free conditioned media, containing metabolites and bioactive compounds including "quorum sensing" molecules, including 2(5H)-furanones, promotes cell differentiation, autolysis and overproduction of specific metabolites [12, 80, 9, 10]. In this way the microbial cultures used in food fermentations can also contribute (by "secondary" reactions and relations) to the formation of flavor and texture [81].

## 4. General steps regarding a virtual fermented food process

In the figure 2, the steps that mainly interest food fermentation are reported. A model virtual fermented food was identified to resume the common denominator of the fermented foods dynamics, particularly focused on the reciprocal influences between environmental fluctuation and LAB fermentation.

Whatever kind of food we want to produce, fermented or not, the first step of the process is the formulation: in this phase the main raw materials (meat, milk, fruit and vegetables or their derivatives) are mixed with other ingredients, that have different roles: salts or sugars to improve taste, spices to give specific sensorial quality and as antimicrobials, additives or other substances able to affect physical and structural properties, preservatives to improve microbial stability and shelf life. The addition of those ingredients can be perceived as stress. In fermented products, proper microorganisms, mainly yeasts and LAB, are also added as starter cultures, in order to start and lead the fermentation and to obtain a stable and standard final product. As a consequence, the microorganisms, naturally occurring or added as starter cultures, have to cope with a completely different system: in particular, naturally occurring microflora have to face the changes induced by the ingredients, while the starter cultures, deriving from growth media or added as lyophilized cultures, have to adapt to a real food system, where different sources of stresses are often present.

In particular, the first sub-lethal stress, which LAB face, regards the difference between the growth medium composition and the real food. Generally, LAB lyophilized cultures can be added to the ingredients after a reactivation and subsequently added to the product. This procedure identify the presence of a stress for the LAB cells. Starter cultures are added to the raw materials in large numbers and incubated under optimal conditions, but the adaptation to substrate or raw material is always necessary [82]. It is very important to consider the physiological state of the LAB before the inoculum. This state strongly depends on the time of harvesting of the culture (whether during the logarithmic or stationary phase of growth), on the conditions leading to transition to the stationary phase, on the treatment of the culture during and after harvesting and on the chemical composition of the environment. Therefore it is important during formulation and technological processes to consider also these factors, mainly for those products where microorganisms are added as starter cultures.

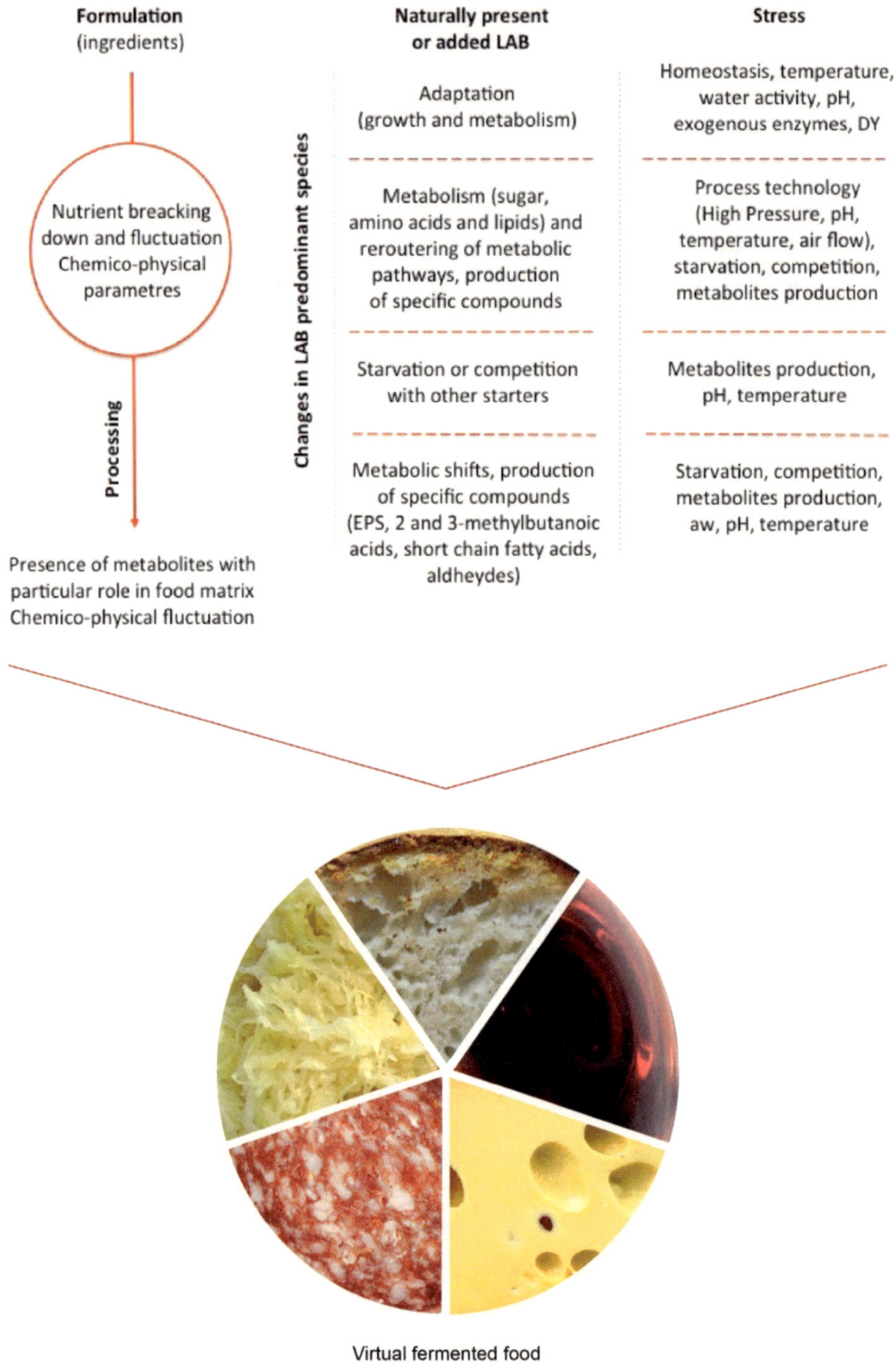


**Figure 2.** Fermented food model: reciprocal influences between environmental fluctuation and lactic acid bacteria fermentation.

The interaction between the starters and the ingredients and between the starters and the naturally present microbial population can trigger few important mechanisms that will influence the quality and the characteristics of the fermented product. Analogously, many food processes and formulations have been tested for safety by challenge test inoculating pathogen bacterial cells at different growth phases, and the results proved that cells grown to the stationary phase or adapted to various stresses have greater resistance than exponential cells [83].

Other ingredients usually added to obtain safe and stable products are food preservatives, including:

a. Antioxidants,
b. Anti-browning agents
c. Antimicrobials.

These latter are arbitrarily classified into two groups: traditional or "regulatory approved" and naturally occurring [84]. The former includes acidifiers such as acetic acid, lactic acid and citric acid and antimicrobials such as benzoic acid and benzoates, propionate, nitrites and nitrates, sorbic acid and sorbates and sulfites. The latter includes compounds from microbial, plant and animal sources that are, for the most part, only proposed for use in foods as antimicrobials (e.g. lactoferrin, lysozyme, nisin). Throughout the ages, food antimicrobials have been used primarily to prolong shelf-life and preserve quality of foods through inhibition of spoilage microorganisms, while only few are used exclusively to control the growth of specific foodborne pathogens (e.g. nitrite, used for hundreds of years to inhibit growth and toxin production of *Clostridium botulinum* in cured meats). In food formulation antimicrobials are part of a multiple intervention system that involves the chemical along with environmental (extrinsic) and food related (intrinsic) stresses and processing steps. Some of these substances (for example lactic acid and citric acid) provoke a direct acidification of a food or food ingredient, and therefore challenge the microflora inducing and increase of acid resistance of the microflora itself. In fermented food the situation can be somewhat different, because the pH is gradually lowered by LAB creating a pH gradient, more likely than a sharp alteration in the pH due to direct acidification.

A good model describing the shock related to the inoculum of LAB in the raw complex material has been described during the production of fermented sausages [85]. The relatively high pH of raw meat rapidly decreases during the initial fermentation phase because organic acids, mainly lactate, are formed by LAB and the water activity is reduced during ripening, because of the addition of salt as well as drying. Furthermore, adjuvants, such as potassium or sodium nitrite and/or nitrate, are mostly added to optimize the fermentation process.

Generally strains used as starter cultures must tolerate these kinds of stresses and exhibit a high ecologic performance in the stressful food environment. Genes related to stress response are induced when *L. sakei* is inoculated in the raw meat system [86]. In fact, ctsR, a gene that coded for a class III heat shock proteins repressor associated with the environmental stress response of Gram positive bacteria, increased its expression when *L.*

*sakei* starts to adapt to the raw environment. This mechanism demonstrated that the sudden changes in the environment conditions are perceived as stress by *Lactobacillus* species. In particular, in the case of *L. sakei*, added to raw meat and spices, the principal stress response regarded high osmolarity and temperature shifts. Moreover, the presence of curing salt is regarded as one of the major hurdles in the initial phase of sausages fermentation. Because nitrite was found to be the effective for growth inhibition of pathogens, nitrite was also hypothesized as a stressor for *L. sakei* [85] and the exposure of this strain to stresses can induce changes in metabolic activities in a food environment [4]. The metabolic changes in *L. sakei* resulted in enhanced exploitation of available nutrients or increased activity of glycolytic enzymes, leading to the accelerated production of lactic acid by stress-treated *L. sakei* cells [85]. However, the exposition of *L. sakei* to low temperature and high osmolarity gives rise to the repression of phosphofructokinase and consequently to a decreased flux through the glycolytic pathway [87].

Moreover, it is important to consider that some ingredients can be also antimicrobials because of their own characteristics: in fact, if the recipe includes herbs and spices (aromatic plants, pepper), garlic and onions, an effect on microorganisms can be exerted by specific compounds characterizing these products, such as essential oils, terpenes and sulfur compounds [88].

Another essential aspect affecting the performances and metabolism of LAB are the intrinsic characteristics of raw materials that sometimes act in a synergic way with other ingredients. Considering for example fermented vegetables, the microflora of the starting fresh vegetables is typically dominated by Gram negative aerobic bacteria and yeasts, while LAB make up a minor portion of the initial population [89] and therefore they would not be able to start and lead a fermentation process. However, if anaerobic conditions are settled and salts are added, LAB can have a competitive advantage and induce spontaneous lactic acid fermentation. The growth of specific LAB is dependent on the chemical (substrate, salt concentration, pH) and physical (vegetable type, temperature) environments. As the environments change during fermentation, so can the dominant organisms, often leading to a specific and reproducible succession of bacteria.

In sauerkraut [89, 90] the presence of 1.8-2.2% of NaCl and a temperature of 18°C inhibits many strains of LAB, with the exception of *Leuconostoc mesenteroides* that initiates the fermentation; however this species is sensitive to acid conditions, so after a few days, when the concentration of lactic acid increases, *L. mesenteroides* is replaced by more acid resistant LAB such as *Lactobacillus brevis* and *L. plantarum*, able to further lower the pH up to 3-3.5, stabilizing the final product.

Considering olives fermentation is possible to outline the characteristics of the product affecting LAB: while the brine provides a good environment for LAB growth, with glucose, fructose and mannitol as the main source of fermentable sugars, the presence of high levels phenols (such as oleuropein) exert an antimicrobial activity, inhibiting some strains and selecting the types of organisms that predominate during the fermentation [91-93]. These LAB have to be resistant not only to phenols, but also to lye treatments and water washes, that can be performed during the processing and increase the initial pH, reducing also the nutrients content on the olive surface. The species able to face these kind of stresses usually belong to the genera *Pediococcus, Leuconostoc* and *Lactococcus;* after the first stage of fermentation, when the pH reaches 6, *L. plantarum* rapidly grows and dominate the fermentation, that goes on until the fermentable sugars are depleted. The viability and vigor of *L. plantarum* can be encouraged also by yeasts that are still present in this stage of fermentation and can produce vitamins [94].

Moreover, the presence of some gases can modify the growth performances of LAB. That is also influenced by the mixing step of the ingredients in some food processes (e.g. dough mixing). In fact, in bread making process, the continuous agitation of the dough can increase the microbes exposure to oxygen, and this can be a source of oxidative stress, mainly for LAB that are usually anaerobic or facultative anaerobic. Also in these cases the bacteria can react in different ways, activating metabolic and transcriptional responses in order to detoxify ROS, as previously described.

For the fermented vegetables ,above reported, the rapid consumption of oxygen due to the presence of yeasts and aerobic bacteria in the first stage of fermentation has a positive effect on LAB. In fact, they are exposed only for a short time to oxidative stress and, due to their competitive advantage, they rapidly and intensively grow in the food system.

After formulation, the technological processes involving LAB include a fermentation process.

It is reported that various beneficial phenotypic traits of LAB in food fermentations such as rapid acidification, selective proteolysis, tolerance of osmotic and stresses, resistance to ROS, and ability to thrive in nutrient poor conditions and at low temperatures are influenced by stress responses in various species of LAB [95, 96]. The knowledge of these mechanisms, and mainly of the stress responses activated by the fermentation process parameters can be useful in order to develop strains with optimal fermentation characteristics [83].

The first metabolic reaction regards the oxidation of carbohydrates (this reaction depends on the hetero-fermentative or homo-fermentative species involved) that give rise to acids, alcohols and $CO_2$. These metabolites are directly involved in flavor, aroma and texture of the product and in a second time can influence the production and the availability of other metabolites such as vitamins and antioxidant compounds [78]. Moreover, the LAB interactions with the ingredients increase also the digestibility and decrease the glycemic index, enhancing the healthy features of the fermented foods [97].

At the same time with carbohydrates oxidation, other metabolic mechanisms interest LAB cells such as proteolysis and lipolysis. The first reaction produces polypeptides with interesting characteristics as antimicrobial compounds, salt substitutes (the oligopeptides are able to increase the palatability of the system), and amino acids deriving aromatic compounds. On the other hand lipolysis produces medium chain fatty acids, with important antimicrobial properties. All these reactions (carbohydrates oxidation, lipolysis and proteolysis) generate precursors for other mechanisms in the cells and in the food matrix that give rise to the dynamic environment characteristics of fermented foods. It is important to outline that the compounds produced by the cells, metabolizing the substrate, can modify the system, producing also compounds that can stimulate the growth of symbiotic species or inhibit the growth of antagonistic microorganisms.

The conversion of carbohydrates to metabolites as acetic acid, lactic acid or $CO_2$ implies the acidification of the system. The contemporary pH decrease and the presence of sugar (osmotic stress) stimulate the exopolysaccharides (EPSs) production. In fact, in sourdough EPSs can be involved in acid tolerance of sourdough LAB [98]. EPSs are long-chain polysaccharides consisting of branched, repeating units of sugars or sugar derivatives. These sugar units are mainly glucose, galactose and rhamnose, in different ratios [99]. The presence of EPSs in the system can create a novel stress to the cells. The inclusion of cells within biofilm can increase their resistance to unfavorable environmental factors such as extreme temperature, low pH and osmolarity, the changes in the texture can induce in LAB also specific stress responses.

For example in yogurt production, the acidification by LAB implies proteins coagulation and thereby changes in the viscosity of the milk. In *L. bulgaricus*, during the acid adaptation present in the fermentation milk to obtain yogurt, some cellular changes were observed: the chaperones GroES, GroEL, HrcA, GrpE, DnaK, DnaJ, ClpE, ClpP and ClpL were induced and ClpC was repressed [100]. Some genes involved in the biosynthesis of fatty acids were induced (*fabH, accC, fabI*), while the genes involved in the mevalonate pathway of isoprenoid synthesis (*mvaC, mvaS*) were repressed [101, 102]. The changes in Aw value are depending not only on EPSs production by LAB after the exposition to acidic and osmotic stress, but also on the ingredients composition and on the step of fermentation.

Considering cheese, the Aw decreases during manufacture and ripening as a result of dehydration, salting, and production of water-soluble solutes from glycolysis, proteolysis, and lipolysis; the cheese Aw values range from 0.70 for extra hard cheeses to 0.99 for fresh, soft cheeses, such as cottage cheese, while semi-hard cheeses have Aw values of around 0.90. The cheese pH also decreases during manufacture and ripening [103]. The effects of different Aw and pH on *L. lactis* simulating cheese ripening have been analyzed [103]. The results evidenced that at low Aw, particularly at low pH, the growth and lactose utilization rates decreased and lactose fermentation to L-(1)-lactate switched to a pathway involving nontraditional saccharide products rather than the traditional lactococcal heterofermentative products.

In *L. plantarum* WCFS1 the addition of 300 mM and 800 mM of NaCl induced mild osmotic stress and osmotic stress respectively. In the presence of 800 mM of NaCl several genes showed an increased expression with respect to the control culture. In particular, those genes were associated with various stress responses in prokariotes, i.e. genes encoding Clp protease, an excinuclease, catalase (peroxide stress) and Dpr-like protein (peroxide stress). These differences in the gene expression were also identified in the presence of acid stress. These results suggest that lactic acid stress in *L. plantarum* WCFS1 also induces a more general stress response (as above described for different *Lactobacillus* species). An overlap between the stimulus for lactic acid and those for peroxide and UV radiation has also been reported for *L. lactis* [104, 66]. The response of *L. sanfranciscensis* to osmotic stress (saccarose 40%) gives rise to the overproduction of 3-methylbutanoic acid and gamma-decalactones when *L. sanfranciscensis* was co-inoculated with yeasts, simulating a sourdough environment. The production of lactones can be indicated as unfavourable environment for microbial growth and metabolism. In fact, these compounds have both particular aromatic and antimicrobial features [13].

The ability of the target strains to dominate the fermentation is related not only to the ingredients (as above described), but also to the fermentation conditions, mainly temperature and atmosphere. If the fermentation is not performed at the optimal growth temperature for the microorganisms, they could be unable to compete with naturally occurring microflora, and consequently the whole process could be compromised. On the contrary, some microbial species have developed specific thermal resistance mechanisms, and they can easily adapt to these unfavorable conditions without implications for the fermentation processes. Moreover, the adaptation to thermal stresses often leads to tolerance to other stresses, in a mechanism usually define "cross protection", as reported for *L. lactis* [105]. The ability of commercial *L. lactis* ssp. *lactis* and *L. lactis* ssp. *cremoris* to withstand freezing at –60°C for 24 h was significantly improved by a prior 25 min heat shock at ~40°C or by a 2 h cold shock at 10°C, opening interesting perspectives for the production on resistant starter cultures, both frozen or lyophilized [105].

Other Authors with regard to different stresses reported the "cross protection" mechanism: for example the mechanisms of multiple adaptations to hops of two different strains of *L. brevis* have been characterized [106]. Hop resistance of lactobacilli requires multiple resistance mechanisms. This is consistent with the stress conditions acting on bacteria in beer, which mainly consist of acid stress and the antimicrobial effect of the hop compounds, in addition to ethanol stress and starvation. The effect of interaction of acid stress and presence/absence of oxygen in the system on *L. helveticus* and *L. sanfranciscensis*, in particular on their cell membrane composition, has been reported [14]. Upon acid stress the level of cyclopropane fatty acids increased at the expense of the level of long-chain unsaturated fatty acids. *L. helveticus* and *L. sanfranciscensis*, exposed to acid sub lethal stress demonstrated the same increase in cyclopropane fatty acids. In particular, *L. helveticus* presented higher concentration of C19cyc11 at pH 4 and pH 3, while *L. sanfranciscensis* presented more C19cyc9 at pH 3 in microaerophilic condition without tween 80, at pH 3.6 in anaerobiosis with tween 80, and at pH 4 in anaerobiosis without tween 80. These results demonstrated the same behavior in front of multiple stresses by LAB membrane [106, 14]

Consider the atmosphere, i.e. the presence or not of oxygen, as another important variable during fermentation, it is known that oxygen can inhibit the growth of LAB, especially in the first stages. However, the food system is usually a consortium of different microorganisms: for example in bakery products and in fermented sausages the fermentation is carried out both by yeasts and LAB; the formers can therefore consume the amount of oxygen present in the mix, allowing the growth of LAB. The same thing happens for fermented vegetables, where naturally occurring Gram negative bacteria and yeast rapidly remove the oxygen, promoting the rapid predominance of Lactobacilli.

Some secondary metabolites such as bacteriocins can play a role in LAB performances and metabolism, affecting also the total population and ecology of fermented foods [107, 108]. Bacteriocins are antimicrobial peptides or proteins produced by bacteria that can be active on different microorganisms, depending on their structure. LAB belonging to the genera *Lactococcus, Pediococcus, Lactobacillus, Leuconostoc, Carnobacterium, Propionibacterium* are known to produce bacteriocins with both narrow and broad inhibitory spectra [109]. The use of functional LAB starter cultures (eg. bacteriocinogenic starter cultures), well adapted to the environment and the process conditions applied, may contribute to the development of better controllable and more efficient production processes [110]. An example can be nisin, a peptide produced by *L. lactis* ssp. *lactis,* that has a narrow spectrum affecting primarily only Gram-positive bacteria and their spores, including lactic acid bacteria, *Bacillus, Clostridium, Listeria,* and *Streptococcus.* However some LAB such as *Streptococcus thermophilus* and *L. plantarum* are able to produce the enzyme nisinase, which neutralizes the antimicrobial activity of the peptide [111]. Therefore these LAB could be suitable for a co-fermentation with *L. lactis.*

Another interesting case of bacteriocin production, as a consequence of oxidative stress and carbon dioxide exposure, has been reported [110]: oxidative stress and carbon dioxide are involved in the production of a specific bacteriocin, amylovorin L, by *Lactobacillus amylovorus,* able to inhibit other LAB species. During traditional sourdough fermentation, a decrease in redox potential of the rather firm mixture occurs. The oxygen initially present is consumed by *Candida* spp. or converted into hydrogen peroxide or water, thereby creating microaerophilic or anaerobic environment in which the growth of the desired LAB is favored. While in a large-scale sourdough type II fermentation currently the use of dough mixture with high dough yield is exploited. This sourdough has to be stirred to liberate part of the carbon dioxide produced to prevent running over. During mixing, oxygen is incorporated into the dough. Also, the development of yeast and hence the production of carbon dioxide is favored in continuously stirred sough mixtures with high water content. Elevation of the airflow rates leading to oxidative stress conditions resulted in an enhanced specific amylovorin L production. Growth in the presence of carbon dioxide also increased the specific bacteriocin production. Mild aeration or a controlled supply of oxygen as well as growth in an environment containing high amounts of carbon dioxide might thus contribute to the competitiveness of *L. amylovorus* DCE471 in a sourdough ecosystem [110]. The production of plantaricin A by *L. plantarum* was also demonstrated in relation to a quorum sensing mechanism [79].

Another example of the influence of the process on LAB metabolism has been widely described [112]. These Authors monitored the evolution of the gene expression of *L. plantarum* IMDO 130201 during a sourdough process. In particular, the genes and the metabolites related to acidic stress were analyzed. It is interesting to highlight that during the pH decrease (production of lactic acid by *L. plantarum*) the genes coding for plantaricin production had higher levels of expression at low pH values, indicating that the bacteriocin production was activated under acid stress conditions by *L. plantarum* IMDO 130201 strain. The presence of the pheromone plantaricin A (PlnA) in a system inoculated with *L. plantarum* DC400 was also reported [79]. Biosynthesis of PlnA was variously stimulated depending on the microbial partner. In fact, *L. sanfranciscensis* DPPMA174 induced the highest synthesis of PlnA, which, in turn, determined lethal conditions for it. The proteome of *L. sanfranciscensis* DPPMA174 responded to the presence of PlnA. The up-regulation of 31 proteins related to stress response, amino acid metabolism, energy metabolism, membrane transport, nucleotide metabolism, regulation of transcription and cell redox homeostasis was found. At the same time, other proteins such as cell division protein (FtsZ), glutathione reductase (LRH_11212) and response regulator (rrp11) were down-regulated. These results demonstrated a hypothetically and interesting waterfall of events all related with stresses response and with the typical fermentation products dynamics (Figure 3). At the same time, the low pH values implied a poor expression of the genes involved in carbohydrate degradation in *L. plantarum* IMDO 130201. The bacterium was directed toward survival at low pH by amino acid conversions rather than by relying on growth [112]. The same behavior was identified in *L. sanfranciscensis* LSCE1 response to pH 3.6 [15]. Under the adopted experimental conditions, which did not produce any decrease in viability of *L. sanfranciscensis* LSCE1, the acid stress, within 2 h, was accompanied by a reduction of the carbohydrate metabolism, as shown by the decrease of ethanol, acetate, and lactate. This mechanism suggests the existence of a switch from sugar to amino acid catabolism that supports survival and growth also in specific and restricted environments, such as sourdoughs, characterized by acid stress and recurrent carbon starvation. Under the acid conditions (pH 3.6) and in the presence of specific nutrients 3-methylbutanoic acid was the predominant metabolite among those detected by solid phase micro-extraction gas chromatographic analysis and mass spectrometry (GC-MS-SPME), released after 2 h of acid stress exposure [15]. The acid stress implied less carbohydrate utilization and ethanol, lactate, and acetate production, but high amino acids catabolism that confers a different and characteristic metabolites pattern. Stress resistance assume great importance as one of the adaptation factors to gastrointestinal tract of probiotic strains as reported in a detailed review [113].

## 5. Stress resistance of probiotic LAB

There are two main categories of factors that contribute to the optimal functioning of probiotic lactobacilli: factors that allow optimal adaptation to the new niches that they temporarily encounter in the host (adaptation factors) and factors that directly contribute to the health-promoting effects (probiotic factors) [113].

Adaptation factors include stress resistance, active metabolism adapted to the host environment, and adherence to the intestinal mucosa and mucus.

In fact, probiotic lactobacilli encounter various environmental conditions upon ingestion by the host and during transit in the gastro intestinal tract (GIT). They need to survive to: 1) the harsh conditions of the stomach secretion generating a fasting pH of 1.5, increasing to pH 3 to 5 during food intake; 2) the bile excreted by liver in small intestine represents another challenge for bacteria entering the GIT. Bile salts also seem to induce an intracellular acidification so that many resistance mechanisms are common for bile and acid stress. Indeed, the protonated form of

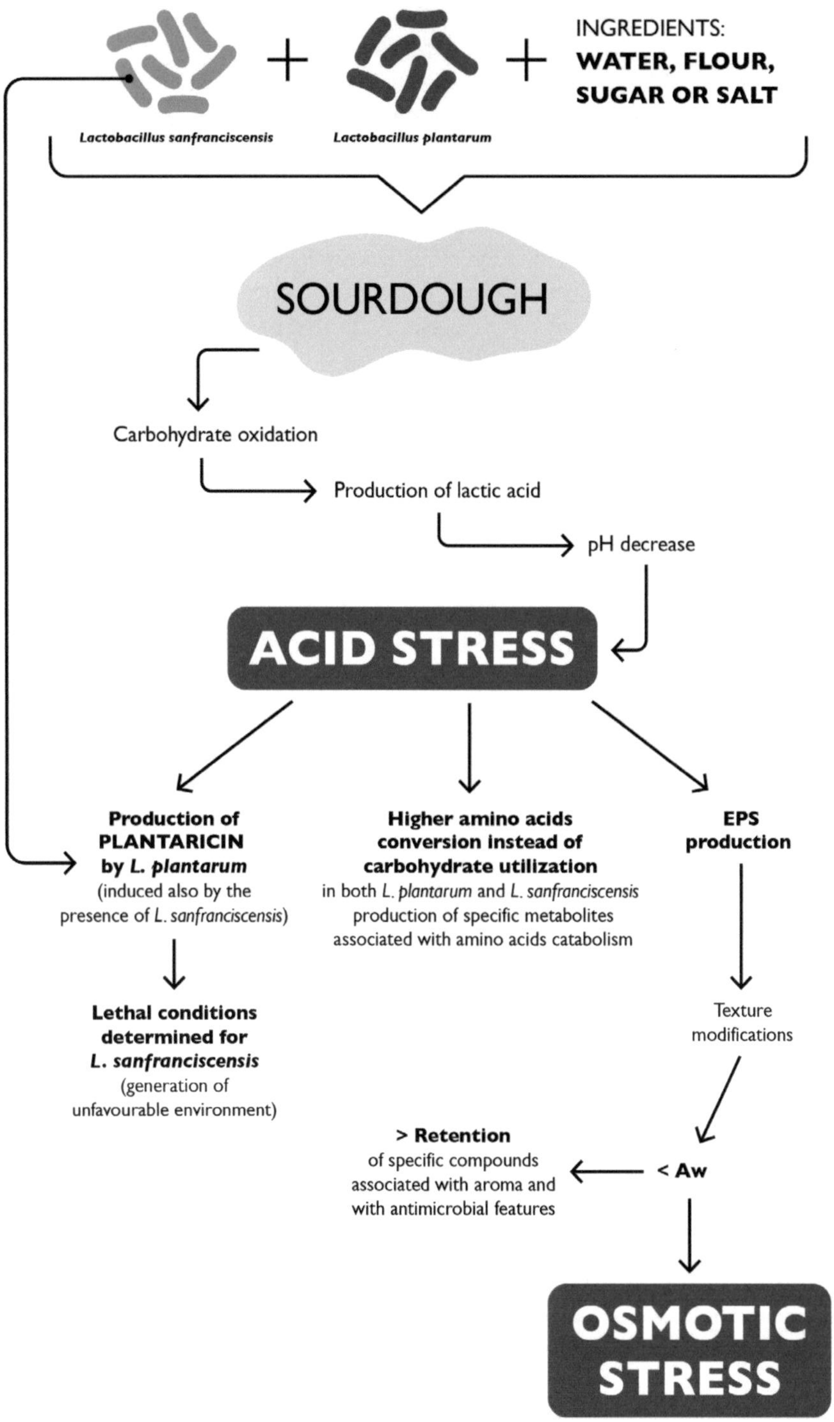


**Figure 3.** Sourdough fermentation dynamics. Case of possible parallel phenomena interesting acid and osmotic stress.

bile salts is thought to exhibit toxicity through intracellular acidification in a manner similar to those of organic acids like the lactic acid produced by the lactobacilli themselves. For a detailed overview of acid, bile, and other stress resistance mechanisms of lactobacilli, the reader is referred to more extensive review [113]. 3) In analogy to the stresses encountered by intestinal pathogens, they also encounter oxidative and osmotic stress in GI tract. 4) Interactions with other microbes and 5) Interactions with cells of the host immune system and the various antimicrobial products that they produce can also impose a serious threat for the probiotic microbes. Analogously to what described in food LAB, the phenomenon of cross-adaptation is often observed, i.e., that adaptation to one stress condition also protects against another stress factor, implying some common mechanisms. In this respect, also for probiotic LAB non-actively-growing stationary-phase cells are generally more resistant to various stressors than early-log-phase cells.

### 5.1. Maintaining integrity of the cell envelope

The different macromolecules constituting the cell membranes and cell walls of lactobacilli have been shown to contribute to maintaining cell integrity during stress to various degrees. For example, low pH caused a shift in the fatty acid composition of the cell membrane of an oral strain of *L. casei*. Similarly, bile salts have been shown to induce changes in the lipid cell membrane of *Lactobacillus reuteri* CRL1098.

The role of EPS in acid and bile resistance is less clear. However, EPS production has not been studied in detail after exposure to bile. In fact, to our knowledge, phenotypic analyses of dedicated *Lactobacillus* mutants affected in EPS biosynthesis genes have not yet been performed. Homopolysaccharides (HoPSs) from *L. reuteri* have been reported to have a more established role in stress resistance by the maintenance of the cell membrane in the physiological liquid crystalline phase under adverse conditions.

### 5.2. Repair and protection of DNA and proteins

A number of proteins that play a role in the protection or repair of macromolecules such as DNA and proteins also seem to be essential for acid and bile resistance. Intracellular acidification can result in a loss of purines and pyrimidines from DNA. Bile acids have also been shown to induce DNA damage and the activation of enzymes involved in DNA repair. Perhaps even more vital in the general stress response are chaperones that intervene in numerous stresses for important tasks such as protein folding, renaturation, protection of denatured proteins, and removal of damaged proteins.

### 5.3. Two-component and other regulatory systems

Mechanisms to specifically sense the presence of certain stress factors and regulate gene expression in response to these stimuli are also crucial for bacterial survival under adverse conditions. Although these mechanisms are not well characterized for lactobacilli, they often involve two-component regulatory systems (2CRSs). 2CRSs allow bacteria to sense and respond to changes in their environment after receiving an environmental signal through transmembrane sensing domains of the histidine protein kinase (HPK).

# 6. Methodological approaches to study the effects of stress on LAB

The study of stress responses by LAB is getting closer and closer to the different "omic" fields: genomic, proteomic and metabolomic. Other traditional approaches regarding the membrane cells composition and modifications, both from a structural (cellular fatty acids composition by gas-chromatographic method) and morphological (membrane and wall modification by electronic microscopy) point of view are still used.

Genes implicated in LAB stress responses are numerous and the levels of characterization of their actual role and regulation differ widely between species. The studies concerning stress responses in LAB sometimes benefit from the knowledge already acquired in other bacteria. For example, parts of the studies on heat response have been focused on specific genes because of their major role demonstrated in other microorganisms [17]. The cheapest and easiest way to study a stress response in LAB is to follow some specific genes related to stresses such as heat shock, salts and acids [114, 115]. This type of study is useful especially if the entire genome sequence of some LAB is still unknown. However, nowadays the study of whole trascriptome (the total set of RNAs) is one of the most exhaustive ways to study modifications of gene expression as a result of a stress condition. The transcriptome of a cell contains information about the biological state of the cell and the genes that play a role under specific circumstances. The principal technique used to study the trascriptome is microarray [116].

DNA microarray technology has been used in numerous experiments to analyze gene expression: one example is the evaluation of the general stress response of *B. subtilis* [117] or the investigation of the transcription profiles of *L. plantarum* grown in steady-state cultures that varied in lactate/lactic acid concentration, pH, osmolarity [66, 104]. This approach is useful also to study the behaviour of bacteria in a real food system. Hüfner et al. [5] studied the global transcriptional response of *L. reuteri* to sourdough environment, showing a significant changes of mRNA levels for 101 genes involved in diverse cellular processes, from carbohydrate and energy metabolism, to cell envelope biosynthesis, exopolysaccharide production, stress responses, signal transduction and cobalamin biosynthesis.

The gene expression dynamics of *L. casei* during fermentation in soymilk when grown up to lag phase, late logarithmic phase, or stationary phase were also studied. Comparisons of different transcripts close to each other revealed 162 and 63 significantly induced genes, in the late logarithmic phase and stationary phase, whose expression was at least threefold up-regulated and down-regulated, respectively. Approximately 38.4% of the up-regulated genes were associated with amino acid transport and metabolism, followed by genes/gene clusters involved in carbohydrate transport and metabolism, lipid transport and metabolism, and inorganic ion transport and metabolism [118].

The study of trascriptome is a good approach that gives a good overview of the changes that can occur inside a stressed bacterium. A limitation of this technique is that it is expensive and requires that the genome sequences of the organisms under study should be available for designing the oligonucleotides for the microarray [119].

A different but, at the same time, related point of view regards the study of proteins and proteome. The most common method to obtain this information is to extract total proteins and separate them by a sodium dodecyl sulphate polyacrylamide gel electrophoresis (SDS-PAGE) followed by a western blotting (in the first case) or a two dimesional electrophoresis (2D-E) analysis (in the second case). Also in this case if the study is focused on a single protein, it is necessary to know before the characteristic of the target protein to optimize the analytical conditions. 2D-Electrophoresis can provide more than 10000 detectable protein spots in a single gel run. Thus, proteins with post-translational modifications (PTMs), such as processing, phosphorylation and glycosylation, can be easily detected as separate spots. A spot separated by 2D-E theoretically consists of an almost homogeneous protein, and thus can be identified following digestion with a sequence-specific protease by peptide mass fingerprinting (PMF) approaches, typically using matrix-assisted laser desorption ionization (MALDI)- time-of-flight (TOF) mass spectrometers. The same level of automation is also available for proteomic approaches involving tandem mass spectrometry (MS-MS) analysis, extremely useful when studying organisms with incomplete or partial genomic information [120].

This kind of approach was used to investigate the cell surface proteins of a typical strain of *L. casei* in response to acidic growth conditions [121]. They demonstrated that growth of *L. casei* under acidic conditions caused molecular changes at the cell surface in order to accomplish an adaptive strategy, resulting in slower growth at low pH. Moreover, the proteomic approach was useful to study the heat shock response respectively on *L. helveticus* PR4 and *L. plantarum* [26, 31]. The cold adaptation of *Lactococcus piscium* strain CNCM I-4031 was studied with the same approach [122]. This analysis could be also performed to compare the effects that new technologies produce on bacteria comparing with the normal stress conditions. In fact, the HHP stress response of *L. sanfranciscensis* was compared with cold, heat, salt, acid and starvation stresses responses [18].

Due to increasingly available bacterial genomes in databases, proteomic tools have recently been used to screen proteins expressed by microorganisms in food, in order to better understand their metabolism *in situ*. While up to now the main objective has been the systematic identification of proteins, the next step will be to bridge the gap between identification and quantification of these proteins [123]. Proteomics has also been used to analyse the proteins released during the ripening of Emmentaler cheese. In an innovative study, proteomics was used to prepare a reference map of the different groups of proteins found in cheese [124]. These authors were able to categorize these proteins into five classes: those involved in proteolysis, glycolysis, stress response, nucleotide repair and oxidation-reduction. In addition, information was obtained regarding the peptidases released into the cheese during ripening process. This study enabled the Authors to differentiate between the various casein degradation mechanisms present, and to suggest that the streptococci within the cheese matrix are involved in peptide degradation and together with the indigenous lactobacilli contribute to the ripening process. Using proteomics these Authors were able to get a greater understanding of the microbial succession involved in the ripening of Emmentaler cheese, which information could not have been obtained using other protein separation techniques. This example illustrates the power of proteomics as a tool for analyzing the composition of a complex mixture of proteins and peptides [119].

The global identification of stress-induced proteins in a given organism has technical limitations. Membrane proteins, for example, are rarely detected by this method. Secondly, it may be that changes in membrane proteins composition result from long-term adaptation processes, while short-term responses may primarily be accounted for the activation (and/or stabilization) of proteins already present. The latter hypothesis is valid especially in the case of transport systems, although for some of the systems studied a transcriptional induction has also been observed [17]. The use of this technique is not as widespread as that of DNA microarrays due to the challenges associated with the purification and separation of complex mixtures of proteins found in cell extracts. At the same time the study of the only transcriptome should take into consideration that a lot of post-transcriptional processes may act on RNA (ex. RNA interference, polyadenilation ecc) [125].

As reported above, the stress responses of LAB are studied also through the analysis of membrane composition, structure and integrity. Not unexpectedly, in fact, the cell membrane plays an important role in stress resistance. First of all, the membrane itself can change in adaptation to environmental conditions and these changes contribute to the protection of the bacteria [17]. The adaptive response to sub-lethal acid and cold stresses in *L. helveticus* and *L. sanfranciscensis* has been analyzed (as described above) [14]. The extraction and identification by GC-MS of lipid fatty acids and free fatty acids could give an overview of the membrane fluidity state. In the same article they developed a gas chromatographic method to separate and quantify the cell cyclopropane fatty acids lactobacillic (C19cyc11) and dehydrosterculic (C19cyc9) demonstrating different responses of the strains tested in terms of cyclopropane fatty acids production, probably due to the different original optimal environment. The comparison between the wild type and the acid-resistant mutant *L. casei* LBZ-2 evidenced in the latter higher membrane fluidity, higher proportions of unsaturated fatty acids, and higher medium chain length. In addition, cell integrity analysis showed that the mutant maintains a more intact cellular structure and lower membrane permeability after environmental acidification [126].

The last but not least approach used to study the stress response of LAB is the metabolic one. The study of the metabolites released, as a consequence of the stress exposure, can contribute to the understanding of the mechanisms that regulate the microbial interactions and the metabolic alterations induced by stress conditions. Moreover, these approaches can be exploited to identify which technological conditions induce microorganisms to produced desirable metabolites [4, 15].

With this perspective the use of GC-MS-SPME as a potent and easy tool to study the generation of volatile metabolite compounds such as flavoring molecules or aroma precursors was widely adopted [9,11-13, 15] and contributed to rationalize the process and optimize the products. In particular, the effects of HPH on different species of *Lactobacillus* involved in dairy product fermentation and ripening, monitoring the changes in volatile compounds as indicators of metabolic profiles has been studied [11].

Analysing the oxidative and heat stresses in *L. helveticus* two new 2[5H]-furanones released by this strain both as a possible signalling molecules and as possible important flavouring compounds has been identified by GC-MS-SPME [9]. On the contrary the study of non-volatile metabolites can be performed by normal chromatographic technique (HPLC), especially for amino acids and sugars [15], or by Fast Protein Liquid Chromatography (FPLC) separation for peptides, followed by a mass spectrometry identification [127]. An NMR approach to evaluate the effects on the growth of *L. plantarum* raising the medium molarity by high concentrations of KCl or NaCl and iso-osmotic concentrations of non-ionic compounds was performed [128].

Since all the techniques described above, if used alone, do not allow a total comprehension of stress responses, a lot of studies are trying to combine two or more approaches together. Combined transcriptomic and proteomic analyses were used to evaluate the glucose-limited chemo-stat in *Enterococcus faecalis* V583 [129] or to study the effect of bile salts in the growth of *L. casei* [130]. A combined physiological and proteomic approach, instead, was followed to unravel lactic-acid-induced alterations in *L. casei* [131].

Therefore it is possible to understand, from the references above, that techniques used to study the stress responses of LAB are taking more and more "omic" approach. This comports an accumulation of a huge number of data that it is not easy to manage and to compare. For this reason the use of new programs of data analysis is required. One of these approaches could be the use of heat maps, a technique born as a tool to understand microarray results [66]. Nowadays it could be useful also to manage the data from other fields: in fact, a heat maps was used to show the correlation between metabolites produced, the relative gene expression of specific genes and stress conditions [15]. The same useful tool, combined with other statistical analyses, has been also applied [132].

## 7. Conclusion

It is known that LAB can adapt to stress with different mechanisms widely studied in model and real systems. An overview of those responses has been described and reported in this chapter.

Stress not only induces changes enabling better survival, but also different performances in a system. In fermented food, the knowledge of the mechanisms that regulate LAB metabolic changes and their effects gain importance especially when those responses can be exploited in order to improve the food properties [4]. In particular, fermented foods are dynamic systems subjected to continuous evolution of their physico-chemical characteristics. The complex fluctuation of the food environment itself, during processing, is stress source for every microorganism involved and the changes that affect the fermented food habitats, can be perceived by LAB as stress.

In this chapter examples of the dynamic fluctuation effect on LAB metabolism have been described in order to outline that every reaction can cause a waterfall of metabolic events influencing the sensorial quality, the shelf-life and the bioactive compounds production of fermented foods.

The subjects of those events are LAB, indicating the importance of metabolism of these microorganisms in food. The cell physiology is crucial to ensure that cells are well suited to survival during downstream processes and that they exhibit high performances.

The production and exploitation of naturally adapted strains can be interesting for companies because of the absence of ethical and legal concerns. The adapted strains are not considered genetically modified microorganisms (GMOs) and therefore they can be applied in food processing without legal restrictions and, more important, without affecting the consumer perception, currently (in Europe) not ready to introduce in his diet foods produced with GMOs.

Individual stresses used in food processing and preservation may render probiotic LAB more resistant to further and different stresses, including those encountered in the human body, e.g. those encountered during gastro-intestinal passage (pH of the stomach, exposure to bile salts in small intestine etc.). A positive correlation has been recently observed between EPS production and resistance to bile salt and low pH stress in *Bifidobacterium* species isolated from breast milk and infant faeces [128].

This knowledge can open interesting perspectives to improve at the same time the performances of LAB, the quality of fermented food and the health-promoting properties of the LAB used.

Moreover, it will be interesting to identify the gastrointestinal tract also as a complex and dynamic system in which LAB need to adapt to adverse conditions, responding with metabolic shifts provided with interesting technological an healthy features.

The "omics" technologies could be particularly useful for identifying the mechanism leading to LAB stress responses. These approaches could also help to identify the mechanisms for cell fitness and stress adaptation that will be needed to develop more generic and science based technologies [7].

## Author details

Diana I. Serrazanetti *, Davide Gottardi, Chiara Montanari and Andrea Gianotti
*Department of Food Science, Alma Mater Studiorum, University of Bologna, Bologna, Italy*

*Inter-Departmental Center of Industrial Agri-Food Research (CIRI Agroalimentare), Cesena, Italy*

## Acknowledgement

We thank Prof.ssa Maria Elisabetta Guerzoni for her enormous scientific support and Luca Vagnini for his graphic abilities (http://www.lucavagnini.com).

* Corresponding Author

## 8. References

[1] Poolman B, Ruhdal J, Gruss A (2011) LAB Physiology and energy metabolism. In: Ledeboer A, Hugenholtz J, Kok J, Konings W, Wouters J. Thirty years of research on lactic acid bacteria.Rotterdam: 24 Media Labs. pp. 77-101.
[2] Ledeboer A, Hugenholtz J, Kok J, Konings W, Wouters J (2011) In: Ledeboer A, Hugenholtz J, Kok J, Konings W, Wouters J. Thirty years of research on lactic acid bacteria.Rotterdam: 24 Media Labs. pp. v-vi.
[3] Marles-Wright J, Lewis R (2007) Stress response of bacteria. Curr. Opin. Struct. Biol. 17:755–760.
[4] Serrazanetti DI, Guerzoni ME, Corsetti A, Vogel RF (2009) Metabolic impact and potential exploitation of the stress reactions in lactobacilli. Food Microbiol. 26:700–711.
[5] Hüfner E, Britton RA, Roos S, Jonsson H, Hertel C (2008) Global transcriptional response of *Lactobacillus reuteri* to the sourdough environment. Syst. Appl. Microbiol. 31:323–338.
[6] Heller KJ (2001) Probiotic bacteria in fermented foods: product characteristics and starter organisms. Am. J. Clin. Nutr. 73:374S–379S.
[7] Lacroix C, Yildirim S (2007) Fermentation technologies for the production of probiotics with high viability and functionality. Curr. Opin. Biotechnol. 18:176–183.
[8] Serrazanetti DI (2009) Effects of acidic and osmotic stresses on flavor compounds and gene expression in *Lactobacillus sanfranciscensis*. Ph.D. thesis. University of Teramo, Teramo, Italy.
[9] Ndagijimana M, Vallicelli M, Cocconcelli PS, Cappa F, Patrignani F, Lanciotti R, Guerzoni ME (2006) Two 2[5H]-furanones as possible signaling molecules in *Lactobacillus helveticus*. Appl. Environ. Microbiol. 72:6053–6061.
[10] Vannini L, Ndagijimana M, Saracino P, Vernocchi P, Corsetti A, Vallicelli M, Cappa F, Cocconcelli PS, Guerzoni ME (2007) New signaling molecules in some Gram-positive and Gram-negative bacteria. Int. J. Food Microbiol. 120:25–33.
[11] Lanciotti R, Patrignani F, Iucci L, Saracino P, Guerzoni ME (2007) Potential of high-pressure homogenization in the control and enhancement of proteolytic and fermentative activities of some *Lactobacillus* species. Food Chem. 102:542–550.
[12] Guerzoni ME, Vernocchi P, Ndagijimana M, Gianotti A, Lanciotti R (2007) Generation of aroma compounds in sourdough: effects of stress exposure and lactobacilli–yeasts interactions. Food Microbiol. 24:139–148.
[13] Vernocchi P, Ndagijimana M, Serrazanetti DI, Gianotti A, Vallicelli M, Guerzoni ME (2008) Influence of starch addition and dough microstructure on fermentation aroma production by yeasts and lactobacilli. Food Chem. 108:1217–1225.
[14] Montanari C, Sado-Kamdem SL, Serrazanetti DI, Etoa FX, Guerzoni ME (2010) Synthesis of cyclopropane fatty acids in *Lactobacillus helveticus* and *Lactobacillus sanfranciscensis* and their cellular fatty acids changes following short term acid and cold stresses. Food Microbiol. 27:493–502.
[15] Serrazanetti DI, Ndagijimana M, Sado SL, Corsetti A, Vogel RF, Ehrmann M, Guerzoni ME (2011) Acid stress-mediated metabolic shift in *Lactobacillus sanfranciscensis* LSCE1. Appl .Environ. Microbiol. 77:2656-2666.

[16] Klaenhammer TR, Barrangou R, Buck BL, Azcarate-Peril MA, Altermann E (2005) Genomic features of lactic acid bacteria effecting bioprocessing and health. FEMS Microbiol. Rev. 29:393–409.

[17] van de Guchte M, Serror P, Chervaux C, Smokvina T, Ehrlich SD, Maguin E (2002) Stress responses in lactic acid bacteria. Antonie Leeuwenhoek. 82:187–216.

[18] Hörmann S, Scheyhing C, Behr J, Pavlovic M, Ehrmann M, Vogel RF (2006) Comparative proteome approach to characterize the high-pressure stress response of *Lactobacillus sanfranciscensis* DSM 20451$^{T}$. Proteomics. 6:1878–1885.

[19] Guerzoni ME, Lanciotti R, Cocconcelli PS (2001) Alteration in cellular fatty acid composition as a response to salt, acid, oxidative and thermal stresses in *Lactobacillus helveticus*. Microbiol. 147:2255–2264.

[20] Pavlovic M, Hörmann S, Vogel RF, Ehrmann MA (2008) Characterisation of a piezotolerant mutant of *Lactobacillus sanfranciscensis*. Z. Naturforsch. 63:791–797.

[21] Burns P, Patrignani F, Serrazanetti DI, Vinderola GC, Reinheimer JA, Lanciotti R, Guerzoni ME. (2008) Probiotic crescenza cheese containing *Lactobacillus casei* and *Lactobacillus acidophilus* manufactured with High Pressure-Homogenized Milk. J Dairy Sci. 91:500–512.

[22] Patrignani F, Burns P, Serrazanetti DI, Vinderola G, Reinheimer J, Lanciotti R, Guerzoni ME (2009) Suitability of high pressure-homogenized milk for the production of probiotic fermented milk containing *Lactobacillus paracasei* and Lactobacillus acidophilus. J. Dairy Res. 5:1-9.

[23] Gouesbert G, Jan G, Boyaval P (2002) Two-dimensional electrophoresis study of *Lactobacillus delbrueckii* subsp. *bulgaricus* thermotolerance. Appl. Environ. Microbiol. 68:1055–1063.

[24] Desmond C, Fitzgerald GF, Stanton C, Ross RP (2004) Improved stress tolerance of GroESL-overproducing *Lactococcus lactis* and probiotic *Lactobacillus paracasei* NFBC 338. Appl. Environ. Microbiol. 70:5929–5936.

[25] Corcoran BM, Ross RP, Fitzgerald GF, Dockery P, Stanton C (2006) Enhanced survival of GroESL-overproducing *Lactobacillus paracasei* NFBC 338 under stressful conditions induced by drying. Appl. Environ. Microbiol. 72:5104–5107.

[26] Di Cagno R, De Angelis M, Limitone A, Fox PF, Gobbetti M (2006) Response of *Lactobacillus helveticus* PR4 to heat stress during propagation in cheese whey with a gradient of decreasing temperatures. Appl. Environ. Microbiol. 72:4503–4514.

[27] Laplace JM, Sauvageot N, Harke A, Auffray Y (1999) Characterization of *Lactobacillus collinoides* response to heat, acid and ethanol treatments. Appl. Microbiol. Biotechnol. 51:659–663.

[28] Schmidt G, Hertel C, Hammes WP (1999) Molecular characterisation of the dnaK operon of *Lactobacillus sakei* LTH681. Syst. Appl. Microbiol. 22:321–328.

[29] Zink R, Walker C, Schmidt G, Elli M, Pridmore D, Reniero R (2000) Impact of multiple stress factors on the survival of dairy lactobacilli. Sci. Aliment. 20:119–126.

[30] Prasad J, McJarrow P, Gopal P (2003) Heat and osmotic stress responses of probiotic *Lactobacillus rhamnosus* HN001 (DR20) in relation to viability after drying. Appl. Environ. Microbiol. 69:917–925.

[31] De Angelis M, Gobbetti M (2004) Environmental stress responses in *Lactobacillus*: a review. Proteom. 4:106–122.

[32] Bucio A, Hartemink R, Schrama JW, Verreth J, Rombouts FM (2005) Survival of Lactobacillus plantarum 44a after spraying and drying in feed and during exposure to gastrointestinal tract fluids in vitro. J. Gen. Appl. Microbiol. 51:221–227.

[33] Castaldo C, Siciliano RA, Muscariello L, Marasco R, Sacco M (2006) CcpA affects expression of the groESL and dnaK operons in *Lactobacillus plantarum*. Microb. Cell Fact. 5:35.

[34] Gardiner GE, O'Sullivan E, Kelly J, Auty MAE, Fitzgerald GF, Collins JK, Ross RP, Stanton C (2000) Comparative survival rates of human-derived probiotic *Lactobacillus paracasei* and *L. salivarius* strains during heat treatment and spray drying. Appl. Environ. Microbiol. 66:2605–2612.

[35] Earnshaw RG, Appleyard J, Hurst RM (1995) Understanding physical inactivation processes: combined preservation opportunities using heat, ultrasound and pressure. Int. J. Food Microbiol. 28:197–219.

[36] Teixeira P, Castro H, Mohacsi-Farkas C, Kirby RJ (1997) Identification of sites of injury in *Lactobacillus bulgaricus* during heat stress. Appl. Microbiol. 83:219–226.

[37] Hansen PJ, Drost M, Rivera RM, Paula-Lopes FF, AI-Katananit YM, Krininger CE, Chase CC (2001) Adverse impact of heat stress on embryo production: causes and strategies for mitigation. Theriogenol. 55:91–103.

[38] Kilstrup M, Jacobsen S, Hammer K, Vogensen FK (1997) Induction of heat shock proteins DnaK, GroEL, and GroES by salt stress in *Lactococcus lactis*. Appl. Environ. Microbiol. 63:1826–1837.

[39] KimWS, Dunn NW (1997) Identification of a cold shock gene in lactic acid bacteria and the effect of cold shock on cryotolerance. Curr. Microbiol. 35:59–63.

[40] KimWS, Khunajakr N, Dunn NW (1998) Effect of cold shock on protein synthesis and on cryotolerance of cells frozen for long periods in *Lactococcus lactis*. Cryobiol. 37:86–91.

[41] Wouters JA, Jeynov B, Rombouts FM, de Vos WM, Kuipers OP, Abee T (1999) Analysis of the role of 7 kDa cold-shock proteins of *Lactococcus lactis* MG1363 in cryoprotection. Microbiol. 145:3185–3194.

[42] Derzelle S, Hallet B, Francis KP, Ferain T, Delcour J, Hols P (2000) Changes in cspL, cspP, and cspC mRNA abundance as a function of cold shock and growth phase in *Lactobacillus plantarum*. J. Bacteriol. 182:5105–5113.

[43] Fonseca F, Béal C, Corrieu G (2001) Operating Conditions That Affect the Resistance of Lactic Acid Bacteria to Freezing and Frozen Storage. Cryobiol. 43:189–198.

[44] Zhang G, Fan M, Li Y, Wang P, Lv Q (2012) Effect of growth phase, protective agents, rehydration media and stress pretreatments on viability of *Oenococcus oeni* subjected to freeze-drying Afr. J Microbiol. Res. 6:1478-1484.

[45] Bâati L, Fabre-Gea C, Auriol D, Blanc PJ (2000) Study of the cryotolerance of *Lactobacillus acidophilus*: effect of culture and freezing conditions on the viability and cellular protein levels. Int. J. Food. Microb. 59:241–247.

[46] Panoff JM, Thammavongs B, Guéguen M. (2000) Cryotolerance and cold stress in lactic acid bacteria. Sci. Aliment. 20:105-110.

[47] Panoff JM, Thammavongs B, Laplace JM, Hartke A, Boutibonnes P, Auffray Y (1995) Cryotolerance and Cold Adaptation in *Lactococcus lactis* subsp. *lactis* IL1403. Cryobiol. 32:516–520.

[48] Grogan DW, Cronan JE (1986) Characterization of *Escherichia coli* mutants completely detective in synthesis of cyclopropane fatty acids. J. Bacteriol. 166:872.

[49] Grogan D, Cronan J (1997) Cyclopropane ring formation in membrane lipids of bacteria. Microbiol. Mol. Biol. Rev. 61:429-441.

[50] Raynaud S, Perrin R, Cocaign-Bousquet M, Loubiere P (2005) Metabolic and Transcriptomic Adaptation of *Lactococcus lactis* subsp. *lactis* biovar *diacetylactis* in Response to Autoacidification and Temperature Downshift in Skim Milk. Appl. Environ. Microbiol. 71:8016-8023.

[51] Wolf G, Arendt EK, Pfähler U, Hammes WP (1990) Heme-dependent and hemeindependent nitrite reduction by lactic acid bacteria results in different N-containing products. Int. J. Food Microbiol. 10:323–329.

[52] Duwat P, Cesselin B, Sourice S, Gruss A (2000) *Lactococcus lactis*, a bacteriamodel for stress responses and survival. Int. J. Food Microbiol. 55:83–86.

[53] Rochat T, Gratadoux JJ, Gruss A, Corthier G, Maguin E, Langella P, van de Guchte M (2006) Production of a heterologous nonheme catalase by *Lactobacillus casei*: an efficient tool for removal of $H_2O_2$ and protection of *Lactobacillus bulgaricus* from oxidative stress in milk. Appl. Environ. Microbiol. 72:5143–5149.

[54] Yamamoto Y, Poyart C, Trieu-Cuot P, Lamberet G, Gruss A, Gaudu P (2005) Respiration metabolism of Group B Streptococcus is activated by environmental haem and quinone and contributes to virulence. Mol. Microbiol. 56:525–534.

[55] Vido K, Diemer H, Van Dorsselaer A, Leize E, Juillard V, Gruss A, Gaudu P (2005) Roles of thioredoxin reductase during the aerobic life of *Lactococcus lactis*. J. Bacteriol. 187:601–610.

[56] Rezaïki L, Cesselin B, Yamamoto Y, Vido K, van West E, Gaudu P, Gruss A (2004) Respiration metabolism reduces oxidative and acid stress to improve long-term survival of *Lactococcus lactis*. Mol. Microbiol. 53:1331–1342.

[57] Lechardeur D, Cesselin B, Fernandez A, Lamberet G, Garrigues C, Pedersen M, Gaudu P, Gruss A (2011) Using heme as an energy boost for lactic acid bacteria. Curr. Opin. Biotechnol. 22:143-149.

[58] Brooijmans RJW, de Vos WM, Hugenholtz J (2009) *Lactobacillus plantarum* WCFS1 Electron Transport Chains. Appl. Environ. Microbiol. 75:3580-3585.

[59] Lushchak VI (2001) Oxidative stress and mechanisms of protection against it in bacteria. Biochem. Moscow. 66:476–489.

[60] Hertel C, Schmidt G, Fischer M, Oellers K, Hammes WP (1998) Oxygendependent regulation of the expression of the catalase gene katA of *Lactobacillus sakei* LTH677. Appl. Environ. Microbiol. 64:1359–1365.

[61] Miyoshi A, Rochat T, Gratadoux JJ, Le Loir Y, Costa Oliveira S, Langella P, Azavedo V (2003) Oxidative stress in *Lactococcus lactis*. Genet. Mol. Res. 2:348–359.

[62] Bruno-Bárcena JM, Andrus JM, Libby SL, Klaenhammer TR, Hassan HM (2004) Expression of a heterologous manganese superoxide dismutase gene in intestinal lactobacilli provides protection against hydrogen peroxide toxicity. Appl. Environ. Microbiol. 70:4702–4710.

[63] Jänsch A, Korakli M, Vogel RF, Ganzle MG (2007) Glutathione reductase from *Lactobacillus sanfranciscensis* DSM20451$^T$: contribution to oxygen tolerance and thiol exchange reactions in wheat sourdoughs. Appl. Environ. Microbiol. 73:4469–4476.

[64] Rallu F, Gruss A, Ehrlich SD, Maguin E (2000) Acid- and multistress-resistant mutants of *Lactococcus lactis*: identification of intracellular stress signals. Mol. Microbiol. 35:517–528.

[65] Cotter PD, Hill C (2003) Surviving the acid test: responses of Gram-positive bacteria to low pH. Microbiol. Mol. Biol. Rev. 67:429–453.
[66] Pieterse B, Leer RJ, Schuren FHJ, van der Werf MJ (2005) Unravelling the multiple effects of lactic acid stress on *Lactobacillus plantarum* by transcription profiling. Microbiol. 151:3881–3894.
[67] Piuri M, Sanchez-Rivas C, Ruzal SM (2005) Cell wall modifications during osmotic stress in *Lactobacillus casei*. J. Appl. Microbiol. 98:84–95.
[68] Piuri M, Sanchez-Rivas C, Ruzal SM (2003) Adaptation to high salt in *Lactobacillus*: role of peptides and proteolytic enzymes. J. Appl. Microbiol. 95:372–379.
[69] Jordan S, Hutchings MI, Mascher T (2008) Cell envelope stress response in Gram-positive bacteria. FEMS Microbiol. Rev. 32:107–146.
[70] Kets EPW, Teunissen PJM, Bont JAM (1996) Effect of Compatible Solutes on Survival of Lactic Acid Bacteria Subjected to Drying. Appl. Environ. Microbiol. 62:259-261.
[71] Ward DE, van der Wejden CC, van der Merwe MJ, Westerhoff HV, Claiborne A, Snoep JL (2000) Branched-chain $\alpha$-keto acid catabolism via the gene products of the bkd operon in *Enterococcus faecalis*: A new, secreted metabolite serving as a temporary redox sink. J Bacteriol. 182:3239–3246.
[72] Considine KM, Kelly AL, Fitzgerald GF, Hill C, Sleator RD (2008) High-pressure processing – effects on microbial food safety and food quality. FEMS Microbiol. Lett. 281:1–9.
[73] Jofré A, Champomier-Vergès M, Anglade P, Baraige F, Martín B, Garriga M, Zagorec M, Aymerich T (2007) Protein synthesis in lactic acid and pathogenic bacteria during recovery from a high pressure treatment. Res. Microbiol. 158:512–520.
[74] Tabanelli G (2011) Use of sub-lethal high pressure homogenization (HPH) treatments to enhance functional properties of Lactic acid bacteria probiotic strains. Ph.D. thesis. University of Bologna, Bologna, Italy.
[75] Chung HJ, Bang W, Drake MA (2006) Stess response of *Escherichia coli*. Compr Rev Food Sci F. 5:52-64
[76] Burns P, Vinderola G, Molinari F, Reinheimer J (2008) Suitability of whey and buttermilk for the growth and frozen storage of probiotic lactobacilli. Int. J. Dairy Technol. 61:156-164.
[77] Sieuwerts S, de Bok FAM, Hugenholtz J, van Hylckama Vlieg JET (2008) Unraveling Microbial Interactions in Food Fermentations: from Classical to Genomics Approaches. Appl. Environ. Microbiol. 74:4997-5007
[78] Caplice E, Fitzgerald GF (1999) Food fermentations: role of microorganisms in food production and preservation. Int. J. Microbiol. 50:131-149
[79] Di Cagno R, De Angelis M, Calasso M, Vincentini O, Vernocchi P, Ndagijimana M, De Vincenzi M, Dessì MR, Guerzoni ME, Gobbetti M (2010) Quorum sensing in sourdough *Lactobacillus plantarum* DC400: induction of plantaricin A (PlnA) under co-cultivation with other lactic acid bacteria and effect of PlnA on bacterial and Caco-2 cells. Proteomics. 10:2175-2190.
[80] Lorenz MC, Cutler NS, Heitman J (2000) Characterization of Alcohol-induced Filamentous Growth in *Saccharomyces cerevisiae*. Mol. Biol. Cell 11:183–199.
[81] Hansen PJ, Drost M, Rivera RM, Paula-Lopes FF, Al-Katananit YM, Kríninger CE, Chase CC (2001) Adverse impact of heat stress on embryo production: causes and strategies for mitigation. Theriogenology. 55: 91–103.

[82] Giraffa G (2004) Studying the dynamics of microbial populations during food fermentation. FEMS Microbiol. Rev. 28:251-260.

[83] Johnson EA (2002) Microbial Adaptation and Survival in Foods. In: Microbial Stress Adaptation and Food Safety. Ed. A.E. Yousef and V.K Juneja. CRC Press

[84] Davidson PM (2001) Chemical preservatives and natural antimicrobial compounds, p. 593–627. In: Food Microbiology: Fundamentals and Frontiers, 2nd ed. M.P. Doyle.

[85] Hüfner E, Hertel C (2008) Improvement of raw sausage fermentation by stress-conditioning of the starter organism *Lactobacillus sakei*. Curr. Microbiol. 57:490-496.

[86] Hüfner E, Markieton T, Chaillou S, Crutz-Le Coq AM, Zagorec M, Hertel C (2007) Identification of *Lactobacillus sakei* genes induced during meat fermentation and their role in survival and growth. Appl. Environ. Microbiol. 73:2522-2531.

[87] Marceau A, Zagorec M, Chaillou S, Mera T, Champomier-Verges MC (2004) Evidence for involvement of at least six proteins in adaptation of *Lactobacillus sakei* to cold temperatures and addition of NaCl. Appl. Environ. Microbiol. 70:7260-7268.

[88] Kamdem S, Patrignani F, Guerzoni ME (2007) Shelf-life and safety characteristics of Italian Toscana traditional fresh sausage (Salsiccia) combining two commercial ready-to-use additives and spices. Food Control. 18: 421 - 429.

[89] Schneider M (1988) Zur mikrobiologie von sauerkraut bei der vergärung in verkaufsfertigen kleinbehältern. Dissertation Hohenheim University, Stuttgart.

[90] Pederson CS, Albury MN (1969) The sauerkraut fermentation. New York State Agricultural Experiment Station Technical Bulletin 824, Geneva, New York.

[91] Ruiz-Barba JL, Garrido-Fernàndez A, Jiménez-Diaz R (1991) Bactericidal action of oleuropein extracted from green olives against *Lactobacillus plantarum*. Let. Appl. Microbiol. 12:65-68.

[92] Ciafardini G, Marsilio V, Lanza B, Possi N (1994) Hydrolysis of oleuropein by *Lactobacillus plantarum* strains associated with olive fermentation. Appl. Environ. Microbiol. 60:4142-4147

[93] Lavermicocca P, Valerio F, Lonigro SL, De Angelis M, Morelli L, Callegari ML, Rizzello CG, Visconti A (2005) Study of Adhesion and Survival of *Lactobacilli* and *Bifidobacteria* on Table Olives with the Aim of Formulating a New Probiotic Food.Appl. Environ. Microbiol. 71:4233–4240.

[94] Ruiz-Barba JL, Jiménez-Diaz R (1995) Availability of essential B-group vitamins to *Lactobacillus plantarum* in green olive fermentation brines. Appl. Environ. Microbiol. 61:1294-1297.

[95] O'Sullivan E, Condon S (1997) Intracellular pH is a major factor in the induction of tolerance to acid and other stresses in *Lactococcus lactis.* Appl. Environ. Microbiol. 63:4210–4215.

[96] Sanders JW, Venema G, Kok J (1999) Environmental stress responses in *Lactococcus lactis*. FEMS Microbiol. Rev. 23:483–501.

[97] Guerzoni ME, Gianotti A, Serrazanetti DI (2011) Fermentation as a tool to improve healthy properties of bread". In V. R. Preedy, R. R. Watson, & V. B. Patel, (Eds.), Flour and breads and their fortification in health and disease prevention (pp.385-393). London, Burlington, San Diego: Academic Press, Elsevier.

[98] Gänzle MG, Schwab C (2009). Ecology of exopolysaccharide formation by lactic acid bacteria: sucrose utilisation, stress tolerance, and biofilm formation. In: Ulrich, M. (Ed.), Bacterial Polysaccharides – Current Innovation and Trends. Horizon Press.

[99] Welman AD, Maddox IS (2003) Exopolysaccharides from lactic acid bacteria: perspectives and challenges. Trends Biotechnol. 21:269-274

[100] Lim EM, Ehrlich SD, Maguin E (2000). Identification of stress-inducible proteins in *Lactobacillus delbrueckii* subsp. *bulgaricus*. Electrophoresis. 21:2557–2561.

[101] Streit F, Corrieu G, Beal C (2007) Acidification improves cryotolerance of *Lactobacillus delbrueckii* subsp. *bulgaricus* CFL1. J. Biotechnol. 128:659–667.

[102] Streit F, Delettre J, Corrieu G, Beal C (2008) Acid adaptation of *Lactobacillus delbrueckii* subsp. *bulgaricus* induces physiological responses at membrane and cytosolic levels that improves cryotolerance. J. Appl. Microbiol. 105:1071–1080.

[103] Liu SQ, Asmundson RV, Gopal PK, Holland R, Crow VL (1998) Influence of Reduced Water Activity on Lactose Metabolism by *Lactococcus lactis* subsp. *cremoris* at Different pH Values. Appl. Environ. Microbiol. 64:2111-2116

[104] Hartke A, Bouche S, Giard JC, Benachour A, Boutibonnes P, Auffray Y (1996) The lactic acid stress response of *Lactococcus lactis* subsp. *lactis*. Curr. Microbiol. 33:194–199.

[105] Broadbent JR, Lin C (1999). Effect of heat shock or cold shock treatment on the resistance of *Lactococcus lactis* to freezing and lyophilization. Cryobiology. 39: 88–102.

[106] Behr J, Gänzle MG, Vogel RF (2006) Characterization of a highly hop-resistant *Lactobacillus brevis* strain lacking hop transport. Appl. Environ. Microbiol. 72:6483-92.

[107] Atrih A, Rekhif N, Michel M, Lefebvre G (1993) Detection and characterization of a bacteriocin produced by *Lactobacillus plantarum* C19. Can. J. Microbiol. 39:1173-1179.

[108] Jiménez-Diaz R, Rios-Sànchez RM, Desmazeaud M, Ruiz-Barba JL, Piard JC (1993) Plantaricin S and T, two new bacteriocins produced by *Lactobacillus plantarum* LPCO10 isolated from a green olive fermentation. Appl. Environ. Microbiol. 59:1416-1424.

[109] Klaenhammer TR (1988) Bacteriocins of lactic acid bacteria. Biochim. 70:337-349.

[110] Neysens P, De Vuyst L (2005) Carbon dioxide stimulates the production of amylovorin L by *Lactobacillus amylovorus* DCE 471, while enhanced aeration causes biphasic kinetics of growth and bacteriocin production. Int. J. Food Microbiol. 105:191–202.

[111] Davidson PM, Harrison MA (2002) Microbial Adaptation to Stresses by Food Preservatives, in Microbial Stress Adaptation and Food Safety. Ed. A.E. Yousef and V.K Juneja. CRC Press

[112] Vrancken G, De Vuyst L, Rimaux T, Allemeersch J, Weckx S (2011) Adaptation of *Lactobacillus plantarum* IMDO 130201, a Wheat Sourdough Isolate, to Growth in Wheat Sourdough Simulation Medium at Different pH Values through Differential Gene Expression. Appl. Environ. Microbiol. 77:3406–3412

[113] Lebeer S, Vanderleyden J, De Keersmaecker SCJ (2008) Genes and Molecules of Lactobacilli Supporting Probiotic Action. Microbiol. Mol. Biol. R. 72:728–764

[114] Jobin MP, Garmyn D, Divies C, Guzzo J (1999) Expression of the *Oenococcus oeni* trxA gene is induced by hydrogen peroxide and heat shock. Microbiol. 145:1245-1251.

[115] Capozzi V, Arena MP, Crisetti E, Spano G, Fiocco D (2011) The hsp 16 Gene of the Probiotic *Lactobacillus acidophilus* Is Differently Regulated by Salt, High Temperature and Acidic Stresses, as Revealed by Reverse Transcription Quantitative PCR (qRT-PCR) Analysis. Int. J. Mol. Sci. 12:5390-5405.

[116] Pieterse B (2006) Transcriptome analysis of the lactic acid and NaCl-stress response of *Lactobacillus plantarum* Ph.D. thesis. Wageningen University, Wageningen, The Netherlands

[117] Petersohn A, Brigulla M, Haas S, Hoheisel JD, Volker U, Hecker M (2001) Global Analysis of the General Stress Response of *Bacillus subtilis*. J. Bacteriol. 183:5617-5631.

[118] Wang JC, Zhang WY, Zhong Z, Wei AB, Bao QH, Zhang Y, Sun TS, Postnikoff A, Meng H, Zhang HP (2012) Transcriptome analysis of probiotic *Lactobacillus casei* during fermentation in soymilk. J. Ind. Microbiol. Biotech. 39:191-206.

[119] Anandan S (2006) Genomic and Proteomic Approaches for Studying Bacterial Stress Responses. In: Advances in Microbial Food Safety. V. Juneja, J.P. Cherry and M.H Tunick eds. ACS Books.

[120] Renzone G, D'Aambrosio C, Arena S, Rullo R, Ledda L, Ferrara L, Scaloni A (2005) Differential proteomic analysis in the study of prokaryotes stress resistance. Ann. Ist. Super. Sanità. 41:459-468.

[121] Nezhad MH, Knight M, Britz ML (2012) Evidence of changes in cell surface proteins during growth of *Lactobacillus casei* under acidic conditions. Food. Sci. Biotech. 21:253-260.

[122] Garnier M, Matamoros S, Chevret D, Pilet MF, Leroi F, Tresse O (2010) Adaptation to Cold and Proteomic Responses of the Psychrotrophic Biopreservative *Lactococcus piscium* Strain CNCM I-4031. Appl. Environ. Microbiol. 76:8011–8018.

[123] Jardin J, Mollé D, Piot M, Lortal S, Gagnaire V (2012) Quantitative proteomic analysis of bacterial enzymes released in cheese during ripening. Int. J. Food Micro. 155:19–28.

[124] Gagnaire V, Piot M, Camier B, Vissers JPC, Gwenael J, Leonil J (2004) Survey of bacterial proteins released in cheese: a proteomic approach. Int. J. Food Microbiol. 94:185-201.

[125] Güell M, Yus E, Lluch-Senar M, Serrano L (2011) Bacterial transcriptomics: what is beyond the RNA horiz-ome? Nat. Rev. Microbiol. 9:658-669.

[126] Wu C, Zhang J, Wang M, Du G, Chen J (2012) *Lactobacillus casei* combats acid stress by maintaining cell membrane functionality. J. Ind. Microbiol. Biotechnol. Accepted

[127] Coda R, Rizzello CG, Pinto D, Gobbetti M (2012) Selected Lactic Acid Bacteria Synthesize Antioxidant Peptides during Sourdough Fermentation of Cereal Flours, Appl. Environ. Microbiol. 78:1087-1096.

[128] Glaasker E, Tjan FSB, Ter Steeg PF, Konings WN, Poolman B (1998) Physiological Response of *Lactobacillus plantarum* to Salt and Nonelectrolyte Stress. J. Bacteriol. 180:4718-4723.

[129] Mehmeti I, Faergestad EM, Bekker M, Snipen L, Nes IF, Holo H (2012) Growth Rate-Dependent Control in *Enterococcus faecalis*: Effects on the Transcriptome and Proteome, and Strong Regulation of Lactate Dehydrogenase. Appl. Environ. Microbiol. 78:170-176.

[130] Alcántara C, Zúñiga M (2012) Proteomic and transcriptomic analysis of the response to bile stress of *Lactobacillus casei* BL23. Microbiol. Accepted

[131] Wu C, Zhang J, Chen W, Wang M, Du G, Chen J (2011) A combined physiological and proteomic approach to reveal lactic-acid-induced alterations in *Lactobacillus casei* and its mutant with enhanced lactic acid tolerance. Appl. Microbiol. Biotechnol. 93:707-722.

[132] Jozefczuk S, Klie S, Catchpole G, Szymanski J, Cuadros-Inostroza A, Steinhauser D, Selbig J, Willmitzer L (2010) Metabolomic and transcriptomic stress response of *Escherichia coli*. Mol. Sys. Biol. 6:364.

Chapter 24

# Lactic Acid Bacteria in Philippine Traditional Fermented Foods

Charina Gracia B. Banaay, Marilen P. Balolong and Francisco B. Elegado

Additional information is available at the end of the chapter

http://dx.doi.org/10.5772/50582

## 1. Introduction

The Philippine archipelago is home to a diverse array of ecosystems, organisms, peoples, and cultures. Filipino cuisine is no exception as distinct regional flavors stem from the unique food preparation techniques and culinary traditions of each region. Although Philippine indigenous foods are reminiscent of various foreign influences, local processes are adapted to indigenous ingredients and in accordance with local tastes. Pervasive throughout the numerous islands of the Philippines is the use of fermentation to enhance the organoleptic qualities as well as extend the shelf-life of food.

Traditional or indigenous fermented foods are part and parcel of Filipino culture since these are intimately entwined with the life of local people. The three main island-groups of the Philippines, namely – Luzon, Visayas, and Mindanao, each have their own fermented food products that cater to the local palate. Fermentation processes employed in the production of these indigenous fermented foods often rely entirely on natural microflora of the raw material and the surrounding environment; and procedures are handed down from one generation to the next as a village-art process. Because traditional food fermentation industries are commonly home-based and highly reliant on indigenous materials without the benefit of using commercial starter cultures, microbial assemblages are unique and highly variable per product and per region. Hence the possibility of discovering novel organisms, products, and interactions are likely.

Various microorganisms are involved in common food fermentation processes. In particular, lactic acid bacteria (LAB) in food is a type of biopreservation system. They not only contribute to the flavor of the food but LAB are also able to control pathogenic and spoilage microorganisms through various ways that include, but are not limited to, production of peroxidases, organic acids, and bacteriocins. Traditionally, identification of LAB in foods is largely dependent on culture-based methods; and properties of each isolate are evaluated under controlled conditions. However, with the advent of molecular techniques, the enumeration of microorganisms missed by culture-dependent methods is now possible. Also, as more LAB metabolites, such as bacteriocins, are being reported, a wider database for identification and comparison with potential novel products are now available.

As the production and consumption of traditional fermented food products become increasingly relevant in the face of rapidly increasing population and food insecurity, more research and development to ensure the safety and nutritional quality of these fermented products is warranted. For a more extensive discussion of the principles and technology of Philippine fermented foods, the readers are directed to Sanchez (2008). This book is a detailed reference based on decades of research. Some data from the book will be presented again here in addition to other data from more recent studies. It is not the intention of this present paper to repeat what has been presented in the book, especially regarding fermentation processes, but only to present, as complete as possible, the data that are available regarding LAB present in indigenous/traditional fermented foods.

This paper aims to briefly review the various lactic acid-fermented indigenous fermented specialties in the different regions of the Philippines. Majority of the discussion will focus on recent data gathered from bacteriocin research and metagenomics studies of Philippine fermented specialties. Lastly, the health applications of the different fermented food products and their development as functional foods will be evaluated.

## 2. Regional fermented specialties in the Philippines

There are various lactic acid-fermented indigenous food products in the Philippines. Table 1 gives a summary of these different fermented specialties found in the different regions. Although a particular product type can be seen throughout the whole country, the texture, taste, and appearance would vary depending on the local taste, materials used, and process employed. For example, bagoong is a common fermented fish paste found all over the Philippines but the characteristic of the product found in Luzon is different from that found in the Visayas and Mindanao regions. Bagoong also takes on different names; there is bagoong na isda, bagoong alamang, bagoong na sisi, and guinamos (Sanchez, 2008). A product that is processed in a similar manner is dayok; it is made of brined fish entrails. Research indicates that this is also a lactic acid-fermented food but the LAB involved have not been identified yet (Besas and Dizon, 2012). Longanisa is sausage made of beef, pork, or chicken. It also takes on many forms depending on where it is made. The more famous ones are Vigan Longanisa in Northern Luzon, Pampanga Longanisa in Central Luzon, Lucban Longanisa in Southern Luzon, and Cebu Longanisa in the Visayas. The tastes vary from spicy, garlicky, sour, to sweet.

In lactic acid-fermented foods, LAB are important in preventing the growth of spoilage organisms, and altering flavor, aroma, and texture of the product. Although LAB are initially present in low numbers in the raw materials used, they soon proliferate as other organisms are inhibited by the initial addition of salt and as the continuous growth of LAB decreases the pH of the food making it less conducive for growth of other organisms. Recent studies, however, have shown that there are a lot more benefits that can be derived from LAB in traditional fermented foods.

| CATEGORY | PRODUCT NAME | REGION | MAJOR INGREDIENTS | LACTIC ACID BACTERIA INVOLVED (as determined from culture-based methods) | APPEARANCE AND/OR USAGE |
|---|---|---|---|---|---|
| Fermented vegetables, fruits | Burong mustasa | Luzon | Mustard leaves, cooked rice and/or rice washings | *Leuconostoc mesenteroides, Enterococcus faecalis, Lactobacillus plantarum* | Side dish |
| | Burong pipino | Whole Phil | cucumber | *Leu. mesenteroides, L. brevis, Pediococcus cerevisiae, L. plantarum* | Side dish |
| | Burong mangga | Whole Phil | Immature mango | *Leu. mesenteroides, L. brevis, P. cerevisiae, L. plantarum* | Side dish |
| | Atchara | Whole Phil | Immature papaya or chayote, or turnip (singkamas) | Unknown | Side dish |
| Cheese | Kesong puti | Luzon, Visayas | Cow or carabao milk | *Lactococcus lactis* | White soft cheese |
| Fermented fish and fishery products | Balao-balao | Luzon | Cooked rice, shrimp, salt | *Leu. mesenteroides, P. cerevisiae, L. plantarum* | Side dish, condiment |
| | Burong-isda | Luzon | Freshwater fish, rice, salt | *Leu. mesenteroides, E. faecalis, P. cerevisiae, L. plantarum, P. acidilactici, Leu. paramesenteroides* | Side dish, condiment |
| | Tinabal | Visayas | Parrot fish (for tinabal molmol) and frigate fish (for tinabal mangko), salt | *P. pentosaceus, S. equinus, Leuconostoc* sp., *Lactobacillus* sp. | Side dish, viand |
| | Burong talangka | Luzon | Small shore crabs (*Varuna litterata*) | *Leu. mesenteroides, E. faecalis, P. cerevisiae, L. plantarum* | Side dish, viand |
| | Patis | Whole Phil | Small fish, salt | *P. halophilus* (in mixed fermentation) | Fish sauce (patis), fish paste (bagoong), used as condiment, sauce, flavoring agent, viand |
| | Bagoong isda | Whole Phil | Small fish, salt | | |
| | Bagoong alamang | Whole Phil | Small shrimps, salt | | |
| | Bagoong na sisi | Visayas | Shell fish, salt | | |
| | Guinamos | Bagoong isda in Visayas, Mindanao | Salt water small fish (dilis/belabid – *Stolephorus* sp.), salt | | Condiment, viand, side dish |
| | Dayok | Visayas, Mindanao | Fish entrails, salt | Unidentified LAB | Condiment, viand, side dish |
| Fermented meat, sausages | Longanisa | Whole Phil | Ground pork, beef, or chicken meat, spices and preservatives | *P. acidilactici, Lactococcus lactis* (together with *Micrococcus aurantiacus*) | Viand |
| | Agos-os | Visayas | Sweet potato and ground pig's head | *E. faecalis* | Viand |
| | Burong kalabi | Luzon | Cooked rice, ground carabao meat | *L. plantarum* | Side dish, viand |
| | Burong babi | Luzon | Cooked rice, ground pork | *L. plantarum* | Side dish, viand |

| | | | | | |
|---|---|---|---|---|---|
| Fermented rice, cassava, sugar cane, coconut, soya | Puto | Whole Phil | Rice, sugar | *L. mesenteroides, E. faecalis, P. cerevisiae* (in mixed fermentation with *Saccharomyces cerevisiae*) | Steamed rice cake |
| | Bibingka | Whole Phil | Rice, sugar | | Baked rice cake |
| | Tapuy | Luzon | Rice, glutinous rice | *Leuconostoc, L. plantarum* (in mixed fermentation with molds and yeasts) | Wine; beer |
| | Pangasi | Mindanao | Rice | Unknown | Wine |
| | Landang | Visayas, Mindanao | Cassava, or buli palm flour | Unknown | Dried jelly pellets pellets, rice substitute |
| | Puto balanghoy | Mindanao | Cassava | Unknown | Steamed cake |
| | Basi | Luzon | Sugar cane | Unknown | Wine |
| | Suka | Whole Phil | Sugar cane juice (for sukang Iloco), palm inflorescence sap (for sukang tuba) | *Leuconostoc, Lactobacillus, Streptococcus* in the initial fermentation phase only | Vinegar, condiment, flavoring |
| | Sinamak | Luzon | Sugar cane juice, spices (chilies, onions, garlic) | Unknown | Spiced vinegar, condiment, flavoring |
| | Pinakurat | Visayas, Mindanao | Coconut sap, chilies, salt, various spices | Unknown | Spiced vinegar, condiment, flavoring |
| | Tuba | Whole Phil | Coconut sap | Unknown | Wine |
| | Lambanog | Whole Phil | Coconut sap | Unknown | Wine |
| | Toyo | Whole Phil | Soybeans | *P. halophilus, E. faecalis, L. delbrueckii* (in mixed fermentation with *Aspergillus sojae* and *Saccharomyces rouxii*) | Condiment, flavoring agent, seasoning |

(Sources: Banaay et al., 2004; Besas and Dizon, 2012; Lee, 1999; Olympia et al., 1995; Sanchez, 2008; Tan et al., 2007)

**Table 1.** Regional Lactic Acid-Fermented Specialties in the Philippines

## 3. Research initiatives on LAB from Philippine fermented foods

### 3.1. Bacteriocin research

Bacteriocins are antimicrobial proteins or peptides produced by certain bacterial strains. Unlike the peptide antibiotics they usually have a narrow spectrum of antimicrobial activity, usually inhibiting growth of closely related bacterial species or strains and lacking lethality to the producer strain (Riley and Wertz, 2002).

The bacteriocins of LAB are small, cationic, hydrophobic, or amphiphilic peptides or small proteins, composed of 20 to 60 amino acid residues (Chen & Hoover, 2003). The bactericidal mode of action and biochemical properties depend on the protein moiety that could be specific to a particular LAB strain, *i.e.* the N-terminal amino acids as determinant of receptors in the cell wall of the susceptible strains/species and C-terminal amino acids for the biochemical properties. LAB bacteriocin must have the following desirable properties: "(1) not active and nontoxic to eukaryotic cells, (2) become inactivated by digestive proteases, having little influence on the gut microbiota, (3) low pH and heat-tolerant, (4) have a relatively broad antimicrobial spectrum, against many food-borne pathogenic and spoilage bacteria, (5) show a bactericidal mode of action, usually acting on the bacterial cytoplasmic membrane: no cross resistance with antibiotics, and (6) have genetic determinants that are usually plasmid-encoded, facilitating genetic manipulation" (Apaga, 2012 as cited from Abriouel et al., 2007).

LAB bacteriocins have attracted attention in recent years because of their generally regarded as safe (GRAS) status and good value as natural biopreservatives which can find applications in the food and cosmetic industries (Cleveland et al., 2001; Daeschel, 1993; Riley and Wertz, 2002). Nisin, produced by strains of *Lactococcus lactis*, has been used in over 50 countries as anti-listerial and anti-clostridium substance. LAB bacteriocins with selective inhibition on food pathogens such as *Listeria monocytogenes*, but no inhibition on important lactic acid bacterial inocula such as the noted probiotic *Lactobacillus paracasei* or *Lactobacillus rhamnosus*; and yogurt-producing *Lactobacillus delbrueckii* subsp. *bulgaricus* and *Lactococcus thermophilus*, may provide advantage over those that have a wider spectrum of antimicrobial activity and would kill these beneficial organisms, including nisin (De Vos, 1993; Jack and Ray, 1995; Nielsen et al., 1990). Hence, efforts on the search for LAB bacteriocins and elucidation of their properties are actively being pursued by several research laboratories. The future holds a wide array of LAB bacteriocins available for various specific applications.

### 3.2. Isolation and identification of bacteriocin-producing LAB

Some efforts on the isolation of bacteriocin-producing LAB had been started for more than a decade now in two major research institutions in the country namely: University of the Philippines Los Banos (specifically, the National Institutes of Molecular Biology and Biotechnology or BIOTECH-UPLB and the Institute of Biological Sciences or IBS-UPLB) and the Philippine Root Crop Research and Training Center, Visayas State University (VSU). These two institutions branched out knowledge on bacteriocin research through affiliate tutorship, as thesis advisers and as trainors to students and staff from a few other academic institutions which also did bacteriocin researches like University of Santo Tomas (UST), University of the Philippines Manila (UPM), De La Salle University (DLSU) and Ateneo de Manila University (ADMU). BIOTECH-UPLB and IBS-UPLB jointly worked on bacteriocins of *Lactobacillus plantarum* or plantaricins and those of *Pediococcus acidilactici* or pediocins. On the other hand, VSU devoted some efforts on the enterocins of *Enterococcus* spp. (Tan et al., 2001). DLSU also tried isolation of bacteriocin-producing LAB for food applications. UST was able to isolate bacteriocin-like inhibitory substances against medically important pathogens like *K. pneumoniae* (Dedeles et al., 2011). UPM and ADMU worked on human and animal health applications of bacteriocins.

Various fermented food products with proteinaceous components were the major sources of isolated LAB for bacteriocin screening. Such fermented food products are home-grown or produced by small enterprises and are still commercially available from public markets in Luzon, Philippines and some parts of the Visayas like Leyte island. Examples of Philippine indigenous fermented foods that were good sources of bacteriocin-producing LAB are fermented rice and shrimp (*balao-balao*), fermented rice and fish mixture (*burong kanin at isda*), fermented pork (*burong babi*) in Central Luzon (Elegado et al., 2003; Gervasio and Lim, 2007) and fermented pork and sweet potato (*agos-os*) in Eastern Visayan region (Samar and Leyte). On the other hand, pickled vegetables like mustard leaf (*burong mustasa*) and green papaya (*achara*), fermenting fruits like pickled green mango, *bignay* or mango wine (Samnang 2010), fermented salted fish (*bagoong*), spicy sausages (*longganisa*) may contain some LAB but often times they are not bacteriocinogenic (Gervasio and Lim, 2007). The obvious reasons are the presence of inhibitory substances like salt, spices, alcohol or acid and of course the dearth of proteinaceous materials in the food material.

In one of the first isolation studies for bacteriocinogenic LAB, various proteinaceous fermented foods native to Central and Southern, Philippines were screened for bacteriocin-producing bacterial isolates. Seventy one out of several hundreds of colony-forming unit isolated by agar plate streaking were found antagonistic to the indicator microorganism, *Lactobacillus plantarum* ATCC 14917, through direct assay. By "spot-on-lawn" assay by pH-neutralized culture supernatant, nine (9) isolates were confirmed to be bacteriocin producers (Elegado et al., 2003). Banaay et al. in 2004 also reported on the isolation of 1,100 putative LAB from indigenous fermented foods in Luzon, Philippines. A strain of *Lactobacillus plantarum* was selected as the best bacteriocin producer. In another study, out of the 160 putative LAB obtained from 19 fermented food products from public markets in Central Luzon, 32 LAB isolates were found to be bacteriocinogenic (Gervasio and Lim, 2007). Santiago et al. (2008) were also able to find two LAB isolates, *Lactobacillus fermentum* LBA-19 and *Lactobacillus casei* LTI-21, screened from among several LAB isolates from various fermented food products from different regions in the Philippines.

Being pleomorphic, identification of LAB is quite challenging. A combination of various microbiological and molecular biology tools would help in finding the real identity. Banaay et al. (2004) did a thorough identification of the bacteriocinogenic LAB isolate using conventional morphological, biochemical and physiological methods, chemotaxonomic methods, as well as molecular methods. This is especially relevant to the identification of *Lactobacillus plantarum* which is a known pleomorphic bacteria. Most other Philippine LAB researchers often times directly apply 16S rRNA gene sequencing and homology search for LAB purified through repeated agar streaking and putatively identified as LAB just after determining its acid–forming, Gram positive and catalase negative properties. (Elegado et al., 2003; Gervasio and Lim, 2007; Santiago et al., 2008). Aside from 16S rRNA genes, other conserved genes were used for identification such as phenylalanyl-tRNA synthase (*pheS*) gene (Dedeles et al., 2011). Detection of bacteriocin genes through PCR may also be helpful in confirming the identity of the bacteriocinogenic LAB as well as the probability of producing the bacteriocin (Table 2).

| ISOLATE/ STRAIN No. | IDENTIFICATION | (primer) HOMOLOGY to *P. acidilactici* type strain | REFERENCE | Bacteriocin gene by PCR; fingerprinting; HOMOLOGY |
|---|---|---|---|---|
| AA-5a | partial 16S rRNA gene ID: *P. acidilactici* | (1492R)98% *P. acidilactici* UL5; 99% *P. acidilactici* DSM20284 (27F) 99% *P. acidilactici* LAB 001; 99% *P. acidilactici* DSM20284 | Elegado et al. 2003 | ped $^{+}$; REP and RAPD |
| 4E2 | partial 16S rRNA gene ID: *P. acidilactici* | (1492R) 98% *P. acidilactici* UL5; 99% *P. acidilactici* DSM20284 (27F) 99% *P. acidilactici* LAB 001; 99% *P. acidilactici* DSM20284 | Apaga (2012) | ped $^{+}$ |
| 4E4 | partial 16S rRNA gene ID: *P. acidilactici* | (1492R) 97% *P. acidilactici* UL5; 99% *P. acidilactici* DSM20284 (27F) 98% *P. acidilactici* 8D2CCH01MX; 99% *P. acidilactici* DSM20284 | Apaga (2012) | ped $^{+}$ |
| 4E5 | partial 16S rRNA gene ID: *P. acidilactici* | (1492R) 99% *P. acidilactici* DSM20284 (27F) 99% *P. acidilactici* DSM20284 | Laxamana et al. (2011) | ped $^{+}$ ; REP |
| 4E6 | partial 16S rRNA gene ID: *P. acidilactici* | (1492R) 98% *P. acidilactici* UL5; 99% *P. acidilactici* DSM20284; (27F) 99% *P. acidilactici* 8D2CCH01MX ; 99% *P. acidilactici* DSM20284 | Apaga (2012) | ped $^{+}$;[ 99% *P. acidilactici* bacteriocin genes ; pSMB74 ] |
| 4E10 | partial 16S rRNA gene ID: *P. acidilactici* | (1492R) 96% *P. acidilactici* UL5; 99% *P. lolii to* NGRI0510Q (27F) 99% *P. acidilactici* LAB 001 ; 99% *P. lolii* NGRI 0510Q | Apaga (2012) | ped $^{-}$ |
| 4BL7 | partial 16S rRNA gene ID: *P. acidilactici* | (1492R) 98% *P. acidilactici* UL5; 99% *P. acidilactici* DSM20284 (27F) 99% *P. acidilactici* 8D2CCH01MX; 99% *P. acidilactici* DSM20284 | Apaga (2012) | ped $^{+}$ |
| 3G3 | API CHL50 ID: *Lactobacillus pentosus*(doubtful) partial 16S rRNA gene ID: *P. acidilactici* | (1492R) 99% *P. acidilactici* IMAU20090 (27F) 98% *P. acidilactici* DSM20284 | Elegado and Perez (2012) | ped $^{+}$ ; REP; ped $^{+}$ |
| 3G8 | partial 16S rRNA gene ID. *P. acidilactici* | (1492R) 99% *P. acidilactici* UL5 (27F) 98% *P. acidilactici* DSM20284 | Elegado and Perez (2012) | ped $^{+}$ |
| 3F3 | partial 16S rRNA gene ID: *P. acidilactici* | (1492R) 95% *P. acidilactici* UL5 (27F) 98% *P. acidilactici* UL5; 99% *P. acidilactici* DSM20284 | Apaga (2012) | ped $^{+}$ |
| 3F8 | partial 16S rRNA gene ID: *P.acidilactici* | (1492R) 98% *P. acidilactici* UL5; 99% *P. acidilactici* DSM20284 (27F) 99% *P. acidilactici* LAB 001; 99% *P. acidilactici* DSM20284 | Apaga (2012) | ped $^{+}$ |
| 3F10 | partial 16S rRNA gene ID: *P. acidilactici* | (1492R) 97% *P. acidilactici* UL5; 99% *P. acidilactici* DSM20284 (27F) 97% *P. acidilactici* LAB 001; 99% *P. acidilactici* DSM20284 | Apaga (2012) | ped $^{+}$ [99% *P. acidilactici* genomic scaffold]; |
| IG7 | partial 16S rRNA gene ID: *P. acidilactici* | (1492R) 97% *P.acidilactici* UL5 99% *P. acidilactici* 8D2CCH01MX; (27F) 98% *P. acidilactici* DSM20284 | Apaga (2012) | ped $^{+}$ [100% pediocin operon;PSMB74]; |

| | | | | |
|---|---|---|---|---|
| $K_2A_2$-3 | API: *Pediococcus pentosaceus* (good) partial 16S rRNA gene ID: *P. acidilactici* | (1492R) 97% *P. acidilactici* UL5; 99% *P. acidilactici* DSM20284 (27F) 99% *P. acidilactici* LAB 001; 99% *P. acidilactici* DSM20284 | Villarante (2011); Elegado and Perez (2012) | ped$^+$ ; plan$^+$ ped$^+$ ; REP |
| $K_2A_2$-1 | API: *P. acidilactici* (doubtful) | - | Abuel (2007) | ped$^+$ ; plan$^+$ ped$^+$ |
| $K_2A_2$-5 | API: *P. acidilactici* (doubtful); partial 16S rRNA gene ID: *P. acidilactici* | (1492R) 97% *P. acidilactici* UL5; 99% *P. acidilactici* DSM20284 (27F) 99% *P. acidilactici* LAB 001; 99% *P. acidilactici* DSM20284 | Apaga (2012) | ped$^+$ [99% *P. acidilactici* genomic scaffold]; plan$^+$ |
| $K_2A_1$-1 | partial 16S rRNA gene ID: *P. acidilactici* | (1492R) 99% *P. acidilactici* L94; 99% *P. acidilactici* DSM20284 (27F) 98% *P.acidilactici* JS-9-4; 99% *P. acidilactici* DSM20284 | Apaga (2012) | ped$^+$ |
| $K_3A_2$-2 | API: *Lactococcus lactis* (good) partial 16S rRNA gene ID: *P. pentosaceus* | - | | ped$^+$ ; plan$^+$; ped$^+$ |
| $K_3A_2$-3 | partial 16S rRNA gene ID: *P. acidilactici* | 100% *P. acidilactici* UL5 | Elegado and Perez (2012) | ped$^+$ |
| S3 | partial 16S rRNA gene ID: *P. acidilactici* | (1492R) 98% *P. acidilactici* UL5; 99% *P. acidilactici* DSM20284 (27F) 97% *P. acidilactici* LAB 001; 99% *P. acidilactici* DSM20284 | Apaga (2012) | ped$^+$ [99% pediocin operon; pSMB72]; |

**Table 2.** Identification and bacteriocin gene determination of putative *Pediococcus acidilactici* through 16S rRNA and pediocin gene PCR amplification and sequencing.

### 3.3. Purification and characterization of bacteriocins

Purification of bacteriocin peptides or small proteins into homogeneity is necessary in order to fully characterize them, particularly the determination of molecular mass, the primary structure or amino acid sequence and secondary structure. For pediocin, it was found that a simple and rapid method is effective for its purification. This method involves adsorption of pediocin onto the cell wall of the producer cell at pH 6 and 0.05 M NaCl and then subsequent desorption at pH 2.0 and 1 M NaCl (Elegado et al., 1997; Yang et al., 1992). This method seemed more applicable to pediocin but not with the lactococcin, nisin or plantaricin. The reason is not clear but it could be related to variation in cell wall properties. The pH-adsorption/desorption method was able to provide materials for pH and temperature tolerance assays, estimation of molecular mass through SDS-PAGE, residual activity determination after protease, amylase and other enzyme actions (Laxamana et al., 2011). Enough amount of semi-purified bacteriocin from pediococci using this method was obtained for further purification through preparative reverse phase HPLC for various characterization studies, including the determination of secondary structures by circular dichroism and confirmation of double bonds through trypsin digestion and electrospray mass spectrometry (Elegado and Kwon, 1998). Other preparative purification methods prior to reverse phase HPLC and spectrometry included ion exchange chromatography and gel filtration chromatography (Elegado et al., 2003), and hydrophobic interaction chromatography (Villarante et al., 2011). This method could also be applied with bacteriocins of pediococci and lactobacilli. The properties obtained from well characterized bacteriocinogenic LAB are shown in Table 3.

| Isolate | Identity | Bacteriocin | Purification mode | Properties |
|---|---|---|---|---|
| AA5a | *Pediococcus acidilactici* | pediocin | pH adsorption/desorption<br>Reversed-phase HPLC | Tolerant to pH 2-9 and 121 °C |
| BS25 | *Lactobacillus plantarum* | plantaricin | Gel filtration chromatography<br>Reversed-phase HPLC | MW = 3,830 Da |
| K2a2-3 | *Pediococcus acidilactici* | pediocin | Hydrophobic interaction and ion-exchange chromatographies<br>Reversed-phase HPLC | MW = 4,626 Da |
| K2a2-1 | *Pediococcus acidilactici* | pediocin | pH adsorption/desorption | Optimum pH = 5-7<br>Resistant to boiling but not to autoclaving |
| 4E5 | *Pediococcus acidilactici* | pediocin | pH adsorption/desorption | Tolerant to pH 2-9; slight loss of activity at 100 °C; loss of activity at 121 °C; tolerates high salt; est. MW = 6,500 Da by SDS-PAGE |

**Table 3.** List of purified and characterized bacteriocins from LAB isolated from Philippine indigenous fermented foods.

## 3.4. Optimization of bacteriocin production through fermentation kinetics

Bacteriocin production is largely dependent on the nutrients and nitrogen content of the fermentation medium. For instance, increased yeast extract concentration and polypeptone amount increases bacteriocin production. Molasses, raw sugar and sago hydrolyzates of amylase digestion were found to be good carbon sources. Other possible substrate base and supplements are cheese whey, coconut water and rice bran extract. Initial sugar concentration of usually 2 to 3% and inoculation rate of 3% by volume of at least $10^8$ cells/mL provides good bacteriocin production (Elegado et al., 2001).

Bacteriocin production is highly dependent on cell or biomass growth. LAB are microaerophilic and most are either mesophilic or slightly thermophilic. The following conditions are applicable to their production: pH= 5.5 to 6.0; temperature = 35 – 40 °C; agitation = 50 rpm; without aeration. Usually, bacteriocin is optimally produced or secreted in the culture broth during the early stationary phase of growth. For *Pediococcus acidilactici*, culturing at 40 °C promotes earlier optimum bacteriocin production of around 10-12 hours. At 37 °C, bacteriocin production is from 14-16 hours (Sagpao et al., 2007).

## 3.5. Applications

Pediocins and plantaricins are the commonly found bacteriocins in Philippine fermented foods so far studied. Their antimicrobial properties have been investigated in several studies (Banaay et al., 2004; Elegado et al., 2003, 2004, 2007; Marilao et al., 2007). Although pediocins and plantaricins show promise, their applications are limited at present because it is a well-known fact that other bacteriocins aside from nisin are not yet approved for food use. For pediocins and plantaricins, the most practical use for now would be dermatological and animal health care use. But since the bacteriocin-producing LAB are of GRAS status, those with probiotic properties such as tolerance to acidic pH (2.0 -3.0) and bile (0.3%) and adhesion properties to intestinal mucosa would be an advantage when used as adjunct inocula in fermented food products (Gervasio and Lim, 2007).

Perhaps another importance of bacteriocin-producing LAB is their effectiveness in biomedical applications. In one study, for example, partially-purified pediocin K2a2-3, through pH-mediated bacteriocin extraction method, was found cytotoxic against human colon adenocarcinoma (HT29) and human cervical carcinoma (HeLa) cells *in vitro* as determined by MTT [3-(4,5-dimethylthiazol-2-yl)-2,5-diphenyltetrazolium bromide] assay (Villarante et al., 2011). Other potential biomedical applications will be discussed in the succeeding section.

## 4. Probiotics and functional foods

An offshoot of the initial research on bacteriocins of LAB isolated from indigenous fermented foods is the emergence of probiotic research towards developing functional foods for biomedical applications. Probiotics refer to microorganisms that, when administered in adequate amounts, confers health benefits to the host. Although there are many microorganisms that can be considered as probiotics, LAB are the most common types because they produce antimicrobial compounds that inhibit other harmful microorganisms, they are able to tolerate acids and bile present in the digestive system, and they are able to adhere and establish themselves in the gut surfaces.

Many benefits have been ascribed to probiotics. For example, *Lactobacillus casei* (Shirota strain in Yakult®) have been shown effective in preventing diarrhea due to enterotoxigenic *Escherichia coli* (ETEC) and choleragenic vibrios (*V. cholerae* biotype E1 Tor and classical *V. cholerae*) using rats (Jacalne et al., 1990). This may be accounted for by its ability to kill the pathogens and inhibit further growth (Consignado et al., 1994). Because the probiotic used in the two studies mentioned is a commercial strain, current research on probiotics progressed to the search for indigenous LAB for use in the development of locally-produced functional food and investigation of their utility for biomedical applications. Metagenomic approaches to investigating LAB present in fermented foods have shown the diversity of potentially beneficial species present other than those that are readily detected by conventional culture-based methods. The development of functional food products shows potential in disease management. Research using metagenomic analysis in searching for microbial markers for use in functional foods to address certain lifestyle diseases as well as malnutrition is on the way.

### 4.1. Metagenomic and diversity studies

Traditional culture-based methods have been used for isolating LAB from fermented foods. These studies form the basis for the starter cultures used in food fermentation technologies employed for commercial production. Sanchez (2008) gives detailed information on the different technologies and cultures used for the production of some traditional as well as developed technologies that have arisen from the culture-based studies conducted in earlier years.

In recent years culture-based approaches in LAB isolation have become more targeted for detection of bacteriocin-producers and those that have potential as probiotics. In one initiative, LAB isolates from fermented foods were screened for bacteriocin production and a PCR-based assay was used to detect specific bacteriocin-encoding genes. Acid and bile tolerance were also determined. Among all the isolates tested, *Lactobacillus fermentum* 4B1 and *Lactobacillus pentosus* 3G3 (later identified as *Pediococcus acidilactici*) have been identified as most promising for the development of new probiotic food products, hence they were chosen for subsequent biomedical application assays (Lim and Gervacio, 2007). In another study, LAB from traditionally fermented wine and vinegar from Visayas and Mindanao were isolated, identified, and tested for inhibitory activity against *Enterococcus faecium, Listeria innocua,* and *Staphylococcus aureus*. Five *Lactobacillus paracasei* and one *Lactobacillus brevis* showed antimicrobial properties against the tester strains (Licaros and Bautista, 2009).

With the advent of molecular techniques, the existence of non-culturable microorganisms has been acknowledged especially since the occurrence of culture-bias is already well-accepted. Culture-independent approaches, therefore, have been gaining popularity in microbial diversity studies and this includes researches on microorganisms found in fermented foods. The microbial populations in selected Philippine fermented foods were assessed through Polymerase Chain Reaction followed by Denaturing Gradient Gel Electrophoresis (PCR-DGGE) in two recent studies (Dalmacio et al., 2011; Larcia, 2010). Food samples tested include *burong mustasa* (fermented mustard), *alamang* (fermented shrimp paste), *burong isda* (fermented rice-fish mixture), *balao-balao/burong hipon* (fermented rice-shrimp mixture), *tuba* (sugar cane wine), and *sinamak* (spiced vinegar). Analysis of the 16S rRNA gene sequences revealed the presence of several LAB that have not been reported in these food products before. *Weissella cibaria, Lactobacillus plantarum, Lactobacillus pontis, Lactobacillus panis,* and *Lactobacllus fermentum* were detected in *burong mustasa* (Larcia, 2010). *L. panis* and *L. fermentum* were present in *alamang; L. pontis* and *L. plantarum* in *burong isda; L. panis, L. pontis,* and *L. fermentum* in burong hipon; and *W. cibaria, L. pontis, L. panis, L. fermentum* and *L. plantarum* in *burong mustasa* (Dalmacio et al., 2011).

The results of the two studies using molecular approaches in defining diversity of LAB in Philippine fermented foods show that culture-independent approaches are efficient tools for the analysis of microbial populations in fermented foods. Majority of the identified bacteria (LAB and other bacterial groups) have not been reported in culture-dependent studies. As such, the isolated bacterial 16S rRNA genes were cloned to have an initial partial 16S rRNA gene library for Philippine fermented foods (Dalmacio et al., 2011).

## 4.2. Biomedical applications

1. Anti-Obesity

Obesity is defined as an abnormal or excessive fat accumulation that presents risks to health. Probiotics can help in fighting obesity by reducing lipid absorption through its action on bile acid metabolism, and by assimilation of cholesterol thus eliminating it from the host's system. Several studies were conducted to examine anti-obesity properties of different probiotic strains.

In one study, oral administration of *Lactobacillus paracasei* K3-4C, isolated from a locally fermented food had significant effect on lowering blood glucose levels (by 46%) and body weight (by 13%) in female BALB/c mice induced to be diabetic and obese through a 28-day high-fat diet (Parungao et al., 2006). In another study, orally administered *L. fermentum* 4B1 reduced adipose cell size, and decreased adipose tissue weight and overall body weight of mice fed with a high-fat diet for 49 days (Bautista et al., 2008). Likewise, oral administration of *P. acidilactici* 3G3 reduced body weight in diet-induced obese female Swiss mice (Parungao et al., 2009). In the last two studies described, the effects of the probiotics were determined to be comparable with the effects of the commercial anti-obesity drug Orlistat based on the parameters measured.

Recently, it has been postulated that the development of obesity may be caused by a shift in the composition of the gut microbiota towards the Firmicutes population (Ley et al., 2005). Firmicutes characterize obese versus lean/non-obese individuals together with a drop or no change in Bacteroidetes (Delzenne and Cani, 2010). Interestingly, Ley et al. (2006) found that a low fat diet had an effect to reverse the shift of Firmicutes/Bacteroidetes proportion. Because of this, dietary manipulation has been seen as a potential means of changing bacterial populations in the colonic microbiota and perhaps treating or at least preventing diseases like obesity. Although the root cause of obesity is excessive caloric intake coupled with a sedentary lifestyle (Blaut and Bischoff, 2010), Ley et al. (2005) proposed in their findings that alteration in the populations of mice gut microflora may have caused or may have been an effect of obesity. Because of this, current researches aim in using probiotics in the treatment of diseases such as obesity.

In two related studies (Arroyo and Fabiculana, 2011; Parungao et al., 2012), the effect of a functional food containing *P. acidilactici* 3G3 on microbial community changes in the gut of obese and non-obese mice was determined through PCR-DGGE. Results of these two preliminary studies showed that obese and non-obese mice had different baseline colonic microbiota. There were also indications that treatment with probiotics shifts the microbiota of obese mice towards the normal non-obese type. As these are preliminary studies, more research is warranted to elucidate the nature of the changes in gut microbiota and how it is related to obesity and the anti-obesity effects of probiotics.

2. Immuno-enhancement

A preliminary *in vitro* study to examine the immune-enhancing properties of viable and heat-killed preparations of two LAB previously isolated from traditional fermented foods (*L. fermentum* 4B1 and *P. acidilactici* 3G3) on murine peritoneal macrophage cells and spleenic T-cells showed that isolate 4B1 was able to induce NO production in murine macrophages but, like 3G3, was unable to stimulate murine T-cell proliferation (Tan et al., 2008). Furthermore, this study showed that preparations of *L. fermentum* 4B1 have the ability to induce NO production in murine macrophage cells and its effects were more potent when it was alive.

The study also showed that isolate 4B1 exhibited better immune-enhancing effect than the probiotic species found in a commercial probiotic drink. T-cell proliferation, however, was not observed in any of the treatments in this study and was attributed to the delayed stimulation in cells responding to a first-time exposure to the different probiotic strain preparations used.

3. Reduction of blood glucose levels

A study by Ngo et al. (2008) showed that oral administration of kefir, a common fermented food consumed by the elderly, significantly decreased blood glucose levels and body weight of diabetic obese male Sprague Dawley rats. The results of the study showed lower blood glucose levels (from 198.5 to 105.6 mg/dL) and clinically lower body weights (from 342.9 to 311.5 g) of the treated diabetic-obese rats than the untreated diabetic-obese control group.

4. Prevention of hypercholesterolemia

The effect of *P. acidilactici* 3G3 administration on hypercholesterolemic Swiss Albino mice was determined (Parungao et al., 2009). This strain was able to assimilate cholesterol in the *in vitro* plate assay and decrease HDL, LDL, and total cholesterol in the *in vivo* assay using mice. Strain 3G3 was also shown to adhere well to the duodenum and middle colon. Results suggest the potential of *P. acidilactici* 3G3 in preventing hypercholesterolemia.

### 4.3. Development of functional foods

The development of functional foods containing known probiotic strains stems from earlier researches on bacteriocins and isolation of potential probiotics from traditional fermented foods. The beneficial effects of probiotic-supplemented chocolate bars (Arroyo et al., 2010; Arroyo and Fabiculana, 2011), fermented mustard leaves (Calapardo et al., 2006), and coffee wine (Parungao, 2007) have been investigated. Initial studies on mango-milk and carrot juice drinks supplemented with probiotic strains have also been conducted (Bugarin et al., 2010; Elegado et al., 2005). These potential functional foods contain probiotic strains, previously isolated from traditional fermented foods such as *P. acidilactici* AA5a (Elegado et al., 2003), *L. plantarum* BS25 (Banaay et al., 2004), and *P. acidilactici* 3G3 (Lim and Gervacio, 2007). Research on functional foods is still in its infancy but this food category shows promise in disease management as well as in contributing to food security in the country. Commercial interest in probiotic food products is increasing due to the growing understanding of its health benefits. This growing industry can derive benefits from the researches conducted on this emerging food category.

## 5. Future perspectives

Aside from the research works presented earlier in this paper as well as on-going follow-up studies related to them, future goals may include research on a variety of other possible biomedical applications of LAB with potential probiotic properties. The effect of probiotics on *Helicobacter pylori* infections (that may cause peptic ulcers) may be determined. Their ability to modulate inflammatory and hypersensitivity responses as well as their effect on irritable bowel syndrome and colitis may be investigated. Further research on possible anti-cancer properties of probiotics is warranted as follow-up studies on the work done by Villarante et al. (2011). These studies are very important as these have the potential to address some of the more serious health concerns of our society.

Much is still to be learned about the existing probiotic strains. The molecular biology and genomics of these isolates may be pursued in order to further elucidate their properties and mechanisms of action.

Determination of factors affecting probiotic viability in foods is also important as these will determine if their survival in the food, and therefore their delivery into the host, is maintained. This will constitute a quality control for functional foods.

The potential physiological effects of multiple prebiotic strains, as opposed to a single strain, are also interesting areas of research. The delivery of multiple probiotic strains may help ensure its effectiveness in an environment that contains high diversity of resident microflora. The potential benefits of synbiotics, (combination of probiotic and prebiotic) which have synergistic interaction, may also be investigated. A good combination will greatly enhance the health benefits to humans.

## Author details

Charina Gracia B. Banaay
*Institute of Biological Sciences, College of Arts and Sciences,*
*University of the Philippines Los Baños, Laguna, Philippines*

Marilen P. Balolong
*Department of Biology, University of the Philippines Manila,*
*Padre Faura Street, Ermita, Manila, Philippines*
*Institute of Molecular Biology and Biotechnology, National Institutes of Health,*
*University of the Philippines Manila, Pedro Gil Street, Ermita, Manila, Philippines*

Francisco B. Elegado
*National Institute of Molecular Biology and Biotechnology,*
*University of the Philippines Los Baños, Laguna, Philippines*

## 6. References

Abriouel H, Galvez A, Lopez RL, Omar NB (2007) Bacteriocin-based strategies for food biopreservation. *International Journal of Food Microbiology* 120: 51–70.

Apaga DLT (2012) Detection of bacteriocin structural genes and bacteriocinogenic activity of several pediococcus isolates. Undergraduate thesis. University of the Philippines Manila. 56 p.

Arroyo KZO, Garcia JAR, Elegado FB, Calapardo M, Parungao MM (2010) Survival of *Lactobacillus pentosus* 3G3 in home-made chocolate bars. 39th Annual Convention and Scientific Meeting,, PSM, Inc. April 29-30, 2010, City of Naga, Camarines Sur.

Arroyo MJJ, Fabiculana PRS (2011) Gut microflora assessment of obese and non-obese mice (*Mus musculus* L.) orally-fed with lp-3g3 chocolate. Department of Biology, University of the Philippines Manila, Manila, Philippines. Unpublished Undergraduate Thesis.

Banaay CGB, Elegado FB, Dalmacio IF (2004) Identification and characterization of bacteriocinogenic *Lactobacillus plantarum* BS25 isolated from balao-balao, a locally fermented rice-shrimp mixture from the Philippines. *The Philippine Agricultural Scientist* 87(4): 427-438.

Bautista RLS, Ecarma NCA, Balolong ECJr, Hallare AV, Parungao MM (2008) The Effects of Orally-Administered *Lactobacillus* sp. 4B1 on the Adipose Tissues of Diet-Induced Obese Mice (*Mus musculus* L.). International Symposium on Probiotic from Asian Traditional Fermented Foods for Healthy Gut Function. August 19-20, 2008. Sari-Pan Pacific Hotel, Jakarta, Indonesia.

Besas JR, Dizon EI (2012) Influence of salt concentration on histamine formation in fermented tuna viscera (Dayok). *Food and Nutrition Sciences* 3:201-206.

Blaut M and Bischoff SC (2010) Probiotics and Obesity. *Ann Nutr Metab* 2010; 57(Suppl.1):20-23.

Bugarin MA, Sison AAD, Elegado FB, Calapardo M, Parungao MM (2010) Survival of *Lactobacillus plantarum* BS25 on Varying Ratio of Mango-Milk Substrates and Storage Temperature. 39th Annual Convention and Scientific Meeting, PSM, Inc. April 29-30, 2010, City of Naga, Camarines Sur.

Calapardo MR, Bueno MOV, Guillermo MKB, Saguibo JD, Parungao MM, Elegado FB (2006) Bile and Acid Tolerance of Bacteriocinogenic *Lactobacillus plantarum* and its Use as Adjunct Inoculum in Pickled Mustard Leaves. Proceedings of the 5th Asia-Pacific Biotechnology Congress and 35th Annual Convention of the PhilippineSociety for Microbiology Inc. (PSM). "Microbiology and Biotechnology: Roadmaps and Milestones for Enhancing Sustainable Productivity in the Asia-Pacific Region." Bohol Tropics Resort. Tagbilaran City, Bohol, Philippines.

Chen H, Hoover DG (2003) Bacteriocins and their food applications. http://www.ift.org/publications/crfsfs (accessed 22 July 2006)

Cleveland J, Chikindas ML, Montville TJ, Nes IF (2001) Bacteriocins: safe, natural antimicrobials for food preservation. *International Journal of Food Microbiology* 71: 1–2 0.

Consignado GO, Pena AC, Jacalne AV (1994) *In vitro* study on the bacterial activity of *Lactobacillus casei* (commercial Yakult drink) against four diarrhea-causing organisms: enterotoxigenic *E. coli, Salmonella enteritidis, Shigella dysenteriae, Vibrio cholerae*. Philipp J Microbiol Infect Dis . 23(2): 50-55.

Daeschel MA (1993) Applications and interactions of bacteriocins of lactic acid bacteria in food and beverages. In: Hoover, D.G., L.R. Steenson, eds. Bacteriocins of Lactic Acid Bacteria. New York: Academic Press Inc. P. 63-70.

Dalmacio LMM, Angeles AKJ, Larcia LLH, Parungao-Balolong MM, Estacio RC (2011) Assessment of microbial diversity in Philippine fermented food products through polymerase chain reaction- denaturing gradient gel electrophoresis (PCR-DGGE). Beneficial Microbes, 2(4): 273-281.

De Vos WM (1993) Future prospects for research and applications of nisin and other bacteriocins. In: Hoover, D.G., L.R. Steenson, eds. Bacteriocins of Lactic Acid Bacteria. New York: Academic Press Inc. P. 249-258.

Dedeles GR, Caranza MAE, Elegado FB (2011) Bacteriocin-like inhibitory substances (BLIS) from lactic acid bacteria and their potential as biopreservatives in foods. Proccedings of the 6th Asian Conference on Lactic Acid Bacteria and XIII International Congress of Bacteriology and Applied Microbiology. September 6-10, 2011. Sapporo, Japan. 120 p.

Delzenne NM and Cani PD (2010) Nutritional modulation of gut microbiota in the context of obesity and insulin resistance: potential interest of prebiotics. *Int. Dairy J.* 20: 277-280.

Elegado FB, Kim WJ, Kwon DY (1997) Rapid purification, partial characterization and antimicrobial spectrum of the bacteriocin, Pediocin ACM from *Pediococcus acidilactici* M. *International Journal of Food Microbiology* 37(1): 1-11.

Elegado FB, Kwon DY (1998) Primary structure and conformational studies of Pediocin AcM, a bacteriocin from *Pediococcus acidilactici* M. *The Philippine Journal of Biotechnology* 9(1): 19-26.

Elegado FB, Sonomoto K, Ishizaki A (2001) Molecular characterization of a bacteriocin of *Pediococcus acidilactici* and its production in sago and sugarcane-based substrate. *Biotechnology for Sustainable Utilization of Biological Resources in the Tropics*. Murooka, Y., T. Yoshida, T. Seki, P. matangkasombut, T.M. Espino, U. Soetisna and M.I.A. Karim. eds. JSPS-NRCT/DOST/LIPI/VCC Joint Seminar. Nov. 7-9, 2001. Bangkok, Thailand. 15:173-180.

Elegado FB, Opina ACL, Banaay CGB, Dalmacio IF (2003) Purification and characterization of novel bacteriocins from lactic acid bacteria isolated from Philippine fermented rice-shrimp or rice-fish mixtures. *The Philippine Agricultural Scientist* 86(1): 65-74.

Elegado FB, Guerra MARV, Macayan RA, Estolas MT, Lirazan MB (2004) Antimicrobial activity and DNA fingerprinting of bacteriocinogenic *Pediococcus acidilactici* through RAPD-PCR. *The Philippine Agricultural Scientist* 87(2):229-237.

Elegado FB, Calapardo MR, Fabregas JF, Ona SEN, Parungao MM (2005) Antilisterial Efficacy of Bacteriocinogenic *Pediococcus acidilactici* AA5a in Carrot Juice Fermentation. Proceedings of the 3rd Asian Conference on Lactic Acid bacteria. August 25-27, 2005. Sanur Paradise Hotel, Bali, Indonesia.

Elegado FB, Abuel BJA, Te JT Jr.,Calapardo MR, Parungao MM (2007) Antagonism against *Listeria* spp. and *Staphylococuus aureus* by bacteriocin-producing lactic acid bacteria screened from the intestine of Philippine carabao using polymerase chain reaction. *The Philippine Agricultural Scientist* 90(4): 305-314.

Elegado FB and Perez MTM (2011) Genetic identification of lactic acid bacteria and bacteriocin structural gene elucidation. National Academy of Science and Technology (NAST) Scientific Meeting. July 11-12, 2012. Manila Hotel, Manila, Philippines.

Gervasio ATR, Lim VMT (2007) Probiotic characterization of bacteriocinogenic lactic acid bacteria isolated from fermented foods of selected areas. Undergraduate thesis. University of the Philippines Manila. 105 p.

Jack RJT, Ray B (1995) Bacteriocin of gram positive bacteria. *Microbiological Reviews*. 59(2): 171–200.

Jacalne AV, Jacalne RR, Hirano H, Suetomi T, Villahermosa CG, Castaneda I (1990) *In-vivo* studies on the use of *Lactobacillus casei* (Yakult Strain) as biological agent for the prevention and control of diarrhea. Acta Med Philipp. 26(2): 116-122

Larcia, LLH (2010) A study of the bacterial profile of Philippine fermented mustard (*burong mustasa*) through culture-independent methods / Levi Letlet Larcia, II. Thesis (MS Biochemistry)--University of the Philippines Manila. 158pp. LG995 2010 B3 L37

Laxamana FLM, Carillo MCO, Elegado FB (2011) Characterization of a bacteriocin of lactic acid bacteria isolated from fermented rice-fish mixture. Proceedings of the 40th Annual Convention and Scientific Meeting of the Philippine Society for Microbiology, Inc. May 10-14, 2011. Manila, Philippines.

Lee C (1999) Cereal fermentations in countries of the Asia-Pacific Region. In: FAO, Fermented Cereals: A Global Perspective, FAO Agricultural Services Bulletin Issue 138. Food and Agriculture Organization 144 p.

Ley RE, Bäckhed F, Turnbaugh P, Lozupone C A, Knight RD, Gordon JI (2005) Obesity alters gut microbial ecology. Proc Natl Acad Sci USA, 102(31): 11070-11075.

Ley RE, Turnbaugh PJ, Klein S, Gordon JI (2006) "*Human gut microbes associated with obesity*." Nature 444:1022-1023.

Licaros A, Bautista A (2009) Culture-Dependent and Culture-Independent Analysis of Vinegars from Visayas and Mindanao. Department of Biology, University of the Philippines Manila, Manila, Philippines. Unpublished Undergraduate Thesis.

Lim V, Gervacio TA (2007) Bacteriocin-producing LAB from fermented foods of Central Luzon. Department of Biology, University of the Philippines Manila, Manila, Philippines. Unpublished Undergraduate Thesis.

Marilao CG, Calapardo MR, Ciron CE, Elegado FB (2007) Antilisterial action of *Pediococcus acidilactici* AA5a in pork sausage fermentation. *The Philippine Agricultural Scientist* 90(1): 40-45.

Ngo RE, Estrada MZ, Balolong ECJr., Parungao MM (2008) Orally-administered kefir on lowering blood glucose levels and body weight in diet-induced diabetic obese rats. Project Terminal Report. National Institutes of Health University of the Philippines Manila.

Nielsen JW, Dickson JS, Crouse JD (1990) Use of a bacteriocin produced by *Pediococcus acidilactici* to inhibit *Listeria monocytogenes* associated with fresh meat. *Applied and Environmental Microbiology* 54: 2349-2353.

Olympia M, Fukuda H, Ono H, Kaneko Y, Takano M (1995) Characterization of starch-hydrolyzing lactic acid bacteria isolated from a fermented fish and rice food, "Burong Isda", and its amylolytic enzyme. *Journal of Fermentation and Bioengineering* 80(2):124-130.

Parungao MM (2007) Probiotic Properties of Philippine Coffee Wine. The 4th Asian Conference on Lactic Acid Bacteria & the 3rd International Symposium on Lactic Acid Bacteria and Health. October 17-19, 2007. Shanghai, China.

Parungao MM, Castillo RL, Mortel MRA (2006) Effects of Oral Administration of *Lactobacillus paracasei* K3-4C on Blood Glucose Levels and Body Weight in Diabetic

Obese Mice (*Mus musculus* L.). December 15-17, 2006. Proceedings of the 11th Biological Sciences Graduate Congress, Chulalongkorn University, Bangkok, Thailand.

Parungao MM, Ong C, Laluces N, De Torres K, Usisa J, Calapardo M, Trinidad L, Elegado FB (2009) The Cholesterol-Reducing Ability and Adhesion Properties of *Lactobacillus pentosus* 3G3. Proceedings of the 5th Conference of Federation for Societies of Lactic Acid Bacteria. July 1-4, 2009. National University of Singapore, Singapore.

Parungao-Balolong M, Libed AAO, Loma K, Villena JPDS, Villafuerte AR, Balolong EC Jr., Dalmacio LMM (2012) Influence of probiotic drinks on the distal gut bacterial flora of mice (*Mus musculus* L.) fed with standard diet or high-fat diet. Terminal report, National Institutes of Health University of the Philippines Manila.

Riley MA, Wertz JE (2002) Bacteriocins: evolution, ecology and applications. *Annual Review of Microbiology* 56:117-137.

Sagpao SMN, Elegado FB, Zamora A (2007) Optimization of production and partial purification of pediocin from *Pediococcus acidilactici* PNCM 10289. Poster paper presented at the 29th National Academy of science and Technology (NAST) Philippines Annual Scientific Meeting, Manila Hotel, July 11-12, 2007.

Samnang (2010) Isolation and identification of lactic acid bacteria from fermenting bignay (*Antidesma bunius* (L.) Spreng) and mango (*Mangifera indica* L.) wines. M.S. thesis. University of the Philippines Los Banos. 242 p.

Sanchez PC (2008) Philippine fermented foods: principles and technology. The University of the Philippines Press. Diliman, Quezon City, Philippines. 516 p.

Santiago MR, Lopez CM, Tenorio EL, de Guzman EE (2007) Detection and partial charcaterization of two bacteriocins produced by *Lactobacillus casei* and *Lactobacillus fermentum* isolated from fermented rice-fish mixtures indigenous in the Philippines. Proceedings of the 37th Annual Convention and Scientific Meeting of the Philippine Society for Microbiology, Inc. May 7-9, 2008. Boracay Philippines. P. 32.

Tan JD, Galvez FCF, Asano K, Tomita F (2001) Isolation and partial purification of bacteriocin produced by microorganisms from *agos-os*. Biotechnology for Sustainable Utilization of Biological Resources in the Tropics. Murooka, Y., T. Yoshida, T. Seki, P. matangkasombut, T.M. Espino, U. Soetisna and M.I.A. Karim. eds. JSPS-NRCT/DOST/LIPI/VCC Joint Seminar. Nov. 7-9, 2001. Bangkok, Thailand. 15: 104-110.

Tan JAS, Yalung PM, Evangelista KV, Parungao MM (2008) *In vitro* Study on the Effect of *Lactobacillus* sp. 4B1 and *Lactobacillus* sp. 3G3 on Murine Macrophage Nitric Oxide Production and Splenic T-cell Proliferation. Proceedings of the International Symposium on Probiotic from Asian Traditional Fermented Foods for Healthy Gut Function.August 19-20, 2008. Sari-Pan Pacific Hotel, Jakarta, Indonesia.

Villarante KI, Elegado FB, Iwatani S, Zendo T, Sonomoto K, de Guzman EE (2010) Purification, characterization and *in vitro* cytotoxicity of the bacteriocin from *Pediococcus acidilactici* K2a2-3 against human colon adenocarcinoma (HT29) and human cervical carcinoma (HeLa) cells. *World Journal of Microbiology and Biotechnology* 27: 975-980.

Yang R, Johnson MC, Ray B (1992) Novel method to extract large amounts of bacteriocins from lactic acid bacteria. *Appl. Environ. Microbiol.* 58: 3355-3359.

Chapter 25

# Lactic Acid Bacteria as Source of Functional Ingredients

Panagiota Florou-Paneri, Efterpi Christaki and Eleftherios Bonos

Additional information is available at the end of the chapter

http://dx.doi.org/10.5772/47766

## 1. Introduction

Lactic acid bacteria (LAB) are widespread microorganisms which can be found in any environment rich mainly in carbohydrates, such as plants, fermented foods and the mucosal surfaces of humans, terrestrial and marine animals. In the human and animal bodies, LAB are part of the normal microbiota or microflora, the ecosystem that naturally inhabits the gastrointestinal and genitourinary tracts, which is comprised by a large number of different bacterial species with a diverse amount of strains [1,2].

Phylogenetically the LAB belong to the *Clostridium* branch of Gram positive bacteria. They are non-sporing, aero tolerant anaerobes that lack catalase and respiratory chain, with a DNA base composition of less than 53 mol% G+C [3,4]. According to their morphology LAB are divided to robs and cocci and according to the mode of glucose fermentation to homofermentative and heterofermentative. The homofermentative LAB convert carbohydrates to lactic acid as the only or major end-product, while the heterofermentative produce lactic acid and additional products such as ethanol, acetic acid and carbon dioxide [5,6]. Thus, the main metabolism of LAB is the degradation of different carbohydrates and related compounds by producing primarily lactic acid and energy. Although many genera of bacteria produce lactic acid as primary or secondary fermentation products, typical lactic acid bacteria are those of the Lactobacillales order, including the following genera: *Lactobacillus, Carnobacterium, Lactococcus, Streptococcus, Enterococcus, Vagococcus, Leuconostoc, Oenococcus, Pediococcus, Tetragonococcus, Aerococcus* and *Weissella* [7].

Many strains of LAB are among the most important groups of microorganisms used in the food and feed industries, although some of the genus Pediococcus cause deterioration of foods, which results in their spoilage [4]. LAB have been used in food preservation and for the modification of the organoleptic characteristics of foods, for example flavors and texture [2]. Various strains of LAB [8] can be found in dairy products (yoghurt, cheese), fermented meats (salami), fermented vegetables (olives, sauerkraut), sourdough bread, etc [9]. The European Food Safety Authority (EFSA) has stated that several LAB strains can be considered to have "Qualified Presumption of Safety" QPS-status [9].

Moreover, nowadays, LAB play an important role in the industry for the synthesis of chemicals, pharmaceuticals, or other useful products (Figure 1). Also, the biotechnological production of lactic acid has recently reported that offers a solution to the environmental pollution by the petrochemical industry [10].

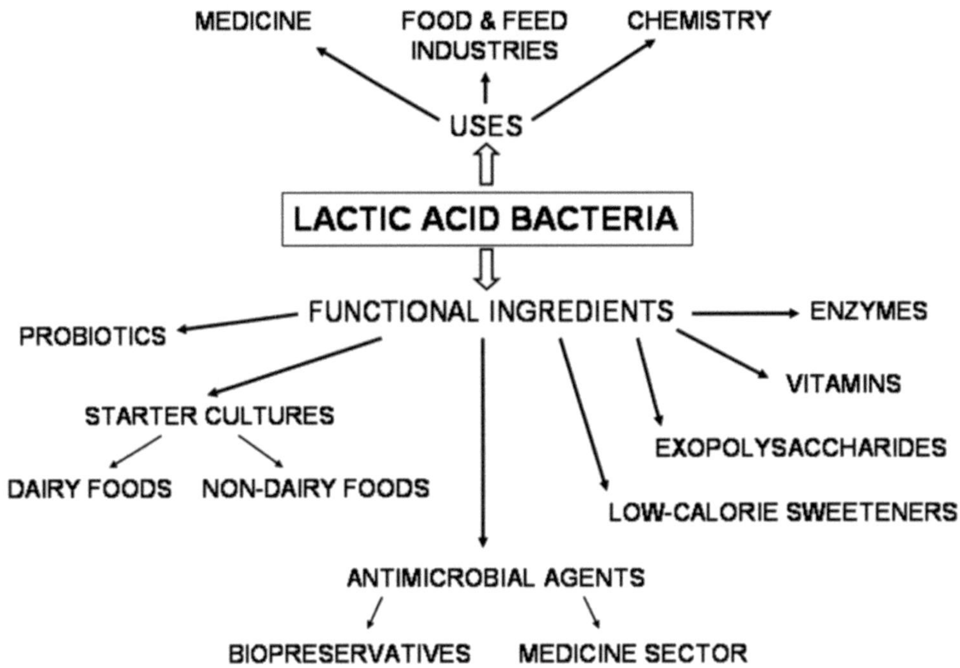


**Figure 1.** Uses and Functional Ingredients of Lactic Acid Bacteria

This chapter will discuss recent applications of LAB as source of probiotics, starter cultures, antimicrobial agents, vitamins, enzymes and exopolysaccharides, especially those that can satisfy the increasing consumer's demands for natural products and functional foods in relation with human health.

## 2. Lactic acid bacteria as source of probiotics

Etymologically the term probiotics is derived from the Greek "probios" which means "for life". In 1974 Parker [11] defined as probiotics "organisms and substances which contribute to intestinal microbial balance". Fuller in 1989 [12] defined as probiotic "a live microbial feed supplement which beneficially affects the host animal by improving its intestinal microbial balance". Later the Food and Agriculture Organization / World Health Organization defined probiotic bacteria as "live microorganisms which when administered in adequate amounts confer a health benefit on the host" [13]. Since probiotics can colonize the gastrointestinal tract and exert their beneficial effect long term, without requiring continuous medical intervention, they have been used for a century to treat a variety of mucosal surface infections (gut, vagina), but their use decreased after the appearance of antibiotics. However, today, probiotics are considered as an alternative solution to antibiotics due to the increasing spread of antibiotic resistance and the need for treatment cost reduction [14].

Microorganisms considered as commercial probiotics are mainly of the Lactobacillus genus with over one hundred species recognized, for example: *L. acidophilus, L. rhamnosus, L. reuteri, L. casei, L. plantarum, L. bulgaricus, L. delbrueckii, L. helveticus* [15-17]. Lactobacilli are Generally Recognized As Safe (GRAS) organisms [18,19].

Probiotic bacteria are very sensitive to many environmental stresses, such as acidity, oxygen and temperature [20,21] and they must fulfill some functional and physiological aspects such as [21,22]: a) Adherence to the intestinal epithelium and colonization of the lumen of the tract. b) Ability to stabilize the intestinal microbiota. c) Counteracting the action of harmful microorganisms. d) Production of antimicrobial substances. e) Stimulation of the immune response.

There are subcategories of the general term probiotic [23,24] which are: a) Probiotic drugs: intended to cue, treat and prevent disease. b) Probiotic foods: food ingredients and dietary supplements. c) Direct-fed microbials: probiotics for animal use. d) Designer probiotics: genetically modified. Generally, foods containing probiotic bacteria fall in the category of functional foods [25].

## 2.1. Mechanism of action of probiotics

Probiotics have multiple and diverse effects on the host. The main mechanisms of action of probiotic bacteria by which they improve mucosal defenses of the gastrointestinal tract include:

a. Antimicrobial activity: The probiotics block the colonization of pathogenic bacteria by decreasing luminal pH, inhibiting bacterial invasion and adhesion to epithelial cells and producing antimicrobial compounds such as bacteriocins and defensins, organic acids and hydrogen peroxide. The interaction of LAB with the mucosal epithelial cells of the gastrointestinal tract and the lymphoid cells in the gut enhance the gut immune response against ingested pathogens [26,27].
b. Enhancement of mucosal barrier function against ingested pathogens: It is achieved with the increasing mucus production through modulation of cytoskeletal and tight junctional protein phosphorylation. The probiotic bacteria compete with pathogenic bacteria for epithelial binding sites, inhibiting the colonization of strains like *Salmonella* and *E. coli* [28,29]. Probiotic bacteria interact with the epithelial cells of the gut, either directly (via cell compounds like DNA, lipoteichoic acids and cell-surface polysaccharides) or indirectly (through production of bioactive metabolites) [30]. The enhancement of mucosal barrier function may be an important mechanism by which probiotics benefit the host in various diseases such as Type 1 diabetes [31,32].

c. Immunomodulation: Specific strains of probiotics might influence the innate and the acquired immune system, thus playing an important role in human diseases. Probiotic bacteria may affect the epithelial cells, the dendritic cells, the monocytes / macrophages and the various types of lymphocytes (Natural killer cells, T-cells and T-cell redistribution) directly or secondarily [33,34]. This action of probiotics could be important for the elimination of neoplastic host cells [22]. Moreover, the effects of probiotics on B-lymphocytes and antibody production resulted in an increase in IgA secretion and the enhancement of response to vaccination [34]. Recently, it was also reported that probiotics can have positive effects on the respiratory system by preventing and reducing the severity of respiratory infections, because of an increase of IgA in the bronchial mucosa [35].

## 2.2. Probiotics and health

Functional properties of probiotics have been demonstrated for various therapeutic applications. Nevertheless, the health benefits provided by probiotics are strain-specific, therefore no probiotic strain will have all proposes benefits, not even strains of the same species [36]. Among the LAB probiotic strains *L. rhamnosus* GG and *L. casei Shirota* have the strongest human health efficacy in the management of lactose intolerance, rotaviral diarrhea and antibiotic associated diarrhea [17]. An optimal single oral dose, based on detection of the bacteria in human feces is $10^9$ bacterial colony forming units (CFU) [37], while in other reports $10^6$-$10^7$ CFU / g of food are considered adequate amounts [13,17,38].

Moreover, in animal nutrition dietary probiotics or direct fed microbials, term which is preferred in the USA, are able to help the maintenance of a healthy intestinal microflora. This microflora may serve to improve performance and health status of the animals, but also to suppress food born pathogens such as *Salmonella* and *Campylobacter*. These conditions are necessary for the production of safe meat and meat products [39]. For instance, the gastrointestinal tract of broilers can be colonized by ingested probiotic bacteria from the first days of their life, which results in shorter period for the achievement of microflora stability. Also, in another example the dietary inclusion of probiotics in young calves' milk replacers may improve their growth performance [40].

Some of the beneficial effects of probiotics are well established as shown in Table 1.

| Lactic Acid Bacteria | Effects on human health | References |
|---|---|---|
| *Lactobacillus rhamnosus* GG | May shorten the course of rotavirus causing diarrhea.<br>Helps to alleviate the symptoms of ulcerative colitis and atopic dermatitis. | 41-45 |
| *Lactobacillus casei* | Reduces the severity and duration of diarrhea. It can stimulate the immune system of the gut and alleviates the symptoms of Crohn's disease | 17, 43, 46 |

| Lactic Acid Bacteria | Effects on human health | References |
|---|---|---|
| *Lactobacillus acidophilus* | Secretes lactic acid which reduces the pH of the gut and inhibits the development of pathogens (*Salmonella spp, E. coli*).<br>Reduces blood cholesterol. | 17, 46, 47 |
| *Lactobacillus johnsonii* | Effective in inhibition of *H. pylori* and against inflammation | 17, 36 |
| *Lactobacillus plantarum* | Produces short-chain fatty acids that block the generation of carcinogenic agents by reducing enzyme activities | 17, 36 |
| *Lactobacillus fermentum* | Effective in restoration of a normal microflora. Effective against bacterial vaginosis flora | 48 |
| *Lactobacillus reuteri* | Reduces the duration of diarrhea | 49 |
| *Enterococcus faecium* | Can reduce blood cholesterol leading to decreased blood pressure | 50-52 |

**Table 1.** Lactic acid bacteria derived probiotics and human health

LAB derived probiotics have potential health benefits in the following situations:

1. Diarrheal diseases:
   a. Infective diarrhea. The most studied gastrointestinal condition treated by probiotics is acute infectious diarrhea in infants. Children represent a main target of studies due to the importance of limiting the spread of diseases and decreasing the need of antibiotics (Aureli et al. 2011). Clinical trials with LAB derived probiotics (*L. rhamnosus* GG; *L. reuteri*; *L. casei*; *L. delbrueckii subsp. Bulgaricus*) support the efficacy of these probiotics in preventing diarrhea [14,49,53], due to their direct or indirect interaction with the enterotoxins [54].
   b. Antibiotic associated diarrhea. A variety of probiotic bacteria, mainly lactobacilli have been used in the treatment and prevention of antibiotic associated diarrhea [55,56]. In a recent study *L. acidophilus* and *L. casei* seemed to be effective in reducing the risk of development of diarrhea [57]. Nevertheless, the results obtained were from pilot studies, so further investigation is needed to evaluate the efficacy of probiotics on such disorders.
   c. Clostridium difficile associated diarrhea. *C. difficile* is an opportunistic pathogen often responsible for diarrhea in vulnerable people. *L. rhamnosus* GG, has demonstrated positive effects on treated patients [14].
   d. Travelers diarrhea. Probiotics with Lactobacilli did not seem to be effective on such diarrhea, which is caused by bacteria, in particular enterotoxigenic E. coli [14,53].
   e. Radiation induced diarrhea. Although there is little research on this subject, probiotics seem to be promising in decreasing radiation diarrhea [53].

2. Inflammatory bowel disease

LAB may affect positively the intestinal mobility and relieve constipation, possibly through a reduction of the intestinal pH [58].

a. Pouchitis. It is a chronic inflammation of the ileal pouch. The treatment with probiotics such as *L. rhamnosus* GG and *L. acidophilus* reduced the risk of pouchitis due to decreased mucosal inflammation [59,60].
b. Crohn's disease. This disease can involve the whole gastrointestinal tract and is characterized by inflammatory processes occurring deeper in the tissues. Among other typical treatments *L. rhamnosus* is used aiming at decreasing the rate and the severity of disease after surgery [61].
c. Ulcerative colitis. It is an acute or chronic disease only affecting the large bowel. LAB probiotics (*L. acidophilus*) provide some promising initial indications [53].

3. Irritable bowel syndrome

This term is used to describe a heterogenous group of gastrointestinal symptoms, like diarrhea, constipation, bloating and abdominal pain. *L. plantarum* strain 299V and *E. faecium* PR88 could be effective treatments against this syndrome [62,63].

4. Prevention of colon cancer

The anticarcinogenic effect of probiotiocs may be attributable to a combination of mechanisms like the induction of pro- or anti-inflammatory and secretary responses that could inhibit carcinogenesis [22]. In vitro studies with lactobacillus strains have shown anti-mutagenic activities. However, there is no evidence yet that probiotics can protect against the development of colon cancer in humans [64,65]. Although, it is hypothesized that the strains tested may have anti-carcinogenic effects by reducing the activity of the enzyme β-glucuronidase.

5. Helicobacter pylori

It is a common chronic bacterial infection in humans, which causes many problems, such as chronic gastritis, septic ulcers and gastric cancer. Probiotics *like L. salivarius, L. casei Shirota* and *L. acidophilus* appear to be promising in inhibiting the growth of H. pylori in vitro [66,67]. Moreover, *L. johnsonii* was also shown effective to inhibit *H. pylori* [68].

6. Lactose intolerance

It is the most common disorder of the intestinal carbohydrate digestion. In both adults and children it has been shown that probiotics can improve the lactose digestion by reducing the intolerance symptoms and slowing orocecal transit [69,70].

7. Blood cholesterol

Recently it has been suggested that some strains of probiotic bacteria, *like L. acidophilus, L. plantarum* and *Enterococcus faecium* could significantly reduce blood cholesterol and increase resistance of low density lipoprotein oxidation, leading to decrease of blood pressure [50-52].

8. Other disorders

The majority of probiotics use has focused on diseases related to the gut, but there are studies that evaluated probiotics, in allergic conditions, including atopic dermatitis, rhinitis, bacterial vaginosis and food allergies [53,71].

a. Atopic Dermatitis. It is the most common of the chronic skin disorders, known as eczema. Investigations have shown that probiotics like *L. rhamnosus* GG, can prevent or reduce the symptoms [42]. Even eczema can be prevented if mothers ingest probiotics during pregnancy and neonatals ingest them during the first 6 months of their life [53].
b. Bacterial Vaginosis. Probiotics (*L. rhamnosus* GR-1 and *L. fermentum* RC-14) are considered to have theraupetic benefits in vaginosis. Probably this is due to the large numbers of lactobacilli in the healthy vaginal microflora [72,73].
c. Other ailments. Probiotics such as *Lactococcus, Pediococcus* and *Leuconostoc* can prevent or limit mycotoxinogenic mould growth [74-78]. Moreover, LAB according to their bacterial strain could bind aflatoxin $B_1$ both in vivo and in vitro [79]. It was reported in studies that *L. paracasei* ST11 reduced body and abdominal fat [80]. These probiotic bacteria seemed to have an anti-obese action. Probably intestinal bacteria may regulate body weight by affecting the host's metabolic neuroendocrine and immune functions [80]. Additionally, probiotics may have anticariogenic effects, preventing and treating dental caries and generally be effective in the oral cavity and the treatment of periodontal disease [81,82].

## 3. Lactic acid bacteria as source of starter cultures

### 3.1. Starter cultures and functional starter cultures in fermentation of foods

LAB for a long time have been applied as starter cultures in fermented foods and beverages, because they can improve nutritional, organoleptic, technological and shelf-life characteristics [83,84]. LAB initiate rapid and adequate acidification in the raw materials, through the production of various organic acids from carbohydrates. Lactic acid is the most abundant, followed by acetic acid, whilst LAB can also produce ethanol, bacteriocins, aroma compounds, exopolysaccharides and some enzymes [85]. Earlier the production of fermented foods and beverages was obtained on a spontaneous fermentation, due to the microflora naturally present on the raw materials. Later on, the direct addition of selected starter cultures to the food matrix was preferred by the food industry. The advantages were the high degree of control over the fermentation process and the standardization of the final product [84].

As starter culture can be defined a microbial preparation of a large number of one or more microorganisms which is introduced to a raw material aiming to produce a fermented food by accelerating and steering its fermentation process [86,87].

The industries of fermented foods mainly utilize commercial starter cultures for the direct inoculation to the food matrix, which are available as frozen and freeze dried concentrates or lyophilized preparations [88].

Recently the use of functional starter cultures in food and beverage fermentation is being explored. These cultures have at least one functional property, contributing in the improvement of the fermentation process, enhancing the quality and of the end safety product and conferring health benefits [84]. Nevertheless, the selection of starter cultures must also eliminate undesirable side effects like the formation of D-lactic acid or a racemate of lactic acid (DL) or the formation of biogenic amines [84,89].

### 3.2. Probiotics as functional starter cultures

A category of successful starter cultures are LAB produced probiotic cultures. Firstly, Metchnikoff [90] discovered the beneficial effects of LAB on human health, through the consumption of yoghurts and fermented milks. Currently probiotic cultures are used for a number of products such as yoghurt, yoghurt drinks, infant formulas, dietary supplements, etc [91]. Yoghurt is manufactured using *Streptococcus thermophilus* and *Lactobaccilus delbrueckii subsp. bulcaricus* as starter cultures [17].

A manufacture in order to choose any probiotic microbial strain to be used as starter culture or better as a blend with a traditionally used starter culture (co-culture), must check the following aspects [92]: 1) The ability of the probiotics to grow in a medium to increase the cells counts. 2) The robustness of the organism to withstand the freezing and drying stages of preparation. 3) The tolerance to acidity of the gastric acid and the bile salts during their passage in the gastrointestinal tract. Thus, the probiotic strains must be stable in order to claim the health benefits [92].

LAB are used as starter cultures either in dairy or non-dairy products (Table 2).

| Genus | Application in dairy foods | Application in non-dairy foods |
|---|---|---|
| *Lactobacillus spp* | Cultured dairy products, cheese, yoghurt, kefir | Sausage, sourdough bread, fermented vegetables |
| *Lactococcus spp* | Cheese, butter milk sour cream, cultured dairy products | - |
| *Leuconostoc spp* | Cheese, cultured dairy products, sour cream, buttermilk | Fermented vegetables |
| *Streptococcus thermophilus* | Cheese, yoghurt | - |
| *Pediococcus* | - | Sausage, fermented vegetables |
| *Tetragenococcus* | - | Soy sauce |
| *Oenococcus* | - | Wine |

**Table 2.** Lactic acid bacteria used as starter cultures in fermented foods

## 3.3. Functional starter cultures in fermented dairy products

Traditionally, LAB have been used in the fermentation of dairy products, as a simple and safe way of preserving such foods. The main species of LAB that can potentially be used as probiotic cultures in dairy products belong to the *Lactobacillus spp* (*L. acidophilus, L. lactis, L. casei, L. plantarum, L. rhamnosus, L. reuteri, L. delbrueckii subsp. bulgaricus*) or to the *Enterococcus spp*. (*E. faecalis, E. faecium*) [17,92].

Dairy products are considered as ideal vehicles for delivering probiotics to the human gut. Yoghurt is considered the most important, followed by cultured buttermilk, kefir, cheeses, ice-cream [17,22,92] or frozen desserts like chocolate mousse [93]. Moreover proteolytic strains of LAB produced probiotics are used to release bioactive peptites called angiotensin I-converting enzyme inhibitors, which are examined for their hypotensive role [94]. Furthermore interaction between probiotics and starter cultures are possible, either as synergism (e.g. yoghurt) or antagonism (e.g. bacteriocins which exhibit antibiotic properties) [20].

A minimum viable LAB count of $10^6$ CFU/g in fermented dairy food is recommended for the claimed health benefits [95].

## 3.4. Functional starter cultures in fermented non-dairy products

### 3.4.1. Fermented meat and meat products

The preservation of meat and meat products by fermentation has been used from ancient times and it was based mainly on natural meat microorganisms. Recently, researchers begun to develop starter cultures for meat products, in order to ensure standard quality for the fermentation process [87]. In 1995 the first LAB meat starter culture used by Niven et al. [96] in the USA was a pure culture of *Pediococcus cerevisiae*. Essential requirements of meat LAB starter cultures are the immediate and rapid production of organic acids at the start of the fermentation, which will result in a pH below 5.1 [97]. Therefore, the original characteristics of the foods are changed, resulting in enhancement of the final products [98].

As commercial meat LAB starter cultures the species more used belong to the Lactobacillus and Pediococcus strains, which can be isolated from dry sausages [97], sauerkraut [20], or smoked salmon [99]. Strains of the above LAB were found to have the best survival activity under acidic conditions and high levels of bile salts [98]. The role of the starter culture as aforementioned is for the safety of foods by inactivating pathogens and spoilage microorganisms via the acid and bacteriosin production. Therefore, the production of biogenic amines is inhibited and microbial growth is suppressed, without the use of antibiotics [97].

Several studies have reported that LAB from meat and meat products can have antibiotic resistance [100]. Thus, before using novel starter cultures or probiotic cultures it is important to check that they do not contain transferable resistance genes [97]. In addition, the selection of LAB starter cultures for sausage production must not have amino decarboxylase activity.

Otherwise, biogenic amines will be produced in foods, such as histamine, tryptamine, tyramine, cadaverine, putrescine and phenylethylamine which have toxic effects [101].

### *3.4.2. Fermented vegetables*

LAB fermentation of vegetables can be achieved due to the presence of carbohydrates. Usually fermented vegetable juices are produced from cabbage, red beet, carrot, celery and tomato [18,102]. Also, LAB play an important role in pickles and table olives fermentation, affecting the final flavour and shelf-life [103,104].

### *3.4.3. Starter cultures in silages*

Ensiling is a traditional method of preserving forages and is widely used all over the world. It is based on natural fermentation, where LAB ferment water-soluble carbohydrates into organic acids, mainly lactic acid or acetic and formic acids, under anaerobic conditions. Inoculation of LAB is often used as silage additive to enhance lactic acid fermentation [18]. This results in decreasing pH, inhibiting detrimental anaerobes and preserving the nutritional value and palatability of the forage [105,106].

Among the LAB genera frequently used are *Lactobacillus plantarum*, Enterococcus *faecium*, *Pediococcus acidilactici*, *Pediococcus pentoseceus* and *Lactobacillus acidophilus*, with usual rates $10^5$-$10^6$ viable cells / g [107]. Feeding ruminants with silages that have been treated with LAB beyond improving their performance, it is believed to induce probiotic effects [108].

## 4. Lactic acid bacteria as source of antimicrobial agents

LAB derived probiotic bacteria display a wide range of antimicrobial activities. Some strains of LAB produce non specific antimicrobial substances (short chain fatty acids, hydrogen peroxide) while others produce toxins (bacteriosins, bacteriosin-like components) [109]. Short chain fatty acids (formic, acetic, propionic, butyric and lactic acids) which are produced during the anaerobic metabolism of carbohydrates, decrease the pH. It has been considered that these acids are responsible for the domination of mucosal ecosystems by LAB [110]. Also, hydrogen peroxide inhibits the growth of pathogens [111].

### 4.1. Bacteriocins

Most of bacteriocins originating by Gram positive bacteria are produced from LAB. They are proteins that have bacteriocidal activity against species closely related to the bacteriocin producing strains, which could be applied in food preservation and health care [112,113]. Traditionally bacteriocin production has been considered an important characteristic in the selection of probiotic strains, while nowadays it is considered that they may function within the gastrointestinal tract [114], perhaps as alternatives to antibiotics for medical and veterinary use [115]. Generally bacteriocins are cationic peptides which display hydrophobic or amphilitic properties and usually the bacterial membrane is the target for their action [116].

The majority of bacteriocin produced by LAB are distinguished from classical antibiotics because: a) They are ribosomally synthesized and have a relatively narrow killing spectrum. b) They can be divided into two main groups, produced by Gram-negative and Gram-positive bacteria [117,118].

Bacteriocins according to their structure and characteristics can be classified mainly in the following classes:

- Class I (lantibiotics), small peptides [119].
- Class II, small heat-stable proteins which are further divided into subclasses such as IIa (pediocin-like bacteriocins) and IIb (two peptite bacteriocins) [119].
- Class III (helveticin) [120].

Bacteriocins mainly produced by *Lactobacillus acidophilus* have strong antimicrobial capacity against various food pathogens [113]. Bacteriocins can act as bactericidal or bacteriostatic, a distinction which is strongly dependent on bacteriocin dose and degree of purification, physiological state of the indicator cells and experimental conditions such as incubation temperature, pH, presence of agents disrupting cell wall integrity, etc [121,122].

## 4.2. Traits of LAB derived bacteriocins

LAB derived bacteriocins are suitable to use as food preservatives due to their characteristics: a) protein nature – they are inactivated by proteases in the gastrointestinal tract. b) Non-toxic and generally non-immunogenic. c) Thermoresistant thus the antimicrobial activity remains after pasteurization and sterilization. d) Affect most of the Gram-positive bacteria. e) Genetic determinants generally located in plasmid facilitating genetic manipulation to increase the variety of natural peptides. f) Usually act on the bacterial cytoplasmic membrane having no cross resistance with antibiotics [122,123].

Some benefits of the use of bacteriocins as food preservatives are: a) extended shelf-life of foods. b) reduction of the risk of transmission of food born pathogenic bacteria. c) Amelioration of economic losses due to food spoilage. d) No addition of chemical preservatives. e) Decrease of the intensity of heat treatments resulting in better preservation of food nutrients and sensory properties of the food. f) Marketing of "novel" foods, less acidic, less salty and with higher water content [123].

Nicin (lantibiotic – class I) is the first bacteriosin produced by LAB (*Lactococcus lactis*) whilst today it is used in many countries as biopreservatives in foods [109]. Nicins have a dual mode of action: a) Binding to lipid II thus preventing correct cell wall synthesis and b) employing lipid II as a docking molecule to initiate a process of membrane insertion and pore formation which leads to rapid cell death [109]. Nisin-producing bacteria can be found in about 30% of human milk samples. This substance may protect mothers from mastitis and infants from toxication by pathogenic skin flora like Staphylococcus aureus [124]. Except from nicin, currently pediocin PA-1/AcH from several Pediococcus strains and enterocin AS-48 from Enterococcus faecalis are used as biopreservatives [119].

There are at least three ways in which bacteriocins can be incorporated into a food to ameliorate its safety: a) By using a purified or semi-purified bacteriocin preparation as food ingredient. b) By introducing an ingredient that has earlier been fermented with a bacteriocin producing strain. c) By using a bacteriocin-producing culture in fermented products to produce the bacteriocin in situ [125].

Additionally, bacteriocin production can contribute to the probiotic functionality of intestinal LAB, while in certain cases may be directly responsible for it, with respect to either beneficially modulating the gut microbiota or inhibiting some gastrointestinal pathogenic bacteria [30].

Consequently bacteriocins, derived from LAB can cover a broad field of applications, including the food industry and the medical sector, mainly in combination with other treatments to increase their effectiveness in humans and animals [126]. In the latter, bacteriocins can be used as growth promoters, instead of antibiotics, which have been banned in the European Union since 2006 [127].

## 5. Lactic acid bacteria as source of vitamins

Human life cannot exist without vitamins, because they are involved in essential functions e.g. cell metabolism and antioxidant activities. Humans cannot synthesize most of these vitamins, although it is well known that some intestinal bacteria like LAB can produce some vitamins (folate, vitamin $B_{12}$ or cobalamin, vitamin $K_2$ or menaguino, riboflavin and thiamine) [128,129]. The gut microbiota has been recognized as a source of some water-soluble vitamins, while such vitamins have also been reported as results of the LAB fermentation in yogurt, cheeses and other fermented foods.

### 5.1. Folate

Folate is the term used to describe the folic acid derivatives, such as the folyl glutamates which are naturally present in foods and folic acid that is the chemically synthesized form of folate, commonly used for food fortification and nutritional supplements. Folate belongs to the B-group of vitamins and participates in many metabolic pathways like the biosynthesis of DNA and RNA and the inter-conversions of amino acids. Moreover, folate possesses antioxidant capacity that protects the genome by preventing free radical hack of DNA [130].

Dietary folate is essential for humans, since it cannot be synthesized by mammalian cells. Folate can be found in legumes, leafy greens, some fruits and vegetables, in liver and fermented dairy products [131], especially in yogurts, where it may be increased depending on the starter cultures used and the storage condition, to values above 200 μg / lt [132]. Epidemiological studies indicated that folate deficiency is associated with a variety of disorders like Alzheinmer's disease, coronary heart diseases, osteoporosis and increased risk of breast and colorectal cancer [130,133].

LAB having the ability to produce folate belong to the *Lactobacillus spp (L. lactis, L. plantarum, L. bulgaricus), Streptococcus spp.* and *Enterococcus spp*. Nevertheless, some lactobacilli strains (*L. gasseri, L. salivarius, L. acidophilus* and *L. johnsonii*) used as both starter cultures and probiotics, cannot synthesize folate due to their lacking in some genes involved in folate biosynthesis [130]. Furthermore, it has been reported that some starter cultures and probiotic lactobacillus strains in non-dairy foods utilize more folate than they produce [128,130]. For this reason nowadays the food industry focuses on the strategy to select and use folate producing probiotic strains, to produce fermented products with elevated amounts of "natural" folate concentrations, without increasing production cost, although increasing health benefits [130,133].

### 5.2. Vitamin $B_{12}$

Vitamin $B_{12}$ or cobalamin is required for the metabolism of fatty acids, amino acids, nucleic acids and carbohydrates [134]. Vitamin $B_{12}$ cannot be synthesized by mammals and must be obtained from exogenous sources like foods or the intestinal microbiota [128]. It has been reported that among the microorganisms some members of the *Lactobacillus spp* have the ability to produce this vitamin. In particular a probiotic strain of *L. reuteri* which exhibits hypocholesterolaemic activity in animals can produce $B_{12}$ [135].

Vitamin $B_{12}$ deficiency can cause various pathological disorders that affect the haematopoietic (pernicious anaemia), nervous and cardiovascular system. Furthermore, this deficiency in male animal models influenced the number of offspring which showed growth retardation and decrease in some blood parameters [136].

### 5.3. Vitamin K

Vitamin K is involved in blood clotting, tissue calcification, atherosclerotic plaque and bones and kidneys function [137]. Vitamin K is present as phylloquinone (Vitamin $K_1$) in green plants and as menaquinone ($K_2$) produced by some intestinal bacteria, like LAB and especially strains of the genera *Lactococcus, Lactobacillus, Enterococcus, Leuconostoc* and *Streptococcus* [128]. Vitamin K deficiency has been involved in some clinical disorders like intracranial hemorrhage in newborn infants and possible bone fracture resulting from osteoporosis [129]. LAB producing menoquinone could be useful to supplement vitamin K requirements in humans [138].

### 5.4. Riboflavin

Riboflavin or vitamin $B_2$ is necessary in cellular metabolism, being the precursor of coenzymes acting as hydrogen carriers in biological redox reactions [129]. Although, riboflavin is present in many foods such as dairy products, meat, eggs, green vegetables, its deficiency occurs with damages in the liver, skin and changes in the brain glucose metabolism [128,129], with symptoms like hyperaemia, sore throat, odema of oral and mucous membranes, cheilosis and glossitis [139].

Currently, riboflavin-producing LAB strains were isolated and used as a convenient biotechnological application for the preparation of bread (fermented sourdough) and pasta to enrich them with vitamin $B_2$ [140].

## 6. Lactic acid bacteria as source of enzymes

LAB possess an extensive collection of enzymes many of which have the potential to influence the composition and the processing, organoleptic properties and quality of foods and feeds. LAB release various enzymes into the gastrointestinal tract and exert potential synergistic effects on digestion and alleviate symptoms of intestinal malabsorption [141]. In other cases these organisms may serve as a source for the preparation of enzyme extracts that are able to function under the environmental conditions of fermentation [142]. The enzymatic activity has been studied mainly in LAB isolated from wine or other fermented foods like cheeses and yoghurt [143,144]. Species of *Lactococcus* and *Pediococcus* are the LAB most commonly associated with fermented foods [143]. The LAB produced enzymes and in particular amylases which are the most stable can be used in sourdough technology for the natural improvement of bread texture [145]. Moreover, LAB contribute to the aroma and flavor of fermented foods. Certain peptidases produced by *Lactococcus lactis subsp. cremoris* improved the sensory quality of cheese [146]. In addition, proteolysis and lipolysis may enhance the flavour of most varieties of cheese [147]. LAB strains isolated from a traditional Spanish Genestoso cheese were evaluated for the enzymatic activity and it was reported that dipeptidase activity of high level was found for *Lactococcus spp*, enterolytic activity was detected for *Enterococcus spp.*, while carboxypeptidase activity was very low or undetectable [147].

Also, enzymes play an important role in winemaking. Wine flavor and aroma apart from aromas originating in grapes and alcoholic fermentation, is derived mainly from the activity of the LAB, through the action of their enzymes. These bacteria grow in wine during malolactic fermentation, following alcoholic fermentation, while a broad range of secondary modifications improve the taste and flavor of wine [144].

## 7. Lactic acid bacteria as source of exopolysaccharides

### 7.1. Definition and classification of exopolysaccharides

A number of LAB can produce a variety of long chain sugar polymers, called exopolysaccharides (EPS) which are mainly employed for the production of fermented dairy products. They are synthesized either extracellularly from sucrose by glycansucrases or intracellularly by glycosyltransferases from sugar nucleotide precursors [148]. These EPS can be classified according to their chemical composition and biosynthesis mechanism as homopolysaccharides, consisting of a single type of monosaccharide and heteropolysaccharides consisting of repeating units of two or more types of monosaccharides, substituted monosaccharides and other units like phosphate, acetyl and glycerol [149-151].

Homopolysaccharides are further divided into fructans including levan and inuline-type and glucans including dextran, mutan, alteran and b-1, 3 glucan [152]. On the other hand, heteropolysaccharides demonstrate little structural similarity to one another. Their production is influenced by the bacterial growth, phase, medium composition (carbon and nitrogen source), pH and temperature [153]. They can be produced by *Lactococcus spp.* and *Lactobacillus spp.* and they play a crucial role in the food industry [151]. Homopolysaccharides can be introduced in sourdough products, influencing the structural quality and backing ability in bakery products, while heteropolysaccharides are used as food additives in dairy products [154]. EPS contribute to the organoleptic quality of the fermented foods, in texture, taste perception, mouth-feel and stability [153,155]. The above researchers reported that there is no information about the effects of bacterial EPS in non-dairy foods, such as meat products, sauerkraut and vinegar. Although EPS are tasteless, they prolong the time that the milk product spends in the mouth, enhancing its delicacy through an improved volatilization of the intrinsic flavors [153,155].

## 7.2. Applications of exopolysaccharides in the industry

In the last years, EPS derived from LAB have received increasing interest because of their GRAS status and their properties. EPS can improve the rheology of fermented foods (viscosity and elasticity) as natural biothickeners, emulsifiers, gelling agents and physical stabilizers to bind water and limit syneresis [153,156]. In particular commercial products like LAB dextran could be utilized apart from foods in gel filtration products, in the pharmaceutical industry, as blood volume expander and flow improver, in chemistry as paper and metal plating processes, in enhanced oil recovery and in chromographic media. Furthermore, levan can find use in the food industry as biothickener, while alteran as low-viscocity factor, extender, etc [128,151,157]. Additionally, EPS may produce oligosaccharides having prebiotic properties that could find important applications in functional foods [158]. The successful application of EPS in the manufacture of fermented milks is determined by the ability to bind water, interact with proteins and increase the viscosity of the milk serum phase [159]. Although, many LAB strains are able to produce EPS, their yield is low [149] and their industrial applications for the improvement of the properties of food products are limited [155].

## 7.3. Potential health benefits of exopolysaccharides

Apart from the technological benefits some EPS derived from LAB are claimed to have beneficial physiological effects on consumer's health. These benefits are detectable at very low concentrations [153]. The EPS due to their increased viscosity in foods may remain for longer time in the gastrointestinal tract and therefore be beneficial to the transient colonization by probiotic bacteria [153,160]. Another health benefit is the generation of short chain fatty acids by colonic microflora degradation in the gut. Several of these fatty acids are possibly involved in the prevention of colon cancer [153,159]. In addition, LAB synthesized EPS appear to have anti-tumor, anti-ulcer, immuno-modulating and cholesterol-lowering activity [155].

## 7.4. Factors limiting the use of exopolysaccharides

The EPS used in the industry represent only a small fraction of the biopolymers used. The reasons are their economical production which needs a global knowledge of their biosynthesis and an adapted bioprocess technology [151]. Moreover, large scale production of LAB derived EPS is low, since LAB being anaerobes, are relatively inefficient in converting energy from carbohydrates, compared to aerobes [149,161]. So this technological barrier must be overcome for cost effective production of EPS. Furthermore, the genetic instability of EPS production is a problem to industrial applications, resulting in loss or reduction of production or change in the composition of EPS [162].

Increasing the knowledge on EPS structure may lead to the production of the "designer" EPS, including the modification primary in structure by altering their physical properties, their function and their production levels. However, for such production of EPS legal approval and the acceptance by the consumers and the food industries are required [151,163]. However, if these biomolecules are to be developed commercially, they must be cost effective.

### 7.5. Negative effects of exopolysaccharides

In some circumstances EPS cause food spoilage. For instance during the fermentation of wine or cider the final products receive undesirable properties. The EPS synthesis is responsible for dental plaque that results in dental caries. Moreover, the accumulation of EPS cause many technical and hygienic problems in the cheese and milk industries [156,164].

## 8. Lactic acid bacteria as source of low-calorie sweeteners

Recently, low calorie sugars produced from LAB have attracted the interest of researchers, industries and consumers, since they can find application as vital food ingredients mainly in foods marketed as "diabetic foods", like sugar-free candies, cookies and chewing gums [165,166]. Manitol, sorbitol, xylitol, tagatose and thehalose are sweeteners produced by LAB. These substances are polyols, i.e. sugar alcohols and can be produced in food fermentation processes. They can be incorporated directly to foods or be produced in the food by LAB, leading to the production of foods containing such sweeteners [165]. *Leuconostoc* and *Lactobacillus spp*. seem to be the most promising producers of these sweeteners [167,168].

A number of health benefits have been attributed to these LAB produced low calories sweeteners, like low glycemic index, osmotic diuretics, weight control, antiplaque, prebiotic. These GRAS substances could be used especially by children, diabetic patients and weight watchers [165,166].

## 9. Conclusion

Lactic acid bacteria are very promising sources for novel products and applications, especially those that can satisfy the increasing consumer's demands for natural products and functional foods. They can be used in the diet of humans and animals, with particular role in their health status. Despite recent advances, the study of LAB and their functional ingredients is still an emerging field of research that has yet to realize its full potential.

## Author details

Panagiota Florou-Paneri and Efterpi Christaki
*Laboratory of Nutrition, Faculty of Veterinary Medicine, Aristotle University of Thessaloniki, Thessaloniki, Greece*

Eleftherios Bonos
*Animal Production, Faculty of Technology of Agronomics, Technological Educational Institute of Western Macedonia, Florina, Greece*

## 10. References

[1] Aureli P, Capurso L, Castellazzi AM, Clerici M, Giovannini M, Morelli L, Poli A, Pregliasco F, Salvini F, Zuccotti (2011) Probiotics And Health: An Evidence-based Review. Pharmacol. res. 63: 366-376.
[2] Barinov A, Bolotin A, Langella P, Maguin E, Van De Guchte M (2011) Genomics Of The Genus *Lactobacillus*. In: Sonomoto K, Yokota A, editors. Lactic Acid Bacteria and Bifidobacteria: Current Progress in Advanced Research. Caister Academic Press, Portland, USA.
[3] Stiles ME, Holzapfel WH (1997) Lactic Acid Bacteria Of Foods And Their Current Taxonomy. Int. j. food Microb. 36: 1–29.
[4] Johnson-Green P (2002) Introduction To Food Biotechnology. CRC Press, Boca Raton.
[5] Halasz A (2009) Lactic Acid Bacteria. In: Lasztity R, editor. Food Quality and Standards (Vol. 3). EOLSS Publishers Co Ltd, UK.
[6] Jay J (2000) Modern Food Microbiology (6th edition). Aspen, Maryland
[7] Hutkins RW (2006) Microbiology and Technology of Fermented Foods. Blackwell Publishing, Iowa, USA. p. 24.
[8] Teitelbaum JE, Walker WA (2002) Nutritional Impact Of Pre- and Probiotics As Protective Gastrointestinal Organisms. Annu. rev. nutr. 22: 107-138.
[9] Korhonen J (2010) Antibiotic Resistance Of Lactic Acid Bacteria. Dissertations in Forestry and Natural Sciences, University of Eastern Finland.
[10] Hamdan AM, Sonomoto K (2011) Production Of Optically Pure Lactic Acid For Bioplastics. In: Sonomoto K, Yokota A, editors. Lactic Acid Bacteria and Bifidobacteria: Current Progress in Advanced Research. Caister Academic Press, Portland, USA.
[11] Parker RB (1974) Probiotics, The Other Half Of The Antibiotic Story. Animal nutr. health. 29: 4-8.
[12] Fuller R (1989) Probiotics In Man And Animals. J. appl. bact. 66: 365-378.
[13] FAO/WHO – Food and Agriculture Organization / World Health Organization (2001) Health And Nutritional Properties Of Probiotics In Food Including Powder Milk With Live Lactic Acid Bacteria, Report Of A Joint FAO/WHO Expert Consultation On Evaluation Of Health And Nutritional Properties Of Probiotics In Food Including Powder Milk With Live Lactic Acid Bacteria, Cordoba. Argentina. http://www.who.int/ foodsafety/ publications/ fs_management/ en/ probiotics.pdf. Accessed 2011 Dec 15.

[14] O'May GA, Macfarlane GT (2005) Health Claims Associated With Probiotics. In: Tamime AY, editor. Probiotic Dairy Products. Blackwell Publishing, Oxford, UK.

[15] Krishnakumar V, Gordon IR (2001) Probiotics: Challenges And Opportunities. Dairy ind. Int. 66: 36-40.

[16] Playne MJ, Bennet LE, Smithers GW (2003) Functional Dairy Foods And Ingredients. Aust. J. dairy technol. 58: 242-264.

[17] Shah NP (2007) Functional Cultures And Health Benefits. Int. dairy j. 17: 1262-1277.

[18] Avall-Jaaskelainen S, Palva A (2005) Lactobacillus Surface Layers And Their Applications. FEMS microbiol. rev. 29: 511-529.

[19] Choi SS, Kang BY, Chung MJ, Kim SD, Park SH, Kim JS, Kang CY and Ha NJ (2005) Safety Assessment of Potential Lactic Acid Bacteria *Bifidobacterium longum* SPM1205 Isolated from Healthy Koreans. J. microbiol. 43: 493-498.

[20] Heller KJ (2001) Probiotic Bacteria In Fermented Foods: Product Characteristics And Starter Organisms. Am. j. clin. nutr. 73(S): 374S-379S.

[21] Parvez S, Malik KA, Kang SA, Kim (2006) Probiotics And Their Fermented Food Products Are Beneficial For Health. J. applied microbiol. 100: 1171-1185.

[22] Soccol CR, Vandenberghe LPDS, Spier MR, Medeiros ABP, Yamaguishi CT, Lindnen JDD, Pandey A, Thomaz-Soccol V (2010) The Potential Of Probiotics: A Review. Food technol. biotechnol. 48: 413-434.

[23] Ahmed F (2003) Genetically Modified Probiotics In Foods. Trends biotechnol. 21: 491-497.

[24] Sanders ME (2009) How Do We Know When Something Called "Probiotic" Is Really A Probiotic? A Guideline For Consumers And Health Care Professionals. Functional Food Rev. 1: 3-12.

[25] Stanton C, Gardiner G, Meehan H, Collins K, Fitzgerald G, Lynch PB, Ross RP (2001) Market Potentials For Probiotics. Am. j. clin. nutr. 73(S): 476-483.

[26] Bourlioux P, Koletzko B, Guarner F, Braesco V (2003) The Intestine And Its Microflora Are Partners For The Protection Of The Host: Report In Danone Symposium The Intelligent Intestine. Am. j. clin. nutr. 78: 675-683.

[27] Mazahreh AS, Ershidat OTM (2009) The Benefits Of Lactic Acid Bacteria In Yoghurt On The Gastrointestinal Function And Health. Pakistan j. nutr. 8: 1404-1410.

[28] Sherman PM, Johnson-Henry KC, Yeung HP, Ngo PSC, Goulet J, Tompkins TA (2005) Probiotics Reduce Enterohemorrhagic Escherichia Coli O157:H7- And Enteropathogenic *E. Coli* O127:H6-Induced Changes In Polarized T84 Epithelial Cell Monolayers By Reducing Bacterial Adhesion And Cytoskeletal Rearrangements. Infect. immun. 73: 5183-5188.

[29] Lin CK, Tsai HC, Lin PP, Tsen HY, Tsai CC (2008) *Lactobacillus Acidophilus* LAP5 Able To Inhibit The *Salmonella Cholerasuis* Invasion To The Human Caco-2 Epithelial Cell. Anaerobe. 14: 251-255.

[30] O'Shea EF, Cotter PD, Stanton C, Ross RP, Hill C (2012) Production Of Bioactive Substances By Intestinal Bacteria As A Basis For Explaining Probiotic Mechanisms: Bacteriocins And Conjugated Linoleic Acid. Int. J. Food Microbiol. 152: 189-205.

[31] Watts T, Berti I, Sapone A, Gerarduzzi T, Not T, Zielke, Fasano A (2005) Role Of The Intestinal Tight Junction Modulator Zonulin In The Pathogenesis Of Type I Diabetes In BB Diabetic prone Rats. Proc. natl. acad. Sci. USA. 102: 2916 –2921.

[32] Meddings J (2008) The Significance Of The Gut Barrier In Disease. Gut. 57: 438–440.

[33] Walker WA (2008) Mechanisms Of Action Of Probiotics. Clinic. inf. dis. 46(S2): S87.

[34] Ng SC, Hart AL, Kamm MA, Stagg AJ, Knight SC (2009) Mechanisms Of Action Of Probiotics: Recent Advances. Inflamm. bowel dis. 15: 300-308.

[35] Perdigon G, Alvarez S, Medina M, Vintini E, Roux E (1999) Influence Of The Oral Administration Of Lactic Acid Bacteria On Iga Producing Cells Associated To Bronchus. Int. j. immunopathol. Pharmacol. 12: 97–102.

[36] Figueroa-Gonzalez I, Quijano G, Ramirez G, Cruz-Guerrero (2011) Probiotics And Prebiotics – Perspectives And Challenges. J. sci. food agric. 91: 1341-1348.

[37] Tannonck GW (2003) Probiotics: Time For A Dose Of Realism. Curr. i. intestinal microbiol. 4: 33-42.

[38] Vaz-Velho M, Todorov SD (2011) Potential Probiotic Evaluation OF Bacteriocin Producing *Lactobacillus Plantarum* ST16PA Isolated From Papaya (*Carica Papaya*). 12th ASEAN Food Conference, Bangkok, Thailand.

[39] Vila B, Esteve-Garcia E, Brufau J (2010) Probiotic Micro-organisms: 100 Years Of Innovation And Efficacy; Modes Of Action. World poult. Sci. 66: 369-380.

[40] Frizzo LS, Soto LP, Zbrun MV, Signorini ML, Bertozzi E, Sequeira G, Rodriguez Armesto R, Rosmini MR (2011) Effect Of Lactic Acid Bacteria And Lactose On Growth Performance And Intestinal Microbial Balance Of Artificially Reared Calves. Livestock sci. 140: 246-252.

[41] Thomas MR, Litin SC, Osmon DR, Corr AP, Weaver AL, Lohse CM (2001) Lack Of Effect Of *Lactobacillus* GG On Antibiotic-associated Diarrhea: A Randomized, Placebo-controlled Trial. Mayo Clinic Proc. 76: 883-889.

[42] Isolauri E, Arvola T, Sutas Y, Moilanen E, Salminen E (2000) Probiotics in the management of atopic eczema. Clin. Experim. Allergy 30: 1604-1610.

[43] Kaur IP, Chorpa K, Saini A (2002) Probiotics: Potential Pharmaceutical Applications. Eur. j. pharmaceut. sci. 15: 1-9.

[44] Kruis W, Fric P, Pokrotnieks J, Lukas M, Fixa B, Kascak M, Kamm MA, Weismueller J, Beglinder C, Stolte M, Wolff C, Schulze J (2004) Maintaining Remission Of Ulcerative Colitis With The Probiotic *Escherichia Coli Nissle* 1917 Is As Effective As With Standard Mesalazine. Gut 53: 1617-1623.

[45] Henker J, Muller S, Laass MW, Schreiner A, Schulze J (2008) Probiotic Echerichia Coli Nissle 1917(EcN) For Successful Remission Maintainance Of Ulcerative Colitis In Children and Adolescents: An Open-label Pilot Study. Z. Gastroenterol. 46: 874-875.

[46] Itsaranuwat P, Shal-haddad K, Robinson PK (2003) The Potential Therapeutic Benefits Of Consuming 'Health-Promoting' Fermented Dairy Products: A Brief Update. Int. j. dairy technol. 56: 203–210.

[47] Gill H, Prasad J (2008) Bioactive Components Of Milk: Probiotics, Immunomodulation, And Health benefits. In: Bosze Z. Advances in Experimental Medicine and Biology. Springer, New York, USA. Pp. 423-464.

[48] Reid G, Charbonneau D, Erb J, Kochanowski B, Beuerman D, Poehner R, Bruce AW (2003) Oral Use Of Lactobacillus Rhamnosus GR-1 And L. Fermentum RC-14 Significantly Alters Vaginal Flora: Randomized Placebo-Controlled Trial In 64 Healthy Women. FEMS immunol. med. microbiol. 35: 131-134.

[49] Shornikova AV, Casas IA, Mykkanen H, Salo E, Vesikari T (1997) Bacteriotherapy With *Lactobacillus Reuteri* In Rotavirus Gastroenteritis. Pediatric inf. dis. j. 16: 1103-1107.

[50] Goel AK, Dilbaghi N, Kamboj DV, Singh L (2006) Probiotics: Microbial Therapy For Health Modulation. Defence sci. j. 56: 513-529.

[51] Liong MT, Shah NP (2005) Roles Of Probiotics And Prebiotics On Cholesterol: The Hypothesized Mechanisms. Nutrafood 4: 45-57.

[52] Cavallini DC, Bedani R, Bomdespacho LQ, Vendramini RC, Rossi EA (2009) Effects Of Probiotic Bacteria, Isoflavoned And Simvastatin On Lipid Profile And Aherosclerosis In Cholesterol-fed Rabbits: A randomized Double-Blind Study. Lipids health dis. 8, Article 1.

[53] Harish K, Varghese T (2006) Probiotics In Humans – Evidence Based Review. Calicut med. j. 4: e3.

[54] Bomba A, Nemcova R, Mudronova D, Guba P (2002) The Possibilities Of Potentiating The Efficacy Of Probiotics. Trends food sci. technol. 13: 121-126.

[55] Cremonini F, Di Caro S, Nista EC, Bartolozzi F, Capelli G, Gasbarini G, Gasbarrini A (2002) Meta-Analysis: The Effect Of Probiotic Administration On Antibiotic-Associated Diarrhoea. Aliment. pharmacol. ther. 16: 1461-1467.

[56] Sazawal S, Hiremath G, Dhingra U, Malik P, Deb S, Black RE (2006) Efficacy Of Probiotics In Prevention Of Acute Diarrhoea: A Meta-Analysis Of Masked, Randomized, Placebo-Controlled Trials. Lancet infect. dis. 6: 374-382.

[57] Yoon SS, Sun J (2011) Probiotics, Nuclear Receptor Signaling, and Anti-Inflammatory Pathways. Gastroenterol. res. pract. Vol: 2011, Article ID: 971938.

[58] Mallett AK, Bearne CA, Rowland IR (1989) The Influence Of Incubation Ph On The Activity Of Rat And Human Gut Flora Enzymes. J. appl. bacteriol. 66: 433–437.

[59] Laake K, Bjorneklett A, Bakka A, MIdtvedt T, Norin K, Eide TJ (1999) Influence Of Fermented Milk On Clinical State, Faecal Bacterial Count And Biochemical Characteristics In Patient With Ileal-Pouch-Anal-Anastomosis. Microb. ecol. health dis. 11: 211-217.

[60] Gosselink MP, Schouten WR, Van Lieshout LM, Hop WC, Laman JD, Ruseler-Van Embden JG (2004) Delay Of The First Onset Of Pouchitis By Oral Intake Of The Probiotic Strain Lactobacillus Rhamnosus GG. Dis. Colon rectum. 47: 876-884.

[61] Prantera C, Scribano ML, Falasco G, Andreoli A, Luzi C (2002) Ineffectiveness Of Probiotics In Preventing Recurrence After Curative Resection For Crohn's Disease: A Randomized Controlled Trial With *Lactobacillus* GG. Gut 51: 405-409.

[62] Hunter J, Lee A, King T, Barratt M, Linggood M, Blades J (1996) *Enterococcus Faecium* Strain PR88 – An Effective Probiotic. Gut 38(S1), A62.

[63] Niedzielin K, Kordecki K, Kosik R (1998) New Possibility In The Treatment Of Irritable Bowel Syndrome: Probiotics As A Modification Of The Microflora Of The Colon. Gastroenterology 114: A402.

[64] Brady LJ, Gallaher DD, Busta FF (2000) The Role Of Probiotic Cultures In The Prevention Of Colon Cancer. The j. nutr. 130(2S): 410S-455S.

[65] Rafter J. (2003) Probiotics And Colon Cancer. Best Pract. Res. Clin. Gastroenterol. 17: 849-859.

[66] Hamilton-Miller JM (2003) The role of probiotics in the treatment and prevention of Helicobacter pylori infection. Int. j. antimicrob. agents. 22: 360-366.

[67] Cats A, Kulpers EJ, Bosschaert MA, Pot RG, Vandenbroucke-Gauls CM, Kusters JG (2003) Effect Of Frequent Consumption Of *Lactobacillus Casei*-Containing Milk Drink In Helicobacter *Pylori*-Colonised Subjects. Pharmacol. therap. 17: 429-435.

[68] Marteau P, De Vrese M, Cellier CJ, Schrezenmeir J (2001) Protection From Gastrointestinal Diseases With The Use Of Probiotics. Am. j. clin. nutr. 73: 430S–436S.

[69] Saltzman JR, Russell RM, Golner B, Barakat S, Dallal GE, Goldin BR. (1999) A Randomized Trial Of *Lactobacillus Acidophilus* BG2FO4 To Treat Lactose Intolerance. Am. j. clin. nutr. 69: 140–6.

[70] Roberfroid MB (2000) Prebiotics And Probiotics: Are They Functional Foods? Am. j. clin. nutr. 71(S): 1682S-1687S.

[71] Pessi T, Sutas Y, Hurme M, Isolauri E (2000) Interleukin-10 Generation In Atopic Children Following Oral *Lactobaccilus Rhamnosus* GG. Clin. experim. Allergy. 30: 1804-1808.

[72] Reid G, Bruce AW, Fraser N, Heinemann C, Owen J, Henning B (2001) Oral Probiotics Can Resolve Urogenital Infections. FEMS immunol. med. microbiol. 30: 49-52.

[73] De Vrese M, Schrezenmeis J (2002) Probiotics And Non-Intestinal Infectious Conditions. Br. j. nutr. 88(S1): S59-66.

[74] Suzuki I., Nomura M., Morichi T. (1991) Isolation Of Lactic Acid Bacteria Which Suppress Mold Growth And Show Antifungal Action. Milchwissenschaft, 46: 635–639.

[75] Florianowicz T (2001) Antifungal Activity Of Some Microorganisms Against Penicillium Expansum. Eur. food res. technol. 212: 282–286.

[76] Mandai V., Sen S.K., Mandai N.C. (2007) Detection, Isolation And Partial Characterization Of Antifungal Compound(S) Produced By Pediococcus Acidilactici LAB 5. Nat. prod. commun. 2: 671–674.

[77] Sathe SJ, Nawani NN, Dhakephalkar PK, Kapadnis BP (2007) Antifungal Lactic Acid Bacteria With Potential To Prolong Shelf-Life Of Fresh Vegetables. J. appl. microbiol. 103: 2622–2628.

[78] Gerez CL, Torino MI, Rollan G, Font de Valdez G (2009) Prevention Of Bread Mould Spoilage By Using Lactic Acid Bacteria With Antifungal Properties. Food Control. 20: 144–148.

[79] Kankaanpaa P, Tuomola E, El-Nezami H, Ahokas J, Salminen SJ (2000) Binding Of Aflatoxin $B_1$ Alters The Adhesion Properties Of *Lactobacillus Rhamnosus* Strain GG In Caco-2 Model. J. food prot. 63: 412–414.

[80] Deshpande G, Rao S, Patole S (2011) Progress In The Field Of Probiotics: Year 2011. Curr. Opin. Gastroenterol. 27: 13-18.

[81] Chen F, Wang D (2010) Novel Technologies For The Prevention And Treatment Of Dental Caries: A Patent Survey. Expert. opin. Ther. Pat. 20: 681–694.

[82] Flichy-Fernandez AJ, Alegre-Diago T, Penarrocha-Oltra D, Penarrocha-Diago (2010) Probiotic Treatment In The Oral Cavity: An Update. Med. oral patol. oral cir. bucal. 15: e677-680.

[83] Wood BJB, Holzapfel WH (1995) The Genera of Lactic Acid Bacteria. London, Blackie Academic & Professional, UK.

[84] Leroy F, De Vuyst L (2004) Lactic Acid Bacteria As Functional Starter Cultures For The Food Fermentation Industry. Trends food sci. technol. 15: 67-78.

[85] De Vuyst L, Leroy F (2007) Bacteriocins From Lactic Acid Bacteria: Production, Purification, And Food Applications. J. mol. Microbiol. biotechnol. 13: 194-199.

[86] Ray B (1992) The Need For Food Biopreservation. In: Ray B, Daeschel M, editors. Food biopreservatives of microbial origin Boca Raton, Florida: CRC Press. pp. 1–23.

[87] Caplice E, Fitzgerald G F (1999) Food Fermentations: Role Of Microorganisms In Food Production And Preservation. Int.j food microbiol. 50: 131–149.

[88] Sandine WE (1996) Commercial Production Of Dairy Starter Cultures. In: Cogan TM, Accolas JP, editors. Dairy Starter Cultures. Wiley-VCH. New York, USA. pp. 191–206.

[89] Joosten HMLJ, Gaya P, Nunez, M (1995) Isolation Of Tyrosin Decarboxylaseless Mutants Of A Bacteriocin-Producing Enterococcus Faecalis Strain And Their Application In Cheese. J food prot. 58: 1222–1226.

[90] Metchnikoff II, Chalmer Mitchell P (1910) Nature Of Man Or Studies In Optimistic Philosophy. Kessinger Publishing, Whitefish, MT, USA.

[91] Hansen EB (2002) Commercial Bacterial Starter Cultures For Fermented Food Of The Future. Int. j. food microbiol. 78: 119-131.

[92] Tamine AY, Saarela M, Korslund Sondergaard A, Mistry VV, Shah NP (2005) Production And Maintenance Of Viability Of Probiotic Micro-organisms In Dairy Products. In: Tamine AY, editor. Probiotic Dairy Products. Blackwell Publishing Ltd, Oxford, UK. pp. 39-72.

[93] Aragon-Alegro LC, Alarcon-Alegro JH, Cardarelli HR, Chiu MC, Sadd SMI (2007) Potentially Probiotic And Symbiotic Chocolate Mousse. LWT-Food sci. technol. 40: 669–675.

[94] Conlin PR, Chow D, Miller ER, Svetkey LP, Lin PH, Harsha DW, Moore TJ, Sacks FM, Appel LJ (2000). The Effect Of Dietary Patterns On Blood Pressure Control In Hypertensive Patients: Results From The Dietary Approaches To Stop Hypertension (DASH) Trial. Am. j. hypertension 13: 949–955.

[95] Karma BKL, Emata OC, Barraquio VL (2007) Lactic Acid And Probiotic Bacteria From Fermented And Probiotic Dairy Products. Sci. Diliman. 19: 23_34.

[96] Erkkila S (2001) Bioprotective And Probiotic Meat Starter Cultures For The Fermentation Of Dry Sausages. Academic Dissertation, Department of Food Technology, University of Helsinki, Finland.

[97] Ammor MS, Mayo B (2007) Selection Criteria For Lactic Acid Bacteria To Be Used As Functional Starter Cultures In Dry Sausage Production: An Update. Meat sci. 76: 138-146.

[98] Zhang W, Xiao S, Samaraweera H, Lee EJ, Ahn DU (2010) Improving Functional Value Of Meat Products. Meat Sci. 86: 15-31.

[99] Todorov SD, Furtado DN, Saad SMI, Tome E, Franco BDGM (2011) Potential Beneficial Properties Of Bacteriocin-producing Lactic Acid Bacteria Isolated From Smoked Salmon. J. appl. microbiol. 110: 971-986.

[100] Gevers D, Danielsen M, Huys G, Swings J (2003) Molecular Characterization Of Tet(M) Genes In Lactobacillus Isolates From Different Types Of Fermented Dry Sausage. Appl. Environm. Microbiol. 69: 1270–1275.

[101] Suzzi G, Gardini F (2003) Biogenic Amines In Dry Fermented Sausages: A Review. Int.j. food microbiol. 88: 41–54.

[102] Buruleanu L, Nicolescu CL, Bratu MG, Manea I, Avram D (2010) Study regarding some metabolic features during lactic acid fermentation of vegetable juices. Romanian biotechnol. Letters. 15: 5177-5188.

[103] Fleming HP (1984) Development In Cucumber Fermentation. Solid State Fermentation Symposium, London, UK. pp. 241-252.

[104] Medina E, Gori C, Servili M, de Castro A, Romero C, Brenes M (2010) Main Variable Affecting The Lactic Acid Fermentation Of Table Olives. Int. j. food sci. tech. 45: 1291-1296.

[105] Weinberg ZG, Ashbell G, Hen Y, Azrieli A (1993) The Effect Of Applying Lactic Acid Bacteria At Ensiling On The Aerobic Stability Of Silages. J. applied bacteriol. 75: 512-518.

[106] Broberg A, Jacobsson J, Strom K, Schnurer J (2007) Metabolite Profiles Of Lactic Acid Bacteria In Grass Silage. Applied environm. microbiol. 73: 5547-5552.

[107] McDonald P, Henderson AR, Heron SJE (1991) The Biochemistry Of Silage, 2nd ed. Chalcombe Publications, Aberystwyth, UK. pp. 184–236.

[108] Weinberg ZG, Muck RE, Weimer PJ, Chen Y, Gamburg M (2004) Lactic Acid Bacteria Used In Inoculants For Silage As Probiotics For Ruminants. Appl. biochem. biotechnol. 118: 1-9.

[109] Gillor O, Etzion A, Riley MA (2008) The Dual Role Of Bacteriocins As Anti- And Probiotics. Appl. microbiol. biotechnol. 81: 591-606.

[110] Lavermicocca P, Valerio F, Evidente A, Lazzaroni S, Corsetti A, Gobbetti M (2000) Purification And Characterization Of Novel Antifungal Compounds From The Sourdough Lactobacillus Plantarum Strain 21B. Appl. environ. microbiol. 66: 4084–4090.

[111] Falagas ME, Betsi GI, Athanasiou S (2007) Probiotics For The Treatment Of Women With Bacterial Vaginosis. Clin. microbiol. infect. 13: 657-664.

[112] Rajaram G, Manivasagan P, Thilagavathi B, Saravanakumar A (2010) Purification And Characterization Of A Bacteriocin Produced By *Lactobacillus Lactis* Isolated From Marine Environment. Adv. j. food sci. technol. 2: 138-144.

[113] Surwase SS, Adsul GG, Jadhav DS (2011) Anti-microbial Activity Associated With Bacteriocin From *Lactobacillus Acidophilus*. J. res. antimicrob. 1: 5-8.

[114] Dobson A, Cotter PD, Ross RP, Hill C (2012) Bacteriocin Production: A Probiotic Trait? Appl. Environ. Microbiol. 78: 1-6.

[115] Kos B, Beganovic J, Jurasic L, Svadumovic M, Pavunc AL, Uroic K, Suskovic J (2003) Coculture-Inducible Bacteriocin Biosynthesis Of Different Probiotic Strains By Dairy Starter Culture *Lactococcus Lactis*. Mljekarstvo 61: 273-282.

[116] Savadogo A, Ouattara CAT, Bassole IHN, Traore AS (2006) Bacteriocins And Lactic Acid Bacteria – A Minireview. African J. Biotechnol. 5: 678-683.

[117] Gordon DM, Oliver E, Littlefield-Wyer J (2007) The Diversity Of Bacteriocins In Gram-Negative Bacteria. In: Riley MA, Chavan M, editors. Bacteriocins: Ecology and evolution. Springer, Berlin, Germany. pp. 5-18.

[118] Heng NCK, Wescombe PA, Burton JP, Jack RW, Tagg JR (2007) The diversity of bacteriocins in Gram-positive bacteria. In: Riley MA, Chavan M, editors. Bacteriocins: Ecology and evolution. Springer, Berlin, Germany. pp. 45-92.

[119] Zendo T, Sonomoto K (2011) Classification And Diversity Of Bacteriocin. In: Sonomoto K, Yokota A, editors. Lactic Acid Bacteria and Bifidobacteria: Current Progress in Advanced Research. Caister Academic Press, Portland, USA.

[120] Dobson AE, Sanozky-Dawes RB, Klaenhammer TR (2007) Identification Of An Operon Inducing Peptide Involved In The Production Of Lactacin B By *Lactobacillus Acidophilus*. J. appl. microbiol. 103: 1766-1778.

[121] Cintas LM, Casaus MP, Herranz C, Nes IF, Hernandez PE (2001) Review: Bacteriocins Of Lactic Acid Bacteria. Food sci. tech. int. 7: 281-305.

[122] Juodeikiene G, Bartkiene E, Viskelis P, Urbonaviciene D, Eidukonyte D, Bobinas C (2012) Fermentation processes using lactic acid bacteria producing bacteriocins for preservation and improving functional properties of food products. In: Petre M, editor. Advances in applied biotechnology. Intech, Croatia. pp. 63-100.

[123] Galvez A, Abriouel H, Lopez RL, Omar NB (2007) Bacteriocin-Based Strategies For Food Biopreservation. Int. j. food microbiol. 120: 51-70.

[124] Beasley SS, Saris PEJ (2004) Nisin-producing *Lactococcus Lactic* Strains Isolated From Human Milk. Appl. environ. microbiol. 70: 5051-5053.

[125] Deegan LH, Cotter PD, Hill C, Ross P (2006) Bacteriocins: Biological Tools For Bio-Preservation And Shelf-Life Extension. Int. dairy j. 16: 1058-1071.

[126] Parada JL, Caron CR, Medeiros ABP, Soccol R (2007) Bacteriocins From Lactic Acid Bacteria: Purification, Properties And Use As Biopreservatives. Brazilian arch. biol. technol. 50: 521-542.

[127] Bonos EM, Christaki EV, Florou-Paneri PC (2010) Performance And Carcass Characteristics Of Japanese Quail As Affected By Sex Or Mannan Oligosaccharides And Calcium Propionate. South African j. animal sci. 40: 173-184.

[128] O'Connor EB, Barrett E, Fitzgerald G, Hill C, Stanton C, Ross RP (2005) Production Of Vitamins, Exopolysaccharides And Bacteriocins By Probiotic Bacteria. In: Tamime AY, editor. Probiotic Dairy Products. Blackwell Publishing, Oxford, UK.

[129] LeBlanc JG, Laino JE, Juarez del Valle M, Vannini V, van Sinderen D, Taranto MP, Font de Valdez G, Savoy de Giori G, Sesma (2011) B-Group Vitamin Production By Lactic Acid Bacteria – Current Knowledge And Potential Applications. J. applied microbiol. 111: 1297-1309.

[130] LeBlanc JG, de Giori GS, Smid EJ, Hugenholtz J, Sesma F (2007) Folate Production By Lactic Acid Bacteria And Other Food-Grade Microorganisms. Commun. curr. res. educ. top. trends appl. microbiol. 1: 329-339.

[131] Eitenmiller RR, Landen WO (1999) Folate. In: Eitenmiller RR, Landen WO, editors. Vitamin Analysis For The Health And Food Sciences. CRC Press LLC, Boca Raton. p. 411-465.

[132] Wouters JTM, Ayad EHE, Hugenholtz J, Smit G (2002) Microbes From Raw Milk For Fermented Dairy Products. Int. dairy j. 12: 91-109.

[133] Rossi M, Amaretti A, Raimondi S (2011) Folate Production By Probiotic Bacteria. Nutrients 3: 118-134.

[134] Quesada-Chanto A, Afschar AS, Wagner F (1994) Microbial Production Of Propionic Acid And Vitamin $B_{12}$ Using Molasses Or Sugar. App. microbiol. biotechnol. 41: 378-383.

[135] Taranto MP, Vera JL, Hugenholtz J, De Valdez GF, Sesma F (2003) *Lactobacillus Reuteri* CRL1098 Produces Cobalamin. J. bacteriol. 185: 5653-5647.

[136] Molina V, Medici M, Taranto MP, Font de Valdez G (2008) Effects Of Maternal Vitamin $B_{12}$ Deficiency From End Of Gestation To Weaning On The Growth And Haematological And Immunological Parameters In Mouse Dams And Offspring. Arch. anim. nutr 62: 162–168.

[137] Olson RE (1984) The Function And Metabolism Of Vitamin K. Ann. rev. nutr. 4: 281-337.

[138] Morishita T, Tamura N, Makino T, Kudo S (1999) Production Of Menaquinones By Lactic Acid Bacteria. J. dairy sci. 82: 1897-1903.

[139] Wilson JA (1983) Disorders Of Vitamins: Deficiency, Excess And Errors Of Metabolism. In: Petersdorf RG, Harrison TR. Harrison's Principles of Internal Medicine. McGraw-Hill Book Co. New York, USA. pp. 461–470.

[140] Capozzi V, Menga V, Digesu AM, De Vita P, van Sinderen D, Cattivelli L, Fares C, Spano G (2011) Biotechnological Production Of Vitamin $B_2$-Enriched Bread And Pasta. J. agric. food chem. 59: 8013–8020.

[141] Naidu AS, Bidlack WR, Clemens RA (1999) Probiotic Spectra of Lactic Acid Bacteria (LAB). Crit. rev. food sci. nutr. 38: 13-126.

[142] Tamang JP (2011) Prospects Of Asian Fermented Foods In Global Markets. 11th ASEAN Food Conference, Bangkok, Thailand.

[143] Matthews A, Grimaldi A, Walker M, Bartowsky E, Grbin P, Jiranek V (2004) Lactic Acid Bacteria As A Potential Source Of Enzymes For Use In Vinification. Appl. environ. mcrobiol. 70: 5715-5731.

[144] Mtshali PS (2007) Screening And Characterisation Of Wine-related Enzymes Produced By Wine-associated Lactic Acid Bacteria. MSc. Thesis, Stellenbosch University.

[145] Mogensen G (1993). Starter Cultures. In: Smith J, editor. Technology of reduced-additive foods London: Blackie Academic & Professional, UK. pp. 1–25.

[146] Guldfeldt LU, Sorensen KI, Stroman P, Behrndt H, Williams D, Johansen E (2001). Effect Of Starter Cultures With A Genetically Modified Peptidolytic Or Lytic System On Cheddar Cheese Ripening. Int. dairy j. 11: 373–382.

[147] Gonzalez L, Sacristan N, Arenas R, Fresno JM, Tornadijo ME (2010) Enzymatic Activity Of Lactic Acid Bacteria (With Antimicrobial Properties) Isolated From A Traditional Spanish Cheese. Food microbiol. 27: 592–597.

[148] Ganzle MG, Schwab C (2009) Ecology Of Exopolysaccharide Formation By Lactic Acid Bacteria: Sucrose Utilisation, Stress Tolerance, And Biofilm Formation. In: Ullrich M, editor. Bacterial Polysaccharides: Current Innovations And Future Trends: Caister Academic Press, Bremen, Germany.

[149] Ruas-Madiedo P, Hugenholtz J, Zoon P (2002) An Overview Of The Functionality Of Exopolysaccharides Produced By Lactic Acid Bacteria. Int. dairy j. 12: 163-171.

[150] Ruas-Madiedo P, Salazar N, de los Reyes-Gavilan (2009) Biosynthesis and Chemical Composition Of Exopolysaccharides Produced By Lactic Acid Bacteria. In: Ullrich M, editor. Bacterial Polysaccharides: Current Innovations And Future Trends: Caister Academic Press, Bremen, Germany.

[151] De Vuyst L, de Vin F, Vaningelgem F, Degeest B (2001) Recent Developments In The Biosynthesis And Applications Of Heteropolysaccharides From Lactic Acid Bacteria. Int. dairy j. 11: 687-707.

[152] Monsan P, Bozonnet S, Albenne C, Jouela G, Willemot RM, Remaud-Simeon M (2001) Homopolysaccharides From Lactic Acid Bacteria. Int. dairy j. 11: 675-685.

[153] Duboc P, Mollet B (2001) Application Of Exopolysaccharides In The Dairy Industry. Int. dairy j. 11: 759-768.

[154] Waldherr F, Vogel RF (2009) Commercial Exploitation Of Homo-exopolysaccharides In Non-dairy Food Systems. In: Ullrich M, editor. Bacterial Polysaccharides: Current Innovations And Future Trends: Caister Academic Press, Bremen, Germany.

[155] Zhang Y, Li S, Zhang C, Luo Y, Zhang H, Yang Z (2011) Growth And Exopolysaccharide Production By *Lactobacillus Fermentum* F6 In Skim Milk. African j. biotechnol. 10: 2080-2091.

[156] De Vuyst, L, Degeest B (1999) Heteropolysaccharides From Lactic Acid Bacteria. FEMS microbiol. rev. 23: 153-177.

[157] Freitas F, Alves VD, Reis AM (2011) Advances In Bacterial Exopolysaccharides: From Production To Biotechnological Applications. Trends biotechnol. 29: 388-398.

[158] Gibson R, Roberfroid MB (1995) Dietary Modulation Of The Human Colonic Microbiota: Introducing The Concept Of Prebiotics. J. nutr. 125: 1401-1412.

[159] Ruijssenaars HJ, Stingele F, Hartmans S (2000) Biodegradability Of Food-associated Extracellular Polysaccharides. Curr. microbiol. 40: 194-199.

[160] German B, Schiffrin E, Reniero R, Mollet B, Pfeifer A, Neeser JR (1999) The development of functional foods: Lessons from the gut. Trends biotechnol. 17: 492–499.

[161] Welman AD, Maddox IS (2003) Exopolysaccharides From Lactic Acid Bacteria: Perspectives And Challenges. Trends biotechnol. 21: 269-274.

[162] Bouzar F, Cerning J, Desmazeaud M (1996) Exopolysaccharide Production In Milk By *Lactobacillus Delbrueckii Ssp. Bulgaricus* CNRZ 1187 And By Two Colonial Variants. J. dairy sci. 79: 205-211.

[163] Sutherland IW (1999) Polysaccharases For Microbial Exopolysaccharides. Carbohydr. polym. 38: 319-328.

[164] Patel S, Majumder A, Goyal A (2012) Potentials Of Exopolysaccharides From Lactic Acid Bacteria. Indian j. microbiol. 52: 3-12.

[165] Patra F, Tomar SK, Arora S (2009) Technological And Functional Applications Of Low-Calorie Sweeteners From Lactic Acid Bacteria. J. food sci. 74: 16-21.

[166] Monedero V, Perez-Martínez G, Yebra MJ (2010) Perspectives Of Engineering Lactic Acid Bacteria For Biotechnological Polyol Production. Appl. microbiol. biotechnol. 86: 1003–1015.

[167] Soetaert W (1990) Production Of Mannitol With *Leuconostoc Mesenteroides*. Med. fac landbouwwet rijksuniv gent. 55: 1549–52.

[168] Kim P (2004) Current Studies On Biological Tagatose Production Using L-Arabinose Isomerase: A Review And Future Perspective. Appl. microbiol. biotechnol. 65: 243–249.

Chapter 26

# The Current Status and Future Expectations in Industrial Production of Lactic Acid by Lactic Acid Bacteria

Sanna Taskila and Heikki Ojamo

Additional information is available at the end of the chapter

http://dx.doi.org/10.5772/51282

## 1. Introduction

Conversion of carbohydrates to lactic acid is one of the most employed fermentation processes in food industry. Applications of lactic acid fermentation are found in dairy industry, production of wine and cider, production of fermented vegetable products and meat industry.

The main markets for lactic acid have been in food, pharmaceutical and cosmetics industry, but presently the main growing application of lactic acid is in the production of biodegradable and renewable raw material based poly lactic acid (PLA) polymers. Production of lactate esters (*e.g.* butyl lactate) is another growing application as environmentally friendly solvents [1]. Lactic acid has two optical isomers, L-(+)-lactic acid and D-(−)-lactic acid. Lactic acid is classified as GRAS (generally recognized as safe) for use as a food additive, although D(-)-lactic acid can be harmful to human metabolism and result in *e.g.* acidosis [2]. The optical purity of lactic acid is required for the production of PLA. The properties of PLA may however be adjusted by the ratio of the L- and D-PLA in a copolymer D-form increasing the melting point of the copolymer [3]. Optically pure L- or D-lactic acid can be obtained by microbial fermentation and presently more than 95 % of industrial production of lactic acid is based on fermentation.

Production figure of 260,000 t as 100 % lactic acid for conventional (excluding PLA) markets in 2008 and forecast over 1 million ton annual production of lactic acid for conventional markets and PLA by 2020 has been presented in 2010 [4]. DuPont patented PLA already in 1954 but it took almost 50 years before first large-scale production was started. The US-based NatureWorks is the largest producer of PLA having lactic acid production capacity of 180,000 t/a. The sustainability of the PLA product Ingeo® from NatureWorks has been evaluated [5]. Greenhouse gas emissions and nonrenewable energy consumption for Ingeo from cradle to factory-gate are 1.3 kg $CO_2$ eq./kg polymer and 42 MJ/kg polymer. These compare favorably with *e.g.* fossil-based PET (polyethylene terephtalate) with 3.2 kg $CO_2$ eq./kg polymer and 80 MJ/kg polymer, respectively. There is a huge potential for biodegradable and renewable raw materials based polymers if and when the economics for these become competitive. It is estimated that altogether 140 million tons of petroleum-based synthetic polymers are produced annually [6]. It should be emphasized that also many petroleum-based synthetic polymers (*e.g.* polyesters) are biodegradable. However at the moment there are only three commercial synthetic polymers replacing petroleum-based ones and produced on renewable raw materials: PLA, PTT (polytetramethylene terephtalate which is partly renewable) and PHA (polyhydroxyalkanoates). Natural polymers such as starches and celluloses are biodegradable and based on renewable raw materials, but their applications are limited by their properties. Reliance Life Sciences is producing copolymers of PLA and glycolic acid mainly for high-value medical applications. Lactic acid in this case is produced by bacterial fermentation.

The price of PLA is ca. 2.2 $/kg, the target being half of that [7]. This means that the price of lactic acid in captive use should be less than 0.8 $/kg. A major cost factor is the raw material used in fermentation medium. This is especially the case with fastidious lactic acid bacteria. Processes based on cheap polymeric waste and side stream materials are indeed widely studied. So far research on alternative fermentation modes and reactor systems has been mainly academic. PLA production requires both optically and chemically pure lactic acid. Optical purity can be guaranteed with several microbial strains under optimized fermentation conditions. Chemical purity is mainly dependent on the constituents in the fermentation medium especially when cheap materials are being used. Contrary to many other fermentation products lactic acid yield on monosaccharides is usually very high (> 90 %) the main impurity being the cell mass itself, which is easily separated from the product. The key economic drivers in the fermentative production of lactic acid are optimization of the production medium, high product yields, productivity, and the concentration of products formed, which influences the down-stream processing costs [8].

Lactic acid bacteria (LAB) are a group of Gram-positive bacteria belonging to genera Aerococcus, Alloiococcus, Atopobium, Bifidobacterium, Carnobacterium, Enterococcus, Lactobacillus (Lb.), Lactococcus (L.), Leuconostoc (Leuc.), Oenococcus, Pediococcus, Streptococcus (S.), Tetragenococcus, Vagococcus and Weissella (W.). LAB are non-sporulating rods or cocci which produce lactic acid as the main fermentation product under suitable substrates. LAB are oxidase and benzidine negative, lack cytochromes, and do not reduce nitrates to nitrite [9]. Most of the LAB are anaerobic, but some of them can shift to oxygen-dependent metabolism in aerobic conditions [10,11]. Lactic acid bacteria have complex nutrient requirements, including specific minerals, B vitamins, several amino acids, and purine and pyrimidine bases.

LAB ferment sugars via homo-, hetero-, or mixed acid fermentation. Homofermentative LAB produce lactic acid as main product from sugars, while hetero- or mixed acid fermentations produce also ethanol and/or acetic acid, formic acid and carbon dioxide.

Although it is a common practice to divide LAB into homo- and heterofermentative strains, the division is not that straightforward as the actual metabolism is dependent on both the nature of the C/energy substrate (*e.g.* hexose vs. pentose sugars) and fermentation conditions (*e.g.* growth rate and availability of the C/energy source). LAB used for lactic acid production are used to be classified as homofermentative (*Lactococcus, Enterococcus, Streptococcus* and some lactobacilli) as their hexose metabolism under non-limiting conditions is entirely via Embden-Meyerhof pathway to pyruvate which is then used to regenerate the reducing power (NADH) in the lactate dehydrogenase (LDH) catalyzed reaction to lactic acid. However at slow growth rate and low glycolytic flux mixed acid fermentation may take place and acetic acid, formic acid and ethanol are formed in addition to lactic acid [12]. The key enzyme in this metabolic shift *e.g.* in *L. lactis* is claimed to be pyruvate-formate lyase (PFL) [13]. There are two types of LDH for both enantiomers D-LDH and L-LDH. In addition some species have a racemase enzyme catalyzing the reaction between the two enantiomers. Thus enantiomerically pure lactic acid is produced by species with only one type of LDH and no racemase. A comprehensive list of different LAB strains used in lactic acid production is available elsewhere [1].

Biotechnical production of lactic acid may be based on several alternative micro-organisms. In addition to lactic acid bacteria filamentous fungi (*e.g. Rhizopus* spp.), other gram-positive bacteria (*e.g. Bacillus coagulans*) and metabolically engineered yeasts have been used also in industrial scale. The advantage of fungi is that they are active at and tolerate low medium pH. Low pH reduces significantly the consumption of neutralizing agent ($Ca(OH)_2$) in the fermentation stage and subsequent formation of gypsum ($CaSO_4$) in the product recovery stage. The advantage of filamentous fungi, *Bacillus* spp. and yeasts compared to lactic acid bacteria is their simple nutrient requirement in the fermentation medium. Filamentous fungi and *Bacillus* spp. are better suited to lignocellulosic fermentation raw materials as they are in general able to utilize pentose sugars in addition to hexoses. Anaerobic fermentation is generally speaking more feasible and this favors yeasts and lactic acid bacteria. When optimized the technical parameters such as product yield, $R_P$ and final product concentration are quite similar for each of these production organisms.

In the wide literature on lactic acid production two examples based on other than lactic acid bacteria should be taken up. The first of them presents results with a thermotolerant *B. coagulans* strain [14]. High lactic acid $Y_{P/S}$ on both glucose and xylose (96 % and 88 %, respectively) were achieved at reasonable $R_P$ (2.5 g/lh) and product concentration (100 g/l). Exceptionally high levels of lactic acid (200 g/l) were produced in fed-batch fermentation. Yeasts have been metabolically engineered aiming at lactic acid production since 1990's [15]. A recent article reported on metabolic engineering of *Candida utilis* having pyruvate decarboxylase deleted and a bovine L-lactate dehydrogenase expressed under the *pdc* promoter resulting in the production of lactic acid with high yield from glucose (95 %) and reasonable $R_P$ (4.9 g/lh) ending up with lactic acid concentration of 103.3 g/l and more than 99.9 % enantiomeric purity [16].

Heterofermentative LAB (*Leuconostoc, Weissella* and some lactobacilli such as *Lb. brevis*) utilize both hexose and pentose sugars via phosphoketolase pathway (PKP). Several LAB possess the genes for PPP as well. The different pathways are presented in Fig. 1. Heterofermentative LAB may be applied for the production of side products such as polyols (mannitol, erythritol) and ethanol or acetic acid. This is only feasible if the markets for the side products are comparable to those of lactic acid and the production more than covers the added down-stream processing costs.

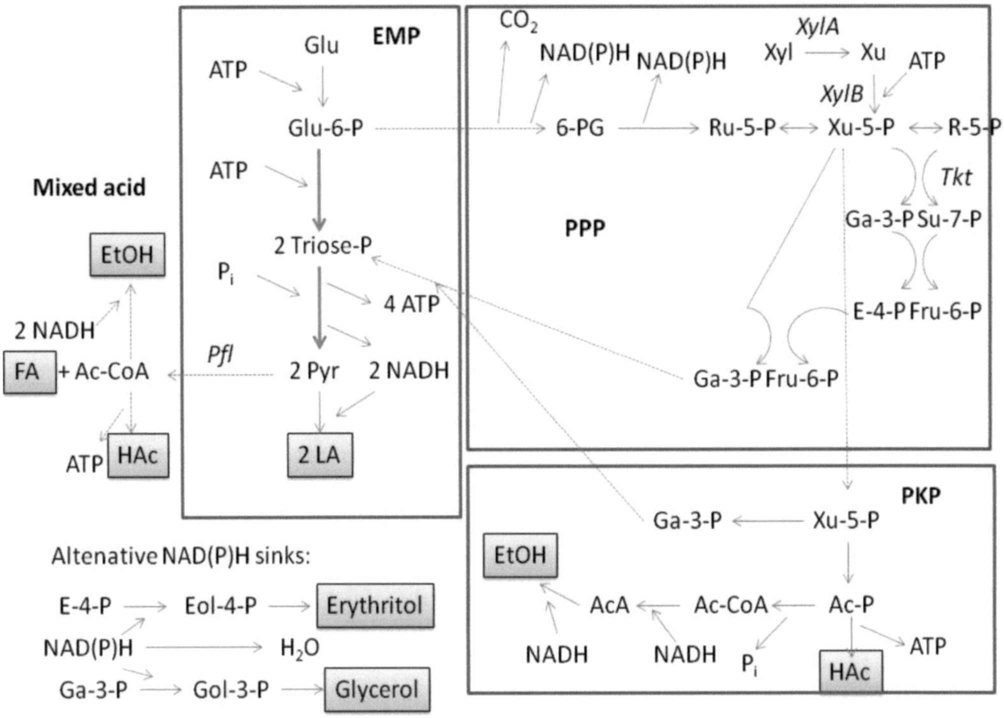


**Figure 1.** The main metabolic pathways in LAB. EMP: Embden-Meyerhof-Parnas pathway. PPP: pentose phosphate pathway. PKP: phosphoketolase pathway. Glu: glucose. LA: lactic acid. HAc: acetic acid. FA: formic acid. EtOH: ethanol. -P: energy-rich phosphate group. Pi: inorganic phosphate. Xyl: xylose. Xu: xylulose. 6-PG: 6-phosphogluconate. Ru: ribulose. R: ribose. Ga: glyceraldehyde. E: erythrose. Su: seduheptulose. Fru: fructose. Ac-CoA: acetyl-coenzymeA. Pyr: pyruvate. AcA: acetaldehyde. Ac-P: acetyl-phosphate. Pfl: pyruvate-formate lyase. XylA: xylose isomerase. XylB: xylulokinase. All carbohydrates are in D-form. Various metabolic end-products are presented with the dark background.

## 2. Future raw materials for production of lactic acid by LAB

The carboxylate platform is comprised of biological and chemical pathways that can be used in order to convert waste to bioproducts, such as lactic acid [17]. Lactic acid is a relatively cheap product, and one of the major challenges in its large-scale fermentative production is the cost of the raw material. This is the situation even in case of so called low-cost substrates [18]. Therefore, development of processes that utilize cheap raw materials at minimal costs have been under extensive studies. These substrates can be roughly classified as starch-based non-processed biomasses, lignocellulosic non-processed biomasses, and waste or side stream feedstocks. The former are nowadays generally considered as non-ideal feedstocks due to ethical reasons, and therefore they are not discussed in this review. Extensive reviews including starch-based feedstocks are available elsewhere [19]. With respect to future applications, the most likely raw materials for the lactic acid production are industrial side-streams and lignocellulosic biomasses. Recent advances in case of both raw material groups are discussed in the following.

As in other bioconversion processes, also in lactic acid production the focus of research has turned towards the use of lignocellulosic feedstocks. The major driving forces are fossil fuel deprivation and general paradigm change to bioeconomy, and the abundancy of lignocellulose materials. Generally, the effective utilization of lignocellulosic biomass for biochemical processes is limited due to seasonal availability, scattered distributions and high logistics cost [20]. The fermentation of lignocellulosic biomasses can also be hampered by inefficient pretreatment, high enzyme costs and end-product inhibition, formation of unwanted by-products under metabolism of pentoses, and carbon catabolite repression caused by the heterogeneous substrates. These challenges are further discussed in a recent review [8].

Paper industry residues and recycled paper products include various possible feedstocks for lactic acid production, which are together with agroindustrial residues discussed in a recent review [21]. Due to economical and ecological reasons, an intensive research interest is currently devoted to complex industrial by-products. In this field the advances presented during the past five years include the utilization of cellulosic biosludges from a Kraft pulp mill [22,23], and recycled paper sludge [24]. In both cases a nutrient supplement has increased the lactic acid productivity. LAB could be used for the bioconversion of hemicellulose fractions, *e.g.* from alfafa processing [25] to lactic acid. Direct conversion of xylan to lactic acid by LAB is already possible by use of genetically modified strains [26].

Food industry residues comprise a large variety of different biomasses and sludges that can be roughly categorized to agricultural wastes and food production wastes. Since the use of agricultural residues for lactic acid production is summarized in a recent review [21], it is not futher discussed here. Food production residues have been tested for bioconversion applications for ages, and the variety of used materials is large. Whey and other dairy industry residues are the prominent raw materials with respect to lactic acid production, reviewed in *e.g.* [27,28]. Whey retains about 55% of total milk nutrients, from which approximately 70% consists of lactose [29]. Availability of the lactose carbohydrate reservoir and the presence of other essential nutrients, such as proteins and phosphates, for the growth of microorganisms make whey and other cheese-making residues potent raw materials for the production of biochemicals.

Other quite often referred raw materials include brewery residues, especially spent grain [30], and winery wastes [31-35]. Additionally, there are various other proposed food industry residues that could fit to the lactic acid fermentation. The recently proposed include *e.g.* apple pomace [36], canned pineapple syrup [37], cashew apple juice [38], Jerusalem artichoke tubers [39], macaroni milk and rice-green pea salad refectory wastes

[40], rice residues [41], sap from palmyra and oil palms [42], and spent coffee grounds [43]. Despite of the large variety of the raw materials, the main conclusions of these studies are that the optimization and control of pH and temperature is critical for the process, and that the supplementation of low-cost substrate with *e.g.* inorganic salts and yeast extract is necessary or at least improves the productivity remarkably. In a recent study the use of mixed cultures of *Lb. casei*, *Lb. helveticus*, and *S. thermophilus* was observed to reduce the demand of supplements compared to single strain cultures [44].

The required supplements and their concentrations depend on the low-cost substrate. Drawbacks of complex supplements are their cost and extensive down-stream processing required for the purification of lactic acid from fermentation broth, especially in applications requiring high purity. Therefore, the optimization of supplement concentration is essential. Although yeast extract is often considered superior to other supplements in terms of efficiency, its major drawback is the relatively high cost, and therefore substitutive supplements have been suggested. Equal productivities may be achieved via use of cheaper alternatives, such as inorganic phosphates [45], and microbial lysates [46,47]. It is notable that the use of lysates in combination to *e.g.* whey proteins could cause unwanted proteolytic activity. Other options for the increased productivity include *e.g.* the addition of manganese, which is a constituent of lactate hydrogenase [48], whey protein hydrolyzate [49], malt combing nuts [49], corn steep liquor [50], fish hydrolyzates and other fishery by-products [51-53], hydrolyzed spent cells [54] or red lentil flour [55]. It is likely that this is one of the future trends in lactic acid production, *i.e.* fermentation media are optimized from mixtures of different low-cost raw materials in order to avoid the use of expensive complex supplements.

The modern biorefineries are looking into oceans in order to find new abundant and less land- or water-using biomasses for the production of commodities. Among the plenty marine biomasses, brown seaweed and especially species *Laminaria japonica*, a common food in Japan, has been recognized as a potential raw material for the production of platform chemicals. *L. japonica* is interesting due to its high carbohydrate content and fast growth. Production of lactic acid from *L. japonica* hydrolyzates was reported in a recent study [56]. Another potential raw material for bioconversion is shrimp shell waste, which is produced in vast amounts as a by-product of food industry. It has been reported that the production of lactic acid can be combined to the recovery of biopolymer chitin, a precursor for largely applied chitosan [57,58]. Since the recovery of chitin is traditionally done via chemical processing, the integrated process offers both economical and ecological advantage. Similar to the previous examples of other food industry residues, also the marine food processing industry generates various different side streams, such as fish waste and shells that could perhaps be combined in biochemical production.

## 3. Novel LAB strains

Metabolic engineering in general is applied when *e.g.* $Y_{P/S}$, $R_P$, substrate flux through a desired pathway in growth phases or resting cells are aimed at. Metabolic engineering studies aiming at increased flux in glycolysis to lactic acid in LAB are fairly scarce. That may be explained by the fact that the metabolism of LAB is already tuned for efficient lactic acid production.

Some of these studies are listed in a review on metabolic engineering for lactic acid production [59]. The overexpression of L-LDH in *Lb. plantarum* can result 13-fold increase in LDH activity, and still show no effect on lactic acid production [60]. It has also been shown by overexpression that glyceraldehyde-3-P dehydrogenase (GAPDH) is not limiting the glycolytic flux either in growing or resting cells of a *L. lactis* strain [61]. Metabolic flux and control analysis (MFA and MCA) combined with the estimation of the kinetic parameters of the enzymes of a pathway are indeed needed in systematic and systemic approach to study and optimize also such seemingly simple - there is always growth and maintenance functions involved as well - metabolic pathway as that from glucose to lactic acid in LAB. An excellent view on topic is available in a review [62], which includes several references also for LAB (*e.g.* [63-66]).

More straightforward work on lactic acid production has been performed to achieve high enantiomeric purity by expressing and deleting respective genes for LDH. There are several examples of these as discussed in the recent review [59], such as construction of two different strains of *Lb. helveticus* for optically pure L-lactic acid production [67]. These strains differed from each other at the level of L-LDH activity (53 and 93 % higher than the wild type strain). Lactic acid production in a fermentation batch was equal to that of the wild type strain. However, at low pH when the growth and production are uncoupled, the strain with higher activity produced 20 % more lactic acid compared to construct with the lower activity.

Another straightforward target for the construction of genetically modified strains is widening of the raw materials for the production of lactic acid especially to lignocellulosic biomass-based materials. There are no reports on work to produce cellulolytic enzymes in LAB. Instead several groups have tried to produce xylanase in LAB [26]. This is however focused on heterofermentative LAB as they are naturally able to utilize pentoses and especially *Lb. brevis* as it has been shown to have endogenous beta-xylosidase activity [68]. Another approach is based on *L. lactis* IO-1 strain being able to metabolize xylose both via PKP and PPP [69]. PPP provides a homolactic fermentation route for pentoses. As the molecular biology tools or protocols for this strain were not available, another strain of *L. lactis* was used as the host. *XylRAB* genes from IO-1 strain were expressed in the host. *XylA* and *XylB* encode genes for xylose isomerase and xylulokinase, respectively. *XylR* is a putative transcriptional activator of the *XylAB* operon. In addition the gene for phosphoketolase was disrupted. Such a strain construct had homolactic fermentation for xylose. The rate of xylose fermentation was further improved by overexpressing the gene for transketolase, one of the enzymes in PPP. Almost theoretical $Y_{P/S}$ of lactic acid (1.58 vs. 1.67 mol/mol xylose) was achieved with lactic acid concentration of 50,1 g/l. Acetic acid concentration was as low as 0.3 g/l. Enantiomeric purity was very high (99.6 %). Similar approach has been applied for the production of D-lactic acid from xylose and L-arabinose [70-71].

Typical LAB fermentations are run at minimum pH of 5 – 5.5, which is much higher than the pKa-value of lactic acid (*i.e.* 3.8). Thus more than 90 % of the product exists as lactate. This is a major cost factor in the product recovery stage as well as the cause of high salt burden and/or gypsum formation. The tolerance to acid and low pH is difficult to explain at genetic level and thus hardly be affected by metabolic engineering methods on specific genes. A successful approach to engineer LAB strains for lower fermentation pH has been genome shuffling. *E.g.* populations from nitrosoguanidine (NTG) mutations and low pH acclimatization in chemostat cultivation have been used for the shuffling [72]. The resultant population grew at pH 3.8 and lowered pH by lactic acid formation down to 3.5. This is a promising result even though the population was not used with realistic sugar concentrations. Similar approach has been reported aiming at improving acid tolerance as well as $R_P$ and glucose tolerance, respectively, with *Lb. rhamnosus* [73,74]. NTG and UV irradiation were used for mutagenesis and lethal mutants were fused from protoplasts. The best strain of [73] lowered pH down to 3.25 and increased average $R_P$ by 60 % compared to the wild type strain. However, average $R_P$ was still moderately low (ca. 1 g/lh). Final lactic acid concentration and $Y_{P/S}$ from glucose were 84 g/l and 82 %, respectively. In [74] higher $Y_{P/S}$ (> 95 %) and $R_P$ (ca. 3.6 g/lh) were reached with the best strains on industrially relevant fermentation medium with 150 g/l glucose. The $Y_{P/S}$ from 200 g/l glucose was still 90 %, but the average $R_P$ decreased to 2 g/lh. In a recent study *Lb. casei* mutants induced by NTG were screened in high glucose concentration (360 g/l) [75]. A mutant strain with highest osmotic tolerance produced 198.2 g/l lactic acid from 210 g/l glucose with increased $R_P$ (5.5 g/lh).

## 4. Novel process technologies

From fermentor design point of view lactic acid production by LAB is quite simple and conventional as the process requires no gassing, gas exchange or gas mass transfer. When the production strain and fermentation conditions are optimized for lactic acid production there is no or little formation of side products (metabolites, cellular mass, exopolysaccharides). Thus *e.g.* the rheology of the fermentation broth is Newtonian and very close to that of water. Power consumption is mainly for the sake of homogeneity and reduction of gradients of pH-controlling agents. The biggest challenges for process technology are to minimize osmotic effects by substrates and the product, to reach high $R_P$ and to minimize the costs and waste formation in the product recovery stage.

Typical fermentation approaches other than simple batch include repeated batch and fed-batch fermentation and continuous fermentation with cell-recycle as solutions with free cells and the use of immobilized cells in different reactor types (fixed or fluidized bed). A novel fed-batch strategy was developed recently by combining pH-control and substrate feeding [76]. The rationale behind the strategy was the linear relationship between the consumption amounts of alkali and that of substrate. Thus these two components were mixed together in the feeding liquid. This resulted in higher efficiency compared to batch fermentation, but the efficiency parameters were not especially high if compared with data from several other reports. By far the most studied method to increase the $R_P$ and/or separate cell growth from product formation is based on the immobilization of the cells. These have also been reviewed [27]. Several immobilization methods have been applied including entrapment within gels such as alginate [77,78], modified alginate [79,80], or pectate [81], adsorption on granulated DEAE-cellulose [82] or porous glass [83], and biofilm formation on solid supports [84-85]. Solid incompressible supports and carrier materials such as granulated cellulose and porous glass may be applied in any scale and in various reactor designs while gels as compressible materials suit less well for larger scale especially in fixed-bed column reactors.

Immobilized cells may be utilized in various fermentation modes and reactor designs such as repeated batch or fed-batch, continuous fermentation with cell retention or recycle, in continuous stirred tank reactors (CSTR), fixed-bed or fluidized-bed reactors. High $R_P$ (19-22 g/lh) have been achieved in a two-stage process with immobilized cells [86]. A special arrangement consisting of a CSTR for pH-control and substrate feeding and a fixed-bed reactor with immobilized cells was used in a concept with intermittent refreshing of the cells in a patent [87]. Short residence time within the column was possible because of the incompressible nature of the carrier material. Chemically pure product was achieved by using a production medium with few nutrients. Once the productivity decreased below a threshold value based on the consumption of alkali the cells were refreshed with nutrients. Incompressible carriers for cell adsorption have obvious advantages. However, new solutions to secure cell adherence on the carrier are required. This would facilitate efficient use of fluidized-bed reactors with minimal pressure losses in the reactor. Biological means for cell adherence may be one solution which could offer a further advantage to selectively keep the productive cells in the reactor.

Another approach to increase $R_P$ is high cell-density fermentation with free cells recycled by membrane separation technique. This has been in use in industrial scale for lactic acid production already in 1980's in Finland. Several academic reports on this approach have been since published demonstrating very high $R_P$ of 26 g/lh [88], 31.5 g/lh [89] and up to 57 g/lh [90].

It should be kept in mind that $R_P$ is affected by the concentration of lactic acid. Thus not all published figures are comparable. Product inhibition may be diminished by in-situ recovery of the product. Electrodialysis [91-92], nanofiltration [93] and ion-exchange [94-95] have thus been coupled with the fermentation system.

Conventional lactic acid recovery from fermentation broth consists of cell and other solids separation, lactic acid precipitation as calcium lactate and precipitate recovery, acidification of the precipitate by sulfuric acid and the separation of the gypsum precipitate formed. The amount of gypsum is usually higher than the amount of lactic acid produced. Lately NatureWorks has reported to have reduced the formation of gypsum significantly. Probably this has been achieved by performing the fermentation at lower pH *e.g.* by using metabolically engineered yeast for the production of lactic acid. The amount of gypsum can be avoided by using electrodialysis for the acidification and separation of the acid and alkali formed with bipolar membranes [96]. The alkali formed may be recycled back to the fermentation. Electrodialysis has been considered too expensive technology for lactic acid recovery [97]. However, specific energy consumption of only 0.25 kWh/ kg lactic acid is presented [96]. Nanofiltration has been used as a pretreatment method to remove Mg- and Ca- and sulfate-ions and color before electrodialysis increasing significantly the capacity in electrodialysis [98]. Alternative techniques for lactic acid recovery are extraction [99] and use of ion-exchange [100-101], neither of which is a proper solution to the salt burden.

## 5. Conclusions

Lactic acid production in LAB has both cell mass and growth dependent portions. Typically LAB require several nutrient components for their growth increasing the fermentation and down-stream processing costs. Down-stream processing is especially important in the production of lactic acid for PLA. As $R_P$ is the a major investment factor affecting costs, the minimization of medium and product purification costs should be accompanied by methods increasing cell mass concentration without excess growth. For this several different strategies have been applied so far mainly in academia (cell immobilization, cell-recycling and cell-retention). As history shows some of these could be applicable in industrial production as well, however pilot and demonstration plant studies and some risk-taking are required.

The main C/energy source spectrum available for LAB has been widened significantly. Reports of new possible substrates are frequently published, and the utilization of industrial side streams is a growing trend. Into this direction major successes have also been achieved with metabolic engineering providing strains for efficient production of lactic acid from pentoses as well, which is to promote sustainable use of renewables.

In an ideal fermentation process product inhibition should be minimized so that high $R_P$ would be achieved even at high lactic acid concentrations resulting in feasible average productivities. For this purpose both acclimatization and mutagenesis has been applied successfully. However, it has to be considered how far can we go in respect to fermentation pH and lactic acid concentration. There are already remarkable alternatives to LAB with naturally better properties in this sense. Some success has been achieved with in-situ product recovery, but also these procedures lack experiences in any larger scale.

Conventional lactic acid production process with LAB is accompanied with the formation of large amounts of gypsum in the product recovery stage. Fermentation at lower pH diminishes this amount, but does not prevent its formation. Electrodialysis has been considered too expensive technique for the recovery of such cheap, bulk products as lactic acid. However, recent reports claim promising results with this technology. Forecasted figures for lactic acid market show up to one million tons per year. The growth would come mainly from the growth of PLA as a biodegradable polymer based on renewable raw materials. Economies of scale should decrease the production costs, but new technical approaches are also needed to reach these figures.

## Author details

Sanna Taskila* and Heikki Ojamo
*University of Oulu, Faculty of Technology, Department of Process and Environmental Engineering, Bioprocess Engineering Laboratory, Oulu, Finland*

* Corresponding Author

## Abbreviations

η % - Efficiency, i.e. the ratio of YP/S to the maximum theoretical value
D-LDH - D-lactate dehydrogenase
LAB – lactic acid bacteria
L-LDH - L-lactate dehydrogenase
NTG – Nitrosoguanidine
$R_P$ - Volumetric productivity g/l*h
SSF – Simultaneous saccharification and fermentation
PLA – poly lactic acid
PPP - Pentose phosphate pathway
PKP - Phosphoketolase pathway
$Y_{P/S}$ – Yield of lactic acid per substrate consumed g/g
$Y_{P/X}$ – Yield of lactic acid per cell mass g/g

## 6. References

[1] Wee YJ, Kim JN, Ryu HW. Biotechnological production of lactic acid and its recent applications. Food Technology and Biotechnology 2006;44 163-172.

[2] Datta R, Tsai SP, Bonsignore P, Moon SH, Frank JR. Technological and Economic-Potential of Poly(Lactic Acid) and Lactic-Acid Derivatives. Fems Microbiology Reviews 1995;16 221-231.

[3] Fukushima K, Sogo K, Miura S, Kimura Y. Production of D-lactic acid by bacterial fermentation of rice starch. Macromolecular Bioscience 2004;4 1021-1027.

[4] Jem JK, van der Pol JF, de Vos S. Microbial lactic acid, its polymer poly(lactic acid), and their industrial applications. Microbiology Monographs 2010;14 323-346.

[5] Vink ETH, Davies S, Kolstad JJ. The eco-profile for current Ingeo® polylactide production. Industrial Biotechnology 2010;6 212-224.

[6] Shah AA, Hasan F, Hameed A, Ahmed S. Biological degradation of plastics: A comprehensive review. Biotechnology Advances 2008;26 246-265.

[7] Nampoothiri KM, Nair NR, John RP. An overview of the recent developments in polylactide (PLA) research. Bioresource Technology 2010;101 8493-8501.

[8] Abdel-Rahman MA, Tashiro Y, Sonomoto K. Lactic acid production from lignocellulose-derived sugars using lactic acid bacteria: Overview and limits. Journal of Biotechnology 2011;156 286-301.

[9] Carr FJ, Chill D, Maida N. The lactic acid bacteria: A literature survey. Critical Reviews in Microbiology 2002;28 281-370.

[10] Murphy MG, Condon S. Correlation of Oxygen Utilization and Hydrogen-Peroxide Accumulation with Oxygen Induced Enzymes in Lactobacillus-Plantarum Cultures. Archives of Microbiology 1984;138 44-48.

[11] Sedewitz B, Schleifer KH, Gotz F. Physiological role of pyruvate oxidase in the aerobic metabolism of *Lactobacillus plantarum*. Journal of Bacteriology 1984;160 462-465.

[12] Zaunmuller T, Eichert M, Richter H, Unden G. Variations in the energy metabolism of biotechnologically relevant heterofermentative lactic acid bacteria during growth on sugars and organic acids. Applied Microbiology and Biotechnology 2006;72 421-429.

[13] Melchiorsen CR, Jokumsen KV, Villadsen J, Israelsen H, Arnau J. The level of pyruvate-formate lyase controls the shift from homolactic to mixed-acid product formation in *Lactococcus lactis*. Applied Microbiology and Biotechnology 2002;58 338-344.

[14] Ou MS, Ingram LO, Shanmugam KT. l(+)-Lactic acid production from non-food carbohydrates by thermotolerant *Bacillus coagulans*. Journal of Industrial Microbiology & Biotechnology 2011;38 599-605.

[15] Porro D, Brambilla L, Ranzi BM, Martegani E, Alberghina L. Development of Metabolically Engineered Saccharomyces-Cerevisiae Cells for the Production of Lactic-Acid. Biotechnology Progress 1995;11 294-298.

[16] Ikushima S, Fujii T, Kobayashi O, Yoshida S, Yoshida A. Genetic Engineering of Candida utilis Yeast for Efficient Production of L-Lactic Acid. Bioscience Biotechnology and Biochemistry 2009;73 1818-1824.

[17] Agler MT, Wrenn BA, Zinder SH, Angenent LT. Waste to bioproduct conversion with undefined mixed cultures: the carboxylate platform. Trends in Biotechnology 2011;29 70-78.

[18] Yadav AK, Chaudhari AB, Kothari RM. Bioconversion of renewable resources into lactic acid: an industrial view. Critical Reviews in Biotechnology 2011;31 1-19.

[19] John RP, Nampoothiri KM, Pandey A. Fermentative production of lactic acid from biomass: an overview on process developments and future perspectives. Applied Microbiology and Biotechnology 2007;74 524-534.

[20] Lin Y, Tanaka S. Ethanol fermentation from biomass resources: current state and prospects. Applied Microbiology and Biotechnology 2006;69 627-642.

[21] Alonso JL, Dominguez H, Garrote G, Gonzalez-Munoz MJ, Gullon B, Moure A, Santos V, Vila C, Yanez R. Biorefinery processes for the integral valorization of agroindustrial and forestal wastes. Cyta-Journal of Food 2011;9 282-289.

[22] Romani A, Yanez R, Garrote G, Alonso JL. SSF production of lactic acid from cellulosic biosludges. Bioresource Technology 2008;99 4247-4254.

[23] Romani A, Yanez R, Garrote G, Alonso JL, Parajo JC. Sugar production from cellulosic biosludges generated in a water treatment plant of a Kraft pulp mill. Biochemical Engineering Journal 2007;37 319-327.

[24] Marques S, Santos JAL, Girio FM, Roseiro JC. Lactic acid production from recycled paper sludge by simultaneous saccharification and fermentation. Biochemical Engineering Journal 2008;41 210-216.

[25] Sreenath HK, Moldes AB, Koegel RG, Straub RJ. Lactic acid production by simultaneous saccharification and fermentation of alfalfa fiber. Journal of Bioscience and Bioengineering 2001;92 518-523.

[26] Hu CY, Chi DJ, Chen SS, Chen YC. The direct conversion of xylan to lactic acid by *Lactobacillus brevis* transformed with a xylanase gene. Green Chemistry 2011;13 1729-1734.

[27] Kosseva MR, Panesar PS, Kaur G, Kennedy JF. Use of immobilised biocatalysts in the processing of cheese whey. International Journal of Biological Macromolecules 2009;45 437-447.

[28] Panesar PS, Kennedy JF, Gandhi DN, Bunko K. Bioutilisation of whey for lactic acid production. Food Chemistry 2007;105 1-14.

[29] Jelen P, Whey processing. In: Encyclopedia of dairy sciences. H.Roginski, J.W.Fuquay, P.F.Fox, (Eds.), Academic Press: London, 2003, pp. 2739-2751.

[30] Aliyu S, Bala M. Brewer's spent grain: A review of its potentials and applications. African Journal of Biotechnology 2011;10 324-331.

[31] Devesa-Rey R, Vecino X, Varela-Alende JL, Barral MT, Cruz JM, Moldes AB. Valorization of winery waste vs. the costs of not recycling. Waste Management 2011;31 2327-2335.

[32] Bustos G, Moldes AB, Cruz JM, Dominguez JM. Production of lactic acid from vine-trimming wastes and viticulture lees using a simultaneous saccharification fermentation method. Journal of the Science of Food and Agriculture 2005;85 466-472.

[33] Bustos G, Moldes AB, Cruz JM, Dominguez JM. Production of fermentable media from vine-trimming wastes and bioconversion into lactic acid by *Lactobacillus pentosus*. Journal of the Science of Food and Agriculture 2004;84 2105-2112.

[34] Bustos G, Moldes AB, Cruz JM, Dominguez JM. Formulation of low-cost fermentative media for lactic acid production with *Lactobacillus rhamnosus* using vinification lees as nutrients. Journal of Agricultural and Food Chemistry 2004;52 801-808.

[35] Bustos G, de la Torre N, Moldes AB, Cruz JM, Dominguez JM. Revalorization of hemicellulosic trimming vine shoots hydrolyzates trough continuous production of lactic acid and biosurfactants by L-pentosus. Journal of Food Engineering 2007;78 405-412.

[36] Gullon B, Yanez R, Alonso JL, Parajo JC. L-lactic acid. production from apple pomace by sequential hydrolysis and fermentation. Bioresource Technology 2008;99 308-319.

[37] Nakanishi K, Ueno T, Sato S, Yoshi S. L-Lactic Acid Production from Canned Pineapple Syrup by Rapid Sucrose Catabolizing *Lactobacillus paracasei* NRIC 0765. Food Science and Technology Research 2010;16 239-246.

[38] Honorato TL, Rabelo MC, Goncalves LRB, Pinto GAS, Rodrigues S. Fermentation of cashew apple juice to produce high added value products. World Journal of Microbiology & Biotechnology 2007;23 1409-1415.

[39] Ge XY, Qian H, Zhang WG. Enhancement of L-Lactic Acid Production in *Lactobacillus casei* from Jerusalem Artichoke Tubers by Kinetic Optimization and Citrate Metabolism. Journal of Microbiology and Biotechnology 2010;20 101-109.

[40] Omay D, Guvenilir Y. Lactic Acid Fermantation from Refectory Waste. Ekoloji 2011;20 42-50.

[41] Lu ZD, Lu MB, He F, Yu LJ. An economical approach for D-lactic acid production utilizing unpolished rice from aging paddy as major nutrient source. Bioresource Technology 2009;100 2026-2031.

[42] Chooklin S, Kaewsichan L, Kaewsichan L. Potential use of *Lactobacillus casei* TISTR 1500 for the bioconversion from palmyra sap and oil palm sap to lactic acid. Electronic Journal of Biotechnology 2011;14 .

[43] Mussatto SI, Machado EMS, Martins S, Teixeira JA. Production, Composition, and Application of Coffee and Its Industrial Residues. Food and Bioprocess Technology 2011;4 661-672.

[44] Secchi N, Giunta D, Pretti L, Garcia MR, Roggio T, Mannazzu I, Catzeddu P. Bioconversion of ovine scotta into lactic acid with pure and mixed cultures of lactic acid bacteria. Journal of Industrial Microbiology & Biotechnology 2012;39 175-181.

[45] Amrane A. Effect of inorganic phosphate on lactate production by Lactobacillus *helveticus* grown on supplemented whey permeate. Journal of Chemical Technology and Biotechnology 2000;75 223-228.

[46] Amrane A. Evaluation of lactic acid bacteria autolysate for the supplementation of lactic acid bacteria fermentation. World Journal of Microbiology & Biotechnology 2000;16 207-209.

[47] Coelho LF, de Lima CJB, Bernardo MP, Contiero J. d(-)-Lactic Acid Production by *Leuconostoc mesenteroides* B512 Using Different Carbon and Nitrogen Sources. Applied Biochemistry and Biotechnology 2011;164 1160-1171.

[48] Fitzpatrick JJ, Ahrens M, Smith S. Effect of manganese on Lactobacillus casei fermentation to produce lactic acid from whey permeate. Process Biochemistry 2001;36 671-675.

[49] Fitzpatrick JJ, O'Keeffe U. Influence of whey protein hydrolysate addition to whey permeate batch fermentations for producing lactic acid. Process Biochemistry 2001;37 183-186.

[50] Kim HO, Wee YJ, Kim JN, Yun JS, Ryu HW. Production of lactic acid from cheese whey by batch and repeated batch cultures of *Lactobacillus* sp RKY2. Applied Biochemistry and Biotechnology 2006;131 694-704.

[51] Gao MT, Hirata M, Toorisaka E, Hano T. Acid-hydrolysis of fish wastes for lactic acid fermentation. Bioresource Technology 2006;97 2414-2420.

[52] Beaulieu L, Desbiens M, Thibodeau J, Thibault S. Pelagic fish hydrolysates as peptones for bacterial culture media. Canadian Journal of Microbiology 2009;55 1240-1249.

[53] Vazquez JA, Montemayor MI, Fraguas J, Murado MA. High production of hyaluronic and lactic acids by *Streptococcus zooepidemicus* in fed-batch culture using commercial and marine peptones from fishing by-products. Biochemical Engineering Journal 2009;44 125-130.

[54] Gao MT, Hirata M, Toorisaka E, Hano T. Study on acid-hydrolysis of spent cells for lactic acid fermentation. Biochemical Engineering Journal 2006;28 87-91.

[55] Altaf M, Naveena BJ, Reddy G. Use of inexpensive nitrogen sources and starch for L(+) lactic acid production in anaerobic submerged fermentation. Bioresource Technology 2007;98 498-503.

[56] Jang S, Shirai Y, Uchida M, Wakisaka M. Production of L(+)-Lactic Acid from Mixed Acid and Alkali Hydrolysate of Brown Seaweed. Food Science and Technology Research 2011;17 155-160.

[57] Adour L, Arbia W, Amrane A, Mameri N. Combined use of waste materials - recovery of chitin from shrimp shells by lactic acid fermentation supplemented with date juice waste or glucose. Journal of Chemical Technology and Biotechnology 2008;83 1664-1669.

[58] Healy M, Green A, Healy A. Bioprocessing of marine crustacean shell waste. Acta Biotechnologica 2003;23 151-160.

[59] Singh SK, Ahmed SU, Pandey A. Metabolic engineering approaches for lactic acid production. Process Biochemistry 2006;41 991-1000.

[60] Ferain T, Garmyn D, Bernard N, Hols P, Delcour J. Lactobacillus-Plantarum Ldhl Gene - Overexpression and Deletion. Journal of Bacteriology 1994;176 596-601.

[61] Solem C, Koebmann BJ, Jensen PR. Glyceraldehyde-3-phosphate dehydrogenase has no control over glycolytic flux in *Lactococcus lactis* MG1363. Journal of Bacteriology 2003;185 1564-1571.

[62] Teusink B, Smid EJ. Modelling strategies for the industrial exploitation of lactic acid bacteria. Nature Reviews Microbiology 2006;4 46-56.

[63] Bai DM, Zhao XM, Li XG, Xu SM. Strain improvement and metabolic flux analysis in the wild-type and a mutant *Lactobacillus lactis* strain for L(+)-lactic acid production. Biotechnology and Bioengineering 2004;88 681-689.

[64] Neves AR, Ramos A, Nunes MC, Kleerebezem M, Hugenholtz J, de Vos WM, Almeida J, Santos H. In vivo nuclear magnetic resonance studies of glycolytic kinetics in *Lactococcus lactis*. Biotechnology and Bioengineering 1999;64 200-212.

[65] Even S, Lindley ND, Cocaign-Bousquet M. Transcriptional, translational and metabolic regulation of glycolysis in *Lactococcus lactis* subsp. cremoris MG 1363 grown in continuous acidic cultures. Microbiology-Sgm 2003;149 1935-1944.

[66] Cocaign-Bousquet M, Even S, Lindley ND, Loubiere P. Anaerobic sugar catabolism in *Lactococcus lactis*: genetic regulation and enzyme control over pathway flux. Applied Microbiology and Biotechnology 2002;60 24-32.

[67] Kylä-Nikkilä K, Hujanen M, Leisola M, Palva A. Metabolic engineering of Lactobacillus helveticus CNRZ32 for production of pure L-(+)-lactic acid. Applied and Environmental Microbiology 2000;66 3835-3841.

[68] Garde A, Jonsson G, Schmidt AS, Ahring BK. Lactic acid production from wheat straw hemicellulose hydrolysate by *Lactobacillus pentosus* and *Lactobacillus brevis*. Bioresource Technology 2002;81 217-223.

[69] Shinkawa S, Okano K, Yoshida S, Tanaka T, Ogino C, Fukuda H, Kondo A. Improved homo l-lactic acid fermentation from xylose by abolishment of the phosphoketolase pathway and enhancement of the pentose phosphate pathway in genetically modified xylose-assimilating *Lactococcus lactis*. Applied Microbiology and Biotechnology 2011;91 1537-1544.

[70] Okano K, Yoshida S, Yamada R, Tanaka T, Ogino C, Fukuda H, Kondo A. Improved Production of Homo-D-Lactic Acid via Xylose Fermentation by Introduction of Xylose Assimilation Genes and Redirection of the Phosphoketolase Pathway to the Pentose Phosphate Pathway in L-Lactate Dehydrogenase Gene-Deficient *Lactobacillus plantarum*. Applied and Environmental Microbiology 2009;75 7858-7861.

[71] Okano K, Yoshida S, Tanaka T, Ogino C, Fukuda H, Kondo A. Homo-D-Lactic Acid Fermentation from Arabinose by Redirection of the Phosphoketolase Pathway to the Pentose Phosphate Pathway in L-Lactate Dehydrogenase Gene-Deficient *Lactobacillus plantarum*. Applied and Environmental Microbiology 2009;75 5175-5178.

[72] Patnaik R, Louie S, Gavrilovic V, Perry K, Stemmer WPC, Ryan CM, del Cardayre S. Genome shuffling of *Lactobacillus* for improved acid tolerance. Nature Biotechnology 2002;20 707-712.

[73] Wang YH, Li Y, Pei XL, Yu L, Feng Y. Genome-shuffling improved acid tolerance and L-lactic acid volumetric productivity in *Lactobacillus rhamnosus*. Journal of Biotechnology 2007;129 510-515.

[74] Yu L, Pei X, Lei T, Wang Y, Feng Y. Genome shuffling enhanced L-lactic acid production by improving glucose tolerance of *Lactobacillus rhamnosus*. Journal of Biotechnology 2008;134 154-159.

[75] Ge XY, Yuan JA, Qin H, Zhang WG. Improvement of L-lactic acid production by osmotic-tolerant mutant of *Lactobacillus casei* at high temperature. Applied Microbiology and Biotechnology 2011;89 73-78.

[76] Zhang Y, Cong W, Shi SY. Application of a pH Feedback-Controlled Substrate Feeding Method in Lactic Acid Production. Applied Biochemistry and Biotechnology 2010;162 2149-2156.

[77] Idris A, Suzana W. Effect of sodium alginate concentration, bead diameter, initial pH and temperature on lactic acid production from pineapple waste using immobilized *Lactobacillus delbrueckii*. Process Biochemistry 2006;41 1117-1123.

[78] Shen XL, Xia LM. Lactic acid production from cellulosic waste by immobilized cells of *Lactobacillus delbrueckii*. World Journal of Microbiology & Biotechnology 2006;22 1109-1114.

[79] Rao CS, Prakasham RS, Rao AB, Yadav JS. Production of L(+) lactic acid by *Lactobacillus delbrueckii* immobilized in functionalized alginate matrices. World Journal of Microbiology & Biotechnology 2008;24 1411-1415.

[80] Rao CS, Prakasham RS, Rao AB, Yadav JS. Functionalized alginate as immobilization matrix in enantioselective L (+) lactic acid production by *Lactobacillus delbrucekii*. Applied Biochemistry and Biotechnology 2008;149 219-228.

[81] Panesar PS, Kennedy JF, Knill CJ, Kosseva MR. Applicability of pectate-entrapped *Lactobacillus casei* cells for L(+) lactic acid production from whey. Applied Microbiology and Biotechnology 2007;74 35-42.

[82] Lommi, H., Swinkels, W., Viljava, T., and Hammond, R. Bioreactor with immobilized lactic acid bacteria and the use thereof. WO patent 94/12614. 1994.

[83] Senthuran A, Senthuran V, Mattiasson B, Kaul R. Lactic acid fermentation in a recycle batch reactor using immobilized *Lactobacillus casei*. Biotechnology and Bioengineering 1997;55 841-853.

[84] Dagher SF, Ragout AL, Sineriz F, Bruno-Barcena JM. Cell Immobilization for Production of Lactic Acid: Biofilms Do It Naturally. Advances in Applied Microbiology, Vol 71 2010;71 113-148.

[85] Qureshi N, Annous BA, Ezeji TC, Karcher P, Maddox IS. Biofilm reactors for industrial bioconversion processes: employing potential of enhanced reaction rates. Microbial Cell Factories 2005;4 .

[86] Schepers AW, Thibault J, Lacroix C. Continuous lactic acid production in whey permeate/yeast extract medium with immobilized *Lactobacillus helveticus* in a two-stage process: Model and experiments. Enzyme and Microbial Technology 2006;38 324-337.

[87] Viljava, T. and Koivikko, H. Method for preparing pure lactic acid. US patent 5,932,455. 1999.

[88] Richter K, Nottelmann S. An empiric steady state model of lactate production in continuous fermentation with total cell retention. Engineering in Life Sciences 2004;4 426-432.

[89] Xu GQ, Chu J, Wang YH, Zhuang YP, Zhang SL, Peng HQ. Development of a continuous cell-recycle fermentation system for production of lactic acid by *Lactobacillus paracasei*. Process Biochemistry 2006;41 2458-2463.

[90] Kwon S, Yoo IK, Lee WG, Chang HN, Chang YK. High-rate continuous production of lactic acid by *Lactobacillus rhamnosus* in a two-stage membrane cell-recycle bioreactor. Biotechnology and Bioengineering 2001;73 25-34.

[91] Gao MT, Hirata M, Koide M, Takanashi H, Hano T. Production of L-lactic acid by electrodialysis fermentation (EDF). Process Biochemistry 2004;39 1903-1907.

[92] Min-tian G, Koide M, Gotou R, Takanashi H, Hirata M, Hano T. Development of a continuous electrodialysis fermentation system for production of lactic acid by *Lactobacillus rhamnosus*. Process Biochemistry 2005;40 1033-1036.

[93] Jeantet R, Maubois JL, Boyaval P. Semicontinuous production of lactic acid in a bioreactor coupled with nanofiltration membranes. Enzyme and Microbial Technology 1996;19 614-619.

[94] Monteagudo JM, Aldavero M. Production of L-lactic acid by *Lactobacillus delbrueckii* in chemostat culture using an ion exchange resins system. Journal of Chemical Technology and Biotechnology 1999;74 627-634.

[95] Senthuran A, Senthuran V, Hatti-Kaul R, Mattiasson B. Lactate production in an integrated process configuration: reducing cell adsorption by shielding of adsorbent. Applied Microbiology and Biotechnology 2004;65 658-663.

[96] Wee YJ, Yun JS, Lee YY, Zeng AP, Ryu HW. Recovery of lactic acid by repeated batch electrodialysis and lactic acid production using electrodialysis wastewater. Journal of Bioscience and Bioengineering 2005;99 104-108.

[97] Akerberg C, Zacchi G. An economic evaluation of the fermentative production of lactic acid from wheat flour. Bioresource Technology 2000;75 119-126.

[98] Bouchoux A, Roux-de Balmann H, Lutin F. Investigation of nanofiltration as a purification step for lactic acid production processes based on conventional and bipolar electrodialysis operations. Separation and Purification Technology 2006;52 266-273.

[99] Yankov D, Molinier J, Kyuchoukov G, Albet J, Malmary G. Improvement of the lactic acid extraction. Extraction from aqueous solutions and simulated fermentation broth by means of mixed extractant and TOA, partially loaded with HCl. Chemical and Biochemical Engineering Quarterly 2005;19 17-24.

[100] Gullon B, Alonso JL, Parajo JC. Ion-Exchange Processing of Fermentation Media Containing Lactic Acid and Oligomeric Saccharides. Industrial & Engineering Chemistry Research 2010;49 3741-3750.

[101] Moldes AB, Alonso JL, Parajo JC. Recovery of lactic acid from simultaneous saccharification and fermentation media using anion exchange resins. Bioprocess and Biosystems Engineering 2003;25 357-363.

# Chapter 27

# Application of Amylolytic *Lactobacillus fermentum* 04BBA19 in Fermentation for Simultaneous Production of Thermostable α-Amylase and Lactic Acid

Bertrand Tatsinkou Fossi and Frédéric Tavea

Additional information is available at the end of the chapter

http://dx.doi.org/10.5772/50456

## 1. Introduction

Lactic acid bacteria (LAB) have diverse applications for both animals and humans. Food, pharmaceutical and chemical industries rely on these microorganisms to produce fermented beverage, foods and other important compounds of industrial interests. In recent years the industrial relevance of lactic acid bacteria is on an increasing trend because of the application of lactic acid as chemical for the production of biodegradable plastics [1]. Typical LAB are Gram-positive, non-sporing, catalase-negative, devoid of cytochromes, anaerobic but aerotolerant cocci or rods that are acid-tolerant and produce lactic acid as the major end product during sugar fermentation [2]. Although most LAB are unable to degrade starch because of the lack of the amylolytic activity, a few exhibit this activity and are qualified as amylolytic lactic acid bacteria (ALAB) which are able to decompose starchy material through the amylases production during the fermentation processes [3]. Regarding the importance and availability of starchy biomass in the world, amylases and lactic acid production from starch appear as two potential industrial applications of ALAB. Amylases play important role in degradation of starch and are produced in bulk from microorganisms and represent about 25 to 33% of the world enzyme market [4]. The spectrum of amylases application has widened in many fields, such as clinical, medical and analytical chemistry as well as in the textile, food, fermentation, paper, distillery and brewing industries [4]. The advantages of using thermostable amylases in industrial processes include the decreased risk of contamination, cost of external cooling and increased diffusion rate [4]. Several thermostable α-amylases have been purified from *Bacillus* sp. and the factors influencing their thermostability have been investigated [5]. However, no study has yet dealt with thermostable amylase from lactic acid bacteria (LAB). The use of thermostable amylases from *Lactobacillus* is of advantage as they are generally non-pathogenic. On the other hand, the major end product of LAB fermentation, lactate, has applications as a preservative, acidulant and flavouring agent in the food industry, because of the tartness provided by lactate and also because lactate is generally regarded as safe (GRAS) [6].

Thus a thermostable amylase producing lactic acid bacterium would be a potential candidate for food industries and especially for the making of high density gruel from starchy raw material as corn or wheat [7]. This would require a good knowledge of the conditions required to optimally produce amylase and lactic acid of good quality. The present study deals with the co-production of thermostable α-amylase and lactic acid from a LAB, *Lactobacillus fermentum* 04BBA19, isolated from a starchy waste of a soil sample from the western region of Cameroon.

## 2. Background and significance

### 2.1. Amylolytic lactic acid bacteria

Amylolytic lactic acid bacteria (ALAB) have been reported from different tropical starchy fermented foods, made especially from roots as cassava and sweet potato or grains as maize sorghum and rice. Strains of *Lactobacillus plantarum* have been isolated from African cassava-based fermented products [8], *L. plantarum* A6 (LMG 18053) have been isolated from retted cassava in the Congo [9] and *Lactobacillus manihotivorans* OND32 have been isolated from cassava sour starch fermentations in Colombia [10]. Amylolytic strains of *Lactobacillus fermentum* were isolated for the first time from Benin maize sourdough (ogi and mawè) by Agati et al. [11]. Sanni et al. [12] described amylolytic strains of *L. plantarum* and *L. fermentum* strains in various Nigerian traditional amylaceous fermented foods. ALAB are generally screened in fermented amylaceous foods. Owing to their relatively high starch content, starchy biomass appears as an important eco-niche for the screening and isolation of ALAB, which can be industrially applied to convert starch into mono- and disaccharides for lactic acid fermentation. The composition of the microbiota and in particular the occurrence of ALAB is determined by the way the raw material is processed [13]. Most ALAB isolated belong to the *Lactobacillus* genus, however few studies reported the existence of amylolytic activity in some strains of *Bifidobacterium* isolated from the human large intestinal tract [14, 15]. The distribution of amylolytic microorganisms in the human large intestinal tract has been investigated in various individuals of different ages using anaerobic cultures techniques. So far, twenty one amylolytic bifidobacteria have been isolated from adult faeces and tested for rice fermentation [16].

Owing to the ability of their α-amylases to partially hydrolyze raw starch, ALAB can ferment different types of amylaceous raw material, such as corn [17], potato [18], or cassava [19] and different starchy substrates [20, 21, 8]. Amylolytic LAB utilize starchy biomass and convert it into lactic acid in a single step fermentation. ALAB are mainly used in food fermentation, they are involved in cereal based fermented foods such as European sour rye bread, Asian salt bread, sour porridges, dumplings and non-alcoholic beverage production.

Few of them are used for production of lactic acid in single step fermentation of starch [1]. The common method to produce lactic acid from starchy biomass involves the pretreatment for gelatinisation and hydrolysis (liquefaction and saccharification). The liquefaction of the starch is carried out at high temperatures of 90–130 °C for 15 min followed by enzymatic saccharification to glucose and subsequent conversion of glucose to lactic acid by fermentation [22, 1]. This two-step process involving consecutive enzymatic hydrolysis and fermentation makes it economically unattractive. The bioconversion of carbohydrate materials to lactic acid can be made much more effective by coupling the enzymatic hydrolysis of carbohydrate substrates and microbial fermentation of the derived glucose into a single step. This has been successfully employed for lactic acid production from raw starch materials with many representative bacteria including *Lactobacillus* and *Lactococcus* species [23, 20, 24, 21].

Because at industrial scale, the use of glucose addition is an expensive alternative, there is interest in the use of a cheaper source of carbon, such as starch, the most abundantly available raw material on earth next to cellulose. This, in combination with amylolytic lactic acid bacteria may help to decrease the cost of the overall fermentation process. Amylolytic lactic acid bacteria can convert the starch directly into lactic acid. Development of production strains which ferment starch to lactic acid in a single step is necessary to make the process economical. Very few bacteria have been reported so far for direct fermentation of starch to lactic acid [1, 25, 26] Approximately 3.5 billion tonnes of agricultural residues are produced per annum in the world [27]. The use of a specific carbohydrate feedstock depends on its price, availability, and purity. Although agro-industrial residues are rich in carbohydrates, their utilization is limited [27]. Different food/agro-industrial products or residues form the cheaper alternatives to refined sugars as substrates for lactic acid production. Sucrose-containing materials such as molasses are commonly exploited raw materials for lactic acid production. Starch produced from various plant products is a potentially interesting raw material based on cost and availability. Laboratory-scale fermentations have been reported for lactic acid production from starch by *Lactobacillus amylophilus* GV6 [20], *L. amylophilus* B4437 [28], *Lactobacillus amylovorus* [29, 23, *Lactococcus lactis* combined with *Aspergillus awamorii* [30] and *Rhizopus arrhizus* [31]. *L. amylophilus* NRRL B4437 [32], *L. amylovorus* [17] and *L. amylophilus* GV6 are exceptions that have been described to actively ferment starch to lactic acid and this may lead to alternative process of industrial lactic acid production [23, 20]. To make the process cost effective in terms of substrate, various groups have worked on acid/enzyme hydrolysis of starchy substrates followed by *Lactobacillus* fermentation or simultaneous saccharification and fermentation by co-culture/mixed culture fermentations. It has been reported that starch is used as substrate in two steps [1].

## 2.2. Thermostable amylases

Amylases are among the most important enzymes and are of great significance in present-day biotechnology. Although they can be derived from several sources, such as plants, animals and microorganisms; enzymes from microbial sources generally meet industrial demands. The spectrum of amylase application has widened in many other fields, such as clinical, medical and analytical chemistries, as well as their widespread application in starch saccharification and in the textile, food, brewing and distilling industries. Thermostability is one of the main features of many enzymes sold for bulk industrial usage. Thermostable $\alpha$-amylases are of interest because of their potential industrial applications. They have extensive commercial applications in starch liquefaction, brewing, sizing in textile industries, paper and detergent manufacturing processes. [33,34, 35]. The advantages of using thermostable amylases in industrial processes include the decreased risk of contamination and cost of external cooling, a better solubility of substrates, a lower viscosity allowing accelerated mixing and pumping [36]. Several thermostable $\alpha$-amylase have been purified from *Bacillus sp.* and the factors influencing their thermostability have been investigated, but the thermostability of amylases from lactic acid bacteria have attracted very few scientific attention. *Lactobacillus amylovorus, Lactobacillus plantarum, Lactobacillus manihotivorans,* and *Lactobacillus fermentum* are some of the lactic acid bacteria exhibiting amylolytic activity which have been studied [37, 10, 38, 5, 39, 40]. However, most of $\alpha$-amylase from these bacteria presented weak thermostability compared to those of genus *Bacillus*. Owing to the important acidification of fermenting medium by most lactic acid bacteria, the production of thermostable amylase by a lactic acid bacterium under submerged or solid-state fermentation can help to reduce the risk of contamination caused by undesirable micro-organisms during the process [41, 42]. Another advantage is the non-pathogen character of the genus *Lactobacillus* that allows their utilization in food fermentation processes.

### 2.3. Lactic acid

Lactic acid a water soluble and highly hygroscopic aliphatic acid is present in humans, animals and microorganisms. It is the first biotechnologically produced multi-functional versatile organic acid having wide range of applications [1], namely as a preservative in many food products. It can be produce by LAB trough fermentation or synthetically from lactonitrile [43]. It is non-volatile, odorless organic acid and is classified as GRAS (Generally Recognized As Safe) for use as a general purpose food additive. The lactic acid consumption market is dominated by the food and beverage sector since 1982 [1]. More than 50% of lactic acid produced is used as emulsifying agent in bakery products [44]. It is used as acidulant/flavoring/pH buffering agent or inhibitor of bacterial spoilage in a wide variety of processed foods, such as candy, breads and bakery products, soft drinks, soups, sherbets, dairy products, beer, jams and jellies, mayonnaise, and processed eggs, often in conjunction with other acidulants. Lactic acid or its salts are used in the disinfection and packaging of carcasses, particularly those of poultry and fish, where the addition of aqueous solutions during processing increased shelf life and reduced microbial spoilage. The esters of calcium and sodium salts of lactate with longer chain fatty acids have been used as very good dough conditioners and emulsifiers in bakery products. The water retaining capacity of lactic acid makes it suitable for use as moisturizer in cosmetic formulations. Ethyl lactate is the active ingredient in many anti-acne preparations. The natural occurrence of lactic acid in human body makes it very useful as an active ingredient in cosmetics [45]. Lactic acid has long been used in pharmaceutical formulations, mainly in topical ointments, lotions, and parenteral solutions. It also finds applications in the preparation of biodegradable polymers for medical uses such as surgical sutures, prostheses and controlled drug delivery systems [45]. Because of ever-increasing amount of plastic wastes worldwide, considerable research and development efforts have been devoted towards making a single-use, biodegradable substitute of conventional thermoplastics. Biodegradable polymers are classified as a family of polymers that will degrade completely – either into the corresponding monomers or into products, which are otherwise part of nature – through metabolic action of living organisms. The demand for lactic acid has been increasing considerably, owing to the promising applications of its polymer, the polylactic acid (PLA), as an environment-friendly alternative to plastics derived from petrochemicals. PLA has received considerable attention as the precursor for the synthesis of biodegradable plastic [46]. The lactic acid polymers have potentially large markets, as they many advantage like biodegradability, thermo plasticity, high strength etc., have potentially large markets. The substitution of existing synthetic polymers by biodegradable ones would also significantly alleviate waste disposal problems. As the physical properties of PLA depend on the isomeric composition of lactic acid, the production of optically pure lactic acid is essential for polymerization. L-Polylactic acid has a melting point of 175–178 °C and slow degradation time. L-Polylactide is a semicrystalline polymer exhibiting high tensile strength and low elongation with high modulus suitable for medical products in orthopedic fixation (pins, rods, ligaments etc.), cardiovascular applications (stents, grafts etc.), dental applications, intestinal applications, and sutures [45].

## 3. Materials and methods

### 3.1. Samples

Twenty-eight samples of soils were collected from main geographic zones of Cameroon in four localities: (Ngaoundere, Yaounde, Bafoussam and Mbouda) at the factories where starchy wastes are frequently submitted to natural fermentation. Four kinds of factories were investigated: "gari" factories, corn and cassava mills, cassava plantation after harvesting and treatment of tubers and flour markets. At the site where degradation of starchy material was remarkable and visible, one to five grams of soils were collected and transferred to polyethylene aseptic bag, the factory age recorded; the samples were finally transported to the laboratory and analyzed in the same week.

### 3.2. Screening of thermostable amylases and lactic acid producing bacteria

The starch degrading amylolytic lactic acid bacterial strains were isolated from different samples of soil. Amylolytic micro-organisms were firstly enriched by introduction of 1 g of soil sample in 100 ml Erlenmeyer flasks containing 50 ml of enrichment liquid medium, composed of (gram per litre): 5 g soluble starch, 5 g peptone, 5 g yeast extract 0.5 g $MgSO_4.7H_2O$, 0.01g FeSO4.7H2O, 0.01 g NaCl. Enrichment of thermostable amylases producing bacteria was carried out by heating Erlenmeyer flasks at 90°C for 5 min followed by incubation in an alternative shaker at 37-40°C and speed of 150 oscillations per minute for 24 h. Amylases producing bacteria strains were screened on agar plate, containing (gram per liter): 10 g soluble starch, 5 g peptone, 5 g yeast extract, 0.5 g $MgSO_4.7H_2O$, 0.01g $FeSO_4.7H_2O$, 0.01 g NaCl, 15 g agar. Incubation at 37-40°C was carried out for 48 h, after which the plates were stained with lugol solution (Gram iodine solution: 0.1% $I_2$ and 1% KI). The colonies with the largest halo forming zone were pre-selected and tested for Gram staining and catalase activity.

Preliminary tests were carried out to determine the heat stability of the amylase of each isolate as we described previously [47, 48]. The gas production from glucose, growth at different temperature (10, 40, 45 °C) as well as the ability to grow in different concentration of NaCl was determined as described by Schillinger and Lucke [49] and Dykes et al [50]. The isolates which were Gram positive and catalase negative, non-motile and producing heat stable amylase and lactic acid were finally selected and identified using API 50 CH test kit (bioMerieux, France). The APILAB PLUS database identification was used to interpret the results.

### 3.3. Microbial growth, amylase and lactic acid production

In order to study microbial growth, amylase and lactic acid production, the microorganism was propagated at 40°C for 70 h in 50 ml of a basal medium containing: soluble starch, 1% (w/v); yeast extract, 0.5 % (w/v) placed in 100 ml Erlenmeyer flask with shaking at 150 oscillations per minute in an alternative shaker (Kotterman, Germany). The initial pH of the medium was adjusted to 6.5 using 0.1 M HCl. After removal of cells by centrifugation (8000xg, 30 min, 4°C) in centrifugator (Heraeus, Germany), the supernatant was considered as the crude enzyme solution and was also used for lactic acid evaluation.

### 3.4. Optimisation of raw starch degrading thermostable amylase and lactic acid production

The amylase and lactic acid production was optimized by studying the effect of cultural and environmental variables (carbohydrate and nitrogen sources, metal salts and surfactants) individually and simultaneously. The effect of carbohydrate sources was studied by replacing soluble starch in basal medium with different sugars, gelatinized and raw natural crude starch sources (glucose, fructose, maltose, amylose, amylopectine, cassava, corn, rice tapioca, and sorghum flours at final concentration of 1% (w/v)). Nitrogen sources were tested by replacing yeast extract with various nitrogen sources (peptone, tryptone, beef extract, soyabean meal, ammonium sulphate, and urea at final concentration of 1.5% (w/v)). The effect of metal salts was studied by adding individually various metal salts ($CaCl_2.2H_2O$, $MgSO_4.7H_2O$, $FeSO_4.7H_2O$, $FeCl_3$, NaCl at concentration of 0.1% (w/v)). Similarly the effect of surfactants was studied by supplementing the culture medium with Tween 80 and Tween 40 at concentration of 1.5% (v/v).

All media containing gelatinized starch sources were autoclaved at 121 °C for 20 min, while for the media containing raw starch flour, starch powder was sterilized by washing in ethanol and added to sterile nutrient broth.

### 3.5. Partial enzyme purification

The culture supernatant was supplemented with solid ammonium sulphate to 65% (w/v) final concentration, with mechanical stirring at 4°C. The suspension was retained for 1 h at 4 °C, and centrifuged at 8000 g for 30 min at the same temperature. The resultant supernatant was brought to 70 % w/v ammonium sulphate saturation at 4°C. 50-70% (w/v) ammonium sulphate precipitate was recovered, dissolved in 0.1 M phosphate buffer and dialysed using Spectra/PorR, VWR 2003 dialysis membrane overnight against the same buffer at 4°C and used as partial purified enzyme solution.

### 3.6. Effect of temperature and pH on activity and stability

The optimal temperature for amylase activity was determined by assaying activity between 30 and 100°C for 30 min in 50 mM phosphate buffer. Measurement of optimum pH for amylase activity was carried out under the assay conditions for pH range of 3.0-10.0, using 50mM of three buffer solutions: Tris-HCl (pH 3.0), $Na_2HPO_4$-Citrate (pH 4.0 – 6.0), and Glycine-NaOH (pH 7.0-10.0).

The temperature stability was determined by incubating the partial purified enzyme solution in water bath for temperature range of 30-100°C for 30, 60, 90, 120, 180 min and then cooled with tap water. The remaining α-amylase activity was measured and expressed as the percentage of the activity of untreated control taken as 100%. The first order inactivation rate constants, $k_i$ were calculated from the equation: $\ln A = \ln A_0 - k_i t$, where $A_0$ is the initial value of amylase activity and A the value of amylase activity after a time t (min).

For the determination of pH stability, the enzyme was incubated in a water bath at 60 °C at varying pH value for 30 min. The residual activity was detected under the same conditions and expressed as the percentage of the activity of untreated control taken as 100%.

### 3.7. Effect of metal salts and chelating agent

The effect of metal salts and EDTA on amylase activity was determined by adding 0.05 to 0.1% (w/v) of metal salts ($CaCl_2.2H_2O$, $MgSO_4.7H_2O$, $FeSO_4.7H_2O$, NaCl, $FeCl_3$, $CuSO_4.5H_2O$) and EDTA to the standard assay. The effect of metal salts and chelating agent on amylase activity were evaluated by pre-incubating the enzyme in the presence of effectors for 30 min at 60°C. The remaining amylase activity was determined and expressed as the percentage of the activity of untreated control taken as 100%.

### 3.8. Analytical methods

Cell growth was evaluated by reading the absorbance of culture medium at 600 nm using a Secoman spectrophotometer and numeration of total colony forming unit by 10-fold serial dilution of fermented broth and pour plating on MRS-starch agar (De Man Rogosa and Sharpe medium in which glucose has been replaced by soluble starch (Prolabo-Merck Eurolab, France)). In order to evaluate the capacity of microorganism to acidify the culture medium, the pH of the fermented broth was measured using an electronic pH meter (Mettler Seven S20, Japan)

The amylolytic power of Lactic acid bacteria was determined using the method of wells by inoculation of 10 µl of microbial strain in 4 mm depth micro-wells on the surface of MRS-starch agar plate. The starch hydrolysis halo was revealed after 48 h of incubation using iodine solution. The amylolytic power was defined as the average diameter (mm) of hydrolysis halo provoked by a strain after its inoculation in micro-well on MRS-starch agar plate for 48 h incubation at optimum temperature of growth for three assays

The activity of amylase both in crude and purified extracts was assayed by iodine method. In a typical run, 5 ml of 1% soluble starch solution and 2 ml of 0.1M phosphate buffer (pH 6.0) were mixed and maintained at a desired temperature for 10 min, then 0.5 ml of appropriately diluted enzyme solution was added. After 30 min the enzyme reaction was stopped by rapidly adding 1ml of 1M HCl into the reaction mixture. For the determination of residual starch, 1 ml of the reaction mixture was added to 2.4 ml of diluted iodine solution and its optical density was read at 620 nm using a spectrophotometer (Secoman). One unit of amylase activity (U) was defined as the amount of enzyme able to hydrolyse 1 g of soluble starch during 60 min under the experimental condition. The lactic acid was determined according to Kimberley and Taylor [51]. The nature of amylase (endo-acting or exo-acting) was determined according to Ceralpha method (Megazyme) which uses a blocked maltoheptaoside as substratre [57].

The affinity of the enzyme preparation from selected LAB toward raw cassava starch was studied by incubating 0.2 g of raw cassava flour with 1ml of the enzyme solution at 60 °C for 15 min. After centrifugation, the α-amylase activity of the supernatant was measured and the adsorption percentage was calculated as follows: $Adsorption\,(\%) = \frac{A - B}{A} \times 100$, $A$ is the original α-amylase activity and $B$ is the α-amylase activity in the supernatant after adsorption on raw potato starch granules.

For the determination of raw starch digestibility, raw cassava was used and the reaction mixture containing 100 U of α-amylase preparation from the selected LAB and 100 mg of raw cassava starch in a final volume of 10ml dispensed in 100ml Erlenmeyer flasks were incubated in alternative water bath shaker at 60°C and 150 oscillations per min. After a time interval of 6 h, the reducing sugars liberated in the reaction mixtures were determined by dinitrosalicyclic acid method [58].

Light microscopy was used for the examination of the effect of enzyme on raw starch granules using Olympus microscope BH-2.

# 4. Results and discussion

## 4.1. Biochemical properties of amylolytic LAB isolated

From the 28 samples of soil collected from different localities of Cameroon, 90 amylolytic isolates were screened but only 9 isolates (04BBA15, 04BBA19, 05BBA22, 05BBA23,

14BYA42, 20BBA60, 17BNG51, 23BYA21, 26BMB81) presented very high amylolytic power (≥15mm) and were qualified as amylase overproducing isolates. The amylolytic power was defined as the average diameter (mm) of starch hydrolysis halo (Fig.1.) provoked by a strain after its inoculation in micro-well on MRS-starch agar plate for 48 h incubation at optimum temperature of growth for three assays. The amylolytic power is an expression of the capacity of an isolate to degrade starch during the culture. Among the amylase overproducing isolates, two (04BBA19, 26BMB81) were aero-anaerobic non spore forming, gram positive and catalase negative bacteria; this characteristic is proper to lactic acid bacteria. Microscopic observation showed rod cells. Biochemical characteristics of these isolates were carried out using API 50 CH kit bioMerieux system, the results are summarized in Table 1, the isolates were tested for their possibility to ferment 50 carbohydrates, and this fermentation profile was use for their numerical identification. According to their biochemical profile, 04BBA19 and 26BMB81 were respectively identified as *Lactobacillus fermentum* and *Lactobacillus plantarum.* The strain 04BBA19 (*Lactobacillus fermentum*) presented a very high amylolytic power, as it was able to cause a starch hydrolysis halo of 45 ±1.5mm on MRS-starch agar plate after 48 h of incubation at 40 °C; consequently it was selected for further studies. The preliminary test of thermostability carried out on its crude extract amylase showed that it produce a very high thermostable enzyme and it was selected for an application on simultaneous production of thermostable amylase and lactic acid from starchy material.

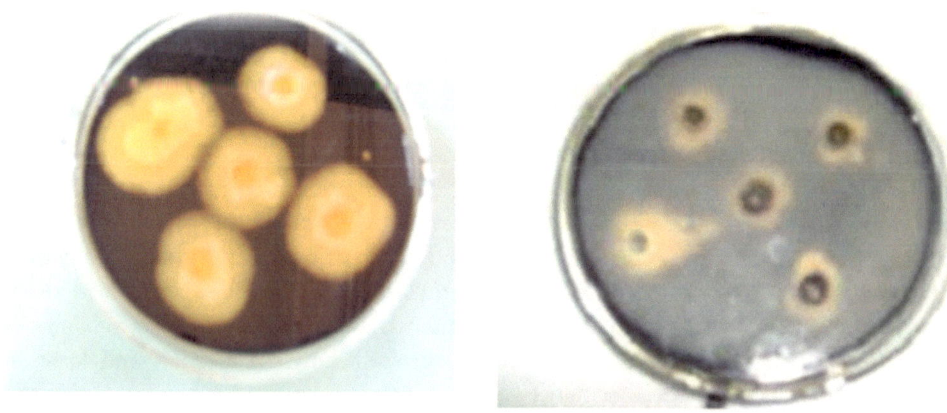

Strain 04BBA19

Strain 26BMB81

**Figure 1.** Plate assays for detection of amylase activity of lactic acid bacteria (04BBA19, 26BMB81) on MRS-starch agar plate medium. The diameter of hydrolysis halo was revealed by flooding the plates with Iodine solution (0.1% $I_2$+1% KI) after 48 h of culture at 40°C

| Test number | 1 | 2 | 3 | 4 | 5 | 6 | 7 | 8 | 9 | 10 | 11 | 12 | 13 | 14 | 15 | 16 | 17 | 18 | 19 | 20 | 21 | 22 | 23 | 24 | 25 | 26 | 27 | 28 | 29 |
|---|---|---|---|---|---|---|---|---|---|---|---|---|---|---|---|---|---|---|---|---|---|---|---|---|---|---|---|---|---|
| Strains cod | Heterofermentative | Gaz production | Optimum temperature of growth | Growth at 10°C | Dextran | Ammonia from Arginine | Nitrate reduction | Glycerol | Erythritol | D-arabinose | L-arabinose | Ribose | D-xylose | L-xylose | Adonitol | ß methyl-D-Xyloside | Galactose | Glucose | Fructose | Mannose | Sorbose | Rhamnose | Dulcitol | Inositol | Mannitol | Sorbitol | α-Methyl-D-mannoside | α-Methyl-D-glucoside | N-Acetyl-Glucosamine |
| 04BBA19 | + | + | 45°C | - | - | + | - | - | - | - | - | + | - | - | - | - | + | + | + | + | - | - | - | nc | - | - | - | - | - |
| 26BMB81 | - | + | 40°C | - | - | - | - | - | - | - | + | + | - | - | - | - | + | + | + | + | - | - | - | - | + | + | - | + | + |

| Test number | 30 | 31 | 32 | 33 | 34 | 35 | 36 | 37 | 38 | 39 | 40 | 41 | 42 | 43 | 44 | 45 | 46 | 47 | 48 | 49 | 50 | 51 | 52 | 53 | 54 | 55 | 56 | |
|---|---|---|---|---|---|---|---|---|---|---|---|---|---|---|---|---|---|---|---|---|---|---|---|---|---|---|---|---|
| Strains cod | Amygdalin | Arbutin | Aesculin | Salicin | Cellobiose | Maltose | Lactose | Melibiose | Sucrose | Trehalose | Inulin | Melezitose | Raffinose | Starch | Glycogen | Xylitol | ß Gentiobiose | D-turanose | D-lyxose | D-tagatose | D-fucose | L-fucose | D-arabitol | L-arabitol | Gluconate | 2-gh-Gluconate | 5-Keto-Gluconate | Identification species (API 50CHL) |
| 04BBA19 | - | - | nc | - | - | + | + | + | + | - | - | - | + | + | - | - | - | - | - | - | - | - | - | - | + | - | - | *Lactobacillus fermentum* |
| 26BMB81 | + | + | + | + | + | + | + | + | + | + | nc | + | + | + | + | - | + | + | - | - | - | - | - | - | + | - | - | *Lactobacillus plantarum* |

+, positive reaction; -negative reaction, nc, non -conclusive

**Table 1.** Biochemical characteristics of amylases overproducing *Lactobacillus* isolated from soils

## 4.2. Amylase and lactic acid production

In the presence of starch as carbon source at 40°C, *L. fermentum* 04BBA19 strain grew, exhibited amylolytic activity and produced lactic acid in the culture medium. The amylase production pattern in *L. fermentum* 04BBA19 (Fig.2) indicates that the induction of amylase took place during the lag phase (after 10 h of incubation) in the presence of starch. The level of amylase production increased significantly during the exponential phase of growth. Lactic acid production became visible around 15 h after incubation and also increased considerably during the exponential phase of growth. Cell growth, amylase and lactic acid production reached maxima values at the same time (40 h of fermentation). The values of those maxima were $1.1x10^9$ cfu/ml, 107.3±0.5 U/ml, 8.7±0.5g/l for cell growth, amylase activity, and lactic acid production respectively. Such coincidence shows that amylase production by *L. fermentum* 04BBA19 was tightly linked to cell growth. These results are in agreement with the report of Goyal et al. [53], Liu and Xu [54] on the relationship between pattern of cell growth and amylase production. The decline of cell growth and amylase production after the peak occurred around 50 h of incubation and could be attributed to the rise of lactic acid concentration in fermented broth [6] or to the rise of protease levels [55]. The acidification was also expressed by the decrease of initial pH of culture broth (Fig. 2). The initial pH of culture broth declined significantly and reached a value of 3.0 around 50 h of incubation and then remained constant.

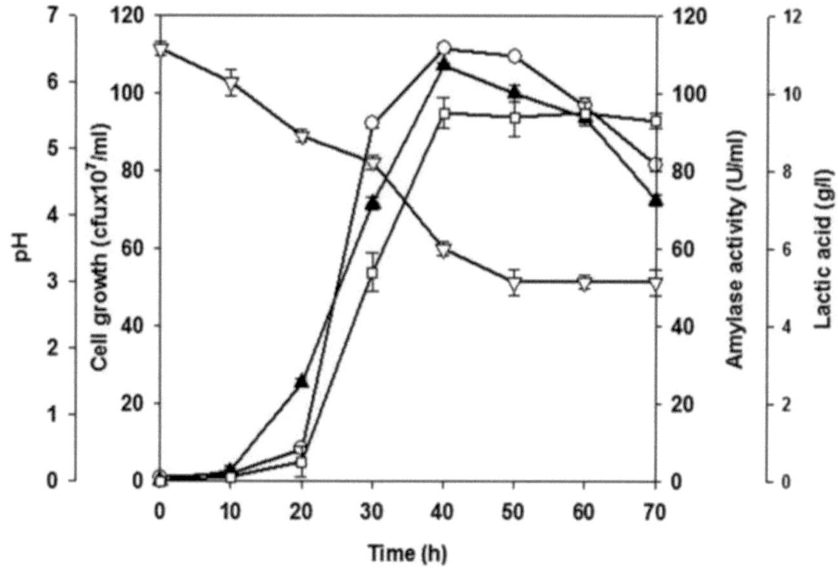


**Figure 2.** Time course of growth (○), pH (▽), α-amylase (▲) and lactic acid (□) production by *L. fermentum* 04BBA19 in 1% (w/v) soluble starch medium at 40°C, pH 6.0. The data shown are averages of triplicates assays within 10% of the mean value.

The study of cell growth and amylase production as a function of temperature (Fig. 3a) showed that *L. fermentum* 04BBA19 exhibited maximal growth and amylase activity at 45°C, confirming thus the strong relationships between cell growth and amylase production. On the other hand the maximum value of lactic acid was produced at the same temperature. Many other investigators reported that maximum amylase production occurred at the optimum growth temperature [56, 53]. These results are contrary to the findings of Chandra et al.[57] who studied the growth and amylase production of *Bacillus licheniformis* CUM 305. They have observed that this microorganism grew very well at 30°C, but did not produce α-amylase at that temperature. In addition, Saito and Yamamoto [58] found α-amylase production at 50°C and cell growth at a temperature lower than 45°C for another strain of *B. licheniformis*.

The amylase and lactic acid production by *L. fermentum* 04BBA19 was influenced significantly by initial pH of culture broth (Fig. 3b). Maximum amylase and lactic acid production was achieved for pH range of 4.0-6.5. These results could be explained by the fact that pH generally act by inducing morphological change in microorganism which facilitate enzyme production [59].

## 4.3. Optimisation of amylase and lactic acid production

Amylase production is known to be induced by a variety of carbohydrate, nitrogen compounds and minerals [60, 61]. In order to achieve high enzyme yield, efforts are made to develop a suitable medium for proper growth and maximum secretion of enzyme, using an adequate combination of carbohydrates, nitrogen and minerals [53, 62].

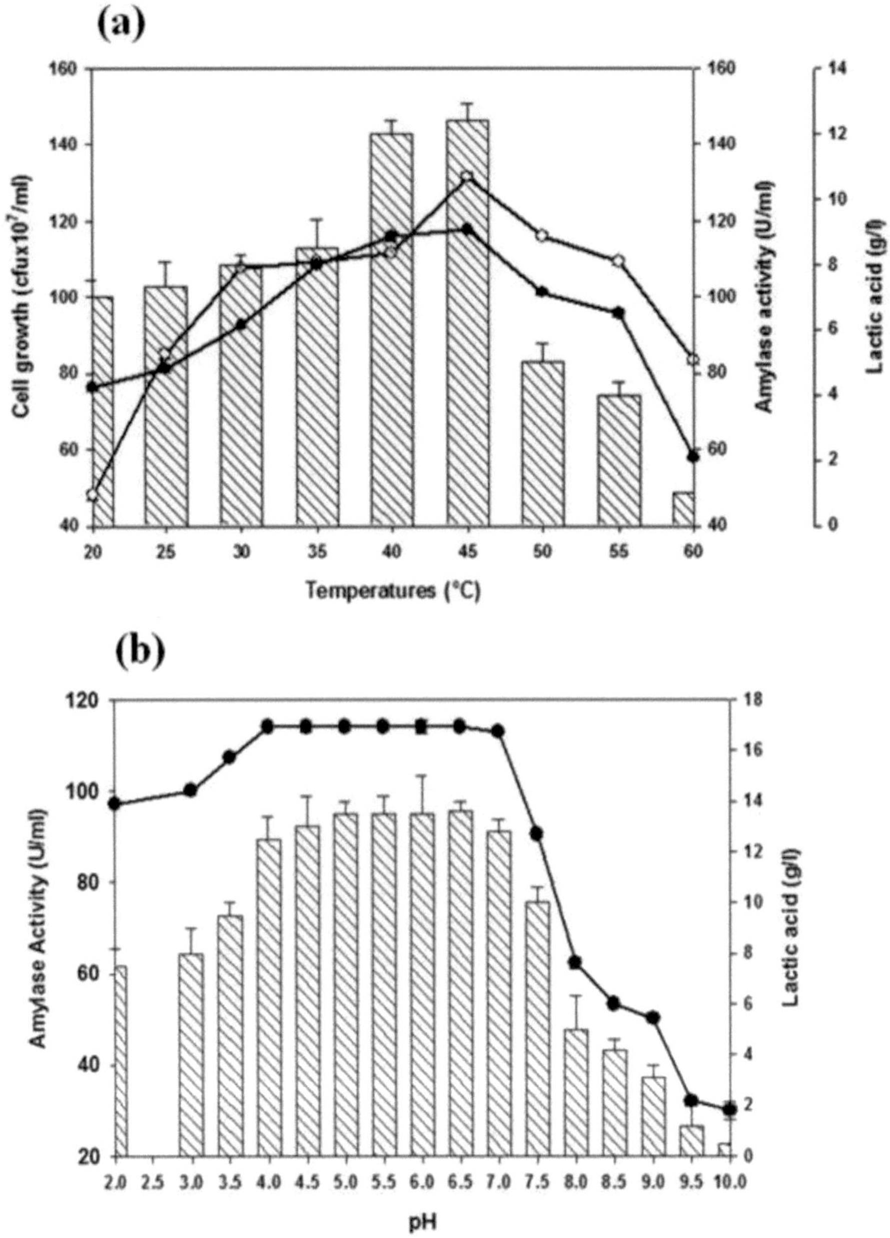


**Figure 3.** (a) Effect of temperature on microbial growth (○), α-amylase (●) and lactic acid (▨) production. (b) Effect of initial pH of culture broth on α-amylase (●) and lactic acid (▨) production. The data shown are averages of triplicate assays within 10% of the mean value.

From the use of different carbohydrate sources in the present study, soluble starch proved to be the best inducer of amylase production (Table 1). In the presence of soluble starch at concentration of 1% (w/v), the enzyme yield reached 107.0±1.2 U/ml after 48 hours of fermentation, while in the presence of raw cassava starch at the same concentration, the enzyme yield was 67.1±0.5 U/ml. These results are in agreement with the reports of Cherry et al. [63], Saxena et al. [4] who reported maximum amylase production when starch was used as carbohydrate source. In the presence of glucose and fructose, amylase production was almost nil; and that was a proof that glucose and fructose repressed amylase synthesis by *L. fermentum* 04BBA19. This observation is in agreement with the reports of Theodoro and Martin [64] showing that synthesis of carbohydrate degrading enzymes in some microbial species leads to catabolic repression by substrate such as glucose and fructose. Similar results were observed by Halsetine et al. [65] for the production of amylase by the hyperthemophilic archeon *Sulfolobus solfataricus.* According to them, glucose prevented α-amylase gene expression and not only secretion of performed enzyme. Since amylase yield is higher with amylose (92.3 U/ml) as carbohydrate source than with amylopectin (50.1 U/ml), the *L. fermentum* 04BBA19 amylase is more efficient for hydrolysis of alpha-1,4 linkages than those of alpha-1,6. The amylase production increased with the soluble starch concentration (Fig. 4), reaching a maximum (180.5 ± 0.3 U/ml) at the concentration range of 8-16 % (w/v). These optimum starch concentrations for amylase production by *L. fermentum* 04BBA19 are higher than that observed for amylase production in *Bacillus* sp. PN5 reported by Saxena et al. [4]. This microorganism presented an optimum soluble starch concentration of 0.6% (w/v) for amylase production. The lactic acid production also increased with the soluble starch concentration, the optimum starch concentration for lactic acid production was achieved at the same range of concentration for amylase production.

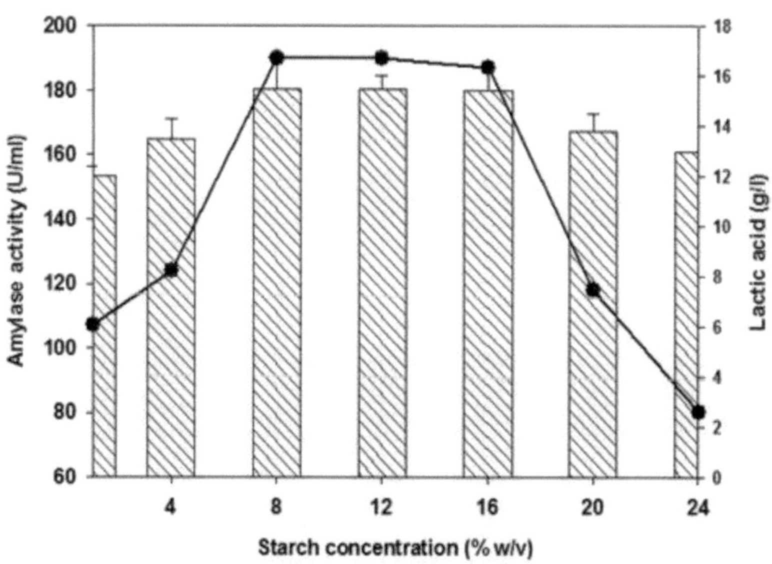


**Figure 4.** Effect of starch concentration on α-amylase (●) and lactic acid production (▨) by *L. fermentum* 04BBA19. The data shown are averages of triplicate assays with SD within 10% of mean value

Among the various gelatinized starchy sources tested, corn and sorghum flour were found to be the most suitable for α-amylase and lactic acid production by *L. fermentum* 04BBA19 while for the raw starchy sources tested, potato starch was most suitable (Table 2). On the other hand the level of lactic acid was more important when corn and sorghum flours were used. The good production of α-amylase and lactic acid when these starchy flours are used is based on their composition; they also contain proteins and vitamins which are required by lactic acid bacteria for their growth, enzymes and acids production [66].

Among nitrogen sources used in the present study, soya bean meal and yeast extract showed significant effect on α-amylase and lactic acid production. Soya bean meal, rich in protein is a potential nutrient for lactic acid fermentation. Similar results were obtained by several authors. Goyal et al. [53] reported that soybean meal presented a positive effect and was the best nitrogen source for raw starch digesting thermostable α-amylase production by the *Bacillus* sp I-3 strain. The yeast extract was also reported to be a potential nutrient for lactic acid fermentation, since it contains vitamins, amino acids [66]. Though all nitrogen sources are positively influencing enzyme production by *L. fermentum* 04BBA19, an inverse behaviour has been observed with other bacterial strains, for instance, Tanyildizi et al. [67] reported zero effect of yeast extract on amylase production by *Bacillus* sp.

All metal salts tested in this study increased amylase and lactic acid production by *L. fermentum* 04BBA19, except $CuSO_4.5H_2O$ that acted as inhibitor. The inhibition of amylase production by CuSO4.5H2O was also reported by Wu et al. [68] for the *Bacillus* sp CRP strain. Copper ion acted as poisonous compound for this strain and consequently inhibited amylase synthesis. The effect of $CaCl_2.2H_2O$ was the most important, and was in agreement with the observation of Gangadharan et al. [61] who described the rise of amylase production by *B. amyloliquefaciens* when $CaCl_2.2H_2O$ was supplemented to the culture medium. The supplementation of metal ions has been reported to provide good growth and also influence higher enzyme production. Most α-amylases are metalloenzymes and in most cases, $Ca^{2+}$ ions are required for maintaining the spatial conformation of the enzyme, thus play an important role in enzyme stability [61].

From the surfactants tested in this study, Tween-80 appeared to be the best surfactant sources for amylase production by *L. fermentum* 04BBA19. Similar results were obtained by Reddy et al. [69]. These authors reported that the supplementation of culture medium with Tween-80 resulted in a marked increase in the yields of thermostable β-amylase and pullullanase by *C. thermosulfurogenes* SV2, and that the stimulation of enzyme production was greater when the surfactants were added after 18 h of incubation of culture. Beside stimulation, the surfactants caused and increased secretion of the enzymes into extracellular fluid [59].

From various environmental factors tested for α-amylase and lactic acid production by L. fermentum 04BBA19, it has been observed that all factors that increase amylase synthesis also positively affect lactic acid production. The optimization of the basal medium by supplementation of all carbohydrate, nitrogen, mineral and surfactant sources (excepted $CuSO_4.5H_2O_4$) in culture medium resulted to a significant improvement of enzyme and lactic acid yield. In the optimized medium, amylase activity and lactic acid content reached 732.4±0.4 U/ml and 53.2±0.4 g/l respectively.

| Parameters | Enzyme yield (U/ml) | Lactic acid(g/l) |
|---|---|---|
| Carbohydrate sources (1% w/v) | | |
| Glucose | 0.1±0.0[d*] | 14.3±0.5[a] |
| Fructose | 0.2±0.0[d] | 12.1±0.5[b] |
| Maltose | 0.1±0.0[d] | 12.8±0.4[b] |
| Amylose | 92.3±0.1[b] | 12.2±0.1[b] |
| Amylopectin | 50.1±0.5[c] | 10.3±0.5[c] |
| Soluble starch | 107.3±0.5[a] | 8.7±0.5[d] |
| Nitrogen sources (1.5% w/v) | | |
| Yeast Extract | 107.3±0.5[b] | 8.7±0.5[b] |
| Beef extract | 92.4±0.5[c] | 7.3±0.2[b] |
| Peptone | 88.3±1.7[d] | 7.1±0.3[b] |
| Tryptone | 75.3±0.5[e] | 6.5±0.5[b] |
| Soya bean meal | 397.3±0.4[a] | 29.2±0.4[a] |
| Ammonium sulphate | 95.4±1.5[c] | 7.2.±0.8[b] |
| Urea | 76.3±0.3[e] | 5.3±0.6[c] |
| Minerals (0.1% w/v) | | |
| $CaCl_2. 2H_2O$ | 412.1±0.6[a] | 33.2±0.1[a] |
| $MgSO_4. 7H_2O$ | 315.1±0.4[b] | 31.2±0.5[a] |
| FeSO4. $7H_2O$ | 237.3±0.7[c] | 20.2±0.4[b] |
| NaCl | 315.2±0.9[b] | 22.1±0.6[b] |
| $CuSO_4.5H_2O$ | 12.2±0.6[e] | 3.2±0.3[d] |
| Surfactants (1.5% w/v) | | |
| Tween-40 | 209.5±0.1[b] | 27.3±0.4[b] |
| Tween-80 | 215.1±0.3[a] | 35.2±0.3[a] |
| Gelatinized starchy sources (1 %w/v) | | |
| Corn flour | 303.5±0.2[a] | 36.3±0.6[a] |
| Cassava flour | 182.3±0.4[c] | 24.2±0.8[d] |
| Sorghum flour | 305.8±0.7[a] | 35.2±0.1[a] |
| Rice flour | 187.3±0.8[c] | 30.1±0.5[b] |
| Tapioca flour | 237.4±0.6[b] | 27.2±0.7[c] |
| Raw starchy sources | | |
| Cassava starch | 67.1±0.5[c] | 21.3±0.4[a] |
| Potato starch | 87.2±0.5[a] | 23.4±0.1[a] |
| Cocoyam starch | 78.6±0.2[b] | 22.7±0.4[a] |
| Media | | |
| Basal medium | 107.5±0.3 | 8.7±0.5[b] |
| Optimized medium | 732.3±0.4 | 53.2±0.4[a] |

**Table 2.** Effect of different parameters on α-amylase and lactic acid production by *L. fermentum* 04BBA19 in submerged state fermentation at 45 °C and initial pH 6.5.

The basal medium contained soluble starch, 1% (w/v); yeast extract, 0.5 % (w/v); while the optimized medium contained all parameters without $CuSO_4.5H_2O$. The data shown are averages of triplicate assays with SD within 10% of mean value. For each group of parameters (Carbohydrate, Nitrogen, Mineral, Starchy sources, media), means with different superscripts within columns are significantly different ($p<0.05$).

### 4.4. Enzyme properties

The amylase produced by *L. fermentum* 04BBA19 showed high affinity toward cassava raw starch granules with 80% adsorption and brought about 79% hydrolysis of 1% (w/v) suspension of raw cassava starch. On the other hand, the enzyme was able to hydrolyze blocked p-nitro phenyl methyl heptaoside, releasing a yellow compound (p-nitro phenol) with maximum absorption at 530 nm. This result was a proof that amylase from *L. fermentum* 04BBA19 is an endo acting amylase (α-amylase), since the blocked p-nitro phenyl methyl heptaoside is known to be hydrolysed only by endo-acting amylases [38].

The enzyme exhibited maximum activity at 60-70°C and maintained 100% of its initial activity at 80°C for 30 min of heat treatment (Fig 5-a). When the enzyme was treated for the same time (30 min.) at 90°C and 100°C, the remaining activities were 90 and 87% respectively. These results showed the thermophilic character and very high thermostability of α-amylase from *L. fermentum* 04BBA19. In general, most of lactic acid bacteria do not produce amylases. However, this property have been observed in some genera of lactic acid bacteria, especially in *L. plantarum* and *L. amylovorus* [37], *L. manihotivorans* [70, 13], *L. fermentum* OGI E1 [38]. But amylases produced by these strains are not thermostable. Traditionally high thermostable and thermophiles amylases are found in *Bacillus* and *Thermococcus* genera as: *B. amyloliquefaciens* [71]; *B. licheniformis* [72]; *B. stearothermophilus* [73]; *B. subtilis* and *T. aggreganes* [74], *T. profundus* [75], *Bacillus* sp PN5 [4], *B. cohnii* US147 [35], *Chromohalobacter* sp. TVSP 101 [76].

Fig. 6 shows the thermostability pattern of α-amylase from *L. fermentum* 04BBA19 at 80°C, 90°C and 100°C when the time of heat treatment is beyond 30 min. Table 3 presents the thermal inactivation rate constant ($k_i$) and half-life (T) at these temperatures. The half-life of this enzyme is higher than that of α-amylase from *B. licheniformis*: 120 min at 70°C [76]. The thermal stability was considerably improved by addition of 0.1% (w/v) $CaCl_2.2H_2O$. Goyal et al. [53] obtained a half-life value of 3.5 h at 80°C with α-amylase from *Bacillus* sp.I-3 in the presence of 0.1 % (w/v) calcium chloride, while under the same conditions; α-amylase from *L. fermentum* 04BBA19 displayed a half-life of 6.1 h.

Due to its high thermostability, α-amylase from *L. fermentum* 04BBA19 could be highly competitive in industrial bioconversion reactions, as compared to α-amylase from *Bacillus*. In addition, this competitiveness is enhanced by the fact that lactobacilli, due to their non-pathogen character, are easily used in food industry [6].

The *L. fermentum* 04BBA19 α-amylase is active and stable in pH range of 4.0 – 7.0 (Fig. 5-b), which is the pH range of many foods. In this respect, this amylase could be used in starch hydrolysis, brewing and baking.

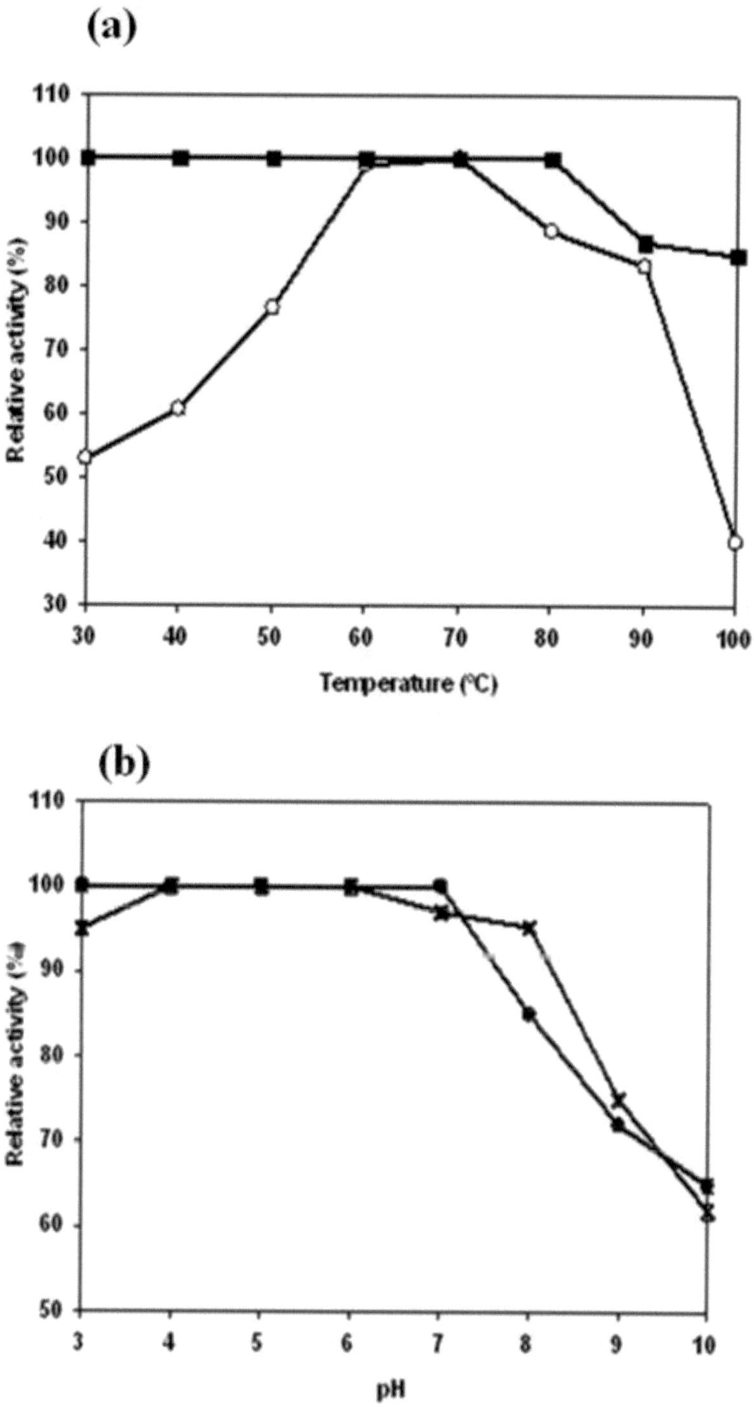


**Figure 5.** (a) Effect of temperature on activity (O) and stability (■) of α-amylase from *L. fermentum* 04BBA19. (b) Effect of pH on activity (×) and stability (■) of α-amylase from *L. fermentum* 04BBA19. The data shown are averages of triplicate assays within 10% of the mean value.

The metal salts generally act on activity of enzyme through their ions. The enzyme activity was highly improved by $Ca^{2+}$, while $Fe^{2+}$, $Fe^{3+}$, $Na^{+}$ and $Mg^{2+}$ had less significant effect. On the contrary $Cu^{2+}$ and EDTA acted as inhibitors (Fig. 7). The behaviour of the enzyme towards metal ions, particularly calcium, indicates its metalloenzyme nature, which is confirmed by the action of EDTA.

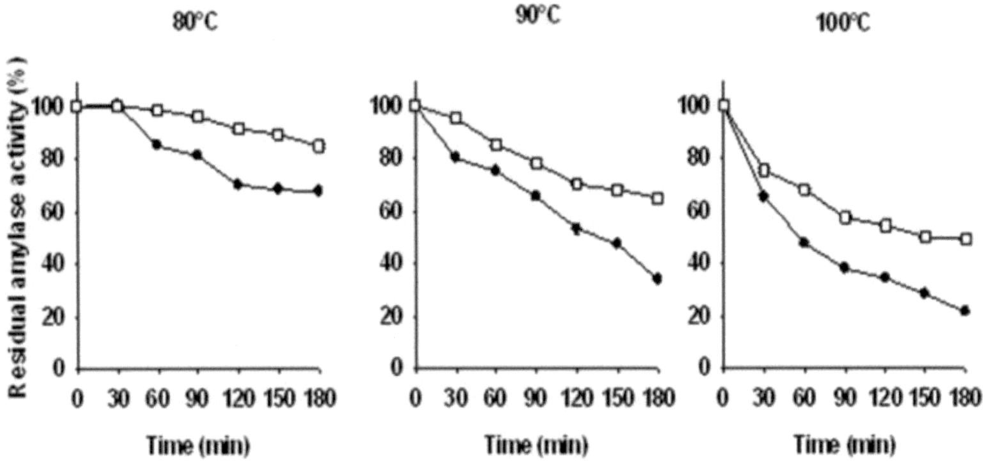


**Figure 6.** Thermostability pattern of α-amylase from *L. fermentum* 04BBA19, at 80, 90, 100°C without $CaCl_2.2H_2O$ (●) and with 0.1% (w/v) $CaCl_2.2H_2O$ (□). The enzyme was pre-incubated at optimum pH, for 30, 60, 90, 120 and 180 min at temperatures (80, 90 and 100°C). The remaining activity was determined incubating the enzyme at optimum temperature, 60°C for 30 min. The data shown are averages of triplicate assays with SD within 10% of mean value.

| | Temperatures | | | | | |
|---|---|---|---|---|---|---|
| | 80 °C | | 90°C | | 100°C | |
| $CaCl_2.2H_2O$ (% w/v) | $k_i$ ($10^{-3}.min^{-1}$) | T (min) | $k_i$ ($10^{-3}.min^{-1}$) | T (min) | $k_i$ .($10^{-3}.min^{-1}$) | T (min) |
| 0 | 3.4 | 204.0 | 5.6 | 123.8 | 7.9 | 87.7 |
| 0.1 | 1.9 | 364.8 | 2.6 | 266.6 | 3.8 | 182.4 |

**Table 3.** Inactivation rate constant ($k_i$) and half-live (T) of amylase from *L. fermentum* 04BBA19 at 80, 90 and 100°C in the absence and the presence of 0.1% (w/v) $CaCl_2.2H_2O$.

The main ALAB that have been isolated for the past decade are summarized in Table 4. No study has dealt with the thermostability of their amylases, except the case reported by Aguilar et al. (2000) concerning the properties of the extracellular amylase produced by *L. manihotivorans* LMG $18010^T$. This strain produced an amylase with a moderate themostability exhibiting maximum activity at 55°C.

The strain *L. fermentum* 04BBA19 appears as the first ALAB producing highly thermostable amylase. The potential industrial application of this strain could be the bioconversion of inexpensive raw material as starch into lactic acid in single step process. On the other hand this strain and its amylase are potential candidates for food industries (making of high density gruels, baking, brewing) and for the production of biodegradable plastic from starchy raw material.

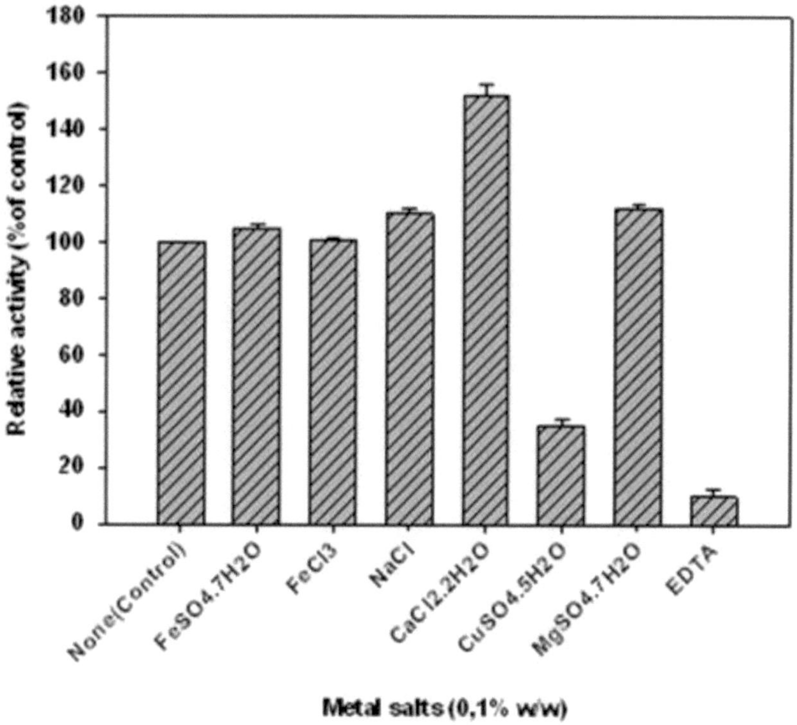


**Figure 7.** Effect of metal salts and EDTA on the activity of α-amylase from *L. fermentum* 04BBA19. The data shown are averages of triplicate assays with SD within 10% of mean value.

| Bacteria | Strains | References |
|---|---|---|
| *L. fermentum* | 04BBA19 | [47, 48] |
| *L. manihotivorans* | LMG18010$^T$ | [69] |
| *L. fermentum* | Ogi E1 | [11] |
| *L. fermentum* | MW2 | [11] |
| *L. fermentum* | K9 | [12] |
| *L. acidophilus* | L9 | [78] |
| *L. amylovorus* | ATCC33622 | [23] |
| *L. amylovorus* | B-4542 | [29] |
| *L. amylovorus* | | [17] |
| *L. manihotivorans* | OND32T | [10, 13] |
| *L. manihotivorans* | | [13] |
| *L. manihotivorans* | LMG 18011 | [79] |
| *L. acidophilus* | | [78] |

| | | |
|---|---|---|
| *L. plantarum* | A6 | [80, 9] |
| *L. plantarum* | LMG18053 | [9] |
| *L. plantarum* | NCIM 2084 | [81] |
| *Streptococus. bovis* | 148 | [82] |
| *Lactobacillus sp* | LEM 220, | [85] |
| *Lactobacillus sp* | LEM 207 | [85] |
| *Leuconostoc sp* | | [86] |
| *Leuconostoc* | St3-28 | [80] |
| *S. macedonicus* | | [87] |
| *L. amylolyticus* | | [88] |
| *L. amylophilus* | JCIM 1125 | [84] |
| *L. amylophilus* | B 4437 | [28, 32] |
| *L. amylophilus* | GV6 | [20] |
| *Bifidobacterium adolescentis* | Int57 | [15] |
| *B. adolescentis* | ZS8 | [16] |

**Table 4.** The main amylolytic lactic bacteria strains isolated during the past two decade

## 5. Conclusion

*L. fermentum* 04BBA19 which is a soil isolate produced very high thermostable α-amylase. This is the first study dealing with high thermostable amylase from a lactic acid bacterium. According to its properties, this enzyme is a good candidate for starch hydrolysis at high temperature. An economical process could be attained through the use of this enzyme at the liquefaction stage at high temperatures.

On the other hand the fact that thermostable amylase and lactic acid production can be combined in single fermentation step would not only provide a way to make gruels with high energy density, but also improve its safety, since lactic acid bacteria fermentation is an efficient way to inhibit food-borne pathogens.

Owing to the importance of this finding, further studies will focus on the development of an accurate method for preparing high energy density complementary food using local starchy sources and the *L. fermentum* 04BBA19 strain.

## Author details

Bertrand Tatsinkou Fossi*
*Department of Microbiology and Parasitology, University of Buea, Buea, Cameroon*

Frédéric Tavea
*Department of Biochemistry, University of Douala, Douala, Cameroon*

* Corresponding Author

## Acknowledgement

We gratefully acknowledge the assistance of the Ministry of Higher Education and Brewing society "Société Anonyme des Brasseries du Cameroun" (SABC) who supported this work through Research-Development Grant Programme.

## 6. References

[1] Reddy G, Altaf M, Naveeana BJ, Ventkateshwar M., Vijay Kumar E (2008) Amylolytic bacterial fermentation. A review. Biotechnology Advanced 26: 22-34.

[2] Axelsson L (2004) Lactic acid bacteria: classification and physiology. In: Salminen S, von Wright A, Ouwehand A, editors. Lactic acid bacteria: microbiological and functional aspects. 3rd rev. and exp. ed.New York: Marcel Dekker, Inc.; 2004. p. 1-66.

[3] Asoodeh A, Chamani J, Lagziana M (2010) A novel thermostable, acidophilic α-amylase from a new thermophilic "*Bacillus* sp. Ferdowsicous" isolated from Ferdows hot mineral spring in Iran: Purification and biochemical characterization. Inter. J. Biol. Macromol. 46: 289-297.

[4] Saxena RK, Dutt K, Argawal L, Nayyar P (2007) A highly thermostable and alkaline amylase from a *Bacillus* sp PN5. Biores. Technol. 98: 260-265.

[5] Haki GD, Rakshit, SK (2003) Developments in industrially important thermostable enzymes: a review. Biores. Technol. 89: 17-34.

[6] Singh SK, Ahmed SU, Pandey A (2006) Metabolic engineering approaches for lactic acid production. Process Biochem. 41: 991-1000.

[7] Nguyen TT, Loiseau G, Icard-Vernière C, Rochette I, Trèche S, Guyot, JP (2007) Effect of fermentation by amylolytic lactic acid bacteria in process combinations, on characteristics of rice/soybean slurries: A new method for preparing high energy density complementary foods for young children. Food Chem. 100: 623-631.

[8] Songré Ouattara, LT, Mouquet-Rivier C., Icard-Vernière C, Humblot C, Diawara B, Guyot JP (2008) Enzyme activities of lactic acid bacteria from a pearl millet fermented gruel (ben-saalga) of functional interest in nutrition. Int. J. of Food Microbiol. 128: 395–400.

[9] Giraud E, Lelong B, Raimbault M (1991) Influence of pH and initial lactate concentration on the growth of *Lactobacillus plantarum*. Appl Microbiol Biotechnol 36: 96–9.

[10] Morlon-Guyot J, Guyot JP, Pot B, Jacobe de Haut I, Raimbault M (1998). *Lactobacillus manihotivorans* sp. Nov., a new starch-hydrolysing lactic acid bacterium isolated during cassava sour starch fermentation. Int. J. Syst. Bacteriol. 48: 1101-1109.

[11] Agati VJP, Guyot J, Morlon-Guyot P, Talamond, Hounhouigan DJ (1998). Isolation and characterization of new amylolytic strains of *Lactobacillus fermentum* from fermented maize doughs (mawe and ogi) from Benin. J Appl Microbiol 85: 512–20.

[12] Sanni A, Morlon-Guyot J, Guyot JP. New efficient amylase-producing strains of *Lactobacillus plantarum* and *L. fermentum* isolated from different Nigerian traditional fermented foods. Int J Food Microbiol 2002;72:53–62.

[13] Guyot JP, Calderon M, Morlon-Guyot JP (2000). Effect of pH control on lactic acid fermentation of starch by *Lactobacillus manihotivorans* LMG 180010T. J. Appl. Microbiol. 88: 176-182.

[14] Ji, GE., Han HK, Yun SW, Rhim SL (1992). Isolation of amylolytic *Bifidobacterium* sp. Int57 . J. Microbiol. Biotechnol. 2: 85–91.

[15] Lee SK, Kim, YB, Ji, GE, (1997). Purification of amylase secreted from *Bifidobacterium adolescentis*. J. Appl. Microbiol. 83: 267–272.

[16] Lee JH, Lee SK,. Parka KH, In K,. Jib GE (1999). Fermentation of rice using amylolytic *Bifidobacterium*. Int J Food Microbiol 50: 155–161.

[17] Nakamura LK (1981). *Lactobacillus amylovorus*, a new starch-hydrolyzing species from cattle waste-corn fermentations. Int J Syst Bacteriol: 31:56–63.

[18] Chatterjee M, Chakrabarty SL, Chattopadhyay BD, Mandal RK (1997) Production of lactic acid by direct fermentation of starchy wastes by an amylase-producing Lactobacillus. Biotechnol Lett: 19:873–4.

[19] Giraud E, Champailler A, Raimbault M (1994) Degradation of raw starch by a wild amylolytic strain of *Lactobacillus plantarum*. Appl Environ Microbiol 60: 4319–23.

[20] Vishnu C, Seenayya G, Reddy G. Direct fermentation of various pure and crude starchy substrates to L(+) lactic acid using *Lactobacillus amylophilus* GV6.World J Microbiol Biotechnol 2002;18:429–33.

[21] Naveena BJ, Altaf Md, Bhadrayya K, Madhavendra SS, Reddy G (2005) Direct fermentation of starch to L(+) lactic acid in SSF by *Lactobacillus amylophilus* GV6 using wheat bran as support and substrate—medium optimization using RSM. Process Biochem 40: 681–90.

[22] Anuradha R, Suresh AK, Venkatesh KV (1999) Simultaneous saccharification and fermentation of starch to lactic acid. Process Biochem 35: 367–75.

[23] Zhang DX, Cheryan M (1994) Starch to lactic acid in a continuous membrane reactor. Process Biochem 29: 145–50.

[24] Naveena BJ, Vishnu C, Altaf Md, Reddy G (2003) Wheat bran an inexpensive substrate for production of lactic acid in solid state fermentation by *Lactobacillus amylophilus* GV6-optimization of fermentation conditions. J Sci Ind Res62: 453–6.

[25] Goel MK. Biotechnology: an overview; 1994 (http://www.rpi.edu/dept/chem-eng/Biotech-Environ/goel.html ).

[26] Altaf M, Venkateshwar M, Srijana M, Reddy G (2007) An economic approach for L-(+) lactic acid fermentation by *Lactobacillus amylophilus* GV6 using inexpensive carbon and nitrogen sources. J Appl Microbiol;103:372–80.

[27] Pandey A, Soccol CR, Rodriguez-Leon JA, Nigam P (2001). Solid state fermentation in biotechnology: fundamentals and applications. New Delhi: Asiatech Publishers

[28] Mercier P, Yerushalami L, Rouleau D, Dochania D (1992). Kinetics of lactic acid fermentations on glucose and corn by *Lactobacillus amylophilus*. J Chem Technol Biotechnol 55:111–21.

[29] Cheng P, Muller RE, Jaeger S, Bajpai R, Jannotti EL (1991) Lactic acid production from enzyme thinned cornstarch using *Lactobacillus amylovorus*. J Ind Microbiol 7: 27–34

[30] Kurusava H, Ishikawa H, Tanaka H. L-lactic acid production from starch by co-immobilized mixed culture system of *Aspergilus awamori* and *Streptococcus lactis*. Biotechnol Bioeng 1988;31: 183–7.

[31] Kristoficova L, Rosenberg M, Vlnova A, Sajbidor J, Cetrik M. Selection of Rhizopus strains for L (+) lactic acid and gammalinolenic acid production. Folia Microbiol 1991;36:451–5.

[32] Nakamura LK, Crowell CD (1979). *Lactobacillus amylophilus*, a new starch hydrolyzing species from swine waste-corn fermentation. Dev Ind Microbiol 20: 531– 40.

[33] Prasana VA (2005). Amylases and their applications. Afr. J Biotechnol 4(13): 1525-1529.

[34] Sivaramakrishnan S, Gangadharan D, Madhavan K (2006) α-Amylase from microbial sources-An overview on recent developments. Food technol. Biotechnol. 44(2): 173-184.

[35] Ghorbel RE, Maktouf S, Massoud EB, Bejar S, Chaabouni SE (2009) New Thermostable Amylase from *Bacillus cohnii* US147 with a Broad pH Applicability. Appl. Biochem. Biotechnol. 157: 50–60.

[36] Lin LL, Chyau CC, Hsu WH (1998) Production and properties of a raw starch degrading amylase from thermophilic and alkaliphilic Bacillus sp TS-23. Biotechnol Appl Biochem, 28 (1): 61-68.

[37] Giraud E, Cunny G (1997) Molecular characterization of the α-amylase genes of *Lactobacillus plantarum* A6 and *Lactobacillus amylovorus* reveals an unusual 3′ end structure with direct tandem repeats and suggest a common evolutionnary origin. Gene 198: 149-157.

[38] Talamond, P, Desseaux V, Moreau Y, Santimone M, Marchis-mouren G (2002) Isolation, characterization and inhibition by acarbose of the α-amylase from *Lactobacillus fermentum* : comparison with *Lb. manihotivorans* and *Lb. plantarum*. Comparative Biochem. and Physiol. 133: 351-360.

[39] Qui and Yao (2007) Lactic acid production, rice straw. Bioresources 2(3): 419-429.

[40] Sawadogo-Lingani H, Diawara B.Traoré AS, Jakobsen M (2008) Technological properties of *Lactobacillus fermentum* involved in the processing of dolo and pito, West African sorghum beers, for the selection of starter cultures. J.Appl Microbiol, 104(3): 873-882.

[41] Champagne C (1998). La production des ferments dans l'industrie laitière Ed. Agriculture et Agroalimentaire, Canada,pp : 32-145.

[42] Edelman S, Westerlund-Wikström B, Leskelä S, Kettunen H, Rautonen N, Apajalahti J., Korhonen TK (2002) In vitro adhesion specificity of indigenous *lactobacilli* within the avian intestinal tract. *Appl Environ Microbiol* 68: 5155-5159.

[43] Hofvendahl K, Hahn-Hagerdal B (1997) L-lactic acid production from whole wheat flour hydrolysate using strains of *Lactobacilli* and *Lactococci*. Enzyme Microbiol Technol 20: 301-307.

[44] Datta RS, Sai PT, Patric B, Moon SH, Frank JR (1993). Technological and economic potential of polylactic acid and lactic acid derivatives. International congress on chemicals from biotechnology, Hannover, Germany p. 1–8.

[45] Wee YJ, Kim JN, Ryu HW (2006). Biotechnological production of lactic acid and its recent applications. Food Technol Biotechnol 44: 163–72.

[46] Senthuran A, Senthuran V, Mattiasson B, Kaul R (1997). Lactic acid fermentation in a recycle batch reactor using immobilized *Lactobacillus casei*. Biotechnol Bioeng 55:843–53.

[47] Tatsinkou, FB, Tavea F, Jiwoua C, Ndjouenkeu R (2009) Screening of thermostable amylase producing bacteria and yeasts strains from some cameroonian soils. Afr. J. Microbiol. Res. 3(9): 504-514.

[48] Tatsinkou, FB, Tavea F, Jiwoua C, Ndjouenkeu R (2011) Simultaneous production of raw starch degrading highly thermostable a-amylase and lactic acid by *Lactobacillus fermentum* 04BBA19. Afr. J. Biotechol. 10(34): 6564-6574.

[49] Schillinger U, LuKe F.K (1987). Identification of lactobacilli from meat and meat products. Food Microbiology 4, 199– 208.

[50] Dykes GA, Britz TJ, von Holy A (1994). Numerical taxonomy and identification of lactic acid bacteria from spoiled, vacuum packaged Vienna sausages. Journal of Applied Bacteriology 76, 246– 252.

[51] Kimberley AC, Taylor, C (1996) A simple colorimetric assay for muramic acid and lactic acid. Appl. Biochem. Biotechnol. 56: 49-58.

[52] Miller GL (1959) Use of dinitrosalicylic acid reagent for determination of reducing sugar. Anal. Chem. 31: 426-429.

[53] Goyal N, Gupta JK, Soni. SK (2005) A novel raw starch digesting thermostable α-amylase from *Bacillus* sp. I-3 and its use in the direct hydrolysis of raw potato starch. Enzyme and Microb. Technol. 37: 723-734.

[54] Liu DX, Xu Y (2008) A novel raw starch digesting α-amylase from a newly isolated *Bacillus* sp YX-1: purification and characterization. Biores. Technol. 99: 4315-4320.

[55] Hiller P, Wase D, Emery, AN (1996) Production of α-amylase by *Bacillus amyloliquefaciens* in batch and continuous culture using a defined synthetic medium. Biotechnol. Lett. 18 : 795-799.

[56] Burhan A, Nisa U, Gökhan, C, Ômer C, Ashabil, Osman, G (2003) Enzymatic properties of a novel thermostable, thermophilic, alkaline amylase and chelator resistant amylase from an alkaliphilic *Bacillus* sp isolate ANT-6. Process Biochem. 38: 1397-1403.

[57] Chandra AK, Medda S, Bhadra, AK (1980) Production of extracellular thermostable α-amylase by *Bacillus licheniformis.* J. Ferment. Technol. 58: 1-10.

[58] Saito N, Yamamoto K (1975) Regulatory factors affecting α-amylase production in *Bacillus licheniformis*. J. Bacteriol. 121: 848-856.

[59] Gupta R, Gigras P, Mohapatra H, Goswami VK, Chauhan, B (2003) Microbial α-amylases: a biotechnological perspective. Process Biochem. 38: 1599-1616.

[60] Muralikrishna G, Nirmala M, (2005) Cereal α-amylase –an overview. Carbohydrate polymer 60: 193-173.

[61] Gangadharan D, Sivaramakrishna, K, Nampoothiri KM, Sukumaran RK, Pandey A (2008) Response surface methodology for the optimisation of alpha amylase production by *Bacillus amyloliquefaciens*. Biores. Technol. 99: 4597-4602.

[62] Sodhi HK, Sharma K, Gupta, JK, Soni SK (2005) Production of a thermostable -amylase from *Bacillus* sp PS-7 by solid state fermentation and its synergistic use in the hydrolysis of malt starch for alcohol production. Process Biochem. 40: 525-534.

[63] Cherry HM, Hussain T, Anwar M.N (2004) Extracellular glucoamylase from isolate of *Aspergillus fumigates*. Pak. J. Biol. Sci. 7(11): 1988-1992.

[64] Theodoro CE, Martin ML (2000) Culture conditions for the production of thermostable amylase by *Bacillus* sp. Brazilian J. Microbiol. 31: 298-302.

[65] Halsetine C, Rolfsmeier M, Blum P (1996) The glucose effect and regulation of α-amylase synthesis in the hyperthermophilic archeon Sulfolobus solfataricus . J.Bacteriol. 178: 945-950.

[66] Gao L, Yang H, Wang X, Huang Z, Ishii M, Igarashi Y, Cui Z (2008) Raw straw fermentation using lactic acid bacteria. Biores.Technol. 99: 2742-2748.

[67] Tanyildizi, MS, Ozer D, Ebiol, M (2005) Optimisation of α-amylase production by *Bacillus* sp using response surface methodology. Processs Biochem. 40, 2291-2296.

[68] Wu WX, Mabinadji J, Tatsinkou FB (1999) Effect of culture conditions on the production of an extracellular thermostable alpha-amylase from an isolate of *Bacillus* sp. J. Zhejiang Univ Agric Life Sci 25: 404-408.

[69] Reddy PR, Reddy G Seenayya G (1999). Enhance production of thermostable β-amylase and pullulanase in the presence of surfactant by *Clostridium thermosulfurogenes* SV2. Process Biochem. 34: 87-92.

[70] Aguilar G, Morlon-Guyot J, Trejo-Aguilar, B, Guyot, JP (2000) Purification and Characterization of an extracellular alpha-amylase produced by *Lactobacillus manihotivorans* LMG 18010T, an amylolytic lactic bacterium. Enzyme Microb. Technol. 27: 406-413.

[71] Underkoffer L (1976) Microbial enzyme In Miler B. Litsky W. (Eds). Industrial Microbiology. Mc Graw-Hill, New York.

[72] Viara NI, Elena PD, Elka IE (1992) Purification and characterisation of a thermostable alpha amylase from *Bacillus licheniformis*. J. of Bacteriol. 28: 277-289.

[73] Vihinen M, Mantsala P (1990). Characterization of a thermostable *Bacillus stearothermophilus* alpha-amylase. Biotechnol. Appl. Biochem. 12: 427-435.

[74] Canganella F, Andrade C, Antranikian G (1994) Characterization of amylolytic and pullulytic enzymes from thermophilic archea and from a new *Ferividobacterium* species. Appl. Microb. Biotechnol. 42 : 239-245.

[75] Kwak Y, Akeba T, Kudo T (1998) Purification and characterization of enzyme from hyperthermophilic archeon *Thermococcus profundus*, which hydrolyses both α-1-4 and α-1 6 glucosidic linkages. J. Ferment. Bioeng, 86: 363-367.

[76] Prakash B, Vidyagar M, Madhukumar MS, Muralikrishna G, Sreeramulu K (2009). Production, purification and characterization of two extrememely halotolerant thermostable, and alkaline stable α- amylases from *Chromohalobacter* sp TVSP 101. Process Biochem. 44: 210-215.

[77] Bayramoglu G Yilmaz M Arica Y (2003) Immobilization of a thermostable a-amylase onto reactive membranes: kinetics characterization and application to continuous starch hydrolysis. Food chem. 57: 83-90.

[78] Lee Hs, SeG, Carter S (2001) Amylolytic cultures of *Lactobacillus acidophilus*: potential probiotics to improve dietary starch utilization. J food Sci. 66:2.

[79] Ohkouchi Y, Inoue Y (2006) Direct production of L(+)-lactic acid from starch and food wastes using *Lactobacillus manihotivorans* LMG18011. Bioresour. Technol 97:1554–62.

[80] Mette Hedegaard Thomsen, Guyot JP, Kiel P. Batch fermentations on synthetic mixed sugar and starch medium with amylolytic lactic acid bacteria. Appl Microbiol Biotechnol 2007;74: 540–6.

[81] Krishnan S, Bhattacharya S, Karanth NG (1998) Media optimization for production of lactic acid by *Lactobacillus plantarum* NCIM 2084 using response surface methodology. Food Biotechnol 12: 105–21.

[82] Junya Narita, Nakahara S, Fukuda H, Kondo A (2004) Efficient production of L-(+)-lactic acid from raw starch by *Streptococcus bovis* 148. J Biosci Bioeng 97: 423–5.

[83] Yumoto I, Ikeda K (1995) Direct fermentation of starch to L(+)-lactic acid using *Lactobacillus amylophilus*. Biotechnol Lett 17: 543–6.

[84] Champ MO, Szylit P, Raimbault M, Abdelker N (1983) Amylase production by three Lactobacillus strains isolated from chicken crop. J Appl Bacteriol 55:487–93.

[85] Lindgren S, Refai O (1984) Amylolytic lactic acid bacteria in fish silage. J Appl Bacteriol 57: 221–8.

[86] Diaz-Ruiz G, Guyot JP, Ruiz-Teran F, Morlon-Guyot J, Wacher C (2003) Microbial and physiological characterization of weakly amylolytic but fast-growing lactic acid bacteria: a functional role in supporting microbial diversity in pozol, a Mexican fermented maize beverage. Appl Environ Microbiol 69:4367–74.

[87] Bohak I, Back W, Richter L, Ehrmann M, Ludwing W, Schleifer KH (1998) *Lactobacillus amylolyticus* sp. nov., isolated from beer malt and beer wort. Syst Appl Microbiol : 21:360–4.